Remote Sensing of Northwest Himalayan Ecosystems

R. R. Navalgund • A. Senthil Kumar
Subrata Nandy
Editors

Remote Sensing of Northwest Himalayan Ecosystems

Springer

Editors
R. R. Navalgund
Indian Space Research Organisation
Bangalore, Karnataka, India

A. Senthil Kumar
Indian Institute of Remote Sensing
Indian Space Research Organisation
Dehradun, Uttarakhand, India

Subrata Nandy
Indian Institute of Remote Sensing
Indian Space Research Organisation
Dehradun, Uttarakhand, India

ISBN 978-981-13-2127-6 ISBN 978-981-13-2128-3 (eBook)
https://doi.org/10.1007/978-981-13-2128-3

Library of Congress Control Number: 2018957836

This Springer imprint is published by the registered company Springer Nature Singapore Pte Ltd.
The registered company address is: 152 Beach Road, #21-01/04 Gateway East, Singapore 189721, Singapore

Foreword

The Himalaya, the youngest mountain ranges of the earth in the north of India, is home to very many high mountain peaks; mighty rivers like the Indus, Ganga and the Brahmaputra; forests very rich in biodiversity; and vast snow-covered regions and glaciers. It is also a very seismically active region. In view of the geotectonic activities, climate change and ever-increasing demographic pressure, there is degradation of natural resources, loss of biodiversity and soil erosion. It is also beset with many natural disasters like the earthquakes, landslides, flash floods, forest fires and snow avalanches. In recent years, spaceborne remote sensing along with other geospatial techniques has emerged as excellent tools in assessment of the current status of various resources, understanding complex interactions and providing information for conservation and development.

I am very happy to note that the Indian Institute of Remote Sensing has over the years carried out detailed investigations in different themes for the northwest Himalaya (NWH) region comprising Himachal Pradesh, Uttarakhand and Jammu and Kashmir. I am also glad to note that a volume describing this work is being brought out. I am sure such a volume would not only be a good documentation of the present understanding of the resource status of NWH and a useful reference material for future work but also would be useful to the development planners of the region.

Chairman, ISRO,
Bangalore, Karnataka, India
September 14, 2017

A. S. Kirankumar

Preface

The Himalayan mountain range is the youngest and the mightiest on the earth. It extends east-west over a 2400 km long arch in the north of South Asia. It is home for more than 100 million people, with a good fraction living at very high altitudes. It influences the South Asian Monsoon, has large orographic precipitation and stores water as snow and ice. A large number of mighty rivers such as the Indus, Ganga and the Brahmaputra originate from here. A billion people living downstream in the plains are dependent on these resources. The Himalayan mountains are among the most biodiversity-rich ecosystems in the world. The northwest Himalaya (NWH) region extends from the foothills of Shivaliks to the greater Himalaya and the three states of India, Himachal Pradesh, Uttarakhand and Jammu and Kashmir form part of this.

This youngest mountain chain of the world is seismically very active and has experienced numerous earthquakes, landslides, snow avalanches, flash floods, forest fire, etc. causing damage to life, property and the natural ecosystems. The climate change and anthropogenic pressures are also exerting tremendous impacts on the status of natural resources and biodiversity of NWH. Hence, there is an urgent need to assess and monitor the NWH ecosystems to take remedial steps to facilitate proper management of resources and sustainable development of this region.

In the recent past, there has been significant advancement in the field of remote sensing. There have been a number of satellites that collect data in visible, infrared, thermal and microwave regions at different resolutions. Geospatial technology, a combination of remote sensing (RS), geographic information system (GIS) and global navigation satellite system (GNSS), plays a crucial role in providing very useful data for inaccessible terrain like the NWH and to understand complex interactions between different components of NWH ecosystems.

Multidisciplinary studies have been carried out in the NWH region in various themes using geospatial technology over the last few years by the Indian Institute of Remote Sensing, ISRO, Dehradun. They are described in various parts of this volume. The first part gives an overview of the issues, challenges and role of geospatial technology with respect to NWH ecosystem. The second part on geology and geodynamics discusses geology and geohazards of NWH. The morphotectonic analysis, debris flow modelling and total electron content (TEC) modelling as an earthquake precursor studies are specifically addressed in this part. The third part focuses on the water resources of NWH. It comprises cryosphere studies, hydrological modelling and mapping of hydrometeorological hazards, monitoring and modelling. The spatio-temporal characteristics of rainfall over the NWH region are also discussed in this part. In the fourth part, various aspects of the forest resources and biodiversity of NWH region have been addressed. It includes biodiversity characterisation, spatial biodiversity information system, Indian bioresource information network and possible status of forests in different climate change scenarios. Forest biomass assessment using various techniques, monitoring carbon exchange through CO_2 flux tower and remote sensing, emissions from biomass burning and wildlife habitat evaluation studies are also included in this part. The fifth part pertains to agriculture and soils of NWH region. Predicting soil erosion and nutrient loss using different models specific to mountain region and climate change impact assessment of mountain agriculture are described in this part. Urban environment, urban settlement pattern and growth dynamics have been discussed in the sixth part. An appraisal on the vector-borne diseases is also provided. Articles discussing the EO data requirements, availability, gaps with respect to the Himalayan region, various geoweb services and online repositories for disaster monitoring and mitigation, and geostatistical and deterministic interpolation methods are included in the seventh part. This part also emphasises on the role of citizen science in disaster mitigation, bioresource inventory and governance.

Overall, the volume summarises various investigations carried out in NWH region in different themes, and it provides a reference material for further work and indicates how the outputs of these studies can be of use in development planning of the NWH region.

Bangalore, India R. R. Navalgund
Dehradun, India A. Senthil Kumar
Dehradun, India Subrata Nandy

Acknowledgements

We sincerely acknowledge all the authors for their contributions in writing state-of-art information on the theme of this book and for their painstaking revisions based on the comments received from unanimous reviewers on time. Enormous work described in each of the chapters of this book has been the result of efforts of many scientists, thanks to ISRO for its continued support to carry out research on this subject over many years. Grateful appreciation is due for all the supervisors, Heads of respective departments, Deans and Directors of the Indian Institute of Remote Sensing (IIRS) and associated scientists from participating organisations. Compiling, organising and editing these many articles and chapters was no simple task. We sincerely acknowledge critical comments received from reviewers, which have brought the contents of the book to this maturity.

As editors, it is our honour and pleasure to sincerely thank Chairman of ISRO for providing necessary encouragement and guidance for embarking on this programme and to his kind words on taking up this initiative of bringing out this volume.

September 14, 2017

Indian Space Research Organisation, Bangalore, Karnataka, India — R. R. Navalgund

Indian Institute of Remote Sensing, Indian Space Research Organisation, Dehradun, Uttarakhand, India — A. Senthil Kumar

Indian Institute of Remote Sensing, Indian Space Research Organisation, Dehradun, Uttarakhand, India — Subrata Nandy

Contents

About the Editors and Contributors

Editors

Ranganath Navalgund obtained his PhD in physics from the Tata Institute of Fundamental Research, Mumbai, and joined the Space Applications Centre (ISRO) in 1977. Since then, he has worked in ISRO till today in different capacities. His scientific contributions are in the broad area of earth observation systems, science and applications. As Director of National Remote Sensing Centre (2001–2005) and the Space Applications Centre (2005–2012), the two major centres of ISRO, he has overseen the formulation and execution of national-level RS application programmes, establishment of EO data reception systems and Decision Support Centre for disaster monitoring and mitigation, the development of electro-optical and microwave sensors for Indian earth observation and planetary science missions and the communication and navigation payloads on-board Indian satellites and has interfaced with several space agencies of different countries in a leadership role. He was the president of the International Society for Photogrammetry and Remote Sensing Technical Commission VII (2000–2004). He is a fellow of the Indian Academy of Sciences, an academician of the International Academy of Astronautics and a fellow of many other professional societies. He has been a recipient of many awards including the Bhaskara Award of the Indian Society of Remote Sensing and the Outstanding Achievement Award of ISRO besides many others. He is also in the editorial board of *Current Science*, a professional Indian journal. He is currently honorary distinguished professor at ISRO Headquarters, Bangalore.

A. Senthil Kumar received PhD from the Indian Institute of Science, Bangalore, in the field of image processing in 1990. He joined ISRO in 1991 and has been serving in Indian satellite programmes in various capacities. His research includes sensor characterisation, radiometric data processing, image restoration, data fusion and soft computing. He is currently the Director of the Indian Institute of Remote Sensing, Dehradun, and also the Director of UN-affiliated Centre for Space Science and Technology Education in Asia and the Pacific. He is the President of ISPRS Technical Commission V on Education and Outreach and Chair of CEOS Working Group on Capacity Building and Data Democracy. He has published about 80 technical papers in international journals and conferences, besides technical reports. He is a recipient of ISRO Team Awards for Chandrayaan-1 mission and Prof. Satish Dhawan Award conferred by the Indian Society of Remote Sensing. He is also the associate editor of *Journal of the Indian Society of Remote Sensing*.

Subrata Nandy received MSc and MPhil in ecology from Assam University, Silchar, Assam. He obtained his PhD in forest geoinformatics from the Forest Research Institute University, Dehradun. He joined the Indian Institute of Remote Sensing, ISRO, Dehradun, as Scientist/Engineer-SC in 2008. His research expertise includes forest biomass/carbon and productivity assessment and LiDAR remote sensing in forestry. He has more than 15 years of research experience in the applications of remote sensing and GIS in forestry and ecology. He worked significantly in the synergistic use of passive optical and LiDAR data for forest biomass/carbon assessment using various techniques. He also contributed immensely in national-level projects, ISRO's earth observation application mission and technology development projects. He has published 33 research papers in peer-reviewed journals. He is a life member of the Indian Society of Remote Sensing, Indian Society of Geomatics and Indian Meteorological Society. He is currently working as Scientist/Engineer-SE in Forestry and Ecology Department of the Indian Institute of Remote Sensing, ISRO, Dehradun.

Contributors

S. P. Aggarwal holds BTech in agricultural engineering from Allahabad University; PG and PhD from IARI, New Delhi; and postdoctoral research from IHE and ITC, Netherlands. Currently, he is Head of Water Resources Department, Indian Institute of Remote Sensing, ISRO, Dehradun. He has more than 20 years' experience in remote sensing and GIS applications in water resources management. He is also programme coordinator of Centre for Space Science and Technology Application in Asia and the Pacific, UN-affiliated Centre, Dehradun. He is associate editor of *Journal of the Indian Society of Remote Sensing*. He is secretary of the International Society for Photogrammetry and Remote Sensing, Technical Commission V. He has been conferred with Eminent Engineers Award in 2014 by the Institution of Engineers of India, Uttarakhand State Centre, and President Appreciation Medal by the Indian Society of Remote Sensing for 2016. He has published 100 research papers in peer-reviewed national/international journals and symposia.

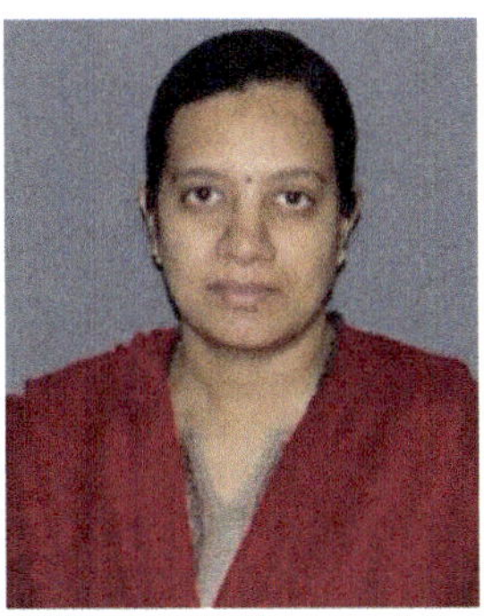

Shefali Agrawal, a physicist working at the Indian Institute of Remote Sensing (ISRO), pursues her academic interest in addressing both fundamental and applied aspects of remote sensing and allied spatial technology with emphasis on satellite photogrammetry, LiDAR remote sensing, UAV remote sensing and advanced image processing related to land use/land cover characterisation and vegetation physics dynamics. She is currently Head of Photogrammetry and Remote Sensing Department, and she has over 25 years of experience and produced over 50 papers in national and international journals.

Joyson Ahongshangbam is a PhD student in Tropical Silviculture and Forest Ecology, University of Göttingen, Germany. He is presently working on measuring tree and oil palm water use in Indonesia as a part of CRC EFForTS project. His areas of interest are carbon and water flux, remote sensing, sap flux and eddy covariance measurements and application of UAV in forest and ecological studies. He holds BSc in forestry from North Eastern Regional Institute of Science and Technology, Nirjuli, Arunachal Pradesh, and MTech in remote sensing and GIS with specialisation in forestry and ecology from the Indian Institute of Remote Sensing, Dehradun, India. He worked as a junior research fellow at the Indian Institute of Remote Sensing, Dehradun, and at Manipur University, Manipur, India. He has published one research paper in a peer-reviewed journal.

Akarsh Asoka is a PhD scholar in the discipline of earth sciences, IIT Gandhinagar. He received MTech in remote sensing and GIS from the Indian Institute of Remote Sensing, Dehradun, and BTech in agricultural engineering from Kelappaji College of Agricultural Engineering and Technology, Kerala Agricultural University. His research interests include remote sensing, climate change impact assessment, food and water security and sustainability. His PhD is supported by Information Technology Research Academy (ITRA), Media Lab Asia under the project 'Measurement to Management: Improved Water Use Efficiency and Agricultural productivity through Experimental Sensor Network (M2M)'. He is a recipient of Water Advanced Research and Innovation Fellowship (WARI) supported by the Department of Science and Technology, Government of India, the University of Nebraska-Lincoln (UNL), the Daugherty Water for Food Institute (DWFI) and the Indo-US Science and Technology Forum (IUSSTF). He has three publications in *Journal of Geophysical Research*.

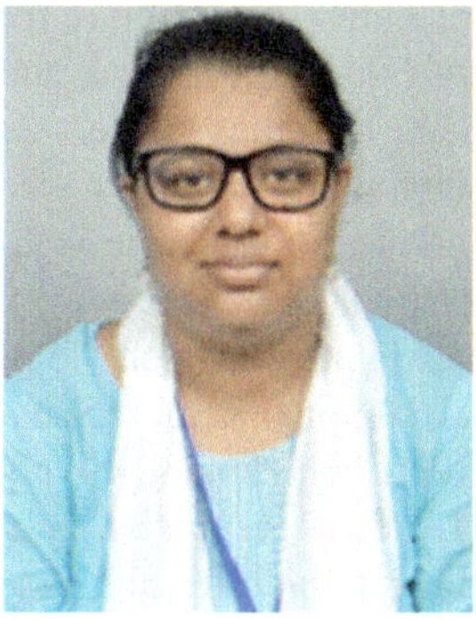

Arifa Begum is currently working as a research fellow at Urban and Regional Studies Department, Indian Institute of Remote Sensing, Dehradun. She has completed her postgraduation from the Department of Geography, Delhi School of Economics, University of Delhi. She is profoundly interested in the study of growth and expansion of spatial entities in urban spaces. She has worked in various projects at Jawaharlal Nehru University, Institute for Studies in Industrial Development, University of Delhi sponsored by various organisations like DST, Finance Commission of Uttarakhand Government and UGC, respectively. Her research interests also include environmental and crime modelling.

B. D. Bharath architect-planner by profession, has a master's degree in city planning (MCP) from the Department of Architecture and City Regional Planning, IIT, Kharagpur. From May 1998 to April 2017, he worked at the Urban and Regional Studies Department, IIRS, Dehradun, and was involved in capacity building activities including training, education and research. Presently, he is working in Urban Studies Department, Urban Studies Group of RSAA, National Remote Sensing Centre (NRSC), Hyderabad. He is involved in national projects like National Urban Information System (NUIS) and Atal Mission for Rejuvenation and Urban Transformation (AMRUT). He is a member of professional bodies including ITPI India, ISRS and ISG and working closely with town planning professionals for national projects. His research is directed towards the use of thermal remote sensing for urban environmental applications with the specific aim to develop planning measures to counter urban heat islands.

Vidhi Bharti is the doctorate student at Monash University, Melbourne, Australia. She is an alumnus of the Indian Institute of Remote Sensing (IIRS), ISRO, Dehradun, and the University of Twente, Netherlands, from where she received her Master of Science (MS) in geoinformation science and earth observation. She has been a recipient of IIRS ISRO gold medal and co-funded MGS (Australia) and IPRS (Australia) scholarships. Her research interests include satellite meteorology, extreme events, precipitation, surface heat fluxes and the Southern Ocean.

Chandra Mohan Bhatt has recently joined Centre for Space Science and Technology Education in Asia and the Pacific (CSSTEAP), Indian Institute of Remote Sensing (IIRS), ISRO Dehradun, India. Prior to this, he has about 10 years of experience in scientific technique development and operational contribution in disaster mitigation and response during his service at Decision Support Centre (DSC), National Remote Sensing Centre (NRSC), ISRO, Hyderabad. He has actively contributed to the widespread use of geospatial techniques for near real-time flood mapping, river blockade and flood hazard and flood mitigation studies, useful for disaster managers at state and central level. He has published about more than 35 peer-reviewed research papers in various scientific journals, seminars and conferences. He is recipient of ISRO-ASI, Young Scientist Award for 2013 and ISRO Team Excellence Awards for 2013 (Decision Support Centre in DMS Programme) and 2014 (BHUVAN online GIS Platform).

P. K. Champati Ray with a postgraduate and PhD degree from IIT Bombay, India, and MS and PDF from the University of Twente, Netherlands, is actively involved in research, education and training in the field of applications of remote sensing, GNSS and GIS in geosciences, geohazards and planetary geology. Currently, he is Head of Geosciences and Disaster Management Studies (GDMS) Group and Head of Geosciences and Geohazards Department of IIRS. His research interests include monitoring and modelling of landslides, active fault mapping, seismic hazard assessment, geodynamics, crustal deformation, earthquake precursor studies, mineral exploration and planetary geology. His professional career spans over 30 years during which he has implemented 14 projects; guided more than 100 students, including 11 PhD students; published around 200 papers, including 54 papers in peer-reviewed national and international journals; and most importantly delivered more than 100 invited presentations at national and international forum.

R. S. Chatterjee is working as a Senior faculty member of Geosciences and Geohazards Department and is Head of Disaster Management Science Department of IIRS (ISRO), Dehradun. The areas of specialisations include microwave remote sensing applications in geosciences, remote sensing-based thermal anomaly detection and modelling, structural geology and geodynamics. He is postgraduate in applied geology and has PhD in geology and geophysics from IIT Kharagpur. He did the professional course DÉSS de Télédétection (master's in remote sensing) from the University of Pierre and Marie Curie (Paris VI University), France, and postdoctoral research for a brief duration in the University of Marne-la-Vallée, France. He was awarded with the Institute Silver Medal from IIT Kharagpur in 1992, P.R. Pisharoty Memorial Award (National Remote Sensing Award) in 2010 and ISRO Team Excellence Award in 2015. He has published more than 50 papers in journals and symposia proceedings, chapters in books and technical reports.

Shovan Lal Chattoraj has been working as a Scientist in Geosciences and Disaster Management Studies Group, Indian Institute of Remote Sensing (ISRO), Dehradun, for the last 5 years. After receiving the University Gold Medal in master's degree, he completed his PhD in sedimentary geochemistry from IIT Bombay, Mumbai. He is honoured with high ranks in Graduate Aptitude Test (GATE) and UPSC Geologist Exam. He was exposed to real-time challenges in the field of engineering geology while being associated with NHPC Ltd. as a geologist for 3 years. He has authored many research articles, published in national and international journals/periodicals and books in the field of sedimentary geology, landslide modelling, hazard assessment and applications of remote sensing in mineral exploration. Till date, he has supervised one PhD, seven MTech and many PG diploma students. His current interest lies in modelling of debris flows/landslides and spectroscopy of minerals.

Arpit Chouksey was born in 1987 and grew up in Jabalpur, Madhya Pradesh, India. He completed BTech in agricultural engineering in 2004. During his 4-year stay at the university, he became interested in fulfilling his career goals in hydrology. After finishing his BTech, he joined IIT Guwahati in 2008 to pursue MTech in civil engineering with specialisation in water resources engineering. His MTech research involved an assessment of the non-point source pollution of small streams at different altitudes which led him to work more on hydrology. His doctoral research, completed recently in 2008, includes runoff and sediment yield modelling of agro-forestry watersheds under climate change. He received scientist position in 2013 at Water Resources Department, Indian Institute of Remote Sensing, Dehradun. He has published more than ten research papers in different fields of water resources and guided eight MTech and MSc candidates so far.

V. K. Dadhwal is a Distinguished Scientist and Director of the Indian Institute of Space Science and Technology Thiruvananthapuram, India, since July 2016. His research interests are crop modelling, remote sensing applications in agriculture, terrestrial carbon cycle, land use/land cover change modelling and land surface processes. He has received ISCA Young Scientist Award, 1987; INSA Young Scientist Medal, 1989; Indian National Remote Sensing Award, 1999; Hari Om Ashram Prerit Dr. Vikram Sarabhai Research Award, 1999; ISRO-Astronautical Society of India Award, 2005; ISRO Merit Award, 2006; and Corresponding Member, International Academy of Astronautics, 2010. He acted as president of ISRS and ISPRS Technical Commission VIII. He has been a project director of National Carbon Project under ISRO Geosphere Biosphere Programme, India. Currently he is editor of *Journal of Indian Society of the Remote Sensing*. He has more than 300 publications in national and international peer-reviewed journals.

Aniruddha Deshmukh is currently working as senior scientific assistant in Geoinformatics Department of the Indian Institute of Remote Sensing, ISRO, Dehradun. He has done MSc in geoinformatics from Savitribai Phule Pune University. He has a professional experience of around 8 years in the field of remote sensing and GIS. His research focus is on RS and GIS applications including advance techniques, urban and regional studies, land use planning, etc. He has six research papers on his name published in peer-reviewed journals, conference proceeding and technical reports.

Pankaj Ramji Dhote is presently working as Scientist/ Engineer 'SD' in Water Resources Department of the Indian Institute of Remote Sensing, Indian Space Research Organisation, Dehradun, India. He has done his MTech in the field of water resources engineering and management from the National Institute of Technology, Surathkal. His research interests are hydrological and hydrodynamic modelling and groundwater modelling. He has participated in many national- and organisation-level research and consultancy projects related to the flood management and conjunctive use of surface water and groundwater. Apart from research activities, he is also involved in education and capacity building activities of the Indian Institute of Remote Sensing (IIRS) and Centre for Space Science and Technology Education in Asia and the Pacific (affiliated to United Nations). Mr. Dhote has around ten publications to his credit including two publications in peer-reviewed journals.

K. N. Ganeshaiah from the University of Agricultural Sciences, Bangalore, worked in the area of evolutionary ecology and plants and insects, on biodiversity mapping and conservation. In the recent past, his interest lied in documenting and mapping biological resources of the country and on developing models for conservation of bioresources. He has published about 240 papers and has written and/or edited 12 books. A series of CDs on databases of the Indian Bioresources was developed by his group. He is a fellow of Indian Academy of Sciences, Indian National Science Academy, National Association for Agricultural Scientists and Current Science Association; honorary senior fellow of Jawaharlal

Nehru Centre for Advanced Scientific Research, Bengaluru, and of Ashoka Trust for Research in Ecology and the Environment, Bengaluru; and adjunct fellow of NIAS Bengaluru. He has been awarded Parisara Prashasthi from the Government of Karnataka; Fulbright fellow, USA; International Radio Hope Award; and Sahitya Academy Datti Award.

Vaibhav Garg is presently working as Scientist at Water Resources Department, Indian Institute of Remote Sensing (IIRS), ISRO, Dehradun, India. His field of interests are large-scale hydrological modelling and application of geospatial technology in water resources problems. He did his doctoral research at the Department of Civil Engineering, IIT, Bombay, India. He had completed his master's in water resources engineering from the Department of Civil Engineering, Malaviya National Institute of Technology Jaipur, India. He has also worked with the National Institute of Hydrology, Roorkee; IIRS, Dehradun; and IIT Bombay, Mumbai, with research fellow position. He has more than 8 years of professional experience in the field of water resources. Till date, he has published 27 refereed journal publications dealing with water resources problems. He is a life member of International Association of Hydrological Sciences, Indian Society of Hydraulics, Indian Society of Remote Sensing and Indian Meteorological Society.

Surajit Ghosh has over 6 years of experience in the field of remote sensing and GIS. In his current job at IORA Ecological Solutions Pvt. Ltd, New Delhi, he is associated with developing forest monitoring and evaluation tools. Prior to joining IORA, he has had experience of working with APRIL Asia, Indonesia, where he has used LiDAR data for forest resource assessment and hydrological modelling. He also had a brief association with International Water Management Institute, Sri Lanka, where his focus area of work was mapping of flood inundation extent in Southeast Asia. During his association with the Indian Institute of Remote Sensing, Dehradun, he worked in different applications area of remote sensing and GIS. Currently, he is pursuing his PhD in engineering from National Institute of Durgapur. He published more than ten peer-reviewed articles in scientific journals.

Kshama Gupta is currently working as Scientist in the Indian Institute of Remote Sensing (Indian Space Research Organisation), Dehradun. She had completed her MTech in urban planning from School of Planning and Architecture, New Delhi, India, after the completion of bachelor's in architecture from Malaviya National Institute of Technology, Jaipur. Since then she is working as researcher in the field of remote sensing and GIS applications for urban management and made contributions in many national level projects and research areas. She has more than 60 publications to her name as research papers in international/national journals and conferences and ISRO technical reports. Her research interest includes smart planning, urban climate and microclimate, urban green spaces and 3D modelling of urban areas.

Stutee Gupta holds a master's in botany from HNB Garhwal University, Srinagar (Garhwal), and PhD in forest informatics from FRI Deemed University, Dehradun. She works in the area of remote sensing application in natural resources and community development. She joined IIRS as a Scientist in Forestry and Ecology Department on 13 September 2012. She is currently working at the Rural Development and Watershed Monitoring Division, National Remote Sensing Centre, Hyderabad. Before joining ISRO, she has also worked with IIFM, Bhopal, and RMSI Pvt. Ltd, Hyderabad. She has a vast experience of more than 15 years during which she contributed in several national and international projects. Her area of interest includes landscape characterisation, biodiversity conservation, ecosystem services, community forestry and rural development. She has authored several publications in peer-reviewed national and international journals and conferences.

Sadhana Jain is presently working as Scientist at Regional Remote Sensing Centre – Central (NRSC) since December 2015. Prior to this, she served as scientist at Urban and Regional Studies Department, Indian Institute of Remote Sensing, Dehradun, during March 1998 to November 2015. She received her PhD degree from IIT Roorkee in 2006. She completed her master's in urban development planning (1997) and Bachelor of Architecture (1995) from Maulana Azad National Institute of Technology (formerly MACT), Bhopal. She is one among the pioneers in the applications of high-resolution satellite data in urban management. She has contributed greatly to the understanding of development patterns of informal settlements in a city. Her expertise in the digital image processing and geographic information system is very well recognised among remote sensing and urban planning fraternity.

R. K. Jauhari joined Janta (postgraduate) College, Bakewar (Etawah), as lecturer in zoology after completing PhD degree in zoology at the Department of Zoology, Banaras Hindu University, Varanasi, and carried out independent research work from February 1980 to September 1982, besides teaching UG and PG classes. For the last 35 years, he is engaged at the Department of Zoology, DAV (PG) College, Dehradun. From the last 6 years, he is actively engaged at the Department as Head. Till date, 25 candidates are awarded PhD under his guidance. He is also acting as principal investigator of research projects in the field of parasitology/mosquito ecology/remote sensing. He has 152 research papers to his credit, published in the journals of national and international repute.

Ashutosh Kumar Jha is scientist in Geoinformatics Department, IIRS. His area of interest has been high-performance spatio-temporal modelling, volunteer GIS and geospatial modelling. He has been involved in the development of ISRO Land Use/Land Cover (ILULC) Modelling software for land use/land cover modelling for large-area simulation and weather forecast system modelling on HPC systems. He has been architect of Smart Nager mobile application for volunteer-supported Clean India Mission. He is currently working on the BigGIS data application. Before joining IIRS, he worked for the British Telecom. He has experience in developing business intelligence solution. He has an MTech in remote sensing, BIT (Mesra), Ranchi, India, and BE in computer science and engineering from VTU, Belgaum, India. He has received the Innovation Award [Asian Association on Remote Sensing (AARS) Foundation]. He has authored different papers on Land Use/Land Cover Modelling, a volunteer GIS application for Swachh Bharat Mission.

Suresh Kannaujiya has been working as a Scientist in Geosciences and Disaster Management Studies Group, Indian Institute of Remote Sensing (ISRO), Dehradun, for the last 4 years. After receiving the master's degree in applied geophysics, he is pursuing his PhD in crustal deformation from IIT Dhanbad. He was exposed to real-time challenges in the field of marine geophysics while being associated with Fugro India Pvt. Ltd. as a processing geophysicist for 5 months. Mr. Kannaujiya has authored many research articles, published in national and international journals/periodicals in the field of landslide modelling, groundwater depletion, TEC modelling, crustal deformation, hazard assessment and integration of remote sensing and geophysics for landslide demarcation. Till date, he has supervised 16 MTech/MSc/PG diploma students. His current interest lies in modelling of GNSS data for total electron content/strain/crustal deformation and active fault mapping through geophysical survey.

Harish Chandra Karnatak is currently working as a Scientist and Head of Geoweb Services, IT & Distance Learning Department at the Indian Institute of Remote Sensing (IIRS), ISRO, Dehradun, India. He received his PhD degree in computer science with specialisation in geoinformatics. He has made significant contributions in various national-level projects and outreach programmes of ISRO on applications of space technology for natural resources and disaster management in India. His area of specialisation includes web/Internet-based GIS, spatial DBMS, online geoprocessing and analysis. He has published more than 65 peer-reviewed research papers in various scientific journals, seminars and conferences. He is the recipient of Indian National Award 2016 by ISRS, two national awards for excellence in training by DOPT and UNDP in 2015, ISRO Team Excellence Award in 2009 and ISRO-ASI Team Achievement Award 2009.

Pramod Kumar is Head of Urban and Regional Studies Department, IIRS, Dehradun, India. He is an alumnus of IIT, Kharagpur, India, and joined Indian Space Research Organisation in 1991. Earlier, he has worked as assistant engineer at CES, New Delhi. He has been involved in more than 50 mission/technology demonstration and research projects using geospatial data and techniques to evolve solutions for natural resources management and brought out technical reports and research publications. He has published more than 40 papers in journals and conference proceedings and many technical reports. He is the recipient of ISRO Team Excellence Awards for two projects. At present, he has research interests in urban hydrology and urban water utilities.

Rakesh Kumar who is a researcher in forestry and ecological domain, has done his postgraduation in environment management from FRI University and MTech degree in remote sensing and GIS technology from Andhra University. He has worked with some of the reputed research institutes of India based in Dehradun, viz. Forest Research Institute, Indira Gandhi National Forest Academy and Indian Institute of Remote Sensing, and contributed in scientific projects related to natural resources monitoring and management. His area of specialisations are applications of geospatial technology in forest resource monitoring and management, carbon flux monitoring using eddy covariance-based flux tower and forest and climate change especially REDD+.

Shashi Kumar received BSc (Hons.) degree in physics from Veer Kunwar Singh University, Arrah, India, in 2002, and MSc degree in physics from Patna University, Patna, India. He completed the MSc geoinformatics course under the joint education programme of the Indian Institute of Remote Sensing (IIRS), Dehradun, India, and ITC, Enschede, Netherlands. He is currently a Scientist in the IIRS, ISRO, Dehradun, India. He has worked as a member of SAR Task Group to develop SAR protocols for forest carbon inventory of Indian forest under the USAID Forest-PLUS Technical Assistance Programme. He is actively involved in NASA-ISRO Synthetic Aperture Radar-related activities. His research interests include SAR remote sensing with special emphasis on polarimetric SAR, polarimetric SAR interferometry and SAR tomography for structural and biophysical characterisation of man-made and natural features.

Shashi Gaurav Kumar is a nature enthusiast with over 3 years of research and corporate experience in geospatial technology in natural resource management and earth science. He holds Master of Remote Sensing and GIS with specialisation in geosciences from the Indian Institute of Remote Sensing, ISRO, and a bachelor's degree in geoscience engineering from the University of Petroleum and Energy Studies. He has research experience in earthquake and geodynamics and has published research papers in peer-reviewed journals. Presently, he is associated with Quantum Asia Pvt. Ltd., Jaipur, in the capacity of Tech Lead-GIS. He has contributed significantly in various projects on mapping and monitoring of natural resources during his association with RMSI Pvt. Ltd., Noida, and commissionerate of Watershed and Soil Conservation, Jaipur.

Suresh Kumar is working as Scientist SG and Head of Agriculture and Soils Department at the Indian Institute of Remote Sensing (IIRS), Government of India, Indian Space Research Organisation (ISRO). He is Bachelor of Science in agriculture, Master of Science in soil science and Doctor of Philosophy in soil science from GB Pant University of Agriculture and Technology, Pantnagar, India. He has vast experience in applications of remote sensing and GIS in soil resource survey, land evaluation, soil carbon assessment, modelling soil erosion processes and watershed management. He did commendable research and published research papers in various international journals (15 nos.) and in national journals (27 nos.). He had carried out several national projects such as National Land Degradation Mapping, Wasteland Mapping, Integrated Mission for Sustainable Development, National Soil Carbon Project, Climate Change Impact on Soil Quality and Land Degradation in Northwest Himalaya.

S. P. S. Kushwaha works at Forest Research Institute, Dehradun, after 35-year career at National Remote Sensing Centre, Hyderabad, and Indian Institute of Remote Sensing, Dehradun (under ISRO). He has PhD in ecology from North-Eastern Hill University, Shillong, and diploma in forest remote sensing and postdoctoral research experience from Albert Ludwigs University at Freiburg, Germany. His research interests are forest resources inventory, ecosystem analysis, species-habitat modelling, biodiversity conservation, microwave and LiDAR sensing, carbon flux modelling and sustainable development planning. He has been involved in 25 projects on remote sensing and GIS applications and has 65 publications in international and 50 in national journals. He has served as member of IUFRO, Vienna Working Group 4.2.2 on Multipurpose Inventories. He is fellow of Alexander von Humboldt Foundation, Germany, and National Academy of Sciences, India.

Sandeep Maithani is working as a Scientist at the Indian Institute of Remote Sensing (Indian Space Research Organisation) since 1996. He holds a bachelor's in civil engineering from National Institute of Technology (NIT) Allahabad in 1992, master's in urban and rural planning from IIT Roorkee in 1995 and PhD in urban growth modelling from IIT Roorkee in 2008. His research work focuses on spatial urban growth modelling, urban risk vulnerability analysis and application of night-time data in urban and regional planning. He has nearly 30 publications to his credit in journals, conferences and book chapters.

Ritwik Mondal is a prominent researcher who worked on CSIR-sponsored project on surveillance of *Aedes* species based on dengue in Dehradun district. He submitted his PhD in HNB Garhwal University which is almost at the brim of award. He published nine research papers. In three academic meets [8th USSTC, held in Doon University, Dehradun, on 26–28 December 2013; in national symposium on 'Perspectives on research in science and health care' held on 29–30 January 2016, at SBS (PG) Institute, Balawala, Dehradun; and national seminar on 'Environmental health vis-à-vis human welfare in present scenario' held on 18–19 April 2016, at DAV (PG) College, Dehradun], he was conferred with Young Scientist Awards. Nowadays, he is engaged as assistant professor in zoology at North Bengal University, West Bengal.

Indranil Nandi is an MTech (Master of Technology) student at Marine and Atmospheric Sciences Department, Indian Institute of Remote Sensing (IIRS), Government of India, Indian Space Research Organisation (ISRO). He received Bachelor of Science in chemistry and Master of Science in marine science from the University of Calcutta. He has actively participated in research conferences/seminars/workshops/symposia of national level in India.

Subrata Nandy is currently working as Scientist/Engineer-SE in Forestry and Ecology Department of the Indian Institute of Remote Sensing, ISRO, Dehradun. He received MSc and MPhil in Ecology from Assam University, Silchar, Assam. He obtained his PhD in Forest Geoinformatics from the Forest Research Institute University, Dehradun. His research expertise includes forest biomass/carbon and productivity assessment, and LiDAR remote sensing in forestry. He has more than 15 years of research experience in the applications of remote sensing and GIS in forestry and ecology. He worked significantly in the synergistic use of passive optical and LiDAR data for forest biomass/carbon assessment using various techniques. He also contributed immensely in National level projects, ISRO's earth observation application mission and technology development projects. He has published 33 research papers in peer-reviewed journals.

Somalin Nath is a research scholar in Geosciences and Disaster Management Studies Group, Indian Institute of Remote Sensing, ISRO, Dehradun. She is pursuing PhD on 'Crustal deformation study in Uttarakhand and Himachal Himalaya' from IIT (ISM) Dhanbad. She holds MTech from ISM Dhanbad and MSc from Sambalpur University. She is gold medallist in MSc and First rank holder in BSc.

Bhaskar Ramchandra Nikam is presently working as Scientist/Engineer 'SE' at Water Resources Department of the Indian Institute of Remote Sensing, Indian Space Research Organisation, Dehradun, India. He has done his PhD in the field of water resources management from the Indian Institute of Technology Roorkee, Roorkee. His research interests are in the field of retrieval of hydrological parameters using remote sensing, irrigation water management, hydrological modelling, climate change studies, etc. He has participated in many national- and organisation-level research and operational projects related to the applications of remote sensing to real-world problems of water sector. Apart from research activities, he is also involved in education and capacity building activities of the Indian Institute of Remote Sensing (IIRS) and Centre for Space Science and Technology Education in Asia and the Pacific. Dr. Nikam has around 80 publications to his credit including 26 publications in peer-reviewed journals and 1 book chapter.

Kapil Oberai is working as Scientist/Engineer 'SE' at the Indian Institute of Remote Sensing (IIRS), Indian Space Research Organisation (ISRO), Dehradun, India. He holds master's (MTech) degree in software engineering from Kurukshetra University, India. He joined IIRS/ISRO in year 2008 as Scientist/Engineer 'SC'. Prior to joining IIRS/ISRO, he was working with American Express as programmer analyst. His main research interest includes web technologies, WebGIS, location-based services and spatial database. He received the Best Paper Award in Map India 2010 Conference (13 annual international conference and exhibition on geospatial information technology and applications) held during January 2010 at Gurgaon, India. He has worked in national and in-house research projects and has over ten scientific publications in journals and conferences.

Hitendra Padalia is Scientist/Engineer 'SF' at Forestry and Ecology Department of the Indian Institute of Remote Sensing, ISRO, Dehradun. He received MSc in forest economics and management and PhD degree in forestry from Forest Research Institute (FRI) University, Dehradun. His research interests are advanced sensors (hyperspectral, microwave remote sensing, fluorescence) and modelling applications in forest and ecological studies. He has contributed to several national operational (Natural Resources Census, SIDDP) and research projects (DOS-DBT Biodiversity Project, ISRO-GBP-National Carbon Project, ISRO-GBP-LULC Dynamics, NNRMS-Mapping NPs and WLS in India). He has 32 research publications in peer-reviewed journals.

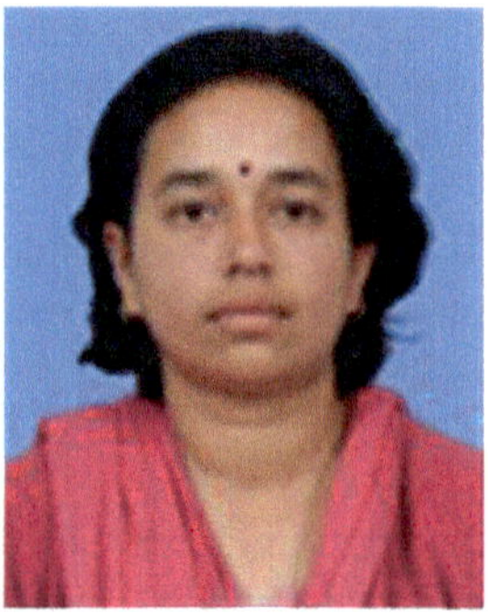

Hina Pande is Scientist and teaching faculty at the Indian Institute of Remote Sensing, ISRO, Dehradun. Her research interest and area of expertise is in the field of high-resolution image analysis for automated feature extraction and 3D modelling. She has over 15 years of teaching and research experience in these domains. She also has about 50 publications in leading journals and conferences. She has a PhD in earth science from IIT Roorkee, India.

N. R. Patel a senior Scientist at the Indian Institute of Remote Sensing, ISRO, has an experience of 20 years on space applications in the field of agriculture. He obtained his master's degree in agronomy and PhD in agrometeorology from Gujarat Agricultural University, Anand (Gujarat). He has received recognition in the field of agriculture and agrometeorology in India and abroad. He has been honoured with Dr. Vikram Sarabhai Research Award for his contribution in the field of space application to agrometeorology. He served as council member/secretary of professional scientific societies (Association of Agrometeorologist, ISPRS WG on agroecosystem and biodiversity, fellow of Earth Science Foundation). He has more than 100 publications with 70 published in peer-reviewed international and national journals. Recently significant contribution is made towards monitoring carbon fluxes over terrestrial ecosystems in India and assessing the vulnerability of agriculture in northwestern Himalaya to climate change.

N. Pemola Devi is currently engaged as assistant professor in the Department of Zoology, DBS (PG) College, Dehradun, since 2008. She availed DPhil degree from HNB Garhwal University. As a research scholar, she had availed JRF, SRF and RA fellowships from various agencies like DST, CSIR and ISRO. She also awarded Women Scientist Fellowship (WOS-B) by DST, New Delhi. She also conferred Young Scientist Award by Zoological Society of India at 14th All India Congress of Zoology 2003 at Kanyakumari. Also conferred two more young scientist awards on USSC during 2006–2007. She has published her research work in 48 papers in national and international reputed journals and books.

A. Ponraj pursued his master's degree in remote sensing and GIS, with specialisation in sustainable agriculture from the Indian Institute of Remote Sensing (IIRS). He worked on 'Climate change and food security' as a junior research fellow at IIRS. Currently, he is working as a research associate at International Maize Wheat Improvement Centre (CIMMYT), New Delhi, and working with climate-smart agriculture and crop insurance schemes in India for Climate Change Agriculture and Food Security (CCAFS) Programme. He has published one research paper in a peer-reviewed journal.

Pooja Rathore is currently a PhD student at the Forestry and Ecology Department, Indian Institute of Remote Sensing, Indian Space Research Organisation, Dehradun, India. She received a master's degree in environmental sciences at Rajasthan University in 2011. She enrolled into her current PhD programme after working for the Arid Forest Research Institute, India. Her current research focuses on the vulnerability assessment of Western Himalayan temperate and alpine species in terms of their spatial variability under projected climate change scenarios.

Arijit Roy is presently a Scientist at the Indian Institute of Remote Sensing, Indian Space Research Organisation, India, and has a PhD in botany with specialisation in ecosystem process and modelling from Banaras Hindu University. He has been working in the fields of geospatial modelling of the impact on the terrestrial ecosystems mainly the structure and functioning (biodiversity, nutrient dynamics) as a result of the climate forcing and anthropogenic influences. His research interests include biodiversity characterisation and spatial landscape modelling, remote sensing and GIS in environment and ecology, terrestrial ecology and ecosystems, network and corridors in ecology and landscape, wildlife habitat and modelling including migration modelling and climate change impacts on ecosystems. He has more than 30 peer-reviewed papers in various international journals and more than 80 publications and reports. Presently he is working in the broad area of climate change impacts on ecosystem structure and functioning.

S. K. Saha presently, is Distinguished Professor of geoinformatics at the University of Petroleum and Energy Studies (UPES), Dehradun, India, and former Dean and Group Director of Earth Resources and System Studies Group, Indian Institute of Remote Sensing (IIRS), Indian Space Research Organisation (ISRO), Dehradun, India. He has more than 30 years of national and 20 years of international postgraduate education and training and research experiences in the field of remote sensing and GIS technology and applications in natural resources and environmental inventory, monitoring and management. He has published about 100 research papers in peer-reviewed international and national journals dealing with geospatial technologies and applications, in addition to other technical publications. He was awarded with the national prestigious award called 'Prof. Satish Dhawan' Award for lifetime research contributions in RS and GIS technology, applications and capacity building in Asia-Pacific region by the Indian Society of Remote Sensing (ISRS).

Raghavendra Sara is working as Scientist in the Indian Institute of Remote Sensing, ISRO, Dehradun, since July 2009. He holds an MTech degree (civil engineering) from the Indian Institute of Technology Kanpur, India, and a bachelor's degree in civil engineering from Osmania University, Hyderabad. His research interests are LiDAR remote sensing, UAV remote sensing and close-range photogrammetry.

Sameer Saran is presently working as Scientist 'SF' and Head of Geoinformatics Department, Indian Institute of Remote Sensing (IIRS), ISRO, Dehradun. He is also course director of Joint Education Programme (JEP) between IIRS and ITC, Netherlands, co-chair of International Society for Photogrammetry and Remote Sensing (ISPRS) Working Group V/3 on Citizen Science, secretary of Indian Society of Remote Sensing and national coordinator of Indian Bioresource Information Network (IBIN) national project. His research interests are distributed GIS, 3D city models and OGC CityGML, citizen science, spatial databases, geospatial Modelling and open-source GIS. He has executed various national projects in conservation and governance. He has published over 40 research papers in peer-reviewed journals and around 50 in proceedings, chapters and technical reports. He is the recipient of Indian National Geospatial Award from ISRS and Team Excellence Award from ISRO and ASI. He has MSc in physics and PhD in geoinformatics.

A. Senthil Kumar received his PhD from the Indian Institute of Science, Bangalore, in the field of image processing in 1990. He joined ISRO in 1991 and has been serving in Indian satellite programmes in various capacities. His research includes sensor characterisation, radiometric data processing, image restoration, data fusion and soft computing. He is currently the Director of the Indian Institute of Remote Sensing, Dehradun, and also the Director of UN-affiliated Centre for Space Science and Technology Education in Asia and the Pacific. He is the President of ISPRS Technical Commission V on Education and Outreach and Chair of CEOS Working Group on Capacity Building and Data Democracy. He has published about 120 technical papers in international journals and conferences, besides technical reports. He is a recipient of ISRO Team Awards for Chandrayaan-1 mission and Prof. Satish Dhawan Award conferred by the Indian Society of Remote Sensing. He is also the associate editor of *Journal of the Indian Society of Remote Sensing*.

Hari Shankar is presently working as Scientist/Engineer 'SD' in Geoinformatics Department at the Indian Institute of Remote Sensing (IIRS), ISRO, Dehradun, India. He had joined IIRS in year 2010 after completing MSc in physics from MJP Rohilkhand University, Bareilly, Uttar Pradesh, India. During his master's degree, he secured gold medal for standing first and honours in the university. His primary area of research interest is transportation GIS and geostatistics. He has published seven papers in peer-reviewed journals and six in conferences/proceedings. He was the Best Paper Award winner at National Symposium on Geomatics for Digital India at JK Lakshmipat University, Jaipur, in 2015.

Gopal Sharma is a Scientist at North Eastern Space Application Centre. He is pursuing his PhD from the Indian Institute of Technology (Indian School of Mines), Dhanbad. His research interest includes various earthquake precursors for understanding of lithosphere-atmosphere-ionosphere coupling, crustal deformation and active tectonics. During his research, he has developed vast experience in GNSS monitoring, surveying and data processing. He is specialised in GNSS total electron content estimation for earthquake precursors, deformation analysis and strain accumulation measurements.

K. Shiva Reddy is faculty in Geoinformatics Department of IIRS. He holds MTech in geomatics and BE in IT from IIT Roorkee and Government Engineering College Bilaspur (CG), respectively. His active research interest is in application of spatial and spatio-temporal data mining in Health GIS. Currently, he is pursuing PhD in the field of geographic data mining from IIT Roorkee. He is an active member of GitHub and contributes in the field of geospatial technologies. His notable open-source contributions are (1) Trivim, a street view mapping software, and (2) SaTSviz, a plugin for QGIS v 1.8 for SatScan analysis.

Asfa Siddiqui is currently working as a Scientist at the Indian Institute of Remote Sensing (Indian Space Research Organisation) since 2014. She holds a bachelor's in architecture from Government College of Architecture, Lucknow, in 2011, and a master's in planning with specialisation in urban planning from School of Planning and Architecture, New Delhi, in 2013. She worked at NIT Kozhikode (Calicut) prior to joining ISRO. She is a double gold medallist and has received government scholarships and awards at college level. Her work focuses on urban and regional areas with emphasis on urban energy and environment. She has 12 publications to her credit in journals, conferences and book chapter.

Charu Singh is a Scientist with Marine and Atmospheric Sciences Department, Indian Institute of Remote Sensing, ISRO, Dehradun. She has been associated with ISRO for the past 10 years, and prior to joining IIRS, she was posted in Space Applications Centre, ISRO, Ahmedabad. She holds postgraduate degree in physics from IIT Roorkee and MSc (Engineering) by research degree in atmospheric sciences from IISc Bangalore. Presently, she is pursuing PhD from IIT Delhi. Her research area includes South Asian monsoon system, extreme rainfall events, aerosol-cloud-precipitation interlink and retrieval of geophysical parameters from remotely sensed data. She has been a recipient of several scholarships/fellowships such as UP state-level scholarship, national scholarship, Indian Academy of Science fellowship and CSIR fellowship for JRF. She has 12 publications in peer-reviewed international journals.

Jyoti Singh received MTech (remote sensing and GIS) from the Indian Institute of Remote Sensing, Dehradun, in 2015. She served as a research associate in Karnataka Knowledge Commission and is currently pursuing PhD from the Centre for Atmospheric Sciences, Indian Institute of Technology, Delhi. The topic of her PhD is 'Impact of climate change and groundwater depletion over India'. Her research interests include agroclimatic suitability, groundwater depletion, food security, remote sensing and data analysis and GIS. She is a lifetime member of *Journal of Agrometeorology*. She has published one paper in a peer-reviewed journal.

Priyanka Singh is working as a senior research fellow in Indian Bioresource Information Network (IBIN) National Project of Geoinformatics Department, Indian Institute of Remote Sensing and also registered for the part-time PhD programme at the Department of Mining Engineering, Indian Institute of Technology (Indian School of Mines), Dhanbad, on the topic 'An interoperability framework for bioresource distributed database'. She has over 3 years of research experience in the field of WebGIS and information technology such as state-of-the-art technology (natural language processing), cloud computing, web application development, semantic web, mobile computing and spatial and non-spatial database management. She has developed various applications such as IBIN mobile app, 3D modelling of a building using terrestrial laser scanner and hydrometeorological information system of flood management, a project in collaboration of the Government of Bihar and World Bank.

Sarnam Singh received PhD from the University of Calcutta, Kolkata, in 1991 and MSc in botany from Kanpur University, Kanpur, in 1977. He was Scientist/ Engineer – G, Dean (Academics), Group Director of Earth Resources & System Studies Group and Head of Forestry and Ecology Department of the Indian Institute of Remote Sensing (IIRS), ISRO, Dehradun. His research expertise includes plant systematics, taxonomy and floristics; application of geospatial technologies for forest management; biodiversity characterisation for conservation prioritisation; vegetation carbon pool assessment; carbon sequestration and productivity; wildlife habitat evaluation; forest monitoring; hyperspectral remote sensing; and digital photogrammetry for forestry. He was also director-in-charge of CSSTEAP (affiliated to United Nations). He has more than 100 publications in journals and proceedings and authored/co-authored more than 50 books/book chapters. He has 48 new plant species to his credit. Two plant species are named after him. He has extensively explored the entire Indian Himalaya to study biodiversity from tropical to alpine ecosystems.

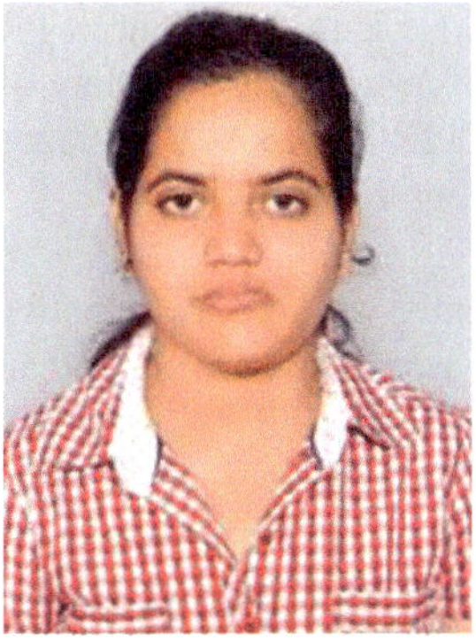

Ritika Srinet is currently pursuing PhD in forestry with specialisation in forest geoinformatics from Forest Research Institute (FRI) University, Dehradun. She has completed her MSc in environment management from FRI University, Dehradun, and postgraduate diploma in remote sensing and GIS with specialisation in forest resources and ecosystem analysis from the Indian Institute of Remote Sensing (IIRS). She received gold medal in her master's. Her area of interest is remote sensing applications in forestry and biomass and productivity assessment.

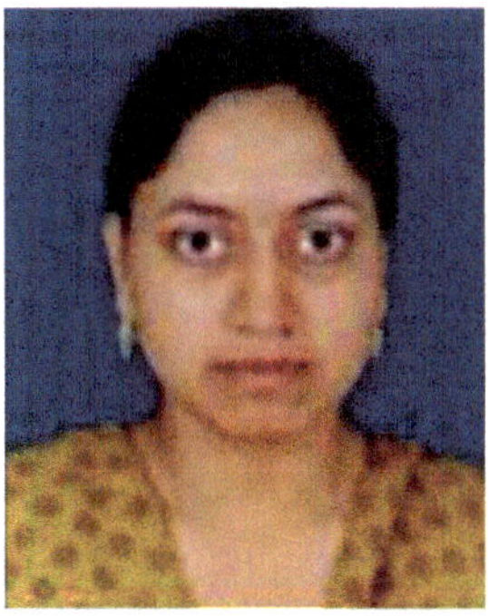

Shuchita Srivastava is Scientist/Engineer 'SE' at Marine and Atmospheric Sciences Department of the Indian Institute of Remote Sensing, ISRO, Dehradun. She received her PhD degree in atmospheric science from Physical Research Laboratory Dehradun. She did commendable research on chemical and dynamical processes affecting the distribution of ozone and its precursors over the Indian subcontinent. She has established a trace gas laboratory at IIRS Dehradun. She is principal investigator of ISRO GBP ATCTM Project. She has published 11 research articles in peer-reviewed journals.

S. K. Srivastav is currently working as Scientist 'G' and Group Director of Geospatial Technology and Outreach Programme (GT&OP) Group at the Indian Institute of Remote Sensing (IIRS), ISRO, Dehradun. He is also the lead of Working Group on Capacity Building of ISRO; deputy project director (training) of ASEAN-India Space Cooperation Programme; associate programme director of Programme Steering Group on Capacity Building of ISRO; co-chair of International Society for Photogrammetry and Remote Sensing (ISPRS) Working Group V/6 on Distance Learning. His research interests are in the fields of geologic remote sensing, groundwater hydrology and land use/land cover change analysis and modelling. He has over 55 scientific publications in journals and proceedings and many technical reports to his credit. He is the recipient of PR Pisharoty Memorial Award from the Indian Society of Remote Sensing and four ISRO Team Excellence Awards. He holds doctorate degree in geology.

Praveen K. Thakur is currently working as Scientist/Engineer 'SF' at Water Resources Department, Indian Institute of Remote Sensing (IIRS), ISRO, Dehradun. He holds BTech in civil engineering from NIT, Hamirpur (2001); MTech in water resources engineering from IIT, Delhi (2002); and PhD from geomatics engineering in IIT Roorkee (2012). He joined IIRS Dehradun in 2004. He has more than 14 years of experience in the usage of remote sensing and GIS in water resources and hydrology. He has published more than 40 research papers in peer-reviewed international/national journals and more than 50 papers/extended abstracts in conferences/symposiums. His current research interests are geospatial technology application in water resources, snow, glacier, flood and groundwater hydrology, hydrological modelling and data assimilation. He also has specialisation in microwave remote sensing for hydrological studies. He is life member of six professional societies. He received ISRO Young Scientist Merit Award in 2014.

Sanjay Kr. Uniyal is a Principal Scientist at the CSIR-Institute of Himalayan Bioresource Technology, Palampur (HP), India. He has more than 25 years of field research experience in the Himalaya, wherein he is involved in research on plant ecology, biodiversity conservation, environmental monitoring, resource use patterns of local people and the development of plant databases. Dr. Uniyal is a member of the National Academy of Sciences, India, Medicinal Plants Specialist Group of the IUCN, International Association for Ecology, International Society for Tropical Ecology and International Society of Ethnobiologists. He has published more than 60 research articles in journals of international repute. He also has three books and six book chapters to his credit.

Prabhakar Alok Verma is presently working as Scientist/Engineer 'SC' in Geoinformatics Department at the Indian Institute of Remote Sensing (IIRS), ISRO, Dehradun, India. He had joined IIRS in year 2013 after completing BTech in physical sciences from the Indian Institute of Space Science and Technology, Trivandrum, India. Till date, he has done research in the field of geostatistics and road transportation network using open-source tools and techniques. Also he has worked in the nation project of ISRO Geosphere Biosphere Programme. He has published three papers in peer-reviewed journals and three in conferences/proceedings and prepared some technical reports.

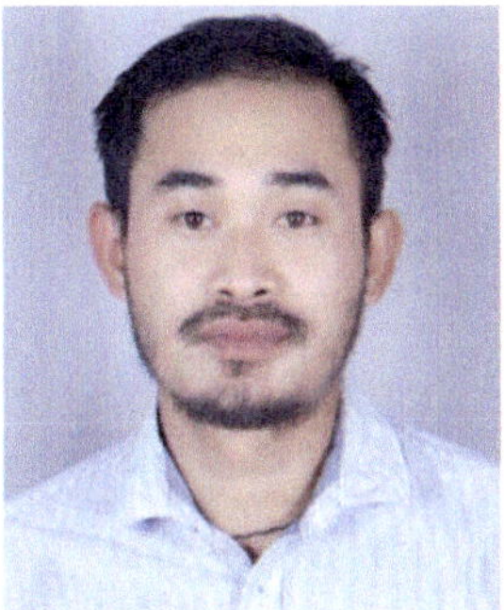

Taibanganba Watham is presently working as a research associate in Wildlife Institute of India, Dehradun. His research interest lies in the field of ecology and environment and understanding evaluation and quantification of ecosystem and its resources using modern techniques. He received PhD in forest ecology and environment from FRI University, Dehradun. During his PhD, he has worked on unifying framework of diverse study domains such as forest inventory, instrument measurement, RS and GIS, ecosystem modelling and ecology. He is experienced on operation and maintenance of various instruments ranging from normal wind anemometer to 3D sonic anemometer, open path Infrared gas analyser and many more slow sensors for micrometeorological observation. He is currently working on e-flow study for conservation of Black-necked crane (*Grus nigricollis*) at Namjang Chu Hydroelectric Power proposed site. He has published nine research papers in peer-reviewed journals.

Yesobu Yarragunta is working as SRF (senior research fellow) in Marine and Atmospheric Sciences Department at Indian Institute of Remote Sensing (IIRS), Government of India, Indian Space Research Organisation (ISRO). He received Master of Science in physics, Master of Technology in remote sensing from Andhra University, Visakhapatnam, India, and pursuing Doctor of Philosophy in physics from Kumaun University, Nainital, India. He has 5 years' experience in air quality modelling and atmospheric science. He has published one research paper in *International Journal*. He has actively participated in more than three research conferences/seminars/workshops/symposia of international/national level in India.

Glossary

AAR	Accumulation Area Ratio
AGB	Aboveground Biomass
AGC	Aboveground Carbon
AGNPS	Agricultural Non-point Source Pollution
AGW	Atmospheric Gravity Waves
AGWB	Aboveground Woody Biomass
AHP	Analytical Hierarchy Process
AI	Artificial Intelligence
AIRS	Atmospheric Infrared Sounder
ALOS	Advanced Land Observing Satellite
AMSR-E	Advanced Microwave Scanning Radiometer on the Earth Observation System
ANN	Artificial Neural Network
ANSWERS	Areal Nonpoint Source Watershed Environment Response Simulation
AOI	Area of Interest
APEX	Agricultural Policy Environmental eXtender
API	Application Programming Interface
AQI	Air Quality Index
ARCTAS	Arctic Research of the Composition of Troposphere from Aircraft and Satellites
ASCAT	Advanced SCATterometer
ASCII	American Standard Code for Information Interchange
ATLAS	Advanced Topographic Laser Altimeter Systems
AWS	Automatic Weather Station
BCLL	Biodiversity Characterisation at Landscape Level
BGB	Belowground Biomass
BIRZ	Bare Ice Radar Zone
BIS	Biodiversity Information System
BMLR	Binomial Multiple Logistic Regression
BR	Biological Richness

BRICs	Bioresource Information Centres
BUE	Built Urban Environment
CA	Cellular Automata
CART	Classification And Regression Tree
CASA	Carnegie-Ames-Stanford Approach
CCI	Climate Change Initiative
CCSM	Community Climate System Model
CERES	Clouds and the Earth's Radiant Energy System
CoK	Co-kriging
COP	Conference of the Parties
CPA	Canopy Projection Area
CRP	Close-Range Photogrammetry
CTS	Coordinate Transformation Service
CWC	Central Water Commission
DBH	Diameter at Breast Height
DEM	Digital Elevation Model
DGVM	Dynamic Global Vegetation Model
DInSAR	Differential Interferometric Synthetic Aperture Radar
DMSP	Defence Meteorological Satellite Programme
DNI	Direct Normal Irradiance
DRR	Direct Radiometric Relation
DSSAT	Decision Support System for Agrotechnology Transfer
DTM	Digital Terrain Model
DVI	Difference Vegetation Index
EC	Eddy Covariance
ECU	Effective Chill Unit
EDGAR	Emission Database for Global Atmospheric Research
ELA	Equilibrium Line Altitude
EML	Ecological Metadata Language
EO	Earth Observation
EOL	Encyclopaedia of Life
EPIC	Erosion Productivity Impact Calculator
ERP	Environmental Resource Potential
ERS	European Remote Sensing
ESA	European Space Agency
ESRI	Environmental Systems Research Institute
ET	Evapotranspiration
ETM	Enhanced Thematic Mapper
EUROSEM	EUROpean Soil Erosion Model
EVI	Enhanced Vegetation Index
FAO	Food and Agriculture Organization
FCC	False Colour Composite
FCD	Forest Canopy Density
FFT	Fast Fourier Transform

*f*PAR	Fraction of Absorbed Photosynthetically Active Radiation
GBH	Girth at Breast Height
GBP	Geosphere-Biosphere Programme
GCM	Global Circulation Model
GCP	Ground Control Point
GDAL	Geospatial Data Abstraction Library
GEPIC	GIS-Based Environmental Policy-Integrated Climate
GFS	Global Forecast System
GHG	Greenhouse Gas
GHI	Global Horizontal Irradiance
GIDS	Gradient plus Inverse Distance Squared
GIS	Geographic Information System
GLAS	Geoscience Laser Altimeter System
GLM	Generalised Linear Model
GLOF	Glacial Lake Outburst Flood
GMB	Glacier Mass Balance
GML	Geography Markup Language
GNSS	Global Navigation Satellite System
GOES	Geostationary Operational Environmental Satellite
GPM	Global Precipitation Measurement
GPP	Gross Primary Productivity
GPR	Ground Penetrating Radar
GPS	Global Positioning System
GPZI	Global Permafrost Zonation Index
GrADS	Grid Analysis and Display System
GRASS	Geographic Resources Analysis Support System
GUI	Graphical User Interface
GVMI	Global Vegetation Moisture Index
HadGEM	Hadley Global Environment Model
HAND	Height Above Nearest Drainage
HEC-HMS	Hydrologic Engineer Centre-Hydrologic Modelling System
HFT	Himalayan Frontal Thrust
HSI	Habitat Suitability Index
HMS	Hydrological Modelling System
HP	Himachal Pradesh
HTAP	Hemispheric Transport of Air Pollution
HTML	Hypertext Markup Language
IBIN	Indian Bioresource Information Network
ICESat	Ice, Cloud and Land Elevation Satellite
ICT	Information and Communication Technology
IDW	Inverse Distance Weighted
IGBP	ISRO Geosphere and Biosphere Programme
IIRS	Indian Institute of Remote Sensing
ILWIS	Integrated Land and Water Information System

IMD	Indian Meteorological Department
IMMS	Integrated Mountain Monitoring System
IMSRA	INSAT Multispectral Rainfall Algorithm
INCCA	Indian National Climate Change Action
InSAR	Interferometric Synthetic Aperture Radar
INTACH	Indian National Trust for Art and Cultural Heritage
IPCC	Intergovernmental Panel on Climate Change
IRS	Indian Remote Sensing
ISM	Indian Summer Monsoon
ISRO	Indian Space Research Organisation
IUCN	International Union for Conservation of Nature
IWMP	Integrated Watershed Management Programme
J&K	Jammu and Kashmir
JAXA	Japan Aerospace Exploration Agency
KED	Kriging with External Drift
KINEROS	Kinematic Runoff and Erosion Model
k-NN	k-Nearest Neighbour
LAI	Leaf Area Index
LCCS	Land Cover Classification System
LiDAR	Light Detection And Ranging
LIS	Lightning Imaging Sensor
LISS	Linear Imaging Self Scanner
LITE	LiDAR In-Space Technology Experiment
LSWI	Land Surface Water Index
LUE	Light Use Efficiency
LULC	Land Use/Land Cover
MAGST	Mean Annual Ground Surface Temperature
MANU	Map the Neighbourhood in Uttarakhand
MBT	Main Boundary Thrust
MCARI	Modified Chlorophyll Absorption Ratio Index
MCE	Multi-criteria Evaluation
MCT	Main Central Thrust
MLR	Multiple Linear Regression
MMF	Morgan, Morgan and Finney
MMI	Modified Mercalli Intensity
MMS	Mobile Mapping Systems
MODIS	Moderate Resolution Imaging Spectroradiometer
MOSDAC	Meteorological and Oceanographic Satellite Data Archival Centre
MPS	Mean Patch Size
MRENDVI	Modified Red Edge Normalised Difference Vegetation Index
MSAVI	Modified Soil Adjusted Vegetation Index
MSI	Moisture Stress Index
MSSDIM	Modified Shi Snow Density Inversion Model
MUSLE	Modified Universal Soil Loss Equation

NAAQS	National Ambient Air Quality Standards
NAPCC	National Action Plan on Climate Change
NASA	National Aeronautics and Space Administration
NBDB	National Bioresource Development Board
NBSS&LUP	National Bureau of Soil Survey and Land Use Planning
NCC	Natural Colour Composite
NCEI	National Centres for Environmental Information
NCEP	National Centre for Environmental Prediction
NCP	National Carbon Project
NDSI	Normalised Difference Snow Index
NDVI	Normalised Difference Vegetation Index
NEE	Net Ecosystem Exchange
NGO	Non-governmental Organisation
NICES	National Information system for Climate and Environmental Studies
NMSHE	National Mission for Sustaining Himalayan Ecosystems
NN	Nearest Neighbour
NOAA	National Oceanic and Atmospheric Administration
NPP	Net Primary Productivity
NREL	National Renewable Energy Laboratory
NRSC	National Remote Sensing Centre
NSRDB	National Solar Radiation Database
NUE	Natural Urban Environment
NUIS	National Urban Information System
NWH	Northwestern Himalaya
NWP	Numerical Weather Prediction
OBIA	Object-Based Image Analysis
OFE	Overland Flow Element
OGC	Open Geospatial Consortium
OGR	OpenGIS Simple Features Reference Implementation
OLS	Operational Linescan System
OSAVI	Optimised Soil Adjusted Vegetation Index
OSM	Open Street Map
PAR	Photosynthetically Active Radiation
PFRZ	Percolation-Freeze Radar Zone
PLAND	Percentage of Landscape
PostgreSQL	Postgres Structured Query Language
PR	Precipitation Radar
PRECIS	Providing REgional Climates for Impacts Studies
PVI	Perpendicular Vegetation Index
RADAR	Radio Detection And Ranging
RAMMS	Rapid Mass Movements Software
RCP	Representative Concentration Pathway
RDBMS	Relational Database Management System
RDVI	Renormalised Difference Vegetation Index

RENDVI	Red Edge Normalised Difference Vegetation Index
RF	Random Forest
RFM	Rational Function Model
RI	Redevelopment Potential Index
RISAT	Radar Imaging Satellite
RPC	Rational Polynomial Coefficients
RS	Remote Sensing
RSPM	Respirable Suspended Particulate Matter
RTM	Radiative Transfer Model
RUSLE	Revised Universal Soil Loss Equation
RVI	Radar Vegetation Index
SAC	Space Applications Centre
SAGA	System for Automated Geoscientific Analysis
SAR	Synthetic Aperture Radar
SAVI	Soil-Adjusted Vegetation Index
SBA	Swachh Bharat Abhiyan
SCA	Snow Cover Area
SCE	Snow Cover Extent
SDM	Species Distribution Model
SEC	Solar Energy Centre
SHP	Small Hydropower Plant
SIR-C	Spaceborne Imaging Radar-C
SLA	Shuttle Laser Altimeter
SLC	Single Look Complex
SLE	Snow Line Elevation
SLR	Simple Linear Regression
SMAP	Soil Moisture Active Passive
SMOS	Soil Moisture and Ocean Salinity
SNAP	Sentinel Application Platform
SOA	Service-Oriented Architecture
SOAP	Simple Object Access Protocol
SOS	Sensor Observation Service
SOx	Sulphur Oxides
SPLAM	Spatial Landscape Model
SPM	Suspended Particulate Matter
SPS	Sensor Planning Service
SQL	Structured Query Language
SRE	Surface Range Envelope
SRM	Snowmelt Runoff Model
SRTM	Shuttle Radar Topography Mission
SVATS	Soil-Vegetation-Atmosphere Transfer Schemes
SVM	Support Vector Machine
SWAT	Soil and Water Assessment Tool
SWE	Snow Water Equivalent

SWM	Stanford Watershed Model
TauDEM	Terrain Analysis Using Digital Elevation Model
TDWG	Taxonomic Databases Working Group
TEC	Total Electron Content
TIR	Thermal Infrared
TJS	Table Join Service
TLS	Terrestrial Laser Scanner
TM	Thematic Mapper
TMI	TRMM Microwave Imager
TML	Transducer Markup Language
TMPA	TRMM Multi-satellite Precipitation Analysis
TOD	Transit-Oriented Development
TRMM	Tropical Rainfall Measuring Mission
TSAVI	Transformed Soil-Adjusted Vegetation Index
TVI	Triangular Vegetation Index
TWI	Topographic Wetness Index
UA	Urban Agglomerations
UAV	Unmanned Aerial Vehicle
UI	Urgency Index
UK	Uttarakhand
UNEP	United Nations Environment Programme
UNFCCC	United Nations Framework Convention on Climate Change
USIS	Urban Spatial Information System
USLE	Universal Soil Loss Equation
VAR	Variable Rain Rate
VARI	Visible Atmospherically Resistant Index
VBD	Vector-Borne Diseases
VGI	Volunteered Geographic Information
VHRS	Very High-Resolution Satellite
VIC	Variable Infiltration Capacity
VIRS	Visible Infrared Scanner
VNIR	Visible Near Infrared
VPD	Vapour Pressure Deficit
VSA	Variable Source Area
VUE	Vulnerable Urban Environment
WCPS	Web Coverage Processing Service
WCS	Web Coverage Service
WDRVI	Wide Range Vegetation Index
WDRVI	Wide Dynamic Range Vegetation Index
WEPP	Water Erosion Prediction Project
WFS	Web Feature Service
WHO	World Health Organisation
WMO	World Meteorological Organisation
WMS	Web Map Service

WPS	Web Processing Service
WRF	Weather Research and Forecasting
WWF	World Wide Fund for Nature
XML	Extensible Markup Language

Part I
Ecosystems of the Northwest Himalaya – An Overview

Chapter 1
Northwest Himalayan Ecosystems: Issues, Challenges and Role of Geospatial Techniques

S. K. Saha and A. Senthil Kumar

1.1 Introduction

Ecosystems of the Northwestern Himalaya (NWH) are fragile and sensitive with respect to topography, geodynamics, geological hazards, soil and land degradation, biogeochemistry, biodiversity, water resources status (snow and glacial) and land use and land cover (LULC) including forest cover and human habitation. This region remained geodynamically active and produced three longest faults on earth surface: MCT (Main Central Thrust), MBT (Main Boundary Thrust) and HFT (Himalayan Frontal Thrust). Topographical diversity, geological complexity, active geodynamic processes, human interference and climatic impact made this region highly prone to various kinds of geological disasters such as earthquakes, landslides, flash flood, etc. Anthropogenic activities such as deforestation, faulty agricultural practices, biomass burning, etc. coupled with ruggedness of terrain contributed to high degree of soil erosion, depletion of soil nutrients and reduced crop and forest productivity of NWH ecosystem. This region is also most prone to ecological degradation because of perturbations in the biogeochemical cycling mainly carbon and nitrogen. Climate change and anthropogenic activities have also affected the water resources in the form of snow and glacial status and conditions. The urban development in the NWH region is a complex process as the human habitation is mainly controlled by natural environment. The unstable nature of terrain, along with heavy rain, soil erosion and mass wasting, constricts the physical distribution of the towns, and it further complicates the situation. This region has also witnessed unprecedented growth in terms of population and development particularly hydropower projects,

S. K. Saha (✉)
University of Petroleum and Energy Studies (UPES), Dehradun, India

A. Senthil Kumar
Indian Institute of Remote Sensing, Indian Space Research Organisation, Dehradun, Uttarakhand, India

R. R. Navalgund et al. (eds.), *Remote Sensing of Northwest Himalayan Ecosystems*,
https://doi.org/10.1007/978-981-13-2128-3_1

infrastructure and urban centres, thereby making it one of the most ecologically vulnerable regions of the country.

1.2 Issues and Challenges in Sustainable Management of Natural Resources and Ecosystem Processes in NWH and Role of Geospatial Technologies

1.2.1 Geodynamics and Seismicity Phenomena

Himalayan region is one of the most seismically active regions of the world due to continuous northward movement of the Indian plate at a rate of 40 mm per year (approx.) that has resulted in three major fault systems (MCT, MBT, HFT) and numerous other faults. Four earthquakes in the recent past, 6.8 Mw Uttarkashi earthquake in 1991, 6.6 Mw Chamoli earthquake in 1999, 7.6 Mw Kashmir earthquake in 2005 and most recently 6.9 Mw Sikkim earthquake in 2011, took place in the Himalayan region. Two mega events also occurred in the past: 1905 Kangra earthquake and 1934 Bihar-Nepal earthquake. High crustal deformation, high upliftment and high incision and erosion rates make it one of the most unstable regions of the world.

In spite of several attempts and ongoing research activities in the country initiated by the Government of India through the Department of Science and Technology and Ministry of Earth Sciences, and many other research organizations, there exist gaps with respect to paleo-seismicity, earthquake precursor studies, activity of faults and spatial variation of deformation in regional context and strain accumulation rates. For example, Quaternary-active fault map for Pakistan and Nepal is available. In Bhutan, active fault mapping is in progress. However, active fault map for Himalaya is yet not complete.

There had been attempts to define the mechanism of lithospheric and atmospheric-ionospheric coupling by ground-based radio-sounding techniques. These observations highlight the intricate relationship between pre-, co- and post-seismic deformations with the ionospheric total electron content (TEC) variations. However, there is a hiatus in explaining the governing physics that correlates seismicity and the TEC variability. How are the ground deformation-related transient signals being transmitted through the atmosphere affecting the electron density in the ionospheric region? Many models have been proposed like anomalous generation of pre-seismic vertical electrical field and its interaction with different ionospheric layers, modulation of electrical signals with seismo-acoustic-gravity waves, etc.

Surface deformation data provide the primary means for estimating inter-seismic strain accumulation that releases during large and damaging earthquakes. It is believed that the Earth's crust behaves intermediate between elastic and brittle materials. The strain changes in the crust observable via differential interferometric synthetic aperture radar (DInSAR) technique can be used to determine stress changes in the crust and can lead to improved earthquake forecasts. Recent developments in DInSAR technique and integrated approach with other geodetic techniques (e.g. GPS observations) have introduced new perspectives in surface deformation studies (Fruneau and Sarti 2000; Chatterjee et al. 2006). DInSAR data processing over a wide swath can hopefully monitor crustal deformation on a regional scale. Internationally, studies have been initiated on the monitoring of active plate boundaries (e.g. Hellenic seismic arc in Europe and San Andreas fault in N. America) using DInSAR and collateral geodetic techniques and ground-based measurements to aim at improved earthquake forecasts.

It is now widely accepted that active faults contribute significantly to the seismic activity; thus mapping and understanding the nature of active fault systems in different segments of Himalaya are of paramount importance. High spatial resolution satellite can be effectively used for studying and mapping active fault by identification of tectonic landforms, fault scarps, tectonic lineaments and secondary features of paleo-earthquakes. Limited studies have been carried out on this problem.

1.2.2 Climate Change and Forest Ecosystem Processes

Due to its unique position and physical features, the Himalayan mountain ranges act as a storehouse of valuable biodiversity. The young and fragile nature and sharp gradients, thus, make the Himalayan mountains vulnerable to climate change and variability. In addition to that, the rapidly growing population pressure is making the natural and socio-economic systems of these mountain regions risky. The rapid change of the ecosystems including the forest, driven by both natural and anthropogenic factors, poses threat to the livelihood of the local people, wildlife and culture and the billions living in the downstream and ultimately to the global environment.

The present efforts by a number of research groups, nongovernmental organizations (NGOs) and institutions of the countries occupying the Himalayan region have initiated work on the information generation and strategy formulation for sustainable resource management and development in the Himalayan region. The Department of Space, Government of India, has been working to generate database on the spatial distribution of the forest and has undertaken numerous work on the characterization of biodiversity-rich areas in the Himalayan region as part of the National-Level Biodiversity Characterization at Landscape level. The Western Himalayan region covering parts of Jammu and Kashmir, Himachal Pradesh and Uttarakhand has been mapped for vegetation type distribution. Using GPS-tagged ground sampling and ancillary data, the biologically rich areas and their extents have been identified. Furthermore, the LULC changes for three decades in the 14 river basins of India as a part of the Indian Space Research Organisation (ISRO) Geosphere and Biosphere Programme (GBP) addressed the human dimension of the impact of the climate change on the river basins, many of which originate from the Himalaya. Yet, there are gaps in information for understanding the forest ecosystems in the mountains such as quantification and estimation of biodiversity of the region, identification of the location and extent of the biodiversity hotspots in the Himalayan region, identification and preservation of the biological corridors in the region and estimation of climate change-induced shift in the vegetation and species loss.

Research studies reveal that integration of remote sensing (RS)-derived long-term LULC change and habitat change maps with other field level species data and socio-economic data is useful for the generation of various anthropogenic and natural pressure gradients to understand habitat change scenario. Based on critical habitat changes and species associated with these habitats, bioclimatic envelopes can be developed using ecological niche models to assess past and projected climate change scenario for selected endemics and invasive to predict potential loss of ecological niches (Thomas et al. 2004a, b). Time series RS data is very effective for the creation of spatial and temporal change database on tree line, while identification of critical and substantial areas of change can be carried out using databases on tree line shifts derived from coarse spatial resolution satellite data in conjunction with climatic and topographic data and climate envelope models.

1.2.3 Sustainable Mountain Agriculture

In the twentieth century, mountain regions have experienced above-average warming (IPCC 2001), which has significant implications for mountain environments and its processes. In the Himalaya, progressive warming at higher altitudes is three times greater than the global average (Eriksson et al. 2009). The second report of Indian Network on Climate Change Assessment (INCCA) reveals that mountains in NWH also experienced three times warming than the whole Indian subcontinent in the last 100 years (MoEF 2010). As per the predictions, temperature will likely increase in mountain areas in the twenty-first century. On the other hand, the number of cold days has been decreasing significantly over NWH. Mountains, in general, have witnessed climate changes, but the knowledge about the impacts of climate change on various sectors, particularly agriculture, is lacking. Agroecosystems of mountain are highly prone to deterioration by various forces of degradation such as water erosion, landslides and frequent occurrence of extreme events. The natural fragility of these ecosystems makes them highly susceptible to small changes in temperature and water availability. Crop growth and development processes are highly sensitive to changes in temperature and water availability, and as a result, the effect of climate change on agriculture in NWH has become a reality. There are evidences of shift in fruit tree belt, increased incidence of pests and diseases, decline in productivity of food and tree crops, etc. Diversity of agro-environments and cropping practices in mountain ecosystem also poses a key challenge to formulate holistic approach of addressing climate change issue.

Soil erosion is one of the major threats to agricultural productivity and environmental quality especially water and soil quality. The Himalayan region is affected largely by soil erosion and sedimentation. These are adversely affecting soil quality and crop yield in the region. Comprehensive field-scale and watershed-scale studies on soil erosion processes and nutrient loss and its impact on crop yield and soil quality of hilly farming system are limited. Understanding of process-based models and simulation of the models for soil erosion, nutrient loss and crop productivity are required for better understanding of land degradation processes and its impact on the ecosystem. It needs to be evaluated for alternative agronomic and management options in hilly farming system. In agricultural watersheds, soil quality assessment is deemed important to understand the long-term effects of conservation practices. These assessments can be used to determine the required soil and water conservation practices and evaluate land management effects or resilience towards natural and anthropogenic forces.

Modern tools such as RS and geographic information system (GIS) technology have shown enormous potential to provide spatial solutions to many problems of mountain agriculture. In the past, many successful applications of these technologies have been made to map and monitor natural resource base and subsequently characterization of agro-environments to improve sustainability of agriculture (Patel et al. 2005). RS and GIS technology has been matured enough to map crop areas, soils and terrain information, which are vital in delineating uniform zones having similar agroecological practices and production prospects. Crop models have been widely used to assess the impact of climate change. In the past few decades, crop models have been widely used to assess climate change impact on crop yields. Such models simulate crop growth and crop yield levels by using variables like daily weather parameters, soil characteristics, crop characteristics and cropping system management options. Climate change impact on crop productivity on a regional or national scale has been realized in the past for plains based on station-/district-level weather inputs and soil as well crop management options. Crop model simulation over mountain agroecosystem is often limited by large heterogeneity in soils/terrain and coarse-resolution climate drivers. Crop model simulation based on uniform agroecological zones is more promising for assessing climate change effects in mountain ecosystem.

The climate change impact on the mountain regions has already started surfacing (Partap and Partap 2002). Mountain ecosystem experiencing shifting of temperate fruit belt upward has adversely affected the productivity of food grains and apples, shifting and shortening of rabi season forward and disrupted rainfall pattern and more severe incidences of diseases and pests over crops. Rana et al. (2009) reported that the apple belt has shifted to higher villages due to warmer temperatures and decreasing chilling periods during November and March.

Crop model predictions at experimental farm scale will be misleading with respect to regional impact of climate change on crop productivity in mountainous region with large heterogeneity in topography, soils and crop management practices. Use of geospatial techniques in delineating agroclimatically uniform zones as simulation unit will improve our ability to assess climate change impact on crop's productivity in mountain agroecosystem. Sophisticated models for fruit and vegetable crops are not yet fully developed as well as tested for regional applications. Development of simple regression models based on historical data of productivity, weather, management and edaphic factors would have strong impact on realizing short-term climate change effects on fruit/vegetables productivity. There is definitely a lack of information on potential shift in production zones of important crops (food and fruit crops) and land suitability classes in NWH. As mentioned above, some studies have an indication of shift in apple cultivation areas in Himachal Pradesh. Information on shift in potential land suitability caused by alteration of agroclimatically potential productivity and length of growing period under changing climate will be a base for formulating adaptation strategy for cropping practices.

A large number of process-based models, e.g. the Water Erosion Prediction Project (WEPP), Agricultural Non-point Source Pollution (AGNPS) (Young et al. 1987), ANSWERS (Beasley et al. 1980), Erosion-Productivity Impact Calculator (EPIC) (Williams 1990), Soil and Water Assessment Tool (SWAT) (Arnold et al. 1998) and Agricultural Policy/Environmental eXtender (APEX) (Williams et al. 2008), are available to study surface runoff, erosion and nutrient loss at watershed scale. Models can provide long-term simulations of various combinations of cropping systems and conservation practices, effects of best management practices and aid in selection of suitable conservation approaches for improved environmental benefits. These models are coming with GIS interfaced, since GIS has emerged as a powerful tool in handling spatial datasets, so models interfaced with GIS. Soil quality can be assessed by determining appropriate indicators and relating them with desired values (critical limits or threshold level) at different time intervals. Such a monitoring system will provide information on the efficiency of the selected farming system, land use practices, technologies and policies.

1.2.4 Water Resources Status and Availability

The NWH has large area under seasonal and perennial snow cover. The seasonal variation in runoff is influenced by the changes in snow and glacier melt, as well as rising snowlines in the Himalayas, causing water shortages during dry summer months. The changes in climate and LULC in NWH will influence the hydrological regime of this area, thus affecting the water availability for drinking, irrigation and hydropower requirement in the region. The physical properties of snow such as snow wetness, which shows the degree of liquid water content in snowpack, snow water equivalent (SWE) and snow density are some of the most significant parameters for many water resources-related studies such as snowmelt runoff and snow avalanche modelling in this area. The traditional survey of these parameters is very expensive and difficult, especially in rugged NWH mountains. Therefore, mapping of snow- and glacier-covered areas using RS techniques is very important for hydropower, irrigation, associated hazard monitoring as well as drinking water needs of this region. A glacial lake is defined as sufficient mass of water blocked in glacial tongue's ablation zone depressions by the end/lateral moraines with the shifting, merging and draining characteristics. The main glacier should be large enough to feed the lake in wet as well as dry spells. The sudden breach in this moraine-dammed lake, known as glacial lake outburst flood (GLOF), can cause enormous destruction in downstream areas. The sudden discharge from the dam contains the huge amount of hustle water and high sediment load, which can endanger the safety of the hydropower projects in downstream. With the changing climate, mainly with increasing temperatures, the numbers of glacier lakes are on the rise in Himalayas. Therefore, GLOF should be modelled accurately so that its discharge is taken into consideration in planning and management of water resources projects.

The optical RS methods have been used successfully to map snow cover area (Dhanju 1983; Rees 2006; Singh and Singh 2001), qualitative snow wetness (Gupta et al. 2005) and snow grain size with hyperspectral data (Dozier and Painter 2004) and glacial mass balance (GMB) studies (Kulkarni 1992). But there is no universally accepted approach for snow cover mapping under dense forested area in SWE and mapping of debris-covered glaciers and crevasses using RS in NWH, except for few recently reported studies (Bhambri et al. 2011; Shukla et al. 2010; Shukla et al. 2009). Traditionally, temperature index model and ELA/AAR methods have been used for the snowmelt runoff and glacier mass balance (GMB) studies. This is a gap area in understanding and quantifying the runoff contribution coming from snow-pack, glacier ice and non-snow areas. Similarly, the roles of topography and thermodynamics variables are the gap areas in GMB and glacier movement studies. Scanty studies have been carried out following energy balance approach (Datt et al. 2008; Takeuchi et al. 2000; Tarboton and Luce 1996) along with other physical parameter-based techniques for estimating the runoff from various sources as well as GMB study, with maximum inputs from RS data and hydrometeorological field stations in NWH.

1.2.5 Temporal and Spatial Growth of NWH Cities

Urban development in NWH region is largely controlled by the natural environment. The unstable nature of terrain and various kinds of geological and hydrometeorological hazards create many problems in the physical distribution of the towns. This region has also witnessed unprecedented growth in terms of population and development particularly hydropower projects, infrastructure and urban centres. It is therefore necessary to understand the causes and dynamics of urban growth and provide models of urban growth to the planning bodies who can utilize it to forecast urban growth patterns and structure the policies in the short and long term to implement the intended plans.

Advances in RS, GIS and system theories are undoubtedly stimulating a new development wave of modelling. Complexity theory brings hopes for re-understanding the systems or phenomena under study. New mathematical methods create new means to represent and quantify the complexity. RS and GIS guarantee the availability of data on various spatial and temporal scales. Scanty research is reported on predictive modelling for spatial growth modelling of complex urban systems using new techniques like multi-criteria evaluation (MCE) and artificial neural network (ANN) and utilization of RS and GIS in such studies as spatial data providers and spatial data handlers, respectively.

1.2.6 Extreme Rainfall Events and Rainfall Retrieval in NWH

Amount of rainfall during monsoon and its spatial coverage is a deciding factor for the heavy erosion and flooding across the Himalayan range (Bookhagen and Burbank 2006; Anders et al. 2006). The relationship between extreme rainfall and Himalayan topography is also not well understood and still needs sincere efforts to address this issue in detail. Spatial patterns of precipitation are greatly affected by topography both at regional and global scales. Mountains act as a barrier by modifying the flow of air and influence the vertical stratification of the atmosphere.

Rain gauges remain the traditional method to determine rainfall at any location, and at present long-term analyses of precipitation pattern over Himalaya are derived from rain gauges. However, the sparse coverage of the rain gauges due to the remoteness of the Himalaya cannot provide essential information about the heavy rainfall events at a finer resolution. Rainfall is associated with large spatio-temporal variability, which offers a great deal of difficulty to retrieve it from satellite measurements (Gairola et al. 2003; Mishra et al. 2009). Rainfall retrieval may be carried out from the variety of methods like visible, infrared and microwave (Barret and Martin 1981; Ferraro et al. 1996). Visible and infrared techniques for rainfall retrieval have its own limitation, because these provide the rainfall estimation based on the cloud top information due to the incapability of visible and infrared radiation to penetrate the clouds. Sometimes it may lead to ambiguous results in terms of rainfall estimation. More recently, the focus has been turned towards the microwave measurements to estimate the rainfall from spaceborne sensors, due to the advantage of microwave frequencies to overcome the above-mentioned inability of visible and infrared radiation. Emission and scattering are the two approaches which are used to estimate rain rate from remotely sensed microwave data (Janowiak et al. 1995). Information from emitted radiation from atmospheric liquid hydrometeors is used to estimate the rain rate in emission-based techniques, whereas scattering technique is based on the measured extinction of microwave caused by the liquid particles or ice. Observed radiation in both the cases is sensitive to the surface emissivity, so emission-based technique is more appropriate to apply on oceanic region, because ocean surface has low microwave emissivity ~0.5. The land surface emissivity being close to unity complicates the signal from the liquid particle and adds difficulties in the rainfall retrieval. Moreover, due to complex terrain of Himalayan region, it is difficult to rely upon the rainfall estimation techniques specifically developed for land and oceanic regions. Nonetheless, over the past three decades, significant progress has been made in the rainfall retrieval techniques over land- and ocean-based on remotely sensed data from space. There is still a scope of considerable improvement to provide better estimation of rainfall specifically over hilly terrain.

Orographic precipitation and its process have been investigated using a variety of RS methods and rain gauges' network in some parts of the Himalayan region at 10–20 km resolution during monsoon season and storm events (Barros et al. 2000; Lang and Barros 2002). Spatio-temporal variability and diurnal cycle of rainfall over Himalayan region have been extensively studied using high-resolution Tropical Rainfall Measuring Mission (TRMM) Precipitation Radar (PR) data (Bhatt and Nakamura 2005). Anders et al. (2006) have examined the spatial patterns of rainfall and effect of topography over the Himalayan region using TRMM PR data for 4 years (1998–2001). They used a simple model of orographic precipitation and found out that the spatial pattern of precipitation is strongly correlated with topography. Since most of the studies have used polar orbiting satellite measurements like TRMM PR data, which actually provides the instantaneous rainfall measurements and cannot provide the continuous coverage of rain events, hence monthly or seasonal rainfall amounts estimated using polar satellite sensors are significantly lower than the true rainfall amounts. Therefore, an extensive calibration of remotely sensed precipitation is required. In addition to this, there is also a need to modify the merged rain product based on IR/MW combined observations, for example, TRMM 3B42 and IMSRA (INSAT Multispectral Rainfall Algorithm (Mishra et al. 2011)), with rain gauge estimates to provide continuous and accurate coverage of rainfall at finer resolution for a remote place like NWH, as the accurate precipitation measurements with a higher spatial resolution are of utmost importance for landslides.

1.3 Indian Research Initiative on Monitoring and Assessment of Ecosystem Processes in NWH Using Geospatial Technologies

Indian Institute of Remote Sensing (IIRS) is in an advantageous position to take up research study on these problems as it is located in the foothill of NWH and has developed excellent in-house expertise of using RS and GIS technology in natural resources inventory and management and also established networking with research institutes in NWH for collaborative research.

Therefore, for sustainable environmental development, making disaster-resilient society and improved livelihood in the NWH region, it is envisaged in an interdisciplinary research programme to study the various aspects of ecosystem processes and services in the NWH using recent advances of earth observation techniques (focusing on ISRO missions) and allied spatial technologies supported by extensive field investigation and field instrumentation in several subthemes. The subthemes are:

- *Geology and geodynamics*
- *Water resources*
- *Forest resources and biodiversity*

- *Mountain agriculture*
- *Urban environment*

Several research organizations/institutes located in NWH participated in the studies. Some of the institutions involved are Wadia Institute of Himalayan Geology, Dehradun; G.B. Pant National Institute of Himalayan Environment and Sustainable Development, Almora; Forest Departments, Uttarakhand and Himachal Pradesh; C.S.K. Himachal Pradesh (HP) Agricultural University, Palampur; Institute of Biotechnology and Environmental Science, Hamirpur, HP; National Institute of Hydrology, Roorkee; Snow and Avalanche Study Establishment, Chandigarh; and Centre for Atmospheric and Oceanic Sciences, Indian Institute of Science, Bangalore.

Details of the work carried out in different themes of this research programme have been summarized in various chapters of this volume.

References

Anders, A.M., Roe, G.H., Hallet, B., Montgomery, D.R., Finnegan, N.J., Adiku, S. G. K., Reichstein, M., Lohila, A., Dinh, N. Q., Aurela, M., Laurila, T., Lueers, J., and Tenhunen, J. D. (2006). PIXGRO: A model for simulating the ecosystem CO2 exchange and growth of spring barley. *Ecol. Model.*, 190(3–4):260–276.

Arnold JG, Srinivasan R, Muttiah RS, Williams JR (1998) Large area hydrologic modeling and assessment: Part I. Model development. Journal of the American Water Resources Association 34(1): 73–89.

Barret E.C., Martin, D.W. (1981) The use of Satellite Data in Rainfall Monitoring. Academic Press, pp. 350.

Barros, A.P., Joshi, M., Putkonen, J., and Burbank, D.W., (2000) A study of the 1999 monsoon rainfall in a mountainous region in central Nepal using TRMM products and rain gauge observations: Geophysical Research Letters, 27, 3683–3686, doi: https://doi.org/10.1029/2000GL011827.

Beasley DB, Huggins LF, Monke A (1980). ANSWERS: A model for watershed planning. *Transactions of the ASAE, 23*(4), 938–0944.

Bhambri, R., Bolch T. and Chaujar R.K. (2011). Mapping of debris-covered glaciers in the Garhwal Himalayas using ASTER DEMs and thermal data. *International Journal of Remote Sensing,* 32:23, 8095–8119.

Bhatt B. C., and K. Nakamura, (2005) Characteristics of Monsoon Rainfall around the Himalayas Revealed by TRMM Precipitation Radar, Monthly Weather Review, 133, 149–165.

Bookhagen, B. and Burbank, D.W. (2006) Topography, relief, and TRMM derived rainfall variations among the Himalaya. Geo. Res. Lett., 33, L08405, doi: https://doi.org/10.1029/2006GL026037,2006.

Chatterjee, R.S., Fruneau, B., Rudant, J.P., Roy, P.S., Frison, P.L., Lakhera, R.C., Dadhwal, V.K., and Saha, R., (2006) Subsidence of Kolkata (Calcutta) City, India during the 1990s as observed from space by differential synthetic aperture radar interferometry (D-InSAR) technique, Remote Sensing of Environment, 102–206: 176–185.

Datt P., Srivastava P.K., Negi P.S., Satyawali P.K., (2008). Surface energy balance of seasonal snow cover for snow melt estimation in N-W Himalaya. *Journal of Earth Systems Sciences,* 117, No.5, pp. 567–573.

Dhanju, M. S. (1983). Studies of Himalayan snow cover area from satellites. Hydrological Applications of Remote Sensing and Remote Data Transmission, *Proceedings of the Hamburg Symposium*, IAHS (145), 401–409.

Dozier, J. and Painter H.T. (2004). Multispectral and hyperspectral remote sensing of alpine snow properties, *Annual Reviews of Earth Planet. Science,* 32, 465–94.

Eriksson, M, Jianchu, X., Shrestha, AB; Vaidya, R.A., Nepal, S., and Sandstroem, K (2009) *The changing Himalayas: Impact of climate change on water resources and livelihoods in the greater Himalayas.* Kathmandu: ICIMOD

Ferraro, R. R., Fuzhong Weng, Norman C. Grody and Alan Basist (1996) An eight –year (1987–1994) time series of rainfall, clouds, water vapor, snow cover and sea ice derived from SSM/I measurements, Bulletin of the American Meteorological Society, pp. 891–905, 1996

Fruneau, B., and Sarti, F., (2000) Detection of ground subsidence in the city of Paris using radar interferometry: isolation from atmospheric artefacts using correlation, Geophysical Research Letters, 27(24): 3981–3984.

Gairola R. M., A. K. Verma and Vijay K. Agarwal (2003) Rainfall estimation using spaceborne microwave radar and radiometric measurements, Mausam, pp. 89–106

Gupta, R. P., Haritashya, U. K., & Singh, P. (2005). Mapping dry/wet snow cover in the Indian Himalayas using IRS multispectral imagery. *Remote Sensing of Environment, 97*(4), 458–469.

IPCC, 2001. Climate Change (2001) The Scientific Basis. Third Assessment Report of the Intergovernmental Panel on Climate Change. Cambridge University Press, Cambridge, United Kingdom.

Janowiak, J.E., Arkin, P.A., Xie, P., Morrey, M.L., Ligates, D.R. (1995) An examination of the east pacific ITCZ rainfall distribution. J. Climate 8, 2810–2823, 1995.

Kulkarni, A.V. (1992). Mass Balance of Himalayan Glaciers Using AAR and ELA Methods. J Glaciology, 38(128), 101–104.

Lang, T.J., and Barros, A.P. (2002) An investigation of the onsets of the 1999 and 2000 monsoons in central Nepal: Monthly Weather Review, v. 130, p. 1299–1316, doi: https://doi.org/10.1175/1520-0493(2002).

Mishra, A., Gairola, R.M., Verma, A.K., Abhijeet Sarkar, Vijay K. Agarwal (2009) Rainfall Retrieval over Indian land and oceanic regions from SSM/I microwave data, Advances in Space Research, 44(7), 815–823.

Mishra, Anoop, R. M. Gairola, Vijay K. Agarwal, (2011) Rainfall Estimation from Combined Observations Using KALPANA-IR and TRMM- Precipitation Radar Measurements over Indian Region, Journal of the Indian Society of Remote Sensing (http://rd.springer.com/article/10.1007/s12524-011-0128-9)

MoEF (2010). Climate Change and India: A 4x4 Assessment. INCCA Report 2, Ministry of Environment and Forests, Government of India.

Partap, Uma and Tej Partap (2002) Warning *Signals from Hindu Kush Himalayas, Productivity Concerns and Pollination Problems,* International Centre for Integrated Mountain Development, Kathmandu, Nepal.

Patel, N. R.; Endang, P.; Suresh Kumar and Pande, L.M. (2005). Agro-ecological zoning using remote sensing and GIS – A case study in part of Kumaon region. In: Sustainable agriculture development, (Eds) B. Bandopadyay, KV Sundaram, M. Moni and M. Zha (Eds), Northen Book Depo, New Delhi. pp. 265–280.

Rana, R. S., Bhagar, R.M., Kalia, V and Lal, Harbans (2009). Impact Of Climate Change On Shift Of Apple Belt In Himachal Pradesh. ISPRS Archives XXXVIII-8/W3 Workshop Proceedings: Impact of Climate Change on Agriculture, pp.131–136.

Rees, W.G. (2006) *Remote Sensing of Snow and Ice*. Taylor and Francis/CRC Press Inc.

Shukla, A., Arora, M.K and Gupta, R.P. (2010). Synergistic approach for mapping debris-covered glaciers using optical-thermal remote sensing data with inputs from geomorphometric parameters. Remote Sensing of Environment, 114, 1378–1387.

Shukla, A., Gupta, R.P. and Arora, M.K. (2009). Estimation of debris cover and its temporal variation using optical satellite sensor data: a case study in Chenab basin, Himalaya. *Journal of Glaciology*, 55, 444–452.

Singh, P. and V.P. Singh (2001). *Snow and Glacier hydrology*. Kluwer Academic Publishers, Dordrecht.

Takeuchi, Y., Kayastha, R.B. and Nakawo, M. (2000). Characteristics of ablation and heat balance in debris-free and debris-covered areas on Khumbu Glacier, Nepal Himalayas, in the pre-monsoon season, *In: Debris-covered glaciers, A. Fountain*, Eds; IAHS: Wallingsford, 264. 53–61.

Tarboton DG, Luce CH (1996) Utah Energy Balance Snow Accumulation and Melt Model (UEB).

Thomas CD, Cameron A, Green RE, Bakkenes M, Beaumont LJ, Collingham YC, Erasmus BF, De Siqueira MF, Grainger A, Hannah L, Hughes L (2004a) Extinction risk from climate change. *Nature*, 427(6970), 145.

Thomas, C.D. et al. (2004b) Extinction risk from climate change. *Nature* 427: 145–148

Williams JR (1990) The erosion-productivity impact calculator (EPIC) model: a case history. Philosophical Transactions of the Royal Society of London B: Biological Sciences 329 (1255): 421–428.

Williams JR, Izaurralde RC, Steglich EM (2008) Agricultural policy/environmental eXtender model: Theoretical documentation version 0604 (draft). BREC report # 2008–17. Texas AgriLIFE Research, Texas A&M University, Temple, TX.

Young RA, Onstad CA, Bosch DD, Anderson WP (1987) AGNPS, Agricultural Non-Point-Source Pollution Model. A Watershed Analysis Tool. US Dept. of Agr. Conservation Research Report 35.

Part II
Geology and Geodynamics

Summary

Ever since the closing and subduction of the Tethyan Ocean, located between India and Asia during the Paleozoic, collision of Asia and Indian landmass has produced the highest and most complex mountain range of the world, the Himalaya. The Tethyan Himalaya is separated by South Tibetan Detachment (STD) from the Higher Himalaya, which is separated by Main Central Thrust (MCT) from the Lesser Himalaya, which in turn is separated by Main Boundary Thrust (MBT) from the Outer or Sub-Himalaya which is finally separated from the Indo-Gangetic Plain (IGP) by most recent Himalayan Frontal Thrust (HFT). Due to immense crustal shortening and faulting, the youngest mountain chain of the world is marked by numerous earthquakes and is prone to other natural disasters such as landslides, snow avalanches and flash floods causing devastation across NWH.

Space observations in conjunction with in-situ geophysical measurements have been providing crucial data to understand geology and geodynamics of the region. While multispectral data in visible, near infrared and thermal regions provides information on the landscape, geomorphological, structural features and lithology, differential interferometric SAR data facilitates the study of surface deformation. In addition, ionospheric disturbances due to impending earthquakes are inferred using total electron content data derived from global navigation satellites. IIRS has been working extensively on many topics that can provide a better understanding on various aspects of Himalayan geology and natural hazards. Some of the studies carried out are preparation of a detailed geological map of Garhwal and Kumaon Himalaya using aerial photo interpretation, highway alignment in Nepal, assessment of many hydro-electric project sites, ground water targeting, landslide mapping, monitoring and modeling, seismic hazard assessment, seismic-induced landslide modelling, and active fault mapping, etc.

Some of the recent studies carried out at the Institute relate to the morpho-tectonic analysis of the Himalayan frontal region of NWH in the light of geomorphic signatures of active tectonics, simulation outputs of major debris flows in Garhwal Himalaya-A geotechnical modelling approach for hazard mitigation and TEC modelling for earthquake precursor studies in western Himalaya using GNSS Data. They have been presented here.

Chapter 2
Morphotectonic Analysis of the Himalayan Frontal Region of Northwest Himalaya in the Light of Geomorphic Signatures of Active Tectonics

R. S. Chatterjee, Somalin Nath, and Shashi Gaurav Kumar

2.1 Introduction

Among the principal thrust belts in the Himalaya such as Main Central Thrust (MCT), Main Boundary Thrust (MBT), and Himalayan Frontal Thrust (HFT), the HFT represents a zone of active deformation between the sub-Himalaya and Indo-Gangetic alluvial plain. The active deformation along the HFT causes tectonic tilting of the terrain and development of different types of tectonic landforms, subtle topographic breaks, and drainage anomalies. The landform development process in the Himalaya is a result of mutual interaction between climate and tectonics (Molnar 2003; Starkel 2003; Srivastava and Misra 2008; Kothyari et al. 2010). In fact, simultaneously operating tectonic and physical processes results in the present-day topography of the terrain (England and Molnar 1990; Bishop 2007). Various tectonic landforms such as fault scarps, stream terraces, back-tilted terraces, relict geomorphic surfaces, alluvial fan offsets, topographic breaks in piedmont-alluvial plain, drainage anomalies, and drainage diversions were observed in and around the HFT (Nakata 1972; Ruhe 1975; Thakur and Pandey 2004; Singh and Tandon 2008; Tandon and Singh 2014). A few-meter-high NW-SE trending scarp of discontinuous nature is observed in the piedmont-alluvial region in front of the HFT at several localities between Pinjore Dun and Dehra Dun (Thakur 2004). Besides, a number of archaeological evidences in the foothill regions of Uttarakhand (Piran Kaliyar in Roorkee district), Uttar Pradesh (Khajnawar and Bargaon in Saharanpur district), and Haryana (Bhirrana and Balu in Fatehabad district) raise the possibility of active tectonic events to cause such extinction and burial. Several factors such as topography, rock types, geological structures, soil, and vegetation cover essentially govern

R. S. Chatterjee (✉) · S. Nath · S. G. Kumar
Geosciences and Disaster Management Studies Group, Indian Institute of Remote Sensing, Indian Space Research Organisation, Department of Space, Government of India, Dehradun, India
e-mail: rschatterjee@iirs.gov.in

R. R. Navalgund et al. (eds.), *Remote Sensing of Northwest Himalayan Ecosystems*,
https://doi.org/10.1007/978-981-13-2128-3_2

the development of a drainage system including channel morphology and drainage pattern. Primarily, active tectonics manifest itself by either steepening or reducing the local valley gradient which in turn changes the existing slope of the channels and introduces disturbance in the natural equilibrium of a drainage system. In the process of restoring the equilibrium, the river tries to adjust to the new conditions by changing its slope, cross-sectional shape, and meandering pattern (Vijith and Satheesh 2006; Pérez-Peña et al. 2010). In general, the phenomena like river incision, asymmetry of the catchment, and river diversion are accelerated by the tectonic processes (e.g., Cox 1994; Jackson et al. 1998; Clark et al. 2004; Salvany 2004; Schoenbohm et al. 2004). Several geomorphic parameters and indices describing tectonic tilting of the catchment, changes in the geometry and slope of the longitudinal and cross profiles of the rivers, anomalous hypsometric curve with high hypsometric integral value, anomalously low (< 1.0) valley floor width to valley ratio, and anomalous mountain front sinuosity index (close to 1.0) can be used to evaluate present-day tectonic activity on drainage basin scale (Bull and McFadden 1977; Rockwell et al. 1985; Keller and Gurrola 2000; Azor et al. 2002; Silva et al. 2003; Molin et al. 2004; Bull 2007; Malik and Mohanty 2007; Ata 2008; Pérez-Peña et al. 2010; Giaconia et al. 2012; Mahmood and Gloaguen 2012).

For evaluating relative tectonic activity in and around the HFT, we carried out drainage basin morphometric analysis at five test sites spread over the region. We studied various geomorphic indices of active tectonics such as drainage basin asymmetry factors, hypsometric integral, valley floor width to valley height ratio, and mountain front sinuosity index. By combining the geomorphic indices, we categorized fifth-order sub-basins into three activity classes such as high, moderate, and low (Bull and McFadden 1977; Rockwell et al. 1985; Silva et al. 2003; Ata 2008; Pérez-Peña et al. 2010; Giaconia et al. 2012; Mahmood and Gloaguen 2012). Subsequently, for identification of the possible active tectonic locations, we used characteristic tectonic landforms, subtle topographic breaks in piedmont region, and drainage anomalies as geomorphic signatures. Drainage anomalies such as rectilinearity of stream segments, anomalous curves or turns (e.g., acute and obtuse angle elbow turns, U-turns), compressed meandering, anomalous pinching and flaring, and abrupt and localized braiding can be used to identify the locations of structural and tectonic features such as faults, lineaments, folds, and warps. Besides, drainage migration, stream deflection, shifting of distributary bifurcation zone, divergence of existing stream, and emergence of new streams can be used to identify the possible locations of active tectonic features. Based on the interpretation of geomorphic anomalies and related structural or tectonic features (e.g., faults, folds, and lineaments), relevant information on post-collision tectonics can be obtained, which bears immense significance to explain the seismicity in the area. In piedmont-alluvial region, to confirm the presence of active tectonic features as inferred from the geomorphic anomalies, it is indeed necessary to unravel the subsurface profile at such locations. This can be accomplished directly by trenching and indirectly by noninvasive geophysical techniques such as ground-penetrating radar (GPR). In the present study, GPR survey was conducted at selective locations to confirm the occurrence of active tectonic features in the subsurface profile.

2.2 Study Area

The study area encompasses the HFT region in parts of NW Himalaya which extends from 75°46′ E to 79°04′ E longitudes and 30°22′ N to 33°12′ N latitudes in Himachal Pradesh state and 77°34′ E to 81°02′ E longitudes and 28°43′ N to 31°27′ N latitudes in Uttarakhand state. The study area is surrounded by Nepal in the east, China in the north, Jammu-Kashmir in the northwest, Haryana and Punjab in the west, and UP in the south. The study area covers the sub-Himalaya tectonic zone within the Kumaun Himalaya (Valdiya 1980) and piedmont-alluvial plain in and around the HFT. For evaluating relative tectonic activity using geomorphic indices of active tectonics, we selected five major sixth-order drainage basins such as Solani, Markanda, Budki Nadi, Khoh, and Dhela basins located along the HFT as test sites (Fig. 2.1).

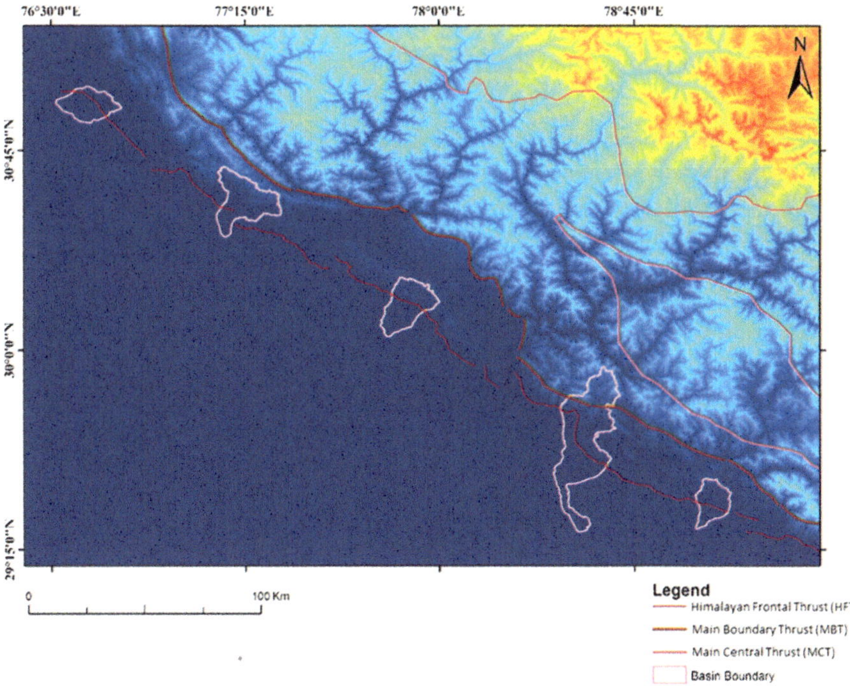

Fig. 2.1 Five major sixth-order drainage basins in and around Himalayan Frontal Thrust (HFT) in parts of NW Himalaya as test sites for assessing relative tectonic activity (SRTM 90 m DEM is used as background): (**a**) Budki Nadi (Rupnagar) basin, (**b**) Markanda basin, (**c**) Solani basin, (**d**) Khoh basin, and (**e**) Dhela basin

2.3 Data Used and Survey Methods

2.3.1 *Satellite Remote Sensing Data*

In the study, we used medium-resolution Resourcesat LISS-III multispectral images with a spatial resolution of 23.5 m acquired during 2007–2010. The LISS-III false color composites were mosaicked to generate a unified FCC image for the entire study area. We also used Cartosat-1 PAN ortho images with a spatial resolution of 2.5 m acquired over the five test sites (in and around the abovementioned sixth-order drainage basins) during 2013 for identification of active tectonic features. We used medium-resolution Landsat TM, ETM+, and OLI multispectral images with a spatial resolution of 30 m acquired during 1985, 1995, 2005, and 2015 for studying dynamic changes in the drainage system possibly due to the presence of active tectonic features.

2.3.2 *Digital Elevation Model (DEM)*

We used Shuttle Radar Topographic Mission (SRTM) 1 arc second (approximately 30 m spatial resolution at equator) DEM available in 1 degree × 1 degree tiles. The DEM is available in geographic (lat/long) projection system with the WGS84 horizontal datum and the EGM96 vertical datum. The vertical error of the DEM is reported to be less than 16 m. We also generated high-resolution DEM for the five test sites with 10 m × 10 m pixels and relative vertical accuracy of 3–5 m with respect to differential GPS-based ground control points (GCPs) from Cartosat-1 optical stereoscopic image pairs (with spatial resolution of 2.5 m) of 2013. We used the DEMs for drainage basin morphometric analysis to evaluate relative tectonic activity in the five major sixth-order drainage basins. Both SRTM 30 m and high-resolution Cartosat-1 DEMs were used for detection of subtle topographic breaks in and around the test sites essentially in the piedmont-alluvial zone those are potentially indicating subsurface active faults or warps.

2.3.3 *Ground-Penetrating Radar (GPR) Survey*

GPR survey is a high-resolution geophysical scanning method that allows investigation of the shallow subsurface based on the dielectric properties of the layers. Ideally, GPR provides high-resolution images of the subsurface over a depth range of few meters to several 10s of meters with a vertical resolution of few 10s of centimeters to a meter (Basson 2000; Knight 2001). However, the quality of the data and the depth of penetration strongly depend on the dielectric properties of the material and the frequency range of the antennae (Davis and Annan 1989). It may be noted that the best results are generally obtained for stratified, clay-free, dry sand or gravel (Smith and Jol 1995).

In the present study, we used IDS GeoRadar GPR system and RIS (IDS GeoRadar) processing software for processing and visualization of GPR data. We conducted GPR survey at selected locations of active tectonic features in and around the HFT. We used 100 MHz and 40 MHz antennae with a common offset bistatic configuration keeping the separation between transmitting (Tx) and receiving (Rx) antennae fixed. The 40 MHz antenna was unshielded having the inherent problem of air reflections or unwanted reflections caused by surrounding features at the survey area (Abdulkareem et al. 2013), whereas the 100 MHz antenna was shielded without interference effect from surrounding areas. In terms of performance, the 100 MHz antenna provided a higher resolution but lower depth of penetration, whereas 40 MHz antenna provides relatively lower resolution but higher depth of penetration.

2.4 Geomorphic Indices of Active Tectonics

Geomorphic indices of active tectonics describe the relative importance of erosional and tectonic forces assuming uniform climate and lithology. The indices are useful for studying active tectonics in the study area. The indices may be categorized into four classes: (a) spatial symmetry/asymmetry of drainage basin such as transverse topographic symmetry factor (T_t) and drainage basin asymmetry factor (A_f), (b) gradient of drainage basin such as shape of hypsometric curve and hypsometric integral (HI), (c) shape of the valley profile such as valley floor width to valley height ratio (V_f), and (d) rectilinearity in the mountain fronts such as mountain front sinuosity (S_{mf}) index (Keller and Pinter 1996). In the present study, geomorphic indices of active tectonics were calculated for the fifth-order sub-basins occurring in five major sixth-order drainage basins spread over the study area in and around the HFT to evaluate relative tectonic activity in the region.

2.4.1 Spatial Symmetry/Asymmetry of Drainage Basin

Spatial symmetry/asymmetry of drainage basin is analyzed to assess tectonic tilting of the drainage basin and therefore degree of tectonic activity in the area. Geomorphic indices such as transverse topographic symmetry factor (T_t) and drainage basin asymmetry factor (A_f) are designed to assess drainage basin asymmetry possibly due to tectonic tilting.

2.4.1.1 Transverse Topographic Symmetry Factor (T_t)

Transverse topographic symmetry factor (T_t) is defined as Da/Dd, where Da is the distance from the midline of drainage basin to the midline of the active meander belt and Dd is the distance from basin midline to the basin divide. It is a reconnaissance tool for inferring lateral tilting in the drainage basin (Cox et al. 2001; Keller and Pinter 2002). T_t was calculated along the midline of the active meander belt at regular intervals and averaged to find the representative T_t value for the drainage basin. T_t was calculated for all the fifth-order sub-basins falling under five major sixth-order drainage basins used as test sites (Table 2.1).

2.4.1.2 Drainage Basin Asymmetry Factor (A_f)

Drainage basin asymmetry factor (A_f) is expressed as (Ar/At)*100, where Ar is the area of right sub-basin on the downstream and At is the total area of the basin. The values of A_f above or below 50% indicate that the basin is asymmetric. It permits to establish the lateral tilting of a drainage basin with respect to the main stream (Hare and Gardner 1985; Cox 1994; Cuong and Zuchiewicz 2001). It also includes the direction of asymmetry suggesting the possible direction of higher tectonic activity and relative uplift or subsidence of discrete blocks (Pinter 2005). A_f can be expressed as the absolute of the value minus 50 and subsequently categorized into four asymmetry classes: < 5% (symmetric), 5–10% (slightly asymmetric), 10–15% (moderately asymmetric), and > 15% (strongly asymmetric) (Giaconia et al. 2012). In the present study, A_f values were calculated for all the fifth-order sub-basins falling under five major sixth-order drainage basins (Table 2.1). It was observed that all the fifth-order sub-basins represent low to high asymmetric classes.

2.4.1.3 Gradient of Drainage Basin

To evaluate the stage of development and thereby to assess relative tectonic activity of the drainage basins, the shape of hypsometric curves and hypsometric integrals were studied. The hypsometric curve of a drainage basin describes the spatial distribution of basin area vs. altitude of the basin (Strahler 1952; Keller and Pinter 2002). The shape of the curve is related to the degree of dissection and the stage of development of the drainage basin. For example, convex hypsometric curves represent relatively young drainage basins and weakly eroded regions; S-shaped curves represent mature drainage basins and moderately eroded regions, whereas concave curves represent old drainage basins and highly eroded regions. The hypsometric integral (HI) is an index which is defined as $\frac{\text{Mean Elevation}-\text{Minimum Elevation}}{\text{Maximum Elevation}-\text{Minimum Elevation}}$ of the drainage basin (Keller and Pinter 2002). It represents the fraction of the area below the hypsometric curve and thus expresses the volume of a drainage basin that has not been eroded (El Hamdouni et al. 2008). It is independent of the basin area and varies

Table 2.1 Geomorphic indices of active tectonics in five test sites

Test site		Drainage basin asymmetry index			Drainage gradient index			Valley profile anomaly index		Mountain front sinuosity index		
					Shape of hypsometric curve and hypsometric integral (HI)			V_f		S_{mf}		
Drainage basin	Sub-basin	T_t	A_f	Magnitude and direction of tectonic tilting	Curve shape	HI value	Tectonic activity	Value	Tectonic activity	Value	Tectonic activity	Overall category of relative tectonic activity
Budki Nadi (Rupnagar)	Fifth order left	0.58	13.72	M (M-M); right	Concave	0.27	L	2.17	L	1.28	H	**L**
	Fifth order right	0.37	7.30	L (L-L); left	Concave	0.26	L	1.60	L	1.22	H	**L**
Markanda	Fifth order left	0.46	17.65	H (M-H); right	Concave	0.31	L	0.70	M	1.16	H	**H**
	fifth order central	0.06	4.8	VL (VL-VL); right	Concave	0.27	L	0.73	M	1.32	H	**L**
	Fifth order right	0.37	6.47	L (L-L); left	Concave	0.28	L	0.51	M	1.68	M	**L**
Solani	Fifth order left	0.19	13.58	M (L-M); right	Concave to S-shaped	0.37	M	0.78	M	1.34	H	**H**
	Fifth order right	0.21	16.10	M (L-H); right	Concave to S-shaped	0.37	M	0.89	M	2.20	M	**M**

(continued)

Table 2.1 (continued)

Test site		Drainage basin asymmetry index			Drainage gradient index			Valley profile anomaly index		Mountain front sinuosity index		
					Shape of hypsometric curve and hypsometric integral (HI)			V_f		S_{mf}		
Drainage basin	Sub-basin	T_t	A_f	Magnitude and direction of tectonic tilting	Curve shape	HI value	Tectonic activity	Value	Tectonic activity	Value	Tectonic activity	Overall category of relative tectonic activity
Khoh	Fifth order left	0.31	7.90	L (L-L); left	Convex to S-shaped	0.48	M	1.16	L	1.64	M	**L**
	Fifth order right	0.38	9.22	L (L-L); right	Convex	0.51	H	0.97	M	1.39	H	**H**
Dhela	Fifth order left	0.41	17.95	H (M-H); right	Concave	0.25	L	0.59	M	1.17	H	**H**
	Fifth order right	0.59	2.32	L (M-VL); right	Concave	0.26	L	0.74	M	1.19	H	**L**

VL, very low; *L*, low; *M*, moderate; *H*, high

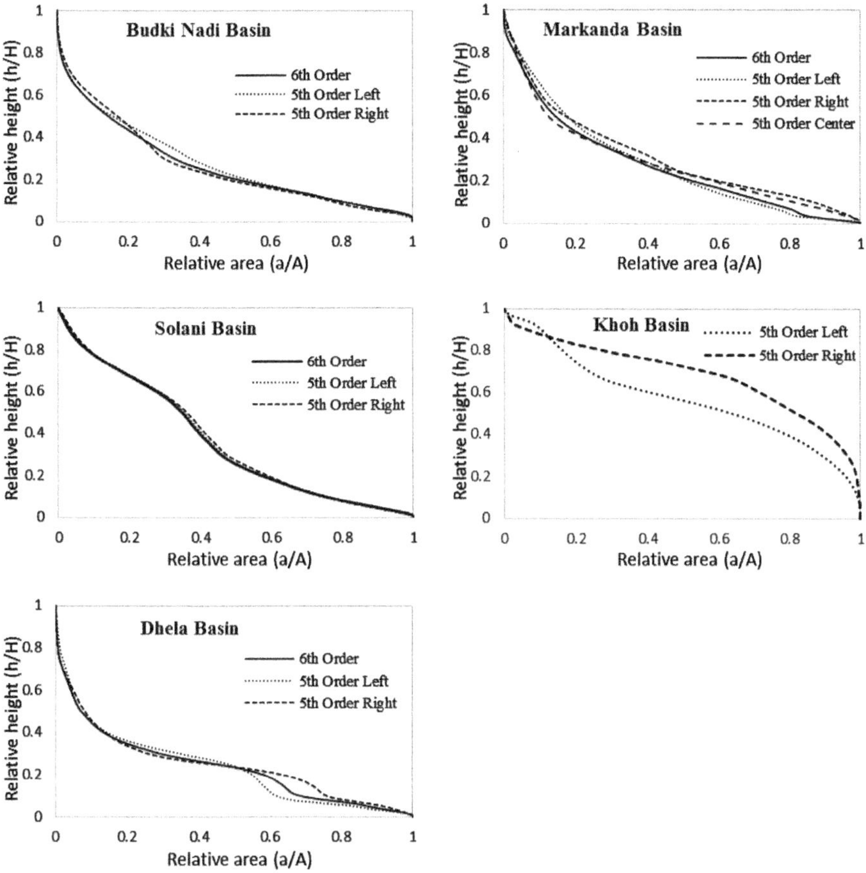

Fig. 2.2 Hypsometric curves of the fifth- and sixth-order drainage basins (as applicable) in the five test sites of sixth-order drainage basins in and around the HFT in the study area

from 0 to 1.0 (in general, from 0.25 to 0.75) for highly eroded to weakly eroded regions. The shape of the hypsometric curves and HI values provide valuable information about the tectonic, climatic, and lithological factors controlling the catchment landscape (Pérez-Peña et al. 2010). Considering the shape of the hypsometric curves (Fig. 2.2) and the HI values, the relative tectonics in and around the five major sixth-order drainage basins were assessed (Table 2.1). For example, in case of Budki Nadi, Markanda, and Dhela basins, both sixth-order basins and fifth-order sub-basins show concave hypsometric curves with low hypsometric integrals (Table 2.1) and therefore represent in general old drainage basins with highly eroded regions. However, there is subtle upward curvature in the middle of the hypsometric curves which is more conspicuous in the fifth-order right sub-basin of Budki Nadi drainage basin and fifth-order left sub-basin of Dhela drainage basin. It infers the possibility of less eroded resistant or tectonically uplifted piedmont-alluvial region in those drainage basins. In case of Solani drainage basin, the shape of the hypsometric curves is overall concave in nature with a sharp upward curvature in the middle part which suggests the possibility of prominent tectonic upliftment of faulted blocks or upwarping in the piedmont-alluvial region. On the other hand, in case of Khoh drainage basin, the hypsometric curves are S-shaped to convex upward representing mature to young drainage basins with moderately to weakly eroded regions. This suggests the possibility of high level of tectonic activity in Khoh drainage basin. In case of Solani and Khoh drainage basins, the HI values are relatively higher (close to 0.4) compared to the rest of the three drainage basins (with HI values close to 0.25).

2.4.2 Shape of the Valley Profile

Deep V-shaped valleys are associated with linear, active downcutting streams characteristically occurring in the areas subjected to active uplift, whereas flat-floored valleys indicate an attainment of the base level of erosion characteristically occurring in the areas of relative tectonic quiescence (Keller and Pinter 2002; bull 2007, 2009). Valley floor width to valley height ratio (V_f) is a geometric index conceived to evaluate the nature of the valley profile in terms of V-shaped and U-shaped valleys. V_f is defined as $\frac{Vfw}{[(Eld-Esc)+(Erd-Esc)]}$ where V_{fw} is the width of the valley floor; E_{ld} and E_{rd} are elevations of left and right valley divides, respectively; and E_{sc} is the elevation of the valley floor. V_f was calculated for the main streams in five sixth-order drainage basins and their fifth-order sub-basins. V_{fw} was measured from high-resolution satellite image (Cartosat image with spatial resolution 2.5 m.), whereas E_{ld}, E_{rd}, and E_{sc} were retrieved from high-resolution DEM (Cartosat DEM with spatial resolution 10 m and vertical accuracy 3–5 m) at regular intervals. Finally, the mean V_f was determined for the fifth-order sub-basins, and relative tectonic activity was assessed (Table 2.1: Silva et al. 2003; Ata 2008; Giaconia et al. 2012; Mahmood and Gloaguen 2012).

2.4.3 Mountain Front Sinuosity (S_{mf})

In general, the mountain fronts represent the thrusted contact between mountain and piedmont-alluvial plain. The plan view of the mountain fronts is essentially straight or gently curved at the time of development. The denudational processes subsequently modify them into curved and wavy mountain fronts. Mountain front sinuosity index (S_{mf}) is defined as $\frac{Lmf}{Ls}$ where L_{mf} is the length of the mountain front along the foothill of the mountain and L_s is the straight line length of the mountain front. S_{mf} describes the degree of denudational modification of the thrusted tectonic contact (Bull and McFadden 1977). The index balances between tectonic uplift and erosional processes. In case of active mountain fronts, tectonic uplift is dominant over the erosional processes, which give rise to straight mountain fronts with S_{mf} values close to 1.0. S_{mf} values less than 1.4 imply tectonically active mountain fronts (Rockwell et al. 1984; Keller 1986). In the present study, all the five sixth-order drainage basins and their fifth-order sub-basins under investigation have well-defined mountain fronts. Based on the S_{mf} values of the five test sites (fifth-order drainage basins), the relative tectonic activity in and around the mountain fronts was assessed (Table 2.1: Silva et al. 2003; Malik and Mohanty 2007; Ata 2008; Pérez-Peña et al. 2010).

2.5 Drainage Anomaly and Active Tectonics

In the sub-Himalaya region, the drainage pattern varies from dendritic to sub-dendritic, trellis, and parallel depending on the slope of the terrain and attitude of the exposed Siwalik rocks. The drainage density is generally high. On the other hand, in the piedmont-alluvial plain, the streams emerging from the Siwalik ranges flow intermittently and show meandered to braided pattern and coarse drainage texture due to high sediment load and gentle topographic slope. Drainage anomalies are defined as local deviations from the overall stream pattern or drainage pattern. Active tectonics primarily manifest itself by either steepening or reducing the local valley gradient which in turn introduces disturbance in the natural equilibrium of a drainage system and produces drainage anomalies. Drainage anomalies may be broadly categorized into two types: (a) anomaly or aberration in some segments of the existing drainage and (b) changes in the stream course and their behavior with time. Geomorphic features such as rectilinearity of stream segments, anomalous curves or turns (e.g., acute and obtuse angle elbow turns, U-turns), compressed meandering, anomalous pinching and flaring, and abrupt and localized braiding may be considered as drainage anomalies to infer the locations of structural and tectonic features such as faults, lineaments, folds, and warps. In the present study, we plotted them as the locations of drainage segment anomaly (Fig. 2.3). Similarly, the changes in part of the drainage system over a period of time (multi-date observations), such as drainage migration, stream deflection, shifting of distributary bifurcation zone, divergence of existing stream, and emergence of new streams, can also be used to identify the possible locations of active tectonics. In the present study, we used multi-temporal satellite images of the last four decades to identify the locations of dynamic changes in the drainage system. In the present study, we identified many locations of well-defined drainage deflections. Some of them probably represent natural deflections in response to hydrodynamic changes of a meandered drainage system. Many of them were deflected/emerged abruptly along straight segments unrelated to the existing drainage courses and follow straight alignments for a long distance which possibly mark the locations of active tectonic faults. In the present study, we plotted them as the locations of drainage shift.

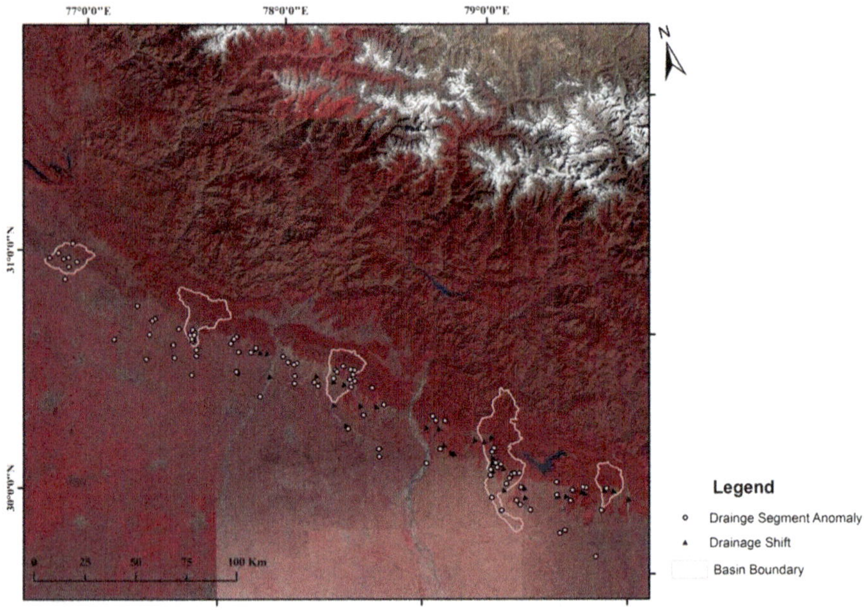

Fig. 2.3 Drainage segment anomalies and drainage shifts in and around five test sites along the HFT in the study area

2.6 Topographic Breaks vs. Active Tectonics

Using subtle topographic break in piedmont-alluvial region, the probable locations of active faults in and around the HFT region can be identified. We found elevated terraces along the rivers and across the mountain fronts which possibly indicate the locations of active tectonics. To identify the locations of subtle topographic breaks, particularly in piedmont-alluvial region where the signatures of active tectonics are quickly obliterated by the physical processes of weathering and erosion, we selected several transverse and longitudinal topographic profiles at regular intervals from the high-resolution Cartosat-1 DEM (spatial resolution, 10 m; vertical accuracy, 3–5 m). Transverse and longitudinal topographic transects were plotted in and around five sixth-order drainage basins (test sites) in piedmont-alluvial region around the HFT. We identified the locations with $\geq$ 10 m topographic breaks along N-S transects and $\geq$ 5 m topographic breaks along E-W transects in and around the five test sites only and plotted them as the possible locations of active tectonic features in the piedmont-alluvial plain region (Fig. 2.4). These locations were further compared in reference to the drainage anomalies and existing structural features. For selective locations, GPR survey was conducted to confirm surface and near-surface active tectonic features. In the study area in between the test sites, similar topographic breaks may be identified and investigated further to confirm the locations of active tectonics.

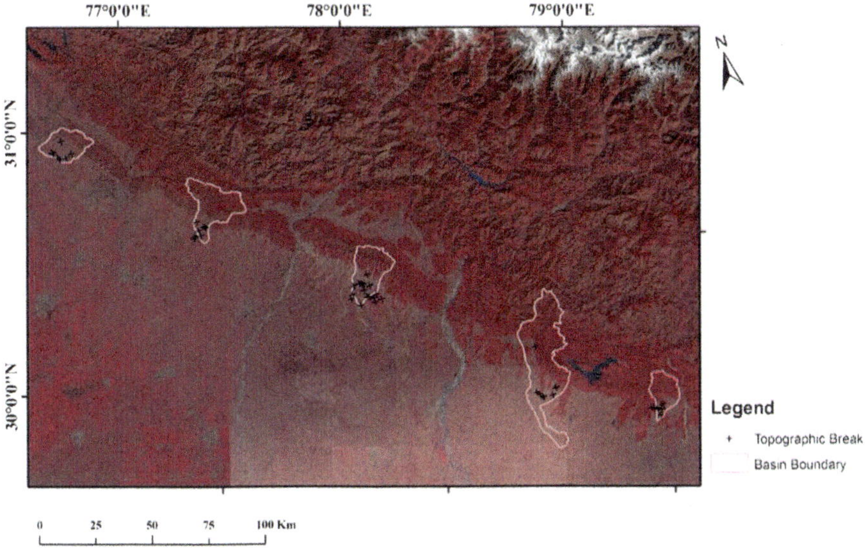

Fig. 2.4 Topographic breaks and possible locations of active tectonics or pre-existing lineaments in and around the five test sites along the HFT in the study area

2.7 GPR Profiles vs. Near-Surface Active Tectonic Features

Geomorphic anomalies and topographic breaks infer the presence of active tectonic features such as near-surface faults, lineaments, and warps in piedmont-alluvial region. In the absence of direct observation by trenching, we conducted ground-penetrating radar (GPR) survey at selective locations to confirm the presence of near-surface active tectonic features such as near-surface faults, lineaments, and warps up to a depth of 10–30 m. In the present study, we used multifrequency (e.g., 40 MHz and 100 MHz antennae) bistatic GPR to obtain 2D profiles at variable depths and resolutions (Fig. 2.5). The result demonstrates that near-surface active tectonic features such as faults, lineaments, and warps produce a variety of signatures in 2D GPR profiles as a function of antenna frequency, relative orientation of the feature, and heterogeneity of materials with respect to the feature.

2.8 Discussion and Summary

In the piedmont-alluvial region in and around the HFT of NW Himalaya covering Himachal Pradesh and Uttarakhand states of India, we studied the geomorphic evidences of active tectonics using geomorphic indices and geomorphic anomalies. Based on the geomorphic indices of active tectonics such as spatial asymmetry of drainage basin (including transverse topographic symmetry factor, T_t, and drainage

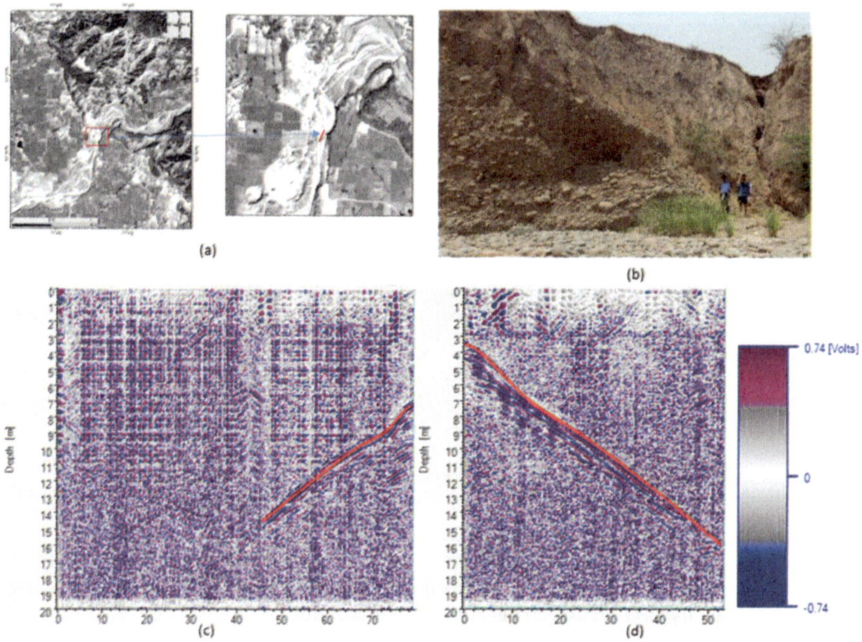

Fig. 2.5 GPR-based 2D radargrams by 40 MHz antenna near Govindpur village (at Thathar ki Nadi) of Haryana, India, around Budki Nadi drainage basin of the study area showing near-surface active fault; (**a**) Location and transect of GPR survey in Cartosat-1 PAN image, (**b**) field photograph of the river section (Thathar ki Nadi), (**c**) NNE-SSW transect 2D radargram, and (**d**) SSW-NNE transect 2D radargram

basin asymmetry factor, A_f), drainage basin gradient (hypsometric analysis), valley profile (valley floor width to valley height ratio, V_f), and mountain front sinuosity (S_{mf}), the relative tectonic activity in the five test sites along the HFT was studied. We considered the fifth-order sub-basins under five major sixth-order drainage basins as the unit for relative tectonic analysis in the study area. In general, all the test sites are found to be tectonically active. However, we classified them into three categories of relative tectonic activity classes: high (Class I), moderate (Class II), and low (Class III) tectonic activity classes to evaluate geodynamic status of the study area in and around the HFT (Fig. 2.6).

Such basic classification is useful in delineating the area into broad relative tectonic activity classes to prioritize detailed large-scale study for identification of active tectonic features and ground-based observation on the rate of active tectonic processes. Thus, morphotectonic analysis based on the geomorphic indices infers the current status of relative tectonic activity in and around the Himalayan Frontal Thrust. On the other hand, based on the locations of geomorphic anomalies such as drainage anomalies and topographic breaks in and around the five test sites along the HFT, the possible locations of surface and near-surface active tectonic features were identified in the study area. The importance of the potential locations of active

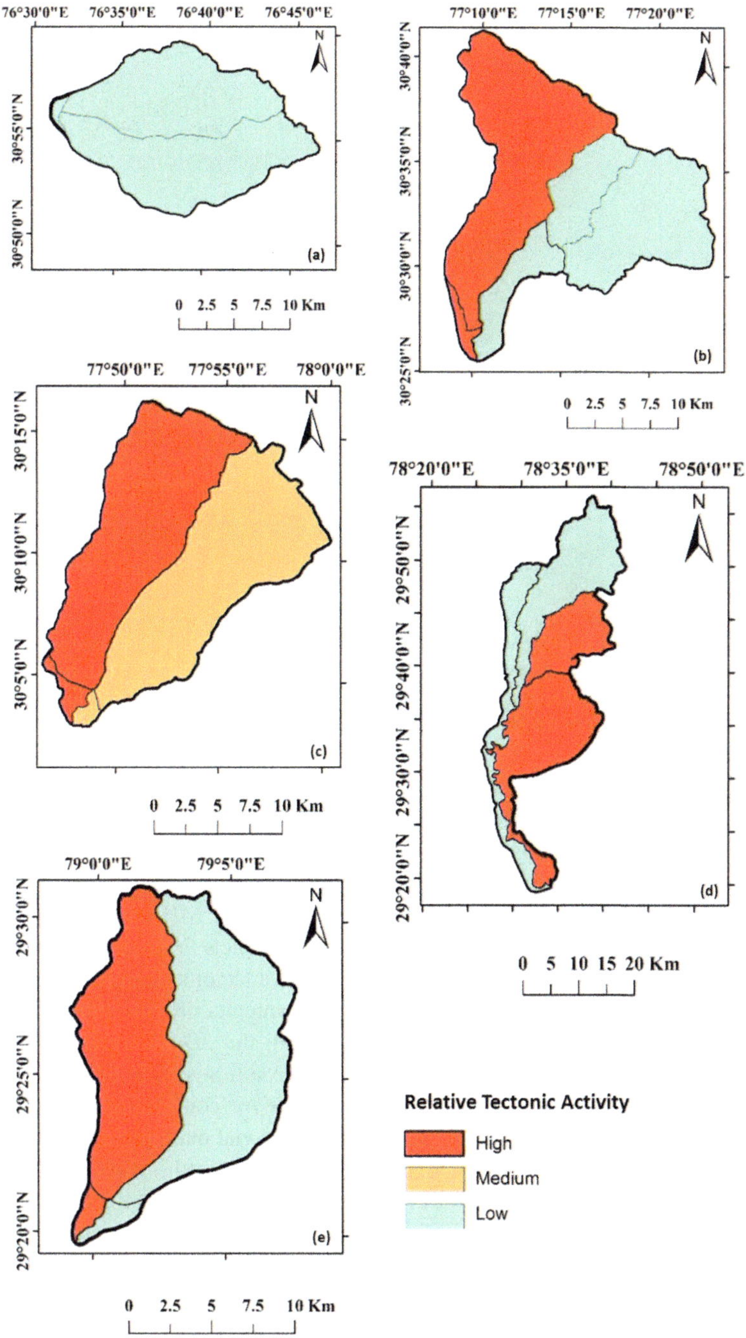

Fig. 2.6 Relative tectonic activity in the five test sites in and around the HFT of the study area (fifth-order sub-basin level relative tectonic activity results are shown): (**a**) Budki Nadi (Rupnagar) basin, (**b**) Markanda basin, (**c**) Solani basin, (**d**) Khoh basin, and (**e**) Dhela basin

Fig. 2.7 Archaeological evidences such as buried houses and prehistoric potteries in vertical topographic sections at Kot Kaliyar Chak village, Haridwar district, Uttarakhand, India (**a** and **b**), and Khajnawar village, Saharanpur district, Uttar Pradesh, India (**c**), occurring at anomaly locations in Solani drainage basin of the study area

tectonics was prioritized based on the number of anomalies coexisting together at the same or nearby locations. We selected a few such locations where two or more than two anomalies coexist together and conducted a multifrequency (40 MHz and 100 MHz) GPR survey for confirming the presence of active tectonic features at such locations. The 2D radargrams obtained from 100 MHz antenna give high-resolution information at shallow depth which depicts near-surface soil layers beautifully but fail to confirm the presence of active tectonic features. On the other hand, the 2D radargrams obtained from 40 MHz antenna provide subsurface information up to 30 m depth. It was observed that in the 2D radargrams, the active tectonic features are manifested beyond the surface soil layers. In the present study, the 40 MHz GPR antenna has been found suitable for confirming the presence of near-surface active tectonic features in piedmont-alluvial plain region in and around the HFT. In and around some of the anomaly locations in the study area, archaeological evidences were observed. For example, in Solani drainage basin test site, archaeological evidences such as buried houses under newly constructed houses and prehistoric potteries were observed in vertical topographic sections at Kot Kaliyar Chak village, Haridwar district, Uttarakhand, and Khajnawar village, Saharanpur district, Uttar Pradesh (Fig. 2.7).

References

Abdulkareem AO, Hameed FS, Muhammed AA, Ali SM (2013) The effect of air reflection on ground penetrating radar (GPR) data. Journal of Babylon University (Engineering Sciences) 21 (5): 1697–1704.

Ata HA (2008) A test of the validity of morphometric analysis in determining tectonic activity from aster derived DEMS in the Jordan-dead sea transform zone. PhD Thesis, University of Arkansas, 220 p.

Azor A, Keller EA, Robert SY (2002) Geomorphic indicators of active fold growth: South Mountain–Oak Ridge anticline, Ventura basin, southern California. Geological Society of America Bulletin 114(6): 745–753.

Basson U, (2000) Imaging of active fault zone in the Dead Sea Rift: Evrona Fault Zone as a case study. PhD Thesis, Tel-Aviv University, Raymond & Beverly Sackler, Faculty of Exact Sciences, Department of Geophysics & Planetary Sciences, 195 p.

Bishop P (2007) Long-term landscape evolution: Linking tectonics and surface processes. Earth Surface Processes and Landforms 32: 329–365.

Bull WB (2007) Tectonic geomorphology of mountains: a new approach to paleoseismology. Blackwell Publishing, Hoboken, 316p.

Bull WB (2009) Tectonically Active Landscapes. Blackwell Publishing, Wiley Online Library, 326p.

Bull WB, McFadden LD (1977) Tectonic geomorphology of north and south of the Garlock fault, California. Geomorphology in arid regions, Proceeding of the 8th annual Geomorphology Symposium, Bingham, NY, pp 115–138.

Clark CD, Evans DJA, Khatwa A, Bradwell T, Jordan C, Marsh SH, Mitchell WA, Bateman MD (2004) Map and GIS database of glacial landforms and features related to the last British Ice Sheet. Boreas 33: 359–375.

Cox RT (1994) Analysis of drainage-basin symmetry as a rapid technique to identify areas of possible Quaternary tilt-block tectonics: An example from the Mississippi Embayment. Geological Society of America Bulletin 106: 571–581.

Cox RT, Van Arsdale RB, Harris JB (2001) Identification of possible Quaternary deformation in the northeastern Mississippi Embayment using quantitative geomorphic analysis of drainage-basin asymmetry. Geological Society of America Bulletin 113(5): 615–624.

Cuong NQ, Zuchiewicz WA (2001) Morphotectonic properties of the Lo River Fault near Tam Dao in North Vietnam. Natural Hazards and Earth System Sciences 1: 15–22.

Davis JL, Annan AP (1989) Ground penetrating radar for high-resolution mapping of soil and rock stratigraphy. Geophys. Prospect 37: 531–551.

El Hamdouni R, Irigaray C, Fernández T, Chacón J, Keller EA (2008) Assessment of relative active tectonics, southwest border of the Sierra Nevada (southern Spain). Geomorphol 96(1): 150–173.

England P, Molnar P (1990) Surface uplift, uplift of rock, and exhumation of rocks. Geology 18: 1173–1177.

Giaconia F, Booth-Rea G, Martínez-Martínez JM, Azañón JM, Pérez-Peña JV, Pérez-Romero J, Villegas I (2012) Geomorphic evidence of active tectonics in the Sierra Alhamilla (eastern Betics, SE Spain). Geomorphology 145: 90–106.

Hare PH, Gardner TW (1985) Geomorphic indicators of vertical neotectonism along converging plate margins, Nicoya Peninsula, Costa Rica. In: Morisawa M., Hack J.T. (Eds.), Tectonic Geomorphology, Allen and Unwin, Boston, pp 75–104.

Jackson MPA, Schultz-Ela DD, Hudec MR, Watson IA, Porter ML (1998) Structure and evolution of Upheaval Dome: a pinch-off salt diaper. Geological Society of America Bulletin 110 (12): 1547–1573.

Keller EA (1986) Investigation of active tectonics: use of surficial Earth processes. In: Active Tectonics: Impact on Society, Chapter 8, pp 136-266, National Academies Press: Washington, D. C.

Keller EA, Gurrola LD (2000) Earthquake Hazard of the Santa Barbara Fold Belt, California. Final report for NEHRP Award #99HQGR0081, 108p. (Available online at http://www.geol.ucsb.edu/faculty/ keller/library/pdf/sbeqh2.pdf, accessed on 15 February, 2017).

Keller EA, Pinter N (1996) Active tectonics: Earthquakes, Uplift and Landscapes. Prentice Hall, New Jersey, 338p.

Keller EA, Pinter N (2002) Active tectonics: Earthquakes, uplift and Landscape (second edition). Prentice Hall, New Jersey, 362p.

Knight J (2001) A geocultural classification of landscape in Northern Ireland: implications for landscape management and conservation. Tearmann 1: 113–124.

Kothyari GC, Pant PD, Joshi M, Luirei K, Malik JN (2010) Active faulting and deformation of Quaternary landform Sub-Himalaya, India. Geochronometria 37: 63–71.

Mahmood SA, Gloaguen R (2012) Appraisal of active tectonics in Hindu Kush: insights from DEM derived geomorphic indices and drainage analysis. Geoscience Frontiers 3(4): 407–428

Malik JN, Mohanty C (2007) Active tectonic influence on the evolution of drainage and landscape: Geomorphic signatures from frontal and hinterland areas along the Northern Himalaya, India. Journal of Asian Earth Sciences 29(56): 604618.

Molin P, Pazzaglia FJ, Dramis F (2004) Geomorphic expression of active tectonics in a rapidly-deforming arc, Sila Massif, Calabria, southern Italy. American Journal of Sciences 304: 559–589

Molnar P (2003) Nature, nurture and landscape. Nature 426: 612–614.

Nakata T (1972) Geomorphic history and crustal movements of the foot-hills of the Himalaya. Tohoku University Science Reports, 7th Series, Japan, 22, pp 39–177.

Pérez-Peña JV, Azor A, Azañón JM, Keller AK (2010) Active tectonics in the Sierra Nevada (Betic Cordillera, SE Spain): Insights from geomorphic indexes and drainage pattern analysis. Geomorphology 119: 74–87.

Pinter N (2005) One step forward, two steps back on U.S. floodplains. Science 308: 207–208.

Rockwell TK, Keller EA, Clark MN, Johnson DL (1984) Chronology and rates of faulting of Ventura river terraces, California, Geol. Soc. Am. Bull. 95(12): 1466–1474.

Rockwell TK, Keller EA, Johnson DL (1985) Tectonic geomorphology of alluvial fans and mountain fronts near Ventura, California. In: Morisawa, M. (Ed.), Tectonic Geomorphology. Proceedings of the 15th Annual Geomorphology Symposium. Allen and Unwin Publishers, Boston, pp 183–207.

Ruhe RV (1975) Review of "Pedology, weathering and geomorphological research." Geoderma 14: 176–177.

Salvany JM (2004) Tilting neotectonics of the Guadiamar drainage basin, SW Spain, Earth Surface Processes and Landforms. 29(2): 145–160.

Schoenbohm LM, Whipple KX, Burchfiel BC, Chen L (2004) Geomorphic constraints on surface uplift, exhumation, and plateau growth in the Red River region, Yunnan Province, China. Geological Society of America Bulletin 116(7/8): 895–909.

Silva PG, Goy JL, Zazo C, Bardají T (2003) Fault-generated mountain fronts in southeast Spain: geomorphologic assessment of tectonic and seismic activity. Geomorphology 250: 203–225.

Singh V, Tandon SK (2008) The Pinjaur dun (intermontane longitudinal valley) and associated active mountain fronts, NW Himalaya: Tectonic geomorphology and morphotectonic evolution. Geomorphology 102(3): 376–394.

Smith DG, Jol HM (1995) Ground penetrating radar: antenna frequencies and maximum probable depths of penetration in Quaternary sediments. Journal of Applied Geophysics 33: 93–100

Srivastava P, Misra DK (2008) Morpho-sedimentary records of active tectonics at the Kameng River exit, NE Himalaya. Geomorphology 96: 187–198.

Starkel L (2003) Climatically controlled terraces in uplifting mountain areas. Quaternary Science Reviews 22: 2189–2198.

Strahler AN (1952) Dynamic basis of geomorphology. Geological Society of America 63 (9): 923–938.

Tandon SK, Singh V (2014) Duns: Intermontane basins in the Himalayan frontal zone. In: V.S. Kale (Ed.) Landscapes and Landforms of India, Springer Science, pp 135–142.
Thakur VC (2004) Active tectonics of Himalayan Frontal Thrust and seismic hazard to Ganga Plain. Current Science 86: 1554–1558.
Thakur VC, Pandey AK (2004) Late Quaternary tectonic evolution of Dun in fault bend/propagated fold system, Garhwal Sub-Himalaya. Current Science 87 (11): 1567–1576.
Valdiya KS (1980) Geology of Kumaun Lesser Himalaya. Wadia Institute of Himalayan Geology, Dehradun, 291p.
Vijith H, Satheesh R (2006) GIS based morphometric analysis of two major upland sub-watersheds of Meenachil River in Kerala. Journal of the Indian Society of Remote Sensing 34 (2): 181–185.

Chapter 3
Simulation Outputs of Major Debris Flows in Garhwal Himalaya: A Geotechnical Modeling Approach for Hazard Mitigation

Shovan Lal Chattoraj, P. K. Champati Ray, and Suresh Kannaujiya

3.1 Introduction

Landslides, one of the major geological hazards, contribute to natural disasters in mountainous region around the globe owing to a wide variety of causative as well as triggering factors like heavy rainstorms, cloudbursts, glacial lake outburst (GLOF), earthquakes, geo-engineering setting, unplanned human activities, etc. In different parts of the Himalaya, landslide has evolved as a frequent problem which severely affects life, property, and livelihood of this mountainous area thriving mainly on pilgrimage, tourism, and agriculture (Anbalagan et al. 2015; Anbalagan 1992; Champati Ray and Chattoraj 2014; Gupta et al. 1993; Kumar et al. 2012; Onagh et al. 2012; Sarkar et al. 1995, 2006; Sundriyal et al. 2007). With the background of higher elevation, rough hilly landscape, scanty cultivated land, strong monsoonal effect, and less industrial growth restricting economic progress, repeated landslide events keep human life and property at stake (Champati Ray et al. 2013a, b, 2015; Ketholia et al. 2015; Paul and Bisht 1993). Landslides in the Himalayan region are on an average smaller in dimension and have shallow depth, but these are more recurring in nature and thereby do not get noticed by authorities but cause higher cumulative losses over a period of time. Landslides, in the Himalaya, are observed particularly in highly fractured and sheared rock mass close to faults and also in weathered hard rocks. The climatic factors play an important role in weathering and disintegration of rock mass that are finally brought down by gravity (Kumar et al. 2007, 2012). Most of these landslides wreak havocked not only on life and property but manifest changes in landform due to large-scale mass wasting, landslide-

S. L. Chattoraj (✉) · P. K. Champati Ray · S. Kannaujiya
Geosciences and Disaster Management Studies Group, Indian Institute of Remote Sensing (IIRS), Indian Space Research Organisation (ISRO), Department of Space, Government of India, Dehradun, India
e-mail: shovan@iirs.gov.in

R. R. Navalgund et al. (eds.), *Remote Sensing of Northwest Himalayan Ecosystems*,
https://doi.org/10.1007/978-981-13-2128-3_3

dammed lake formation, and breaching leading to large-scale landform modification (Champati Ray 2013; Champati Ray et al. 2015).

Among many enormous landslides that have been witnessed by the Garhwal part of Uttarakhand Himalaya, the Malpa landslide in the year 1998 is responsible for taking 300 human lives alone. Ukhimath town, one important stop before reaching Kedarnath Hindu shrine, was damaged by many landslides which in total claimed 37 lives again in 1998 (Naithani 2002; Bist and Sah 1999). To exemplify the quantum of huge property loss, a portion of road leading to Gangotri Hindu shrine, with a length of about 42 km from Uttarkashi to Bhatwari, were demolished by landslides triggered by the 1991 Uttarkashi earthquake. Similarly, seismicity-induced landslides wreak havoc on Chamoli district in the year 1999 taking the death toll to more than a hundred (Kimothi et al. 2005). On the other hand, Bhagirathi basin has seen Varunavat hill landslide in 2003 causing huge property damage (Gupta and Bist 2004; Sarkar et al. 2006, 2010). This landslide was also analyzed taking cues from earth observation data by investigators from the Indian Space Research Organization (ISRO) which revealed higher hazard potential of the region (Sati et al. 1998). Ukhimath area subsequently seen recurrence of the debris flows in 2012 killing 33 persons and partially wiped two villages called Chunni and Mangli (Islam et al. 2013; Martha and Kumar 2013). More recently in June 2013, several interrelated phenomena like cloudburst associated with glacial lake outburst floods, river blockade and breaching, etc. caused devastating floods and landslides in almost all major river basins of Uttarakhand which became the country's worst natural disaster since the 2004 tsunami. Kedarnath area and its downstream became the worst affected region (Champati Ray et al. 2015; Chattoraj et al. 2014; Chattoraj 2016; Dobhal et al. 2013). Apart from these events of national importance, there are many other events which have put a catastrophic effect (Gupta et al. 2008; Bist and Sah 1999). Pertinently, most of the landslides in Uttarakhand have a major debris flow component that travels some distance causing enormous damage en route (Chattoraj 2016; Chattoraj et al. 2014, 2015a; Chattoraj and Champati Ray 2015; Champati Ray et al. 2015). However, most of the works mentioned report either the geo-engineering aspects of landslides or hazard/susceptibility mapping leading to damage assessment. Holistic analysis of landslide hazard which demands physical modeling using mathematical simulation techniques is in the budding stage in this part of the Himalaya. Rainfall-triggered landslide models are abundant in literature and hold tremendous opportunity in the implementation of a successful strategy for landslide hazard mitigation (Brand 1995; Chattoraj et al. 2015b; Deganutti et al. 2000; Hungr et al. 1987; Scott 2000). The work is focused to grout this knowledge gap by analyzing and simulating major landslides/debris flow events in the Garhwal Himalaya. This study leads to derivation of the important physical flow parameters taking cues from earth observation techniques to understand the root cause of the devastation, which is essential for effective mitigation measures.

3.2 Debris Flow Modeling: Garhwal Himalaya

Modeling of debris flows is a process-based approach in principle; considers avalanches, flows (mud and debris), and falls; and hence is important in disaster management and mitigation (Cruden and Varnes 1996; Iverson et al. 1997). Particularly, debris flows can be defined as gravity-induced flows comprising of inhomogeneous materials mixed with a liquid phase resulting in a devastating event. With the availability of globally accepted empirical equations employed to characterize kinematics of a flow, there is an increased demand which come up with sophisticated mathematical simulations which can be utilized to mimic flow paths and analyze the process of entrainment (Tsai et al. 2011; Quan Luna et al. 2011). This work utilized the RAMMS (*Ra*pid *M*ass *M*ovements *S*oftware) model, presented by the WSL Institute of Snow and Avalanche, Switzerland, to model the natural flow of a dislodged geophysical mass in three dimensions from source (release) to base (deposition). A high-resolution digital elevation model supported by ancillary ground truth data is an important input to the model, reinforced with various geo-mechanical parameters. To understand the failure behavior, the Voellmy rheological model has been employed to take care and physically characterize the entrainment of debris material. This gives rise to a physical-based model showing spatial variation of flow, height, velocity, pressure, and momentum along the torrent (Christen et al. 2010; Ayotte and Hunger 2000; Rickenmann 2005). Outputs of these process-based 3-D models take us closer to understand the cause of the event and help in designing suitable remedial engineering structures (Iverson et al. 2000). Huge flows often recur on seasonal basis, and hence time series analysis and change detection become equally important. Temporal variation of important output parameters can be assessed by deriving longitudinal profiles along such flows.

With an aim to check the applicability of the model and for validation of the results, the physical process of important debris flows in Uttarakhand Himalaya was attempted. Four major flows, namely, the Varunavat landslide (Uttarkashi), the Ukhimath landslide, the Kedarnath landslide (Rudraprayag), and the Maithana landslide (Chamoli), were considered in this study. All involved stakeholders are thus enabled to have an access to the actual insight of these events and related disasters. Large-scale mapping of these important landslides is often, in practice, capable of complimenting the process of three-dimensional modeling, as portrayed in this work. This, holistically, will deliver adequate and important time-bound information for prima facie assessment and monitoring of the event and cater to planning of the mitigation measures in case of future events (Champati Ray et al. 2013b; Herva'set et al. 2003).

3.3 Study Area: Regional Geology and Geomorphology

Ukhimath and Kedarnath area belong to the Rudraprayag district, whereas the Varunavat Parvat landslide is in Uttarkashi district of Garhwal division, on the banks of the Mandakini and the Bhagirathi River, respectively. The Maithana landslide, close to Nandaprayag, lies in Chamoli district. The main rock types of the areas where debris flow models have been applied include quartzites intercalated with schist, granite gneisses, migmatites, slates, and limestones/dolostones (Fig. 3.1). Most of the rock types have suffered polyphase deformation (Valdiya et al. 1999). The Kedarnath and Ukhimath area consist of lithological units belonging to Central Crystalline of Himalaya consisting of undifferentiated Proterozoic formations like Jutogh and Vaikrita Group, etc. of the Higher Himalayan zone. Rocks in the vicinity of the Varunavat Parvat, Uttarkashi, belong mainly to Berinag Formation (Thakur and Rawat 1992). The Maithana landslide area, on NH 58 on the bank of Alaknanda, is lithologically clubbed under Berinag Formation (Chaturvedi et al. 2014). Both Uttarkashi and Maithana rocks belong to the Lesser Himalayan zone (Thakur and Rawat 1992; Valdiya et al. 1999). All four places are in close vicinity to the Main Central Thrust (MCT) which renders these areas tectonically more fragile and unstable. These areas show overall rugged topography with dissected hilly regions with deep incised channels of major rivers of Uttarakhand like the Mandakini, the Bhagirathi, and the Alaknanda (Fig. 3.1).

3.4 Methodology and Input Data

3.4.1 Source Area Characterization

3.4.1.1 Kedarnath

The targeted debris flow initiation zones were identified based on field observations and visual analysis of satellite data. Most of the source region lies at the upper reaches of avalanche chutes. The source is relatively steeper with a variation in slope angle from 30° to 60°, having an average height of 5000 m. The estimated depth of debris dislodged from the source varies from 1 to 1.5 m as analyzed from Cartosat-1 DEM and GeoEye-1 images available in Google Earth by visual and simple digital image enhancement techniques. The field observations revealed that the modeled landslides were initiated in an avalanche chute and then flowed downward which at times bifurcated before reaching the lower elevation (Chattoraj et al. 2014). There were two debris flow zones identified in Kedarnath area: the large one is near the temple and the other one is just downstream of the temple location, all on the left bank of the Mandakini River (Fig. 3.2).

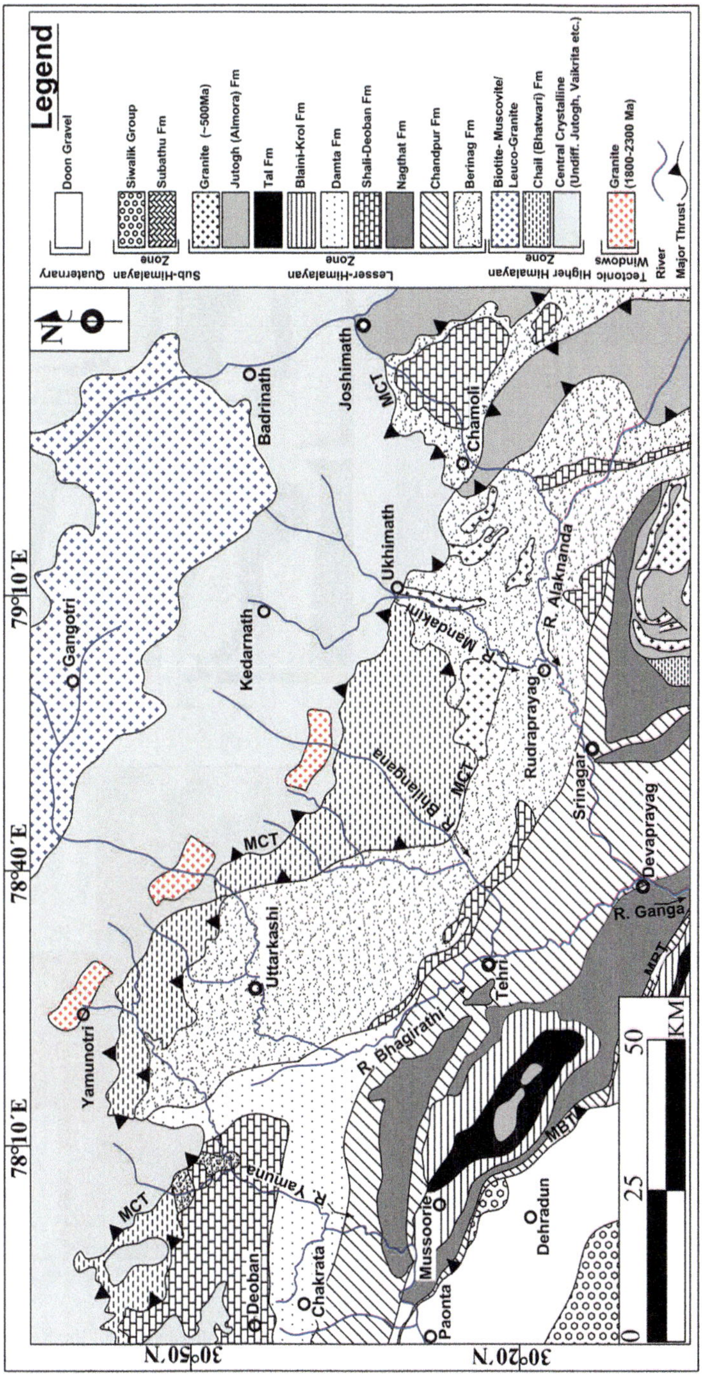

Fig. 3.1 Schematic geological map of Garhwal Himalaya. (After Thakur and Rawat 1992)

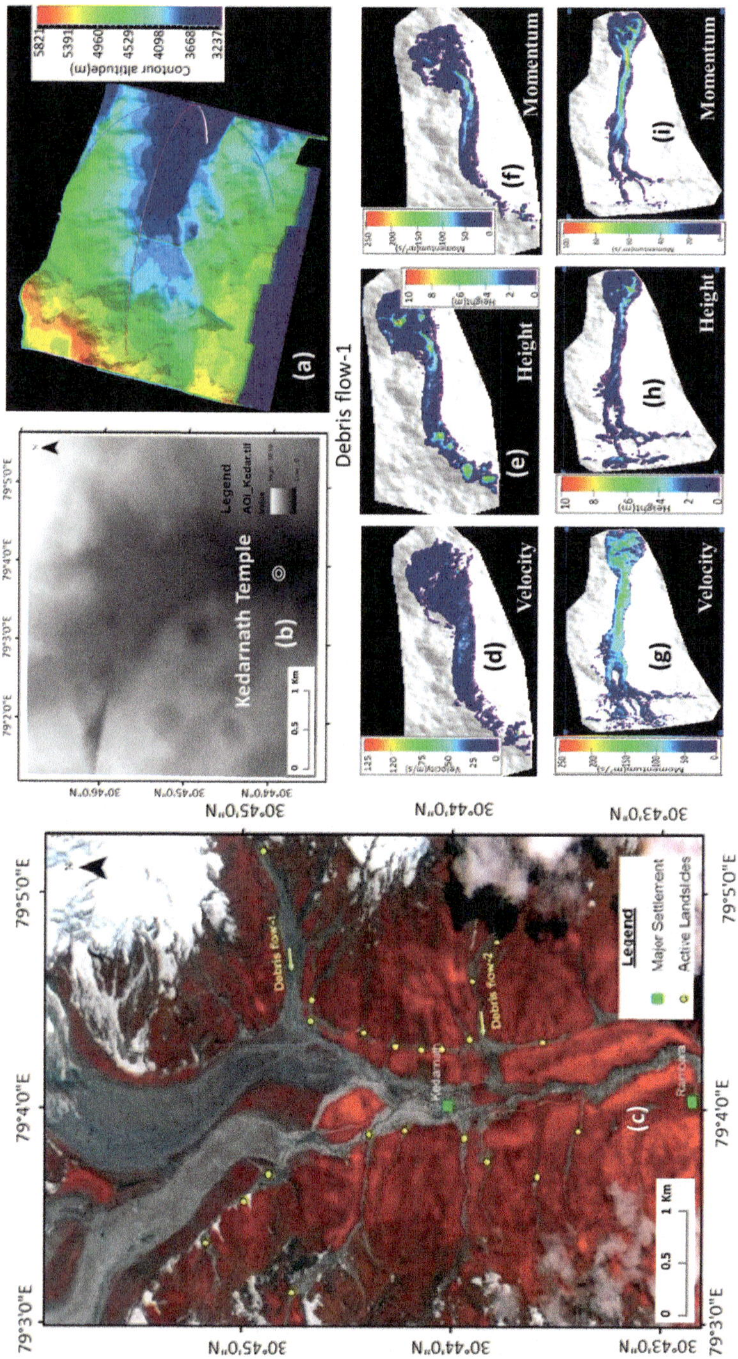

Fig. 3.2 Modeled outputs of debris flow at Kedarnath. (**a**) 3-D contour map of the Kedarnath area; (**b**) Subset of Cartosat DEM; (**c**) Satellite image of the area [IRS P6 LISS IV of 21 June 2013 (RGB: 321)]; (**d–e**) Spatial variation of velocity, momentum, and height along the runout path of debris flow-1; (**g–i**) Spatial variation of velocity, momentum, and height along the runout path of debris flow-2

3.4.1.2 Uttarkashi

The initiation zone lies at the top of Varunavat Parvat at an elevation of 1780 m and a slope of approximately 40–60 degrees. This particular area has been dissected by number of joints, and moreover a major fault passes through this zone (Bist and Sinha 1980). The average length of the initiation zone is approximately 113.2 m. From the field observation, it was clear that the modeled landslide was initiated with a rock fall and then flowed downward in the form of torrent, which was eventually divided into three channels through which most of the debris flowed downstream. Bulk density of 2450 kg/m^3 used quartzite and phyllites and their weathered derivatives which are highly shattered, fragmented, and thinly jointed.

3.4.1.3 Ukhimath

At Ukhimath, the initiation zone lies at an elevation of 1560 m with an average slope of 60^0. The length of the initiation zone is approximately 92 m, and the width varies from 18 to 58 m. From the field observation, it was clear that the modeled landslide was initiated as a rock fall and then flowed downstream in a semi-channelized torrent form. An average height of 2 m was selected to be the release height. A small torrent with a length of 78 m merged with the main flow path near Mangali village. A bulk density of 2350 kg/m^3 was chosen for schist with quartzite band and their weathered counterparts at the release area.

3.4.1.4 Maithana

The release area lies at an elevation of 1120 m with a slope of 35°. The whole slide mass is about 390 m long (till it reaches the river) and 260 m wide (at source region) which suggests it to be a moderately big slide. The direction of movement is toward northwest. The rocks are well jointed having a bedding dip of 37° toward the north. The slope is continuous and is inclined at about 35° above road level and 42° below road level. The right flank of the slide consists of jointed and fractured quartzite intercalated with chlorite schist, while the left flank consists mostly of loose soil and debris material.

3.4.2 Satellite Data Used

Emphasis was put to use mainly Indian Remote Sensing Satellite data products. Consequently, Resourcesat-2 LISS-IV data (21 June 2013) and Cartosat-2 data (20 June 2013) were processed for analyzing debris flow analysis at Kedarnath. Among the others, Cartosat-1 stereo-pair-derived DEM of 2008 and LISS IV images from Resourcesat-1, SPOT images, CNES, and DigitalGlobe data available in Google Earth with complementary SRTM DEM (Ver. 4) for terrain information were also referred. Cartosat-1 stereo-pair-derived DEM (6 March 2010) was analyzed in combination with GeoEye-1 multispectral image and LISS-IV Image of 4 April 2009 for Uttarkashi area. In case of Ukhimath, DEM (Cartosat-1 stereo-pair of 21 January 2010) and GeoEye-1 multispectral image were used. For Maithana area, LISS-IV of 25 October 2011 and 23 May 2013 and DEM from Cartosat-1, 2008, were used. The high-resolution multispectral or pan-sharpened images were used mainly for delineation of release area at the source of debris flows.

3.4.3 Model Input Data

3.4.3.1 Digital Elevation Model

Basic input datasets required for RAMMS simulation comprise of topographic data like elevation and slope from a high-spatial-resolution digital elevation model, release area, and release mass to characterize the head zone and crucial information about friction and related geo-mechanical properties. Topographic information lays an important role for a successful simulation because flow movement path and zone of deposition will be controlled by the elevation, slope, etc. In this context, a precise digital elevation model is required with high spatial resolution defining the release area. In this case, Cartosat-1 DEM (spatial resolution, 10 m) was used. RAMMS can process only the ESRI ASCII Grid and ASCII X, Y, Z data cluster. Contours from topographical maps were also simultaneously digitized to develop a DEM in ArcGIS 10.0 (© ESRI) using Topo to Raster tool in spatial analyst. Both the contours generated from Cartosat DEM and derived from topographical maps were compared to get outputs on numerical simulation.

Further, in debris flow modeling, two inputs were provided to define the initial condition, i.e., release information of the simulation, (a) release area (or block release) and (b) input hydrograph (or simply hydrograph). The starting conditions of a simulation can be selected depending on the type of debris flow expected in the region. In the present case, block release was preferred. Generally, it is also useful to distinguish between channelized and unchannelized debris flows. However, RAMMS use the unchannelized debris flow condition for hillslope debris flows and shallow landslides. The present model consists of both channelized and unchannelized flow path which were validated on satellite images. In RAMMS, for small unchannelized debris flows, it is important to know the release area with a given initial height, which will be released as a block (block release of Rickenmann et al. (2006)). In the present study, landslide-specific release areas were identified which have been demarcated over the DEM. Approximately corresponding calculation domains (within which a debris flow is assumed to be restricted to) were also delineated over DEM considering the possible maximum spatial extent of debris flow runout (Figs. 3.2, 3.3, and 3.4).

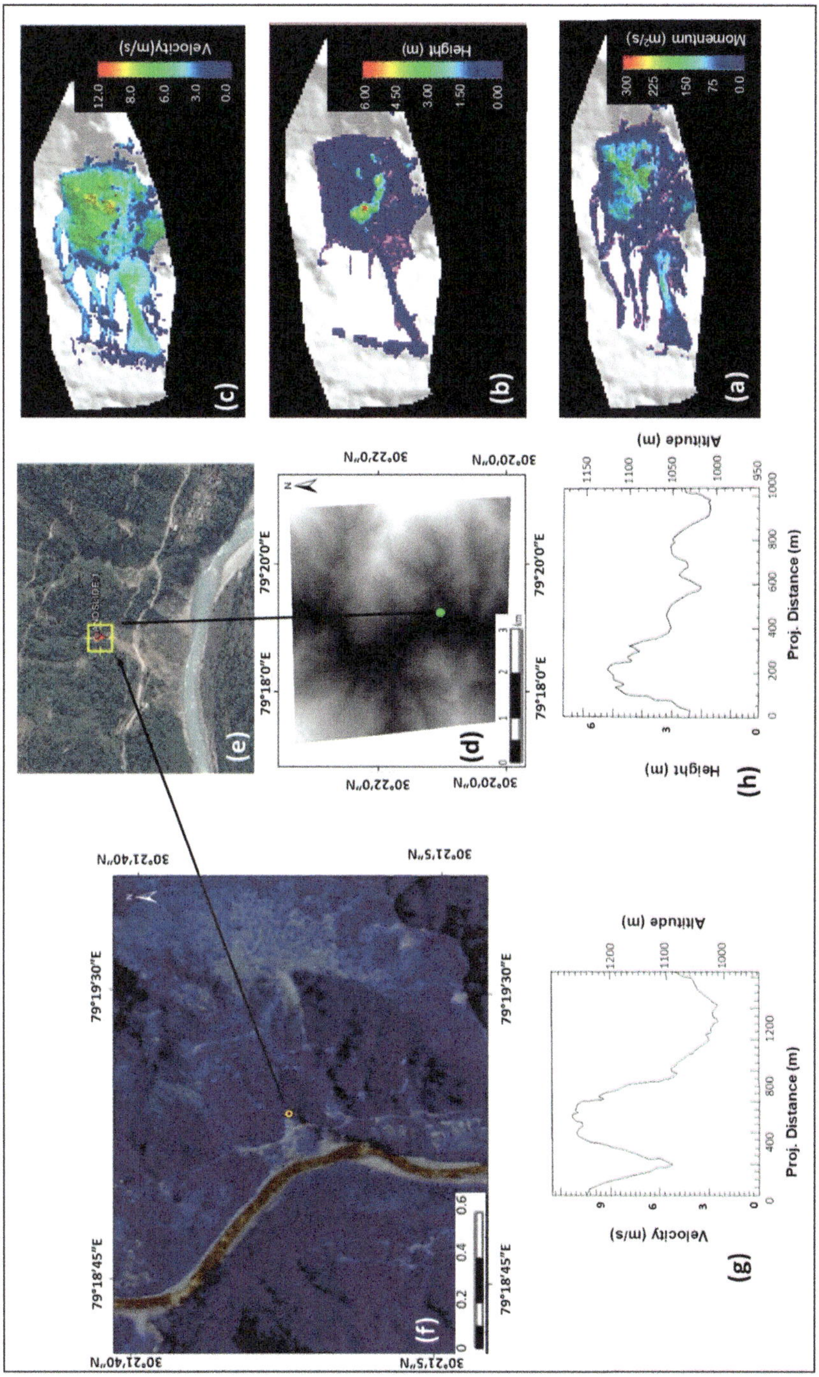

Fig. 3.3 (**a–c**) Modeled outputs of debris flow at Maithana; (**d**) Subset of Cartosat DEM of the study area; (**e**) SPOT image of the Maithana landslide of October 9 2014 © Google Earth; (**f**) LISS-IV (23 May 2013) FCC of study area; (**g–h**) Velocity and height profile along runout path of debris flow

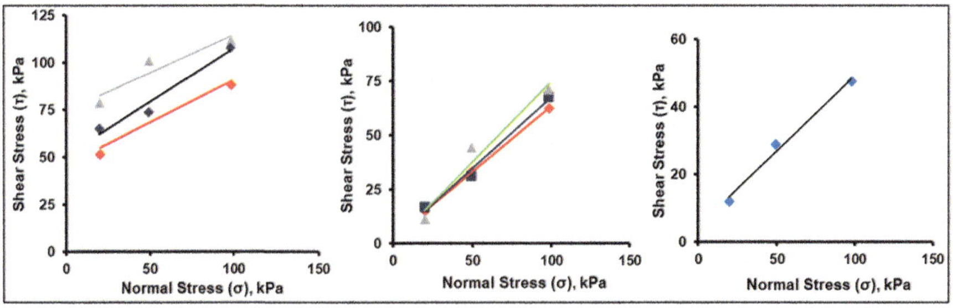

Fig. 3.4 Cross-plots of shear stress and normal stress of soil/debris samples collected from field

3.4.3.2 Frictional and Shear Strength Parameters and Calibration of the Model

The RAMMS mathematical simulation relies upon rheological characteristics of the physical mass resting on unstable slopes and hence gives maximum importance to shear strength parameters. This uses the Voellmy friction law (Salm et al. 1990). RAMMS considers total resistance to flow as a sum of the frictional resistance arising out of a dry-Coulomb-type friction (coefficient, μ) that is related to applied normal stress and a velocity-squared drag or viscous-turbulent friction (coefficient, ξ). The dry friction is imparted on the slope mostly by rock chunks (in situ and weathered/drifted), soil, river-borne material, and associated debris, while the liquid phase or water takes care of the viscous-turbulent friction. The total resistance due to friction S (Pa) is then calculated as:

$$S = \mu \rho H g \cos\ (\varphi) + \left(\rho g U^2\right)/\xi$$

where ρ represents the density of the debris, g the gravitational acceleration, φ the slope angle, H the flow height, and U the initial flow velocity. μ and ξ are considered as the two main geo-mechanical inputs representing frictional resistance.

The representative debris materials sampled from depositional zone at the base of the flow were studied for characterization of their shear strength parameters in electronic direct shear testing equipment (Model No. AIM 104-2kN, Make Aimil Ltd., New Delhi) at the Indian Institute of Remote Sensing, Dehradun. Though the analysis was carried out at different levels of saturation, the ones with maximum-induced saturation were considered best. Samples were examined at 0.25, 0.50, and 1 kgf/cm^2 impounding normal load, and resultant shear strength parameters at respective load failure were considered.

The major problem for carrying out a debris flow simulation lies in its varying constituents, which influence the choice of the friction parameters. RAMMS debris flow uses a single-phase model, and it cannot distinguish between fluid and solid phases, and the entire mass is modeled as a bulk flow. Therefore, the friction parameters should be varied to match the observed flow paths in case of known debris flow events. It is quite possible that different events in the same torrent may show differences in composition. This fact makes the calibration of the friction parameters much more difficult. Therefore, a number of simulations with different values for each input parameters were carried to get the desired results. The results were validated with field data, and the best fitted simulation outputs were adopted for final analysis (Sosio et al. 2008).

Simulation output is required to be verified with previously modeled event as a part of validation strategy. In this context, the input parameters are important because these parameters would affect the simulation results. But unfortunately, in the present study, we did not have such opportunity for validation due to lack of earlier model information. Rather validation was carried out on collected field data in terms of their shear strength parameters and flow characteristics. In this regard, to get real field data, it is always recommended to collect such data at the earliest after an event.

This is why a number of simulations were tried with different combinations of frictional and other geo-mechanical parameters. Optimal friction values were zeroed from standard range for the concerned type of debris. Numerous simulations were carried out with the dry friction varying from 0.05 to 0.2 and viscous turbulent flow from 100 to 2000 m/s^2 (Sosio et al. 2008). Nevertheless, this exercise was done when values of other input parameters, namely, density of materials, release height, earth pressure coefficient (lambda), and the percent of momentum, were kept unchanged. Out of these simulations, the one which approximated the real debris flow most closely, as cross-checked from satellite images and field verification, was selected as the best model. Subsequently, outputs of simulation were validated by comparing the total length of runout distance and their spatial coverage vis-à-vis real flow path visible on the ground.

While carrying out the simulations with varying frictional parameters, pixel-wise spatial matching of not less than 90% with real event visible on satellite image was categorically chosen for the best simulation. The frictional and other parameters of this particular simulation representing the best model were noted. It is observed that an enhancement in the frictional coefficient μ (Mu) causes a shortening of the runout distance due to increase in the basal friction, resisting the flow movement. The variation of ζ (Xi) value, however, did not have any significant effect on the length of flow runout. However, in general parties, an increase in ζ (Xi) value should increase the runout distance as it results in a relatively smoother flow. In the present case, frequent subtle breaks in slopes in the long runout path, and the type of material dislodged has posed a hindrance to this.

Among RAMMS model outputs, momentum is not absolute as it simply considers momentum as a product of flow height and velocity. Thus, the unit is m^2/s. To get real momentum in (kg*m/s), this value is multiplied by density of debris and area under consideration. Additionally, this numeral simulation model does not provide (1) en route erosion and (2) side channel contribution to the main flowing mass along runout. In most of the cases, variation in output geophysical parameters is reported due to the above reason. Therefore, maximum valuation of parameters has been provided with error values. The output bound within error limits ensures that runout is restricted within the real debris flow channel as verified in field and/or satellite image.

3.5 Results and Discussion

3.5.1 Interpretation of Simulation

Numerical simulation, adopted in this work, provides four vital physical outputs, viz., velocity, height, pressure, and momentum of debris flow. Longitudinal profiling of runout and point data collection of a specific location are also permitted. Debris flow height is of major concern due to the fact that the financial costs of clearing huge debris can be very high, and debris of large quantity cut off the road and ultimately disrupt the lifeline of these hilly areas. Therefore, velocity and momentum are very important to specify the type and nature of any remedial structures which can withstand the initial thrust of the flow and arrest further movement of flow and reduce damage. Variation of vital physical parameters of simulations is discussed for each flow.

3.5.1.1 Kedarnath

Immediate upstream of Kedarnath temple, the debris flow-1 appears as an unbranched torrents until deposition. The length of total runout was about 1.5 km which is long enough but with intermittent slope breaks. In the downstream of temple, another flow, namely, debris flow-2, has an even longer runout of a length around 2 km. However, near the depositional zone, a bifurcation happened. Debris flow-1 revealed a maximum velocity of 5–7 m/s and a flow height of 4–6 m near the base. Though debris flow-2 is more in length compared to debris flow-1, it showed lower flow height (1–3 m) and velocity (3–5 m/s) at the base mainly owing to slope breaks. Thus, in terms of flow velocity and height cum depositional thickness, debris flow-1 becomes mightier than debris flow-2. Intriguingly, the high momentum remained concentrated symmetrically in the middle part of both the flows (Fig. 3.2). Spatial variation of pressure appeared quite similar to height and thus became redundant to be presented separately.

In a nutshell, it is important to note that the debris flows, close to Kedarnath shrine, achieved a runout distance of 1.5–2 km and emptied their load into Saraswati River which in turn would reach the Mandakini River. The simulated outputs showcased that the debris flows are capable of contributing sufficient sediment and debris load to the Saraswati River (Champati Ray et al. 2015). In addition to this, satellite image analysis revealed that on the right bank of Mandakini, important tributary streams, viz., Dudhaganga and Madu Ganga, also presumably dumped debris into the Mandakini River. As a result of which, the carrying capacity of the Mandakini (Saraswati) had reduced significantly resulting in diversion of flow to

Kedarnath temple area and easternmost Paleochannel of Saraswati when GLOF occurred on 17 June 2013.

3.5.1.2 Uttarkashi

The simulated model of Uttarkashi region reveals that the total release volume was 27,444 m^3. The flow height varies from 1.69 to 0.28 m considering all three branches. But it can be seen that the maximum flow height (1.69–1.41 m) is in the right channel of debris flow. On the other hand, a sudden change in topographic slope was observed in the left channel, and this could be a possible reason of sudden change in the path of the material. The material started flowing from the release area and went largely to the middle and right channel. So the average range of flow height seen in the middle and right paths varies from 1.13 to 0.56 m. But a lower flow height was observed in the left channel. The base of the slope was covered by the finer materials of the debris flow, and most of the coarse fragments got deposited on relatively gentler upslope areas (Fig. 3.5).

Model result provides a maximum velocity of 10.19 m/s (± 0.56 m/s) at the initiation zone and also in the right channel. From the longitudinal profile, it is evident that velocity of 10–6 m/s continued almost in every channel up to 400 m and then suddenly decreases to 2 m/s at a distance of 430 m in the left channel. This could be due to topographic flattening in that section; afterward it increased slightly till 4 m/s and continued till the end. In other two channels, velocity of flow appears to be intermediate ranging from 4 to 9 m/s which gradually decreases with height. Pressure substantially decreases as the debris flow continues in the torrent from its initiation zone. The highest pressure recorded at the initiation zone, i.e., 254.46 Kpa (± 2.3 Kpa), and the maximum momentum of 14.72 m^2/s (± 0.76 m^2/s) were observed in the right channel. Subsequently, there was a gradual decrease in momentum from 400 m altitude, and then it reduces with a consistent value up to the end of the channel. In the left channel, it was found that there was a decreasing momentum from the beginning due to a flattening in the topography (Fig. 3.5).

3.5.1.3 Ukhimath

In the Ukhimath area, the total release volume was 37,837 m^3. The maximum height value was recorded to be 3.29 m (±0.33 m) near Mangali village and its connecting road with Ukhimath bazar. According to the field information and photographs from the secondary source, it is considered to be quite a good match. A sudden change in topographic slope was observed near the hill edge of Mangali village which could be the reason of sudden velocity decrease and consequent increase of height. However, the longitudinal profile of the whole runout zone indicates that the height (0.5–0.7 m) was relatively constant up to Chunni village, and the rest of the path shows very less value ranging from 0.1 to 0.2 m with slight increase near the Mandakini River where the height was approximately 0.3 m.

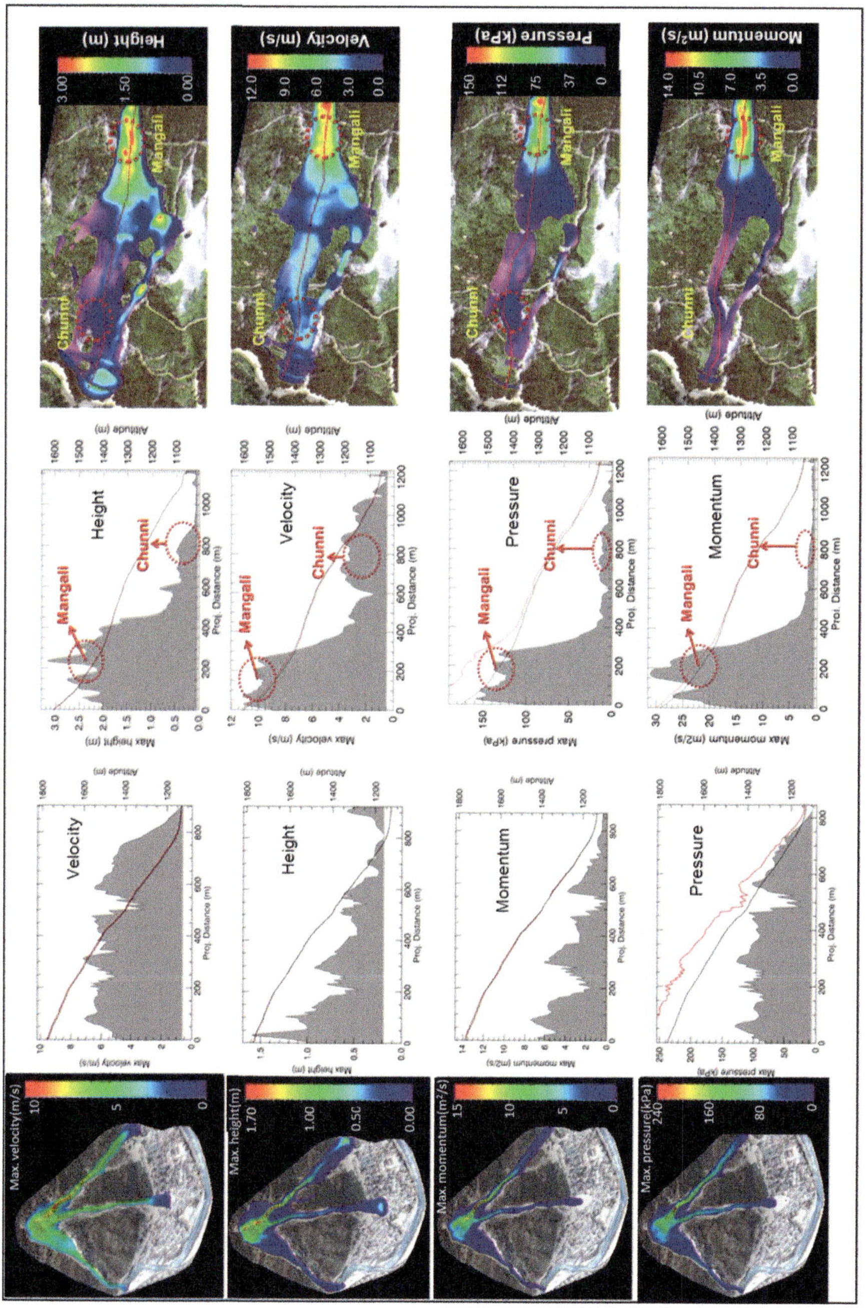

Fig. 3.5 Modeled outputs of debris flow at Uttarkashi (**a–d**) and Ukhimath (**e–h**)

Model result shows a maximum velocity of 12.9 m/s (±0.34 m/s) at the initiation zone (Fig. 3.5). In view of close proximity to the source area, relatively higher flow velocity was observed at Mangali village. From the longitudinal profile, it is evident that a velocity of 10–12 m/s continued up to 400 m and then suddenly reduces to 2 m/s at a distance of 600 m. This could be due to the decrease in topographic slope. Afterward, velocity again increased slightly near the Chunni village reaching 4 m/s.

Pressure substantially decreases as the debris flow continues in the torrent from its initiation zone. But we have observed a sudden drop of pressure from a distance of 400 m up to the Chunni village (Fig. 3.5). Before the Mangali village, pressure varies from 126 kPa to 189 kPa, and at Chunni village, it was 80–85 kPa. Estimation of material pressure in a torrent is an essential for construction of check dam or retaining wall in high-risk areas.

Maximum momentum of 30m^2/s (±0.23 m^2/s) was found near Mangali village due to high velocity (Fig. 3.5). Subsequently, a gradual decrease in momentum was observed up to 380 m and then reduced up to Chunni village and the distal part of the flow. The maximum flow momentum recorded at Chunni village was 2–3 m^2/s (Fig. 3.5).

3.5.1.4 Maithana

From the altitude and velocity plots for the landslide, it was observed that the release area of the sliding material is at 1120 m altitude, and the maximum velocity attained by the sliding material was around 10.5 m/s (±0.78 m/s). From the two-dimensional animations, it was observed that the velocity at the time of sliding ranged between 3 and 10.5 m/s; the height values were between 2.2 and 6.2 m. It was also observed that there is a break in slope. Thus, for the portion of the slide below the slope break, model shows maximum velocity.

3.5.2 Instrumental Validation of Shear Strength Parameters

RAMMS numerical simulation-derived models require cohesion (c) and frictional coefficient for dry and liquid phases (μ and ξ, respectively) for soil/debris as inputs. Cohesion is independent of stress systems and is dependent more on geochemical properties of the material. Frictional coefficient (static) for dry debris phase (μ) is related to the topographic slope by the rule of friction: $\tan \varphi = \mu$ (considering angle of sliding equal to angle of repose). Thus, theoretically, the instrument derived and modeled inputs of shear strength parameters have to be similar if the simulation is correct and assumptions are within the range of error. To get the cohesion (c) and angle of internal resistance (φ), direct shear instrument was utilized. Samples were saturated to the maximum considering the then in situ saturation state in the field. Results derived from direct shear instrument followed a Mohr-Coulomb failure behavior, i.e., $\tau = \sigma \tan \varphi + c$, depicting a straight line in the cross-plot of normal

Table 3.1 Shear strength parameters: input to model vs. instrument-derived outputs

Flow characteristics				Shear strength parameters			
Flow location	Total runout length (Km)	Simulated flow height at base (m)	Simulated velocity at base (m/s)	Inputs provided to model		Outputs derived from direct shear instrument	
				Cohesion (c) (kPa)	Internal shear angle (φ) (°)	Cohesion (c) (kPa)	Internal shear angle (φ) (°)
Kedarnath	2	1–6	5–7	20–30	25–30	40.6	24
Uttarkashi	0.7	5.5	5–25	35–40	25–30	57.13	25
Ukhimath	1.2	6.2	2–13	25–30	28–35	20.12	33
Maithana	0.6	2.2–6.2	3–10.5	20–30	15–25	25.6	16

vis-à-vis shear stress (Fig. 3.4). As each model is frozen once, it approximates the real debris flow and its μ and φ are cross-checked with the instrumentally derived c and φ values from the soil sample (Table 3.1). It is to be noted that when shear strength model inputs in RAMMS model and instrument-derived outputs are comparable, then it is considered that simulation model validates well with the real-world situation.

3.6 Conclusions

Two important outputs were derived by modeling of debris flow events in three dimensions utilizing satellite images and other geo-engineering parameters. Firstly, this work showcased successful simulation of selected debris flow events. The results, in terms of spatial variation of important geophysical parameters, namely, velocity, height, pressure, and momentum, are enriched by remotely sensed and ancillary earth observation data products. This perhaps demonstrates the vast extent of utilization of space technology-derived products. Secondly, it caters stakeholders to the critical insight of the events and related consequences. This work reveals that the modeled flow is capable of depositing enormous debris, which partially blocks and brings in a change in flow capacity and/or path of Mandakini and the Saraswati rivers. With additional water made available to the channels due to breaching of Chorabari Tal, the original course of Mandakini was blocked by debris, allowing the lake discharge and heavy rainwater flow along a Paleochannel, a rather easy alternate path through the town. With a modified flow capacity/path owing to partial blockade of channels by the debris, water and debris caused a maximum devastation inside the Kedarnath town. Considerable velocity, height, pressure, and momentum were also obtained in case of debris flows that happened in Ukhimath, Uttarkashi, and Maithana.

Therefore, the following significant conclusions can be drawn from the study:

- Numerical simulation and flow modeling can be used for calculation of vital physical flow parameters to help mitigation. The height of check dams should always be more than the height of the estimated debris flow. The width of such dam should consider the momentum and pressure of flow estimated at that location.
- Modeled height, momentum, pressure, and velocity of debris flow can be used to assess blockade of rivers and possible diversion and resulting inundation.
- Shear strength parameters used for a numeral simulation model can be validated by laboratory instrumentation techniques.
- Future potential debris flows at nearby vulnerable locations can be modeled with similar technical properties. Change detection in case of repeated flows can be assessed with the help of such models.
- RAMMS-derived models neither consider side channel contribution nor assimilation of mass due to en route erosion of the flow which presumably will increase the volume. The simulated height and momentum thus are to be considered as lower limit of the parameters, while the flow may have been more powerful in reality.

References

Anbalagan R (1992) Landslide hazard evaluation and zonation mapping in mountainous terrain. Engineering Geology 32:269–277

Anbalagan R, Kumar R, Lakshmanan, K, Parida S, Neethu S (2015) Landslide hazard zonation mapping using frequency ratio and fuzzy logic approach. A case study of Lachung Valley, Sikkim. Geoenvironmental Disasters 2015 2:6 DOI: https://doi.org/10.1186/s40677-014-0009

Ayotte D, Hunger O (2000) Calibration of a runout prediction model for debris flows and avalanches. Paper published in Proceedings of the Second International Conference on Debris-Flow Hazards Mitigation: Mechanics, Prediction, and Assessment, Taipei, Taiwan, August 16–18, Rótterdam, pp 505–514

Bist KS, Sah MP (1999) The devastating landslide of August 1998 in Ukhimath area, Rudraprayag district, Garhwal Himalaya. Current Science 76 (4):481–484.

Bist KS, Sinha AK (1980) Some observations on the geological and structural setup of Okhimath area in Garhwal Himalaya. Himalayan Geology 10: 467–475.

Brand EW (1995) Keynote Paper: Slope instability in tropical areas: in Bell (ed.,), Proceedings of the Sixth International Symposium on Landslides, 10–14 February 1992, Christchurch, New Zealand, A.A. Balkema, Rotterdam 3:2031–2051

Champati Ray P K, Chattoraj SL, Bisht M P S, Kannaujiya S, Pandey K, Goswami A (2015) Kedarnath disaster 2013: causes and consequences using remote sensing inputs. Nat Hazards (2016) 81:227–243. DOI https://doi.org/10.1007/s11069-015-2076-0

Champati Ray PK, (2013). A tale of two lakes from Uttarakhand. Indian Landslides, 6 (2): 1–8

Champati Ray PK, Chattoraj SL (2014) Sunkoshi landslide in Nepal and its possible impact in India: a remote sensing based appraisal. International Archive of ISPRS, Commission VIII (WG VIII/1), pp 1345–1351

Champati Ray PK, Chattoraj SL, Chand DS, Kannaujiya S (2013a) Aftermath of Uttarakhand disaster 2013: an appraisal on risk assessment and remedial measures for Yamunotri shrine using satellite image interpretation. Indian Landslides 6 (2):61–70

Champati Ray PK, Chattoraj SL, Kannaujiya S (2013b) Uttarakhand Disaster 2013: Response and Mitigation measures using remote sensing and GIS. Pre workshop full publication In: National work shop on Geology and Geo-heritage sites of Uttarakhand with special reference to geo-scientific development of the region organized by Indian geological Congress (IGC), Roorkee, jointly with L.S.M. Govt. PG College, Pithoragarh, Nov 11 and 12, pp 37–45

Chattoraj SL, Champati Ray PK (2015) Simulation and modelling of debris flows using satellite derived data: A case study from Kedarnath area. International Journal of Geomatics and Geosciences 6(2):1498–1511

Chattoraj SL, Champati Ray PK, Bandopadhyay S (2014) Debris Flow Simulation and Modeling: A Case Study from Kedarnath Area. In Abstract Proceedings: Geo-Environmental Hazards and Neo-Tectonic Activities in Himalaya, being held at HNB Garhwal University Campus Badshahi Thaul, Tehri Garhwal, October 28–30, 2014: 26

Chattoraj SL, Ketholia Y, Champati Ray PK, Kannaujiya S (2015a) Debris flow modelling and risk assessment of selected landslides from Uttarakhand. All India Seminar on slope stability issues in opencast mining and civil engineering (SSIOME), NIT- Rourkela, 25–26 July, pp 90–95

Chattoraj SL, Ketholia Y, Champati Ray PK, Pardeshi P (2015b) 3-Dimensional modeling of 2014-Malin Landslide, Maharashtra using satellite derived data: A quantitative approach by numerical simulation technique Abstract Volume of ISPRS WG VIII/1 Workshop on Geospatial Technology for Disaster Risk Reduction, 17th December 2015, Jaipur, India, pp. 7–8

Chattoraj, SL (2016). Debris Flow Modelling and Risk Assessment of Selected Landslides from Uttarakhand- Case Studies using Earth Observation Data, In: Santra, A. and Mitra, S., (Eds.), Remote Sensing Techniques and GIS Applications in Earth and Environmental Studies. IGI Global Publication, Hershey, Pennsylvania, pp. 111–121. ISBN: 978-1-5225-1814-3.

Chaturvedi P, Jaiswal B, Sharma S, Tyagi N (2014) Instrumentation Based Dynamics Study of Maithana Landslide near Chamoli, Uttarakhand. International Journal of Research in Advent Technology 10:127–132

Christen MK, Walski J, Bartelt P (2010) RAMMS: Numerical simulation of dense snow avalanches in three-dimensional terrain. Cold Regions Science and Technology 63 (1/2):1–14

Cruden DM, Varnes DJ (1996) Landslides types and processes, In: Landslides Investigation and Mitigation, in Turner, A. K., and Schuster, R. L., eds., Transport Research Board, Washington, D.C, Special Report 247:36–71

Deganutti AM, Marchi L, Arattano M (2000) Rainfall and debris-flow occurrence in the Moscardo basin (Italian Alps): *in* Wieczorek, G.F., and Naeser, N.D., eds., Debris-Flow Hazards Mitigation: Mechanics, Prediction, and Assessment: Proceedings of the Second International Conference, Taipei, Taiwan, August 16–18, 2000, A.A. Balkema, Rotterdam, pp. 67–72

Dobhal DP, Gupta AK, Mehta M, Khandelwal DD (2013) Kedarnath disaster: facts and plausible causes. Current Science 105(2):171–174

Gupta RP, Kanungo DP, Arora MK, Sarkar S (2008) Approaches for comparative evaluation of raster GIS-based landslide susceptibility zonation maps. International Journal of Applied Earth Observation and Geoinformation 10:330–341

Gupta V, Bist KS (2004) The 23 September 2003 Varunavat Parvat landslide in Uttaranchal township, Uttaranchal. Current Science 87:119–131

Gupta V, Sah MP, Virdi NS, Bartarya SK (1993) Landslide hazard zonation in the upper Satlej Valley, District Kinnaur, Himachal Pradesh. J Himal Geol 4:81–93

Herva'set J, Barredo JI, Rosin PL, Pasuto A, Mantovani F, Silvano S (2003) Monitoring landslides from optical remotely sensed imagery: the case history of Tessina landslide, Italy. Geomorphology 54:63–75

Hungr, Oldrich, Morgan, GC, VanDine, DF, Lister DR (1987) Debris flow defenses in British Columbia, in Costa, J.E., and Wieczorek, G.F., eds., Debris flows/avalanches: Process, recognition and mitigation, Geological Society of America. Reviews in Engineering Geology 7:201–222

Islam Md Ashraful, Chattoraj SL, Champati Ray PK (2013) Ukhimath landslide 2012: causes and consequences. International journal Geoinformatics and Geosciences 4 (3):544–557

Iverson RM, Denlinger RP, LaHusen RG, Logan M (2000) Two-phase debris flow across 3-D terrain: Model predictions and experimental tests. Paper published in Proceedings of the Second International Conference on Debris-Flow Hazards Mitigation: Mechanics, Prediction, and Assessment, Taipei, Taiwan, August 16–18, Rótterdam, pp 521–529.

Iverson RM, Reid ME, La Husen RG (1997) Debris-flow mobilization from landslides. The Annual Review of Earth and Planetary Sciences 25:85–138

Ketholia Y, Chattoraj SL, Kannaujiya S, Champati Ray PK (2015) Role of Earth Observation Data in determination of Slope Stability in parts of Uttarakhand Himalaya. Full paper in Proceedings of International Conference on Engineering Geology in New Millennium (EGNM), Indian Society of Engineering Geology, IIT Delhi, October 27–29, pp.223

Kimothi MM, Garg JK, Ajay, Joshi V (2005) Slope Ancient religious Uttarkashi town (Garhwal Himalayas, Uttaranchal, Observation from IRS-P6 (Resourcesat-1) high resolution LISS-IV data. Map India, pp 1–11

Kumar K, Devrani R, Kathait A, Aggarwal N (2012) Micro-Hazard Evaluation and validation of landslide in a part of North Western Garhwal Lesser Himalaya, India. International Journal of Geomatics and Geosciences 2:3

Kumar VK, Lakhera RC, Martha TR, Chatterjee RS, Bhattarcharya A (2007) Analysis of the 2003 Varunavat Landslide, Uttarkashi, Indian using Earth Observation data. Environmental Geology 55(4):789–799

Martha T, Kumar VK (2013) September, 2012 landslide events in Okhimath, India-an assessment of landslide consequences using very high resolution satellite data. Landslides 10:469–479

Naithani AK (2002) The August, 1998 Okhimath tragedy in Rudraprayag district of Garhwal Himalya, Uttaranchal, India. GAIA 16:145–156

Onagh M, Kumra VK, Rai PK (2012) Landslide susceptibility mapping in a part of Uttarkashi District (India) by multiple linear regression method. International Journal of Geology, Earth and Environmental Sciences 2(2):102–120

Paul D, Bisht MPS (1993) Pravatiya vikas me bhuskhalan ek paryavaryaniya samasya. Himalayan Geology 14:157–170

Quan Luna B, Blahut J, van Westen CJ, Sterlacchini S, van Asch TWJ, Akbas SO (2011) The application of numerical debris flow modelling for the generation of physical vulnerability curves. Natural Hazards and Earth System Sciences 11:2047–2060

Rickenmann D (2005) Runout prediction methods. In: M. Jakob & O. Hungr (eds.), Debris-flow Hazard and Relation Phenomena, Chichester. Springer pp 305–324

Rickenmann D, Laiglec D, Mc Ardell BW, Huebl J (2006) Comparison of 2d debris-flow simulation models with field events. Computers & Geosciences 10:241–264

Salm B, Burkhard A, Gubler HU (1990) Berechnung von Fliesslawinen: Eine Anleitungfuer Praktiker; mit Beispielen. Mitteilungen des Eidgenoessischen Instituts fuerSchnee- und Lawinenforschung 47:1–37

Sarkar S, Kanungo DP, Chauhan PKS (2010) Varunabat landslide disaster in Uttarkashi, Garhwal Himalaya, India. Quaternary journal of Engineering geology and Hydrogeology 44:1–8

Sarkar S, Kanungo DP, Mehrotra GS (1995) Landslide Hazard Zonation: A Case Study in Garhwal Himalaya, India. Mountain Research and Development 15:301–309

Sarkar S, Kanungo DP, Patra AK (2006) Landslides in the Alaknanda Valley of Garhwal Himalaya, India. Quarterly Journal of Engineering Geology and Hydrogeology 39:79–82

Sati SP, Naithani A, Rawat GS (1998) Landslides in the Garhwal Lesser Himalaya, UP, India, 18 (3): 149–155.

Scott KM (2000) Precipitation-triggered debris-flow at Casita Volcano, Nicaragua: Implications for mitigation strategies in volcanic and tectonically active steeplands. Paper published in

Proceedings of the Second International Conference on Debris-Flow Hazards Mitigation: Mechanics, Prediction, and Assessment, Taipei, Taiwan, August 16–18, Rótterdam, pp 3–13
Sosio R, Crosta GB, Hungr O (2008) Complete dynamic modeling calibration for the Thurwieser rock avalanche (Italian Central Alps). Engineering Geology 100:11–26
Sundriyal YP, Tripathi JK, Sati SP, Rawat GS, Srivastava P (2007) Landslide-dammed lakes in the Alaknanda Basin, Lesser Himalaya: Causes and implications. Current Science 93(4)
Thakur VC, Rawat BS (1992) Geologic Map of Western Himalaya, 1:1,000,000, Dehra Dun, India Wadia Institute of Himalayan Geology
Tsai MP, Hsu YC, Li HC, Shu HM, Liu KF (2011) Application of simulation technique on debris flow hazard zone delineation: a case study in the Daniao tribe, Eastern Taiwan. Natural Hazards and Earth System Sciences 11:3053–3062
Valdiya KS, Paul SK, Chandra T, Bhakuni SS, Upadhyay RC (1999) Tectonic and lithological characterization of Himadri (Great Himalaya) Between Kali and Yamuna rivers, central Himalaya. Himalayan Geology 20(2):1–17

Chapter 4
Ionospheric Total Electron Content for Earthquake Precursor Detection

Gopal Sharma, P. K. Champati Ray, and Suresh Kannaujiya

4.1 Introduction

Understanding earthquake precursory phenomena based on ionosphere perturbation is a fairly new field in geoscience today and has achieved promising success. Scientists across the globe are now trying to learn insight about the physical and chemical processes involved in the upper atmosphere and beyond during the earthquake preparatory period. One of such studies is based on global navigation satellite system (GNSS) observations. Global Positioning System (GPS) is currently one of the most popular global navigation satellite positioning systems widely available for such society application. GPS has led to technical revolutions in the field of applications like navigation as well as in upper atmospheric/ionospheric studies. GPS signals from the satellites encountered the ionosphere before it is captured by the receiver on the ground. In this process, the free electrons in the ionosphere affect the propagation of the signals by changing their velocity and direction of travel. A number of recent investigations have suggested that satellites and ground-based facilities like that of GNSS may detect earthquake precursors a few hours or days prior to the main event due to ionospheric perturbations induced by initiation of earthquake process. The typical phenomenological features of ionospheric precursors of strong earthquakes are summarised by Pulinets et al. (2003). The parameter of ionosphere that produces most of the effects on radio signals is the total electron

G. Sharma
North Eastern Space Applications Centre (NESAC), Department of Space, Government of India, Umiam, India

P. K. Champati Ray (✉) · S. Kannaujiya
Geosciences and Disaster Management Studies Group, Indian Institute of Remote Sensing (IIRS), Indian Space Research Organisation (ISRO), Department of Space, Government of India, Dehradun, India
e-mail: champati_ray@iirs.gov.in

R. R. Navalgund et al. (eds.), *Remote Sensing of Northwest Himalayan Ecosystems*,
https://doi.org/10.1007/978-981-13-2128-3_4

content (TEC). The TEC is defined by the integral of electron density in a 1 metre square column along the signal transmission path. The ionosphere causes GPS signal delays to be proportional to the TEC along the path from the GNSS satellite to a receiver. The TEC measurements obtained from dual frequency GNSS receivers are one of the most important parameters to characterise Earth's ionosphere. The changes in the Earth's ionosphere can be used to derive the information about an impending earthquake. Therefore, it is very important to monitor the TEC variation due to tectonic deformation prior to an earthquake and its validation in real-world situation.

4.2 Theory of Earthquake Preparation Mechanism (Lithosphere–Ionosphere–Magnetosphere Coupling Mechanism)

In the earthquake preparation zone, when rocks are under stress, electronic charge carriers are activated, which are known as positive haloes. These positive haloes leave the electrons and move to the surface as well as unstressed part of the rock. These, in turn, ionise the lower atmosphere, and a lot of positive ions are generated which are free to move through the troposphere up to the lower ionosphere where they join with electrons. Depending upon the process of ionisation, the electrons either deplete or increase as these are initially pulled by positive ions, thereby leading to first decrease and then increase due to continuous flow of electrons. This theory has been described in detail by a number of publications dealing with hypothesis and experimental results (Grant et al. 2015; Freund 2011; Freund et al. 2009). In addition to this mechanical process, active geochemical processes can also contribute. This includes release of radon and several other gaseous components. The ionospheric precursor initiates with the formation of ion clusters resulting from ion–molecular reactions and water molecule attachment to the finally formed ions in the near-ground layer of the atmosphere (Pulinets 2004). Further, quasi-neutral clusters are formed due to Coulomb attraction of positive and negative ion clusters. This is called coagulation (Kikuchi 2001; Horanyi and Goertz 1990). The next stage in the ionosphere perturbation is the generation of electric field. Before an earthquake, intensive gas is released as discussed earlier from the crust (mainly CO_2) in the earthquake preparation zone (Voitov and Dobrovolsky 1994). These gases play a dual role; firstly by generating air motion, they create instabilities to stimulate acoustic gravity waves, and secondly these air motions destroy neutral clusters because of the weakness in Coulomb interaction force. As a result, the near-ground layer of the atmosphere becomes richer in ions within the short time period.

The deviation in the total electron content (TEC) may be created in the ionosphere depending on the electric field generated on the ground surface (Pulinets et al. 1998). These deviation in electron concentration may be either negative or positive vis-à-vis generated electric field direction. Additionally, the electric field penetrates without any decay to the higher levels of the ionosphere (D, E, F1 and F2) in the order of increasing height due to the equipotentiality of geomagnetic field lines. In F-region two main effects take place. Firstly, acoustic gravity waves are generated giving rise to the small-scale density irregularities within the ionosphere (Hegai et al. 1997). This occurs in the area of maximal conductivity. The other probable effect is the intense irregularities of electron concentrations in the F2 region of the ionosphere (Pulinets et al. 2003). These irregularities in the ionosphere can be detected by the navigation satellites and ground-based GNSS receivers and ionosondes (Liu et al. 2004). Due to the complexity of particle motion and large-scale anomalies in the F-region, propagation may be registered not just over the impending earthquake epicentre but also may shift towards equatorial direction. At further heights (magnetosphere), one can expect irregularities along the geomagnetic field lines into the magnetosphere where VLF emissions of different origins remain scattered which results in increased levels of emission within the magnetic duct and change in the shape of magnetospheric area due to plasma drift (Sorokin et al. 2000; McCormick et al. 2002; Kim and Hegai 1997; Shklyar and Nagano 1998). The shape of the area at magnetospheric heights will be elongated in the zonal direction proportionally in the ratio of 1:3 for meridional and longitudinal sizes of the modified volume of the magnetosphere (Kim and Hegai 1997; Larkina et al. 1989). Finally, this chain of processes in the atmosphere and magnetosphere that originated from the near-ground phenomena results in ionisation of the lower ionosphere which increases the electron content in the D-region (Pulinets 2004).

4.3 Recent Advancement in GNSS TEC-Based Precursor Study

Since the great Alaskan earthquake in 1964, many investigations on ionospheric perturbations due to earthquakes have been presented (Blanc 1985). Numerous studies show the TEC derived from GNSS/GPS modelling varies between 0 and 15 days in most of the cases before the occurrence of large-magnitude earthquake (Sharma et al. 2017a, b; Singh and Chauhan 2008; Pulinets 2009; Liu et al. 2011). These anomalies were interpreted either due to the influence of solar flare, geomagnetic storm or earthquakes. The depletions and enhancements in TEC gradient obtained were reported to be induced due to an earthquake in a number of events such as M = 7.0 + global earthquakes (Yao et al. 2012), 12 January 2010 Mw 7.0 Haiti earthquake (Liu et al. 2011), 11 March 2011 Mw 9 Tohoku earthquake (Dimitar et al. 2011), 21 January 2003 Mw 7.8 Colima earthquake (Dogan et al. 2011; Jhuang et al. 2010; Pulinets et al. 2005), 12 May 2008 Mw 8.0 Wenchuan Earthquake and 17 August 1999 Mw 7.6 Izmit earthquake. Some of the recent studies include statistical characteristics of seismo-ionospheric GPS TEC disturbances prior to global Mw $\geq$ 5.0 earthquakes (1998–2014) by Shah and Jin (2015), GNSS modelling for earthquake precursor studies for 25 April 2015 Nepal earthquake by Sharma et al. 2017a and investigation of ionospheric TEC precursors related to the Mw 7.8 Nepal and Mw 8.3 Chile earthquakes in 2015 based on spectral and statistical analysis by Oikonomou et al. (2016). These observed anomalies were considered to be related to the seismic activity with no external effect from solar flare or geomagnetic storm. The results from such analyses indicate lithosphere–ionosphere coupling within the earthquake preparation zone, and it is possible to monitor using space-based observation system.

4.4 GNSS TEC Measurement for Precursory Signal Detection

The slant TEC (TEC along the slant ray paths between a satellite and a ground station) can be obtained from the code pseudo-range and frequency measurements using dual frequency receivers (Blewitt 1990). For TEC applications for earthquake precursor detection, slant TEC measurement is converted into vertical TEC represented by vTEC, in el/m^2, with assumption of thin shell of ionosphere at fixed height (Kersley et al. 2004; Pulinets et al. 2005; Ma and Maruyama 2003) as

$$\mathrm{vTEC} = \mathrm{sTEC} \times \cos z', \tag{4.1}$$

where

$$\mathrm{sTEC} = \frac{2(f1f2)^2}{K\left(f1^2 - f2^2\right)}(P2 - P1), \tag{4.2}$$

sTEC is slant TEC, P1 and P2 are the code pseudoranges, $f1 = 1.57542$ and $f2 = 1.2276$ are the GPS frequencies and $K = 80.62$ (m^3/s^2) is a constant of the plasma frequency to the electron density.

The zenith angle, z', is expressed as

$$z' = \arcsin\left(\frac{\mathrm{RE}\ \cos\ \alpha}{\mathrm{RE} + h}\right), \tag{4.3}$$

where RE is the mean radius of the Earth, α is the elevation angle of the satellite and h is the height of the ionospheric layer, which is 350 km in the present case. Incorporating the satellite and receiver biases, vTEC can be obtained as

$$\mathrm{vTEC} = (\mathrm{STEC} - \mathrm{bs} - \mathrm{br})\cos z' \tag{4.4}$$

where bs and br are the estimated satellite and receiver biases, respectively.

Various statistical methods can be used for identification of anomalous TEC due to the earthquake preparation phenomenon; the behaviour of vTEC in most of the cases has been analysed by statistical method using boundary limit of 15 days running median ± 1.34 running standard deviation (Liu et al. 2004; Sharma et al. 2017a). TEC values crossing these limits are considered as anomaly for impending earthquake.

The TEC in the ionosphere is known to vary with solar activity, geomagnetic storm, user location and pseudo-range elevation angle and has been studied widely since decades (Buonsanto and Fuller-Rowell 1997; Fuller-Rowell et al. 1996; Mendillo 1971). These factors result in deviation of electrons in ionosphere and cause ionospheric delays in the GNSS/GPS signals which can be used to quantify the electron count in the ionosphere. Therefore, while analysing the TEC as a precursor, it is very important to analyse the effect of magnetic as well as solar flare to rule out their possible contributions.

4.5 TEC Application for Earthquake Precursors in the Himalayan Region

In this section, four case examples from Himalayan region are presented where precursors from ground-based GNSS receiver have been detected prior to the earthquake events. With GPS-derived time series, TEC variations over the time window of 31 running days (15 days before and 15 days after the event) were estimated around the earthquake within the earthquake preparation zone. Table 4.1 gives a catalogue which summarises the date of occurrence, epicentre, depth and magnitude of the earthquakes considered for demonstrating the application of TEC precursor in the Himalaya.

Table 4.1 Details of the earthquakes analysed and its relationship with TEC variation

Date of occurrence	Epicentre region	Latitude (degree)	Longitude (degree)	Magnitude (Mw)	Radius of EQ prep zone (km)	Consistent anomaly observed before (days)
08 October 2005	Kashmir	34.539	73.588	7.6	1853.53	6, high values crossing upper bound
07 December 2015	Murghob	38.211	72.78	7.2	1247.38	4–5, low values cross-ing lower bound
25 April 2015	Lamjung (Gorkha)	27.771	86.065	7.8	2259.43	0,1, 8, all high values cross-ing upper bound
11 May 2015	Kodari	27.808	86.065	7.3	1377.20	1, low values crossing lower bound

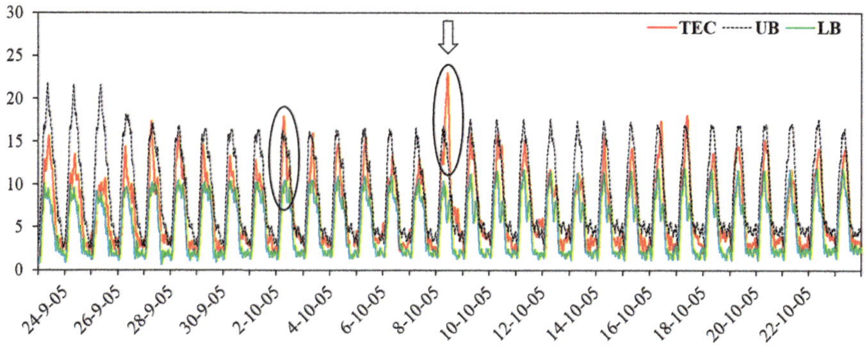

Fig. 4.1 TEC observation during Kashmir earthquake (2005) from IGS station KIT3 situated at a distance of 786 km from the epicentre. Anomaly is detected when TEC value crosses upper or lower bound as shown by dashed circles. Arrow indicates the earthquake day

4.5.1 *Kashmir Earthquake, Mw 7.6*

The Mw 7.6 Kashmir earthquake occurred on 8 October 2005. The vertical TEC variations analysed from data at station KIT3 at a distance of 786 km from the epicentre reveal strong positive pre-earthquake anomalies in 2 and 8 October (6 days before the event and on the day of the event, respectively) as shown in Fig. 4.1, highlighted by dashed circles. The details of the event are shown in Table 4.1. Variations in 25 September and 2 and 8 October are observed as slightly negative and positive pre-earthquake ionospheric anomaly as there is an abnormal decrease and increase in GPS TEC values (Liu et al. 2001, 2002). These anomalies were analysed vis-à-vis solar flare and geomagnetic storm activities with the aid of data from NOAA which reveals no significant role on TEC variation. Thus, it could be inferred that these negative and positive anomalies (enhancements) are ionospheric precursors related to the earthquake. Negative anomalies observed in 12 and 13 October and positive anomalies (enhancements) observed on 16–17 October 2005 may be attributed to 20 intermediate ($4.5 \leq$ Mw ≤ 5.6)-sized shallow earthquakes which occurred between 17 October 2005 and 29 October 2005 in the SE direction of KIT3 station at a distance ranging from 720 to 800 km from earthquake epicentres (USGS Earthquake catalogue).

4.5.2 *Murghob Earthquake, Mw 7.2*

The Mw 7.2 Murghob earthquake occurred on 7 December 2015 in Tajikistan. In this case, low TEC value crossing the lower limit is observed in 2 and 3 December 4–5 days prior to the event from the data at IGS station TASH (Fig. 4.2). Additionally, a pronounced peak (positive anomaly) in the vTEC was also observed on the earthquake day during late evening. It is also inferred that the enhancement observed in 22 December is due to the high-magnitude ($Kp \geq 5$) geomagnetic activity, hence could not be considered as ionospheric precursors.

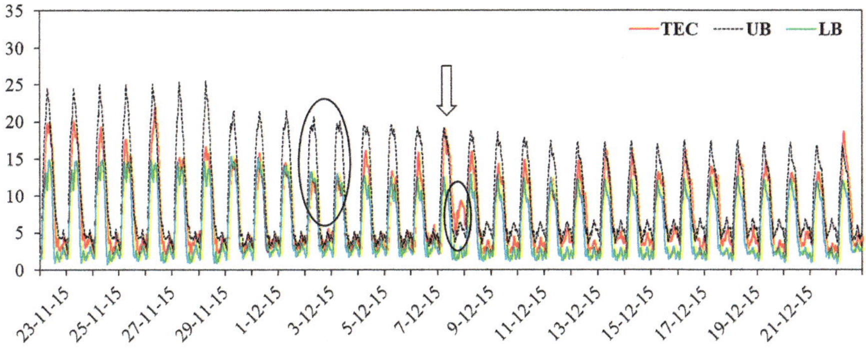

Fig. 4.2 TEC observation during Murghob earthquake estimated from IGS station TASH. Anomalies are detected when TEC value crosses upper or lower bound as shown by dashed circles. Arrow indicates the earthquake day

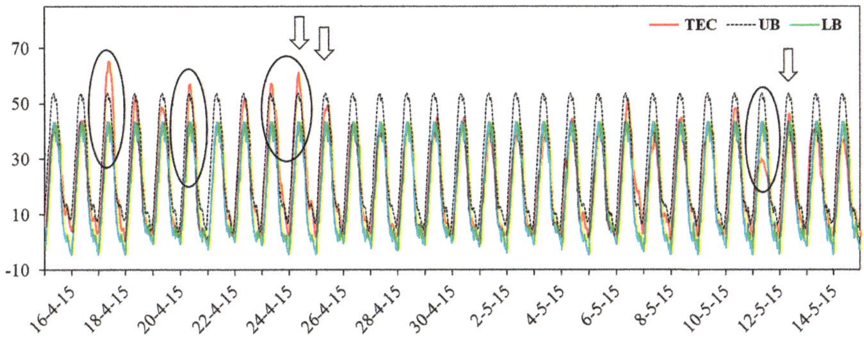

Fig. 4.3 TEC variation during Lamjung (Gorkha) and Kodari earthquakes computed from CORS station of NRSC/ISRO situated in Dehradun at a distance of 870 km. Anomalies are detected when TEC value crosses upper or lower bound as shown by dashed circle. Arrow indicates the earthquake day

4.5.3 Nepal Earthquakes

Continuous TEC observations were made using GNSS (GPS) data observed at a distance of 870 km from the epicentre, within the earthquake preparatory zone (Pulinets et al. 2005) to analyse the TEC variation prior to Mw 7.8 and Mw 7.3 2015 Nepal earthquakes. The details of the earthquake are summarised in Table 4.1. GPS data for the period of 1 month acquired at Dehradun CORS of NRSC/ISRO was analysed to observe TEC variations, Fig. 4.3.

Figure 4.3 shows the TEC time series for the duration of 1 month from 15 April to 14 May 2015 at Dehradun station. TEC values crossing the upper limit were observed on the day of the main event, i.e. 25 April 2015. Besides this, TEC were also observed to cross the upper limit on 17 and 24 April 2015. Consistently, TEC values were exceeding the limits in 24 to 25 April, which could be inferred as precursor to the main event on 25 April 2017 of Mw 7.8. In respect to Mw 7.3 event on 12 May 2015, low TEC values crossing the lower limit were observed on 11 May, and high TEC values crossing the upper limit on 10 May 2015 were observed. These observations are also considered to be attributed to the seismic event as a precursor which were unaffected by both geomagnetic and solar flare activities.

4.6 Conclusions

Numerous modelling methods have already been attempted by various authors in order to interpret anomalous state of the ionosphere during earthquakes. It is assumed that there occurs a lithosphere–atmosphere–ionosphere coupling during the seismic preparation time frame before large earthquakes. The potential mechanisms which lead to seismo-ionospheric perturbations are the penetration of anomalous vertical electric field present above the earthquake hotspots into the upper atmosphere and ionosphere, which further causes abnormality of electron density in terms of total electron count. Although various scientists suggested these ionospheric perturbations as pre-earthquake anomalies and suggest deeper research to understand the pattern and coupling mechanisms of earth's lithosphere and ionosphere, the accuracy of time and place of its probable occurrence still remains a subject of further research on ionospheric studies. The analysed ionospheric anomalies in this chapter associated with different (Mw > 7) earthquakes in Himalayan region by examining the GPS-inferred total electron content variations also support the theory of lithosphere–ionosphere coupling mechanism. Within the time window of 31 running days around the earthquake, anomalous fluctuations in TEC were observed in the form of enhancements and depletion within a range of 0–8 days prior to all the earthquakes. These anomalies were cross-checked with the global solar data in order to ascertain their causative effects. It reveals that TEC variation was unaffected by these factors, thereby indicating that these variations in TEC values were due to seismogenic causes. The enhancement observed for Kashmir earthquake on 8 October 2005 and Nepal earthquake on 25 April 2015, i.e. on the day of the main event, calls for a better approach to investigate the GPS TEC perturbations on an hourly basis in the Himalayan terrain.

References

Blanc, E., 1985. Observations in the upper atmosphere of infrasonic waves from natural or artificial sources: a summary, Ann. Geophys., 3, 673–668

Blewitt, G., 1990. An automatic editing algorithm for GPS data. Geophys. Res. Lett., 17, 199–202.

Buonsanto, M. J., Fuller-Rowell, T. J., 1997. Strides made in understanding space weather at Earth. EOS Transactions American Geophysical Union, 78, 1–7.

Dimitar, O., Pulinets, S., Alexey, R.A,. Konstantin, T., Dimitri, D., Menas, K., Patrick, T. Atmosphere-ionosphere response to the M9 Tohoku earthquake revealed by multiinstrument space-borne and ground observations: Preliminary results., 2011. Earthquake Science, 24, 1–7.
Dogan, U., Ergintav, S., Skone, S., Arslan, N., Oz, D., 2011. Monitoring of the ionosphere TEC variations during the 17th August 1999 Izmit earthquake using GPS data. Earth Planets Space, 63, 1183–1192.
Freund, F. 2011. Pre-earthquake signals: Underlying physical processes. Journal of Asian Earth Sciences, 41(4–5), 383–400.
Freund, F. T., Kulahci, I. G., Cyr, G., Ling, J., Winnick, M., Tregloan-Reed, J., Freund, M. M. 2009. Air ionization at rock surfaces and pre-earthquake signals. Journal of Atmospheric and Solar-Terrestrial Physics, 71(17–18), 1824–1834.
Fuller-Rowell, T. J., M. V. Codrescu, H., Rishbeth, R. J., Moffett, Quegan, S., 1996. On the seasonal response of the thermosphere and ionosphere to geomagnetic storms, J. Geophys. Res., 101, 2343–2353.
Grant, R. A., Raulin, J. P., Freund, F. T. 2015. Changes in animal activity prior to a major (M= 7) earthquake in the Peruvian Andes. Physics and Chemistry of the Earth, Parts A/B/C, 85, 69–77.
Hegai, V. V., V. P. Kim, and L. I. Nikiforova, 1997: A possible generation mechanism of acoustic-gravity waves in the ionosphere before strong earthquakes. J. Earthquake Predict. Res., 6, 584–589.
Horanyi, M., and C. K. Goertz, 1990: Coagulation of dust particles in a plasma. Astrophys. J.,361, 155–161.
Jhuang, H.K., Ho, Y.Y., Kakinami, Y., Liu, J.Y., Oyama, K.-I., Parrot, M., Hattori, K., Nishihashi, M., Zhang, D., 2010. Seismo-ionospheric anomalies of the GPS-TEC appear before the 12 May 2008 magnitude 8.0 Wenchuan Earthquake. Int. J. Remote Sens. 31, 3579–3587.
Kersley, L.,Malan, D., Eleri Pryse, S., Ljiljana. R., Cander., Bamford R. A., Belehaki, A., Leitinger R., Radicella, S.M., Mitchell C, N., Spencer, P.S.J., 2004. Total electron content – A key parameter in propagation: measurement and use in ionospheric imaging. Annals of Geophysics, 47, 2/3, 1067–1091.
Kikuchi, H., 2001: Electrodynamics in dusty and dirty plasmas, Kluwer Academic Publishers.
Kim, V. P., and V. V. Hegai, 1997: On possible changes in the midlatitude upper ionosphere before strong earthquakes. J. Earthq. Predict. Res., 6, 275–280.
Larkina, V. I., V. V. Migulin, O. A. Molchanov, I. P. Khar'kov, A. S. Inchin, and V. B.Schvetcova, 1989: Some statistical results on very low frequency radiowave emissions in the upper ionosphere over earthquake zones. Phys.EarthPlanet.Inter., 57, 100–109.
Liu, J. Y., Le, H., Chen, Y. I., Chen, C. H., Liu, L., Wan, W., Su, Y. Z., Sun Y. Y., Lin, C. H., Chen M. Q., 2011. Observations and simulations of seismoionospheric GPS total electron content anomalies before the 12 January 2010 M7 Haiti earthquake. Journal of Geophysical Research, 116, A04302, 1–9.
Liu, J. Y., Y. J. Chuo, S. J. Shan, Y. B. Tsai, S. A. Pulinets, and S. B. Yu, 2004: Pre-earthquake ionospheric anomalies monitored by GPS TEC. An. Geophys., 22, 1585–1593.
Liu, J.Y., Chen, Y.I., Chuo, Y.J., Tsai, H.F., 2001. Variations of ionospheric total electroncontent during the Chi-Chi earthquake. Geophysical Research Letters 28, 1383–1386.
Liu, J.Y., Chuo, Y.J., Pulinets, S.A., Tsai, H.F., Xeng, X.P., 2002. A study on the TEC perturbations prior to the Rei-Li, Chi-Chi and Chia-Yi earthquakes. In:Hayakawa, M., Molchanov, O.A. (Eds.), Seismo Electromagnetics: Lithosphere–Atmosphere–Ionosphere Coupling. TERRAPUB, Tokyo, pp. 297–301.
Ma, G., Maruyama, T., 2003. Derivation of TEC and estimation of instrumental biases from GEONET in Japan, Ann. Geophys., 21, 2083–2093
McCormick, R. J., C. J. Rodger, and N. R. Thomson, 2002: Reconsidering the effectiveness of quasi-static thunderstorm electric fields for whistler duct formation. J Geophys. Res., 107, 1396, doi: https://doi.org/10.1029/2001JA009219.
Mendillo, M., 1971. Ionospheric total electron content behaviour during geomagnetic storms, Nature, 234, 23–24.

Oikonomou, C., Haralambous, H., and Muslim, B., 2016. Investigation of ionospheric TEC precursors related to the M7.8 Nepal and M8.3 Chile earthquakes in 2015 based on spectral and statistical analysis. Nat Hazards, Volume 83, Supplement 1, pp 97–116

Pulinets, S. A., 2009. Physical mechanism of the vertical electric field generation over active tectonic faults. Advances in Space Research, 44, 767–773.

Pulinets, S. A., Legen'ka, A. D., Gaivoronskaya, T. V., and Depuev, V. Kh., 2003. Main phenomenological features of ionospheric precursors of strong earthquakes, J. Atmos. Solar-Terr. Phys., 65, 1337–1347.

Pulinets, S. A., Leyva, A., Contreras, G., Bisiacchi, G., Ciraolo, L., 2005. Total electron content variations in the ionosphere before the Colima, Mexico, earthquake of 21 January 2003. Geofísica International, 44, 4, 369–377.

Pulinets, S. A., V. V. Khegai, K. A. Boyarchuk, and A. M. Lomonosov, 1998: Atmospheric electric field as a source of ionospheric variability. Physics-Uspekhi, 41, 515–522.

Pulinets, S.A., 2004. Ionospheric Precursors of Earthquakes; Recent Advances in Theory and Practical Applications. TAO, 15, 3, 413–435.

Shah, M and Jin, S., 2015. Statistical characteristics of seismo-ionospheric GPS TEC disturbances prior to global Mw $\geq$ 5.0 earthquakes (1998–2014), Journal of Geodynamics 92, 42–49.

Sharma, G., Champatiray, P, K., Mohanty, S and Kannaujiya, S., 2017a. Ionospheric TEC modelling for earthquakes precursors from GNSS data. Quaternary International, 462, 65–74

Sharma, G., Champatiray, P, K., Mohanty, S., Gautam, P, K, R and Kannaujiya, S., 2017b. Global navigation satellite system detection of preseismic ionospheric total electron content anomalies for strong magnitude (Mw > 6) Himalayan earthquakes. Journal of Applied. Remote Sensing. 11 (4), 046018.

Shklyar, D. R., and I. Nagano, 1998: On VLF wave scattering in plasma with density irregularities. J. Geophys. Res., 103, 29515–29526.

Singh, O.P., Chauhan, V., 2008.Animalous Behaviour of GPS Based Total Electron Content (TEC) Associated with Earthquakes. 12th International Conference of International Association for Computer Methods and Advances in Geomechanics (IACMAG), 1–6 October. Goa, India

Sorokin, V. M., V. M. Chmyrev, and M. Hayakawa, 2000: The formation of ionosphere magnetosphere ducts over the seismic zone. Planet. Space Sci., 48, 175–180.

Voitov, G. I., and I. P. Dobrovolsky, 1994: Chemical and isotopic-carbon instabilities of then active gas flows in seismically active regions. Izvestiya Earth Science, 3, 20–31.

Yao, Y. B., Chen, P., Zhang, S., Chen, J. J., Yan F., Peng, W. F., 2012. Analysis of pre-earthquake ionospheric anomalies before the global M = 7.0+ earthquakes in 2010. Nat. Hazards Earth Syst. Sci., 12, 575–585

Part III
Water Resources

Summary

The Himalayan region, which is also known as the third pole, is regarded as the most fragile hydrosphere particularly in view of climate change. Therefore, it is necessary to assess its water resources, so that, policies may be drawn to preserve it for present and future. The best way to carry out such assessment is to study hydrological cycle, its processes and emulate them by adapting suitable modelling approach. The advent of geospatial technology and improvement in computing capabilities are facilitating the assessment of water resources under various scenarios for any region on the globe. This chapter covers the application of remote sensing and geographical information system in cyrosphere studies, hydrological modelling parameterization, hydrological modelling approach and identification of hydro-meteorological extremes.

The precipitation is considered as most important input to any hydrological system and its modelling. The Northwest Himalayan (NWH) region, geographical and topographical settings receive precipitation in both the forms snow and rainfall which varies spatio-temporally in the region. The snow fall and its physical properties, are again very important from availability of water during lean period for all the sectors (domestic, agriculture or Industry). Moreover, a large number of important glaciers located in this region, those needs to be studied for their retreat and shrinkage. However, the field measurements of these parameters and the maintenance of field instruments installed in NWH region are very difficult due to its rugged terrain and accessibility constraints. The region due its hydro-geological setting is very prone to various disasters such as avalanches, flash floods and glacial lake outburst flood. Recently, the region has experienced a number of such hydro-meteorological disasters, which have cost human lives and economy. These conditions in NWH region, even refrained researchers to develop or setup models for the region before the advent of geospatial technology.

The present section highlights the application of remote sensing and GIS in assessing each of the hydrological process of the hydrological cycle and their modeling. The region has unique cyrosphere, initially, the studies carried out on snow physical parameters retrieval through optical and microwave remote sensing data; snow melt modelling; and glacier dynamics using geospatial tools were discussed. In the second article of the chapter, hydrological models, their types, input requirements and capability has been provided with emphasis to NWH region through case studies. The article stresses on model parameterisation and retrieval of each parameter using remote sensing data. The case studies of the chapter highlights water resources assessment through a physically based regional scale hydrological model and probable impact of climate change on water resources of the region. The next article discusses how precipitation can be retrieved using RS and GIS and their validation using field data. The article mainly focuses on extreme rainfall events and their spatio-temporal variation in the NWH region. These extreme events usually turned to many hydro-meteorological disasters in the region. Therefore, in the last article, the disasters that usually take place in the region are discussed. Further, their modelling and mapping using the geospatial data have been illustrated through case studies. Such modelling and mapping are very useful from mitigation and damage assessment point of view. Moreover, various cause and effect kind of scenarios for the region can be generated using these approaches. Overall, the chapter is a comprehensive demonstration of water resources assessment and its management in NWH region.

Chapter 5
Cryosphere Studies in Northwest Himalaya

Praveen K. Thakur, Vaibhav Garg, Bhaskar R. Nikam, and S. P. Aggarwal

5.1 Introduction

Himalaya is also known as the "third pole" of the Earth due to the presence of the largest area and volume of seasonal snow and glacier ice outside the polar regions. Most visible effects of climate change are found in this region (Immerzeel et al. 2010; IPCC-5 2014 WGII AR5 Section 28.3). The Indian states of Uttarakhand (UK), Himachal Pradesh (HP) and Jammu and Kashmir (J&K) together constitute a unique and rugged terrain of Northwest Himalaya (NWH). Cryosphere components of NWH mainly consist of snow cover (SC), glacier ice (GI), frozen lakes and rivers in winters and high-altitude water lakes and rivers in summer, glacier lakes (GL) and a few stretches of permafrost regions. These cryospheric components provide fresh water to Perennial Rivers of NWH such as the Beas, Ganges, Satluj and Indus by melting of seasonal snow cover and glacier ice. This area has significant variations in minimum to maximum elevation and temperature ranging from 74 m to 8611 m of K2 and −40° in cold glacier areas to +40 °C in low hills and Tarai area. The major accumulation of snow mass occurs in winter period during October to March due to precipitation from NW disturbances of winter monsoon, and main melting takes place during spring and late summer time during April to July. The last 2 weeks of August to first week of September, generally, represents end of ablation season, and during this time traditional snow line elevation (SLE), showing the minimum snow line, coincides with equilibrium line altitude (ELA) of major glaciers (Huang et al. 2011, 2013). The inaccessibility, high relief and remoteness of these mountainous regions make it very difficult to map and monitor cryospheric components using traditional ground-based field instruments and surveying and mapping techniques. In

P. K. Thakur (✉) · V. Garg · B. R. Nikam · S. P. Aggarwal
Water Resources Department, Indian Institute of Remote Sensing (IIRS), Indian Space Research Organisation (ISRO), Department of Space, Government of India, Dehradun, India
e-mail: praveen@iirs.gov.in

R. R. Navalgund et al. (eds.), *Remote Sensing of Northwest Himalayan Ecosystems*,
https://doi.org/10.1007/978-981-13-2128-3_5

this scenario, remote sensing with its large area coverage, multi-resolution spatial and temporal scale, offers a unique opportunity to regularly map and monitor cryospheric components as shown by many studies (SAC 2010; National Remote Sensing Centre 2013). This chapter highlights remote sensing (RS) and geographical information system (GIS)-based studies of major components of NWH cryosphere such as seasonal SC, GI, GL and permafrost. This also includes subsections for mapping, monitoring and quantification of SC, GI in NWH along with retrieval and modelling of snowpack properties, snowmelt and glacier mass balance.

5.2 Northwest Himalayan Cryosphere

Himalayan cryosphere including that of NWH holds a very important role in Asia, as it provides meltwater to major rivers of this region and also plays an important role in modifying or controlling seasonal weather and climate. High albedo of snow and glacier ice provides a direct feedback to overall weather system of this region and is also most sensitive to any changes in seasonal weather patterns or long-term climate change or external anthropogenic forcing such as forest fires, land use land cover change and dust storms (Minnett 2014). This section is subdivided in two parts: in the first subsection, description of snow cover and permafrost of NWH is given, and in the second subsection, glaciers and glacier lakes of NWH are described. Figure 5.1a, b shows the snapshot of NWH snow cover, permafrost and glacier regions along with glacier lakes of Beas-Chenab river basin area, as seen in an

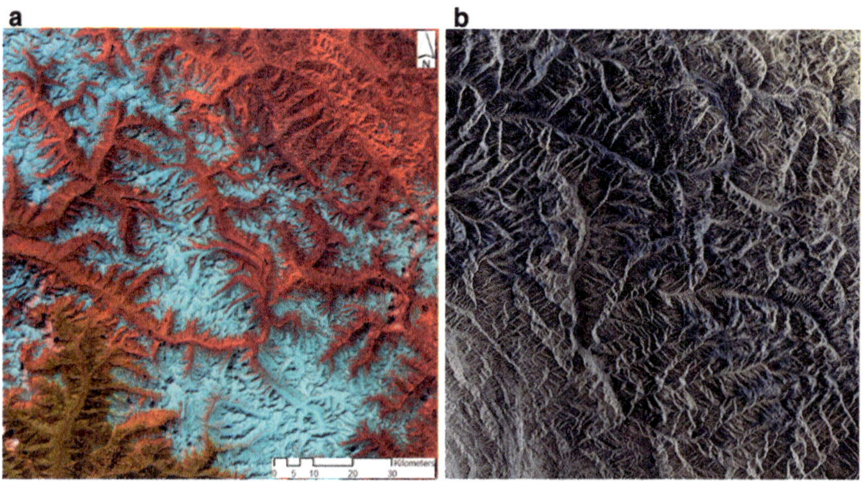

Fig. 5.1 (**a**) Parts of NWH in Beas-Chenab areas of Himachal Pradesh as seen in colour composite (SWIR: Red; NIR: Green and Red: Blue) of Resoursesat-1 AWIFS image (04 Oct 2012), (**b**) Beas-Chenab basins area as seen in RISAT-1 MRS SAR image (27 Sep. 2012)

optical image of Resourcesat-1 AWIFS and radar image of RISAT-1 Medium Resolution scan-SAR mode (MRS) of 2012.

This chapter highlights the major research work of IIRS Dehradun which has been done in this field over the last 15 years. The entire chapter has been divided into three main parts: in first part, snow cover area (SCA) and snow physical parameter estimation using various remote sensing sensors and techniques such as optical-based NDSI method for SCA mapping, SAR-based inversion models for SCA and snow parameters, and hyperspectral data-based GI for snow grain size mapping have been reported. In the second part, mapping, monitoring, glacier ice velocity and thickness estimation have been reported. This section discusses the use of SAR-based glacier classification, DInSAR and feature tracking-based glacier velocity estimation and laminar flow-based glacier ice thickness estimation in Gangotri glacier. In the last and third section of this chapter, the hydrological modelling approach for SCA and snowmelt runoff estimation and AAR-based glacier mass balance methods are discussed.

5.2.1 *Snow Cover and Permafrost of NWH*

Snow cover area usually termed as SCA or snow cover extent (SCE) represents the seasonal snow cover which falls during winters and melts during spring and summer season and also permanent snow on high elevation and glacier areas of mountainous regions in NWH. The minimum SLE can range from 1500 to 1800 m and varies yearly. Remote sensing is essential and invaluable tool for mapping SCA in high relief and rugged areas of NWH due to its large area coverage, temporal revisit and data continuity. The basin-wise area along with long-term annual minimum and maximum SCA of each basin is given in Table 5.1. Details of SCA mapping using remote sensing are given in Sect. 5.3.1.

Permafrost can be defined as any ground- or sub-surface material (excluding glaciers) that remains at or below 0 °C for at least two consecutive years (ACGR 1988; van Everdingen 1998). As both the atmospheric and ground temperatures are closely coupled, permafrost warming is the effect of global climate change in many

Table 5.1 NWH major river basins along with minimum and maximum SCA

River basin with outlet point	Total area (km^2)	Maximum SCA in km^2 (% of basin area)	Minimum SCA in km^2 (% of basin area)
Chenab Basin up to Akhnoor	22,493	15,590 (70%), varies from 65% to 72%	5400 (24%), varies from 15% to 25%
Ganga Basin up to Devprayag	23,177	9080 (46%), varies from 45% to 52%	3800 (19%), varies from 12% to 20%
Satluj Basin up to Bhakra Dam	22275	14498 (65%), varies from 56% to 68%	4528 (20%), varies from 7% to 15%
Beas Basin up to Pandoh Dam	5278	2700 (51%), varies from 40% to 54%	780 (14%), varies from 10% to 20%

parts of the world (Vaughan et al. 2013). Limited estimates and studies on permafrost are done in NWH. Globally, Gruber (2012) has derived and analysed global digital elevation model (DEM) along with global air temperature data derived from global climate reanalysis (NCAR-NCEP) and from interpolated station measurements (CRU TS 2.0) datasets to create a high-resolution estimate of global permafrost extent. Schmid et al. (2015) have made first-order assessment of permafrost distribution maps in the Hindu Kush Himalayan region using rock glaciers mapped in Google Earth with formulation of the ~1 km resolution Global Permafrost Zonation Index (GPZI). Ishikawa et al. (2001) made an attempt to differentiate rock glaciers from the discontinuous mountain permafrost zone in Kanchenjunga Himalaya, Eastern Nepal. In NWH, recently in the study by Allen et al. (2016) in mountainous areas of Kullu district, mapping of permafrost conditions is done by applying a condition on modelled mean annual ground surface temperature (MAGST) below 0 °C. This initial study showed that "permafrost may extend down to 4200 m amsl in favourable instances, which compares favourably with the observed lower elevation limit from the ca. 60 mapped rock glaciers in Kullu district (interquartile range 4280–4560 m amsl), and approximate lower limits to permafrost distribution established in relation to the local 0°C (Mean Annual Air temperature) MAAT isotherm" (Allen et al. 2016). Westermann et al. (2013) completed the transient thermal modelling of permafrost conditions in Southern Norway, and similar techniques can be applied in addition to MAGST-based GPZI method in parts of NWH.

5.2.2 *Glaciers and Glacier Lakes*

Glacier and glacier lakes cover a large area of NWH, with total number of glaciers and the total glacier ice-covered area in these three basins are 32,392 and 78,040 km^2, respectively, as per 2010 report by Space Application Centre (SAC) given to the Ministry of Environment and Forest (SAC 2010). This Himalayan glacier inventory was completed for the Indus, Ganga and Brahmaputra river basins at 1: 50,000 scale using images of Indian Remote Sensing Satellite for the period 2004–2007 (SAC-2010) and is updated from previous glacier inventory by Raina and Srivastava (2008). As per estimates of Central Water Commission (CWC) sponsored project report by the National Remote Sensing Centre (National Remote Sensing Centre 2013), there are about 2028 glacial lakes and water bodies having water spread area more than 10 hectare (Ha) in the Himalayan region catchment which contributes to river flowing in India. Out of these 503 are glacial lakes and 1525 are water bodies. After the 2013 Uttarakhand floods and Kedarnath GLOF, glacier lakes with area of 3 ha are also mapped using high-resolution remote sensing datasets of LiSS-4 sensor.

5.3 Mapping and Monitoring of NWH Cryosphere Using Remote Sensing

The traditional remote sensing working in optical imaging principles, which measures the reflected electromagnetic waves in visible and infrared regions, can provide spatio-temporal maps of snow cover and glacier ice at regional and global scale (Kulkarni et al. 2010). These snow and glacier ice maps derived from optical remote sensing are limited mainly to map surface extent only and have major limitation of mapping Earth surface under cloudy and night-time conditions. These limitations of optical imaging sensors during cloud cover and night-time and non-penetration in snowpack can be solved by the use of microwave remote sensing using both passive microwave radiometers and active microwave radars. The passive microwave provides near global and daily temporal coverage, but at the same time, it has low spatial resolution (10–50 km), which makes it less useful for Himalayan region.

5.3.1 Snow Cover Mapping, Monitoring and Retrieval of Snowpack Properties

5.3.1.1 Snow Cover Mapping Using Optical and SAR Remote Sensing

Time series of SCA maps is used for creating the snow depletion curves or SDCs, which are further used in snowmelt runoff and other hydrological models for estimation of river flow and flood disaster-related events (Thakur et al. 2016b). Conventional maps of SCA in the Indian Himalayas have been prepared using optical remote sensing data of Indian Remote Sensing Satellite (IRS), Resourcesat-1 (Kulkarni et al. 2006 and 2010) and Moderate Resolution Imaging Spectroradiometer (MODIS) (Jain et al. 2010). Recently, 16-day snow cover fraction at $3' \times 3'$ snow product has been made available from Indian Earth observation portal, Bhuvan (National Remote Sensing Centre 2013), under the National Information System for Climate and Environmental Studies (NICES). In majority of SCA products, Normalized Difference Snow Index (NDSI) was first created by Dozier and Marks (1987) and Dozier (1989) using the Landsat Thematic Mapper (TM) and Enhanced Thematic Mapper (ETM+) spectral bands as:

$$\mathrm{NDSI} = (\mathrm{R_G} - \mathrm{R_{SWIR}})/(\mathrm{R_G} + \mathrm{R_{SWIR}}), \tag{5.1}$$

where $\mathrm{R_G}$ and $\mathrm{R_{SWIR}}$ are the reflectance in the green band 2 and short-wave infrared SWIR band 5 (centre wavelengths, 0.57 and 1.65 μm), respectively (Dozier 1989), of Landsat TM. Snow is usually present and mapped if the NDSI value exceeds 0.40 (Dozier 1989; Hall et al. 2002), but as shown in the study by Vogel (2002), optimum threshold value of NDSI can have seasonal variations.

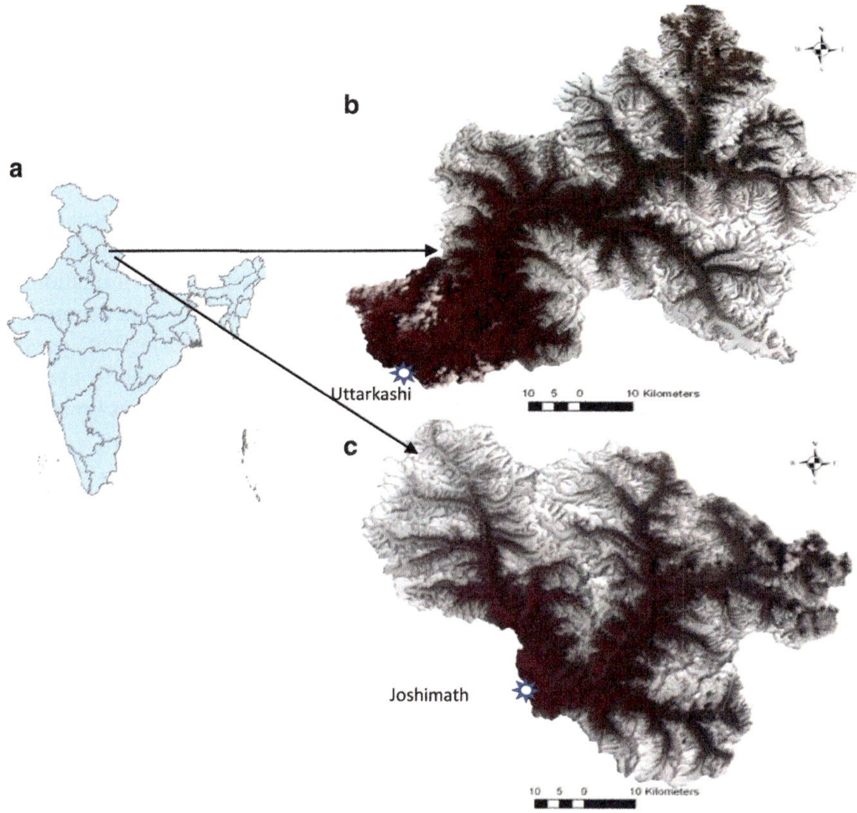

Fig. 5.2 (**a**) Location map of Uttarakhand, India, with AWIFS image-based colour composite, (**b**) Bhagirathi River Basin up to Uttarkashi, (**c**) Alaknanda River basin up to Joshimath

The test basins of Alaknanda and Bhagirathi, shown in Fig. 5.2a, b, are used to estimate SCA using AWiFS and MODIS. Based on the SCA mapping of the Alaknanda River Basin, first maximum snow coverage is found on February 26, 2008 (3106 km^2) and second maximum SCA in April 10, 2008 (2733 km^2). The minimum snow cover was estimated on November 11, 2008 (272 km^2). In the year 2000, maximum accumulation was in April as 3766 km^2. As the SRM model needs daily snow cover area, the available 8 daily SCA was utilized for generating daily SCA with the help of linear interpolation techniques. This approach can give wrong estimates of the SCA by computing under- or overestimated SCA, especially in case of fresh snowfall or sudden snowmelt in the intervening weeks. But this interpolation method maintains good accuracy in matching overall trend of SCA at seasonal timescale. Similarly, SCA mapping was done for the Bhagirathi basin during years 2002 to 2007, with maximum SCA observed in March 2002 i.e., 4200 km^2 and minimum SCA in September 2007 and SCA varied from 1600 to 1800 km^2 during October to December. Beas River basin with Thalot as an outlet

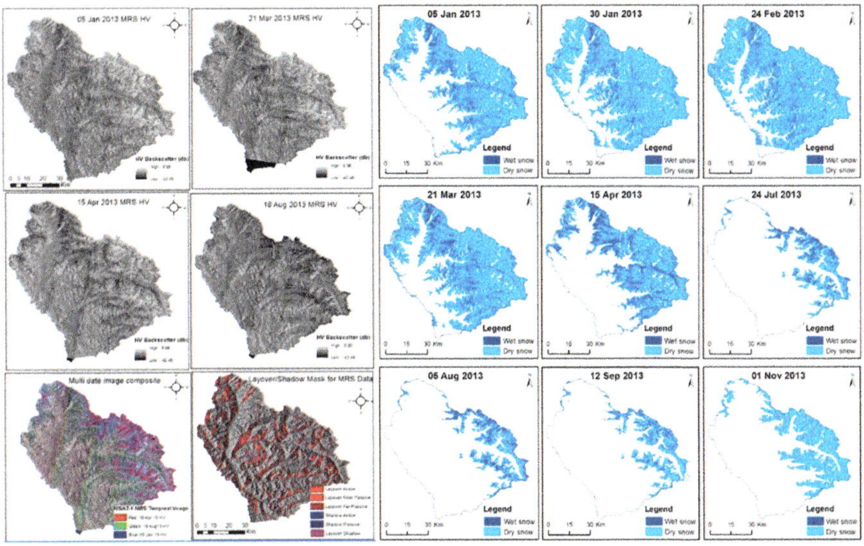

Fig. 5.3 RISAT-1 MRS HV sample backscatter data, image composite, layover/shadow mask and derived SCA for Beas River basin from MRS data for multiple dates of the year 2013

covering an area of 4883 km^2 is the main study area for SCA mapping using both SAR and optical datasets. As per MODIS SCA data, minimum and maximum SCA area during 2003–2006 varied from 7% to 91% for the entire Beas Basin and 3.89 to 90.61% for the Bhagirathi Basin from 2004 to 2006.

The SAR-based SCA mapping techniques have its roots in the knowledge-based image classifiers (Pierce et al. 1994; Al Momani et al. 2007; Tso and Mather 2009; Thakur et al. 2013), wherein SCE is calculated by applying thresholds to ratio of two SAR images, one image of wet snow season and other reference image of October or November time representing minimum snow. Further refinement in SCE is carried out by applying using snow line elevation mask, which is derived from near date or week MODIS SCA maps. The Radar Imaging Satellite-1 (RISAT-1) MRS data was used in SCE mapping of Beas River basin up to Thalot and Bhagirathi River Basin up to Uttarkashi (IIRS-2014, Thakur et al. 2016b).

Thakur et al. (2016b) showed that MRS data in HV polarization has better backscatter range in snow/non-snow areas as compared to the RISAT-1 MRS HH polarization data for the same area. It has been estimated that backscatter threshold for HV MRS data sets varies from −8.0db to −18.0db as per conditions of snowpack and snow season (winter, spring and monsoon). The SAR-based SCE maps of the year 2013 for the Beas River basin were prepared using this approach (Fig. 5.3).

The original SAR image threshold based on SCE contained 11 to 38% wet snow, and pixels were classified as non-wet snow or dry snow pixels. As the local time of RISAT-1 MRS descending overpass is 6 A.M., this early morning overpass can cause less area under wet snow category mainly during October to April time as snowpack can still be under freezing conditions, similarly during May–September

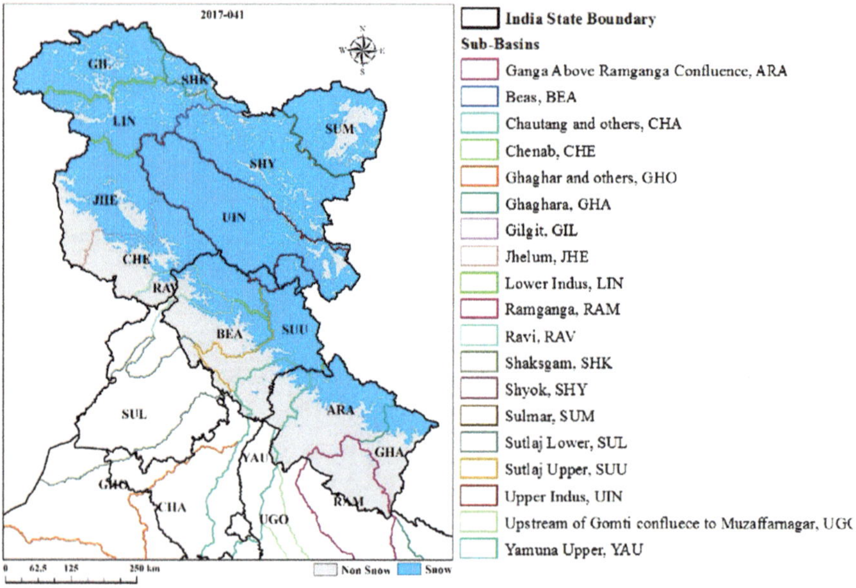

Fig. 5.4 Eight-day maximum SCA in second week of February 2017 derived from MOD10A2 product

time with more area can come under wet snow due to initial melting of surface snow layer. As SAR data threshold based on SCE maps mainly gives only wet snow cover, the large area covered by dry snow is not classified as snow cover. To overcome this limitation, the MODIS-/Landsat-based SCA at 8 daily or 15-day interval data was utilized along with DEM to get snow line elevation (SLE) at the time of SAR data overpass. This SLE mask is used to discard non-snow pixels and also to classify the remaining dry snow pixels as final and total SCE of the study area. This approach is used for total SCE (wet and dry snow) from RISAT-1. Overall SCE accuracy of >95% with error of +/− 10% was achieved using this approach when compared with nearest date or week SCE derived from MODIS optical image (Thakur et al. 2016b).

Recently, the Indian Institute of Remote Sensing (IIRS), Dehradun, has prepared complete SCA map using 8 daily SCA product from MODIS MOD10A2 and AWIFS 15-day SCA products for the entire NWH from 2001 to 2017 (Nikam et al. 2017). Additionally, detailed SCA maps were analysed for all subbasins of NWH (subbasins shown in Figs. 5.4 and 5.5) to derive snow depletion curves from 2001 to 2017. The unusual heavy snowfall of 2017 spring in many subbasins was also analysed with long-term mean SCA data (Fig. 5.6). The results of this work are also hosted in IIRS website (http://iirs.gov.in/content/sca-mapping-activity).

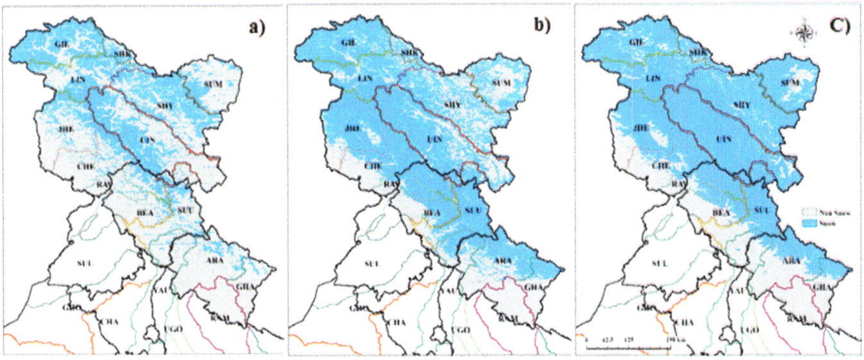

Fig. 5.5 Temporal change in SCA during winter 2016–2017: (**a**) November 25–December 02 2016, (**b**) January 09–16, 2017, (**c**) February 10–17, 2017

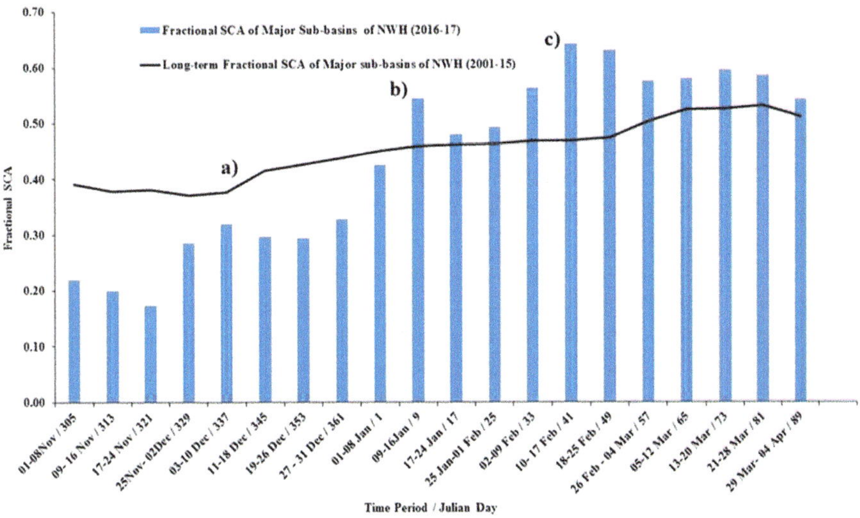

Fig. 5.6 Comparison of fractional snow cover in NWH

5.3.1.2 Snow Physical Properties Retrieval Using Optical and SAR Data

"The mountain shadow, clouds and retrieval of physical properties of snow using optical data remains major gap areas, and in these cases, Synthetic Aperture Radar (SAR) offers a better alternative to estimate the snow and glacier dynamics parameters" (Thakur et al. 2012). SAR-based backscatter resulting from a snowpack is generally modelled as a sum of backscatter coming from the air-snow surface boundary, the volume scatter contribution of the snow layer and the scattering component of the snow-ground interface (Fig. 5.7). To simplify the backscattering from a snowpack, seasonal approach to snowpack can be used. In this approach, the

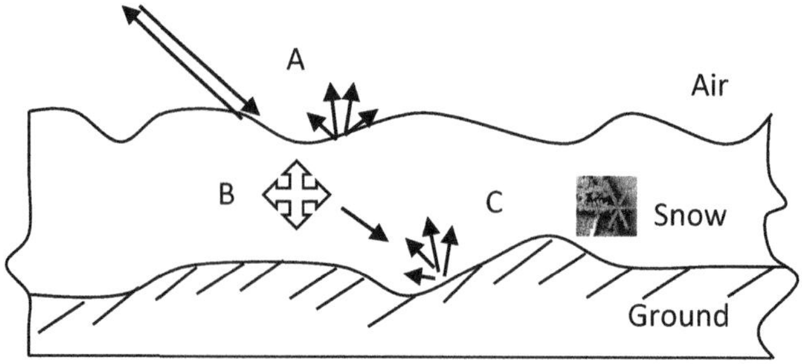

Fig. 5.7 Backscatter from a typical snowpack

total backscatter from the wet snow or spring-summer season comes mainly from surface and volume scattering only, and during dry snow or winter seasons, snow-ground interface scattering dominates in the total backscatter from snowpack. The backscattering terms A, B and C of Fig. 5.7 are defined as (A) top surface backscatter contribution of the snow-air interface, (B) volume backscatter contribution from the snowpack and (C) backscatter component of the snow-ground interface.

The Beas River watershed up to Manali town, with an area of 350.21 km^2, and nearby Rohtang area are the study areas for the comprehensive calibration and validation of RISAT-1-derived snow parameters and snow grain size estimation using hyperspectral Hyperion datasets as described in Sects. 5.3.1.2 and 5.3.1.3, respectively (Fig. 5.8).

The main methodology used for the snowpack physical parameter retrieval using SAR data is given in Fig. 5.9. Shi and Dozier (1995) inversion model was used for retrieval of snow wetness using C-band SAR data of ENVISAT-ASAR, RISAT-1 and RS2 quad-pol data. This inversion model consists of the theoretical integral equation model (IEM)-based backscatter models for simulating surface scattering at snow-ground and air-snow interface, radiative transfer theory-based volume scattering models for main snowpack and finally SAR-based total backscatter-based inversion models for dielectric constant inversion. More detailed description of the models and used coefficients is given in Shi and Dozier (1995), and its application in Manali watershed of Beas Basin is given in Thakur et al. (2012). The original inversion model of Shi and Dozier (2000) was used for estimation of dry snow density using L-band SAR data, and this model was modified for the C-band RISAT-1 quad-pol data by changing the wave number, k, for C-band wavelength and updating inversion model coefficients. This modified snow density model has been named as modified Shi snow density inversion model (MSSDIM) (Thakur et al. 2012). Shi and Dozier (2000) give the full description of the original model and its equations and its usage using ASAR data to part of Himalaya as given in Thakur et al. (2012).

Shi and Dozier (1995) model was used for the calculation of wet snow permittivity, and this model was tested with C-band SAR data of ASAR and RISAT-1 by Thakur et al. (2012, 2016b). Snow wetness has a direct relationship with wet snow permittivity that can be estimated using the following relations. The incremental

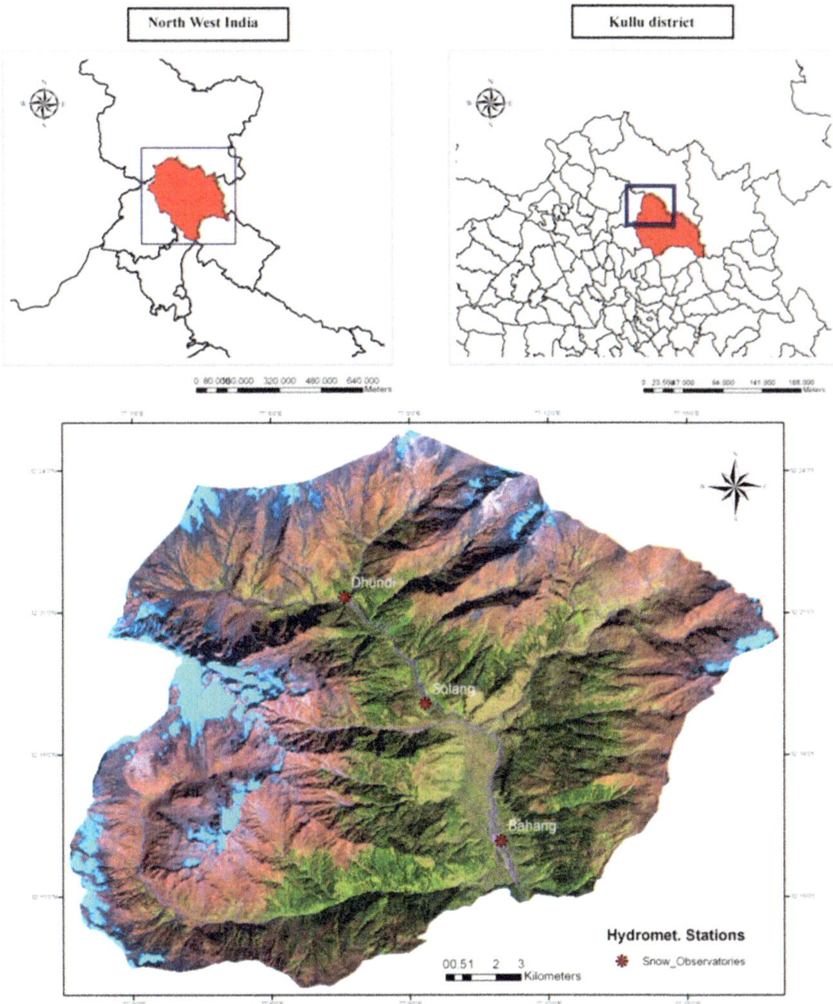

Fig. 5.8 Study area location map with Manali sub-watershed of Beas River basin

increase in snow permittivity due to the presence of the water layer, $\Delta\varepsilon'_s$, in the permittivity was first given by (Hallikainen et al. 1986):

$$\Delta\varepsilon'_s = 0.02m_v^{1.015} + \frac{0.073m_v^{1.31}}{1 + \left({}^{f_s}/_{f_w}\right)^2} \tag{5.2}$$

In this equation, f_s is the frequency at which the permittivity is measured and $f_w = 9.07$ GHz is the relaxation frequency of water at 0 °C and is expressed in percent. If the incremental increase $\Delta\varepsilon'_s$ is known (Eq. 5.3), an accurate estimate of

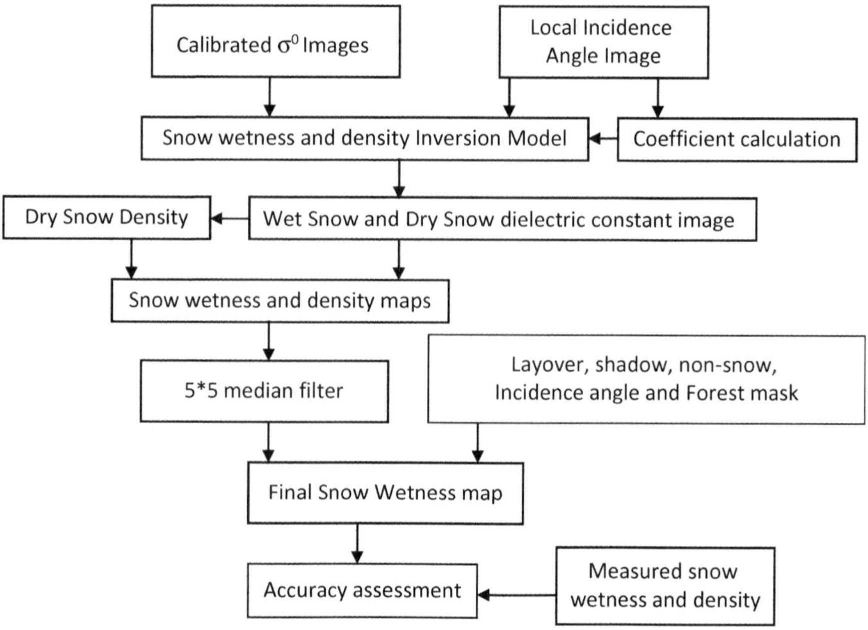

Fig. 5.9 Flow chart for snow wetness and density retrieval using SAR-based inversion models

snow wetness, snow wetness m_v, can be obtained by inverting Eq. 5.2 (Kendra et al. 1998). If the density of snow is known, then the base contribution to ε'_s represented by the dry snow component ε'_{ds} can be calculated; thus, the incremental increase in wet snow permittivity is given by:

$$\Delta\varepsilon'_s(m_v) = \varepsilon'_s(m_v, \rho) - \varepsilon'_{ds}(\rho) \tag{5.3}$$

SAR-based inversion model was used for dry snow permittivity retrieval, and this dry snow permittivity was further used in the calculation of dry snow density using Looyenga's semiempirical dielectric formula (Looyenga 1965), which provides a good fit to Polder and van Santen's physical formula (Mätzler 1996):

$$\varepsilon_s = 1.0 + 1.5995\rho_s + 1.861\rho^3{}_s \tag{5.4}$$

The radiometric and geometric calibration of the SAR multi-looked images was completed using ASTER Global DEM (Fig. 5.10a) for ENVISAT-ASAR datasets. The local incidence angle map (Fig. 5.10c) and calibrated backscatter image (Fig. 5.10d) were generated during the calibration step. The area under snow wetness classes 5 and 6 (410–15%) is very high in case of 30 March 2008 image (Fig. 5.11a), and this gradually decreases from March to February. Similarly, the snow wetness classes 1–2 (dry snow to 2% wetness) are decreasing from January to March images.

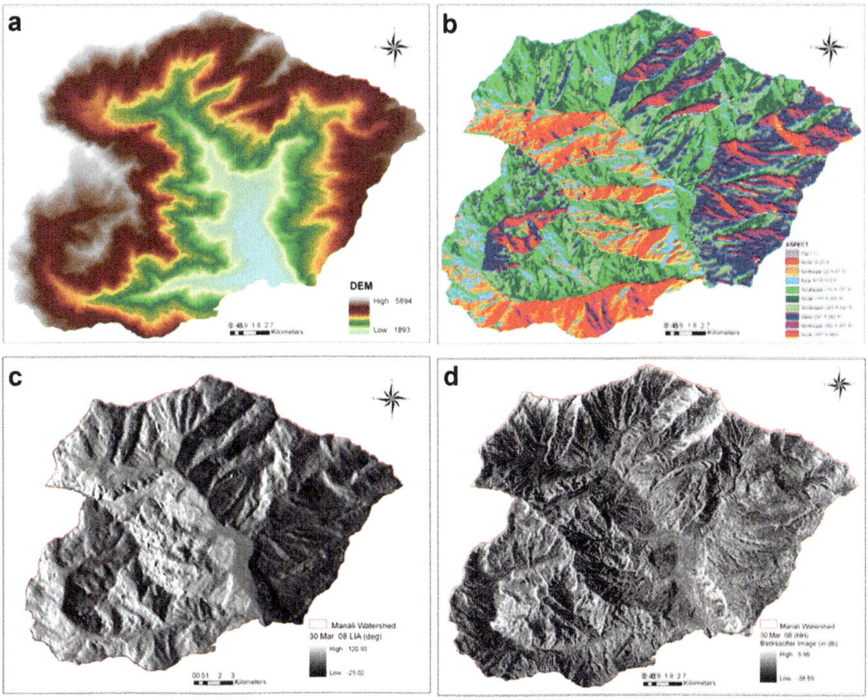

Fig. 5.10 (**a**) DEM map of Manali watershed. (**b**) Aspect map of Manali watershed. (**c**) Local incidence angle map. (**d**) Calibrated Sigma0HH image (both for March 30, 2008)

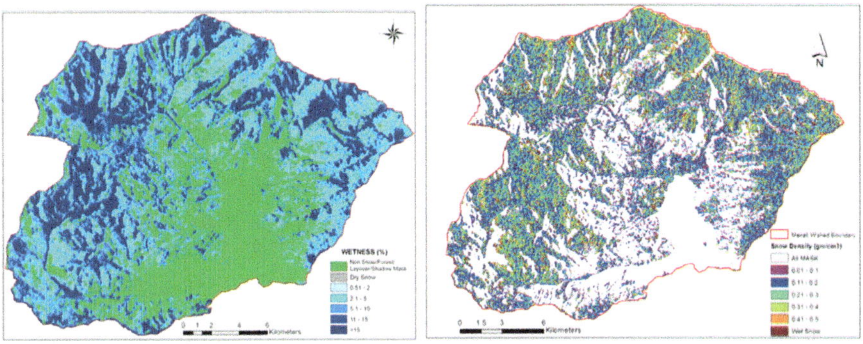

Fig. 5.11 (**a**) Snow wetness map derived from March 30, 2008 ASAR-APS (HH/VV) data; (**b**) snow density map for January 20, 2009

The increasing wetness values and decreasing dry snow area (January to March) match well with the increasing trend of air temperatures from winter to spring period. Snow density map of 20 January 2009 is shown in Fig. 5.11b. The ground observations of the mean snow density at Dhundi, Manali, during January 20 and

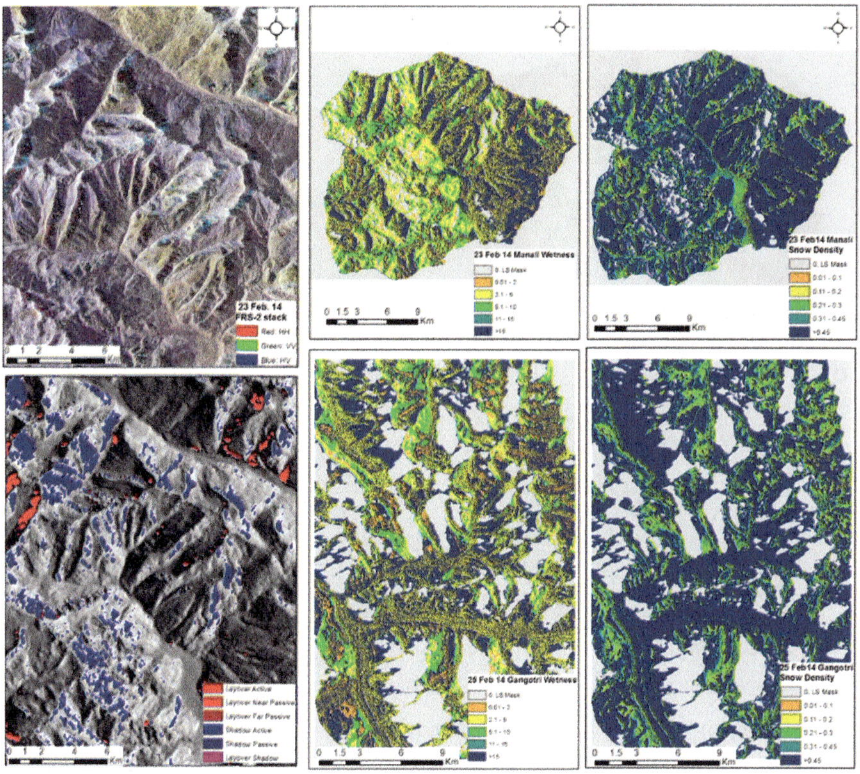

Fig. 5.12 RISAT-1 FRS-2 quad-pol data with RGB stack and with layover/shadow mask (left side panel), derived snow parameters for Manali watershed (top right) and Gangotri area (bottom right) for February 23 and 25 2014

25, 2008, and January 20, 2009, varied from 0.09 to 0.15 g/cm^3 and snow depth varied from 235, 196 and 120 cm, respectively (Thakur et al. 2012).

The RISAT-1 quad-pol data based on snow wetness and snow density maps are shown in the left side of Fig. 5.12. These inversion models are applicable to incidence angle range of 25–70° for snow wetness and 10–70° for snow density. The validation of derived snow density and wetness could be done for Manali watershed along accessible points, and Gangotri glacier area validation could not be done, as during winter access to this area is not feasible. Overall accuracy in terms of R^2 for part of Manali watershed is found to be 0.72 and 0.74 for snow density and snow wetness (Thakur et al. 2016b). The moderate values of R^2 can be attributed to underlying forest and glacier debris/ice which affects the overall snow-ground and volume scattering components. In addition to the use of dual pol and quad-pol C-band datasets, full polarization SAR data was tested for snow physical parameter retrieval (Surendar et al. 2015a, b; Leinss et al. 2014 and 2015), but these datasets cannot be used for regular hydrological studies such as daily or weekly estimates of snow parameters, mainly due to limited temporal scale of fully pol data and

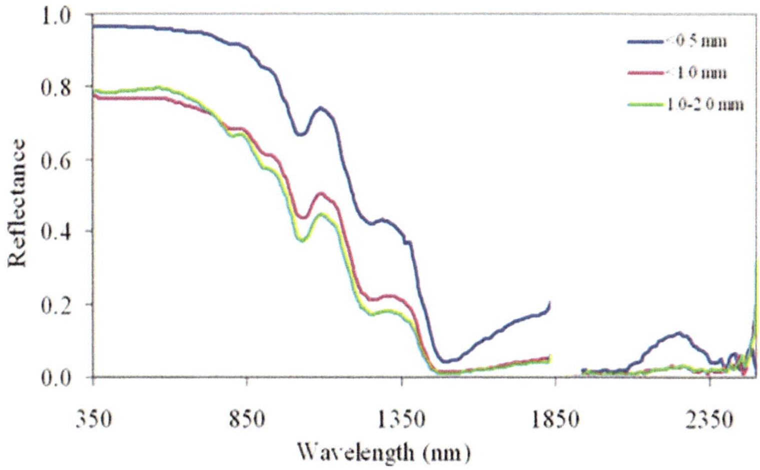

Fig. 5.13 Snow reflectance at varying grain sizes (Painter et al. 1998, 2003)

sub-daily to daily spatio-temporal variations of hydrological processes and parameters such as snow cover, snow depth, density, SWE and finally snowmelt. Therefore, well-calibrated process-based land surface and hydrological models are required to simulate these processes, and remote sensing-based snow parameters can be used in data assimilation mode (Thakur et al. 2016b).

5.3.1.3 Snow Grain Size Mapping Using Hyperspectral Remote Sensing

Literature shows that reflectance decreases with grain size increase as snow ages as shown in Fig. 5.13 (Painter et al. 1998, 2003). Also, the seasonal melting and refreezing of snow causes ice grain to grow in clusters and behave as a large single grain of snow. In this scenario, the traditional optical remote sensing can be updated with hyperspectral remote sensing data and techniques, which has been used to map snow grain size spatially very easily (Negi et al. 2010; Garg et al. 2014).

It has been seen that the snow grain size changes with snow age; simultaneously, the reflectance decreases with increase in grain size. It can be seen in Fig. 5.13 that with the increase in snow grain size, the reflectance at wavelength 1050 nm drops very steeply. Using this information, Negi et al. (2010) developed a band rationing technique for snow grain size mapping, named as grain size index (GI):

$$\mathrm{GI} = \frac{\mathrm{Relectance}\ (590\ \mathrm{nm}) - \mathrm{Reflectance}\ (1050\ \mathrm{nm})}{\mathrm{Relectance}\ (590\ \mathrm{nm}) + \mathrm{Reflectance}\ (1050)}$$

Using this GI approach, grain size mapping of part of Beas Basin, Himachal Pradesh, has been carried out by utilizing Hyperion data. The grain size map of the

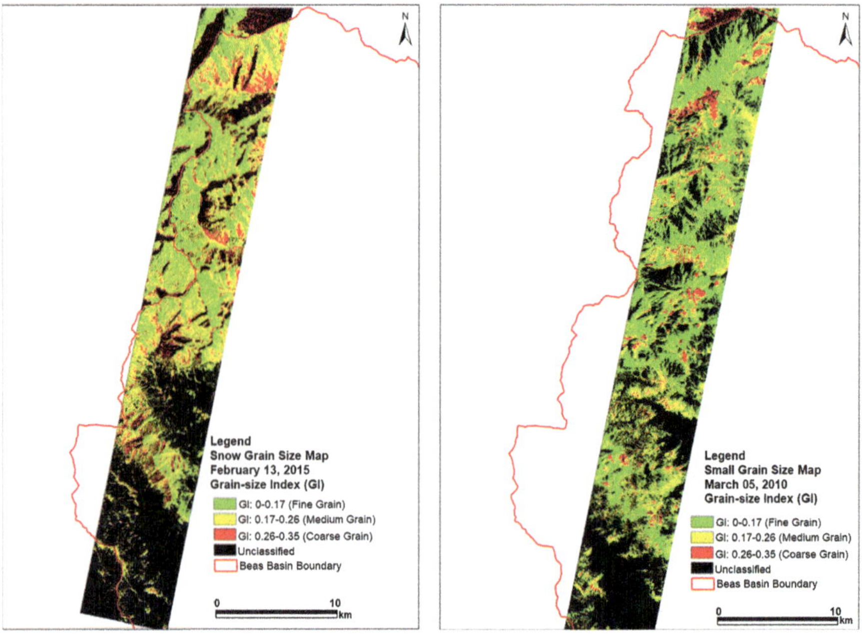

Fig. 5.14 Snow grain size mapping of part of Beas Basin, near Manali, Himachal Pradesh. (**a**) February 13, 2015, (**b**) March 05, 2010

region is shown in Fig. 5.14. The important snow physical property, snow grain size, is very important for hydrological modelling especially snowmelt runoff modelling and climate change studies. With the advent of hyperspectral remote sensing, most of these parameters can be derived at spatial extent temporally. It can be said that the traditional methods of snow physical parameter measurements are difficult and time consuming, whereas, the emerging hyperspectral remote sensing technique is powerful tool for such studies especially for snow grain size.

5.3.2 Mapping and Monitoring of Glaciers, Glacier Velocity and Glacier Depth

5.3.2.1 Glacier Mapping Using SAR Data

Mapping and monitoring of glacier are traditionally done with optical remote sensing data and techniques, but SAR-based images can contribute significantly in mapping and differentiating glacier facies as compared to optical-based VIR images, mainly due to its all-weather, day-night and dry snow/ice penetration capability (Rees 2006).

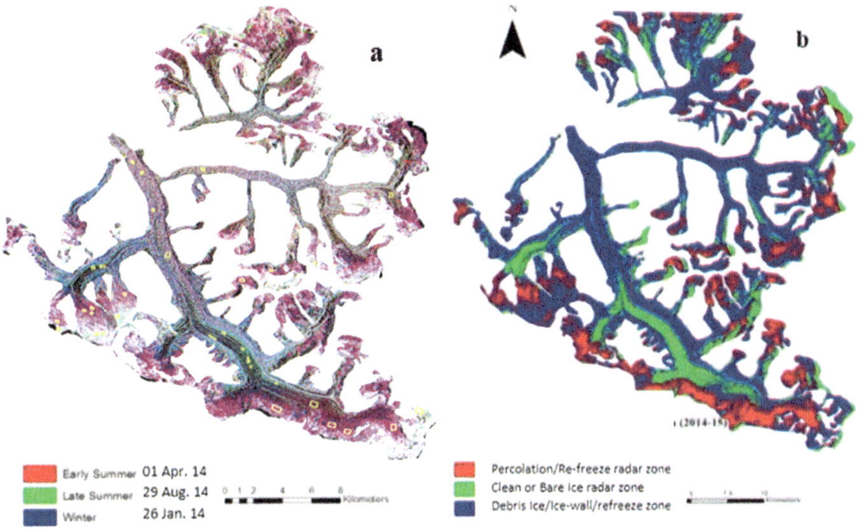

Fig. 5.15 (**a**) MRS data temporal composite for Bhagirathi Basin with sample signature locations and (**b**) SVM-based classified glacier radar zones

The first use of SAR data for glacier facies mapping in India was reported by Kundu and Chakraborty (2015), but this approach did not try temporal SAR composite for making glacier facies or radar zone maps. This multi-temporal seasonal SAR data image composite-based glacier classification approach was tested and proven by Thakur et al. (2016b). This study first created a multi-date co-registered MRS image colour composite, in which three images of melt period (April), late summer or peak ablation (August) and peak winter (January) are assigned to red, green and blue colours, respectively (Fig. 5.15a). This SAR data colour composite is able to identify and distinguish various snow and glacier radar zones (Fig. 5.15a, b). The image interpretation of this colour composite is done with reference to as reported in the literature by Patrington (1998), Rau et al. (2000) and Rees (2006), i.e. dry-snow zone will have shade of dark grey (low backscatter throughout the year, not observed in the present case of Gangotri glacier); the percolation and wet-snow zones look white or light grey in the upper elevation region of glacier and purple in the middle elevation zone and show blue colour at the low elevation zone of glacier (these colours are depended on relative melting/refreezing cycles at each elevation zone and season); and the bare ice or clean ice zone looks greenish in colour (Rau et al. 2000; Rees 2006). This colour composite image was used to create signature sets (given in Table 5.2) for classification of glacier radar zones using support vector machine (SVM)-based supervised image classifier (Thakur et al. 2016b).

The RISAT-1 FRS1 data in hybrid polarization (RH/RV mode) data of February 18, 2014, for the main part of Gangotri glacier was utilized for creating Raney m-chi (Raney 2007; Raney et al. 2012) decomposition (Fig. 5.16a) in PolSARpro 5.0

Table 5.2 Signature set for select glacier radar classes of facies as mapped using MRS data

Signature stats (unit in db)	Min	Max	Range	Mean	Std
Glacier radar class	Early summer, April 2014				
Bare ice	−11.43	−6.34	5.09	−8.80	1.12
Middle percolation zone	−7.58	−0.21	7.37	−3.61	1.36
Lower percolation zone	−6.81	1.63	8.43	−2.17	1.61
Debris glacier	−9.28	−1.15	8.13	−5.28	1.52
Higher percolation zone	−5.81	1.42	7.23	−1.80	1.36
	Late summer, Aug 2014				
Bare ice	−10.91	−5.06	5.85	−7.69	1.18
Middle percolation zone	−11.78	−5.09	6.69	−7.89	1.23
Lower percolation zone	−12.04	−3.84	8.20	−7.74	1.59
Debris glacier	−9.96	−1.46	8.50	−5.36	1.54
Higher percolation zone	−9.01	−2.81	6.21	−5.49	1.22
	Winter, Jan 2014				
Bare ice	−10.89	−4.85	6.05	−7.74	1.23
Middle percolation zone	−7.59	−0.10	7.49	−3.43	1.40
Lower percolation zone	−9.36	−0.35	9.01	−4.96	1.82
Debris glacier	−6.23	1.66	7.90	−1.87	1.41
Higher percolation zone	−6.05	1.15	7.21	−1.99	1.44

software. The Raney m-chi decomposition gave three major scattering components, namely, surface or odd bounce, volume and double bounce scattering components from glaciated terrain, which were further used for glacier radar zone identification using supervised image classifiers (Thakur et al. 2016b). The glacier radar zone classes and facies identified for creating signature sets (Fig. 5.16d) and SVM-based image classification were debris-covered glacier ice, clean or debris-free bare glacier ice, percolation/refreeze area, moraine ice, bare ice overlain by snow, wet snow and supraglacial lakes (Fig. 5.16b, Table 5.3). The entire group of Gangotri and its tributary glacier could not be covered for classification, as previously done with MRS time series images, mainly due to the less swath and extent of FRS1 hybrid pol data. The initial data from MRS-based composite and FRS-1 decomposition maps were clipped with glacier extent boundary of the year 2014 to remove non-glacier area and improve its overall classification accuracy. The RISAT-1 MRS or FRS-1 data based derived glacier radar zones are very useful in marking the peak ablation snow line elevation or equilibrium line altitude, which further can be used for both summer as well as winter, remote sensing based glacier mass balance methods, as such information cannot be easily derived using optical remote sensing due to persistence snow and cloud cover during peak winter and monsoon season (Fig. 5.16c). The classified glacier radar zone accuracy assessment was completed with the help of field ground truth collected during September 2014 near the snout of Gangotri glacier and using cloud-free Google Earth images, resourcesat-1, 2 LiSS-III and LiSS-IV images. RISAT-1 MRS-based glacier radar zone classification accuracy was calculated at 0.92 and 0.86 for FRS-1-based classification. The main

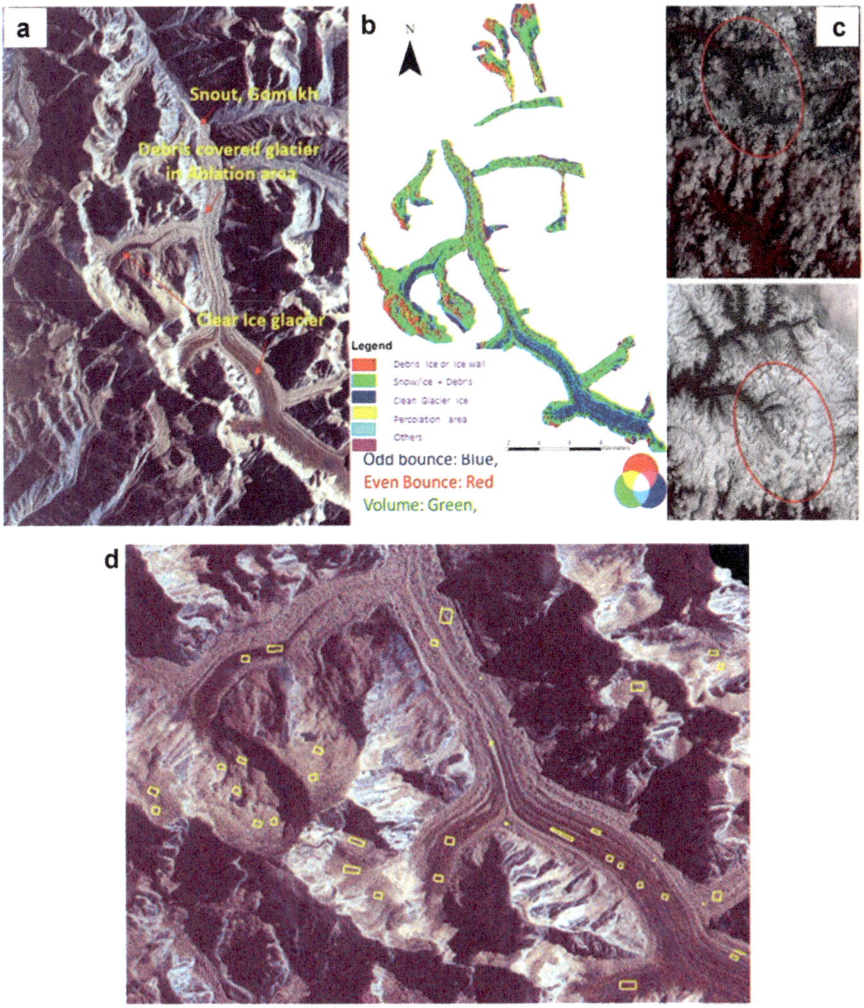

Fig. 5.16 Hybrid polarimetry (RH and RV)-based decomposition results for Gangotri glacier; (**a**) Raney hybrid composition as RGB composite; (**b**) classified Glaciers facies using hybrid data; and (**c**) sample image of same glacier as seen by optical remote sensing during peak winter and peak summer ablation season in late monsoon (**d**) sample signature locations overlaid over hybrid FRS1-based RGB

reasons for difference in accuracy of MRS- and FRS-1-based glacier radar zones are the number of glacier radar zones for each classified map and associated misclassification of debris-covered glacier ice, wet snow and bare ice wall to other classes in case of FRS1 image (Thakur et al. 2016b). The classified glacier radar zones can be used in glacier mass balance studies as they contain information about the equilibrium line altitude (ELA) or the firn line (Engeset and Weydahl 1998; Engeset and Ødegard 1999; König et al. 2001, 2002; Engeset et al. 2002). The

Table 5.3 RISAT-1 FRS1 image (18 Feb 2014) based sample signatures for Raney hybrid decomposition scattering components for various glacier classes

FRS-1 Raney Stats (db)	Min	Max	Range	Mean	Std
Glacier radar class	Even scattering, 18 Feb 2014				
Bare ice	−21.95	−1.30	20.65	−6.37	2.04
Moraine ice	−26.54	−1.56	24.98	−6.08	2.02
Bare ice snow	−20.38	2.00	22.38	−2.84	2.08
Percolation refreeze	−22.26	0.26	22.52	−4.55	2.04
Debris glacier	−22.79	0.94	23.73	−4.73	2.38
Wet snow	−24.20	−2.24	21.96	−7.42	2.36
Supra glacial Lake	−21.35	−1.64	19.71	−6.86	2.35
	Diffuse scattering, 18 Feb 2014				
Bare ice	−6.64	−0.03	6.61	−2.97	0.89
Moraine ice	−5.73	−0.13	5.60	−2.75	0.85
Bare ice snow	−4.01	3.48	7.49	0.43	0.95
Percolation refreeze	−5.16	1.58	6.74	−1.36	0.87
Debris glacier	−5.52	2.68	8.20	−1.25	1.30
Wet snow	−6.76	−0.70	6.06	−3.45	0.92
Supraglacial lake	−6.48	0.22	6.71	−3.02	1.10
	Odd scattering, 18 Feb 2014				
Bare ice	−27.05	−1.54	25.52	−6.87	2.17
Moraine ice	−31.43	−1.71	29.71	−6.63	2.19
Bare ice snow	−26.14	2.28	28.42	−3.15	2.22
Percolation refreeze	−24.37	0.08	24.45	−5.13	2.05
Debris glacier	−22.62	1.31	23.93	−4.65	2.40
Wet snow	−19.89	−2.31	17.58	−6.69	2.01
Supraglacial lake	−21.71	−1.05	20.66	−6.49	2.19

hypothetical line between the bare ice radar zone (BIRZ) and the adjoining wet snow or Percolation-Freeze Radar Zone (PFRZ) is generally considered as the actual transient snow line of a glacier (Rau et al. 2000). This well-marked line at the end of the ablation period is often marked as an approximation of the ELA (Rau et al. 2000; Bindschadler 1998), and it is also considered as an indicator of climatic variations and glacier mass changes. As per Rau et al. (2000), "In the years with a positive mass balance of the glacier, the firn line at the end of the summer coincides with the position of the ELA, and in years which are characterised by a negative mass balance, firn from previous accumulation seasons might be exposed, leading to the assumption of a too low ELA position." Therefore, if we mark ELA from SAR image-based classified glacier radar zone maps, it may lead to an overestimation of glacier mass balance, mainly during years of negative mass balance. This presence of a superimposed ice zone can cause uncertainties in the accurate determination of the ELA from SAR imagery (Thakur et al. 2016b); in any case, the firn line position derived from either optical- or SAR-based data sets provides climatological information and can be considered as an approximation of the ELA (Rau et al. 2000). The transient boundary between PFRZ and BIRZ as discussed above shows the approximate position of the firn line, which can be seen in Figs. 5.15b and 5.16b. In case of the MRS-based glacier radar zones, only three radar glacier zones are created, where some white or light grey areas which are visible in the MRS-based composite images at higher elevations (Fig. 5.15a), which are originally part of percolation/refreeze zone at high elevation, are clubbed with debris ice, ice wall classes in this study (Thakur et al. 2016b). This approach provides an alternate, easy, fast and better method to find glacier radar zones as compared to the earlier reported studies by Kundu and Chakraborty (2015). The signature set for FRS1 Raney hybrid decomposition scattering components for various glacier classes are shown in Table 5.3.

5.3.2.2 Glacier Velocity, Ice Depth and DEM Estimation Using Remote Sensing and Modelling

Interferometric SAR (InSAR) and differential Interferometric SAR (DInSAR) techniques have proven record of estimating glacier velocity (Joughin et al. 2010). Interferometric methods were initially developed for measuring surface topography using airborne systems (Graham 1974; Zebker and Goldstein 1986) and later developed under Shuttle Imaging Radar (SIR) missions, where spaceborne topographic mapping with InSAR-based methods was completed for majority of the land area of the Earth, except the polar regions (Gabriel and Goldstein 1988) and Seasat (Li and Goldstein 1990). In the year 1991, the European Space Agency (ESA) launched, European Remote Sensing, ERS-1 satellite in which one of the sensors was C-band SAR. Goldstein et al. (1993) used this ERS-1 SAR data in InSAR mode study of ice sheets and glaciers. InSAR techniques use the difference in range (line-of-sight distance), Δ, from each SAR antenna to the point which is imaged on the ground. InSAR techniques rely on the ability of SAR sensors to measure phase, ϕ, accurately. This phase of an individual complex SAR image is proportional to the line-of-sight range from the antenna to the ground (Thakur et al. 2016a). Therefore, with the availability of repeat-pass interferometric pair of images at radar wavelength, λ, the product of the first SAR image in single-look complex (SLC) format with the complex conjugate of the second SAR SLC image yields an interferogram with phase $\phi = 4\pi\Delta/\lambda$. This phase, however, ranges from $+\pi$ to $-\pi$, which gives rise to the fringes visible in raw interferograms (Thakur et al. 2016a). In case the fringes are well defined, phase unwrapping process is used to remove the modulo-2π ambiguity (e.g. Goldstein et al. 1988). The interferometric fringe obtained after removing the topographic effect (Fig. 5.17a) gives a clear identification of the horizontal displacement into the flow direction of the Gangotri glacier and its tributaries. A mean coherence of 0.36 (Fig. 5.17b) was recorded for the glaciated area when using 25–26 March ERS-1/ERS-2 tandem pairs. Such values observed at the period of a single day may be due to glacier movement, snowfall or to wet snow and may even happen within a few hours (Strozzi et al. 1999; Negi et al. 2012). Mean line-of-sight displacements into the range direction were equal to 9 cm day 1 and 14 cm day 1 during March 25–26, 1996 (Fig. 5.17c).

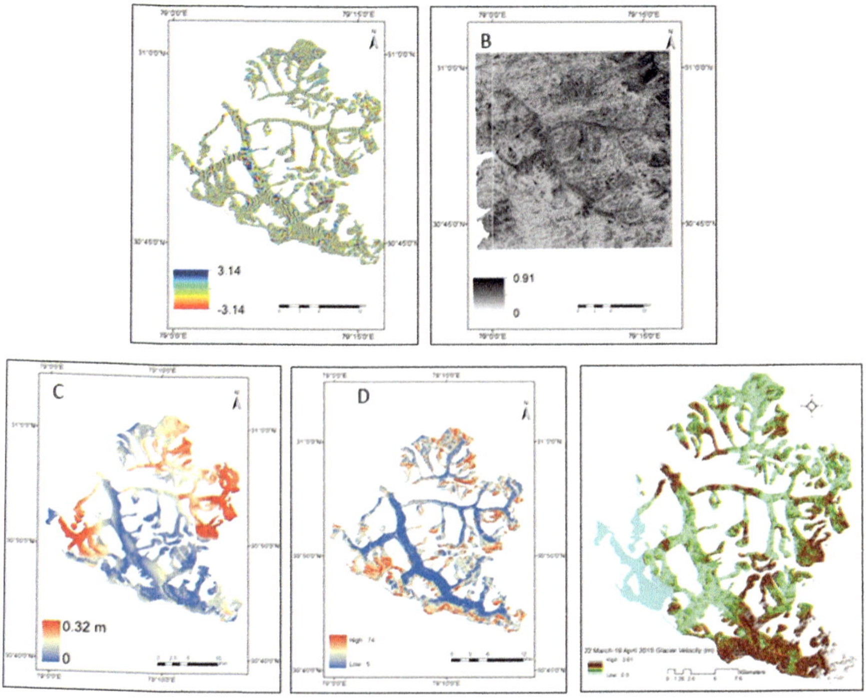

Fig. 5.17 (**a**) Interferogram, (**b**) coherence image for ERS-1/ERS-2 inSAR pair of March 25–26, 1996, (**c**) displacement from ERS-1/ERS-2 for March 25–26, 1996, (**d**) slope map of Gangotri glacier obtained from SRTM DEM, (**e**) displacement from PALSAR-2 inSAR pair of 22 March–19 April, 2015

ALOS-PALSAR-2 polarimetric InSAR data from March 22 to April 19, 2015, was used for deriving the line-of-sight (LOS) velocity in Gangotri glacier, and its mean value varies from 5.4 to 7.4 cm day^{-1} (Fig. 5.17e) during 28 daytime interval for full glacier and main trunk glacier, respectively (Thakur et al. 2016a). This study again highlighted the use of InSAR/DInSAR data for glacier velocity estimation, nearly after 20 years from the last reported results from ERS-1/ERS-2 tandem pairs (Thakur et al. 2016a). In this study, SRTM-x band 30 m DEM (Fig. 5.17d) was utilized for initial co-registration and removal of topographic phase (Thakur et al. 2016a). The velocity from the earlier reported papers varies from 3.8 to 23.3 cm/day in the accumulation region to 5.5–8.21 cm/day near the snout (Kumar et al. 2008; Bhambri et al. 2012; and Gantayat et al. 2014).

Ice thickness calculation for Gangotri glacier is done using velocity values derived from the COSI-CORR software (Fig. 5.18a, b). The basic premise behind the model for thickness estimation follows the logic of basal shear stress and velocities in "laminar flow" as described in Cuffey and Paterson (2010). Here the model of a glacier is a parallel-sided slab of ice of thickness H on a rough plane of slope α. No sliding of the slab is assumed on the plane, and the thickness of the slab

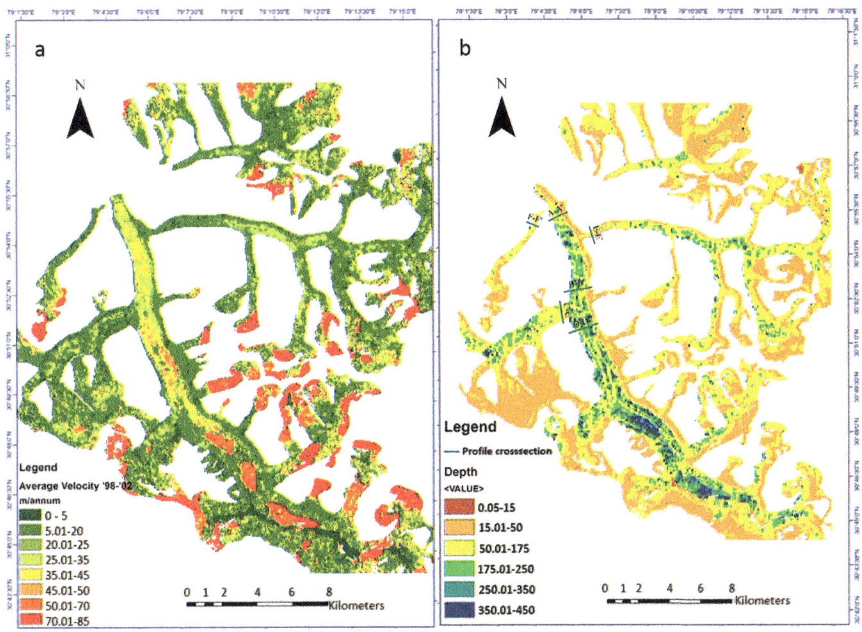

Fig. 5.18 (**a**) Average magnitude of velocity for the years 1998–2002; (**b**) mean ice depth profile of Gangotri glacier from time of analysis (1992–2014)

is much less than its length and width. The slab is perpendicular to the plane and of unit cross section. The weight of the slab is ρgH, where ρ is the density of ice, g is the acceleration due to gravity and H is the height of the slab. The weight of the slab along the surface will be countered by basal stress which will be equal to:

$$\tau_b = \rho g H sin\alpha \tag{5.8}$$

This is a very simple model for glacier movement when the layers of ice do not move over each other. But in real-world scenarios, ice moves over each layer, and hence velocity varies with depth.

Ice thickness was validated by TLS measurement at snout for nine points. These were kept at a distance of 30 m to 100 m apart from each other (Fig. 5.19). The scan of the area was taken and overlaid over the model output. This is then used to measure the thickness of the ice from a stable rock visible in the scan and detached from the main glacier, lying in the river bed. This rock was used as reference for the thickness measurement which varied from 58 m to 67 m at snout validating the model estimation of ice thickness at ~60 m (Bisht et al. 2015). Gantayat et al. (2014) reported the maximum ice thickness as 540 m to 50–60 m as minimum at snout. This is slightly different from the present study, where minimum ice thickness at snout is estimated as 60–67 m and 650 m as maximum ice thickness in central part of main Gangotri glacier. This difference can be attributed to different data sources with

Fig. 5.19 Correlation between TLS ground measurements and laminar flow model at snout and the actual scan of the snout during field visit (September 15–17, 2014)

high-resolution data such as IRS 1C/1D Pan Data and 90 m SRTM DEM-based slope.

In the present study, velocity uncertainty on snow-free ground for IRS 1C/1D is in the range of 0 to 0.351 m and a maximum of 1.5 m. Values for uncertainty of surface velocity were fixed at 3.5 m/year based on observed values by Swaroop et al. (2003). Scaling factor uncertainty was set at 0.1 (Gantayat et al. 2014). Creep factor uncertainty was set at 8.24*10–25 (Farinotti et al. 2009a). Uncertainty over ice density accuracy is taken at 90 kgm-3, i.e. 10% of the defined density used in the study. Uncertainty in the slope estimation (Fig. 5.17d) using SRTM DEM is calculated to be at 0.001. All the reported uncertainties are then added for uncertainty in the volume estimation of the glacier which is reported to be 11.4% for the current study, which shows a significant decrease of uncertainty by 7 %, as reported in an earlier study (Gantayat et al. 2014). This is possible due to the use of high-resolution imagery (5 m to 15 m as compared to 30 m) (Farinotti et al. 2009b) used for velocity estimation and better vertical relative accuracy achieved from STRM DEM (± 10 m) as compared to ASTER GDEM (± 20 m).

In addition, TANDEM-X SAR-based co-registered experimental mode datasets acquired in simultaneous mode provided an opportunity to obtain a high-resolution DEM at a high accuracy with low phase ambiguity, which was caused in the past by repeat-pass interferometry and atmosphere data. Figure 5.20 shows 2012 time series of TanDEM-X-based DEMs for subset of Gangotri glacier area. An interferometric coherence 0.80 is retained in most of this study region. It helps to avoid unnecessary phase jumps, whereas the low coherence in areas with high penetration depth leads to volume decorrelation.

The DEM obtained under such conditions introduces phase unwrapping errors in noisy areas or contain ramps and other artifacts. Areas should therefore be excluded from DEM generation if a specific threshold was reached for the coherence. The obtained DEM of Gangotri glacier for the 9th of June, 1st of July, 23rd of July and 5th of September represents different height information during these periods (Fig. 5.20). To obtain the height changes, differences between a day 1 DEM, a DEM at the 22nd day and a DEM at the 44th day were analysed. Such differences could be due to a mass change or to a vertical displacement between the acquisition dates. The processed TanDEM-X DEM can provide an absolute vertical accuracy of 10 m with a maximum relative height error of 2 m (Kosmann et al. 2010); in our case, the RMSE was estimated to be at 11.32 m using nine numbers of DGPS

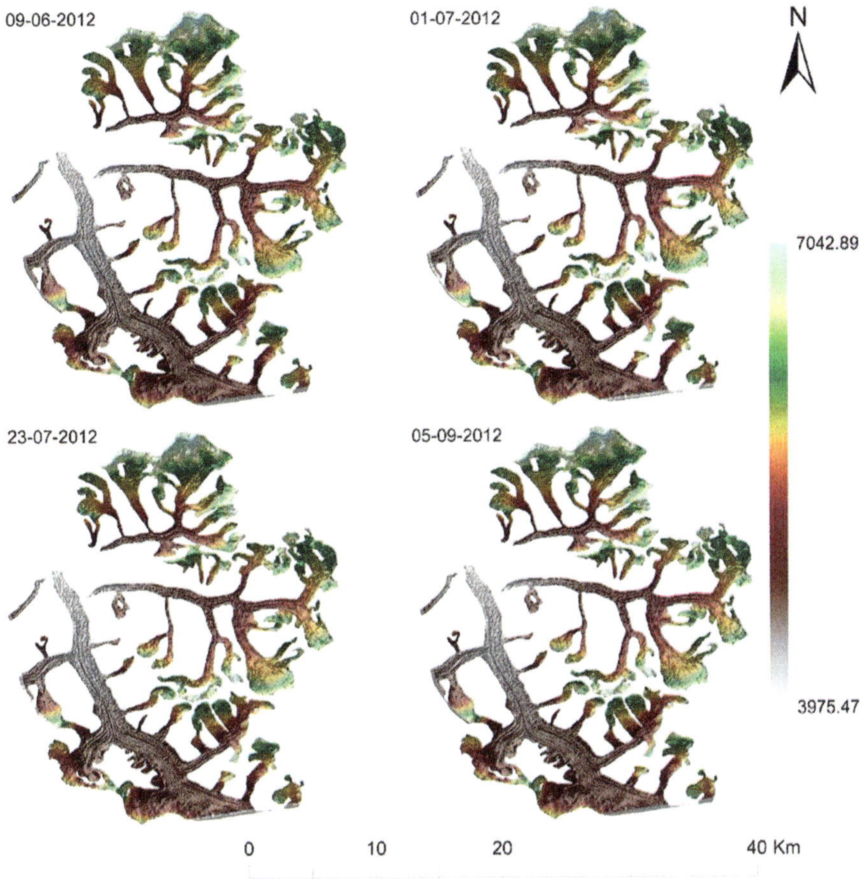

Fig. 5.20 TanDEM-X DEM for different period of 2012

observations near the glacier snout area during the 2012 field survey. In the present study, an elevation change of +0.7 to −1.12m was estimated with mean value of −0.12 m during 09 June to 01 July 2012 time period. During July 2012, The elevation change ranges from 0.58 and − 1.06 m with a mean of −0.042 m, and for July–September 2012, it is in a range of 0.83 and − 0.32 m with a mean of 0.09 m. As we observed from the analysis, the snow-covered area of glacier ice is affected by phase decorrelation which also has considerable influence in error propagation.

5.4 Mapping and Modelling of Snow Cover, Snowmelt and Glacier Mass Balance in NWH

This section highlights the advantages and limitations of remote sensing (RS)-based methods for snow cover, snow parameters and glacier ice and shows the use of land surface or hydrological models for simulation of these parameters. The advantages of RS-based methods are large area and repeat coverage at daily to weekly timescale and are discussed in detail in the previous sections. The major disadvantages of RS-based methods are temporal resolution of such products, as snow cover, snow depth, SWE and glacier ice, as these components can change at sub-daily to few day's time period, and accuracy of such RS products especially in mountainous Himalayan terrain. In this scenario, well-calibrated land surface or hydrological models can be used for simulation of these parameters at higher spatial and temporal scales. The traditional hydrological models, which can be lumped, semi-distributed or fully distributed physical models (Moradkhani and Sorooshian 2008), may or may not have snowmelt or glacier melt runoff modules. If some of these models are able to simulate the snowmelt, it can be done using either the energy balance models or the temperature index models. In case of the energy balance-based models, the net energy of a snowpack governs the production or freezing of melt water. This technique accounts for a given time period, total incoming energy, outgoing energy and the net change in the energy storage in a snowpack. This energy balance gives an estimate of the net available energy, which can be used as the heat equivalent required for snowmelt, or if it is negative, it will further freeze the snowpack (Anderson 1976). The presence or absence of the cloud and vegetation cover also affects the overall energy balance of a snowpack (Anderson 1976). In case of the full energy balance models, we need well-distributed hydrometeorological data over the entire study area to set up model and simulate the snowmelt runoff. The energy balance models use the input hydrometeorological data such as rainfall, minimum and maximum temperature and wind speed to simulate the snow cover, snow density, snow depth, snow albedo and finally snowmelt at point, grid or watershed scale. Whereas, the temperature index models requires less data, such as daily air temperature and rainfall data along with basin or elevation zone wise degree day factor and snow cover information to simulate the daily snowmelt runoff. As the upper Ganga river basin has very few ground-based hydrometeorological data, the present study of estimating snowmelt runoff has used temperature index modelling approach. The brief overview of these snowmelt models is given in the section. Snowmelt runoff model (SRM), developed by Martinec in 1975, is one of the most popular temperature index-based snowmelt runoff models, which is used in upper Ganga basin snowmelt study (Aggarwal et al. 2014). This model was initially developed for small European basins but has been used in other parts of the world in mountain basins of size varying from 0.76 to 122,000 km^2 and any elevation range (Martinec et al. 2007). The details of model structure are given in Martinec et al. (1983) and its implementation for upper Ganga basin in Aggarwal et al. (2014).

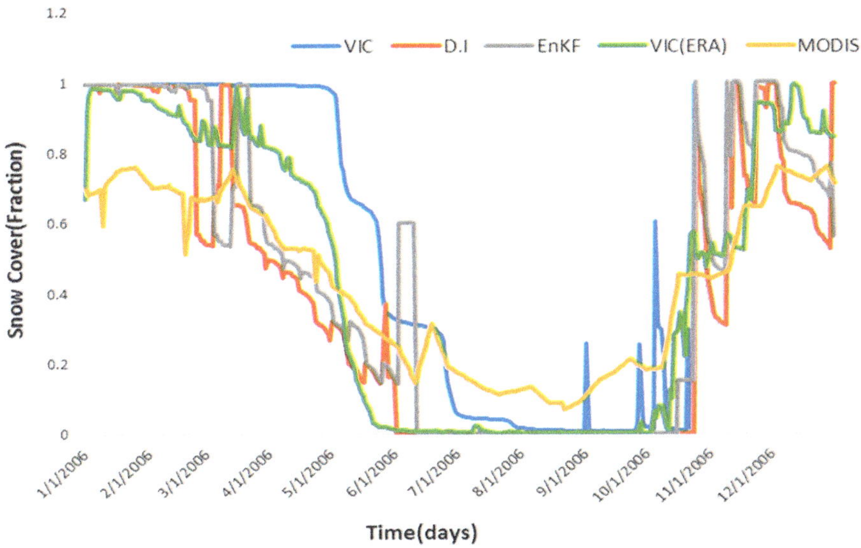

Fig. 5.21 Comparison of Beas basin daily average SCA from VIC modes and MODIS SCA

5.4.1 Mapping and Simulation of Snow Cover Using Remote Sensing, Energy Balance Models and Data Assimilation

In this section, time series data from optical sensors such as MODIS, LANDSAT, AWIFS/LISS-III and SAR data from RISAT-1 data for the years 2004–2006 and 2013–2015 are used to create remote sensing-based SCA maps, and these maps are further used in snowmelt runoff models. This study has also used a grid-based land surface model and variable infiltration capacity (VIC) model in energy balance model to simulate SCA at daily 1-km grid scale for the Beas River basin. Detailed SCA analysis is done with respect to its variations in elevation zones, aspects and slope for test sites. The maximum and minimum snow cover has been estimated in the months of March–April and August–September, respectively, for test river basins of Beas and Bhagirathi basins. The SCA derived from remote sensing (RS) based (optical and SAR) was compared based on VIC-simulated SCA maps, and R^2 of 0.58 was achieved (Naha et al. 2016). The current estimates of SCA with VIC global meteorological forcing data have given more SCA as compared to RS-based SCA. The major advantage of modelling approach is that if meteorological forcing data is correct, it will give accurate SCA and other snow products such as snow depth and water equivalent at daily scale without any interruptions.

Main SCA mapping results from VIC model are shown in Fig. 5.21 as SCA depletion curve from year 2003 to 2006. The input datasets used for simulation are given in flowchart form in Naha et al. (2016), and some key parameter values for simulation snow processes are as follows: minimum temperature at which rain can fall in liquid form is taken as −0.5 °C and maximum temperature at which snowfall can occur is taken as +0.2 °C. Snow density is simulated using formulation given by traditional VIC algorithm taken from Bras (1990). The main snow algorithm of VIC model (Gao et al. 2009) is briefly described here. The snow model in VIC represents the snowpack as a two-layer medium and solves an energy and mass balance for the ground surface snowpack in a manner similar to other cold land process models (Anderson 1976; Wigmosta et al. 1994; Tarboton et al. 1995; Gao et al. 2009). Energy exchange between the atmosphere, forest canopy and snowpack occurs only within the surface layer. The energy balance of the surface layer is (Andreadis et al. 2009):

$$\rho_w c_S \frac{dWT_S}{dt} = Q_r + Q_S + Q_l + Q_p + Q_m \quad (5.4)$$

where c_s is the specific heat of ice (J kg-1 K-1, ρ_w is the density of water (kg/m^3), W is the water equivalent (mm), T_s is the temperature of the surface layer (°C), Q_r is the net radiation flux (W/m^2), Q_s is the sensible heat flux (W/m^2), Q_l is the latent heat flux (W/m^2), Q_p is the energy flux advected to the snowpack by rain or snow (W/m^2) and Q_m is the energy flux given to the pack due to liquid water refreezing or removed from the pack during melting (W/m^2) (Gao et al. 2009). The detailed processes were described in Andreadis et al. (2009) and Gao et al. (2009). VIC considers a grid under snow only when it has some value of SWE else entire grid as non-snow grid (Sheffield et al. 2003; Gao et al. 2009).

The correlation coefficient, R^2, between remote sensing-based MODIS 8-day SCA product and VIC-simulated SCA, after applying the direct insertion (DI) DA technique, is calculated as 0.73 with RMSE of +0.194 (19.4% SCA), as compared to MODIS and original VIC-based SCA without any DA, which resulted in R^2 of 0.58 and RMSE of 0.29 (29% SCA). The snow cover simulation accuracy in terms of R^2 between MODIS SCA and VIC-based SCA, after applying the ensemble Kalman filter (EnKF) DA technique, is 0.76, but it resulted in more RMSE of +0.231 (23%). Further work is needed in the field of DA techniques to improve the model background error and increase the number of ensemble members in case EnKF DA techniques. Apart from this, the relationship between SCA and SWE needs to be established using ground-based data, for proper DA implementation in VIC model.

5.4.2 Simulation of Snowmelt Runoff Using Temperature Index and Energy Balance Models

The upper Ganga River Basin, with two subbasins of Alaknanda and Bhagirathi rivers (as shown in Fig. 5.2a), is chosen as test site for testing and validation of the temperature-based method of snowmelt runoff estimation. SRM model was used for this study, with inputs from RS-based SCA, and the hydrological and meteorological parameters from CWC and IMD were integrated for the snowmelt runoff and overall

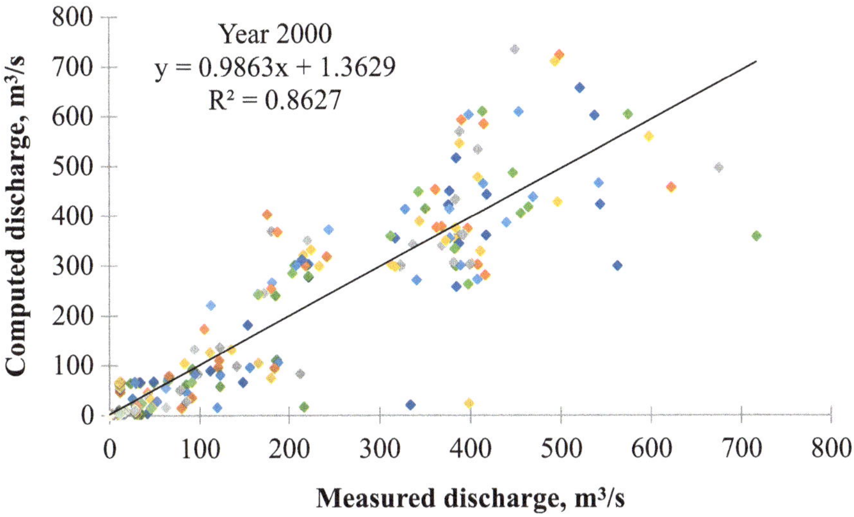

Fig. 5.22 Measured and simulated discharges of the Alaknanda River for the year 2000

discharge hydrograph computation from January 1 to December 31 of the year 2000 and 2008 (Aggarwal et al. 2014). In the SRM model, the main calculated and calibrated parameters were lapse rate, critical temperature (T_{CRIT}), degree-day factors (a_n), lag time, snow runoff coefficient (C_S), rainfall runoff coefficient (C_R), rainfall contribution area (RCA) and X coefficient and Y coefficient, which are part of recession coefficients (Martinec et al. 2007; Aggarwal et al. 2014). The calibrated value of lapse rate varies from 0.65 to 0.75, critical temperature of 2.0°C, degree-day factors estimated to be in the range of 0.35 to 0.65, lag time at 18.0 h, C_S and C_R varied from 0.10 to 0.80, RCA from 0 to 1, X coefficient ranged between 0.90 to 1.02 and Y coefficient varied between 0.80 to 0.88 for the Bhagirathi river basin (Aggarwal et al. 2014). The plot of the observed vs. estimated river flow, along with linear regression equations for the year 2000 (Fig. 5.22), shows a coefficient of determination, R^2, as 0.86 with volume difference of 0.14% for the year 2000. The snowmelt simulation for the year 2008 was done as the validation of selected SRM model parameters, where same set of parameters were applied. The validation simulation of the year 2008 gave correlation coefficient (R^2) of 0.84 which indicates the validity of selected model and parameters (Aggarwal et al. 2014).

5.4.3 Glacier Mass Balance (GMB) Studies Using Remote Sensing-Based Method

Bamber and Rivera (2007) completed a review of remote sensing-based methods for GMB determination and classified remote sensing-based GMB method under three classes, namely, component approach, proxy measures of mass balance and geodetic approach. The accumulation area ratio (AAR) method comes under proxy measures of mass balance. AAR is considered to be one of the indicators of variations in glacier mass balance and also climatic conditions. To estimate the glacier mass balance by AAR, accumulation and ablation area needs to be mapped. The accumulation and ablation zones of the glacier can be mapped by applying various techniques on optical remote sensing data. Remote sensing data in multispectral mode can be used to determine the end of summer snow line by differentiation between (wet) snow and ice (Bindschadler et al. 2001). "Excluding the influence of superimposed ice on the net mass balance, the transient snowline altitude (SLA) at the end of the ablation season is a reasonable proxy for the ELA and can, therefore be used to determine the AAR of the glacier" (Bamber and Rivera 2007). Kulkarni (1992) has used field-based estimates of mass balance and AAR available from 1977–1978, 1982–1983 and 1976–1977 to 1983–1984 for Gara and Gor-Garang glaciers, located in the catchment of Sutlej River, to develop the regression analysis between specific mass balance (SMB) and AAR. This suggests a high correlation coefficient, i.e. 0.88 and 0.96 with AAR values of 0.47 and 0.43, representing zero mass balance for Gara and Gor-Garang glaciers, respectively. The same relation was used along with remote sensing-based Landsat data to find AAR in Gara glacier. Kulkarni (1992) estimated AAR values for the years 1986–87 and 1987–1988 as 0.57 and 0.16, respectively, for the Gara glacier, which showed that Gara glacier had positive mass balance for the year 1986–1987 and negative mass balance for 1987–1988. Kulkarni et al. (2004) used AAR method for monitoring of glacial mass balance in the Baspa basin using relation of AAR and specific glacier mass balance of Gara and Gor-Garang glaciers developed by Kulkarni (1992). However, the extrapolation of this AAR relation to other glaciers not sampled in the field is problematic because this relationship is different from one glacier to another and other uncertainties associated with this method of measuring glacier mass balance from space (Berthier et al. 2007). This error in GMB can be due to variable climate regime, different aspects of glacier, uncertainty due to getting optical- or SAR-based remote sensing image at the time of maximum ablation time period, varying degrees of debris depth and composition (Pratap et al. 2016).

Therefore, based on the latest field investigations for one of the bench mark glacier, Chota Shigri glacier, specific mass balance between year 2002–2010 is estimated by Wagnon et al. (2007); Ramanathan (2011) which is reported in Himalayan Glaciology Technical Report as "Status report on Chhota Shigri glacier" (Himachal Pradesh) (Ramanathan 2011). Chhota Shigri glacier is one of the long-term and well-monitored glaciers (Dhobal et al. 1995). In 2002, the International association of cryospheric science, previously named as the International Commission on Snow and Ice, selected this glacier as one of the benchmark glaciers in whole Hindu Kush Himalayas. Chhota Shigri glacier (Fig. 5.23) is a compound valley-type glacier which extends over 32.19°–32.28°N latitude and 77.48°–77.55°E longitude and is located in the Chandra River basin of Lahaul and Spiti districts of Himachal Pradesh in Western Himalayas. Its drainage basin consists of four tributary glaciers and other small attached glaciers with total area of 34.7 km^2. The length of the

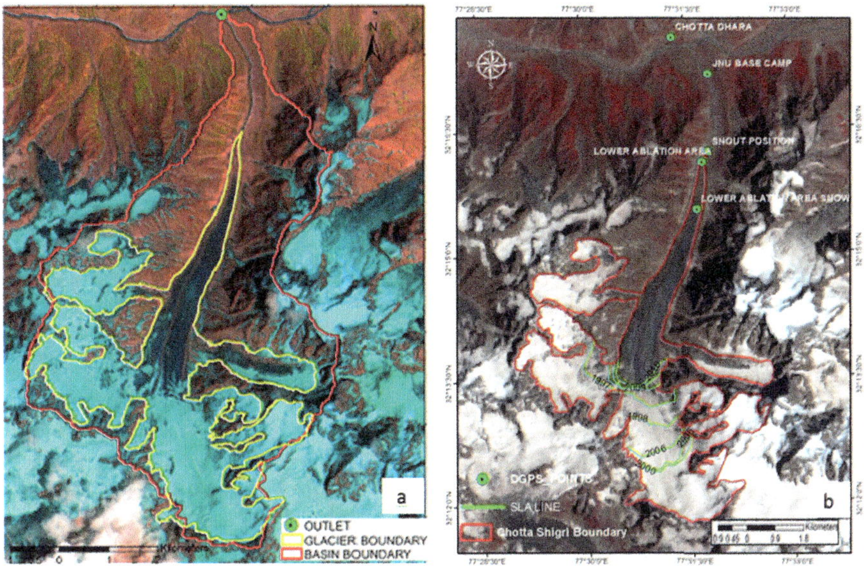

Fig. 5.23 (**a**) The Chhota Shigri Glacier in Lahaul and Spiti district of Himachal Pradesh as seen in the LiSS-III image of IRS P-6 dated 27 July 2012. (**b**) Snow line altitude (SLA) for Chhota Shigri Glacier for 1997, 1998, 2000, 2005, 2006, 2008, 2009, and 2010 derived from remote sensing and DGPS points location

glacier is about 9 km which covers mainly with ice, debris, snow and debris along with other glacier features such as supra-glacier streams and moulins are also visible in lower ablation area.

The study uses this data and established a mathematical model between specific mass balance (Y) and AAR as a following regression equation (Fig. 5.24)

$$\mathrm{Y} = 0.0386 * AAR - 2.50 \text{ with } R^2 = 0.95 \tag{5.6}$$

This mathematical model was then used with the AAR derived from the remote sensing dataset using NDSI and band ratio approach for estimation of specific mass balance of study area. This model helps in the validation of the remote sensing-based-derived AAR and mass balance with the ground data. The average mass balance calculated from field data from 2005 to 2010 comes out to be −0.354 m w. eq. AAR calculated with NDSI techniques estimated specific mass balance from 2005 to 2010 as −0.376 m w. eq. and AAR based on band ratio method estimated SMB of −0.379 m w. eq. The average mass balance estimated by Eq. 5.6 using NDSI and band ratio methods based AAR come out to be −0.47 m w. eq. The correlation coefficient (R^2) and RMSE between observed and estimated GMB comes out to be 0.99 and 0.09 m w.eq. The AAR results are comparable with standard glaciological based method used by Wagnon et al. (2007), where they have reported annual specific mass balance of Chhota Shigri Glacier as −1.4, −1.2, +0.1

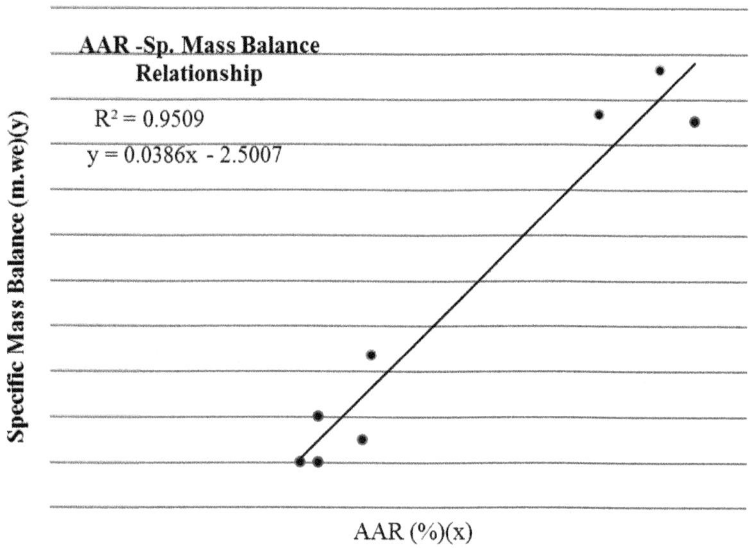

Fig. 5.24 Mathematical model showing relationship between SMB and AAR

and − 1.4mw.e. and AAR of 0.31, 0.31, 0.74, and 0.29 in 2002/2003, 2003/2004, 2004/2005 and 2005/06, respectively. This percentage error can further be reduced if some more subpixel-based classification or higher resolution image or SAR image during minimum snow line of ablation season is used in this area. The snowline altitude (SLA) for these time series has been evaluated using DEM (Fig. 5.23). SLA also has shown high variation during the period of 1997–2010. Figure 5.23 shows the temporal variation of SLA on Chhota Shigri glacier for the given dataset. The minimum SLA, i.e. 4783 ± 43, is obtained from the year 2010 which shows a highly positive mass balance. The requirement of the model used is based on the availability of cloud-free satellite imagery and field data during minimum snow line time period. The SLA used here is calculated based on the availability of cloud-free image, which is the main limitation in this case study as most of the images are in the middle of August time, whereas minimum snow line may have occurred 1–2 weeks after this time. This has resulted in SLA to be less than the reported ELA (Wagnon et al. 2007) for this glacier during common study time period of 2005 to 2010 with this study estimating mean SLA of 4876 ± 33 m and published reports (Wagnon et al. 2007; Ramanathan 2011) giving mean ELA of 5033 m. Therefore, the use of SAR data becomes very useful in finding SLA and ELA of glaciers as shown in Thakur et al. (2016b).

5.5 Conclusions

The cryosphere components of NWH form an integral part of this area as they store and provide water for large number of population living in this area. Remote sensing and GIS is one of the best tools for regular mapping and monitoring of these cryosphere components due to its wide coverage and repeativity. Highlights of this chapter are (a) SCA mapping including dry and wet snow from RISAT-1 MRS datasets; (b) estimation of SCA for the entire NWH using MODIS and AWIFS data from 2001 to 2017; (c) retrieval of the snow physical parameters such as snow density and snow wetness using SAR-based inversion models; (d) classification of major glacier radar zones such as bare ice zone, percolation-refreeze zone, debris glacier, ice wall and supraglacial lakes using time series of SAR data and hybrid polarimetric-based decomposition methods; (e) estimation of glacier ice velocity and thickness using DInSAR and feature tracking and laminar flow-based methods; (e) simulation of SCA using energy balance approach in hydrological models; and (f) estimation of SMR for select subbasins and specific mass balance using AAR method for bench mark Chhota Shigri glacier.

This information, derived from RS-based glacier radar zones, can be utilized for estimating the glacier's accumulation and ablation area, firm line and equilibrium elevation line altitude of a glacier, and can further be used as input to annual summer or winter glacier mass balance models (Thakur et al. 2016b). The results from the research on glacier velocity show various effects of variation in glacier displacement values during a single season and between independent seasons caused by aspect, slope and elevation changes. Snowmelt runoff modelling for the river basins of NWH is often hindered due to non-availability of detailed and accurate hydrometeorological data. This can be partially avoided by integration of the remote sensing-based SCA and DEM data and traditionally limited hydrometeorological data in temperature index-based snowmelt runoff models. The use of elevation zone and aspect map of the basin further improves the quality of SRM simulations. Additionally, with calibrated reanalysis meteorological datasets, grid-based energy balance snowmelt runoff models can be used to simulate the SCA and other snow parameters to get full temporal and spatial coverage over NWH. Overall, the results shown in this chapter can be used for snowmelt and glacier runoff modelling studies, and base data and methodologies created in these works can also be utilized for various cryosphere-related hazards such as snow avalanche modelling, glacier zones and crevasse detection and monitoring in selected areas. There is an urgent need to improve the overall field instruments in NWH, so as to collect timely and spatially well-distributed data on weather, snow and glacier components, which will further help in validating the remote sensing-based map products and also improve accuracy of hydrological models.

Acknowledgements The authors are thankful to NRSC, USGS, BBMB, CWC, IMD and SASE for providing satellite and hydrometeorological data required for this study. Funding for this work is provided by ISRO under various TDP and EOAM projects.

References

ACGR (1988) Glossary of Permafrost and Related Ground-Ice Terms, National Research of Canada, Technical Memorandum No. 142: 64.

Aggarwal, S.P., Thakur, P.K., Nikam, B.R. and Garg, V. (2014). Integrated approach for snowmelt run-off estimation using temperature index model, remote sensing and GIS. Current Science, 106 (3), 397–407.

Al Momani, B., Morrow, P., McClean, S. (2007) Knowledge-based semi-supervised satellite image classification. In 9th international symposium on signal processing and its applications, 2007, ISSPA, February 12–15, 2007, pp. 1–4.

Allen SK, Fiddes J, Linsbauer A, Randhawa SS, Saklani B, Salzmann N (2016) Permafrost studies in Kullu district, Himachal Pradesh. Curr Sci, 11: 257–260.

Anderson, E. A. (1976). A point energy and mass balance model of a snow cover, Tech. Rep. 19, NOAA, Silver Spring, Md.

Andreadis, K., Storck, P. and Lettenmaier, D.P., (2009), Modeling snow accumulation and ablation processes in forested environments, Water Resour Res., 45, W05429, pp. 1–13, doi:https://doi.org/10.1029/2008WR007042.

Bamber, J.L. and Rivera, A. (2007). A review of remote sensing methods for glacier mass balance determination, Global and Planetary Change, 59, 138–148.

Berthier, E., Arnaud, Y., Kumar, R., Ahmad, S., Wagnon, P. and Chevallier, P. (2007). Remote sensing estimates of glacier mass balances in the Himachal Pradesh (Western Himalaya, India), Remote Sensing of Environment, 108, 327–338.

Bhambri, R., Bolch, T., Chaujar, R.K. 2012. Frontal recession of Gangotri Glacier, Garhwal Himalayas, from 1965 to 2006, measured through high-resolution remote sensing data. Curr Sci, 3

Bindschadler, R. (1998). Monitoring ice sheet behaviour from space. Reviews of Geophysics, 36 (1), 79–104.

Bindschadler, R., Dowdeswell, J.A., Hall, D. and Winther, J.G. (2001). Glaciological applications with Landsat-7 imagery: early assessments. Remote Sensing of Environment, 78(1–2), 163–179.

Bisht, S. M., Thakur, P. K., Chouksey, A., & Agarwal, S. P. (2015). Ice thickness estimation using geospatial technology, HYDRO- 2015. In 20th international conference on hydraulics, water resources and river engineering, December 17–19, 2015, p. 12. India: IIT Roorkee.

Bras, R. A. (1990), Hydrology, an Introduction to Hydrologic Science, 643 pp., Addison-Wesley.

Cuffey, K.M. and Paterson, W.S.B. (2010). The Physics of Glaciers. Elsevier Science.

Dhobal, D.P., Kumar, S. and Mundepi, A.K. (1995). Morphology and glacier dynamics studies in monsoon-arid transition zone: An example from Chhota Shigri glacier, Himachal-Himalaya, India. Current Science, 68.

Dozier, J. (1989). Spectral signature of alpine snow covers from the Landsat thematic mapper. Remote Sensing Environ., 28, 9–22.

Dozier, J. and Marks, D. (1987). Snow mapping and classification from Landsat Thematic Mapper (TM) data. Ann. Glaciol., 9, 97–103.

Engeset, R. V., & Ødegard, R. S. (1999). Comparison of annual changes in winter ERS-1 SAR images and glacier mass balance of Slakbreen, Svalbard. International Journal of Remote Sensing, 20(2), 259–271.

Engeset, R. V., & Weydahl, D. J. (1998). Analysis of glaciers and geomorphology on Svalbard using multitemporal ERS-1 SAR images. IEEE Transactions on Geoscience and Remote Sensing, 36(6), 1879–1887.

Engeset, R. V., Kohler, J., Melvold, K., & Lunden, B. (2002). Change detection and monitoring of glacier mass balance and facies using ERS SAR winter images over Svalbard. International Journal of Remote Sensing, 23(10), 2023–2050.

Farinotti, D., Huss, M., Bauder, A. and Funk, M. (2009b). An estimate of the glacier ice volume in the Swiss Alps. Global Planet. Change, 68(3), 225–231, doi: https://doi.org/10.1016/j.gloplacha.2009.05.004.

Farinotti, D., Huss, M., Bauder, A., Funk, M. and Truffer M.A. (2009a). Method to estimate ice volume and ice-thickness distribution of alpine glaciers. J. Glaciol., 55(191), 422–430, doi: https://doi.org/10.3189/002214309788816759.

Gabriel, A.K. and Goldstein, R.M. (1988). Crossed orbit interferometry: theory and experimental results from SIR-B. Int. J. Remote Sens., 9(5), 857–872.

Gantayat, P., Kulkarni, A.V. and Srinivasan, J. (2014). Estimation of ice thickness using surface velocities and slope: case study at Gangotri Glacier, India, Journal of Glaciology, 60(220), pp. 277–282.

Gao, H., Tang, Q., Shi, X., Zhu, C., Bohn, T., Su, F. (2009). Water Budget Record from Variable Infiltration Capacity (VIC) Model Algorithm Theoretical Basis Document.

Garg V, Aggarwal SP, Thakur PK, Nikam BR (2014) Snow and its grain size mapping using hyperspectral remote sensing data. In: Interactive session in ISPRS TC VIII international symposium on operational remote sensing applications: opportunities, progress and challenges, Annual convention of ISRS and ISG and joint sessions with ISPRS TC IV and TC VI, hosted by National Remote, Sensing Centre, Indian Space Research Organisation, Hyderabad, India, Dec 09–12, 2014

Goldstein, R.M., Engelhardt, H., Kamb B. and Frolich R.M. (1993). Satellite radar interferometry for monitoring ice sheet motion: application to an Antarctic ice stream. Science, 262(5139), 1525–1530.

Goldstein, R.M., Zebker H.A. and Werner C.L. (1988). Satellite radar interferometry: two-dimensional phase unwrapping. Radio Sci., 23(4), 713–720.

Graham, L.C. (1974). Synthetic interferometer radar for topographic mapping. IEEE Proc., 62(6), 763–768.

Gruber, S. (2012). Derivation and analysis of a high-resolution estimate of global permafrost zonation. The Cryosphere, 6, 221–233, 2012, www.the-cryosphere.net/6/221/2012/ doi: https://doi.org/10.5194/tc-6-221-2012.

Hall, D. K., Riggs, G. A., Salomonson, V. V., DiGirolamo N. E. and Bayr, K. J. (2002). MODIS snow cover products, Remote Sensing Environ., 83, 181–194.

Hallikainen, M. T., Ulaby, F. T. & Abdelrazik, M. (1986). Dielectric properties of snow in the 3 to 37 GHz range. IEEE Transactions on Antennas and Propagation, AP-34, 1329–1339.

Huang, L., Li, Z., Tian, B., Chen, Q., & Zhou, J. (2013). Monitoring glacier zones and snow/firn line changes in the Qinghai–Tibetan Plateau using C-band SAR imagery. Remote Sensing of Environment, 137, 17–30. doi:https://doi.org/10.1016/j.rse.2013.05.016.

Huang, L., Li, Z., Tian, B.-S., Chen, Q., Liu, J.-L., & Zhang, R. (2011). Classification and snow line detection for glacial areas using the polarimetric SAR image. Remote Sensing of Environment, 115(7), 1721–1732. doi:https://doi.org/10.1016/j.rse.2011.03.004.

Immerzeel, W.W., van Beek, L.P.H., Bierkens, M.F.P. (2010). Climate change will affect the Asian water towers. Science 328(5984): 1382–1385. doi:https://doi.org/10.1126/science.1183188,

IPCC (2014). Climate Change 2014: Impacts, Adaptation, and Vulnerability, Part B: Regional Aspects. Contribution of Working Group II to the Fifth Assessment Report of the Intergovernmental Panel on Climate Change [Barros, V.R., C.B. Field, D.J. Dokken, M.D. Mastrandrea, K.J. Mach, T.E. Bilir, M. Chatterjee, K.L. Ebi, Y.O. Estrada, R.C. Genova, B. Girma, E.S. Kissel, A.N. Levy, S. MacCracken, P.R. Mastrandrea, and L.L. White (eds.)]. Cambridge University Press, Cambridge, United Kingdom and New York, NY, USA, 688 pp,.

Ishikawa, M.,Watanabe, T., and Nakamura, N. (2001). Genetic differences of rock glaciers and the discontinuous mountain permafrost zone in Kanchanjunga Himal, Eastern Nepal, Permafrost Periglac. Process., 12, 243–253, doi:https://doi.org/10.1002/ppp.394.

Jain SK, Goswami A, Saraf AK (2010) Snowmelt runoff modeling in a Himalayan basin with the aid of satellite data. Int J Remote Sens 31(24):6603–6618

Joughin, I., Smith, B.E. and Abdalati, W., (2010). Glaciological advances made with interferometric synthetic aperture radar, Journal of Glaciology, 56(200), pp. 1026–1041.

Kendra, J. R., Sarabandi, K. & Ulaby, F. T. (1998). Radar measurements of snow: experiment and analysis. IEEE Transactions on Geoscience and Remote Sensing, 36(3), 864–879.

König, M., Wadham, J., Winther, J.-G., Kohler, J., & Nuttall, A.-M. (2002). Detection of superimposed ice on the glaciers Kongsvegen and midre Love'nbreen, Svalbard, using SAR imagery. Annals of Glaciology, 34, 335–342.

König, M., Winther, J.-G., & Isaksson, E. (2001). Measuring snow and glacier ice properties from satellite. Reviews of Geophysics, 39(1), 1–27.

Kosmann, D., Wessel, B. Schwieger, V. (2010). Global Digital Elevation Model from TanDEM-X and the Calibration/Validation with worldwide kinematic GPS Tracks. XXIV FIG International Congress 2010, 11–16 April, 2010, Sydney, Australia.

Kulkarni, A. V. (1992) Mass balance of Himalayan glaciers using AAR and ELA methods, Journal of Glaciology, 38(128), 101–104.

Kulkarni, A. V., Rathore, B. P. and Suja, A. (2004). Monitoring of glacial mass balance in the Baspa basin using accumulation area ratio method. Curr. Sci., 86, 101–106.

Kulkarni, A. V., Singh, S. K., Mathur, P. and Mishra, V. D., (2006). Algorithm to monitor snow cover using AWiFS data of Resourcesat-1 for the Himalayan region. Int. J. Remote Sensing, 27, 2449–2457.

Kulkarni, A., Rathore, B. P., Singh, S. K. and Ajai, A. (2010). Distribution of seasonal snow cover in central and western Hima laya. Ann. Glaciol., 51(54), 121–128.

Kumar K, Dumka RK, Miral MS, Satyal GS, Pant M. 2008. Estimation of retreat rate of Gangotri glacier using rapid static and kinematic GPS survey. Curr Sci 94(2): 258–262

Kundu, S., Chakraborty, M. (2015). Delineation of glacial zones of Gangotri and other glaciers of Central Himalaya using RISAT-1 C-band dual-pol SAR. International Journal of Remote Sensing, 36(6), 1529–1550.

Leinss, S., Parrella, G., & Hajnsek, I. (2014). Snow height determination by polarimetric phase differences in X-band SAR data. IEEE Journal of Selected Topics in Applied Earth Observations and Remote Sensing, 7(9), 3794–3810.

Leinss, S., Wiesmann, A., Lemmetyinen, J., & Hajnsek, I. (2015). Snow water equivalent of dry snow measured by differential interferometry. IEEE Journal of Selected Topics in Applied Earth Observations and Remote Sensing, 8(8), 3773–3790.

Li, F.K. and Goldstein, R.M. (1990). Studies of multi-baseline spaceborne interferometric synthetic aperture radars. IEEE Trans. Geosci. Remote Sens., 28(1), 88–97.

Looyenga, H. (1965). Dielectric constant of heterogeneous mixtures. Physica, 21, 401–406.

Mätzler, C. (1996). Microwave remote sensing of dry snow. IEEE Transactions on Geoscience and Remote Sensing, 34(2), 573–581.

Martinec, J., Rango A. & Roberts R. (2007). Snowmelt Runoff Model, User's Manual, Updated Edition 2007, Edited by Enrique Gómez-Landesa, Windows Version 1.11.

Martinec, J., Rango, A. and Major, E. (1983). The Snowmelt-Runoff (SRM) User's Manual, NASA Reference Publ. 1100, NASA, Washington DC.

Minnett, P. J. (2014). Cryosphere, Climate Change Feedbacks, Encyclopedia of Remote Sensing, Part of the series Encyclopedia of Earth Sciences Series, pp 101–104.

Moradkhani, H. and Sorooshian, S. (2008). General review of rainfall - runoff modeling: model calibration, data assimilation, and uncertainty analysis. Hydrological modeling and the water cycle.Springer. 291 p. ISBN 978-3-540-77842-4.

Naha, S., Thakur, P.K. and Aggarwal, S.P. (2016). Hydrological Modelling and data assimilation of Satellite Snow Cover Area using a Land Surface Model, VIC. Paper published in the International Archives of the Photogrammetry, Remote Sensing and Spatial Information Sciences, Volume XLI-B8, 2016, XXIII ISPRS Congress, 12–19 July 2016, Prague, Czech Republic, DOI:https://doi.org/10.5194/isprsarchives-XLI-B8-353-2016.

National Remote Sensing Centre, NRSC, (2013). Glacier lakes in Uttarakhand—a remote sensing based inventory. Geosciences Group, RSA-Area, National Remote Sensing Centre (NRSC), ISRO, Hyderabad, India.

Negi HS, Singh SK, Kulkarni AV, Semwal BS (2010) Field based spectral reflectance measurements of seasonal snow cover in the Indian Himalaya. Int J Remote Sens 31(9):2393–2417.

Negi, H.S., Thakur, N.K. and Ganju, A. (2012). Monitoring of Gangotri glacier using remote sensing and ground observations. Journal of earth system science 121 (4):855–66.

Nikam, B.R., Garg, V., Gupta, P.K., Thakur, P.K., Kumar, A.S., Chouksey, A., Aggarwal, S.P., Dhote P. and Purohit, S. (2017). Satellite-based mapping and monitoring of heavy snowfall in North Western Himalaya and its hydrologic consequences, Current science, 113(12): 2328–2334.

Painter, T. H., Dozier, J., Roberts, D. A., Davis, R. E., & Green, R. O. (2003). Retrieval of subpixel snow-covered area and grain size from imaging spectrometer data. Remote Sensing of Environment, 85(1), 64–77. DOI: https://doi.org/10.1016/S0034-4257(02)00187-6.

Painter, T. H., Roberts, D. A., Green, R. O., & Dozier, J. (1998). The effect of grain size on spectral mixture analysis of snow-covered area from AVIRIS data. Remote Sensing of Environment, 65 (3), 320–332. DOI: https://doi.org/10.1016/S0034-4257(98)00041-8.

Patrington, K. C. (1998). Discrimination of glacier facies using multi-temporal SAR data. Journal of Glaciology, 44(146), 42–53.

Pierce, L. E., Ulaby, F. T., Sarabandi, K., & Dobson, M. C. (1994). Knowledge-based classification of polarimetric SAR images. IEEE Transactions on Geoscience and Remote Sensing, 32, 1081–1086.

Pratap B, Dobhal DP, Bhambri R, Mehta M, Tewari VC (2016) Four decades of glacier mass balance observations in the Indian Himalaya. Reg Environ Change 16:643–658.

Raina, V.K. and Srivastava, D. (2008). Glacier atlas of India. Geological Society of India, Bangalore. p 316.

Ramanathan, A.L. (2011). Status Report on Chhota Shigri Glacier (Himachal Pradesh), Department of Science and Technology, Ministry of Science and Technology, New Delhi, Himalayan Glaciology Technical Report No.1, 88.

Raney, R. K. (2007). Hybrid-polarity SAR architecture. IEEE Transactions on Geoscience and Remote Sensing, 45(1), 3397–3404.

Raney, R. K., Cahill, J. T., Patterson, G., Bussey, D. B. J. (2012). The m-chi decomposition of hybrid dual-polarimetric radar data with application to lunar craters. Journal of Geophysical Research: Planets, 117(E12).

Rau, F., Braun, M., Friedrich, M., Weber, F., & Gobmann, H. (2000). Radar glacier zones and its boundaries as indicators of glacier mass balance and climatic variability. EARSeL eProceedings, 1, 317–327.

Rees, W. G. (2006). Remote sensing of snow and ice. Boca Raton: CRC Press, Taylor & Francis Group.

Schmid, M.-O., Baral, P., Gruber, S., Shahi, S., Shrestha, T., Stumm, D., and Wester, P. (2015). Assessment of permafrost distribution maps in the Hindu Kush Himalayan region using rock glaciers mapped in Google Earth, The Cryosphere, 9, 2089–2099

Sheffield, J., Pan, M., Wood, E.F., Mitchell, K.E., Houser, P.R., Schaake, J.C., Robock, A., Lohmann, D., Cosgrove, B., Duan, Q. and Luo, L., 2003. Snow process modeling in the North American Land Data Assimilation System (NLDAS): 1. Evaluation of model-simulated snow cover extent. Journal of Geophysical Research: Atmospheres, 108(D22).

Shi, J. and Dozier, J. (1995). Inferring snow wetness using C-band data from SIR-C's polarimetric synthetic aperture radar. IEEE Transactions on Geoscience and Remote Sensing, 33, 905–914.

Shi, J. and Dozier, J. (2000). Estimation of snow water equivalence using SIR-C/X SAR, Part I: inferring snow density and subsurface properties. IEEE Transactions on Geoscience and Remote Sensing, 38, 2465–2474.

Space Application Centre, SAC, (2010). Final Technical Report - Snow and Glacier Studies, A joint Project of Ministry of Environment and Forests and Department of Space, Govt. of India, SAC/RESA/MESG/SGP /TR / 59 /2010. 268 pages.

Strozzi, T., Wegmuller, U. and Matzler, C. (1999). Mapping wet snowcovers with SAR interferometry. International Journal of Remote Sensing, 20 (12): 2395-620 403. doi: https://doi.org/10.1080/014311699212083.

Surendar, M., Bhattacharya, A., Singh, G., & Venkataraman, G. (2015a). Estimation of snow density using full-polarimetric synthetic aperture radar (SAR) data. Physics and Chemistry of the Earth, 83–84, 156–165.

Surendar, M., Bhattacharya, A., Singh, G., Yamaguchi, Y., & Venkataraman, G. (2015b). Development of a snow wetness inversion algorithm using polarimetric scattering power decomposition model. International Journal of Applied Earth Observation and Geoinformation, 42, 65–75.

Swaroop, S., Raina, V.K. and Sangeswar, C.V. (2003). Ice flow of Gangotri glacier. In Srivastava D, Gupta KR and Mukerji S eds. Proceedings of the Workshop on Gangotri glacier, 26–28 March 2003, Lucknow, India. (Spec. Publ. 80) Geological Survey of India, Kolkata.

Tarboton, D. G., Chowdhury, T. G. and Jackson, T. H. (1995). A spatially distributed energy balance snowmelt model, in Biogeochemistry of Seasonally Snow Covered Catchments, vol. 228, edited by Tonneson, K. A., Williams, W. and Tranter, M. pp. 141–155, Int. Assoc. of Hydrol. Sci., Wallingford, U. K.

Thakur, P. K., Aggarwal, S.P., Arun, G., Sood, S., Kumar, A.S., Snehmani, Dobhal, D.P. (2016b). Estimation of snow cover area, snow physical properties and glacier classification in parts of Western Himalayas using C-band SAR data. Springer's Journal of the Indian Society of Remote Sensing (JISRS), DOI: https://doi.org/10.1007/s12524-016-0609-y.

Thakur, P.K., Aggarwal, S.P., Garg, P.K., Garg, R.D., Snehmani, Pandit A and Kumar, S. (2012). Snow physical parameter estimation using space based SAR. Geocarto International, 27 (3):263–288, DOI:https://doi.org/10.1080/10106049.2012.672477.

Thakur, P.K., Dixit, A., Chouksey, A., Aggarwal, S.P. and Kumar, A.S. (2016a). Ice sheet features identification, glacier velocity estimation and glacier zones classification using high resolution optical and SAR data, paper published in SPIE Asia-Pacific Remote Sensing conference during April 4-7, 2016 at New Delhi, India, Proc. of SPIE Vol. 9877, 987719-1-16, DOI: https://doi.org/10.1117/12.2224027.

Thakur, P.K., Garg, P.K., Aggarwal, S.P., Garg, R.D. and Snehmani (2013). Snow Cover Area Mapping Using Synthetic Aperture Radar in Manali watershed of Beas River in the Northwest Himalayas. Journal of the Indian Society of Remote Sensing (JISRS), 41(4), 933–945. DOI: https://doi.org/10.1007/s12524-012-0236-1.

Tso, B., Mather, P. (2009). Classification methods for remotely sensed data (2nd ed.). Boca Raton: CRC Press, Taylor & Francis Group.

van Everdingen, R. (ed.), (1998). Multi-Language Glossary of Permafrost and Related Ground-Ice Terms, National Snow and Ice Data Center, Boulder, CO, USA,

Vaughan, D. G. et al., Observations: Cryosphere, in Climate Change 2013: The Physical Science Basis. Contribution of Working Group I to the Fifth Assessment Report of the Intergovernmental Panel on Climate Change (eds Stocker, T. F. et al.), Cambridge University Press, Cambridge, United Kingdom, 2013, pp. 317–382.

Vogel, S. W. (2002) Usage of high-resolution Landsat-7 band-8 for single band snow cover classification. Ann. Glaciol., 34, 53–57.

Wagnon, P., Linda, A., Arnaud, Y., Kumar, R., Sharma, P., Vincent, C., Pottakkal, J. G., Berthier, E., Ramanathan, A., Hasnain, S. I. and Chevallier, P. (2007). Four years of mass balance on Chhota Shigri Glacier, Himachal Pradesh, India, a new benchmark glacier in the western Himalaya, J. Glaciol., 53, 603–611.

Westermann, S., Schuler, T. V., Gisnas, K. and Etzelmuller B. (2013). Transient thermal modeling of permafrost conditions in Southern Norway, The Cryosphere, 7, 719–739, www.the-cryosphere.net/7/719/2013/ doi: 10.5194/tc-7-719-2013.

Wigmosta, M. S., Vail, L. W. and Lettenmaier, D. P. (1994). A distributed hydrology-vegetation model for complex terrain, Water Resour. Res., 30, 1665–1679.

Zebker, H.A. and R.M. Goldstein. 1986. Topographic mapping from interferometric synthetic aperture radar observations. J. Geophys. Res., 91(B5), 4993–4999.

Websites

http://bhuvan.nrsc.gov.in/data/download/index.php accessed latest on September 2016,

https://tandemx-science.dlr.de/ accessed latest on September 2016,

http://www.moef.nic.in/downloads/public-information/Snow-and_Glaciers-of-the-Himalayas-ISRO-SAC%20-%2013th%20June.pdf accessed latest on September 2016,

http://www.eorc.jaxa.jp/ALOS/en/ accessed latest on September 2016,

http://www.portal.gsi.gov.in/portal/page?_pageid=127,711653&_dad=portal&_schema=PORTAL accessed latest on September 2016,

https://www.ipcc.ch/report/ar5/ accessed latest on September 2016,

Chapter 6
Hydrological Modelling in North Western Himalaya

S. P. Aggarwal, Vaibhav Garg, Praveen K. Thakur, and Bhaskar R. Nikam

6.1 Introduction

The Himalayas are one of the largest reservoirs of freshwater in the form of glaciers and snow outside the Polar region (Mani 1981). There are around 32,392 glaciers, covering an area of about 71,182 km^2 in the Indian part of the Himalaya (SAC 2011). Among all, North Western Himalaya (NWH) has the largest area under seasonal and perennial snow cover. This snow/glacier melt contributes significantly to perennial rivers like the Ganga and the Indus during lean time. The Indus Basin is comprised of Chennab, Jhelum, Rawi, Satluj and Beas River subbasins, whereas Upper Ganga Basin is comprised of Bhagirathi, Alaknanda, Mandakini, Dhauliganga and Pindar subbasins. Moreover, these basins have huge hydropower potential, which is a matter of concern during lean period (Kasturirangan et al. 2013).

It has also been reported that most of Himalayan glaciers are in general state of recession under changing climate condition as shown in Fig. 6.1 (Dobhal et al. 2004; Kulkarni et al. 2007; Thayyen and Gergan 2010). Moreover, if these glaciers continue to recede at the alarming rate, the availability of this huge freshwater will be a matter of grave concern. In view of such changes, the development and rational management of the water resources, an assessment of the quantity and quality of available water are necessary. However, it is practically impossible to measure all the components of the water balance or hydrological system due to large heterogeneity and the limitations of measurement techniques. The only way to overcome these limitations is hydrological modelling by extrapolating information from available measurements.

S. P. Aggarwal (✉) · V. Garg · P. K. Thakur · B. R. Nikam
Water Resources Department, Indian Institute of Remote Sensing (IIRS), Indian Space Research Organisation (ISRO), Department of Space, Government of India, Dehradun, India
e-mail: spa@iirs.gov.in

R. R. Navalgund et al. (eds.), *Remote Sensing of Northwest Himalayan Ecosystems*,
https://doi.org/10.1007/978-981-13-2128-3_6

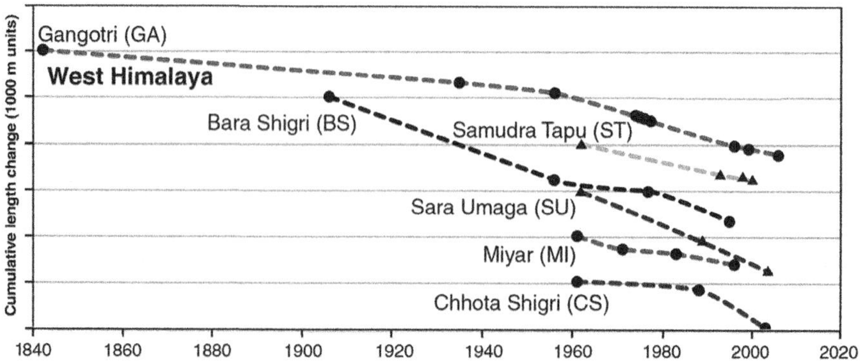

Fig. 6.1 The retreat of NWH glacier since the mid-nineteenth century (Bolch et al. 2012)

6.2 Hydrological Modelling Approach

Hydrological modelling is important from water resources planning, development and management point of view. Hydrological modelling is a mathematical representation of hydrological processes that are generally defined in terms of parameters and states. The parameters represent physical characteristics of surface and subsurface, whereas states represent fluxes and storages of water and energy which generally vary with time. Based on the problem to be addressed, the hydrological modelling may have the following objectives:

- To carry out spatio-temporal extrapolation of field-based point measurement
- To enhance the fundamental understanding of existing hydrological systems
- To assess the impact of climate and land use land cover (LULC) change on water resources
- To improve the existing models or to develop new models for current and future water resource management

6.2.1 Types of Hydrological Models

The hydrological models can be classified based on their structure, spatial distribution, stochasticity and spatio-temporal applications. Based on the model structure, the hydrological models may be classified as metric, conceptual, physics-based and hybrid models. The very basic metric models characterise the system response from the available historic data or observations (Wheater et al. 1993). These kinds of models are also known as empirical models, and the best example for these types of model is unit hydrograph theory developed for event-based catchment-scale simulations of ungauged catchments (Sherman 1932). The most interesting development in the field of metric models is the application of data-based mechanistic approach such as artificial neural networks (ANNs) and evolutionary algorithms in hydrological modelling. ANNs were used extensively to study the behaviour of rainfall-runoff processes just from available rainfall and runoff data. As these kinds of empirical models are developed basically on historically observed input and output, without considering the process of conversion explicitly, on the contrary, the conceptual models are developed based on knowledge of the pertinent natural processes that affect the input to generate the output (Wheater 2002). The structure of these models is defined prior to actual simulation being undertaken (Wheater et al. 1993). The best examples for conceptual model are Stanford Watershed Model and the kinematic-wave runoff model (Crawford and Linsley 1960; Subramanya 2008). Physics-based models represent the component hydrological processes through the governing equations of motion based on continuum mechanics. These kinds of models are considered as most accurate; on the other hand, their data requirement is high.

The models can further be classified based on space discretization criteria as lumped or distributed. Lumped models consider the catchment as a single unit, taking average of state variables over the entire catchment area (Beven 2001). These types of models do not account for spatial variability in physical processes and characteristics of the basin (Singh 1995). On the other hand, in the distributed models, the basin is generally discretized into a number of elements (or regular grids), and then the mathematical equations representing the physical processes are solved for the state variables associated with each element (Singh and Frevert 2006). As in such models, the state variables represent local average and the predictions are distributed spatially. However, due to constraint of spatial scale of input data available, generally the parameters are lumped over the element or grid; such type of models are usually called semi-distributed hydrological models (Beven 2001). As semi-distributed model requires less data and considers at least important features of the catchment as compared to fully distributed model, these models are easy to realise and may require less computational time (Orellana et al. 2008). The best example of this kind of model is variable infiltration capacity (VIC) model.

Further, if the model simulates a single storm, it is regarded as event-based models. The storm duration may range from few hours to few days. The widely used Hydrologic Engineering Centre-Hydrologic Modelling System (HEC-HMS) is one of the event-based models. On the contrary, a continuous model runs at longer period, estimating basin response for entire time series comprised of both precipitation and dry periods. There are many other types of models; the further details of each type of models and their advantages and disadvantages can be found at Wheater et al. (1993), Beven (2001), Singh and Woolhiser (2002), Wagener et al. (2004), Singh and Frevert (2006) and Pechlivanidis et al. (2011).

The input data requirement and the accuracy/reliability of the model output vary from model to model. The distributed physically based models are considered to be more accurate among all. However, the data requirement of such models is huge with higher accuracy in a distributed manner. On the other hand, it is nearly impossible to practically collect the data on field required for the water resources assessment. In such cases, geospatial techniques play a vital role to collect or extract the information at required temporal or spatial scale with reasonably good accuracy. The data collected through remote sensing in regular spatial and temporal domain can easily be incorporated into the hydrological models. These models can further be used to retrieve or estimate the components of hydrological cycle. This helps in achieving more accurate estimates of the hydrological cycle. In this chapter, the hydrological components required to carry out water resources assessment of NWH region have been discussed along with their retrieval from remote sensing data. A table of most widely used hydrological models with their capabilities is shown in Table 6.1 below. Further, the capability of VIC hydrological in studying water balance and impact of climate change on hydrology of a basin is also investigated as shown in case studies in subsequent sections.

6.2.2 *Hydrological Model Input Parameterization Using Remote Sensing Data*

Precipitation/Rainfall

The precipitation means water in its all forms such as rainfall, snowfall, hail, frost and dew, sleet, etc. that reaches the Earth from the atmosphere. The rainfall and snowfall are the predominant form of precipitation generating stream/flood flow in majority of rivers. Based on topography, physiography and climatology, the magnitude of precipitation varies with time and space. Due to this variation, different geographical regions face hydrological problems, either floods or droughts. Precipitation in terms of rainfall is usually measured as the depth, however, in the case of snowfall, an equivalent depth of water (in mm). The precipitation is usually measured through rain gauge infield; however, it is recommended that the catchment area per gauge should be small WMO (1974). The India Meteorological Department (IMD) is making a lot of efforts to instal a large number of rain gauge across the country as shown in Fig. 6.2.

It can be seen that the density of the rain gauge is less in NWH region of the India as compared to other parts due to rugged terrain and difficult conditions. However, in case of basin level water resources assessment study, it is required to have as many as possible rain gauge in the basin. Therefore, in such a case, the remote sensing-derived rainfall products may be best suited as remote sensing can provide spatio-temporal rainfall data of any part on the globe.

Remote sensing can provide better spatial estimate of rainfall than conventional point information of limited observations. The cloud movement and its pattern can easily be observed by satellite images in visible (VIS) and infrared (IR) regions. The useful information such as cloud depth and cloud droplet size, cloud-top temperature and height from and cloud phase can easily be extracted from VIS, IR and water vapour channels, respectively. Moreover, the brightness, pattern, texture, area, shape, shadow and movement of cloud in these images can be correlated to rain and its movement Gruber (1973). At present the following IR sensors, namely, the

Table 6.1 Most widely used hydrological models and their capability

Model	Main advantage	Main disadvantage	Runoff	Baseflow/ groundwater	Watershed scale	Climatic regime	Snow/ glacial Melt	Outputs
HEC-HMS	Focus on runoff, channel routing and water control structure	Suitable only for events not for long-term hydrological simulations	Empirical	Empirical	Small to large	Rain or snow	Yes	FH, AY, PF, LF, SW, ET, WB, SM, IF, OF, SF, GF, RO
SWAT	Focus on water quantity and quality and representation of groundwater	Snow process representation requires improvement	Empirical	Physical	Small to large	Rain or snow	Yes	FH, AY, PF, LF, SW, ET, WB, SM, IF, WT, OF,SF, GF,RO, SE, NF, WQ
Mike-SHE	Simulates complete land phase of hydrologic cycle	Simplified representation of forest cover; high purchase cost	Physical	Physical	Small to large	Rain or snow	Yes	FH, AY, PF, LF, SW, ET, WB, SM, IF, WT, OF,SF, GF,RO, WQ
VIC	Sub-grid variability, macro-scale model; large-scale effects	Large grid size (10–200 km)	Physical	Physical	Medium to large	Rain/ snow/ mixed	Yes	FH, AY, PF, LF, SW, ET, WB, SM, IF, OF,SF, GF, RO

FH full hydrograph, *AY* annual yield, *PF* peak flow, *LF* low flow, *SW* snow water equivalent, *ET* evapotranspiration, *WB* water balance, *SM* soil moisture, *IF* infiltration, *WT* water table, *OF* overland flow, *SF* shallow subsurface flow, *GF* groundwater flow, *RO* basin total runoff, *SE* sediment soil erosion, *NF* nutrient fluxes, *WQ* water quality

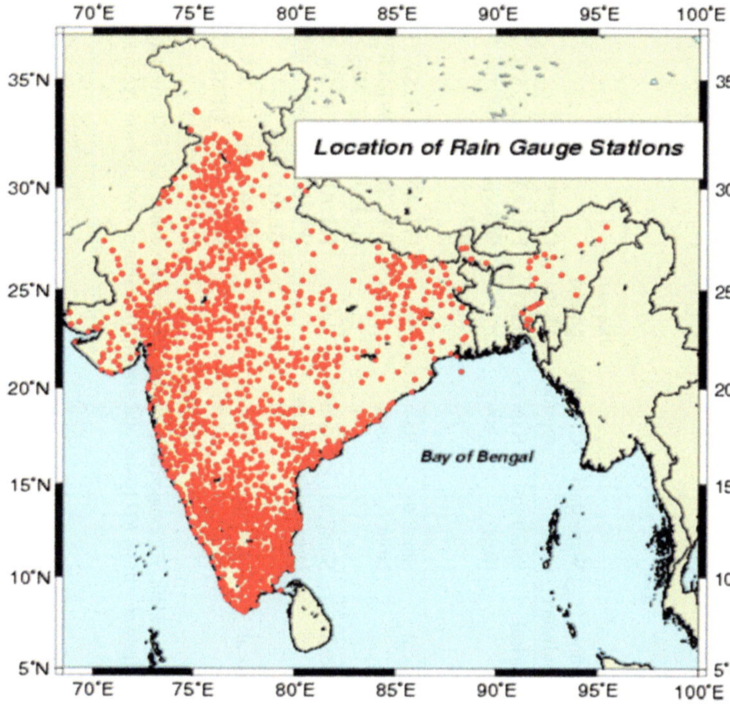

Fig. 6.2 Location of 1803 IMD rain gauges in India and their distribution. (Source: Rajeevan et al. 2005)

National Oceanic and Atmospheric Administration (NOAA), Geostationary Operational Environmental Satellite (GOES), European Meteosat, Russia's Elektro-L and India's INSAT series, are currently operating. The IR image of globe from INSAT 3D along with water vapour image of India and its surroundings is shown in Fig. 6.3.

The simplest cloud indexing method was developed by Arkin (1979). In this method each cloud type identified in the satellite image is assigned a rain rate level. It was initially developed for National Oceanic and Atmospheric Administration (NOAA) polar-orbiting satellites by the University of Bristol, which later found its applicability for geostationary satellite data also. The techniques use the satellite thermal IR measurements (10.5–12.5 μm) of cloud-top temperature; a threshold is applied which is typically around −40 °C, below which it can be regarded as 'rain'. The cloud-top temperature and cloud motion vectors can easily be measured through INSAT series imager and sounder as shown in Fig. 6.3.

The microwave technique has physically been considered as more direct technique in rainfall estimation as compared to VIS/IR method, as at these frequencies, water droplets attenuate the upwelling radiation. The presence of hydrometeors changes the magnitude of emission or scattering which is measured by microwave

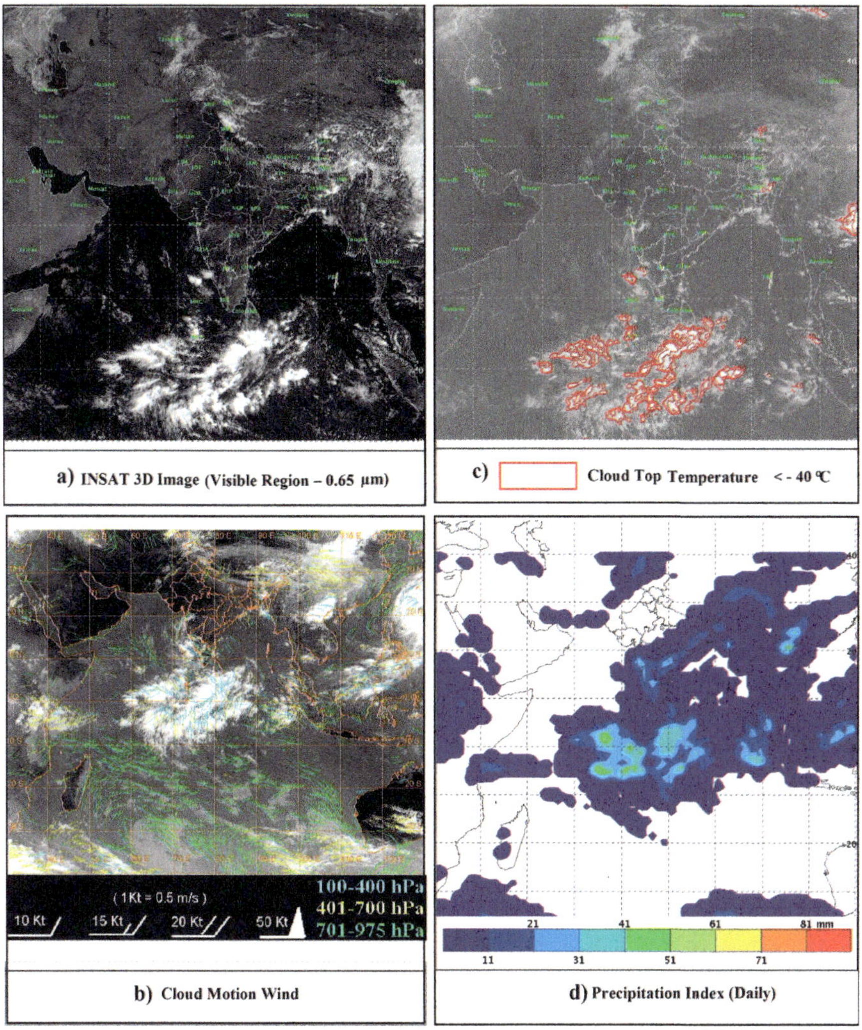

Fig. 6.3 (**a**) INSAT 3D IR image, (**b**) cloud vector motion of India, (**c**) cloud-top temperature image, (**d**) estimated precipitable water images of October 6, 2016, for India and its environment. (Source: http://satellite.imd.gov.in/insat.htm)

sensors, either passive or active as compared to dry atmosphere. Based on the frequency of upwelling radiation, the size and type of hydrometeors can be detected. The common passive microwave imager frequency range is between 19.3 and 85.5 GHz, at which radiation interacts with the hydrometeors and water particles/droplets either in liquid or frozen form. The sensor's field of view (FOV) observes

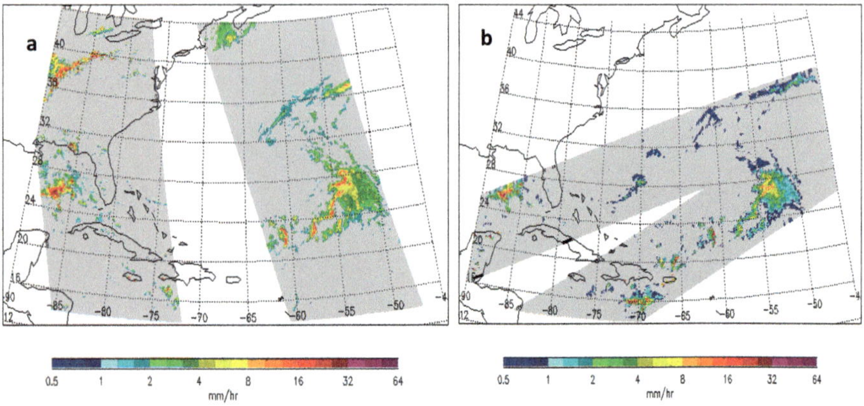

Fig. 6.4 (**a**) Precipitation over the Eastern United States and vicinity, from AMSR-E and (**b**) the TRMM Microwave Imager (TMI), June 5, 2002

cumulative scattering and emission taken place simultaneously with radiation undergoing multiple transformations while interacting with the cloud column. At different frequencies the passive microwave radiometers record brightness temperature of cloud column which is then related to rain (McVicar and Bierwirth 2001; Nagarajan 2009; Khanbilvardi et al. 2015).

At present, the following passive microwave sensors on eight platforms are usually used for precipitation estimation:

- Advanced microwave sounding unit (AMSU)-B (NOAA-15, NOAA-16, NOAA-17, NOAA-18)
- The Special Sensor Microwave Imager (SSM/I) (DMSP 13, 14, 15) – NOAA/NESDIS
- TRMM Microwave Imager
- The Advanced Microwave Scanning Radiometer for EOS (AMSR-E)
- WindSat (ocean only)

The precipitation estimates derived from the passive microwaves sensors, such as AMSR-E and TMI, on-board National Aeronautics and Space Administration (NASA) Aqua and TRMM satellites, respectively, are shown in Fig. 6.4. For generating these precipitation estimates, the algorithms of Ferraro (1997) for SSM/I, Ferraro et al. (2000) for AMSU-B and Kummerow et al. (2001) for TMI are generally used.

Rainfall retrieval using SSM/I data adapting (Ferraro 1997) algorithm is given below:

$$\mathrm{SI} = a_0 + a_1 T_{19V} + a_2 T_{22V} + a_3 {T_{22V}}^2 - T_{85V} \tag{6.1}$$

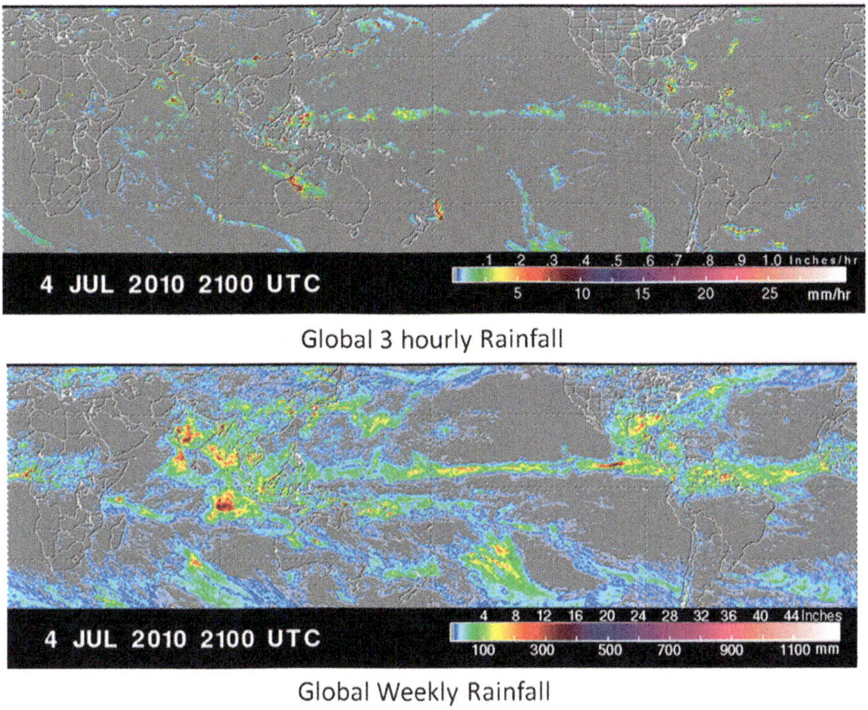

Fig. 6.5 Global rainfall estimates TRMM

$$R = a^{*}\mathrm{SI}^{*}b \tag{6.2}$$

where SI is scattering index, R is rainfall rate, T_{XP} is brightness temperature in given frequency 'x' and polarisation 'p' and a and b are the empirical constants.

The Tropical Rainfall Measuring Mission, TRMM, is a joint mission between the NASA and the Japan Aerospace Exploration Agency (JAXA) targeted to study the tropical and subtropical rainfall launched in the year 1997. TRMM measurements provide information on where the rainfall is occurring with its intensity including 3D structure of storm and clouds. It is a non-sun synchronous satellite carries several sensors, which are useful for landscape monitoring. The TRMM has Precipitation Radar (PR), TRMM Microwave Imager (TMI), Visible Infrared Scanner (VIRS), Clouds and the Earth's Radiant Energy System (CERES) and Lightning Imaging Sensor (LIS) payloads. The global three-hourly and weekly rainfall products by TRMM are shown in Fig. 6.5; however, Fig. 6.6 shows accumulated rainfall in states of NWH region by TRMM.

Built on legacy of TRMM, the Global Precipitation Measurement (GPM) mission is an international partnership co-led by the NASA and JAXA. The mission aims on the deployment of additional satellites in constellation of the GPM Core Observatory. Together the constellation of these satellites provides next-generation

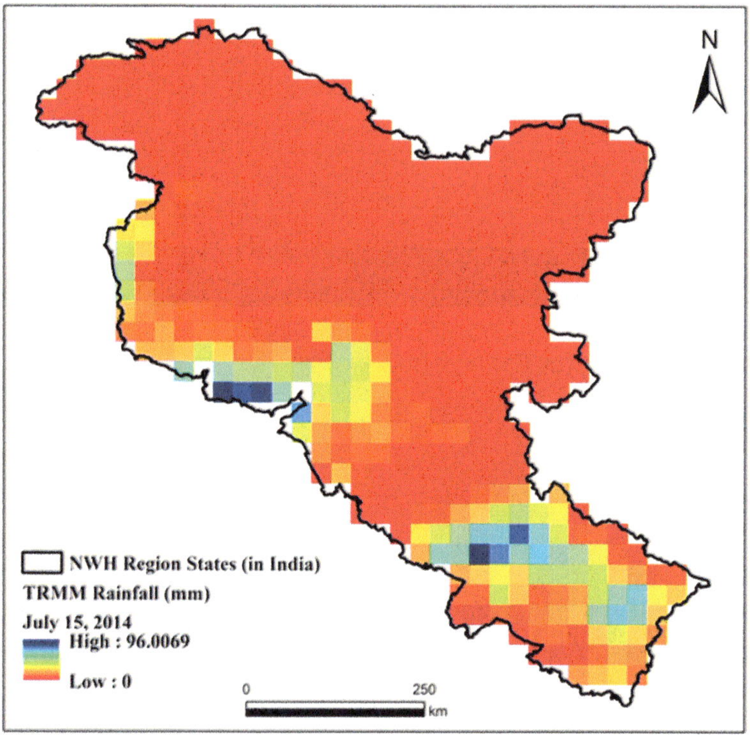

Fig. 6.6 TRMM-based rainfall estimate of NWH states

observations of global precipitation from the space. The objectives of the mission are to understand the horizontal and vertical structure of rainfall, its physical (micro and macro) nature and latent heat associated with it; to develop and calibrate retrieval algorithms for the radiometers in constellation; to provide sufficient number of samples at global level to reduce uncertainties in short-term rainfall accumulations significantly; and to extend its scientific and societal applications. The GPM Core Observatory is carrying an advanced dual-frequency radar (Ku-Ka ranging from 13.6–35 GHz) and multi-frequency (10.7, 19, 22, 37, 85 GHz V&H) radiometer system. It acts as a reference to unify precipitation measurements from all satellites within the GPM constellation. The coverage of the GPM is shown in Fig. 6.7.

The GPM Core Observatory provides greater coverage of precipitation measurements approximately between 65° north latitude and 65° south latitude including both the Arctic and Antarctic Circle. The constellation of these satellites provides global precipitation measurements at approximately every 3 h. With the help of this integrated approach and unified dataset, the scientists will understand Earth's water and energy cycle better. It is expected that the observations from the GPM constellation, combined with land surface data, will improve weather forecast models; climate models; integrated hydrologic models of watersheds; and forecast of

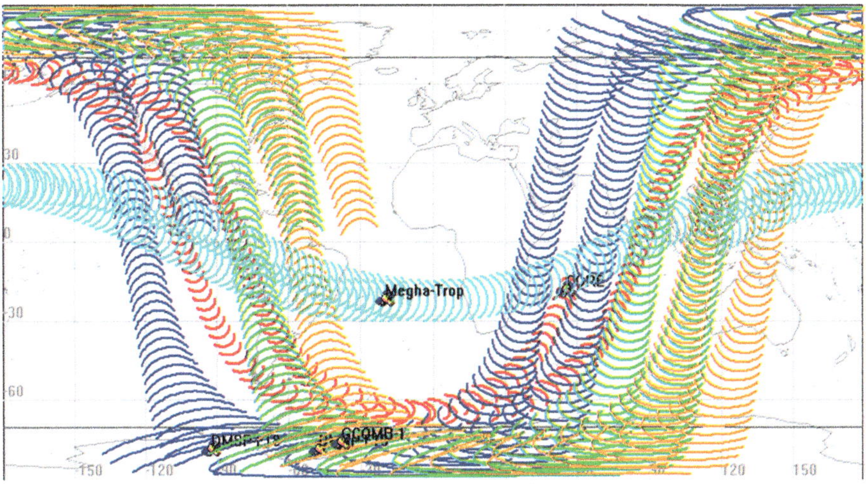

Fig. 6.7 GPM era coverage/3 h, CORE, DMSP-F18, DMSP-F19, Megha-Tropiques, GCOM-B1

hurricanes, landslides, floods, droughts, etc. Further details on various precipitation retrieval algorithms can be found at Thakur et al. (2017).

Soil Moisture

Soil moisture plays a significant role in the energy exchange between the ground surface and the atmosphere through evapotranspiration process. This root zone water storage has important implications especially on agriculture sector, as it partitions the precipitation water into infiltration and runoff. On field, it can be measured through theta probe and different handheld sensors, but these instruments will provide only point information. However, due to the capability of remote sensing technique to detect small spatial variation over a large extent at frequent time interval, this technique overcomes most of all field limitations. The surface soil moisture estimation can be done using optical and thermal bands of electromagnetic radiation (EMR). In the optical region, the changes in spectral signatures of the surface need to be regularly studied with change in soil moisture. One could analyse the changes in the spectral properties due to change in moisture or identify the bands sensitive to moisture; the soil moisture can indirectly be measured. As the emerging hyperspectral sensors provide spectral information at large contiguous narrow bands, very recently, it has been exploited to simulate soil moisture (Finn et al. 2011; Haubrock et al. 2008; Yanmin et al. 2010). The fine spatial resolution (~30 m) is a major advantage of hyperspectral remote sensing data, but it is constrained by availability of the continuous data. On the other hand, the change in thermal inertia between dry and wet soil can easily be detected by thermal sensors. This information can be interpreted to retrieve soil moisture of the land surface. However, the use of this technique has the following limitations such as the coarser spatial resolution and poor capability to penetrate through vegetation.

As the soil moisture estimation requires subsurface information, therefore, it is mostly being estimated using both passive and active microwave remote sensing data. The presence of water content in soil changes its dielectric property, which in turn affects brightness temperature and surface scattering properties. Various inversion models have been developed to map soil moisture using these variations (Dobson et al. 1985). It has been reported that the microwaves L-band (wavelength 15–30 cm), C-band (wavelength 3.8–7.5 cm) and X-band (wavelength 2.5–3.8) are the most sensitive bands for soil moisture estimation (Wagner et al. 2007).

Passive microwave radiometers have large potential among other remote sensing methods for the soil moisture measurement. The advantages of microwave measurements are (1) directly sensitive to changes in surface soil moisture and (2) least affected by clouds and (3) can penetrate moderate amounts of vegetation. Depending on wavelength and soil wetness, these sensors can measure surface moisture up to depths of 2–5 cm. The effect of soil moisture on the measured passive microwave radiometer signal dominates over that of surface roughness; however, on the contrary, it is true for active microwave radars. The following high-frequency Earth-imaging passive microwave radiometers, namely, the Scanning Multichannel Microwave Radiometer launched on the Seasat and Nimbus 7 satellites, Special Sensor Microwave Imager launched on the DMSP satellite series, TRMM Microwave Imager (TMI), Advanced Microwave Scanning Radiometer on the Earth Observation System (AMSR-E) and Soil Moisture and Ocean Salinity (SMOS), have been used extensively for soil moisture estimation. The capabilities of these higher-frequency instruments are limited to soil moisture measurements over predominantly bare soil and in a very shallow surface layer (<5 cm). The major disadvantage of passive microwave soil moisture estimation is coarse spatial resolution. However, the advantage is wider swath which helps to get frequent temporal images at every 4–6 days. The sample images of AMSR-E and SMOS global soil moisture products are given in Figs. 6.8 and 6.9, respectively. The AMSR-E soil moisture (in g/cm^3) for September 10–12, 2009 is shown with few gaps in the data; even on merging ascending and descending passes of AMSR-E, one cannot get soil moisture information for the entire globe. To overcome these issues, ESA launched SMOS mission with an aim of providing global maps of soil moisture, with accuracy better than 0.04 m/m every 3 days and with a space resolution better than 50 km.

The active sensors like synthetic-aperture radar (SAR) and scatterometers measure backscatter signals. A large number of algorithms have been used to derive soil moisture from data acquired by SAR such as Seasat, Spaceborne Imaging Radar-C (SIR-C), European Remote-Sensing (ERS) Satellite, RADARSAT, Environmental Satellite (Envisat) and Advanced Land Observation Satellite (ALOS). Similarly, soil moisture has been derived using scatterometer data of ERS, Advanced SCATterometer (ASCAT) and QuikSCAT.

Later, researchers realised to exploit advantages of both the techniques' passive and active remote sensing through blending. For example, under the ESA's Climate Change Initiative (CCI), the ASCAT active microwave data were blend with the passive microwave remote sensing data of Nimbus 7 SMMR, DMSP SSM/I, TRMM

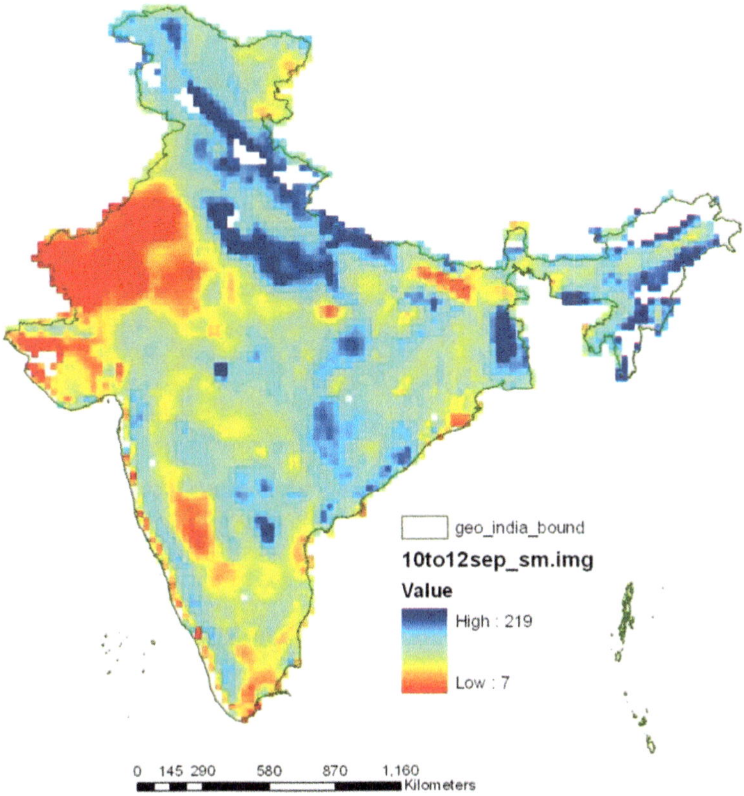

Fig. 6.8 AMSRE-based soil moisture map of India (September 10–12, 2009)

TMI and Aqua AMSR-E. The blended product of CCI soil moisture is available at near global scale at 0.25° spatial resolution.

On a similar guideline, NASA recently launched Soil Moisture Active Passive (SMAP) mission intended to provide global soil moisture at fine spatial resolution with drought monitoring and prediction as its targeted application (Wardlow et al. 2012). As SMAP carries both L-band radar and L-band radiometer, it allows global mapping of soil moisture at 10 km spatial resolution with 2–3 day revisit time. It can provide soil moisture information under clear and cloudy sky conditions even under a wide range of vegetation cover. The SMAP 3-day composite soil moisture products cantered on April 15, 18 and 22, 2015, are given in Fig. 6.10. The soil moisture is provided in volumetric terms (cm^3/cm^3). The blue colour in the map indicates wet condition; however, red indicates dry.

However, major limitation of the microwave remote sensing data in soil moisture estimation include depth over which soil moisture estimates are valid, consisting of the top few centimetres at most. Exposed to the atmosphere, these upper few centimetres of the soil moisture vary rapidly with response to rainfall and

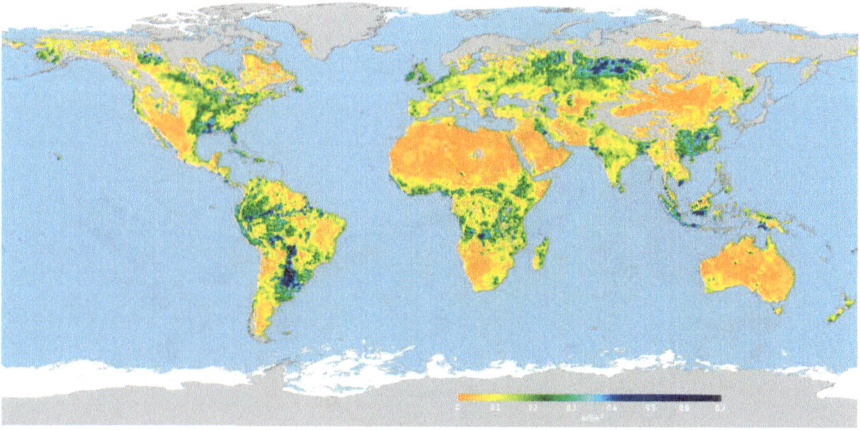

Fig. 6.9 SMOS global root zone soil moisture (in m^3/m^3) product of May 2016. (From European Space Agency, http://www.esa.int/spaceinimages/Images/2016/05/Root-zone_soil_moisture_May_2016)

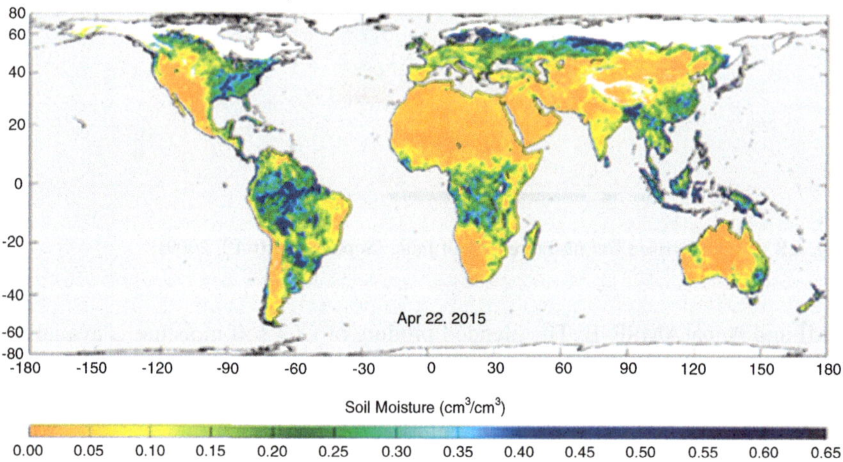

Fig. 6.10 SMAP-derived global soil moisture of April 22, 2015 (https://podaac.jpl.nasa.gov/sites/default/files/content/SmapSSS_SoilMoistureComposite.png)

evaporation, whereas for hydrologic, climatic and agricultural studies, such observations of surface soil moisture must be available for the complete soil moisture profile in the unsaturated zone. In recent years, in order to overcome these issues and to simulate the root zone soil moisture, attempts have been made to assimilate the remote sensing-derived surface soil moisture with physically based distributed land surface models (LSM). Nikam et al. (2015) assimilated AMSR-E-derived soil moisture in physically based semi-distributed VIC LSM at 25 × 25 km grid scale for simulating the root zone soil moisture. A similar attempt to assimilate SMOS soil moisture to simulate soil moisture at multiple depths using VIC LSM is done by Lievens et al. (2015).

Evapotranspiration (ET)

ET represents the exchange of mass and energy between the soil-water-vegetation system and the atmosphere (Senay et al. 2012). The long prevailing weather conditions affect reference ET (ET_o) through numerous variables such as radiation, temperature, wind and relative humidity. Moreover, actual ET (ET_a) is also affected by land cover class and soil moisture at that particular duration. ET is one of the most important parameters which govern the energy and mass transfer between the Earth's surface and atmosphere. It can be measured on field using instruments such as lysimeter and eddy covariance, surface renewal and flux variance systems (French et al. 2012). However, the measurement of ET using these techniques is constrained to a very limited spatial coverage; moreover, they are expensive and require high maintenance. In order to overcome these limitations, various ET estimation methods based on climatological data have been developed, namely, FAO-56 Penman-Monteith (FAO-56PM), Turc radiation, Priestley-Taylor, FAO radiation, Hargreaves and FAO Blaney-Criddle (Kashyap and Panda 2001). The FAO-56PM has been widely used and accepted; however, it requires huge climatic data at high spatio-temporal scale for accurate estimation of ET. On other hand, remote sensing can provide useful information on surface, radiation and climate at regular interval at large spatial scale. Therefore, this technology has widely used for estimation of ET at varying spatial and temporal scale. The ET estimation methods using remote sensing data can be grouped into three: (1) vegetation indices based, (2) surface energy balance based, and (3) the scatterplot between land surface and vegetation indices based (Allen et al. 1998; Bastiaanssen et al. 1998).

Vegetation Indices (VI)-Based Method

The methods discussed above use certain multiplicative factor such as the crop coefficient (derived based on particular crop and its condition) to estimate actual ET from reference ET. The vegetation condition type, health, density and phenology can easily be studied using remote sensing data adapting simple digital image processing techniques. These information have further been used to derive multiplicative coefficients of indirect methods discussed above. The vegetation indices such as normalised difference vegetation index, soil-adjusted vegetation index, leaf area index, enhanced vegetation index and vegetative cover fraction are being used for deriving vegetation-based ET coefficients using regression approach. As these indices provide actual crop/vegetation condition and health, this approach has been operationalized to generate global ET product using MODIS data (Mu et al. 2011). This ET product is freely available and can be used for any region on globe.

Surface Energy Balance-Based Methods

The energy balance models use land surface temperature (LST) as a primary constraint in partitioning of available surface radiant energy between heat and water flux (Senay et al. 2012, 2015). ET is highly dependent on energy available to help these processes (evaporation and transpiration). The estimation of energy balance or available energy can be used to estimate ET. Various surface energy balance methods have been developed and used for this purpose such as Surface Energy Balance Algorithm for Land (SEBAL), Mapping EvapoTranspiration at high Resolution with Internalized Calibration (METRIC), Surface Energy Balance System (SEBS), Atmosphere-Land Exchange Inverse (ALEXI), etc. Among these, the SEBAL methods are the widely accepted and used method as represented by Eq. (6.3):

$$\lambda\text{ET} = \Lambda(Rn - G) \tag{6.3}$$

where λET is the latent heat flux (W/m^2), Rn is the net radiation (W/m^2), G is the soil heat flux (W/m^2) and Λ is the evaporative fraction and can be written as

$$\Lambda = \lambda\text{ET}/(Rn - G) \quad = \lambda\text{ET}/(\lambda\text{ET} + H) \tag{6.4}$$

where H is the sensible heat (W/m^2) and (Rn − G) or (λET + H) is called net available energy.

These relationships can be implemented even instantaneous moment to monthly time scale. However, the assumption in estimation of Λ remains constant during the entire time step.

Scatterplot-Based Methods

The scatterplot-based methods of ET estimation are a combination of both the LST- and VI-based methods. It has been reported that a triangular or trapezoidal shape is formed when remotely sensed LST and VI or albedo (α) derived from heterogeneous areas is plotted in two-dimensional feature space (Petropoulos et al. 2009). These methods identify relative theoretical boundary lines in the observed inverse relationship between LST and VI or α, to estimate the Λ of energy balance or the ETo fraction of VI-based methods. Further details on these types of models can be found at Senay et al. (2012) and Thakur et al. (2017).

These parameters derived using geospatial techniques can easily be incorporated in hydrological models as input to generate fluxes and states. Further, these states and fluxes can be analysed to water resources assessment of the region of interest. Apart from this, these parameters can also be derived adapting suitable hydrological models taking inputs derived from remote sensing observations as discussed in subsequent section.

6.3 Hydrological Modelling: Case Studies

6.3.1 *Water Balance Study of Entire NWH Region*

In earlier days, the discharge in rivers, which is influenced by a large number of basin characteristics, was calculated using statistical approaches due to limitation of land surface parameters on spatial scale. With the advancement in remote sensing and geographical information system (GIS) technique, it showed great potential in providing relevant spatial data and parameters at the appropriate scale for the development of physically based distributed models. These sophisticated hydrological models can provide hydrological fluxes (e.g. runoff, ET, multilayer soil moisture, groundwater) at any desired scale from hydrological component estimation point of view. The improved computing capabilities helped to transform hydrological modelling into finer scale soil-vegetation-atmosphere schemes, where one can incorporate atmospheric components to improve upon estimation of hydrological fluxes (Wood 1991; Abdulla et al. 1996).

The VIC, macro-scale hydrological model, is a semi-distributed model that works at regular grid scale. It has been developed as soil-vegetation-atmosphere transfer schemes (SVATS) by Liang et al. (1994). It has the following advantages over other models: the sub-grid variability in soil moisture storage capacity, sub-grid variability of LULC and consideration of base flow as a non-linear recession (Zhao et al. 1980; Dumenil and Todini 1992). It can be run in different modes such as energy balance, water balance and combination of both. Moreover, it has its own routing module to route flow at desired outlet as simple linear transfer functions (Lohmann et al. 1996, 1998b). The details on VIC model may be found in Liang (1994), Liang et al. (1994, 1996), Nijssen et al. (1997), Lohmann et al. (1998a, b), Cherkauer and Lettenmaier (1999), Maurer et al. (2001a, b), Lettenmaier (2001), Liang and Xie (2003), Aggarwal et al. (2013) and Garg et al. (2017). The VIC model with two soil layers was set up for entire NWH region using geospatial inputs. The analysis was carried out at 5 × 5 km regular grid and the model was run in water balance mode. The model was forced by daily meteorological forcing on rainfall, maximum temperature, minimum temperature and wind speed to estimate daily runoff, ET, etc. on each grid cell independently. The derivation of input parameters for modelling in geospatial domain is defined in the subsequent section.

One has to define land surface characteristics of each grid in the model as four input files, namely, soil parameter, vegetation parameter, vegetation library and meteorological forcings. The unique soil properties of each grid are described in the soil parameter file. For the NWH region analysis, the primary soil properties for particular grid were taken from the National Bureau of Soil Survey and Land Utilisation Planning (NBSS&LUP) for corresponding/representative soil texture of that grid. The NBSS&LUP soil map of the region was analysed in GIS platform. The soil map generated for NWH region from NBSS&LUP database is shown in Fig. 6.11. The other parameters such as field capacity, wilting point, saturated hydraulic conductivity, porosity, etc. were calculated using pedo-transfer functions

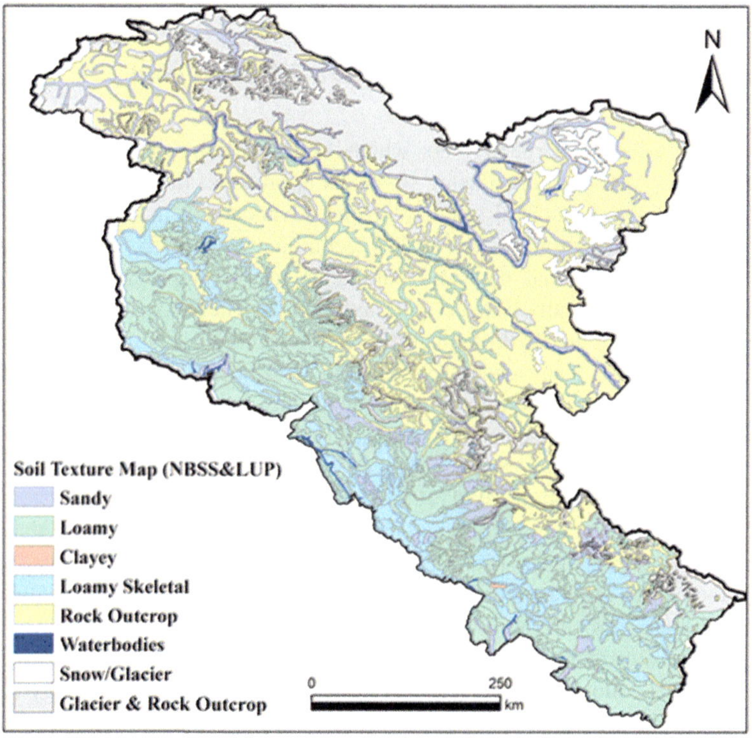

Fig. 6.11 Soil map of the NWH region

(Cosby et al. 1984; Rawls et al. 1998; Reynolds et al. 2000). The runoff generation will also governed by topography; therefore, in the present study, the topographic features (elevation and slope) of the region were derived from GTopo30 digital elevation model available at 90 m resolution (http://edc.usgs.gov/products/elevation/gtopo30/gtopo30.html) as shown in Fig. 6.12.

Vegetation parameter file generally defines the LULC classes present in the grid and fraction coverage of each LULC class along with corresponding rooting depth as suggested by Maurer et al. (2001a, b). The rooting depth is an important parameter as shorter root crops and grasses will draw water from the upper soil layer for transpiration as compared to deeper root trees that draw moisture from deeper soil layers (Garg et al. 2017). The LULC database generated under Indian Space Research Organisation- International Geosphere-Biosphere Programme (ISRO-IGBP) Project on "Land Use and Land Cover Dynamics and Impact of Human Dimensions in Indian River Basins" available at 1:250,000 scale has been used to identify the LULC classes present under the study region as shown in Fig. 6.13. The properties such as LAI, albedo, roughness, displacement, etc. correspond to each LULC class which vary with time; the seasonal variations in these parameters are specified at monthly time scale as vegetation library file in the model domain. The other parameters such as roughness length, displacement height, architectural

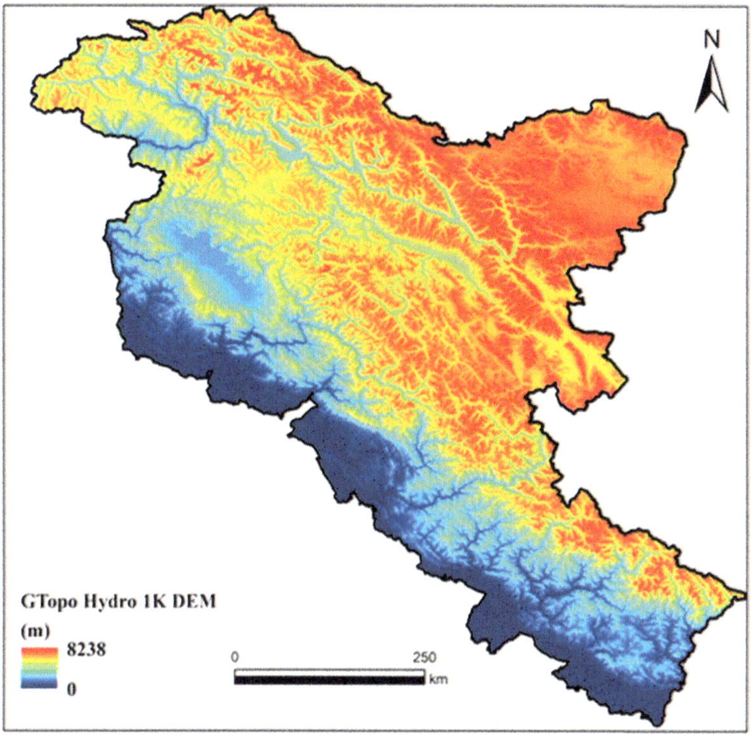

Fig. 6.12 Digital elevation model of NWH region

resistance and minimum stomatal resistance for each LULC class were taken from Global Land Data Assimilation System database (http://ldas.gsfc.nasa.gov/gldas/GLDASmapveg.php).

Later, the VIC2L model was forced with observed meteorological data on daily precipitation, daily minimum and maximum temperature and daily wind speed for 30 years (1985–2014) at daily time step. The forcing file for each year for each grid has been generated from Indian Meteorological Department (IMD) gridded data (Rajeevan et al. 2005; Pai et al. 2014) on rainfall (0.25° × 0.25°) and temperature (1° × 1°). The long-term average rainfall map of NWH region is shown in Fig. 6.14. The model results in flux files correspond to each grid with output on precipitation, evaporation from canopy, evaporation from bare soil, transpiration from canopy, baseflow and soil moisture for each soil layer. The methodology adapted in the present study is presented in Fig. 6.15.

The grid-wise results were aggregated to evaluate the basin average hydrological components as shown in Fig. 6.16. In this way, spatial information with respect to hydrological components may be generated using suitable remote sensing-derived input in geospatial environment. Such information can be utilised for better water resource management, distribution among various sectors and policy making. India is a developing country; therefore, the quantification of these hydrological

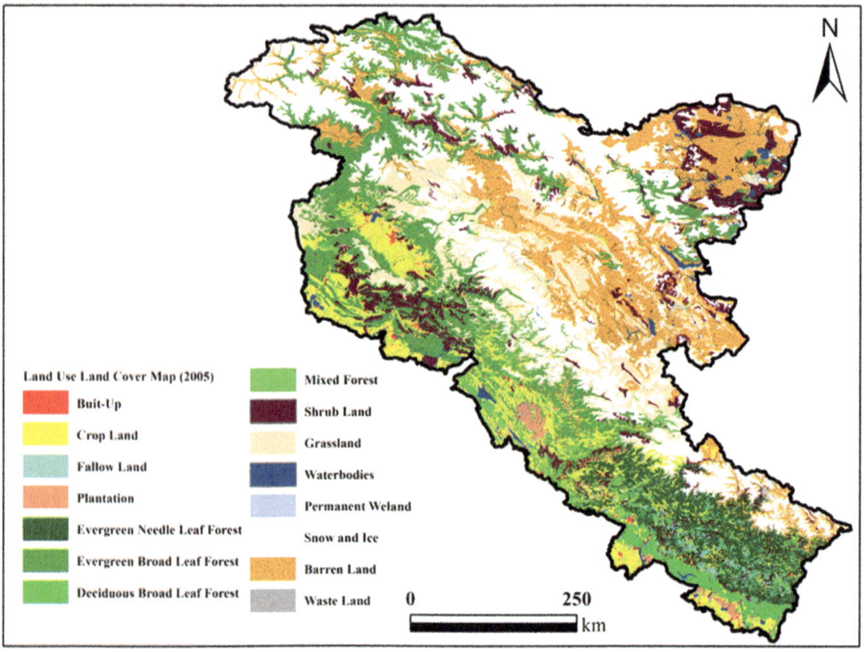

Fig. 6.13 Land use land cover map of NWH region. (Source: ISRO-GBP LULC Project)

components is also subjected to change in LULC pattern and change in climatic condition. The hydrological modelling approach can be best suited for these kinds of assessments. The further attempts were made to analyse the impact of changing climate on basins in NWH region which are discussed in subsequent sections.

6.3.2 *Hydrological Modelling of a NWH Basin Under Future Climate Change Scenario*

The basin scale water resources assessment has been done for Beas River basin which is part of major Indus Basin and located in NWH region using hydrological modelling approach. The Beas River originates from Beas Kund near Rohtang Pass at an elevation of 3960 m, and its basin covers an area of 20,303 km^2 up to its confluence with Satluj River near Harike Wetland in Punjab, India, with total traverse of 460 km. However, in the present study, the part of basin up to Pong Dam, with an area of around 12,417 km^2 which is located between longitudes 75°84′ E to 77°88′ E and latitudes 31°41′ N to 32°46′ N and lies entirely in NWH region, was analysed as shown in Fig. 6.17. Approximately, an area of 777 km^2 in this part of basin is covered under permanent snow and glaciers (http://india-wris.nrsc.gov.in). The remaining basin area receives water in the form of rainfall with maximum

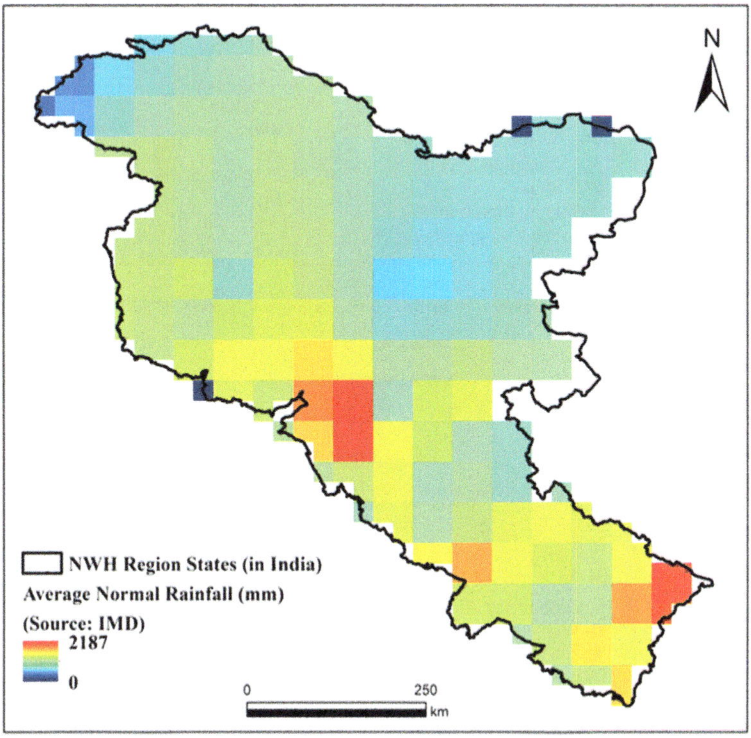

Fig. 6.14 Long-term average rainfall in the states falling in NWH region. (Source: IMD)

contribution during monsoon period (June–October); corresponding to it, maximum discharge in the basin has also been observed.

In this study the VIC hydrological model as discussed above has been used to simulate the impact of climate change on hydrology of Beas River basin. As it works on grid basis, the general circulation model (GCM)/regional circulation model (RCM) outputs can easily be incorporated in the model. For the calibration of the model, the meteorological forcing on precipitation, maximum temperature, minimum temperature and wind speed has been prepared using National Centers for Environmental Prediction and the National Center for Atmospheric Research (NCEP/NCAR) reanalysis data of period 1977–2006. However, for future climate scenarios, the forcings on these parameters for period 2006–2100 were generated from the ensemble of the Geophysical Fluid Dynamics Laboratory (GFDL) GCMs downscaled for Southeast Asia region under Coordinated Regional Climate Downscaling Experiment by the Swedish Meteorological and Hydrological Institute (WAS_GFDL_ESM2M_SMHI_RCAV2_DAILY) for radiation concentration pathways (RCP) scenarios of RCP4.5 and RCP 8.5 available at 0.44° resolution. The data can be found at Earth System Grid Federation, ESGF's data node at the National Supercomputer Centre (NSC), Linköping, Sweden (https://esg-dn1.nsc.liu.se/

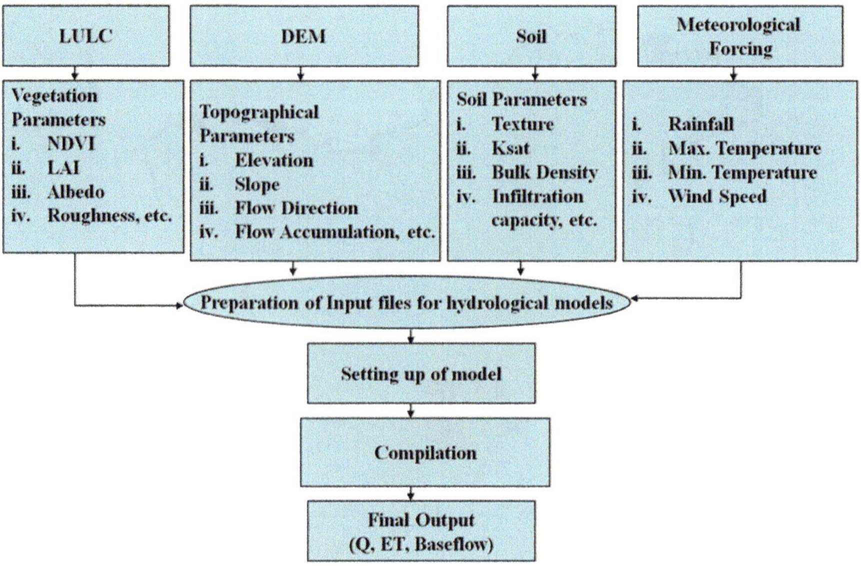

Fig. 6.15 Overall methodology for adapting VIC hydrological modelling

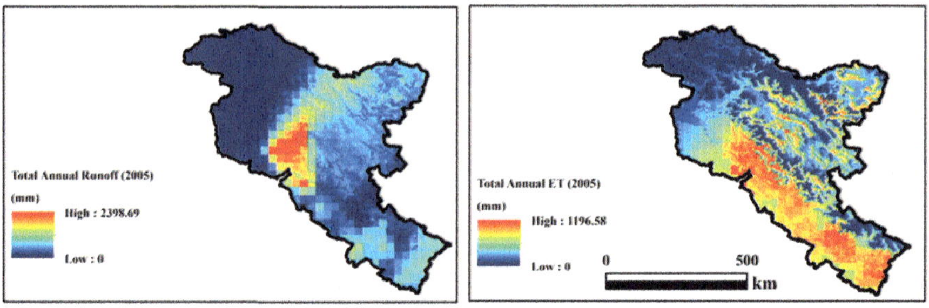

Fig. 6.16 Estimated hydrological components such as annual runoff and ET for NWH region

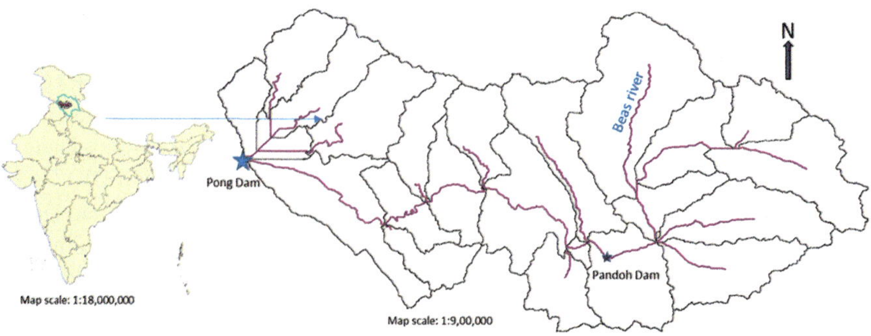

Fig. 6.17 Study area, Beas Basin up to Pong Dam in Himachal Pradesh state of India

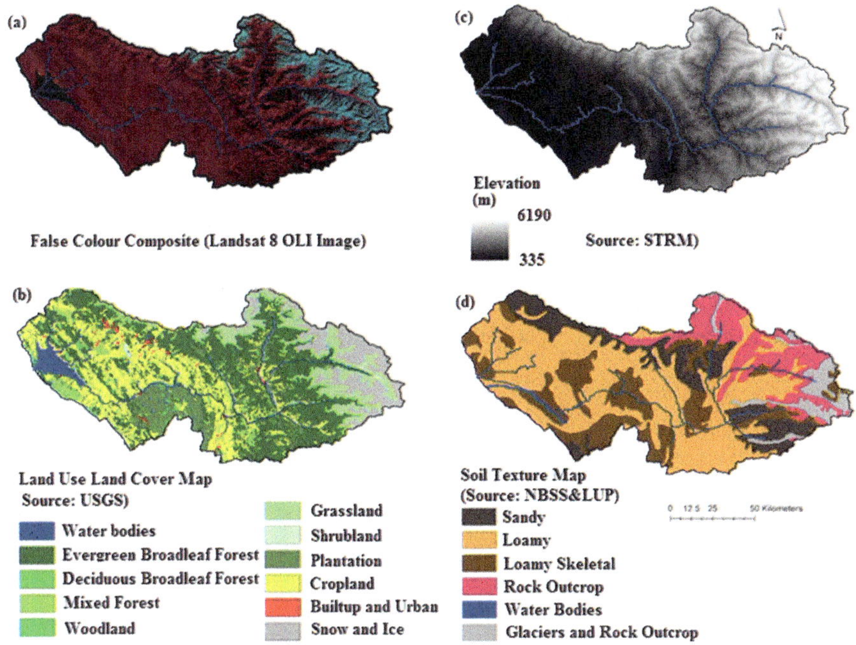

Fig. 6.18 (**a**) False colour composite image of Beas Basin, (**b**) LULC map, (**c**) DEM map and (**d**) soil map of study area

projects/esgf-liu/). The topographic features of the basin were extracted from the 30 m resolution Cartosat-1 DEM available at (http://bhuvan.nrsc.gov.in/data/download/index.php?c=s&s=C1&p=cdv3r1&g=) as shown in Fig. 6.18c. The important soil information has been derived for basin from the NBSS&LUP data as discussed in the previous section (Fig. 6.18d). Similarly, the vegetation parameter and library files were prepared based on ISRO-GBP LULC map of the year 2005 (Fig. 6.18b). In the present analysis, the model was run in energy balance model as basin has large snow cover area.

The climate data of future time period has been analysed, and it was observed that temperature is showing an ever-increasing trend under both RCP 4.5 and RCP 8.5 scenarios. However, the trend of the precipitation is increasing under RCP 8.5, whereas it is decreasing in RCP 4.5 scenario. It has already been studied that slight change in these climatic parameters alters the hydrology of the entire basin. The estimated water balance components (runoff, ET, baseflow) under each scenario are shown in Fig. 6.19. It can be seen that almost all the parameters are showing increasing trend. As precipitation is showing an increasing trend, the estimated runoff is also showing an increasing trend; on the other hand, due to increasing temperature, ET is showing an increasing trend in all the years under consideration.

As major part of the basin is snow fed and has large permanent snow cover area, the trend in model estimated snow cover, snow depth, snowmelt and snow water equivalent (SWE) has also been studied as shown in Figs. 6.20, 6.21, and 6.22. As temperature throughout the period of analysis is showing increasing trend in both the

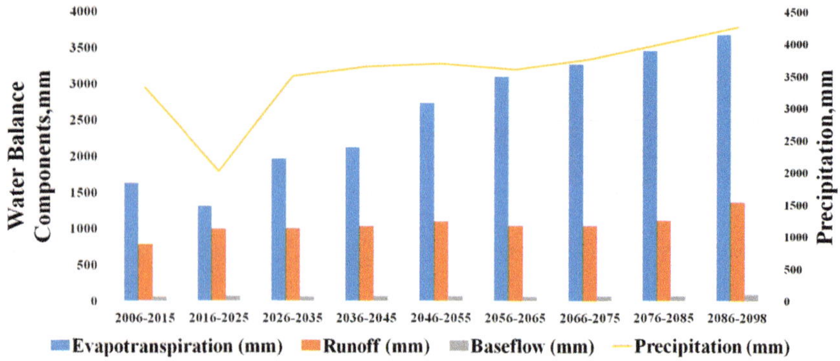

(a) RCP 8.5

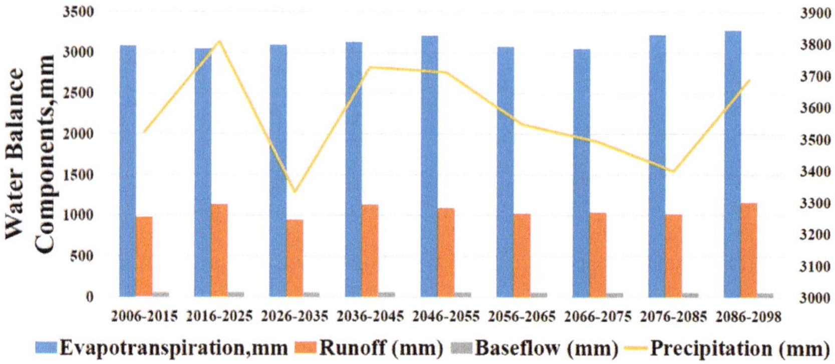

(b) RCP 4.5

Fig. 6.19 Trend of each water balance component under each climatic scenario

scenarios under consideration, it will reduce the amount of snowfall in the region leading to reduction in snow cover. Therefore, in the study, a decreasing trend in the snow parameters was found. It is estimated that around −4.5 m by the end of the century. It is predicted that snowmelt runoff amount may decrease by around 350 mm in the years to come, whereas SWE is showing a decrease of around 280 mm. The hydrological modelling results show that remote sensing data and GIS tools can provide platform to analyse water resources status on any part of the globe even under future climate change scenario.

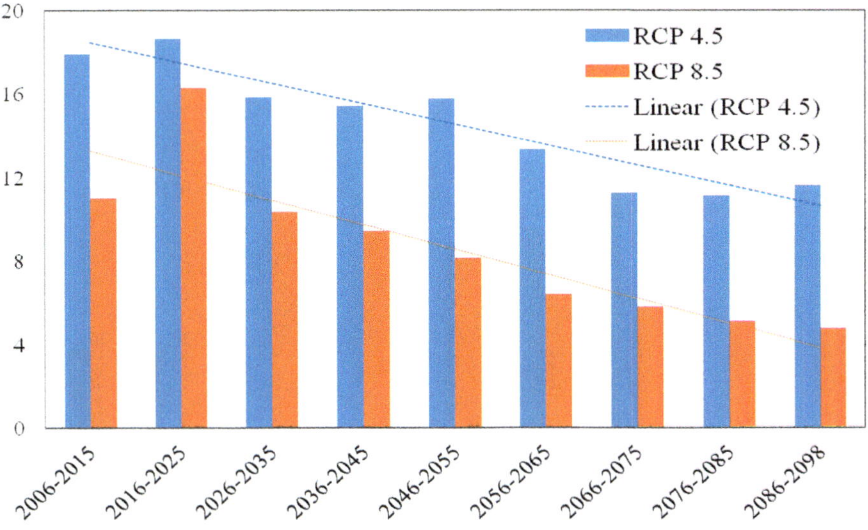

Fig. 6.20 Change in snow depth under each scenario for Beas Basin

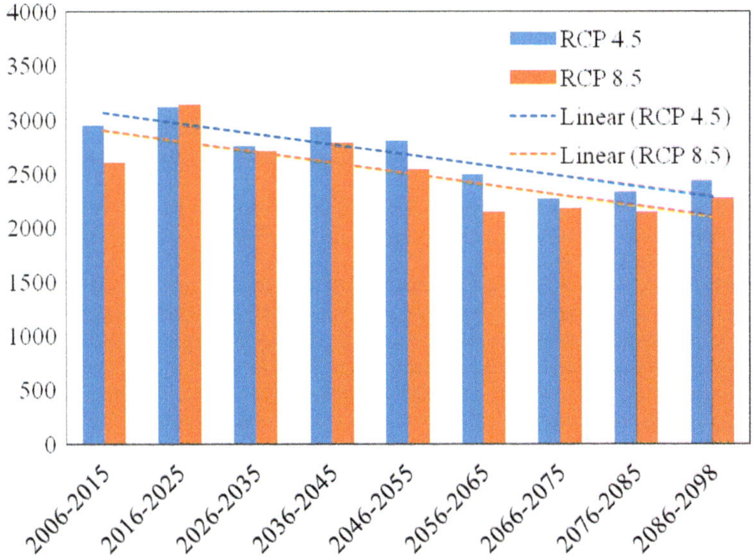

Fig. 6.21 Change in snowmelt under each scenario for the Beas Basin

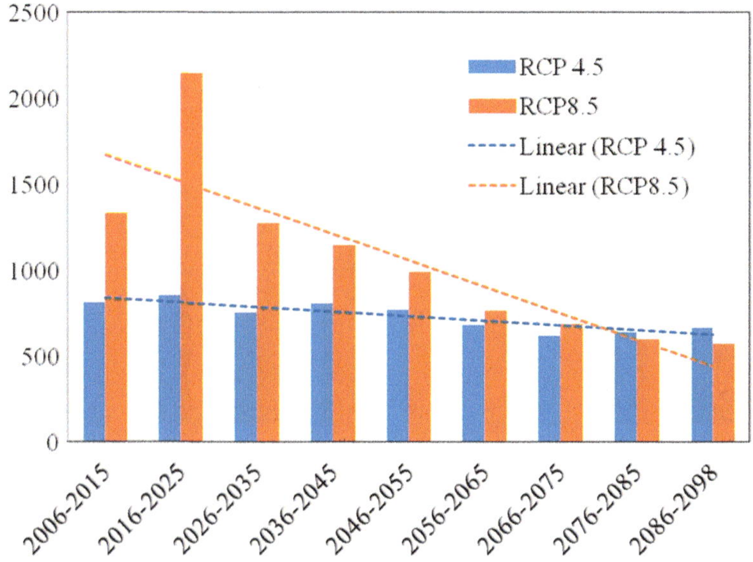

Fig. 6.22 Change in snow water equivalent under each scenario for Beas Basin

References

Abdulla FA, Lettenmaier DP, Wood EF, Smith JA (1996) Application of a macroscale hydrologic model to estimate the water balance of the Arkansas-Red river basin. J Geophys Res 101 (D3):7449–7459

Aggarwal SP, Garg V, Gupta PK, Nikam BR, Thakur PK and Roy PS (2013) Runoff potential assessment over Indian landmass: A macro-scale hydrological modeling approach. Current Science 104(7):950–959

Allen RG, Pereira L, Raes D, Smith M (1998) Crop Evapotranspiration. Food and Agriculture Organization (FAO) Publication 56. Rome, Italy: FAO of the United Nations

Arkin PA (1979) The relationship between fractional coverage of high cloud and rainfall accumulations during GATE over the B-scale array. Mon. Wea. Rev., 106: 1153–1171.

Bastiaanssen WGM, Menenti M, Feddes RA, Holtslag AAM (1998). A remote sensing surface energy balance algorithm for land (SEBAL): 1. Formulation. Journal of Hydrology 212(213): 213–229

Beven K (2001) Rainfall-Runoff modelling. The Primer. John Wiley and Sons, Chichester, UK, pp 1–360.

Bolch T, Kulkarni A, Kääb A, Huggel C, Paul F, Cogley JG, Frey H, Kargel JS, Fujita K, Scheel M, Bajracharya S, Stoffel M (2012) The State and Fate of Himalayan Glaciers. Science 336 (6079):310–314.

Cherkauer KA, Lettenmaier DP (1999) Hydrologic effects of frozen soils in the upper Mississippi River basin. J Geophys Res 104(D16):19599–19610

Cosby BJ, Hornberger GM, Clapp RB et al (1984) A statistical exploration of the relationships of soil moisture characteristics to the physical properties of soils. Water Resour Res. 20 (6):682–690

Crawford NH, Linsley RK (1960) Computation of a synthetic streamflow record on a digital computer. Int. Assoc. Sci. Hydrol. Publication 51:526–538

Dobhal DP, Gergan JT, Thayyen RJ (2004) Recession and morphogeometrical changes of Dokriani glacier (1962–1995), Garhwal Himalaya, India Current Science 86(5):692–696
Dobson MC, Ulaby FT, Hallikainen MT, Elrayes MA (1985) Microwave dielectric behavior of wet soil, Part II: Dielectric mixing models. IEEE Transactions on Geoscience and Remote Sensing GE-23(1):35–46
Dumenil L, Todini E (1992) A rainfall-runoff scheme for use in the Hamburg climate model. In: O'Kane P (ed) Advances in Theoretical Hydrology, A Tribute to James Dooge, , European Geophysical Society Series on Hydrological Sciences 1, pp 129–157
Finn MP, Lewis M, Bosch DD, Giraldo M, Yamamoto K, Sullivan DG, Kincaid R, Luna R, Allam GK, Kvien C, Williams M(2011) Remote Sensing of Soil Moisture Using Airborne Hyperspectral Data. GIScience & Remote Sensing 48(4):522–540. https://doi.org/10.2747/1548-1603.48.4.522
Ferraro RR (1997) Special Sensor Microwave Imager Derived Global Rainfall Estimates For Climatological Applications. J. Geophys. Res., 102:16715–16735.
Ferraro RR, Weng F, Grody NC, Zhao L (2000) Precipitation Characteristics Over Land From the NOAA-15 AMSU Sensor.Geophys. Res. Lett., 27(17):2669–2672.
French AN, Alfieri JG, Kustas WP, Prueger JH, Hipps LE, Chávez JL, Evett SR et al (2012) Estimation of surface energy fluxes using surface renewal and flux variance techniques over an advective irrigated agricultural site. Adv Water Resour 50:91–105
Garg, V., Aggarwal, S.P., Gupta, P.K., Nikam, B.R., Thakur, P.K., Srivastav, S.K. and Senthil Kumar, A. (2017). Assessment of Land Use Land Cover Change Impact on Hydrological Regime of a Basin. Environmental Earth Sciences, 76, DOI: https://doi.org/10.1007/s12665-017-6976-z.
Gruber A (1973) Estimating rainfall in regions of active convection. J. Appl. Meteorol., 12:110–118.
Haubrock S, Chabrillat S, Kuhnert M, Hostert P, Kaufmann H (2008) Surface soil moisture quantification and validation based on hyperspectral data and field measurements. J Appl Remote Sens 2(1):023552-023552-26
Kashyap PS, Panda RK (2001) Evaluation of evapotranspiration estimation methods and development of crop-coefficients for potato crop in a sub-humid region. Agricultural Water Management 50(1):9–25
Kasturirangan K, Navalgund RR and Ajai (2013) Observed changes in the Himalaya-Tibetan glacier. In Working group on Fate of Mountain Glaciers in the Anthropocene. Vatican City. Pontifical Academy of Sciences, Scripta Varia 118: 1–28 www.pas.va/content/dam/accademia/pdf/sv118/sv118-kasturirangan.pdf
Khanbilvardi R, Lakhankar T, Krakauer N, Nazari R, Powell A (2015) "Remote Sensing Data and Information for Hydrological Monitoring and Modeling" in Eslamian, S. (eds), Handbook of Engineering Hydrology, pp. 501–516.
Kulkarni AV, Bahuguna IM, Rathore BP, Singh SK, Randhawa SS, Sood RK, Dhar S (2007) Glacial retreat in Himalaya using Indian remote sensing satellite data. Current Science 92 (1):69–74
Kummerow C, Hong Y, Olson WS, Yang S, Adler RF, McCollum J, Ferraro R, Petty G, Shin D-B, Wilheit TT (2001) The Evolution of the Goddard Profiling Algorithm (GPROF) for Rainfall Estimation from Passive Microwave Sensors. J. Appl. Meteor. 40:1801–1820.
Lettenmaier DP (2001) Present and future of modeling global environmental change: toward integrated modeling. In: Matsuno T, Kida H (ed) Macroscale Hydrology: Challenges and Opportunities, pp 111–136
Liang X (1994) A two-layer variable infiltration capacity land surface representation for general circulation models. Water Resour. Series TR140, Univ. of Washington, Seattle
Liang X, Xie ZH (2003) Important factors in land-atmosphere interactions: surface runoff generations and interactions between surface and groundwater. Global Planet Change 38 (1–2):101–114

Liang X, Lattenmaier DP, Wood EF, Burgess SJ (1994) A simple hydrologically based model of land surface, water, and energy flux for general circulation models. J Geophys Res 99 (D7):14415–14428.

Liang, X., Lettenmaier, D.P., and Wood, E.F. 1996. One-dimensional statistical dynamic representation of subgrid spatial variability of precipitation in the two-layer variable infiltration capacity model. J. Geophys. Res. 101(D16):21403–21422.

Lievens, H., Tomer, S.K., Al Bitar, A., De Lannoy, G.J.M., Drusch, M., Dumedah, G., Hendricks Franssen, H.-J., Kerr, Y.H., Martens, B., Pan, M., Roundy, J.K., Vereecken, H., Walker, J.P., Wood, E.F., Verhoest, N.E.C., Pauwels, V.R.N. 2015. SMOS soil moisture assimilation for improved hydrologic simulation in the Murray Darling Basin, Australia. Remote Sensing of Environment, 168, 146–162, ISSN 0034-4257, https://doi.org/10.1016/j.rse.2015.06.025.

Lohmann, D., et al. (1996), A large scale horizontal routing model to be coupled to land surface parameterization schemes, Tellus (48A), 708–721.

Lohmann, D., Raschke, E., Nijssen, B., and Lettenmaier, D.P. 1998a. Regional scale hydrology: I. Formulation of the VIC-2L model coupled to a routing model. Hydrological Sciences Journal 43(1):131–141.

Lohmann, D., Raschke, E., Nijssen, B., and Lettenmaier, D.P. 1998b. Regional scale hydrology: II. Application of the VIC-2L model to the Weser river, Germany. Hydrological Sciences Journal 43(1):143–158.

McVicar TR, Bierwirth PB (2001) Rapidly assessing the 1997 drought in Papua New Guinea using composite AVHRR imagery. International Journal of Remote Sensing 22:2109–2128.

Mani, A. (1981). The Climate of the Himalaya, in: The Himalaya - Aspects of Change, edited by: Lall, J.S. and Moddie, A.D., Oxford University Press, Delhi, 3–15.

Maurer, E.P., O'Donnell, G.M., Lettenmaier, D.P., and Roads J.O. 2001b. Evaluation of NCEP/NCAR reanalysis water and energy budgets using macroscale hydrologic model simulations. In: (Lakshmi V, Albertson J, Schaake J (eds) Land Surface Hydrology, Meteorology, and Climate: Observations and Modeling. Water Sci. Appl. 3:137–158, AGU, Washington, D. C., doi:https://doi.org/10.1029/WS003p0137.

Maurer, E.P., O'Donnell, G.M., Lettenmaier, D.P., and Roads, J.O. 2001a. Evaluation of the Land Surface Water Budget in NCEP/NCAR and NCEP/DOE Reanalyses using an Off-line Hydrologic Model. J Geophys Res. 106(D16):17,841-17,862.

Mu Q, Zhao M, Running SW (2011) Improvements to a MODIS global terrestrial evapotranspiration algorithm. Remote Sens Environ 115:1781–1800.

Nagarajan R (2009) Drought Assessment. Copublished by Springer with Capital Publishing Company, New Delhi, India.

Nijssen, B.N., Lettenmaier, D.P., Liang, X., Wetzel, S.W., and Wood, E.F. 1997. Streamflow simulation for continental-scale river basins. Water Resour Res. 33(4):711–724.

Nikam, V., Nikam, B.R., Garg, V., Thakur, P.K. and Aggarwal, S.P. 2015. Assimilation of Remote Sensing Derived Soil Moisture in Macroscale Hydrological Model. In proceedings of HYDRO-2015, IIT Roorkee, Uttarakhand, India during December 2015.

Orellana B., Pechlivanidis I.G., McIntyre N., Wheater H.S. and Wagener T., (2008), A toolbox for the identification of parsimonious semi-distributed rainfall-runoff models: Application to the Upper Lee catchment, in iEMSs 2008: International Congress on Environmental Modelling and Software, 1, 670–677, 7–10 July, Barcelona, Spain.

Pai D.S., Latha Sridhar, Rajeevan M., Sreejith O.P., Satbhai N.S. and Mukhopadhyay B., 2014: Development of a new high spatial resolution (0.25° X 0.25°)Long period (1901–2010) daily gridded rainfall data set over India and its comparison with existing data sets over the region; MAUSAM, 65(1), 1–18.

Pechlivanidis I.G., Jackson, B.M., Mcintyre, N.R. and Wheater, H.S. (2011). Catchment scale hydrological modeling a review of model types, calibration approaches and uncertainty analysis methods in the context of recent developments in the technology and applications. Global NEST Journal, 13(3), 193–214.

Petropoulos G, Carlson TN, Wooster MJ and Islam S (2009) A review of Ts/VI remote sensing based methods for the retrieval of land surface energy fluxes and soil surface moisture. Progress in Physical Geography 33(2): 224–250.

Rajeevan, M., Bhate, J., Kale, J.D. and Lal, B. (2005). Development of a high resolution daily gridded rainfall data for the Indian region. National Climate Centre, India Meteorological Department, Government of India Report, Met. Monograph Climatology No. 22/2005, Pune.

Rawls WJ, Gimenez D, Grossman R (1998) Use of soil texture, bulk density, and slope of the water retention curve to predict saturated hydraulic conductivity. Transactions of the ASABE 41 (4):983–988.

Reynolds CA, Jackson TJ, Rawls WJ (2000) Estimating soil water-holding capacities by linking the Food and Agriculture Organization soil map of the world with global pedon databases and continuous pedotransfer functions. Water Resour Res. 36(12):3653–3662.

Senay, G.B., Bohms, S. and Verdin, J.P. 2012. Remote sensing of evapotranspiration for operational drought monitoring using principles of water and energy balance. In Remote Sensing of Drought: Innovative Monitoring Approaches, Eds. Wardlow, B.D., Anderson, M.C. and Verdin, J.P., pp 123–144, CRC Press, Taylor & Francis Group, Boca Raton.

Senay, G.B., Velpuri, N.M., Bohms, S., Budde, M., Young, C., Rowland, J. and Verdin, J.P. 2015. Drought monitoring and assessment: Remote sensing and modeling approaches for the famine early warning systems network. In Hydro-Meteorological Hazards, Risks, and Disasters, Eds. Paron, P and Baldassarre, G.D., pp. 233–262, Elsevier.

Sherman LK (1932) Streamflow from rainfall by the unit hydrograph method. Eng. News-Record, 108:501–505.

Singh V.P., (1995), Computer models of watershed hydrology, Water Resources Publications, LLC, USA.

Singh V.P. and Frevert D., (2006), Watershed models. Boca Raton, Taylor & Francis.

Singh V.P. and Woolhiser D.A., (2002), Mathematical modeling of watershed hydrology, Journal of Hydrologic Engineering, 7(4), 270–292.

Space Application Centre (SAC) (2011). *Snow and Glaciers of the Himalayas*. Space Applications Centre (ISRO), Ahmedabad, India.

Subramanya, K. (2008), Engineering Hydrology 3rd Edition, Tata McGraw-Hill Publishing Company Ltd., New Delhi.

Thakur, P.K., Nikam, B.R., Garg, V., Aggarwal, S.P., Chouksey, A., Dhote, P.R. and Ghosh, S. (2017). Hydrological Parameters Estimation using Remote Sensing and GIS for Indian Region - A Review. Proceedings of the National Academy of Sciences, India Section A: Physical Sciences, 87(4), 641–659.

Thayyen, R.J. and Gergan, J.T. (2010). Role of glaciers in watershed hydrology: a preliminary study of a “Himalayan catchment”. The Cryosphere, 4, 115–128.

Wagener T., Wheater H.S. and Gupta H.V., (2004), Rainfall-Runoff Modelling in Gauged and Ungauged Catchments. Imperial College Press, London, UK, 1–306 pp.

Wagner, W., Bloschl, G., Pampaloni, P., Calvet, J.C., Bizzarri, B., Wigneron, J.P. and Kerr, Y. 2007. Operational readiness of microwave remote sensing of soil moisture for hydrologic applications. Nordic Hydrology, 38(1), 1–20.

Wardlow, B.D., Anderson, M.C., Sheffield, J., Doorn, B.D., Verdin, J.P., Zhan, X. and Rodell, M. 2012. Future opportunities and challenges in remote sensing of drought. In Remote Sensing of Drought: Innovative Monitoring Approaches, Eds. Wardlow, B.D., Anderson, M.C. and Verdin, J.P., pp 123–144, CRC Press, Taylor & Francis Group, Boca Raton.

Wheater H.S., Jakeman A.J., Beven K.J., Beck M.B. and McAleer M.J., (1993), Progress and directions in rainfall-runoff modelling, Modelling change in environmental systems, New York, pp. 101–132.

Wheater HS (2002) Progress in and prospects for fluvial flood modelling. Phil. Trans. R. Soc. Lond. A 360(1796):1409–1431.

WMO (1974) Guide to hydrological practices. WMO, Geneva 2, Switzerland.

Wood, E. F.., Global scale hydrology: advances in land surface modeling. Reviews of Geophysics Supplement, 1991, 29, 193–201.

Yanmin, Y., Na, W., Youqi, C., Yingbin, H. and Pengqin, T. 2010. Soil moisture monitoring using hyper-spectral remote sensing technology. In Goscience and Remote Sensing (IITA-GRS), 2010 Second IITA International Conference on , vol.2, no., pp.373–376, 28–31 Aug. 2010. doi: https://doi.org/10.1109/IITA-GRS.2010.5604219.

Zhao, R.J., Zhang, Y.L., Fang, L.R., LIu, X.R. and Zhang, Q.S., The Xinanjiang model. In Hydrological Forecasting Proceedings Oxford Symposium, IASH, 129, 1980, pp. 351–356.

Chapter 7
Hydrometeorological Hazards Mapping, Monitoring and Modelling

Praveen K. Thakur, S. P. Aggarwal, Pankaj Dhote, Bhaskar R. Nikam, Vaibhav Garg, C. M. Bhatt, Arpit Chouksey, and Ashutosh Jha

7.1 Introduction

Northwest Himalaya (NWH) has unique topographical and climate settings which makes this area prone to various types of hydrometeorological hazards such as flash floods, hail storms, glacier lake outburst floods, avalanches and mudflows. These hazards have high probability of turning into natural disasters if proper planning of natural resources, infrastructure and man-made structures is not done. Floods of June 2013 in Uttarakhand (Dobhal et al. 2013; Thakur et al. 2014) and 2014 floods of Srinagar (Bhatt et al. 2016) are prime examples of such hazards turning into the major disasters. Northwest Himalayan states in the last few years have experienced large number of hydrometeorological disasters such as high-intensity precipitation, cloud burst and subsequent flash flooding in downstream areas, snow avalanches, glacier lake outburst floods (GLOF), hail storms, drought and rainfall-induced mudflows (Kumar et al. 2015; Gupta et al. 2013; Kumar et al. 2012; Rana et al. 2012). This chapter gives an overview of various hydrometeorological hazards which are reoccurring in NWH and provides insights in few such hazards by providing some actual case studies related to such hazards.

P. K. Thakur (✉) · S. P. Aggarwal · P. Dhote · B. R. Nikam · V. Garg · A. Chouksey
Water Resources Department, Indian Institute of Remote Sensing (IIRS), Indian Space Research Organisation (ISRO), Department of Space, Government of India, Dehradun, India
e-mail: praveen@iirs.gov.in

C. M. Bhatt
Centre for Space Science and Technology Education in Asia and the Pacific (CSSTEAP), Dehradun, India

A. Jha
Geoinformatics Department, Indian Institute of Remote Sensing (IIRS), Indian Space Research Organisation (ISRO), Department of Space, Government of India, Dehradun, India

R. R. Navalgund et al. (eds.), *Remote Sensing of Northwest Himalayan Ecosystems*,
https://doi.org/10.1007/978-981-13-2128-3_7

7.2 Snow Avalanches

"Avalanche, landslide, slumps, mudflows, rock falls – these are some of the terms which are used to described movement of snow, soil and rocks under the influence of gravity" (Bromhead 1986). In case of avalanche, it is a special case of movement of snow under the influence of gravity. *"Avalanche" (taken from the French) has a broader meaning, standing for "a sudden and rapid movement of ice, snow, earth or rock down a slope"* (Ghinoi 2003). Snow avalanche can occur due to the failure of a snowpack and presence of critical snow depth and topographical slope in a given area (Malik 2009). Snow avalanche can start with the failure of snowpack at critical slopes and failure occurring due to a cohesionless or cohesive material (Mears 1992). Two components of gravity force acting on block (parallel and perpendicular to slope), physical properties of block, angle of slope and surface properties determine how shelf will move before block slides (McClung and Schaerer 2006). If force parallel to slope exceed, friction created by perpendicular force leads to failure of blocks.

Snowpack goes through three types of deformation, i.e. tension, compression and shear. Tension comprises of two deformation components, i.e. creep and slide. Movement of some snow layers downslope is called as creep, while movement of whole snowpack along downslope is called as glide (Fig. 7.1). Compression acts in a direction perpendicular to snow surface resulting in densification. Shear deformation produces when snow grains move over each other. All these deformations induce shear stress which is nonuniform over the snowpack (McClung and Schaerer 2006; Logen 2005).

Avalanche area can be demarked using avalanche paths. Fixed locality along which avalanche starts, moves and ends is called as avalanche paths. Based on slope angle, avalanche paths can be divided into three main zones (Malik 2009).

Starting zone is located at a place where unstable snow fails and proceeds to move further called as zone of origin or formation zone. Upper limit of avalanche origin is defined by fracture line (catchment area) of slab avalanche and the point at which snow starts to loose (Fig. 7.2). Lower limit of origin avalanche is nor well defined. Different factors affecting formation zones are:

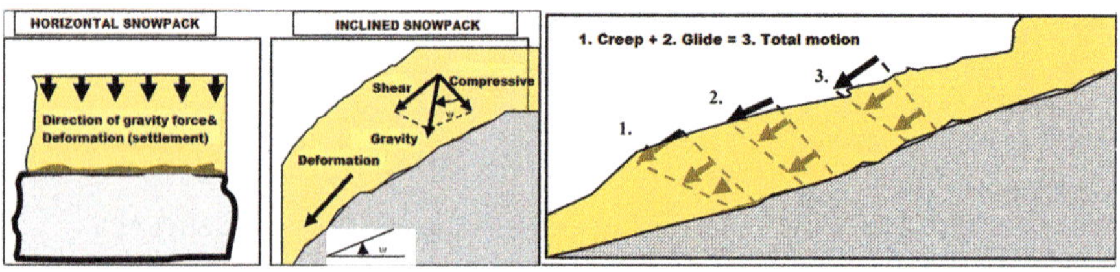

Fig. 7.1 Deformation components of the snowpack. (McClung and Schaerer 2006)

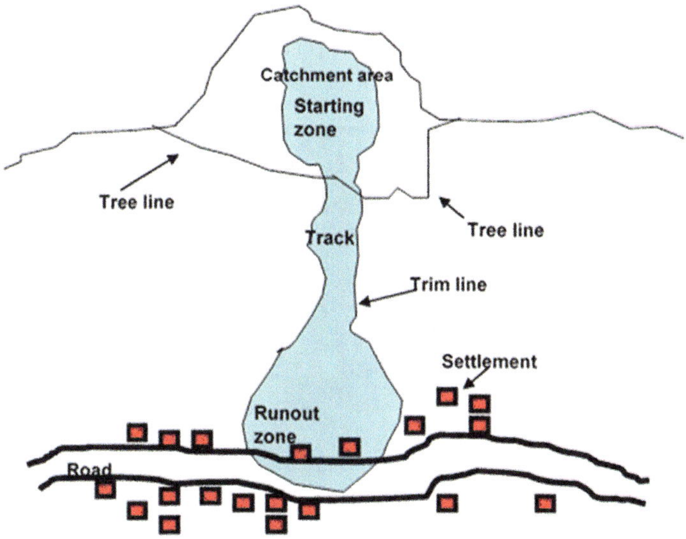

Fig. 7.2 A generalized avalanche path. (Malik 2009)

- Ground/forest cover: The type of forest (open/closed) will affect the wind, precipitation, air temperature and radiation. The effect of forest cover is well reported in Frey and Salm (1990).
- Slope dimensions and inclination (McClung and Schaerer 2006).
- Altitude.
- Crown and flank locations.

Transition zone between starting zone and runout zone where debris comes to rest is called as zone of transition. It is also called as track. There are two types of tracks according to the International Commission on Snow and Ice, i.e.:

Open Slope – consist of no lateral boundaries
Channel – includes gullies and other depressions

When the static friction angle and slope angle become equal, debris comes to rest, the zone called as runout zone. Wide bench, valley bottom and a talus slope are some of the characteristics of runout zone. Correlation is being made between avalanche magnitude and runout distance with underconsideration of water content of snow as an important parameter (Weir 2002). Terrain configuration, roughness coefficient of avalanche path, mechanical properties of snow flow and mass of snow are important determinants of extreme avalanche event runout distances (Mears 1992).

Fig. 7.3 Avalanche-prone areas from Solang-Dhundi road in Manali watershed of Beas River. (Malik 2009)

7.2.1 Mapping of Avalanche-Prone Areas

In one of the studies done at IIRS Dehradun, major avalanche-prone areas were mapped using geospatial tools for part of Manali sub-watershed from Solang to Dhundi (Fig. 7.3). In this work total of ten numbers of avalanche-prone site (Table 7.1) were demarcated, and complete geodatabase was created (Malik 2009).

As it is seen in Fig. 7.3 and Table 7.1, two avalanche sites with name MSP 3 and MSP 7 have the largest formation area. The MSP 7 site has also been selected for construction of snow gallery, designed by SASE, Chandigarh, in which snow avalanche passes over the snow gallery, vehicle traffic through the gallery and stream flows below the gallery. In terms of steepest slope and more frequent avalanche, MSP 10 has more frequent avalanches as compared to other sites (Malik 2009). In addition, there have been reported studies by SASE where

Table 7.1 Avalanche sites with name prefix as Manali South Portal (MSP) of new Rohtang Tunnel

Site name	Area (km^2)
MSP10	0.207
MSP9	0.105
MSP8	0.157
MSP7	3.381
MSP6	0.229
MSP5	0.256
MSP4	0.349
MSP3	**2.916**
MSP2	0.174

Complied by Malik (2009)

geospatial tools were used to integrate various RS-GIS databases such as DEM, land use/land cover (LULC) and published avalanche-prone sites to make final avalanche hazard map for Gangotri Glacier using analytical hierarchy decision rules (Snehmani et al. 2013).

7.2.2 Avalanche Modelling

The studied avalanche tracks are located on the proposed all-weather Manali-Leh National Highway (NH-21) between Solang and Dhundi (see Sect. 7.2.1). The study area lies between 32° 13′ 06″ to 32° 24′ 00″ N latitude and 77° 01′ 35″ and 77° 17′ 00″ E longitudes. All these avalanche tracks have southeast-facing slope. These tracks have been given name "Manali South Portal (MSP) 2 to Manali South Portal (MSP) 10". MSP is the name given to these avalanche sites because these sites are laid on the south side of the proposed 9 Km tunnel from Dhundi to Koksar. Among all the nine avalanche sites, MSP 7 was investigated in depth, as the site is involved in frequent avalanche activities (Malik 2009). Another reason for study is that MSP 7 (19 Km from Manali Town on 32° 20′ N latitude and 77° 08′ 11″ E longitude) has the largest area among all other sites situated on the corridor (Malik 2009). It consists of two main gullies (avalanche tracks) left and right side gullies, which are further divided into three sub-gullies as shown in Fig. 7.4.

The annual precipitation (1990–2005) and mean annual temperature (1990–2005) both are measured by the Snow Avalanche Study Establishment (SASE) at Bahang (2192 m.s.l.), Solang (2480 m.s.l.) and Dhundi (3050 m.s.l.). The climate of the study area is of lower Himalayan zone, which is typically a climate of subtropical zone or a zone classified as the zone of warm temperature, high precipitation and short winter period of 4 months, i.e. December to March. It has been observed by SASE that avalanche activities in this area are due to the unique topographical and climatic settings of this area, which causes heavy snowfall in the critical slopes of

Fig. 7.4 Overview of MSP-7 avalanche site in Solang-Dhundi highway. (Malik 2009)

this corridor, which results in the formation and release of avalanches. The snowpack at the formation zone can develop weak layers due to melt-refreeze cycles, and some weak layers can also form near the ground, which can give way or break at a critical value of overburden pressure. Literature has shown that if snow depth ranges from 150 to 200 cm, the snowpack at formation zones can fail at critical slopes. In general, the avalanche warning bulletin is issued if snow depth of more than 100 cm occurs in a snow storm for the given region (Malik 2009). It has been reported that there are around 13 avalanche-prone sites along the Dhundi-Solang transportation corridor. The terrain along the corridor is generally rocky, rugged and scree field. The corridor generally has 100–150 cm of standing snow during the season. It is also reported that the snow at formation zone is around 30% more than the observed at Dhundi Observatory, which is at an elevation of 3050 m.s.l. The avalanche activity usually starts when the standing snow crosses 250 cm mark at the observatory. It is considered as 350 cm standing snow as critical. It is also observed that the temperature remains in the range of 15 to −15 °C. It is noticed that in this region the

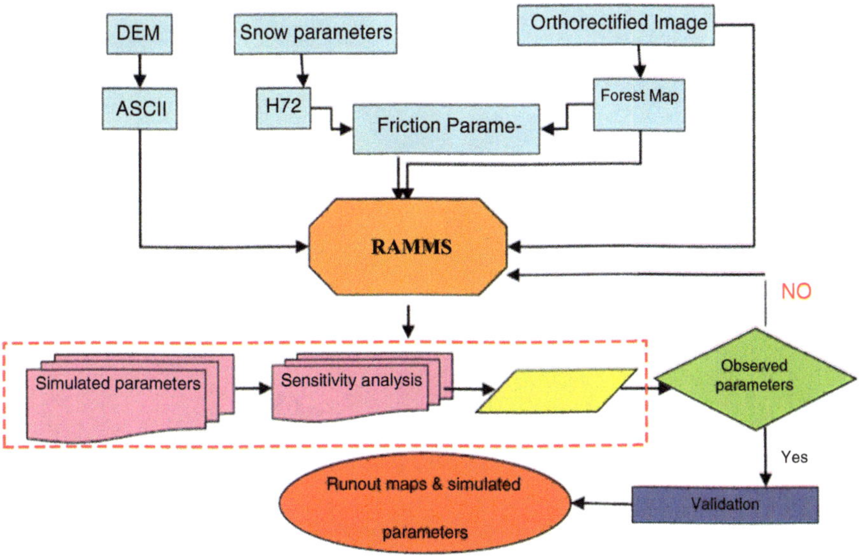

Fig. 7.5 Methodology adopted for the simulation of snow avalanches. (Malik 2009)

overburden and precipitation intensity are the major causes of triggering avalanches. In this region generally, the total snowfall varies from 12 to 18 m.

During the field investigation, it was found that study area between Solang and Dhundi is densely forested and different types of trees and vegetation are present at different elevation zones. The forest line on the mountains in the study area was found between 3000 and 3500 m (Malik 2009). This study uses a simple methodology as shown in the Fig. 7.5.

The main input parameters required to run the RAMMS model are generated using high-resolution remote sensing and elevation data, and in addition sensitivity analysis is done for the case of elevation and friction parameters. Most sensitive parameters, such as friction and critical snow depth, were calibrated with the help of available ground information (Malik 2009), as the entire process of calibration relies on good quality of historical and field information, and this was collected from SASE for the present study. The runout modelling for the rest of snow avalanche sites was then calculated with RAMMS by using the calibrated parameters (Malik 2009). The final calibrated input parameters such as Coulomb and turbulent friction, curvature and velocity results for MSP7 are shown in Fig. 7.6.

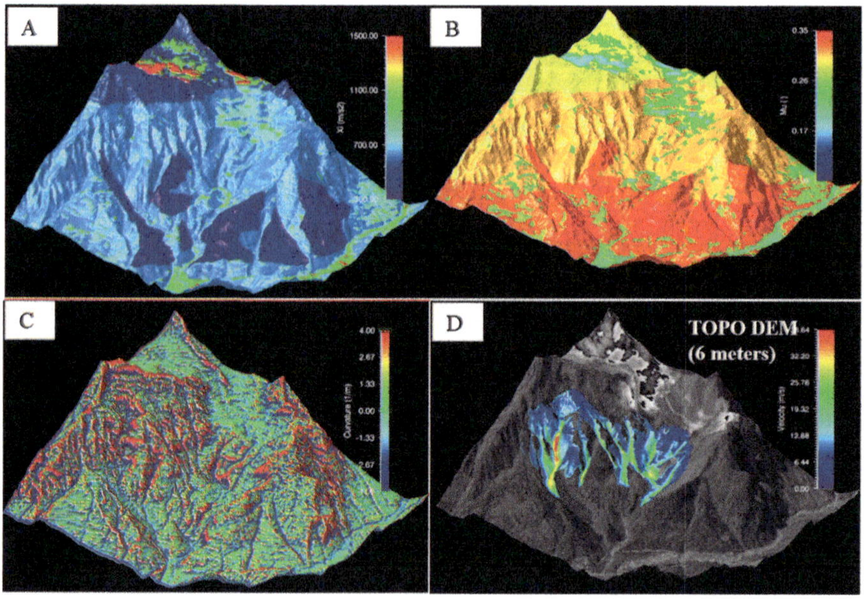

Fig. 7.6 (**a**) Turbulent friction in m/s^2 (xi); (**b**) Coulomb friction coefficient (Mu); (**c**) Topo 6 m DEM-based curvature; and (**d**) snow avalanche results (initial velocity) for MSP7 with 300-year return period. (Malik 2009)

7.3 Flood

Floods in NWH are mainly caused by heavy rainfall or cloud burst and glacier lake outburst or landslide blocked dammed water and subsequent dam break floods. As it is evident from Table 7.2 and Fig. 7.7, NWH is one of the most hazardous areas in terms of its susceptibility for avalanches, cloud burst-related flash floods, floods due to extreme rainfall event during monsoon, landslide or glacier moraine-dammed lake outburst floods, including some spells of drought and hail storms. In India, Indian Meteorological Department (IMD) is the nodal agency for monitoring of weather parameters and prediction of weather, monsoon and weather warnings.

7.3.1 Flood Inundation Mapping

Flood inundation in hilly areas is mainly restricted to nearby channels, and its water spread is constrained by steep topography on both sides of river. Only in few cases where floodplain lies in low areas there is possibility of deriving the flood inundation from remote sensing data. Jhelum River in Jammu and Kashmir State of India is one such river basin where we get frequent floods and water generally exceeds its banks and inundates the surrounding areas as has been reported in floods of September 2014, March 2015 and April 2017 (Bhatt et al. 2016). Such areas can be easily

Table 7.2 List of some of the major flood events, cloud burst and landside/GLOF-related events in NWH

S. No.	Event type	Date/duration	Location
1	Monsoon-related extreme rainfall	September 1952	Various parts of HP
2	January 19, 1975, earthquake-induced land-slide-dammed lake burst	March 1975	Spiti valley of Satluj, HP
3	Cloud burst-induced flash flood	September 29, 1988	Soldang, Satluj Basin, HP
4	Cloud burst-induced flash flood	August 11, 1997	Chirgaon, Rohru tehsil, H.P
5	Cloud burst-induced flash flood	July 22, 2001	Parts of Kullu dist., HP
6	Monsoon-related extreme rainfall	July 31–August 2,1991	Soldang, Satluj Basin, HP
7	Monsoon-related extreme rainfall	September 4–5, 1995	Parts of Kullu valley, HP
8	Monsoon-related extreme rainfall	July 31–August 1, 2000	Satluj Valley, HP
9	Cloud burst-induced flash flood	July 23, 2001	Sainj Valley, Kullu, HP
10	Monsoon-related extreme rainfall	July 17–19, 2001	Mandi District, HP
11	Monsoon-related extreme rainfall	July 29–30, 2001	Chhota Bhangal and Baijnath, Kangra District, HP
12	Monsoon-related extreme rainfall	August 9–10, 2001	Moral-Danda peak, Shimla district, HP
13	Monsoon-related extreme rainfall	August 21–22, 2001	Ani subdivision, Kullu, HP
14	Cloud burst-induced flash flood	July 16, 2003	Kullu district, HP
15	Cloud burst-induced flash flood	August 7, 2003	Kullu district, HP
16	Cloud burst-induced flash flood	August 15, 2007	Ghanvi, Shimla, HP
17	Cloud burst-induced flash flood	August 7, 2009	Dharampur, Mandi, HP
18	Cloud burst-induced flash flood	September 12, 2010	Kharahal valley, HP
19	Monsoon-related extreme rainfall	July 19–20, 2011	Manali, HP
20	Monsoon-related extreme rainfall	August 15, 2014	Hamirpur, Mandi Dist., HP
21	Cloud burst-induced flash flood	2002	Ketgaon, UK
22	Cloud burst-induced flash flood	2004	Ranikhet, UK

(continued)

Table 7.2 (continued)

S. No.	Event type	Date/duration	Location
23	Monsoon-related extreme rainfall	July 5–6, 2004	Chamoli, UK
24	Cloud burst-induced flash flood	2007, July 01, 2016	Bastari, Pithoragarh, UK
25	Monsoon-related extreme rainfall	August 6–7, 2009	Pithoragarh district, UK
26	Cloud burst-induced flash flood	August 18, 2010	Kapkot, Bageshwar, UK
27	Cloud burst-induced flash flood	July 21, 2010	Almora district, UK
28	Monsoon-related extreme rainfall	September 14–15, 2010	Almora district, UK
29	Monsoon-related extreme rainfall	August 2–8, 2012	Bhatwari, Uttarkashi, UK
30	Cloud burst-induced flash flood	September 13, 2012	Ukhimath, Rudraprayag, UK
31	Pre-monsoonal extreme rainfall + Kedarnath GLOF + cloud burst	June 15–17, 2013	Kedarnath, entire UK,
32	Cloud burst-induced flash flood	August 1, 2013	Kapkot, Bageshwar, UK
33	Monsoon-related extreme rainfall	August 07–08, 2015	Dharampur, Mandi, HP
34	Cloud burst-induced flash flood	July 1, 2016	Singhali, Pithoragarh, UK
35	Cloud burst-induced flash flood	May 25, 2017	Bijrani, Almora, UK
36	Monsoon-related extreme rainfall	June 23–24, 2005	Leh Nalla (Ganglass), J&K
37	Monsoon-related extreme rainfall	July 31–August 1, 2006	Leh Nalla (Ganglass), J&K
38	Cloud burst-induced flash flood	August 5–6, 2010	Leh cloudburst, J&K
39	Cloud burst-induced flash flood	June 08, 2011	Baggar, Doda, J&K
40	Late monsoon-related heavy rainfall	02–26 September, 2014	Jhelum and Chenab rivers, Srinagar, J&K and neighbouring downstream areas of Pakistan

Source: (IMD 2013; Thayyen et al. 2013; Kumar et al. 2015; media reports)

mapped using medium- to high-resolution synthetic aperture radar (SAR) data, as in SAR images water causes specular reflection, giving less backscatter and dark appearance in SAR.

Such study is shown in Fig. 7.8 where flood inundation map was created using SAR data during 2014 excess late monsoon rains in Jhelum River (Bhatt et al. 2016). The state of Jammu and Kashmir in North India experienced one of the worst floods

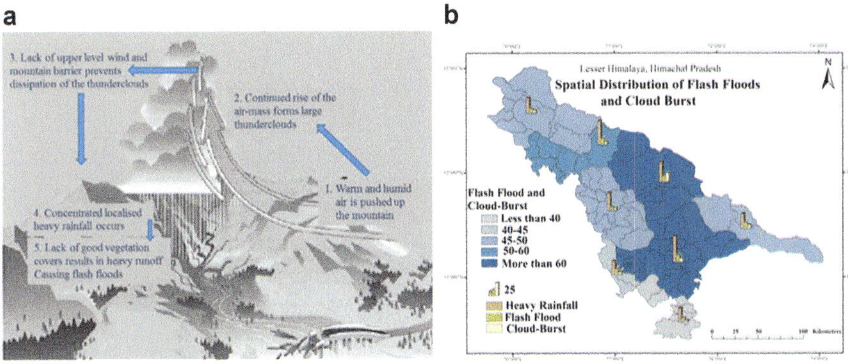

Fig. 7.7 (**a**) Schematic of cloud burst or extreme-intensity rainfall events. (**b**) Spatial distribution of flash flood and cloud burst. (Sources: Indian Meteorological Department (IMD) Centre, Shimla)

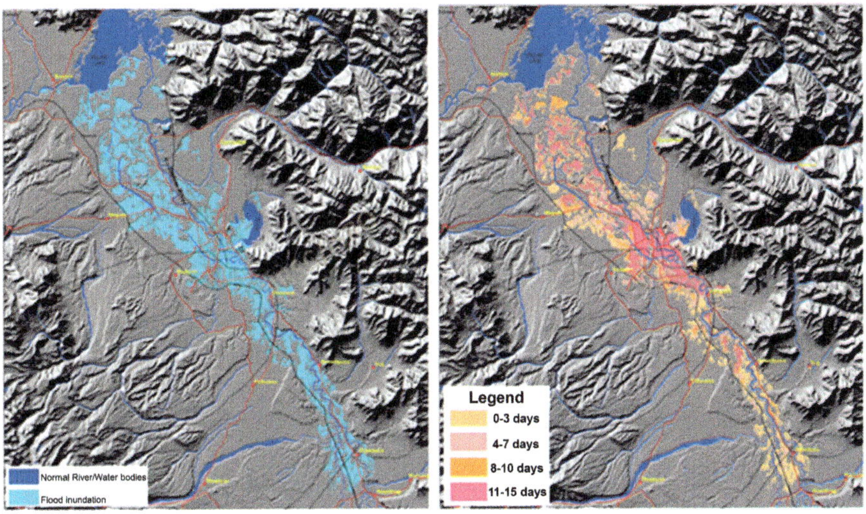

Fig. 7.8 Cumulative flood inundation spatial extent (cyan colour) and flood persistence observed in Kashmir Valley between September 8 and 23, 2014. (Bhatt et al. 2016)

in the past 60 years, during the first week of September 2014. The incessant heavy rainfall in the valley during 3rd–6th of September 2014 has led to the unprecedented flood situation. This river can also get flooded due to heavy snowfall or rainfall in early spring due to sudden melting and rain on snow-induced melting from winter snowpack, as was reported in March 2015 and April 2017 floods.

Flood mapping and monitoring of the floods within the valley was done using synthetic aperture radar (SAR) satellite images operating in C-band (5.35 GHz) acquired from Indian Remote Sensing (IRS) RISAT-1 and Canadian Radarsat-2 satellites operating in C-band (Fig. 7.9). The study shows that all the districts lying within the valley, i.e. Anantnag, Bandipora, Baramulla, Budgam, Ganderbal, Kulgam, Pulwama and Srinagar, are vulnerable to flooding (Tables 7.3 and 7.4).

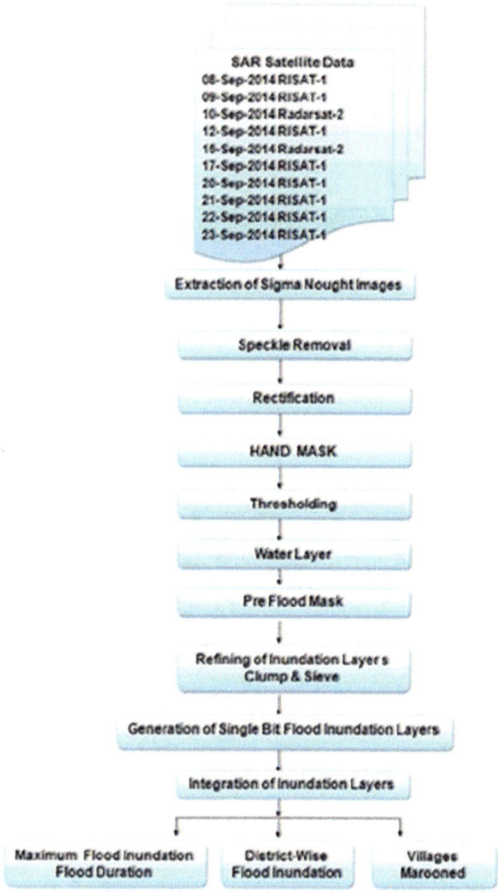

Fig. 7.9 Overall flowchart used for flood mapping, monitoring and damage assessment. (Bhatt et al. 2016)

Table 7.3 District-wise inundated area

S.No	District	Area (km^2)
1	Bandipora	148
2	Pulwama	102
3	Srinagar	100
4	Baramulla	89
5	Budgam	54
6	Anantnag	43
7	Kulgam	15
8	Ganderbal	6
	Total	557

The flat topography of the valley does not allow quick draining of floodwaters. The floodwaters persistence is observed to be greater in the central and northern part of the valley, especially in Srinagar area which was inundated for more than 2 weeks (Bhatt et al. 2016).

Table 7.4 LULC classes under flood inundation

S.No	Land use/land cover	Area (km^2)
1	Agriculture	444
2	Horticulture	20
3	Built-up	67
4	Forests	3
5	Wastelands	21
6	Other	2
	Total	557

7.3.2 *Flood-Prone Area Mapping*

Flood-prone area mapping in NWH remains one of the highest priorities of various water agencies such as the Central Water Commission (CWC) and state disaster management agencies. Flood-prone areas can be assessed using four methods, (a) ground-based flood reports, (b) long-term time series of flood inundation maps derived from remote sensing, (c) flood inundation simulation using 1-D/2-D hydrodynamic (HD) models of historical flood and various flood return periods (Horritt and Bates 2002) and (d) flood plain and upland delineation by use of topographical data. Original assessment of flood-prone area by Rashtriya Barh Ayog (RBA) – in 1980 – gave figure of 40 million hectares, mha, which has been revised to 49.815 mha as per the database maintained by CWC based on the flood damage data reported by states for the period from 1953 to 2010. Remote sensing-based flood inundation maps and DEM-based flood-prone areas also play an important role in delineation of flood-prone areas (Rao et al. 1998). National Remote Sensing Centre (NRSC) is the nodal agency in India for operational flood inundation mapping using remote sensing. NRSC has released flood hazard or flood-prone area maps for some states like Assam, Bihar and Orissa utilizing time series of flood inundation maps from 1995 to 2015 (NRSC 2017). 1-D HD modelling for flood inundation mapping and its comparison with RS-based flood maps has been used in earlier studies (Thakur and Sumangala 2006; Hasan 2012; Dasgupta 2015). Topographical approach-based flood-prone area assessment has gained popularity in recent times as it provides an alternate approach for such studies by utilizing freely available DEM data, is less time-consuming, can work in hydrological data-scarce situations and can be applied to any geographical areas. In literature, three topographical index-based methods have been reported to delineate floodplains and uplands. DEM derived secondary topographical attributes such as Topographic Wetness Index (TWI), Height Above Nearest Drainage (HAND) and Topographic Position Index (TPI) (Beven and Kirkby 1979; Manfreda et al. 2014).

These methods have various advantages and disadvantages, and out of these three methods, HAND methods have given better results for some of flood-prone basin of NHW, and initial results of HAND-based flood-prone areas for part of Beas and Ganga river basin are discussed here. Floodplain in literature corresponds to area under flood with a particular return period. Return period takes account of variations in flood volume and flood stage. Discharge and stage in a stream share a very complex relationship with major influences from local topography. Efforts have been made to simplify the complexity of flood stage with the help of topographic data available in the form of DEM. In HAND method, for each subbasin elevation value of corresponding outlet cell (drainage cell) is subtracted from elevation value of each cell of subbasin above the outlet and thus generates raster containing height above nearest drainage values for all pixels. HAND value remains lower near valleys and goes on increasing towards ridgeline (Nobre et al. 2011). Final flood-prone areas are estimated by giving a suitable threshold for HAND value. Figures 7.10 show results of this method for Ganga basin up to Haridwar and Beas basin up to Thalot, with areas near Rishikesh, Uttarkashi, Kullu and Bhuntar highlighted separately.

7.3.3 Flood Flow and Inundated Area Modelling

7.3.3.1 Integrated Hydrologic/Hydraulic Flood Modelling in Upper Beas River

The Landsat ETM+ data was used for preparing LULC map, and this LULC map was used along with soil map (taken from the National Bureau of Soil Survey and Land Utilisation Planning (NBSSLUP), Nagpur, with 1:5,00,000 scale) in GeoHMS (HEC 2013) extension of ArcGIS for the curve number, CN map generation. This CN method was used to calculate the initial loss from the total rainfall. The LULC was also used to assigning the roughness coefficient along and across the surveyed cross-section with the help of literature and field verification (Acrement and Schneider 1989). In the HE-HMS model, the runoff was estimated using SCS unit hydrograph and time lag method, and flow routing was done using Muskingum-Cunge method (Thakur et al. 2008). The ortho-rectified image of Cartosat-1 was used for mapping the elements at risk with the help of visual interpretation techniques for the study area between Solang and Manlai town (Fig. 7.11). The ASTER ortho-image and DEM were utilized to create the basin characteristics for Geo-HMS model and also for the flood inundated modelling (hydraulic model) as simulated by 1-D MIKE 11 (DHI 2014a, b; Horritt and Bates 2002) and are shown in Figs. 7.11 and 7.12. Hydro-processing of this DEM is done in HEC-Geo-HMS to create drainage lines, small subbasins, total subbasin area and other HEC-HMS model input elements for the Beas River watershed up to Manali (Fig. 7.11).

The Beas River flood inundation maps were generated using MIKE 11 one-dimensional hydrodynamic (HD) model-based highest water level at each cross-section along with the river reaches and DEM, and then the flood inundation-affected area is calculated from Cartosat-1 ortho-images. This flood-inundated area has less areal extent in 1999 as compared to the year 2000. The heavy rainfall during monsoon of 1995 has caused one of the largest flash floods in this part of river, where left bank of river side like part of SASE campus, BRO campus, adjoining roads, etc. were fully inundated with a few (0.1–0.4 m) metres of water level, depending upon the relative elevation of these sites. The highest water

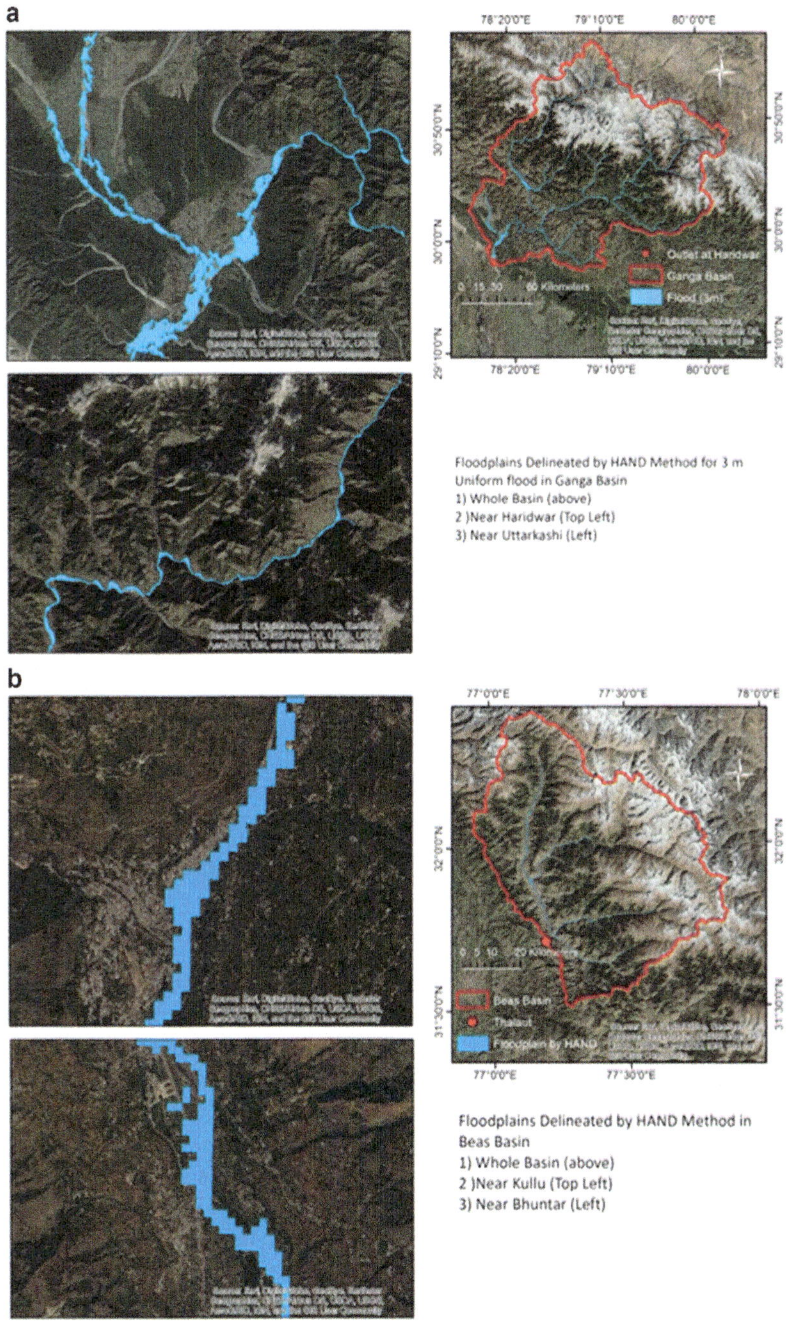

Fig. 7.10 (**a**) Floodplains by HAND method in Ganga basin. (**b**) Floodplains by HAND method in Beas basin

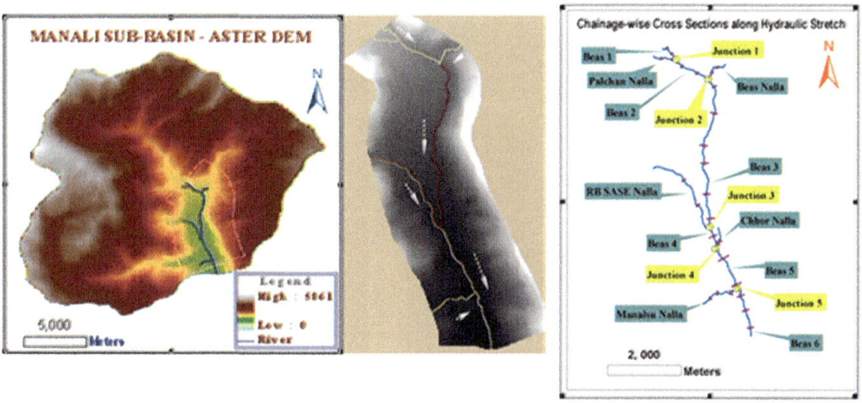

Fig. 7.11 DEM along with location map of flood simulation stretch and river network overlaid with surveyed river cross-sections

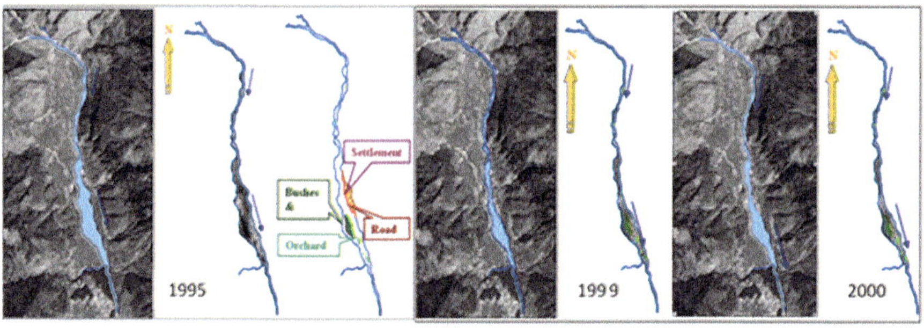

Fig. 7.12 Inundated areas in 1995, 1999 and 2000 (Palchan to Manali portion of Beas River)

level is taken for flood inundation area identification, and they were computed for September 4, August 12 and July 1 in the years 1995, 1999 and 2000, respectively (Fig. 7.12). Most of the flood events reported in this area are mainly due to heavy monsoonal rainfall and cloud burst-related extreme rainfall, which results in flash floods, landslides and damage to roads and other infrastructures. The total flood-inundated area estimated for the year 1995 is 1.39 Km^2, out of which 1.03 Km^2 area is river channel area (Maiti 2009). In the year 1995, a sudden cloud burst of September 4 (103.2 mm rainfall in few minutes) resulted in large flood inundation area, even outside the main river channel; the field campus of SASE and BRO, other settlements (0.24 Km^2), adjoining road (938.42 m), the bush (0.09 Km^2) and orchard (0.03 Km^2) are shown to be affected (Table 7.5). The simulated flood inundation and affected area is similar to that given by SASE officially at Bahang, Manali (as interviewed during field work). This major flood of the year 1995 has resulted in construction of 300 dykes/embankments along the road before (from head ward) up to SASE campus. Additionally, one more 200 m long embankment (about 3 m

Table 7.5 Flood-affected area and elements at risk derived from satellite-based LULC maps for the flood of the years 1995, 1999 and 2000 (Maiti 2009)

Elements at risks	Flood-inundated area (Km^2) 1995	Flood-inundated area (Km^2) 1999	Flood-inundated area (Km^2) 2000
Bushes and trees	0.09	0.09	0.09
Orchards	0.03	0.03	0.03
Settlement	0.24	0	0
River channel	1.03	1.03	1.0
Total area	1.39	1.15	1.12
Road length	938.42 m	0	0

high) and concrete wall has been constructed beside the road and left bank of the Beas River (Maiti 2009) (Table 7.5).

This study highlighted the use of hydrological models in the case of ungauged/gauged Himalayan watersheds for flood peak estimation and use of this simulated flood peak as boundary condition for 1-D HD flood models. The 1-D HD model simulates the water level and discharge at every defined river cross-section and which is combined with DEM to get the flood or water inundation at various flood peaks.

7.3.3.2 Uttarakhand Floods and Kedarnath GLOF Modelling of 2013

The time period of June 14–18 of year 2013 witnessed one of the most devastating floods in state of Uttarakhand and some parts of adjoining states of Himachal Pradesh and Nepal (Rao et al. 2013.). The IMD-TRMM-based rainfall for this duration is shown in Fig. 7.13a–d. The total rainfall during this June 2013 was much more than normal by IMD (2013), and this amount precipitated mainly during June 15–17, 2013. Total of 575 mm of rainfall was recorded at IMD Karanpur and Dehradun Automatic Weather Station (AWS) during June 15 (00:30 h) to June 17 (08:30 h), 2013 (Fig. 7.14). The highest hourly rainfall of 54 mm was reported at 08:30 h on June 16, 2014. The initial study by Dobhal et al. (2013) has reported rainfall of 326 mm from their AWS at Chorabari Lake, Kedarnath, during June 15–16, 2013. This heavy precipitation (rainfall + snowfall in higher areas) caused one of the worst floods in state of Uttarakhand. Some of the devastated sites (Rudraprayag to Kedarnath) were visited by faculty and students of IIRS in September–October 2013 (Fig. 7.15).

This sub-section presents the hydrological and GLOF simulation of Kedarnath area during June 15–17, 2013. The continuous high-intensity precipitation in the entire state of Uttarakhand during this time period caused devastating flash floods. First reported flood damages at Kedarnath and Rambara occurred on June 16 evening. This event was simulated using hydrological modelling system (HMS) model, with input 3 hourly rainfall data, kinematic wave method for runoff transformation

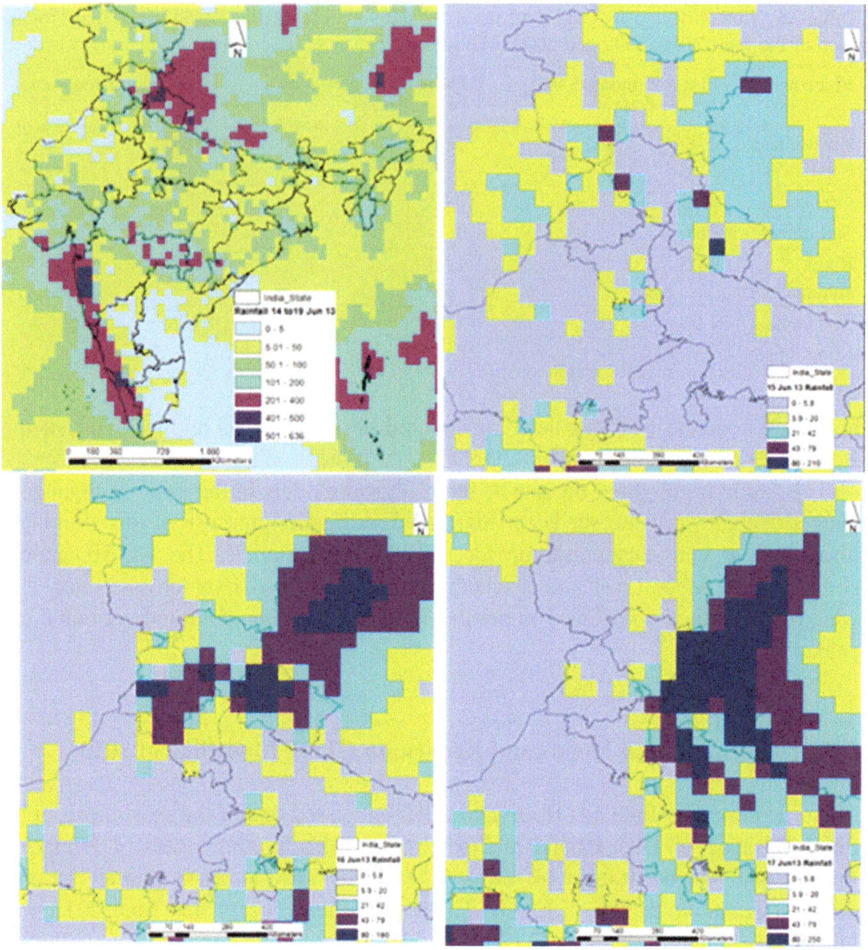

Fig. 7.13 IMD 0.5° daily gridded rainfall data during June 14–19, 2017, with highlight on NWH

and flow routing (Fig. 7.16). Mandakini River watershed up to Sonprayag was taken for this simulation. The peak flood of 610 m^3/s and 950 m^3/s was estimated at Rambara and Sonprayag, respectively, for Mandakini River. The second flood event due to Chorabari Lake breach or GLOF was simulated for June 17, 2013, morning. The resulting dam breach flood has been routed through Mandakini River channel downstream covering Kedarnath, Rambara and Gauri Kund up to Sonprayag. River channel was represented in the model by network file and cross-sections, which were developed from high-resolution Cartosat-1 digital elevation model (DEM) with vertical accuracy of 10 m in GIS environment using HEC-GeoRAS. Cross-section interval has been taken as per the field readings. The altitude of the Chorabari Lake measured in the field using differential GPS was 3855 m, and volume of the lake was calculated using field observations of lake depth and area and comes out to be

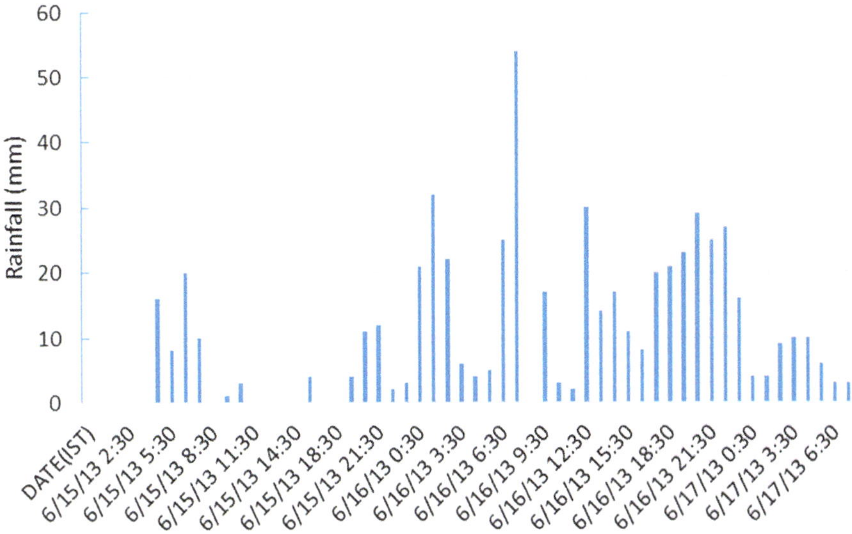

Fig. 7.14 IMD Karanpur and Dehradun Automatic Weather Station (AWS) during June 15 (00:30 h) to June 17 (08:30 h)

Fig. 7.15 Sites visited by faculty and students of IIRS in Sep–Oct 2013 (Rudraprayag to Kedarnath)

4.35 Mm3. The measured breach width for the lake was 46.90 m. The side slope of the lake was calculated as 0.76. The Manning's roughness coefficient was taken as 0.040 considering the boulder beds and hilly terrain of Himalayan Rivers and large debris flow associated with GLOF. The entire dam break simulation is done in DHI's MIKE 11 flood modelling software. MIKE 11 estimates the flood hydrograph of discharge and water level/discharge time series at different cross-sections downstream of lake up to Sonprayag. Its hydrodynamic (HD) module is used for dam break modelling. The peak discharge of 1127 m3/s and 1080 m3/s was estimated at site 715 m and 1714.5 m d/s of lake. The site at 1714.5 m is near Kedarnath Temple. The results of both the simulations (Fig. 7.17) were validated with post-flood field survey and comparison of water/flood marks at various sites.

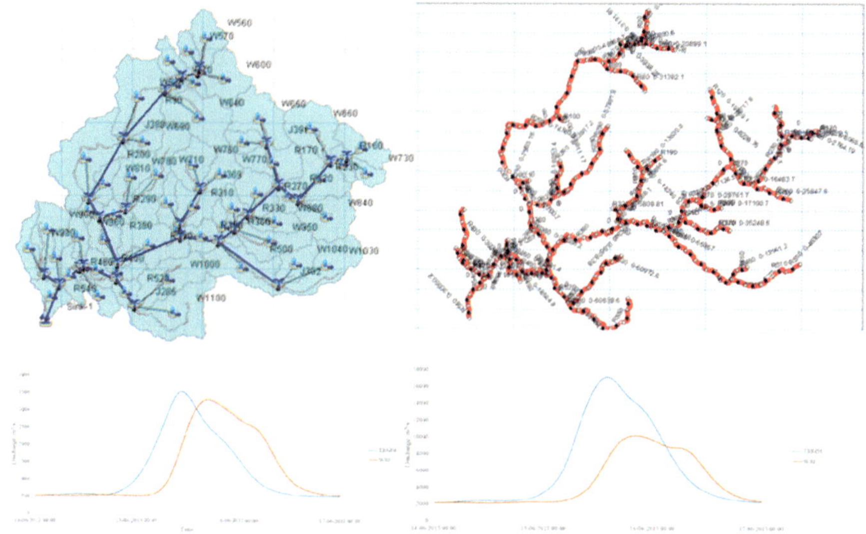

Fig. 7.16 Hydrographs at Uttarkashi and Devprayag for June 14–17, 2013, with rainfall inputs from TRMM 3B42 v.7 and WRF-simulated rainfall

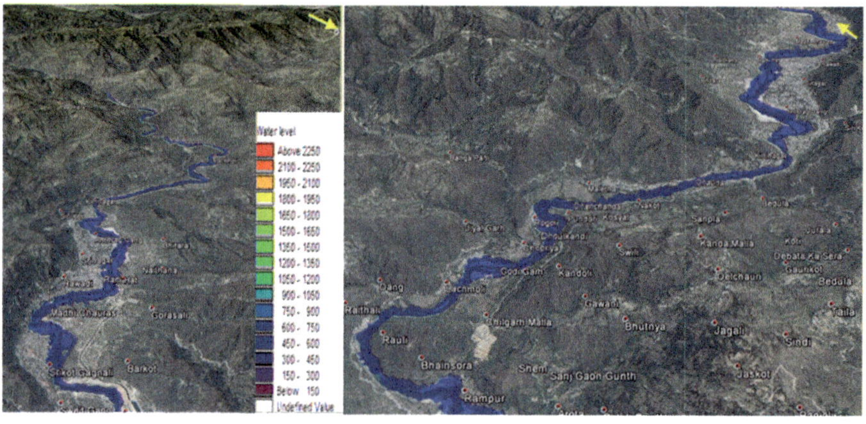

Fig. 7.17 Sample of simulated river flow and flood inundation scenario with water levels for stretch of Alaknanda River near Srinagar, Uttarakhand, during calibration stage July 2007

7.3.3.3 GLOF Scenario for Dhauliganga River and Impact on Hydropower Projects

This sub-section highlights the use of the 1-D HD modelling for the simulation of glacier lake outburst flood (GLOF) in Dhauliganga River of Alaknanda basin where six hydroelectric (HE) power projects are under construction downstream of a major glacier lake. This work used ArcGIS with HEC-GeoRAS extension for creating the

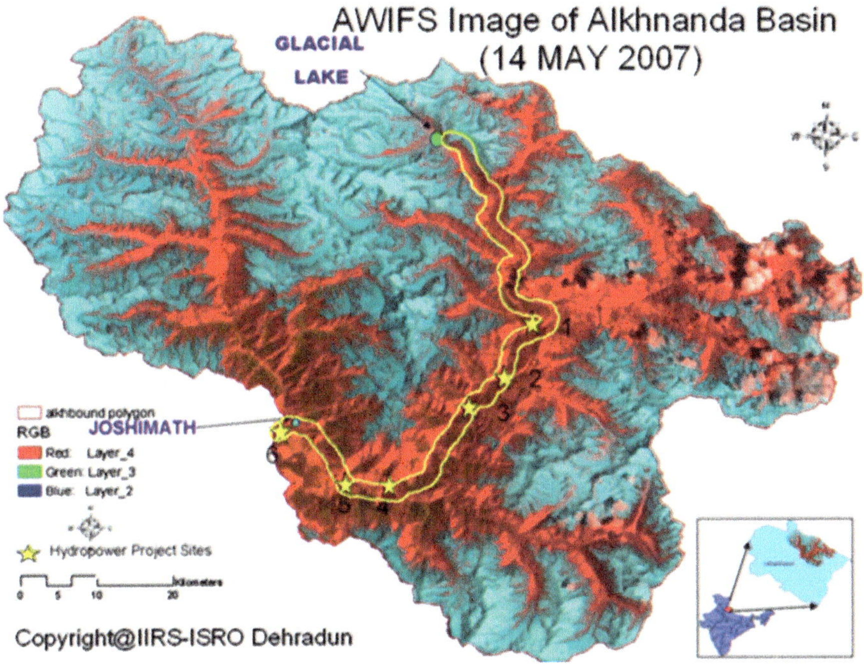

Fig. 7.18 Location of the biggest glacial lake and hydropower projects over Dhauliganga River

river GIS database. Longitudinal and cross-section elevation values were derived from the ASTER global digital elevation model and DEM created from interpolated contour maps (Thakur et al. 2016).

Dhauliganga River, which is a major tributary of Alaknanda river basin, was selected as the study area for estimating the glacier lake outburst flood hazard. The head water catchment of the Dhauliganga River has many glaciers as well as few glacier lakes. One of these glacier lakes has a moraine-dammed lake, which is formed at snout of two merging glaciers, Raikhana Glacier and East Kamet at altitude of 4663 m (Thakur et al. 2016). The total basin area of Dhauliganga River is 4508 km^2 up to the last hydro pilot project, which is taken as outlet of this subbasin (Jain et al. 2012). The potential risk lake of this ablation zone is popularly known as Basudhara Tal and is also known as Kamet Base Camp with altitude of 4663 m, and it forms the source point of the Dhauliganga River (Fig. 7.18, Thakur et al. 2016). Figure 7.19 shows the main flowchart of the methodology adopted to complete this work. The river flow data taken from the Central Water Commission (CWC) is used as inflow hydrographs for defining the upstream boundary condition in MIKE 11 model. The LULC map and Manning's roughness coefficient "N" values were prepared using combination of LANDSAT TM and CARTOSAT-1 ortho-rectified datasets and validated by the field observation. River cross-sections are updated using the field observed actual river cross-sections with corrections for lateral width and depth. Next, the 1-D HD simulations were carried in MIKE 11 model, which

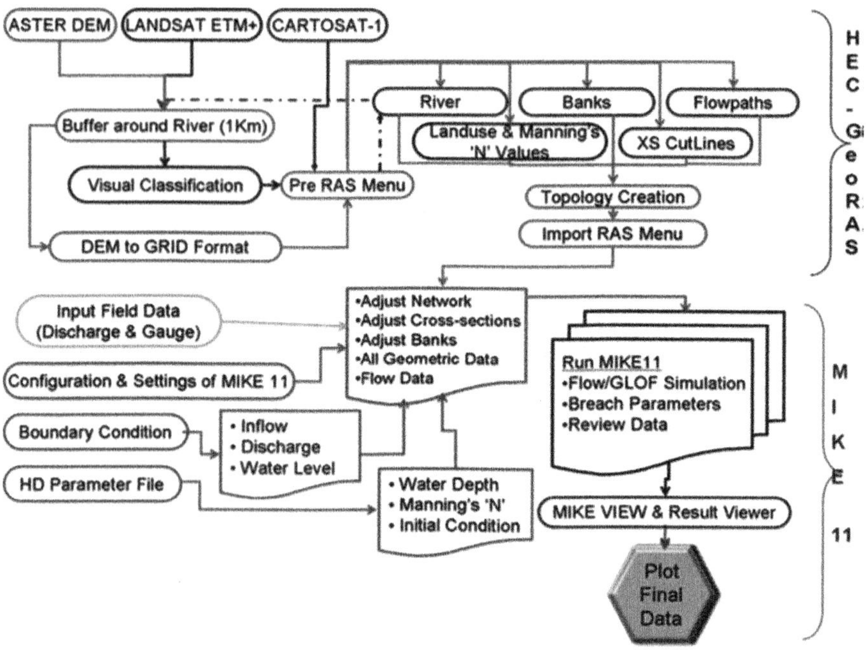

Fig. 7.19 Overall flowchart of GLOF modelling work. (Thakur et al. 2016)

gave output of water level and discharge at defined cross-sections of the selected river (Thakur et al. 2016). The daily discharge, water level and rainfall data from June 1, 2001, to May 31, 2002, were provided by the CWC, Dehradun, India, and this data was used in the setting up of model simulation and calibration as the upstream and downstream boundary conditions. Finally, the discharge data for the year 2004–2005 at Joshimath site was used for the validation of the 1-D HD model (Thakur et al. 2016).

The calibration of 1-D HD model was done from June 1, 2001, to May 31, 2002, and validation was done for the year 2005 with 95% accuracy (Thakur et al. 2016). The present work has used the 100-year return period flood ordinates of Joshimath gauging site and used area distribution method for redistributing the lateral peak flood flow in five sub-catchments of Alaknanda basin (Thakur et al. 2016). The simulated GLOF flow peaks at downstream of the lake site, with the final lake breach width of 40, 60 and 80 m, are 1394.28, 1552.04 and 1898.04 m^3/s, respectively. It is estimated from this study that the GLOF with 80 m final breach width can produce the peak flood of 1575 m^3/s at the project site 1 for GLOF only scenario and 3220 m^3/s at the site 6 for the scenario of GLOF plus 100-year flood (Fig. 7.20). The maximum water level and depth as simulated by 1-D HD model for the CWC gauging site at Joshimath were 1381.49 and 4.329 m, respectively, for the 100-year return period flood peak plus the GLOF event of 80 m final breach (Thakur et al. 2016). This information on peak water level or depth and discharge is very

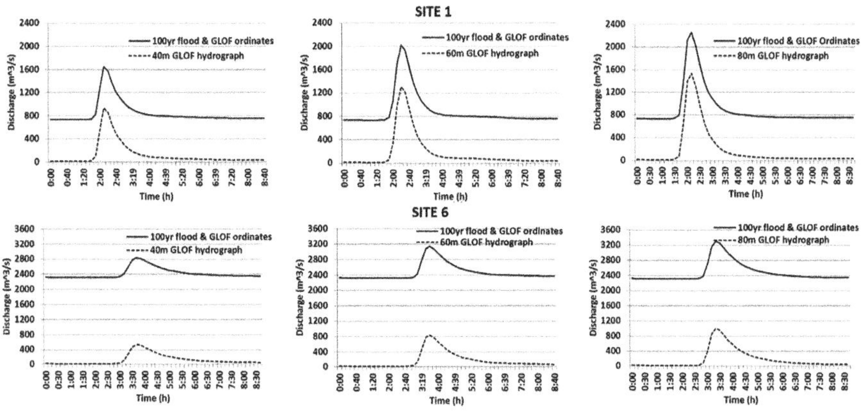

Fig. 7.20 100-year flood peak and GLOF discharge at six project sites for different final breach widths of GLOF scenarios (Thakur et al. 2016)

important for the hydropower project design discharge calculations (Thakur et al. 2016). As per the published report of the Tehri Hydro Development Company, THDC, the calculated design flood value for the standard project flood, SPF, is 6700 m^3/s, and probable maximum flood (PMF) is 10,840 m^3/s at site 6 (Thakur et al. 2016). These designed SPF and PMF discharge values are much higher than the flood peak values calculated in the present study, where a combination of 100-year return period flood (2300 m^3/s) was considered along with GLOF scenarios. It is further recommended that the approach used in the present case study can be extended to include SPF and PMF values and combined with the flood hydrographs from GLOFs and better resolution materials (DEM and river cross-sections), which will help in creating better and accurate flood scenarios (Thakur et al. 2016).

7.4 Early Warning System for Hydrometeorological Extremes in NWH

The flood early warning for any country is very important due to possible saving of human life, minimizing economic losses and devising possible mitigation strategies (Navalgund 2014; Thielen et al. 2009; Jorgensen and Host-Madsen 1997). IIRS and ISRO have developed an integrated methodology to understand the process of rainfall prediction and its resultant flooding in NWH region (Fig. 7.21). The methodology can be divided into three steps. *In the first step*, present study uses 3-day advance weather forecast in NWH using double nested domain of Weather Research and Forecasting (WRF) model (9 km for outer domain and 3 km for inner domain) for the entire monsoon of 2015 and 2016. The WRF-ARW model was run using the National Centre for Environmental Prediction (NCEP), Global Forecasting System

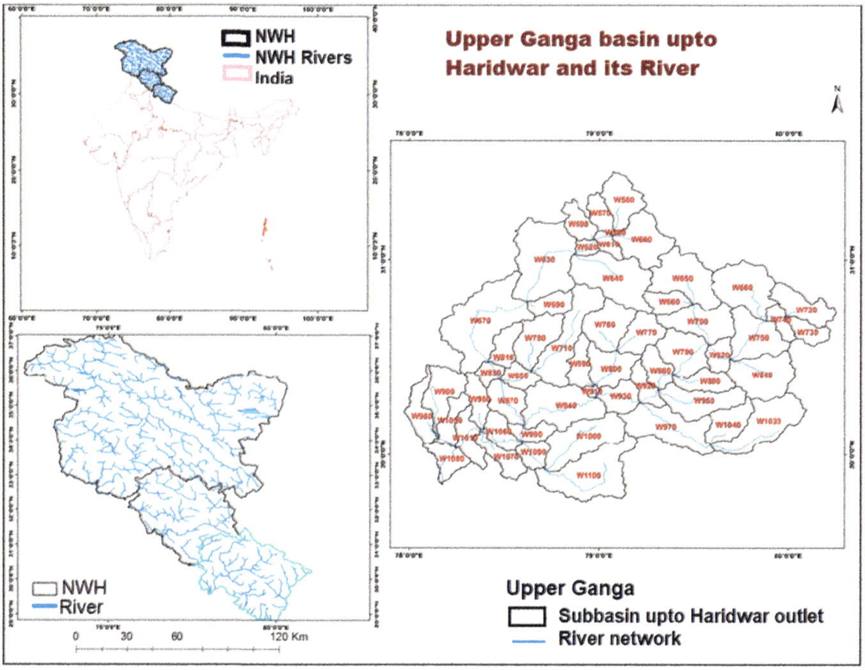

Fig. 7.21 Study area map for flood early warning with initial test subbasin as Ganga up to Haridwar

(GFS) 0.25 degree and hourly data as initialization state files. The WRF outputs were taken at hourly time scale and hosted on 3 hourly scale on IIRS website (http://dms.iirs.gov.in). Based on historical and current Indian Meteorological Department (IMD) 0.5 degree gridded daily rainfall data, validation of few events of 2013–2015 monsoon has been done. The simulation accuracy in prediction of rainfall above 100 mm is found to be 60%, but overall pixel-wise correlation coefficient R^2 is as low as 0.2–0.3.

In the second step, the forecasted precipitation of every 3 h is used in hydrological modelling system (HMS) for flood hydrograph generation at various outlets of study area. Currently, parts of Beas river basin, Upper Ganga basin up to Haridwar and Yamuna basin up to Faizabad near Paonta Sahib are tested for near real-time flood forecasting. Limited validation of simulated river flow hydrograph for Uttarkashi and Joshimath sites has been done using historical data of 2005–2007 monsoon provided by the Central Water Commission (CWC), and R2 of 0.6–0.7 was achieved for two stations.

In the third step, the output of hydrological model in terms of flood hydrograph is used for flood inundation scenarios along the river reaches using 1-D hydrodynamic (HD) modelling to see which are under risk and come under flood inundation.

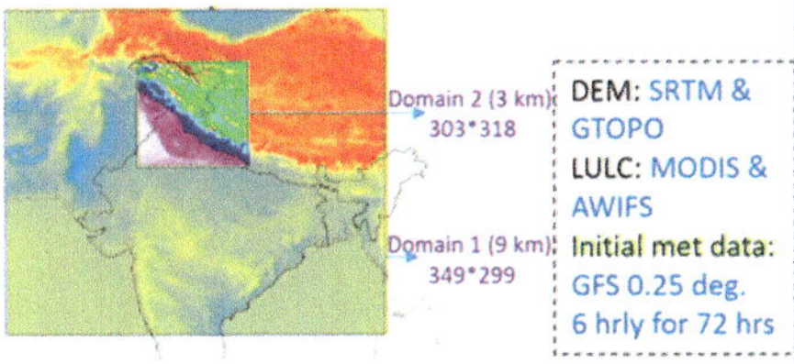

Code	Physics Option	Used Scheme
6	Microphysic (mp_physics)	WRF Single Moment 6-class Scheme
1	Longwave Radiation (ra_lw_physics)	Rapid Radiative Transfer Model Scheme
2	Shortwave Radiation (ra_lw_physics)	Dubia Scheme
2	Land Surface (sf_surface_physics)	Noah Land Surface Model
2	Cumulus Parameterization (cu_physics)	Betts-Miller-Janjic Scheme, Operational Eta Scheme

Fig. 7.22 Domains for WRF simulations and Table 7.6: WRF physics schemes used for simulations

7.4.1 Success Stories

In 2015 monsoon, the flood event of August 05–08 in part of Beas river basin, Mandi District of H.P. near Dharampur, and flood and landslide event at Pithoragarh and Chamoli area during June 30, 2016, have caused very high damages. These events were well forecasted in advance of 3 days by WRF-ARW model. Summary of WRF-ARW options used in present study is given in Table 7.6 (Hong and Lim 2006), and domain details are given in Fig. 7.22. Figure 7.23 shows overall flowchart of methodology used in this project and its implementation for one of flash flood events of 2015 for part of Beas River watershed in Dharampur area of Mandi, H.P., India. Figure 7.24 shows WRF-ARW model-based forecasted accumulated rainfall for two recent successful heavy to very heavy rainfall events of 2016 in Uttarakhand and Himachal Pradesh. IMDs and IIRS rain gauge data (Table 7.6) were used to validate August 05–06, 2016, event of HP, and it shows overall accuracy of 0.748 and difference of 30 mm with rain gauge data (Fig. 7.25).

7.4.2 Hydrometeorological Instrument Network in NWH

In the last 5 years, i.e. 2013–2018, IIRS have played a major role in installing automatic weather stations (AWSs, 27 nos.), digital water level recorders (DWLRs, 6 nos., 4 at NWH; 1 planned at Maitri, Antarctica; 1 at Maithon Dam, West Bengal), snowpack analyser (SPA, 1 at Dhundi, Manali), pressure-based SWE gauge-now scale (SSG, 1 at Kothi, Manali), solar radiation sensors (27 at NWH and 1 at IARI, Delhi), snow precipitation and snow depth gauges in various parts of Northwest Himalaya (NWH) (Fig. 7.26), North India, and few other sites in India. A dedicated rainfall simulator for hill slope hydrology experiments has also been established in IIRS campus (Fig. 7.26). The data from few of these sites is available via mobile-based telemetry systems on IIRS website (http://aws-dwlr.iirs.gov.in/), and this data can be accessed via username and password for AWSs at every 5–15 min and DWLRs at every half hour. Similar sites provide real-time data from SPA and SSG. These datasets are essential for any water balance, hydrological modelling

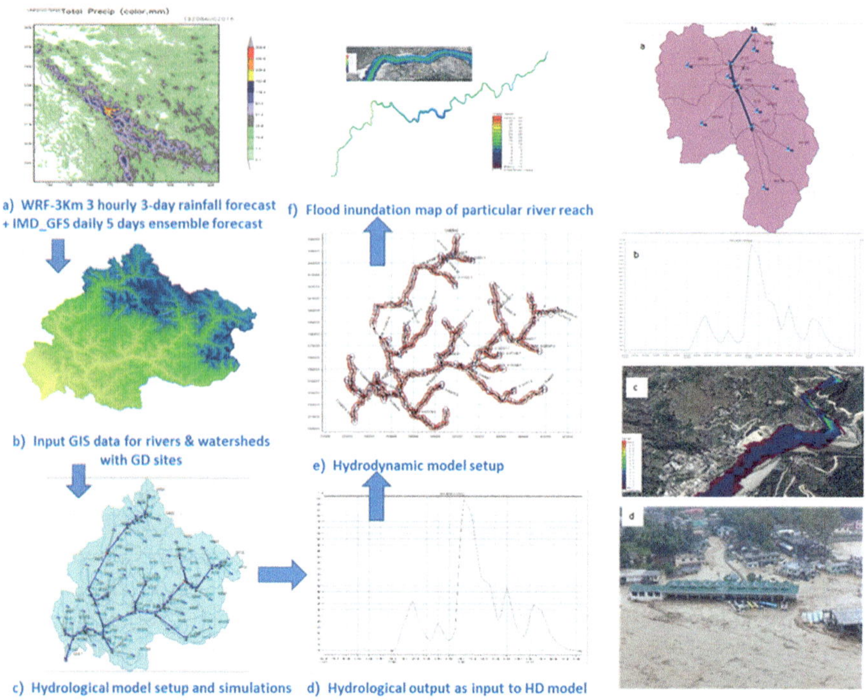

Fig. 7.23 Overall methodology pictorial flowchart and results of one of very heavy rainfall events and subsequent simulated and actual floods in Dharampur, Mandi, H.P. (**a**) Hydrological model setup for HMS; (**b**) simulated flood hydrograph from HMS as boundary condition to 1-D hydrodynamic model (MIKE 11); (**c**) simulated inundation map from HD model; and (**d**) actual flood inundation situation in Dharampur, showing the first floor of bus stand fully under floodwaters

and RS-based hydrological parameter calibration and validation studies. Some of hydrological parameter maps generated using RS data and validated with such ground data are shown in Fig. 7.26. In addition to above-mentioned permanent field instruments, the department also utilizes the portable field instruments for measurement of hydrological parameters such as digital current or flow metre for river velocity, ThetaProbe for soil moisture, multi-parameter water quality kit, turbidity metre, handheld GPS, snow probe for snow wetness and snow density and laser distance metre with 2 km range.

7.5 Conclusions

The chapter highlights the integrated use of remote sensing and weather forecast and hydrological and hydrodynamic models for mapping, monitoring and modelling of climatic extremes in Northwest Himalaya. Avalanche-prone sites were demarcated

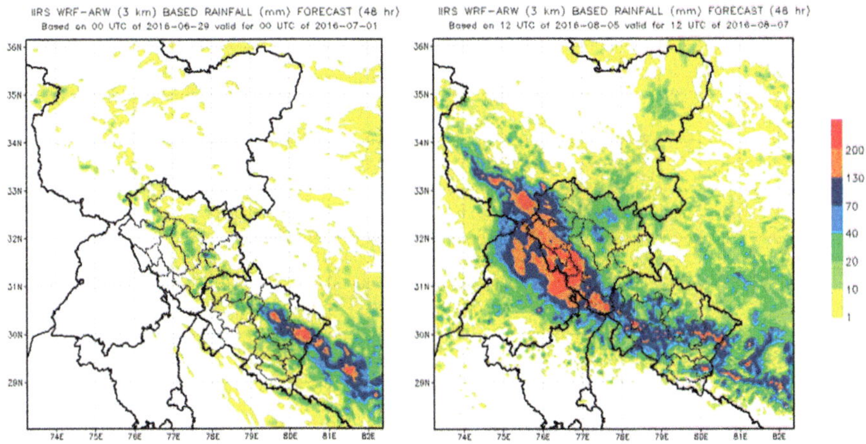

Fig. 7.24 Recent successful rainfall forecast (June 30–July 01, 2016, for Pithoragarh and Chamoli area of UK and August 05–06, 2016, for lower and middle hills of HP) using WRF model for NWH, causing landslides and flash floods in these areas of UK and HP

Table 7.6 IMD and IIRS AWS rainfall data for August 05–06, 2016; 08:30 IST

Station	Rainfall (mm)	Station	Rainfall (mm)
Gaggal	165.8	Kheri	98.8
Jogindernagar	157.6	Baijinath	80
Sujanpur Tira	155	Hamirpur	75
Nagrota Surian	145	Amb	71
Nadaun	135	Sunder Nagar	64.8
Dehra Gopipur	118	Bangana	64
Palampur	111.2	Ghamroor	63
Guler	105.5	Sarahan	57
Gohar	104.6	Bhoranj	52.5
Dharamshala	98.8	Pandoh	52
Bijahi	45.2	Bhota (IIRS)	65.8
Shimla	40.5	Kheri	98.8

Source: CWC, Chandigarh, via personal communication and IIRS hydrometeorological data reception and management website: http://aws-dwlr.iirs.gov.in

using RS&GIS technique, and RAMMS model was being tested in Solan-Dhundi, Manali area, for practical avalanche problems. Results of RAMMS depend upon accuracy of input data which reflects topographical as well as meteorological condition of area. Study shows that RAMMS can be applied in Himalayas with wise selection of input parameters and ground validation. Flood inundation maps were generated for Jhelum River using SAR imageries. Application of topographic feature-based methods was discussed over conventional remote sensing and hydrological/hydrodynamic modelling-based methods to delineate flood-prone areas in case of hydrological scarce data situation. GLOF for Kedarnath flood 2013,

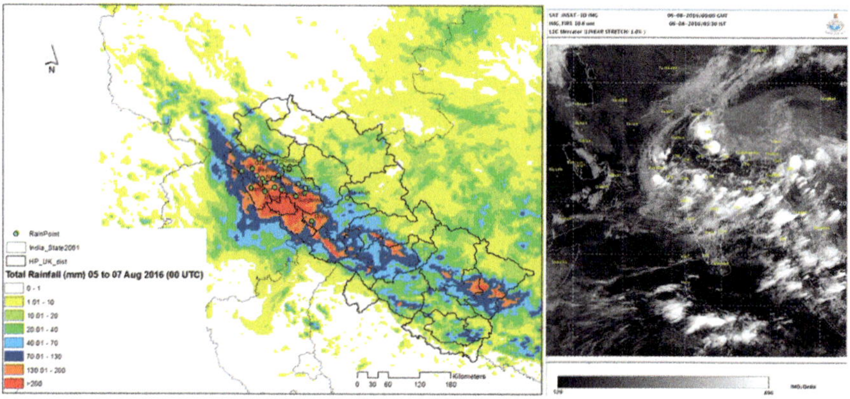

Fig. 7.25 Predicted rainfall for August 05–07, 2016, at 00 UTC and actual INSAT-3D IR image of Indian region. (Source: http://satellite.imd.gov.in/insat.htm)

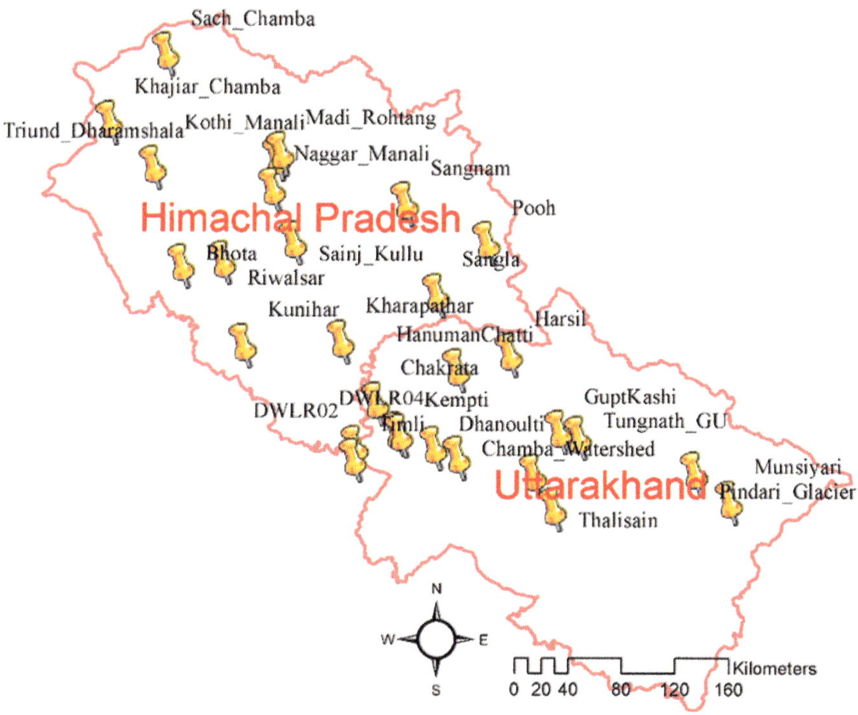

Fig. 7.26 Location map of installed AWSs by IIRS in parts of HP and UK

Dhauliganga River scenarios and its impact on hydropower projects were simulated using hydrodynamic modelling. Flood early warning system experimented in monsoon 2015 and 2016 for parts of Beas basin, Yamuna basin and Upper Ganga basin was discussed. The study showed that, despite the overall pixel-wise low accuracy of weather forecast model, WRF-ARW in this case is able to correctly pick the high rainfall events having more than 100 mm rainfall at medium time range of 3 days, but comparison with other rainfall data products such as from IMD, INSAT or CPC showed a poor pixel to pixel correlation as well as time lag of few hours to 1 day in picking heavy rainfall. Results of these studies can be improved if near real-time hydrometric data is provided by CWC or by state agencies. The accuracy of flood inundation maps is highly dependent on input flow hydrograph as well as input DEM. The actual river cross-section data is also needed to improve results of flow routing and flood inundation. To support the modelling activities in NWH, IIRS has also installed network of AWS, DWLR and snow sensors in various parts of NWH. Overall this chapter gives glimpse of prefeasibility study, methodology and result analysis procedure for hydrometeorological hazards mapping, monitoring and modelling in NWH region.

Acknowledgements WRF-ARW model was provided by NCAR, WRF initialization data was provided by NCEP-GFS and data of rainfall products was provided by MOSDAC, TRMM and CPC sites. Input DEM were provided by USGS EarthExplorer, and hydrometric data was provided by CWC and IMD. Authors acknowledge the special thanks to Bhuvan, NRSC and Uttarakhand state administrative authorities for providing necessary data and their support during field work in October 2013.

References

Acrement, G.J. and Schneider, V.R., 1989. Guide for selecting Manning's roughness coefficients for natural channels and floodplains. Report No. FHWA-TS-84-204. Federal Highways Administration US Department of Transportation, Washington. http://www.fhwa.dot.gov/BRIDGE/wsp2339.pdf, Accessed on 02.02.2018.

Beven, K. J. and Kirkby, M. J. (1979). A physically based, variable contributing area model of basin hydrology, Hydrological Sciences Journal, 24:1, 43–69.

Bhatt C.M., Rao G.S., Farooq, M., Manjusree P., Shukla, A., Sharma, S V S P, Kulkarni, S S, Begum, A., Bhanumurthy, V., Diwakar, P.G. and Dadhwal, V.K (2016). Satellite Based Assessment of the Catastrophic Jhelum Floods of September 2014, Jammu & Kashmir, India, Geomatics, Natural Hazards and Risks. DOI: https://doi.org/10.1080/19475705.2016.1218943.

Bromhead, E.N. 1986. The stability of slopes, Surrey University Press, New York.

Dasgupta A (2015). Reduction of uncertainties in a 1D-2D coupled hydrodynamic model using remote sensing data, IIRS-ITC MSc thesis report in Natural Hazard Disaster and Risk Management, 61 pages.

DHI (2014a). MIKE 11, A Modeling System for Rivers and Channels, User Guide. Danish Hydraulic Institute, Horsholm, Denmark.

DHI (2014b). MIKE 11, A Modelling System for Rivers and Channels, Reference Manual, Danish Hydraulic Institute, Horsholm, Denmark.

Dobhal D. P., Gupta, A.K., Mehta, M. and Khandelwal, D.D. (2013). Kedarnath disaster: facts and plausible causes, Current Science, 105(2), pp. 171–174.

Frey, W. and Salm, B. (1990). Snow properties and movements in forest of different climatic regions.In: Proc. XIX IUFRO World Congress, Montreal, Canada, Div. 1, Vol. 1, p.328–339.

Ghinoi, A., 2003. A new contribution to the assessment of snow avalanche susceptibility: Applications in Tyrol (Austria) and in Alta Val Badia (Dolomites, Italy). Ph. D. Thesis, Modena University, Modena, Italy.

Gupta, V., Dobhal, D.P. and Vaideswaran, S. C. (2013). August 2012 cloudburst and subsequent flash flood in the Asi Ganga, a tributary of the Bhagirathi river, Garhwal Himalaya, India, Current science, 105(2), 249–253.

Hasan N (2012). Flood peak estimation and inundation mapping using modeling approach in Mahanadi river delta. IIRS-ITC post graduate diploma thesis on Natural Hazard Disaster and Risk Management, 41 pages.

HEC (2013). HEC-HMS User's Manual, Hydrologic Modeling System HEC-HMS. US Army Corps of Engineers: Hydrologic Engineering Center, 10. CA, USA, CPD-74A.

Hong SY, Lim JOJ (2006) The WRF single-moment 6-class microphysics scheme (WSM6). Asia-Pac J Atmos Sci 42(2):129–151.

Horritt, M.S., and Bates, P.D., (2002) .Evaluation of 1D and 2D numerical models for predicting river flood inundation. Journal of Hydrology. 268, pp. 87–99.

IMD (2013). A preliminary report on heavy rainfall over Uttarakhand during 16–18 June 2013. Government of India, Ministry of Earth Sciences, India Meteorological Department. http://imd.gov.in/doc/uttrakhand_report_04_09_2013.pdf.

Jain SK, Lohani AK, Singh RD, Chaudhary A, Thakural N (2012) Glacial lakes and glacial lake outburst flood in a Himalayan basin using remote sensing and GIS. Nat Hazards 62:887–899. https://doi.org/10.1007/s11069-012-0120-x

Jorgensen, G. H., and J. Host-Madsen. (1997). Development of a Flood Forecasting System in Bangladesh. In Proceedings of Conference on Operational Water Management, pp. 137–148. 3–6 September 1997, Copenhagen. AA Balkema.

Kumar, M.S. Shekhar, M.S. Rama Krishna, S.S.V.S. Bhutiyani, M.R. and Ganju, A. (2012). Numerical simulation of cloud burst event on August 05 2010, over Leh using WRF mesoscale model. Nat. Hazards, doi: https://doi.org/10.1007/s11069-012-0145-1.

Kumar, P., Shukla, B.P., Sharma, S., Kishtawal, C.M. and Pal, P.K. (2015).A high-resolution simulation of catastrophic rainfall over Uttarakhand, India. Nat Hazards, pp. 1–16, DOI https://doi.org/10.1007/s11069-015-2013-2.

Logen S.C., (2005). Temporal changes in spatial pattern of weak layer shear strength and stability on uniform slopes. Earth Science, Bozman, Montana, Montana State University.

Maiti, D (2009). Hydrological and 1-D Hydrodynamic Modelling in Manali Sub-Basin of Beas River, Himachal Pradesh, India, Master of Science in Geo-information Science and Earth Observation, Specialisation: (Geo-Hazards), Joint MSc Thesis, ITC Netherlands & IIRS, India.

Malik A, (2009). Snow avalanche runout modelling in Solang-Dhundi area of Himachal Pradesh, India using a numerical model (RAMMS). Master of Science in Geo-information Science and Earth Observation, Specialisation: (Geo-Hazards), Joint MSc Thesis,, ITC Netherlands & IIRS, India.

Manfreda, S., Nardi, F., Samela, C., Grimaldi, S., Taramasso, A. C., Roth, G., & Sole, A. (2014). Investigation on the use of geomorphic approaches for the delineation of flood prone areas. Journal of Hydrology, 517, 863–876. https://doi.org/10.1016/j.jhydrol.2014.06.009

McClung, D.M., and P. Schaerer. 2006. The Avalanche Handbook, The Mountaineers Books, Settle, WA.

Mears, A.I., (1992). Snow - Avalanche Hazard Analysis for Land - use Planning and Engineering, Department of Natural Resources, Denver Colorado.

Navalgund, R. (2014). Need for developing effective early warning systems for natural disasters using space technology. Guest Editorial in Current Science, 107(6), 935–936.

Nobre, A. D., Cuartas, L. A., Hodnett, M., Rennó, C. D., Rodrigues, G., Silveira, A., Saleska, S. (2011). Height Above the Nearest Drainage – a hydrologically relevant new terrain model. Journal of Hydrology, 404(1–2), 13–29.

NRSC, 2017. https://nrsc.gov.in/floods (accessed July 27th 2017)

Rana, N., Sundriyal, Y. P. and Juyal, N. (2012). Recent cloud burst-induced landslides around Okhimath, Uttarakhand. Curr. Sci., 2012, 103, 1389–1390.

Rao, D. K. H. V., Rao, V.V., Dadhwal, V.K. and Diwakar, P.G. (2013). Kedarnath flash floods: a hydrological and hydraulic simulation study. Current Science, 106(4), 598–603.

Rao, D.P, Bhanumurthy, V., Rao, G.S., and Manjusri, P. (1998). Remote sensing and GIS in Flood management in India. Memoir Geological Society of India, 41: pp. 195–218.

Snehmani, Bhardwaj, Pandit, A., & Ganju, A. (2013). Demarcation of potential avalanche sites using remote sensing and ground observations: A case study of Gangotri glacier. Geocarto International. https://doi.org/10.1080/10106049.2013.807304.

Thakur P.K. and Sumangala, A. (2006). Flood Inundation Mapping and 1-D Hydrodynamic Modeling Using Remote Sensing and GIS Techniques. ISPRS Orange Book Publications during ISPRS/ISRS commission IV symposium on: "Geospatial Database for Sustainable Development" at GOA from 27–30 September 2006.

Thakur, P.K., Duishonakunov, M., Prasad, V.H., Garg, P.K. and Garg, R.D. (2008). Hydrological Study in Manali Sub-Basin of Beas River using HEC-HMS. Hydro2008, National conference on hydraulics and water resources, at MNIT Jaipur, 15–16 December 2008. pp. 681–693.

Thakur, P. K., Aggarwal, S. P., Nikam, B.R., Garg, V. and Chouksey, A. (2014). Hydrological and Glacier Lake Outburst Flood Modeling of 2013 Kedarnath Floods. In poster presentation in ISPRS TC VIII International Symposium on Operational Remote Sensing Applications: Opportunities, Progress and Challenges, Annual Convention of ISRS and ISG & Joint Sessions with ISPRS TC IV & TC VI, hosted by National Remote, Sensing Centre, Indian Space Research Organisation, Hyderabad, India, December 09–12, 2014.

Thakur, P.K, Aggarwal, S., Aggarwal, S.P., and Jain, S.K. (2016). One dimensional hydrodynamic modeling of GLOF and impact on hydropower projects in Dhauliganga River using remote sensing and GIS applications, Springer's Natural Hazards Journal, pp. 1–19, DOI https://doi.org/10.1007/s11069-016-2363-4 .

Thayyen, R.J., A. P. Dimri, P. Kumar and G. Agnihotri (2013). Study of cloudburst and flash floods around Leh, India, during August 4–6, 2010, Nat Hazards, 65, pp-2175–2204

Thielen, J., J. Bartholmes, M.-H. Ramos, and A. de Roo. (2009). The European Flood Alert System—Part 1: Concept and Development. Hydrology and Earth System Sciences 13 (2), pp. 125–40.

Weir, P. (2002). Avalanche management in forested terrain Ministry of Forests, British Columbia, Canada.

Chapter 8
Rainfall Characteristics over the Northwest Himalayan Region

Charu Singh and Vidhi Bharti

8.1 Introduction

The Northwest Himalayan (NWH) region constitutes a unique geographical setting with a complex interaction between atmosphere and topography. The region is also the birthplace of many perennial rivers like the Ganges, Yamuna, Indus, Chenab, etc. and their various tributaries. The precipitation processes in the region dominate not only the geomorphological processes and terrestrial hydrological cycle but are also highly relied upon for food and water availability, thus, playing a decisive role in the socio-economic survival of millions of people inhabiting the basins of these rivers. The precipitation pattern in this region is mainly controlled by two major atmospheric circulations: Indian summer monsoon (ISM) lasting from June to September and Western disturbances during the winter season from December to March. While the Eastern Himalayan ranges receive precipitation mainly through monsoon rains, the NWH receives both rainfall (liquid) and snow (solid) in summer and winter season, respectively. The monsoon rains contribute roughly 75–80% to the annual precipitation of the NWH and closely control the river discharge in the basins.

There are two major branches of ISM, namely, the Arabian Sea branch and the Bay of Bengal branch. The western Himalayas primarily receive rainfall caused by

C. Singh (✉)
Marine and Atmospheric Sciences Department, Indian Institute of Remote Sensing (IIRS), Indian Space Research Organisation (ISRO), Department of Space, Government of India, Dehradun, India
e-mail: charu@iirs.gov.in

V. Bharti
Marine and Atmospheric Sciences Department, Indian Institute of Remote Sensing (IIRS), Indian Space Research Organisation (ISRO), Department of Space, Government of India, Dehradun, India

Monash University, Melbourne, VIC, Australia

R. R. Navalgund et al. (eds.), *Remote Sensing of Northwest Himalayan Ecosystems*,
https://doi.org/10.1007/978-981-13-2128-3_8

moist air currents coming from the Bay of Bengal which has been deflected westwards off of the Eastern Himalaya range. The monsoon onset occurs at around first week of July at the foothills of the state of Uttarakhand. The north-western part of the country is the last to receive the monsoon rains particularly Jammu and Kashmir as the monsoon strength decreases from east to west along the path of its travel (Basistha et al. 2007). The rainfall decreases westwards due to the increasing distance from the moisture source and weakening of monsoon winds. The official duration of monsoon season is considered 122 days starting from June 1st (Das 2002), although the date of actual monsoon onset may differ annually.

The strength of the monsoon does not remain uniform throughout the period; instead, it undergoes strong intra-seasonal variations known as active and break spells. Most parts of the Indian subcontinent receive good rainfall in an active phase. However, these wet spells are interspersed with dry spells or break phases. During the break spells, the axis of the monsoon trough shifts to the foothills of the Himalayas resulting in excess rainfall over the NWH, while most of the country becomes rainfall deficient (Kelkar 2007; Malik et al. 2011; Rajeevan et al. 2008b). During the northern hemispheric winter, low-pressure systems originating over the Mediterranean Sea travel all the way to the North and North-West India leading to winter precipitation. These migratory systems are known as western disturbance as they enter the country from the west (Pant and Kumar 1997). They are less intense but are capable of reaching higher altitudes in the orogenic interiors resulting in heavy snowfall in Jammu and Kashmir, Himachal Pradesh and Uttarakhand triggering cold waves in the North and Central India (Wulf et al. 2010). These are less intense than convective monsoon rainfall and rarely induce heavy rainfall events.

The thermodynamics and orographic uplifting together are considered responsible for heavy rainfall events over the Himalayas. A paucity of precipitation data due to the inaccessibility and topographical intricacy of the region have been the major hindrances in the precise understanding of the precipitation characteristics of the NWH. Satellite-based precipitation measurements were intended to circumvent the spatial and temporal difficulties related to ground-based precipitation measurements. However, measuring precipitation through satellites is a challenging task in itself as satellite observations are temporally sparse (if measured by polar orbiting satellites) and offer coarser spatial resolution hindering the detection of short-lived storms. Additionally, satellites also have several errors related to observation, bias, data quality and rainfall retrieval algorithms. The current satellite-based techniques for precipitation estimation are inherently indirect measurements. Precipitation estimation techniques predominantly revolve around cloud identification schemes and schemes based on separating raining from non-raining clouds (Kelkar 2007; Kidder and Vonder Haar 1995; Sumner 1988).

Tropical Rainfall Measuring Mission (TRMM), an international project of NASA and JAXA, was designed to primarily observe rain structure, rate and distribution in tropical and subtropical regions. TRMM was the first ever satellite mission dedicated to measuring rainfall in order to understand the global climate system and also carried the first space-borne precipitation radar (PR). The PR operates at 13.8 GHz frequency and has the ability to record information on rain rate above 0.5 mm per hour. It is also able to separate out rain echoes for vertical sample sizes. It has a swath width of 215 km with horizontal resolution at and off nadir 4.3 km and 5 km, respectively. Alongside PR, TRMM has TMI (TRMM microwave imager) which operates at 10.65 GHz, 19.35 GHz, 21.3GHz, 37 GHz and 85.5 GHz microwave frequencies (Kummerow et al. 1998). The outline presented here is based on the results derived from the analysis of TRMM 2B31 and 3B42V7 precipitation products. 2B31 is the merged product of level 1B of TMI and PR. This product provides surface rainfall rate at a spatial resolution of 4.3 × 5 km, but its temporal resolution is poor due to revisit time of 46 days. The satellite passes over the study region four times a day but not precisely above the same location. 3B42V7 is the level 3 version 7 (released in 2012) product of TRMM Multi-satellite Precipitation Analysis (TMPA) constructed by merging estimates from several passive microwave sensors onboard low-Earth orbit satellites including the TRMM Microwave Imager, Special Sensor Microwave/Imager, Advanced Microwave Scanning Radiometer-EOS, Advanced Microwave Sounding Unit-B and NOAA (National Space Development Agency of Japan Earth Observation Center 2001).

TRMM 3B42V7 has been prepared using high-quality (HQ) microwave estimates which are calibrated and combined and then used in the generation of variable rain rate (VAR) infrared (IR) estimates, followed by combining both the estimates and providing the 'best' estimate of precipitation in each grid box at each observation time. Further, the algorithm uses rain gauges indirectly to rescale the estimates to monthly data (Huffman 2013 and references therein). The newly released version 7 of TRMM 3B42 data offers better precipitation measurements with substantially lower bias even over the mountainous terrain and surpasses its predecessor versions (Huffman et al. 2010; Prakash et al. 2015; Qiao et al. 2014; Zulkafli et al. 2014).

In this chapter, the various characteristics of monsoon rainfall over the NWH are discussed. The main focus of the work is rainfall extremes which are frequent over the topographically complex Northwest Himalayan region. The rainfall pattern during monsoon months (JJAS: June, July, August and September) is discussed here using TRMM 2B31 data set. The discussion on extreme rainfall events is based on TRMM 3B42V7 rainfall data. Using the validation study of TRMM 3B42V7 with rain gauge data, the limitations of satellite products have also been discussed in the end. The ground-based gridded daily rainfall dataset prepared and released by India Meteorological Department (IMD) has been used in conjunction with satellite data for the validation purpose. IMD gridded data is available at very fine spatial resolution (0.25 × 0.25°) and has been developed using 6955 rain gauge stations in India for the period of 1901–2013. Rajeevan et al. (2005) and Pai et al. (2014) can be referred to for further details about the product. Digital elevation model (DEM) version 4 of NASA's Shuttle Radar Topography Mission (SRTM) has been employed for the elevation data at spatial resolution of 3 arc second (~90 m).

In a climate change scenario, this study aims to contribute to the ongoing research of extreme weather events over the topographically rich Northwest Himalayan region. Spatial maps and temporal trends of extreme rainfall events would be useful for the policymakers for disaster preparedness and management in the Himalayas.

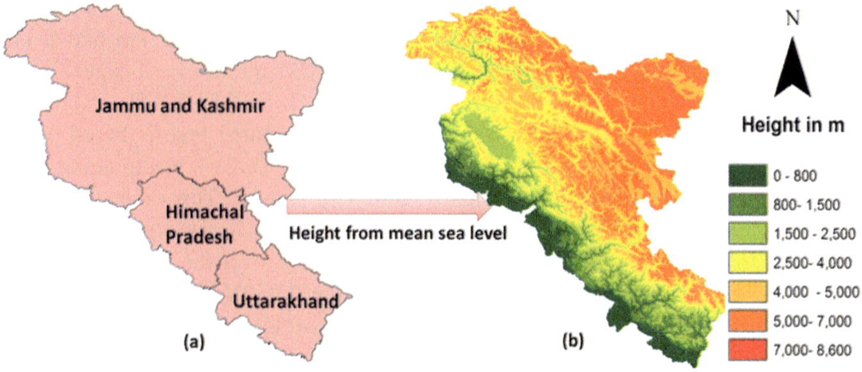

Fig. 8.1 (**a**) Study region comprises three states: Jammu and Kashmir (JK), Himachal Pradesh (HP) and Uttarakhand (UK). (**b**) Elevation map (legend) of the study area resampled to 5 × 5 km resolution as obtained from Shuttle Radar Topography Mission (SRTM)

8.2 Study Region and Methods

8.2.1 Study Region

The study area encompasses the northwestern part of the Himalaya mountain ranges extending roughly from 28° N to 37° N and 72° E to 82° E. The political boundary delineates three states – Uttarakhand (UK), Himachal Pradesh (HP) and Jammu and Kashmir (J&K) – as shown in Fig. 8.1 along with the topographical variability of the region. The NWH is home to some of the world's highest peaks with altitude ranging from ~170 to ~7861 m. The mountain ranges of the Himalayas can be geologically divided into three major fold axes: the Greater Himalaya, the Outer Himalaya and the Lesser Himalaya (Pant and Kumar 1997).

8.2.2 Methodology

High-resolution climatological rainfall pattern during monsoon season for the NWH region has been prepared using TRMM 2B31 data set. It is available as per the satellite passes. Orbital passes of 2B31 during the past 16 years (1998–2013) over the NWH region have been processed for the monsoon season and are further interpolated using nearest neighbour interpolation technique onto an equally spaced grid of 5 × 5 km. Extreme rainfall events (EREs), the diurnal cycle of rainfall and their relationship with elevation have been studied using TRMM 3B42V7 data set. The detailed approach devised using 3B42V7 is illustrated in the following flowchart (Fig. 8.2).

EREs are identified based on the percentile method. The linear correlation analysis has been carried out to determine the relationship between EREs and

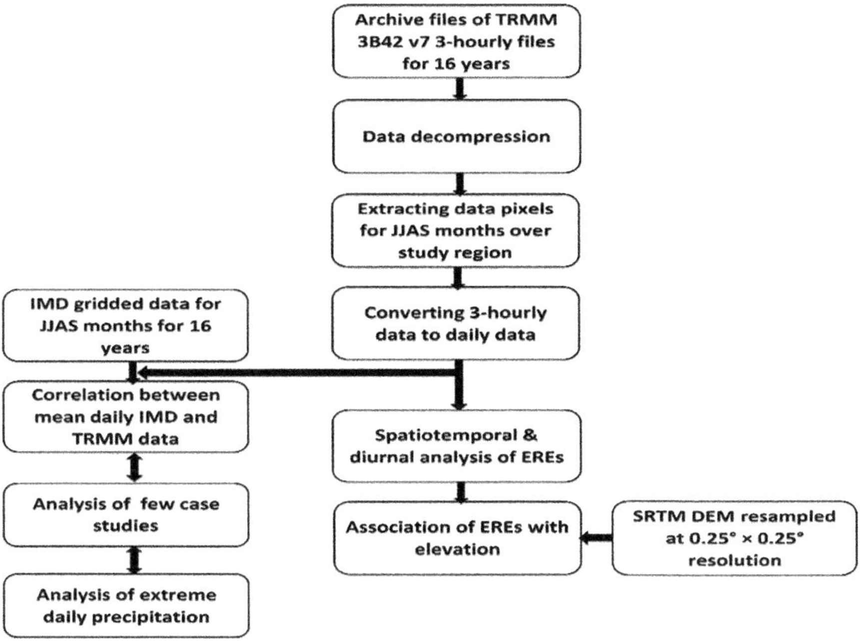

Fig. 8.2 Flowchart of the methodology for ERE and diurnal analysis

elevation using Karl Pearson's coefficient of linear correlation and scatter plot. The Pearson's coefficient of linear correlation denoted as r is calculated as:

$$r = \frac{\sum xy}{N\,\sigma_x\ \sigma_y}$$

where

$x = X - \bar{X}$; $y = Y - \overline{Y}.$
σ_x= standard deviation of series X
σ_y= standard deviation of series Y
N = number of pairs of observations

The statistical significance of the linear trend is estimated using the student's t-test. The null hypothesis states the absence of trend (i.e. slope of the linear regression is zero) with the alternate hypothesis stating that the slope is not zero.

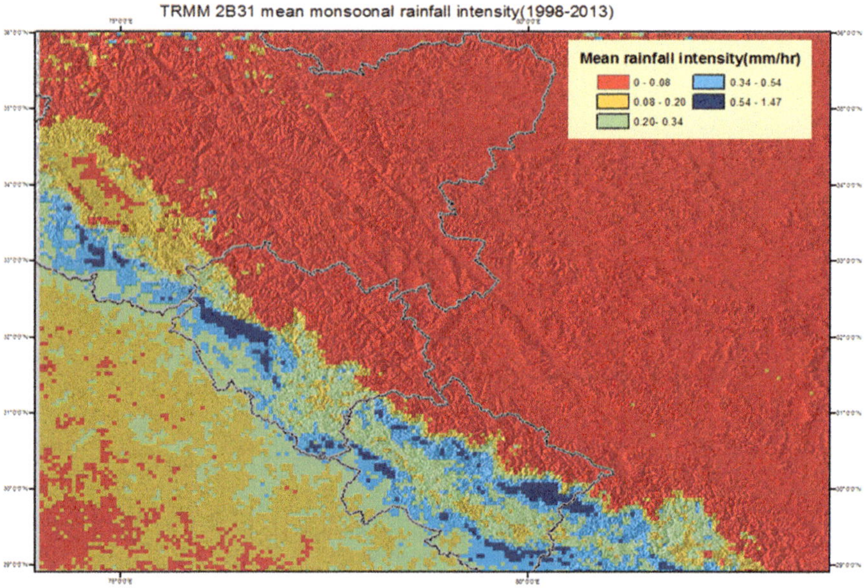

Fig. 8.3 Spatial distribution of the climatological rainfall over the study region. All the orbital passes corresponding to TRMM 2B31 data have been integrated for the monsoon season from 1998 to 2013 to generate the high-resolution (5 × 5 km) rainfall map for the seasonal mean. The legend shows the rainfall intensity in mm/h

8.3 Rainfall Patterns

A high-resolution map that can resolve the fine-scale structure of rainfall over the Northwest Himalayan region is developed using TRMM 2B31 surface rain data (at 5 × 5 km) for 1998–2013 and is shown in Fig. 8.3. The highest amount of rainfall during the monsoon season is observed over UK, followed by HP and J&K. The high rainfall over UK is characterized by a distinct dual-band structure, which is not seen over HP or J&K. This dual-band structure has been reported by previous studies over the Himalayan region (Anders et al. 2006; Bookhagen and Burbank 2006; Shrestha et al. 2012). It is worth noting that such fine-scale structure over UK cannot be resolved by the coarser resolution rainfall data such as TRMM 3B42 (0.25 degrees) or IMD gridded data. Based on IMD gridded rainfall data set, the mean season rainfall is approximately 1500 mm, 1200 mm and 800 mm for UK, HP and J&K, respectively. However, the rainfall variability is highly heterogeneous within the state of J&K as Jammu district receives up to 1200 mm average annual rainfall as compared to Ladakh district which receives approximately 45 mm annual rainfall. Overall, the study region receives almost 80% of its annual total rainfall during the South-West monsoon season.

8.4 Extreme Rainfall Events (EREs)

The NWH region is highly susceptible to EREs and cloudbursts due to its geography and steep topography. The unpredictable weather fluctuations are typical in different sectors of mountains leading to sudden outbursts of heavy rainfall (Nandargi and Dhar 2011) followed by various hazards like flash floods, landslides, debris flows and glacial lake outbursts (e.g. Leh in 2010; Manali in 2011; Rudraprayag in 2012; Kedarnath 2013). Leh (2010) and Kedarnath (2013) disasters were one of the most calamitous natural disasters in the history of India.

An ERE is a rare event, and the lack of a standard definition has produced many definitions. The Intergovernmental Panel on Climate Change (IPCC) defines an extreme weather event as 'an event which is rare within its statistical distribution at a particular place, usually as rare or as rarer than the 90th percentile' (Tank et al. 2009). Researchers use different thresholds for the identification of EREs from the statistical distribution of rainfall. Francis and Gadgil (2006) defined 150–-200 mm day^{-1} as threshold for their study on Indian west coast, whereas Goswami et al. (2006) and Rajeevan et al. (2008a) classified 100 mm day^{-1} as a heavy and 150 mm day^{-1} as a very heavy rainfall event, respectively, for their studies over Central India and all India. Goswami and Ramesh (2007) identified 250 mm day^{-1} as the threshold for the study of EREs over India, though excluding the NWH from their study. Guhathakurta et al. (2011) defined 125–245 mm day^{-1} as a heavy rainfall and > 245 mm day^{-1} as a very heavy rainfall event over India. Nandargi and Dhar (2012) characterized rainfall >200 mm day^{-1} as a heavy rainfall event over the NWH. The quantiles corresponding to the tail of the rainfall probability have also been used to quantify EREs (e.g. May 2004; Krishnamurthy et al. 2009; Bookhagen 2010; Goswami et al. 2010; Malik et al. 2011). The World Meteorological Organization (WMO) advises a standard approach on the analysis of weather extremes and defines extreme precipitation indices such as RX1day (maximum 1-day precipitation), RX5day (maximum 5-day precipitation), R95pTOT (precipitation due to very wet days >95th percentile) and R99pTOT (precipitation due to very wet days >99th percentile) for systematic analysis (Tank et al. 2009). In most of the aforementioned studies, either the NWH region was completely excluded or the studies featured results based on limited datasets derived from in-homogeneously distributed rain gauges. This section discusses the spatiotemporal patterns of EREs in detail based on TRMM 3B42V7 dataset over the NWH. For the analysis presented here, the three percentiles, namely, 98th, 99th and 99.99th, are defined as thresholds for the identification of EREs.

8.4.1 Spatiotemporal Trend of EREs

The frequency distribution of seasonal daily rainfall for the NWH region (shown in Fig. 8.4), as expected, portrays a (right) skewed distribution, i.e. the frequency of

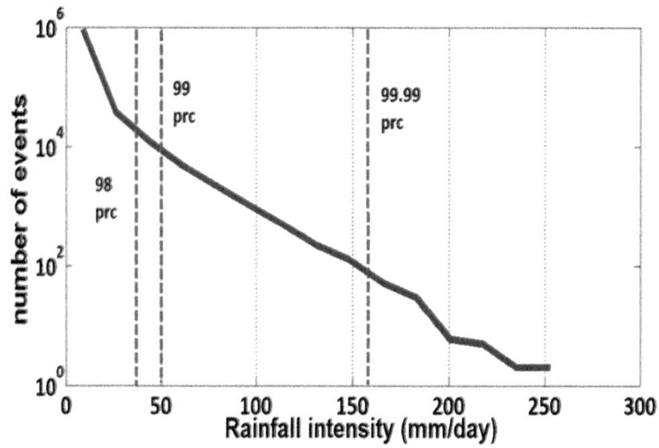

Fig. 8.4 Histogram of TRMM-derived daily rainfall for monsoon season of 1998–2013 for NWH region. The number of events is shown in natural log scale. The rainfall amounts corresponding to 98th, 99th and 99.99th percentiles which are 36.6, 50.7 and 157.1 mm day^{-1} are shown in dashed lines. (Figure and caption are from Bharti et al. 2016)

Table 8.1 Rainfall intensities (mm/day) associated with EREs for NWH and the states of Uttarakhand, HP and J&K

Region	98th percentile (mm/day)	99th percentile (mm/day)	99.99th percentile (mm/day)
NWH	36.5	50.7	157.1
Uttarakhand	59	74.7	179.9
HP	46.2	61.3	160.8
J&K	17.5	26.5	123.4

From Bharti et al. (2016)

rainfall events decreases gradually with increasing rainfall intensity. The magnitude of daily rainfall for 98th, 99th and 99.99th thresholds has been evaluated as 36.6, 50.7 and 157.1 mm day^{-1}, respectively, for the NWH. Table 8.1 highlights the large fluctuations in rainfall intensities associated with different percentiles for the states separately. The maximum mean seasonal rainfall along with the maximum rainfall intensities associated with all three percentiles is reported for Uttarakhand.

Spatially averaged maps (Fig. 8.5) depict the distribution of frequency of EREs spatially that is associated with three categories: (1) 98th $\leq$ daily rainfall <99th percentile, (2) 99th $\leq$ daily rainfall <99.99th percentile and (3) $\geq$99.99th percentile. Typically, the frequency of EREs decreases with the increasing elevation and becomes very low over high-altitude regions (>3000 m). The spatial distribution of EREs is higher over the mountains of Uttarakhand as compared to the other two states. This may correspond to the comparatively higher strength of the monsoon winds in Uttarakhand which steadily decreases towards the northernmost state of J&K. As expected, the frequency distribution of EREs has an inverse relation with the higher percentiles. The easternmost district of Champawat (Uttarakhand) has

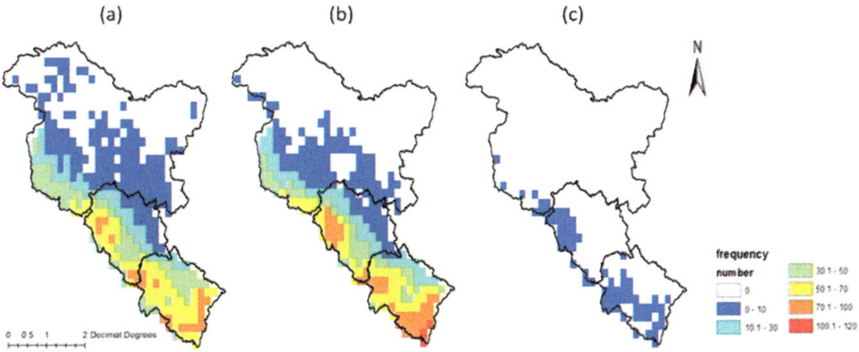

Fig. 8.5 Spatial distribution of frequency of EREs over NWH for (**a**) frequency of EREs with intensity > = 98th and <99th percentile, (**b**) same as (**a**) but for > = 99th and <99.99th percentile and (**c**) same as (**a**) but for > = 99.99th percentile. (Figure and caption are from Bharti et al. 2016)

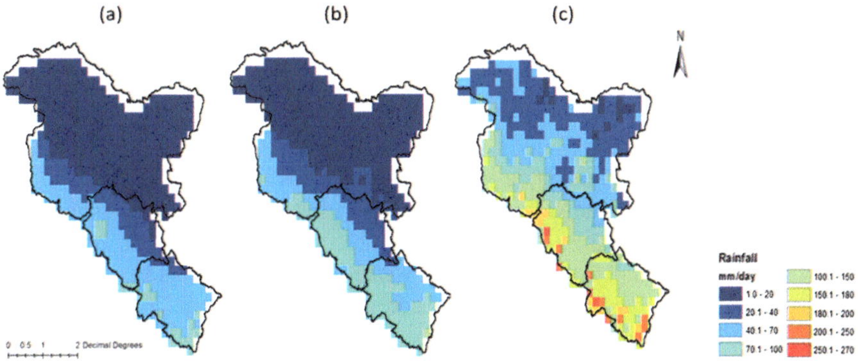

Fig. 8.6 Rainfall intensities for each pixel/grid of NWH for (**a**) 98th percentile, (**b**) 99th percentile and (**c**) 99.99th percentiles during the monsoon season of the period 1998–2013. (Figure and caption are from Bharti et al. 2016)

notably witnessed the maximum frequency of EREs (total 5) exceeding 99.99th percentile during the time period 1998–2013. In HP and J&K, EREs associated with 99.99th percentile have only been observed over low-altitude regions with <2000 m elevation, unlike Uttarakhand. The leeward side of J&K receives very little rainfall; consequently, the frequency of EREs associated with 99.99th percentile is also insignificant. The large spatial distribution of frequency of EREs is mostly observed over the foothills of the Himalaya.

Figure 8.6 depicts the spatial pattern of the rainfall corresponding to 98th, 99th and 99.99th percentiles for the NWH region. As evident in above discussion, the low-altitude regions feature stronger EREs. The elevation again displays a strong association with the spatial distribution of the magnitude of EREs, especially for 98th and 99th percentiles. However, some discrepancies are realized. While some regions having low mean seasonal rainfall experience EREs, other regions receiving steady seasonal rainfall rate almost never experience an ERE. For instance, cold arid regions of the Himalayas like Kargil (J&K) and Ladakh (J&K), featuring mean seasonal rainfall as low as 1.8 mm day^{-1} and 1.3 mm day^{-1}, have witnessed rainfall extremes of intensities 120 mm day^{-1} and 101 mm day^{-1}, respectively. Further, no noteworthy relation could be recognized at this point between EREs exceeding 99.99th percentile and mean seasonal rainfall of the corresponding year, e.g. the highest frequency of EREs exceeding 99.99th percentile was registered in the year 2006 (out of all 16 years) when the average seasonal rainfall was below mean. Besides, in the year 2010, very few EREs exceeding 99.99th percentile were registered when the highest mean seasonal precipitation was recorded.

The presence or absence of any monotonic trend in interannual frequency variations of EREs cannot be established at this stage due to the limited sample size of the satellite precipitation product. While the *t*-test results for all three cases combined with linear regression analysis indicate an increasing trend of EREs at 1% significance level (p-value ~ 0), the non-parametric Mann-Kendall test yields no momentous trend for any of the categories of EREs even at 5% significance level.

8.4.2 Relationship of EREs with Altitude

The formation and movement of localized deep convective systems due to the intricate interaction of moist flow with the windward ridges of mountains are anticipated as the key reason for heavy orographic rainfall (Chow et al. 2013). The heavy monsoon rainfall reception in the western Himalayan region is primarily attributed to the deep and wide convective cores (Romatschke et al. 2010). Elevation or mountain height has been found to be the most influential factor in rainfall distribution out of various topographic parameters like slope, aspect, relief, etc. (Duckstein et al. 1973; Whiteman 2000; Anders et al. 2006; Barry 2008; Houze 2012). Dhar and Rakhecha (1981) proposed a relation of a fourth-degree polynomial between elevation and mean monsoon rainfall; however, the rain gauge distribution was considerably heterogeneous. Based on the simulations study, Alpert (1986) proposed a third rainfall maximum apart from two observed maxima in the Himalayas – primary in the foothills, secondary located between 2 and 2.4 km and a tertiary predicted at ~ 4 km in the Great Himalayas.

An inverse linear correlation between elevation with frequency (Fig. 8.7) and intensity (Fig. 8.8) of EREs is extracted over the NWH. A strong negative correlation of elevation with the frequency of EREs exceeding 98th and 99th percentile thresholds and a weak negative correlation with EREs exceeding 99.99th percentile can be observed. As observed in Fig. 8.7a and b, an inverse relationship between the frequency of EREs and elevation further indicates the highest frequency of EREs at foothills, i.e. <500 m altitude. EREs exceeding 99.99th percentile are nearly absent over high-altitude regions with >3000 m elevation. The higher concentration of EREs is at <3000 m altitude, which is consistent with the spatial distribution of EREs (Fig. 8.5). However, a strong negative correlation between rainfall intensities

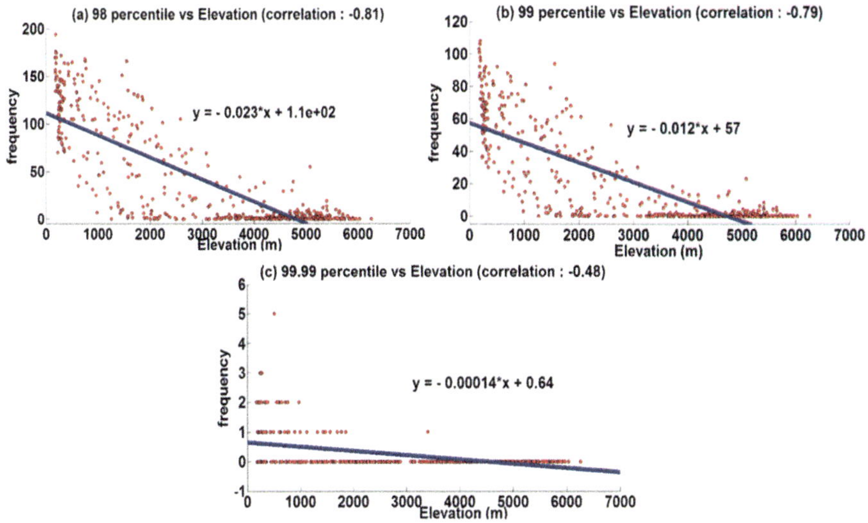

Fig. 8.7 Correlation analysis between elevation and frequency of EREs for (**a**) ≥98th, (**b**) ≥99th and (**c**) ≥99.99th percentiles over NWH region for the past 16 years from 1998 to 2013. (Figure and caption are from Bharti et al. 2016)

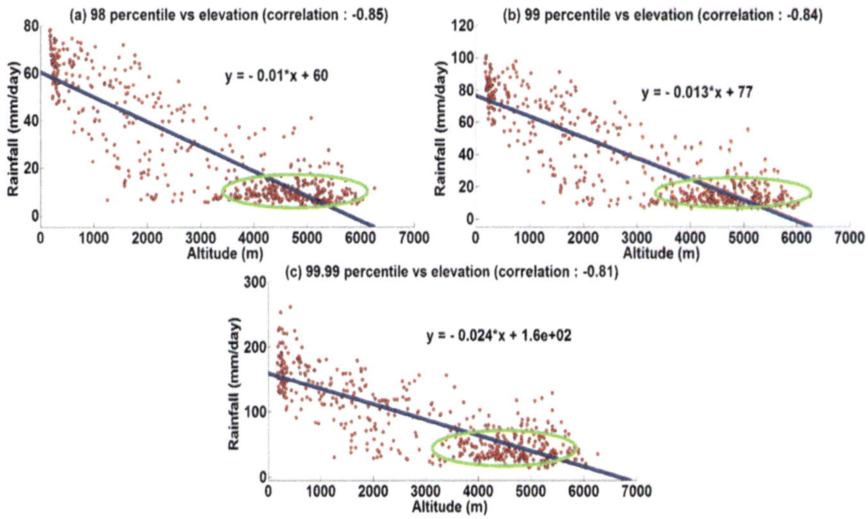

Fig. 8.8 Correlation analysis between elevation and rainfall intensities associated with (**a**) 98th, (**b**) 99th and (**c**) 99.99th percentiles for each pixel of NWH region for the period 1998–2013. The ellipse marks the data that cannot be defined with the linear fit. (Figure and caption are from Bharti et al. 2016)

associated with 98th, 99th and 99.99th percentiles and elevation suggests a distinct inverse relationship between both. A strong negative correlation is noted between rainfall intensity ranging from 20 to 80 mm day^{-1} and up to 3000 m altitudes for all three cases. But the same is not true with low rainfall intensity corresponding to <20 mm day^{-1} and elevation greater than 3000 m which do not correlate well (marked in Fig. 8.8).

Further, an analysis of frequency distribution of EREs based on different elevation ranges of the NWH was performed. The plains and foothills (up to 500 m) witness maximum frequency which decreases, though non-gradually, as one advances towards the high-altitude ranges. The distributions of the frequency of EREs exceeding 98th and 99th percentile are analogous for mountain ranges up to 2000 m altitude. However, the frequency of EREs exceeding 99.99th percentile threshold declines steadily with elevation up to 2000 m with almost no occurrence beyond >2000 m altitude, as only one such event is noted over 3000–4000 m altitude ranges.

8.4.3 Diurnal Variations of Rainfall and Its Extremes

Diurnal variation of precipitation plays a significant role in modulating the mountain climate systems. The diurnal cycles of mountain winds and solar insolation play a major role in the set-up of necessary atmospheric conditions for precipitation in mountains. Solar radiative heating produces thermally driven mountain winds which are considered to be responsible for triggering convective systems over southern slopes of the Himalayas (Zardi and Whiteman 2012 and references therein). Many researchers have reported late night/early morning rainfall peaks over the Himalayan foothills, while some suggested afternoon peaks (Barros et al. 2000, 2004; Basu 2007; Bhatt and Nakamura 2005, 2006; Houze 2012; Lang and Barros 2002; Varikoden et al. 2012). Local circulations due to surface heating, triggering of gravity waves, mountain-valley breeze circulations and propagation of mesoscale convective systems are some of the processes that have been related to daily cycle of precipitation over land (Biasutti et al. 2011; Dai 2001; Jr. Houze 2012; Liu et al. 2009; Nesbitt and Zipser 2003; Yin et al. 2011). The intensity and frequency of precipitation vary considerably within a day over the Himalaya region due to heterogeneous terrain characteristics and dynamic atmospheric conditions (Mao and Wu 2012; Varikoden et al. 2012). Apart from synoptic scale circulations, mountains experience a change in wind directions twice in a day. Therefore, as the diurnal circulation becomes an intrinsic factor in the local climate system, it is important to comprehend the diurnal variation of intensity and frequency of precipitation over the NWH region.

The spatial distribution of the climatology of 3-hour rainfall (Fig. 8.9) suggests rainfall peaks over southwest and southernmost regions with <1000 m elevation during early morning (0530) and midnight (0230) hours. This rain band becomes weaker and shifts to higher-altitude regions (1000–3000 m) later in the day.

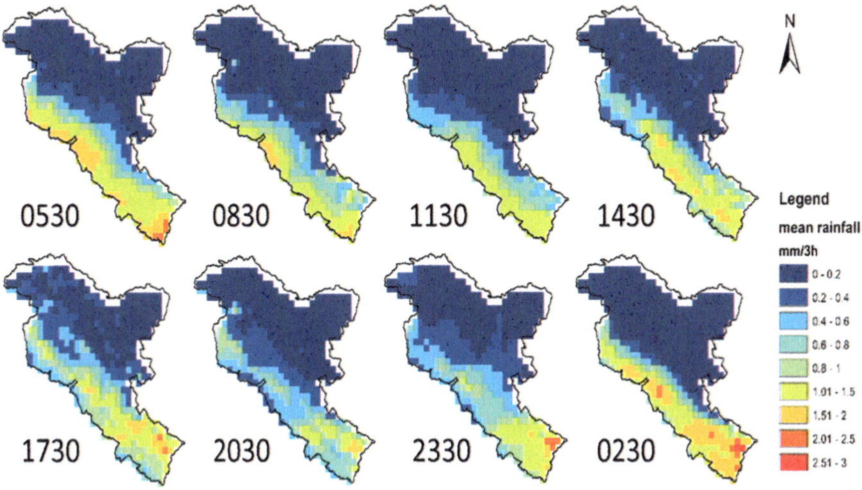

Fig. 8.9 Spatial distribution of 3-hour (Local time) precipitation variations over the NWH

However, the rainfall peaks during the afternoon (1430) and evening hours (1730) are less intense than late night – early morning peaks. Past studies attribute this secondary peak to orographic rainfall, whereas the early morning to late night peaks are thought to be due to the convergence of katabatic winds and warm-moist monsoon winds coming from the southeast during monsoon (Barros et al. 2000; Basu 2007; Sahany et al. 2010; Shrestha and Deshar 2014; Singh and Nakamura 2010; Varikoden et al. 2012). In the study of Nepal Himalayas, Barros (2004) observed that this nocturnal peak is maximum only during monsoon season and practically absent for the rest of the year. The rainfall peaks are strictly restricted to altitude up to 3000 m in HP, but over Uttarakhand, afternoon peaks in rainfall were identified even over few regions having >3000 m altitude (Bharti et al. 2014; Bharti and Singh 2015a, b). Romatschke and Houze (2011) found the rainfall type to be mostly convective over the western Himalayas, in contrast to the eastern ranges of Himalaya (not a part of this study) where rainfall type is predominantly stratiform during monsoon season.

The Himalaya mountain range, due to its steep topography, acts as a barrier and assists in the mechanical uplifting of warm moist air within the atmosphere, thereby promoting free convection and generating cumuliform clouds. This process often occurs during the afternoon when mountain surface becomes warm enough to give rise to atmospheric instability resulting in deep convective cores (Mao and Wu 2012). However, as the night progresses and the mountain surface cools down, this convergence zone is shifted towards lower elevations and foothills. To summarize, the primary rainfall maxima are observed at foothills (<1 km) during the early morning hours (0530 LT), while the secondary maxima are observed during afternoon hours (1430–1730 LT) for elevations ranging from 1 to 4 km. The intensity and frequency of EREs typically follow the mean rainfall distribution pattern; however, a higher frequency of very heavy rainfall events (intensity ≥99.99th percentile) is reported at night rather than during the morning hours (Bharti and Singh 2015a). These results indicate that though EREs can occur at any time of the day, there is a greater probability of occurrence during late night and early morning hours. It is also speculated that regional urbanization may have significant impacts on the patterns of rainfall and its extremes as observed in a case study of Dehradun city of Uttarakhand (Bharti and Singh 2015c), thus, further altering precipitation patterns at the regional scale.

8.5 Satellite vs. In Situ Measured Rainfall

Although rain gauges are the standard method of measuring rainfall, their network is sparse and uneven in the Himalayas. There have been regular efforts to expand and maintain a homogeneous rain gauge network in the region by IMD. Another recommended approach to study rainfall is through weather radars which are extremely capable in providing precipitation estimates with high spatial and temporal resolution. Yet they are not suited over mountainous terrain due to limitations like ground clutter and beam blockage. Satellites provide an effective way of estimating rainfall with high space-time resolution, especially over the rugged topography where installation and maintenance of ground-based measurement systems are challenging. Generally, rain gauges are installed on valley floors in the Himalayas. Therefore, the arithmetic average harbouring large biases may not be representative of accurate rainfall distribution. Present remote sensing algorithms are not instrumental to provide reliable precipitation estimates at finer resolutions over the mountainous region. Available satellite data with poor spatial resolution may not resolve a mountain completely, particularly inept at distinguishing windward and leeward slopes. Thus, missing the signatures of local transient storms and leading to significantly large biases in the precipitation distribution. Errors due to the algorithm, instrument, spatial sampling or geo-referencing are often introduced in satellite-based products when converting received signals into rainfall measurements (e.g. Krajewski 2007). Generally, these errors are challenging to eliminate but can be accepted if they are insignificant in comparison to the estimates (Ebert 2007). Hence, given various uncertainties in the satellite precipitation estimates, their validation with in situ data is required before being employed in climate studies. But their validation with rain gauges is not direct and is mired by several issues. IMD rain gauges record precipitation at daily scales in general with hourly rainfall data available only for few selected stations (Srivastava 2008). The satellite estimates, however, are based on instantaneous snapshots captured from their orbits at certain intervals in a day. Thus, spatiotemporal averaged satellite estimates are preferred, and their reliability degrades at finer resolutions (Levizzani 2009). As both in situ and remotely sensed estimates have their respective drawbacks, rain gauges are still considered as the most reliable way for precipitation measurement, hence, is used as the ground truth for the validation of satellite data in the present study.

TRMM 3B42V7 is validated by monthly accumulated rain gauge measurements in the algorithm itself. As discussed above, merging of satellite-gauge estimates on any other sub monthly scale is limited by sparsely distributed gauge network and biases in observational data itself (Huffman 2013). 3B42V7 is a merged microwave-IR product and has shortcomings in both IR and passive microwave techniques. The IR-based methods estimate the cloud top properties as the proxy for rainfall. Hence, it has a high probability of overestimation by misidentifying cold non-precipitating high clouds as rainy clouds (Kidder and Vonder Haar 1995; Lensky and Levizzani 2008; Petty and Krajewski 1996). Contrarily, although a passive microwave sensor senses the raindrops more directly, yet it often underestimates total rainfall since it can only detect convective rainfall from cumulonimbus clouds while overlooking low stratiform rainfall (Levizzani 2009). Additionally, passive microwave sensors cannot recognize warm rain and localized storm events (Zulkafli et al. 2014 and references therein). Topography-based corrections are integrated into TRMM PR estimates which also act as important input for the TRMM 3B42V7 precipitation data (TRMM Precipitation Radar Team 2011).

Previous studies have proposed numerous validation and bias adjustment techniques (Dinku et al. 2008; Gebregiorgis and Hossain 2015; Müller and Thompson 2013), yet most of them do not focus on reducing the biases in the tails of the distribution. Huang et al. (2013) reported the inability of 3B42V7 to accurately capture the spatiotemporal pattern of rainfall as well as extreme precipitation events due to its tendency to underestimate rainfall peaks, in their detailed study of Beijing extreme rainfall event. Further, Jamandre and Narisma (2013) noted that the accuracy of TRMM 3B42V6 estimates is driven by the geographical location and the amount of rainfall. However, they concluded TRMM as a promising source for the study of extreme rainfall events through their study over the Philippines. Based on a comprehensive comparative study, Chen et al. (Chen et al. 2013) concluded that 3B42V7 performs better than its predecessor (3B42V6) over China region. They also found 3B42V7 underestimating low-intensity rainfall (<24 mm) and overestimating high-intensity rainfall (>100 mm). Xia et al. (2015) incorporated topography-based precipitation correction in monthly and annual 3B43V7 estimates for mountainous regions of China.

The importance of quantifying uncertainties and a deep understanding of error properties for different elevation regions has been well acknowledged concerning future improvements of satellite-based precipitation retrieval algorithms (Gebregiorgis and Hossain 2015; Sorooshian et al. 2011). The spatial bias between TRMM 3B42V7 and rain gauge- based estimates over the region is illustrated in Fig. 8.10 for HP and UK. The satellite usually overestimates rainfall on elevation ranges <3000 m but underestimates rainfall for altitudes >3000 m. However, the range of overestimation is comparably smaller than that of underestimation.

A small positive correlation of ~0.23 is found between 3B42V7 estimates and rain gauge-based IMD data for individual rainfall events, although, as would be expected, the correlation becomes better (~0.67) for area-averaged precipitation estimates. Further, 3B42V7 underestimates the frequency of actual rainfall events with ~75% hit rate in the NWH. It is poor at identifying correct rain and no-rain events with an accuracy of merely ~66%. Figure 8.11 shows the actual mean rainfall

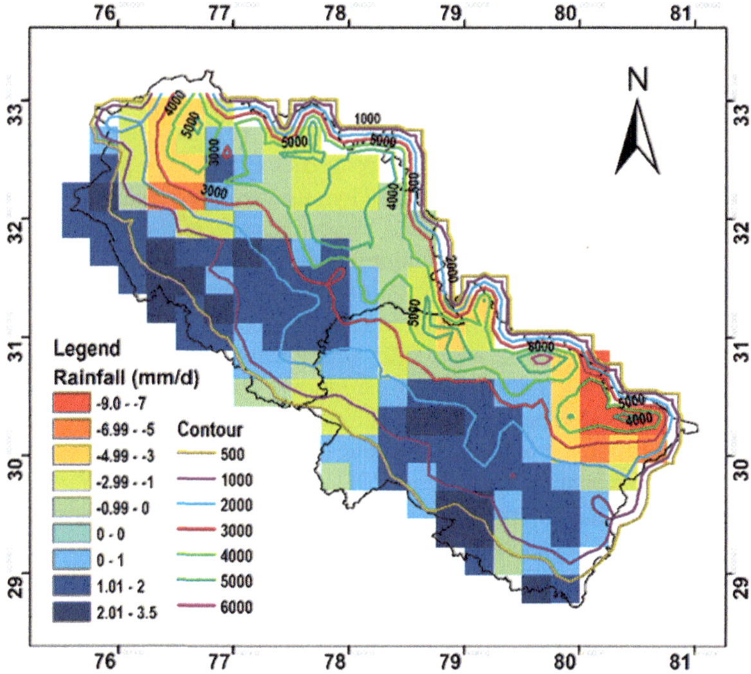

Fig. 8.10 Spatial bias of seasonal mean rainfall over Himachal Pradesh and Uttarakhand. (Figure and caption are from Bharti and Singh 2015d)

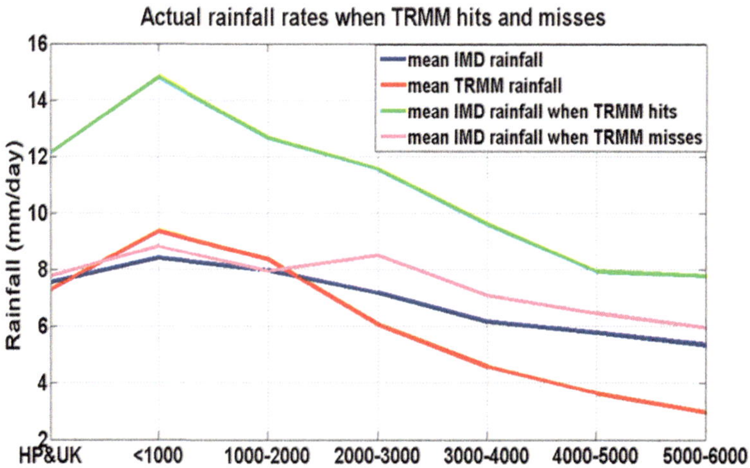

Fig. 8.11 Average IMD rainfall rates when TRMM 3B42V7 hits and misses. (Figure and caption are from Bharti and Singh 2015d)

rate as captured by IMD rain gauges when 3B42V7 hits (both satellite and gauge capture rain) and misses (the only gauge captures the rainfall). Overall, 3B47V7 performs reasonably better for 1000–2000 m altitude-range but is poor for high-elevation regions (>3000 m) in the NWH. Further analysis demonstrates unacceptably large biases in 3B42V7 estimates of EREs corresponding to the 99.99th percentile.

The rapid ascent of monsoonal branches originated from the Arabian Sea and the Bay of Bengal due to steep terrain causes massive adiabatic liquid water production in the Himalayas. The larger surface area and diurnal heating of the slopes of the mountains further accelerate this process. The satellite easily misses the rainfall caused by the convective clouds developed from the collision-coalescence process having a warm base (>10°C), leading to an underestimation of rainfall in the NWH. In addition to this, microwave radiometer algorithms have been found to underestimate heavy rainfall events caused by the shallow orographic systems (Shige et al. 2013).

The cirrus, cirrostratus and deep convective clouds usually dominate during the South Asian monsoon (Tang and Chen 2006). Deep convective clouds are associated with the highest optical thickness and produce heavy rainfall out of these three. However, non-precipitating cirrus clouds also occur at high altitudes up to 18 km and consist of ice crystals with cloud tops of −20° to −30 °C (Liou 2005). These cirrus clouds can sometimes be misidentified by the sensors as raining clouds causing false alarms and precipitation overestimation.

Although the type of precipitation whether it is convective or stratiform precipitation also dictates the actual rain rate, its determination using remotely sensed data sets is still a challenging task (Levizzani 2009). In the relationship of radar reflectivity and rain rate, the coefficients define different types of rain, e.g. convective, stratified or orographic (Sálek et al. 2004). The monsoon winds primarily cause convective rainfall in the mountains which are usually governed by latent-heat-driven vertical circulations. It is noted by Medina et al. (2010) that the convective systems containing intense convective echo which are originated from the Arabian Sea region during monsoon season are released over the foothills of the Himalayas due to orographic forcing. It is proposed that deep and wide convective cores bring intense monsoonal rainfall in the western Himalayas (Mao and Wu 2012; Romatschke et al. 2010; Romatschke and Houze 2011;); nonetheless, all precipitation events are not of convective nature during the rainy season. The undulating terrain of Himalayas is more of a sinusoidal type which also acts as an elevated heat source. This heat source is believed to be responsible of producing various types of convective systems ranging from small cumulus clouds to large mesoscale convective systems. Improvement in existing passive microwave radiometer algorithms by incorporating orographic/non-orographic rainfall classification scheme (Shige et al. 2013) and inclusion of cloud microphysics in VIS/IR rainfall estimation techniques will supplement better rainfall measurements over the NWH region.

8.6 Conclusions

Being a region of intricate topography, the Northwest Himalayas exhibit a complex spatial pattern of rainfall dominated by its geographical heterogeneity during Indian summer monsoon. Since it has a poor network of ground-based rainfall measurement systems, the remote sensing techniques have proven to be highly useful in the precipitation study of inaccessible regions. A comprehensive synopsis of precipitation characteristics over the region using NASA's TRMM 2B31 and TRMM 3B42 provides a detailed insight of the rainfall structure over the region. A very high-resolution rainfall dataset TRMM 2B31 unfolds a dual-band rainfall structure over Uttarakhand in contrast to the other two states, viz., Himachal Pradesh and Jammu and Kashmir. The in-depth analysis of rainfall with elevation and slope for selected swaths reveals a strong interdependence. It is further speculated that these erratic features of topography deprive high-altitude regions of rainfall and are largely responsible for cold dry conditions. The absence of this peculiar dual rainfall band in Himachal Pradesh and Jammu and Kashmir is attributed to non-existent two-step topography and the presence of steep slopes closer to the mountain fronts. The dual-band structure of rainfall is not evident in coarser resolution rainfall products such as TRMM 3B42 and ground-based IMD gridded datasets. Numerical weather prediction (NWP) models hardly resolve the orographic rainfall features; therefore, it is suggested to run the NWP model at high resolution (e.g. 5 km or less) for the development of better understanding of orographic rainfall processes in the rugged mountainous terrain. Apart from this, elevation of mountains is suggested to be one of the most influential factors in affecting the intensity of rainfall. Coarse resolution rainfall datasets hardly resolve the mountain ridges correctly, and therefore it is advisable to carry out sensitivity experiments of varying resolution which can properly resolve the mountain for different elevation ranges. Such sensitivity experiments with NWP models would also be useful in the improved understanding of physics of orographic rain processes and prediction skills for extreme rainfall events over the complicated topographic region.

Further, 3100 m elevation zone acts as the optimal breakpoint (5% significance level) in the linear correlation between elevation and rainfall pattern. The satellite has been found to usually under- and overestimate precipitation above and below this altitude range. The analysis of rainfall extremes using TRMM 3B42 also highlights the proclivity of the Northwest Himalayan region to extreme rainfall events. The extreme rainfall events are identified using three thresholds equivalent to 98th, 99th and 99.99th percentiles of the monsoon rainfall distribution over the NWH region. It is also suggested that the thresholds corresponding to 98th and 99th percentile may be taken as extreme and very extreme rainfall events, respectively, whereas rainfall amount corresponding to 99.99th percentile may be considered as the cloudburst event. The interannual trend analysis of frequency and intensity of extremes does not yield any significant results. A considerable small sample size of rainfall data is held accountable for the uncertainties in the trend results. The spatiotemporal analysis further reveals a higher frequency of rainfall extremes commonly over the plains and foothills. However, the maxima are observed at different altitude zones at state-level analysis.

TRMM 3B42V7 is a gauge-corrected product at monthly scale, yet the large magnitude of biases implies a need for further improvements in the satellite algorithm. The analysis also deemed TRMM 3B42V7 unfit for the study of very heavy rainfall events corresponding to 99.99th percentile. Nevertheless, it can be used for moderate to heavy rainfall studies after bias correction over the Northwest Himalayan region. The advent of Global Precipitation Mission promises improved satellite precipitation estimates ameliorating our understanding of precipitation characteristics over the Himalayan region in the near future. It is further suggested to develop region-specific satellite-derived rainfall retrieval algorithms for the study of the region-specific rainfall variations, as most of the available rainfall algorithms are developed for the global region or for a larger domain. Therefore, they do not represent the region-specific rainfall signatures accurately. Improved rainfall products developed using region-specific coefficients may be used to initialize NWP models for better forecasting of the extreme rainfall events over the NWH.

Acknowledgements Work presented in this chapter is a part of the EOAM project. Authors thankfully acknowledge the collaborator of this project Dr. V. Venugopal (CAOS, IISc Bangalore). Thanks are due to Head MASD, Dean (Academics) and Director IIRS for the support and encouragement. Ms. Sudeshna Purakait and Mr. Abhisek Das are thankfully acknowledged for generating Fig. 8.1 and Fig. 8.3, respectively, of this chapter. The TRMM 3B42V7 and TRMM 2B31 data used in this effort were acquired as part of the activities of NASA's Science Mission Directorate and are archived and distributed by the Goddard Earth Sciences (GES) Data and Information Services Centre (DISC). SRTM DEM is a product of CGIAR-CSI (http://srtm.csi.cgiar.org/). We thank IMD for making the rainfall data set available for the research community.

References

Alpert P. 1986. Mesoscale indexing of the orographic precipitation over high mountains. Journal of Climate and Applied Meteorology 25: 532–545.

Anders AM, Roe GH, Hallet B, Montgomery DR, Finnegan NJ, Putkonen J. 2006. Spatial patterns of precipitation and topography in the Himalaya. Geological Society of America Special Pa: 39–53. DOI: https://doi.org/10.1130/2006.2398(03).

Barros AP. 2004. On the Space-Time Patterns of Precipitation in the Himalayan range: a Synthesis. GAME CD ROM Publ. No. 11, T8APB19Oct04100318.

Barros AP, Kim G, Williams E, Nesbitt SW. 2004. Probing orographic controls in the Himalayas during the monsoon using satellite imagery. Natural Hazards and Earth System Science 4(1): 29–51. DOI: https://doi.org/10.5194/nhess-4-29-2004.

Barros AP, Joshi M, Putkonen J, Burbank DW. 2000. A study of the 1999 monsoon rainfall in a mountainous region in central Nepal using TRMM products and rain gauge observations. Geophysical Research Letters 27(22): 3683–3686. DOI: https://doi.org/10.1029/2000GL011827.

Barry RG. 2008. Mountain Weather and Climate, 3rd edn. Cambridge University Press: New York, NY. ISBN: 13-978-0-511-41367-4

Basistha A, Arya DS, Goel NK. 2007. Spatial Distribution of Rainfall in Indian Himalayas – A Case Study of Uttarakhand Region. Water Resources Management 22(10): 1325–1346. DOI: https://doi.org/10.1007/s11269-007-9228-2.

Basu BK. 2007. Diurnal Variation in Precipitation over India during the Summer Monsoon Season: Observed and Model Predicted. Monthly Weather Review 135(6): 2155–2167. DOI: https://doi.org/10.1175/MWR3355.1.

Bharti V, Singh C. 2015a. Diurnal variations of seasonal rainfall over the state of Himachal Pradesh as observed by satellite data. ACRS 2015 - 36th Asian Conference on Remote Sensing: Fostering Resilient Growth in Asia, Proceedings.

Bharti V, Singh C. 2015b. Influence of Elevation on Spatial and Diurnal Patterns of Orographic Rainfall : A Case Study of Uttarakhand. Journal of Basic and Applied Engineering Research 2 (3): 202–205.

Bharti V, Singh C. 2015c. Characteristics of the spatiotemporal pattern of Extreme Rainfall event over the state of Uttarakhand , India. ICUC9 - 9th International Conference on Urban Climate jointly with 12th Symposium on the Urban Environment Characteristics.

Bharti V, Singh C. 2015d. Evaluation of error in TRMM 3B42V7 precipitation estimates over the Himalayan region. Journal of Geophysical Research: Atmospheres 120: 12458–12473. DOI: https://doi.org/10.1002/2015JD023779.

Bharti V, Singh C, Damen MCJ, Turkington TAR. 2014. Some aspects of spatial distribution of rainfall over Uttarakhand using satellite data. ISRS Proceedings 2014 :ISPRSTCVIII Mid-Symposium.

Bharti V, Singh C, Ettema J, Turkington TAR. 2016. Spatiotemporal characteristics of extreme rainfall events over the Northwest Himalaya using satellite data. International Journal of Climatology. DOI: https://doi.org/10.1002/joc.4605.

Bhatt BC, Nakamura K. 2005. Characteristics of Monsoon Rainfall around the Himalayas Revealed by TRMM Precipitation Radar. Monthly Weather Review 133: 149–165.

Bhatt BC, Nakamura K. 2006. A climatological-dynamical analysis associated with precipitation around the southern part of the Himalayas. Journal of Geophysical Research D: Atmospheres 111(2): D02115. DOI: https://doi.org/10.1029/2005JD006197.

Biasutti M, Yuter SE, Burleyson CD, Sobel AH. 2011. Very high resolution rainfall patterns measured by TRMM precipitation radar: seasonal and diurnal cycles. Climate Dynamics 39 (1–2): 239–258. DOI: https://doi.org/10.1007/s00382-011-1146-6.

Bookhagen B. 2010. Appearance of extreme monsoonal rainfall events and their impact on erosion in the Himalaya. Geomatics, Natural Hazards and Risk 1(1): 37–50. DOI: https://doi.org/10.1080/19475701003625737.

Bookhagen B, Burbank DW. 2006. Topography, relief, and TRMM-derived rainfall variations along the Himalaya. Geophysical Research Letters 33(8): L08405. DOI: https://doi.org/10.1029/2006GL026037.

Chen S, Hong Y, Gourley JJ, Huffman GJ, Tian Y, Cao Q, Yong B, Kirstetter PE, Hu J, Hardy J, Li Z, Khan SI, Xue X. 2013. Similarity and difference of the two successive V6 and V7 TRMM multisatellite precipitation analysis performance over China. Water Resources Research 49(12): 8174–8186. DOI: https://doi.org/10.1002/2012WR012795.

Chow F K, Wekker S F D and Snyder B J (eds.) (2013) Mountain Weather Research and Forecasting, Recent Progress and Current Challenges. Springer Atmospheric Sciences. ISBN 978-94-007-4098-3. DOI: https://doi.org/10.1007/978-94-007-4098-3.

Dai A. 2001. Global Precipitation and Thunderstorm Frequencies. Part I: Seasonal and Interannual Variations. Journal of Climate 14(6): 1092–1111. DOI: https://doi.org/10.1175/1520-0442(2001)014<1092:GPATFP>2.0.CO;2.

Das PK. 2002. Monsoons. National Book Trust. ISBN: 9788123711232.

Dhar ON, Rakhecha PR. 1981. The effect of elevation on monsoon rainfall distribution in the Central Himalayas. Proc. Int. Symp. on Monsoon Dynamics. Cambridge University Press, pp. 253–260.

Dinku T, Chidzambwa S, Ceccato P, Connor SJ, Ropelewski CF. 2008. Validation of high-resolution satellite rainfall products over complex terrain. International Journal of Remote Sensing 29(14): 4097–4110. DOI: https://doi.org/10.1080/01431160701772526.

Duckstein L, Fogel MM, Thames JL. 1973. Elevation effects on rainfall: A stochastic model. Journal of Hydrology 18(1): 21–35. DOI: https://doi.org/10.1016/0022-1694(73)90023-1.

Ebert E. 2007. Methods for verifying satellite precipitation estimates. In: Measuring Precipitation From Space—EURAINSAT and the Future, edited by V. Levizzani, P. Bauer, and F. J. Turk, pp. 345–356, Springer, Dordrecht, Netherlands

Francis PA, Gadgil S. 2006. Intense rainfall events over the west coast of India. Meteorology and Atmospheric Physics 94(1–4): 27–42. DOI: https://doi.org/10.1007/s00703-005-0167-2.

Gebregiorgis AS, Hossain F. 2015. How well can we estimate error variance of satellite precipitation data around the world? Atmospheric Research. Elsevier B.V. 154: 39–59. DOI: https://doi.org/10.1016/j.atmosres.2014.11.005.

Goswami BB, Mukhopadhyay P, Mahanta R, Goswami BN. 2010. Multiscale interaction with topography and extreme rainfall events in the northeast Indian region. Journal of Geophysical Research 115(D12): D12114. DOI: https://doi.org/10.1029/2009JD012275.

Goswami BN, V. Venugopal, Sengupta D, Madhusoodanan MS, Xavier PK. 2006. Increasing Trend of Extreme Rain Events Over India in a Warming Environment. Science 314(December): 1442–1445.

Goswami P, Ramesh K V. 2007. Extreme Rainfall Events : Vulnerability Analysis for Disaster Management and Observation System Design. Current Science 94(8):1037–1044.

Guhathakurta P, Sreejith OP, Menon PA. 2011. Impact of climate change on extreme rainfall events and flood risk in India. Journal of Earth System Science. **120**(3): 359–373.

Huang Y, Chen S, Cao Q, Hong Y, Wu B, Huang M, Qiao L, Zhang Z, Li Z, Li W, Yang X. 2013. Evaluation of Version-7 TRMM Multi-Satellite Precipitation Analysis Product during the Beijing Extreme Heavy Rainfall Event of 21 July 2012. Water 6(1): 32–44. DOI: https://doi.org/10.3390/w6010032.

Huffman G. 2013. 3B42 Version 7 Web Description. [Available at http://trmm.gsfc.nasa.gov/3b42.html.]

Huffman GJ, Bolvin DT, Nelkin EJ,Adler RF. 2010. Highlights of Version 7 TRMM Multi-satellite Precipitation Analysis (TMPA). In: Proceedings of the 5th International Precipitation Working Group Workshop, Hamburg, Germany, 11–15 October 2010.

Jamandre C A, Narisma GT. 2013. Spatio-temporal validation of satellite-based rainfall estimates in the Philippines. Atmospheric Research. Elsevier B.V. 122: 599–608. DOI: https://doi.org/10.1016/j.atmosres.2012.06.024.

Houze RA. 2012. Orographic effects on precipitating clouds. Reviews of Geophysics 50(RG1001): 1–47. DOI: https://doi.org/10.1029/2011RG000365.1.

Kelkar RR. 2007. Satellite meteorology. B S Publications ISBN: 81-7800-137-3.

Kidder SQ, Vonder Haar TH. 1995. Satellite Meteorology An introduction, Academic Press, ISBN: 0-12-406430-2.

Krajewski W. 2007. Ground networks: Are we doing the right things?, in Measuring Precipitation From Space—EURAINSAT and the Future, edited by V. Levizzani, P. Bauer, and F. J. Turk, pp. 403–417, Springer, Dordrecht, Netherlands.

Krishnamurthy CKB, Lall U, Kwon H-H. 2009. Changing Frequency and Intensity of Rainfall Extremes over India from 1951 to 2003. Journal of Climate 22(18): 4737–4746. DOI: https://doi.org/10.1175/2009JCLI2896.1.

Kummerow C, Barnes W, Kozu T, Shiue J, Simpson J. 1998. The Tropical Rainfall Measuring Mission (TRMM) sensor package. Journal of Atmospheric and Oceanic Technology **15**(3): 809–817. DOI: https://doi.org/10.1016/0273-1177(94)90210-0.

Lang TJ, Barros AP. 2002. An Investigation of the Onsets of the 1999 and 2000 Monsoons in Central Nepal. Mon.Wea. Rev. 130: 1299–1316.

Lensky I, Levizzani V. 2008. Estimation of precipitation from space-based platforms. Precipitation: advances in measurement, estimation, and prediction 195–217. DOI: https://doi.org/10.1007/978-3-540-77655-0.

Levizzani V. 2009. Satellite Clouds and Precipitation Observations for Meteorology and Climate. In: Sorooshian S et al (ed) Hydrological Modelling and the Water Cycle. Springer Science +Business Media B.V., 49–68.

Liou KN. 2005. Cirrus clouds and climate. Yearbook Sci. Technol., McGraw-Hill. 51–53.

Liu X, Bai A, Liu C. 2009. Diurnal variations of summertime precipitation over the Tibetan Plateau in relation to orographically-induced regional circulations. Environmental Research Letters 4 (4): 045203. DOI: https://doi.org/10.1088/1748-9326/4/4/045203.

Malik N, Bookhagen B, Marwan N, Kurths J. 2011. Analysis of spatial and temporal extreme monsoonal rainfall over South Asia using complex networks. Climate Dynamics 39(3–4): 971–987. DOI: https://doi.org/10.1007/s00382-011-1156-4.

Mao JY, Wu GX. 2012. Diurnal variations of summer precipitation over the Asian monsoon region as revealed by TRMM satellite data. Science China Earth Sciences 55(4): 554–566. DOI: https://doi.org/10.1007/s11430-011-4315-x.

May W. 2004. Variability and extremes of daily rainfall during the Indian summer monsoon in the period 1901–1989. Global and Planetary Change 44: 83–105. DOI: https://doi.org/10.1016/j.gloplacha.2004.06.007.

Medina S, Houze R A., Kumar A, Niyogi D. 2010. Summer monsoon convection in the Himalayan region: Terrain and land cover effects. Quarterly Journal of the Royal Meteorological Society 136(648): 593–616. DOI: https://doi.org/10.1002/qj.601.

Müller MF, Thompson SE. 2013. Bias adjustment of satellite rainfall data through stochastic modeling: Methods development and application to Nepal. Advances in Water Resources 60: 121–134. DOI: https://doi.org/10.1016/j.advwatres.2013.08.004.

Nandargi S, Dhar ON. 2011. Extreme rainfall events over the Himalayas between 1871 and 2007. Hydrological Sciences Journal 56(6): 930–945. DOI: https://doi.org/10.1080/02626667.2011.595373.

Nandargi S, Dhar ON. 2012. Extreme Rainstorm Events over the Northwest Himalayas during 1875–2010. Journal of Hydrometeorology 13: 1383–1388. DOI: https://doi.org/10.1175/JHM-D-12-08.1.

National Space Development Agency of Japan Earth Observation Center (Feb 2001), TRMM Data Users Handbook. [Available at www.eorc. jaxa.jp/TRMM/document/text/handbook_e.pdf.]

Nesbitt SW, Zipser EJ. 2003. The Diurnal Cycle of Rainfall and Convective Intensity according to Three Years of TRMM Measurements. Journal of Climate 16: 1456–1475.

Pai DS, Sridhar L, Rajeevan M, Sreejith OP, Satbhai NS, Mukhopadyay B. 2014. Development of a new high spatial resolution (0 . 25 ° × 0 . 25 °) Long Period (1901–2010) daily gridded rainfall data set over India and its comparison with existing data sets over the region data sets of different spatial resolutions and time period. 1(January): 1–18.

Pant GB, Kumar KR. 1997. Climates of South Asia. John Wiley & sons. ISBN: 0-471-94948-5.

Petty GW, Krajewski WF. 1996. Satellite estimation of precipitation over land. Hydrological Sciences - Journal des Sciences Hydrologiques 41(4)(August): 433–452.

Prakash S, Mitra AK, Momin IM, Pai DS, Rajagopal EN, Basu S. 2015. Comparison of TMPA-3B42 Versions 6 and 7 Precipitation Products with Gauge-Based Data over India for the Southwest Monsoon Period. Journal of Hydrometeorology 16(1): 346–362. DOI: https://doi.org/10.1175/JHM-D-14-0024.1.

Qiao L, Hong Y, Chen S, Zou CB, Gourley JJ, Yong B. 2014. Performance assessment of the successive Version 6 and Version 7 TMPA products over the climate-transitional zone in the southern Great Plains, USA. Journal of Hydrology. Elsevier B.V. 513: 446–456. DOI: https://doi.org/10.1016/j.jhydrol.2014.03.040.

Rajeevan M, Bhate J, Jaswal a. K. 2008a. Analysis of variability and trends of extreme rainfall events over India using 104 years of gridded daily rainfall data. Geophysical Research Letters 35 (18): L18707. DOI: https://doi.org/10.1029/2008GL035143.

Rajeevan M, Bhate J, Kale JD, Lal B. 2005. Development of a High Resolution Daily Gridded Rainfall Data Set for the Indian Region. Met. Monograph Climatology No. 22/2005. National climate centre, India Meteorological Department.

Rajeevan M, Gadgil S, Bhate J. 2008b. Active and Break Spells of the Indian Summer Monsoon. NCC Research Report 7. National climate centre, India Meteorogical Department.

Romatschke U, Houze R A. 2011. Characteristics of Precipitating Convective Systems in the South Asian Monsoon. Journal of Hydrometeorology 12(1): 3–26. DOI: https://doi.org/10.1175/2010JHM1289.1.

Romatschke U, Medina S, Jr. Houze RA. 2010. Regional , Seasonal, and Diurnal Variations of Extreme Convection in the South Asian Region. Journal of climate 23(2004). DOI: https://doi.org/10.1175/2009JCLI3140.1.

Sahany S, Venugopal V, Nanjundiah RS. 2010. Diurnal-scale signatures of monsoon rainfall over the Indian region from TRMM satellite observations. Journal of Geophysical Research 115 (D02103). DOI: https://doi.org/10.1029/2009JD012644.

Sálek M, Cheze J, Handwerker J, Delobbe L, Uijlenhoet R. 2004. Radar techniques for identifying precipitation type and estimating quantity of precipitation. Document of COST Action 717, WG 1 Task WG 1-: 1–51.

Shige S, Kida S, Ashiwake H, Kubota T, Aonashi K. 2013. Improvement of TMI rain retrievals in mountainous areas. Journal of Applied Meteorology and Climatology 52(1): 242–254. DOI: https://doi.org/10.1175/JAMC-D-12-074.1.

Shrestha D, Deshar R. 2014. Spatial Variations in the Diurnal Pattern of Precipitation over Nepal Himalayas. Nepal Journal of Science and Technology 15(2): 57–64. DOI: https://doi.org/10.1007/s00704-008-0045-1.

Shrestha D, Singh P, Nakamura K. 2012. Spatiotemporal variation of rainfall over the central Himalayan region revealed by TRMM Precipitation Radar. Journal of Geophysical Research: Atmospheres 117(D22): n/a–n/a. DOI: https://doi.org/10.1029/2012JD018140.

Singh P, Nakamura K. 2010. Diurnal variation in summer monsoon precipitation during active and break periods over central India and southern Himalayan foothills. Journal of Geophysical Research: Atmospheres 115(12). DOI: https://doi.org/10.1029/2009JD012794.

Sorooshian S, Aghakouchak A, Arkin P, Eylander J, Foufoula-Georgiou E, Harmon R, Hendrickx JMH, Imam B, Kuligowski R, Skahill B, Skofronick-Jackson G. 2011. Advanced concepts on remote sensing of precipitation at multiple scales. Bulletin of the American Meteorological Society 92(10): 1353–1357. DOI: https://doi.org/10.1175/2011BAMS3158.1.

Srivastava G. 2008. Surface Meteorological Instruments and Measurement Practices, Atlantic Publishers Ltd., New Delhi.

Sumner G. 1988. Precipitation Process and analysis. John Wiley & sons, ISBN: 0-471-90534-8

Tang X, Chen B. 2006. Cloud types associated with the Asian summer monsoons as determined from MODIS / TERRA measurements and a comparison with surface observations. Geophysical Research Letters 33(February 2000): 2–5. DOI: https://doi.org/10.1029/2006GL026004.

Tank AMGK, Zwiers FW, Zhang X. 2009. Guidelines on Analysis of extremes in a changing climate in support of informed decisions for adaptation.

TRMM Precipitation Radar Team. 2011. Tropical Rainfall Measuring Mission (TRMM) precipitation radar algorithm, Instruction manual for version 7, Tech. Rep., JAXA/NASA, 170 pp. [Available at http://www.eorc.jaxa.jp/TRMM/documents/PR_algorithm_product_information/pr_manual/PR_Instruction_Manual_V7_L1.pdf.]

Varikoden H, Preethi B, Revadekar JV. 2012. Diurnal and spatial variation of Indian summer monsoon rainfall using tropical rainfall measuring mission rain rate. Journal of Hydrology. Elsevier B.V. 475: 248–258. DOI: https://doi.org/10.1016/j.jhydrol.2012.09.056.

Whiteman CD. 2000. Mountain Meteorology: Fundamentals and Applications, Oxford Univ. Press, Inc., New York.

Wulf H, Bookhagen B, Scherler D. 2010. Seasonal precipitation gradients and their impact on fluvial sediment flux in the Northwest Himalaya. Geomorphology 118(1–2): 13–21. DOI: https://doi.org/10.1016/j.geomorph.2009.12.003.

Xia T, Wang Z-J, Zheng H. 2015. Topography and Data Mining Based Methods for Improving Satellite Precipitation in Mountainous Areas of China. Atmosphere 6(8): 983–1005. DOI: https://doi.org/10.3390/atmos6080983.

Yin S, Li W, Chen D, Jeong JH, Guo W. 2011. Diurnal variations of summer precipitation in the Beijing area and the possible effect of topography and urbanization. Advances in Atmospheric Sciences 28(4): 725–734. DOI: https://doi.org/10.1007/s00376-010-9240-y.

Zardi D, Whiteman CD. 2012. Diurnal mountain wind systems. In: Mountain Weather Research and Forecasting, F. K. Chow, S. F. J. De Wekker, and B. Snyder Eds. Springer Atmospheric Sciences. DOI https://doi.org/10.1007/978-94-007-4098-32.

Zulkafli Z, Buytaert W, Onof C, Manz B, Tarnavsky E, Lavado W, Guyot J-L. 2014. A Comparative Performance Analysis of TRMM 3B42 (TMPA) Versions 6 and 7 for Hydrological Applications over Andean – Amazon River Basins. Journal of Hydrometeorology 15: 581–592. DOI: https://doi.org/10.1175/JHM-D-13-094.1.

Part IV
Forest Resources and Biodiversity

Summary

Forests play a paramount role in maintaining ecological and environmental security. Manifested as a diverse set of vegetation communities distributed across the elevation gradient, the forests in NWH cover the largest geographical extent. They house a great variety of floral and faunal species with high degree of endemism. NWH is recognised as one of the global biodiversity hotspots. The forests in NWH were largely cleared earlier for agriculture and human habitation. In the recent decades, the forests in NWH have been affected by chronic disturbances resulting from overexploitation of forest resources, intensive grazing, landslide and landslips etc. Forest degradation and fragmentation are causing impoverishment of species populations and incidences of human-wildlife conflicts. Reports of early flowering/fruiting in certain species, densification and upward expansion of vegetation and increasing incidences of forest fires are being attributed to altered climatic regime in the region. Robust scientific approaches blended with technological inputs are the key to innovate strategies for conservation and sustainable utilization of forest resources and biodiversity in the region. Space-based observations in conjunction with *in-situ* observations and geospatial modeling techniques are crucial for the assessment and monitoring of forest resources and biodiversity in the region.

Over the years, researchers at the Indian Institute of Remote Sensing have applied earth observation data and geoinformatics for scientific studies on different aspects of forest resources and biodiversity in NWH. This chapter covers articles related to assessment of biodiversity patterns including the changes expected in biodiversity distribution due to climate change, web-based portals developed for biodiversity data organisation and dissemination, wildlife habitat evaluation and inventory of forest carbon pools and fluxes including emissions from biomass burning. The first article describes an approach for characterizing and prioritizing biodiversity rich areas at landscape level combining satellite remote sensing, field-surveyed data, and landscape ecological principles in geospatial domain. How web-based information system caters to the need of systematic archival, visualization and dissemination of biodiversity data and associated knowledge specific to NWH is presented in the second article. It also features web-based distributed server technology which allows organisation involved in survey of bioresources and biodiversity in NWH to effectively share their data with each other and to the users. Fourth article discusses ecological niche modelling using climatic projections offers valuable insights into the probable consequences of the climate change on forest species of region. Both active and passive remote sensing technologies alongwith the in-situ measurements offer promise for reliable estimation of standing forest biomass is discussed in the next article. The source and sink status of climax sal forests and associated species has been determined using eddy covariance measurements in conjunction with satellite remote sensing. The last two articles discuss how the uncontrolled forest fires are causing higher levels of emissions from the excessive biomass burning and the utility of space data to study wildlife habitation.

Chapter 9
Forest Landscape Characterization for Biodiversity Conservation Planning and Management Gaps in Northwestern Himalaya Using Geospatial Technology

Sarnam Singh

9.1 Introduction

Himalaya is one of the four biodiversity hotspots in India with high level of endemism. The Himalayan arc, commonly referred to as 'Water Tower', spreads from east of Afghanistan (Hindu Kush Region) to west of Myanmar through Pakistan, India, Nepal, Bhutan, Tibet and China and is broadly divided into Hindu Kush, Western, Central and Eastern regions representing nine of ten world's highest and youngest mountains. The sudden rise in height and its terrain complexity make Himalaya environmentally, climatically, ecologically, physically, physiographically as well as biologically unique fostering myriads of life forms, niches and habitats with host of microclimatic conditions across 2500 km and up to 400 km in length and breadth, respectively. The Himalaya supports a great variety of forest ecosystems and other life forms, which vary ecologically, phytosociologically, architecturally, physiognomically, functionally, etc. from east to west and from low to high elevations. For millions of years, these forests have traditionally played a vital role in conserving the environment and have provided the much-needed long-term ecological security to the Indian subcontinent. The forest ecosystems are highly significant to protect wildlife, soil development, erosion and conservation to sustain the livelihood of mankind, to stabilize climate, to optimize water yield and to purify water (Singh and Singh 1992). Northwestern Himalaya (NW Himalaya) is the western part of the Himalayan arc. It is the farthest from sea and latitudinally much higher than the rest of the Himalaya spreading NW to SE directions and has very diverse phytoclimatic patterns quite dissimilar to the areas lying eastwards. Being a part of the youngest mountain ranges of the world, these are very fragile ecosystems due to

S. Singh (✉)
Forestry and Ecology Department, Indian Institute of Remote Sensing (IIRS), Indian Space Research Organisation (ISRO), Department of Space, Government of India, Dehradun, India

School of Ecology and Environment Studies, Nalanda University, Rajgir, Nalanda, Bihar, India

R. R. Navalgund et al. (eds.), *Remote Sensing of Northwest Himalayan Ecosystems*,
https://doi.org/10.1007/978-981-13-2128-3_9

physiographic, edaphic and climatic conditions. Shivalik ranges, frontal Himalaya, the youngest hill range in the world, are one of the biodiversity micro-endemic centres in India. Present communication also includes frontal Himalaya (Shivalik ranges) and Trans-Himalayan regions, ecosystems, biodiversity and their conservation issues.

The significance of biodiversity is well documented as it provides the essential ecological services of maintaining our atmosphere, creating and maintaining soils, sustaining hydrological cycles and controlling the world's climate pattern (Sharma et al. 1997). India being a party to Convention on Biological Diversity (CBD), Ministry of Environment, Forest and Climate Change (MoEF&CC) is the nodal ministry of the Government of India for all matters related to biodiversity conservation. In 2014 MoEF&CC brought out the 5th national report to the CBD updating status, trends, threats to biodiversity and implications for human well-being (Mathur et al. 2014a, b). Special habitats and threatened plants of Ladakh in NW Himalaya India have been discussed by Rawat (2008).

Article 7 of the United Nations Convention of Environment and Development and Agenda 21 requires signatory countries to identify and monitor components of biodiversity vital for conservation and sustainable use. India has long and unique history of institutionalizing inventory of national biological resources especially their geographical distribution through National Biodiversity Authority (NBA), Botanical Survey of India (BSI), Zoological Survey of India (ZSI), Forest Research Institute (FRI) and Wildlife Institute of India (WII) for in situ conservation along with the host of other university systems and other institutions and for *ex situ* conservation such as National Bureau of Plant Genetic Resources (NBPGR) for plants, National Bureau of Animal Genetic Resources (NBAGR) for animals, National Bureau of Agriculturally Important Microorganisms (NBAIM) for cultures of microorganism and National Bureau of Agriculturally Important Insects (NBAII) for insects. India became the first country in the world to bring legislation through Biological Diversity Act (BDA) in 2002 and Biological Diversity Rules (BDR) in 2004 for implementing the provisions of CBD (NBA 2004). National Biodiversity Strategy and Action Plan (NBSAP) from 2000 to 2004 provided macro-level policies and strategies for biodiversity conservation and sustainable development. The mechanism of conservation has further been strengthened, and the responsibility lies with the State Biodiversity Boards (SBB) and the State Forest Departments in the country, steered by MoEF&CC. National Biodiversity Action Plan (NABP) in 2008 has proactively formulated action plans for 2011–2020 CBD Strategic Plan for Biodiversity. India has also made significant progress in achieving 20 Aichi Biodiversity Targets (ABT) and Millennium Development Goals (MDG). MoEF&CC has also updated NABP and identified the targets for 25 ministries/departments to achieve and has emphasized the role of geoinformatics/geospatial technologies.

Traditionally and culturally societies in India understand the need and practice of biodiversity conservation through sacred groves (Ramakrishnan 1998; Jaryan et al. 2010). Rodger and Panwar (1988) emphasized the planning of protected area (PA) network. Since then India has shown immense commitment towards biodiversity conservation through a good network of protected areas (PA) ~735 (July 2017). These are in the form of 103 national parks, 543 wildlife sanctuaries, 73 conservation reserves and 45 community reserves throughout the country covering about 4.93% of forests in addition to 18 biosphere reserves (ENVIS-WII 2017). It is being further strengthened by the identification of Eco-Sensitive Zone (ESZ) around each of the protected areas by the MoEF&CC. The ESZ is a buffer area around all PAs to work as 'shock absorbers' outside buffer zones, and the MoEF&CC would regulate and prohibit activities by law. The PA network also includes species-specific areas like National Citrus Gene Sanctuary-cum-Biosphere Reserve, Baghmara Pitcher Plant Sanctuary, etc., and also the declaration of 50 tiger reserves and 16 elephant reserves is an example of intensive conservation efforts (ENVIS-WII 2017). NW Himalaya has 65 PAs (16 national parks, 57 wildlife sanctuaries and 41 conservation reserves). However, it is important to realize that each species, irrespective of its physical size and position in trophic level or biological hierarchy, requires minimum/optimum area for survival, reproduction (gene exchange), adaptation, evolution, etc. The area of habitat required for each species is very specific and varies greatly, and our knowledge is very limited and may be even scanty. To decide on the size of the conservation area based upon the requirement of each species is a herculean task. Therefore, of late, it has been realized that landscape level conservation planning is more suited for long-term protection of biodiversity rather than single-species conservation. BDA 2002 has now objectively focused on biodiversity conservation in India. This was a historic event to regulate access to biological resources of the country for equitable benefit sharing arising out of the use of biological resources and also to conserve and use sustainably, and was institutionalized in the form of National Biodiversity Authority (NBA), State Biodiversity Boards (SBB) and Biodiversity Management Committees (BMCs) in addition to State Forest Departments and at national level MoEF&CC. The Act also emphasizes on the protection and rehabilitation of rare, threatened and endangered species.

9.2 Climatological Diversity

The arc-shaped orientation of Himalaya from north-west to south and then to north-east covering latitude from 26° 45′ to 37° 30′ N provides different phytoclimatic and transition zones between very cold and xeric conditions with -45°C temperature and 10–70 mm annual precipitation in the NW Himalaya fostering cold desert to very high temperatures, rainfall and humidity in the Eastern Himalaya supporting evergreen forests. NW Himalaya experiences good amount of precipitation during summer from south-east monsoon as well as in winter from the 'Western disturbances'. The climate in the NW Himalaya is quite temperate – often referred to as the extratropical mountain climate. Rainfall distribution varies from 331 mm in Srinagar (Kashmir) to 4000 mm in Kumaon-Garhwal region. Rainfall and humidity reduce from east to west and south to north with large annual and diurnal range of temperature, which are more prominent towards the west than the central or eastern parts of the Himalaya (Pangti and Joshi 1987).

9.3 Ecosystem Diversity (ED)

NW Himalaya has very high ecosystem diversity with more than 34% of the 221 types/subtypes of the forests described in India by Champion and Seth (1968). Sharma and Singh (2000) discussed in detail the ED in India. Major states like Uttarakhand (UK), Himachal Pradesh (HP) and Jammu and Kashmir (J&K) have 65, 60 and 51 forest types, respectively, and of these 43 are common, clearly suggesting that climatic conditions such as rainfall and moisture regulate the ED. In this context, HP is unique as it is the meeting point of high rainfall in south-east with moist deciduous forest to least rainfall in north and north-west with dry cold desert (Lahaul-Spiti). Champion and Seth (1968) have not mentioned about the occurrence of *Lantana* scrub in the region, but these impenetrable thickets can now be seen throughout Shivalik and the degraded southern slopes of the outer Himalaya as well as a replacement of local biodiversity. However, one may find positive aspect in it, has the highest soil organic carbon (SOC) 1.37% in fully invaded/degraded sites and 0.69% in uninvaded sites; and high nitrogen (420.5 kg/ha) in invaded to 201.0 kg/ha in not invaded sites (Panwar et al., 2016). It is an excellent soil binder, leaving aside very good habitat for wild cats, fodder for ungulates, charcoal as minor forest produce, raw material for local furniture, etc. Sal (*Shorea robusta* Gaertn. f.) is one of the timbers as well as ecologically important species in NW Himalaya. Dyer (1872) reported its occurrence from Assam up to Sutlej (River). Brandis (1874) and Duthie (1905) reported its distribution from Assam to the Kangra Valley ascending up to 915 m (3000 ft.) and '... terminate near the Beas River'. The latest surveys by Kapur and Sarin (1990) and Uniyal (2002) have reported an occurrence to further west in Gahar Jito (Aghaar Baba Jito), an extension from river Ravi to Trikuta Hills in J&K. It has also been recorded beyond the river Sutlej near Amb in Himachal Pradesh (Roy et al. 2002b, Roy et al. 2006a). Sal is a very resilient species and grows in moist/hot-humid to dry/hot conditions to very cold and freezing conditions (Brandis1874). It will be interesting to study past and present distribution shift in view of perceived climate change. There is a need to study these patches and climatic impacts on these. The initial assessment of economics of ecosystems and biodiversity has been done by Parikh et al. (2012), Arisdasan and Lakshminarasimhan (2016) in India and by Verma (2000) in HP. Singh (2007) highlighted the role of biodiversity and ecosystem services by the Himalayan forest.

9.4 Floristic Diversity

NW Himalaya is among the regions with high floristic diversity, which is represented at various levels in biological hierarchy such as familial, generic, species, sub-species, varieties, forma, alleles, genotypes, phenotypes, etc. Early part of nineteenth century drew the attention of British army officers, doctors, foresters and explorers giving huge impetus to botanical explorations in India by N. A. Wallich (1820–1829) in Eastern India and in NW India Dr. Stewart 1860–1861, Dr. Jameson (forester and surgeon), D. Brandis during 1863–1870 and J.D. Hooker (1856–1917), Col. R.H. Beddome (1865–1892), Col. C.B. Clarke (1880–1889), C.W. Hope (1889–1904), J.D. Hooker (1850–1897), etc., and the history is summarized by Burkill (1965). In post-independence era, in spite of dozens of expeditions and hundreds of botanical and zoological explorations in Himalaya, still there is a need to understand the biodiversity distribution patterns, abundance, quantification and wilderness vis-à-vis human interventions, particularly in inaccessible areas of NW Himalaya with the help of satellite Earth observation (EO) data. Singh and Panigrahi (2005a, b) reported 48 new taxa of pteridophytes from 1 district alone from Arunachal Pradesh, and discovery of dozens of new records in Himachal Pradesh (Singh et al. 2003a, b) suggests the need for intensive and planned surveys in Himalaya. About 47,513 (~11.4% of the world) plant species have been reported from India (Singh and Dash 2014; Arisdasan and Lakshminarasimhan 2016); and 19,395 taxa (~7% of the described species in the world) provide maximum direct benefit to mankind (Singh and Hajra 1996; Arisdasan and Lakshminarasimhan 2016). In India angiosperms are represented by 17,527 species (Arisdasan and Lakshminarasimhan 2016), and NW Himalaya has ~5000 species (Singh 1997a). Gymnosperms in NW Himalaya are represented by 23 species (48%) out 58 species reported from India (Singh and Mudgal 1997).

The maximum diversity is in genus *Ephedra* (eight species). The genus *Juniperus* is represented by all (five) species reported from India. The genus-to-species ratio in flowering plants in J&K is 1:3.5 against 1:5.6 in India, indicating high biological richness (Singh and Uniyal 2002, Karthikeyan 2000). The microclimatic conditions in NW Himalaya include four micro-endemic centres out of 26 in India. These are facing threat for their very existence from developmental activities and for 'good life'. Singh et al. (2004, 2006) and Roy et al. (2006a) provided detailed account of floristic diversity of Punjab Shivalik. In Himachal Pradesh Chowdhary (1999), in Haryana by Kumar (1999), J&K by Singh et al. (1999), Punjab by Singh (1999) and Uttar Pradesh by Uniyal et al. (1999).

9.5 Biodiversity Endemicity

NW Himalaya is one of the four mega-centres of endemism (E. Himalaya, W. Ghats, W. Himalaya and Sundaland) in India. India has about 171 endemic genera of plant, and nearly half (71) are found in Himalaya indicating the great influence of orographic, climatic and edaphic conditions. Nayar (1996) enumerated and regularly updated status of the endemic plants in different biodiversity hotspot regions of India (Sharma and Singh, 2000; Arisdasan and Lakshminarasimhan 2016). The region has three microcentres of endemism: (a) Kumaon-Garhwal, (b) Lahaul-Himachal Pradesh and (c) Kashmir-Ladakh. NW Himalaya is represented by 12 out of 71 genera endemic to Himalaya. It also has 125 wild relatives of crop plants and wild genetic resource for bioprospecting (Arora and Nayar 1991). Other than angiosperms, NW Himalaya also has very high representation of other plant forms such as conifers, pteridophytes, bryophytes, liverworts, fungi, algae and lichens. The region has vast stretches of nearly pure formations of commercially very important trees, viz. *Cedrus deodara*, *Pinus roxburghii* and *Pinus gerardiana* (whose seeds are edible and which, with very restricted distribution, occurs in cold desert conditions), and mixed coniferous forest of *P. wallichiana*, *Taxus contorta*, *Abies pindrow*, *Picea smithiana*, etc. The genera like *Juniperus* and *Ephedra* have maximum genetic diversity in this region. Out of eight species of *Ephedra* reported from India, seven species and one variety are found in this region (Singh and Mudgal 1997). Similarly *Taxus contorta* Griff. (Western Himalayan yew), well-known plant with anticancerous properties, is found in this region. Nearly 215 genera and 751 species of bryophyte (mosses) *sensu stricto* occur in NW Himalaya out of the total 342 genera and 2000 species in India and 3 genera and 144 species are endemic to the region (Vohra and Aziz 1997). The representation of liverworts in NW Himalaya is about 235 species in 77 genera, which is about 28% of the species in India (Singh and Hajra 1996; Singh 1997b). NW Himalaya is considered as the centre of origin of the family Rebouliaceae (Kachroo 1954). Similarly lichens, the bio-indicators of air pollution, out of 2021 species in 248 genera reported in India, about 550 (27%) species are endemic to NW Himalaya (Singh and Sinha 1997). Several important species of lichens have medicinal and other economic uses in aromatic and cosmetic industry, and are harvested from the wild. The region also harbours varied wealth of edible mushrooms (Arisdasan and Lakshminarasimhan 2016). In Marchantiophyta (liverworts) and Anthocerotophyta (hornworts), 930 species and infraspecific taxa belong to 140 genera in the country of which 1 genus and 200 species are endemic to the country. Of these 306 species belonging to 84 genera occur in NW Himalaya, and 1 genus and 53 species are endemic to the region.

9.6 Genetic Wild Relatives and Bioprospecting Germplasm

NW Himalaya is a storehouse of wild plant genetic resource for bioprospecting for agriculture (cereals, millets, pulses, oil seeds, etc.), horticulture (fruits, vegetables, flowers, etc.) and economic (fibres, timber, spices, condiments, etc.) and medicinal values (Anonymous 1991; Kala 2010). Some of the important bioprospecting germplasm in J&K listed are the species of *Avena*, *Hordeum*, *Pennisetum*, *Lathyrus*, *Vicia*, *Cicer*, *Vigna*, *Malus*, *Prunus*, *Pyrus*, *Rubus*, *Brassica*, *Allium*, *Foeniculum*, etc. (Arora 1994) and Singh and Uniyal (2002). More than 675 wild plant species, representing 384 genera and 149 families, are used by locals as food (Singh and Uniyal (2002). The region shows the highest diversity (50.96%) of edible plants than East Himalaya. The region also has maximum number of endemics (18 taxa) and wild relatives of economic plants (39 taxa) (Samant and Dhar 1997). About 266 and 167 plants with economic and medicinal values, respectively, encountered during stratified random sampling approach in J&K have been reported by Roy et al.

(2011). These grow in normal to very harsh conditions, e.g. cold desert of Leh and Ladakh and Lahaul and Spiti. Other important wild genetic resources of plants with bioprospecting potential are species of *Allium*, *Avena*, *Brassica*, *Cicer*, *Ephedra*, *Taxus*, *Aconitum*, *Dioscorea*, *Podophyllum*, *Acorus calamus*, *Stevia rebaudiana* Bertoni, etc. and several species of *Artemisia*. Family Asteraceae is represented by the highest number of medicinal plants. Genera *Artemisia* and *Polygonum* have the highest number species of medicinal importance. Habitat-wise sandy-moist forest, grassland and dry areas represent highest species richness of medicinal plants (Singh et al. 2009).

Bioprospecting of germplasm is significant in view of biotechnological interventions for societal benefits and sustainable product formulation in a growing economy with high population like in India. The germplasm of the cold and tress tolerant species such as *Caragana jubata* (Pall.) Poir. is being used for bioprospecting to grow cereal crop like wheat in the higher Himalaya, and mass propagation of *Podophyllum hexandrum* Royale (CSIR-IHBT 2017). Traditionally plant taxonomists and biotechnologists have their own domain. However, the correct species identification is of prime importance for any reporting on leads, active agents/molecules, etc. The joint studies carried out by the Department of Space and Department of Biotechnology from 1998 to 2010 (Roy et al. 2002a, b, c, d; Roy et al. 2011, 2012) in India have collected geotagged data on plants (trees, shrubs, herbs, pteridophytes, etc.) including medicinally and economically important plants from 16,000+ locations, and are organized as Biodiversity Information System (BIS) (http://bis.iirs.gov.in/) and Indian Bioresource Information Network (IBIN) (http://www.ibin.gov.in/). This facilitates users to find the location of the plant(s) *a priori* to collect the desired material. It saves time, cost and efforts to collect the germplasm. It also provides information where else this is occurring. Additional information on the forest types and known medicinal and economic uses are also compiled and provided.

9.7 Indicators of Climate Change in NW Himalaya

The ecosystem in Trans-Himalaya region and Shivalik ranges (Lower Himalaya) are very fragile due to unconsolidated soil conditions, mostly sandy and steep slopes with rugged terrain. It is not suitable for intensive agriculture. Normal and terrace cultivation is being practiced. The growth of *Betula* plantations in Spiti valley is not so good/healthy, but in the long run, it will contribute to the local environment and alter local soil ecology and environmental set up and could be disastrous in view of climate change. Mountainous range of Greater Himalaya acts as barriers not only for the movement, seed dispersal or exchange of biodiversity or gene but also clouds from south-west monsoon. Trans-Himalaya receive nil to scanty rainfall. In recent past it has been observed that rainfall in Trans-Himalaya has increased. The soil in this region is mostly sandy, and rainfall will lead to high soil erosion and change the local ecosystem processes in the cold desert. The occurrence of isolated population of relics of tall coniferous trees in desert near Gue settlement is implausible.

Growing of apple orchards in Spiti valley is another indicator of changes. Growth and exceptionally large size of vegetables in Leh region is worth noticing (Singh et al. 2011).

9.8 Speciation: New Species at High Altitude

Global warming caused due to enhanced emissions of greenhouse gas from the pre-industrial era of the mid-nineteenth century, is reported to cause fluctuations in the ecotone between alpine pastures and barren land below permanent snow cover, and alpine scrub pasture and tree line (IPCC 2007; Chaudhary and Bawa 2011; Shrestha et al. 2012; Singh et al. 2012; Agnihotri et al. 2017). Since climate is one of the governing factors of species presence/absence, occurrence and endemicity, the change in climatic conditions and environmental setup will allow migration, adaptation as well as speciation. Genera belonging to advance and highly evolved families such as Asteraceae, Poaceae, Fabaceae, Orchidaceae, Rosaceae, Boraginaceae, etc. have adapted to grow in very harsh conditions (extreme coldness, very dry, etc.) in Greater Himalaya as well as Trans-Himalaya. There is urgent need to mark permanent sample plots near the mouth of receding glaciers, tree lines, close to permanent snow cover, etc. for taxonomic inventory and phytosociological characterization and monitoring the migration of species due to potential climate change. The primitive group of plants such as certain algae, bryophytes and lichens are sensitive and resistant to extreme temperatures and high UV. Cold desert regions of J&K have highest UV radiation, and one might find plant which may be grown on Mars in low temperature and pressure and high UV radiation and wind, the most life limiting factors.

9.9 Issues for Biodiversity Conservation

In India the population explosion (human as well as animal), urban and rural sprawl and industrialization after independence have resulted in large-scale changes in land use/land cover (LULC) and therefore, the challenge of biodiversity conservation has increased manifold. The population density of human in HP has increased from 109 to 123 persons/km^2; in Uttarakhand from 159 to 189 persons/km^2; in Haryana from 478 to 573 persons/km^2; 1.74 person/annum, in Punjab from 484 to 551 persons/km^2; and in Jammu and Kashmir from 46 to 56 persons/km^2 from 2001 to 2011 (Census, 2011). This increase is despite a large out-migration of hill population to the plains especially in Uttarakhand (Mamgain and Reddy 2015) and J&K (Dabla 2014). The natural landscapes have been modified which have resulted in the fragmentation of forests accompanied by poor species composition and invasion of non-native and alien species. Hence, the resultant landscape mosaics are a mixture of human-managed and natural patches, which vary in size, shape and arrangement.

Therefore, it has been realized that one must move from single-species conservation to the scales beyond individual sites and levels of organization, i.e. landscape. Understanding spatial pattern of landscape is vital as it covers various levels of biological hierarchy, from ecosystems to species and genes, which are crucial for biodiversity conservation. Disturbance and/or deforestation are expected to increase with time for want of more and more arable land, for industries and other infrastructure development. It will further cause anthropogenic pressures, resulting in the fragmentation of landscape. Fragmentation of forests or habitats is considered as the biggest threat to species loss (Hunter 1995). It is the greatest threat to biodiversity and is the chief cause of species extinction (Wilcox and Murphy 1985). Degradation in the Himalayan forests and its adverse repercussions on the environment have attracted considerable attention recently (Singh and Singh 1985). Frequent fires in NW Himalaya in chir pine forest cause loss of nitrogen which makes site unsuitable for other species to grow (Singh et al. 1984).

It will be interesting to study the impact on local biodiversity in abandoned fields and ghost villages especially in terms of invasive and alien species. Local Eco Task Force can contribute significantly in the forestation of these lands. It may have positive impact on biodiversity since dependency on forest resources may reduce because source of income becomes 'money order'. However, it needs to be studied further. However, it can be an alert to anthropogenic pressure on natural resources for less local requirement. However, ever-increasing demand for more road network and other infrastructural requirement; pilgrimage, e.g. Char Dham Yatra in Uttarakhand, and site degradation, over-exploitation of medicinal plants from the wild, etc. are causes of worry. Tourism is one of the important income-generating activities for local population in the NW Himalaya. Development of infrastructures such as resorts, hotels, widening of the road, new roads, new settlement, etc. puts pressure on the local resources and increasing man-animal conflict in the region. A very large number of micro- to macro-hydel projects have caused extinction of local population of several species. There is selective mass harvesting of local plants, e.g. *Polygonum* sp., for fodder from wild from plateau Kibber village in Spiti valley of HP.

Ownership issue in Punjab Shivalik is proving to be detrimental to biodiversity conservation. More than 95% of the Punjab Shivalik forested land has private ownership. Efforts of the government to maintain vegetation cover in the private lands have been successful to a large extent, but the right to harvest forest lies with the owner. Large stretches of wilderness have been replaced with nearly impenetrable *Lantana camara* thickets, not allowing to regenerate indigenous species. There is a lot of demand of land for industrial requirement. It has also been reported that *L. camara* is extending to subtropical and subtemperate regions also. Being unaware of the consequences of uncontrolled tourism is creating havoc to the local biodiversity in and around the shrines in Kedarnath (Uttarakhand), Amarnath (J&K), etc. After Kedarnath tragedy in June 16–17, 2013, large-scale clearing of alpine scrub communities like *Berberis* (a highly medicinal shrub), *Cotoneaster* (tufted with horizontal branches protects soil), *Lonicera*, *Rhododendron* and the host of other wild genetic resources representing climatic climax has been done to pitch tents for pilgrims, civil administration, hospital, shelter huts, new path, hotels and to build new helipads between Bhim Vatika, Lincholi, and Kedarnath without analysing the consequences or environmental impact assessment. This is in addition to vegetation loss due to landslides. It is strange why this has not caught the attention of conservationist and agencies responsible for protecting the biodiversity such as MoEF&CC, State Forest Department, State Biodiversity Board and host of NGOs working in the area. However, in spite of such activities the State Forest Departments and Central Agency of MoEF&CC deserve applaud for having biodiversity from underdeveloped and poverty-ridden society in a highly populous country.

9.10 Biodiversity Conservation Efforts in NW Himalaya

Although a number of studies on biodiversity of NW Himalaya were carried out at community level (Saxena and Singh 1982; Tiwari 1998; Joshi and Tiwari 1990), however, studies using landscape ecological parameters as basis for diversity analyses are few. The first landmark study to objectively characterize terrestrial biodiversity and prioritize forest ecosystems/landscapes for conservation planning at landscape level in NW Himalaya was carried out jointly by the Department of Space and Department of Biotechnology, Government of India (Roy et al. 2000; Roy et al. 2002a, b, c, d; Singh et al. 2003a,b; Chandrashekhar et al. 2003; Chandrashekhar et al. 2004; Singh et al. 2004; Roy et al. 2006a, 2011; Thakur et al. 2011; Roy et al. 2013). NW Himalaya has very good, unique and threatened faunal diversity as well. Recently a wild goat from Central Asian region has been sighted in Uttarakhand. Efforts have been made for habitat characterization for *in situ* conservation such as hangul in J&K (Ahmad et al. 2009), tigers, co-predators and prey species (Jhala et al. 2011) and management effectiveness evaluation (Mathur et al. 2011). Focus has always been on angiosperms and 'glamorous' animal species.

9.11 Gaps in Conservation Management

PA area network aims to conserve high-quality, unique and representative habitats across the country. Areas in PA network are in Jammu and Kashmir (19), Punjab Shivalik (2), Uttarakhand (13), Himachal Pradesh (33) and Haryana Shivalik (10). Majority of these are animal-centric and are delineated without detailed scientific or biodiversity surveys, not to mention about modern tools, technology and methods. And it is because of these we are faced with man-animal conflict issues in many regions. In NW Himalaya the forest landscape has overall good connectivity despite non-recommended land-use practices such as agriculture on more than 30° slopes. Bird's eye view of satellite data overlaid by PA network boundaries indicates that there are gaps in the conservation management. NW Himalaya has a good network of about 119 PAs of the 735 in the country. Excellent PA network exists in tropical forest ecosystems. An analysis of the PAs indicates that tropical forest ecosystems dominated by moist and dry Sal forests and mixed moist and mixed dry deciduous forests are well represented in low to high mountainous ecosystem and terrain complexities. Due to high terrain complexity, orography, drainage and springs coupled with the variability in microclimatic conditions create very unique habitats and niches. Ecosystems in subtropical and temperate climate are not optimally represented. This is 'gap' in biodiversity conservation and management. This calls for scientific surveys and decision based on recent scientific data/documentation. Despite poaching of wild animals, it is heartening to note that the population of 'star species' like tiger, elephant, leopard, bear, etc. has increased as is indicated by Animal Census reports and increasingly more animals are straying outside the protected areas for finding their own territory or food, and thus coming in conflict with human population.

9.12 Current Status and Way Forward for Biodiversity Conservation

The biome level mapping of forests/vegetation and land use was carried out using IRS WiFS data (Roy et al. 2004, 2006b; Joshi et al. 2001a, b, 2002, 2006) whereas, at landscape level, landmark studies on 'Biodiversity characterization at landscape level using satellite Remote Sensing and Geographic Information System' were carried out during 1998–2002 in NW Himalaya (Himachal Pradesh, Uttarakhand (earlier Uttaranchal) and Uttar Pradesh, J&K part, Punjab Shivalik) and 2008–2010 in J&K and Haryana Shivalik and Chandigarh Shivalik carried out jointly by the Department of Space and Department of Biotechnology (Roy et al. 2002a, b, c, d; Roy et al. 2011; Roy et al. 2012). The satellite remote sensing data for LULC mapping is well established since the 1970s, starting from Landsat series of satellites and since 1989 IRS series of data. The studies used two-seasons (dry and growing) IRS LISS III satellite data for forest/vegetation cover type and land-use/land cover mapping. The supervised, unsupervised and hybrid classification approaches were followed to classify forest/vegetation and LULC using satellite data. These maps were further used for stratification of forests for sampling design. For inventory of trees, shrubs, herbs, etc., more than 2525 samples were laid (835, 150, 594 and 109,168, 678 samples in Himachal Pradesh, Haryana Shivalik, J&K, Punjab and Chandigarh Shivalik, Uttarakhand and Uttar Pradesh, respectively). A stratified random sampling approach based on the satellite remote sensing data of forest/vegetation and land-use maps, covered almost all forest types and majority of the climatic conditions for laying sample plots for biodiversity inventory. For conservation prioritization, the forested landscapes were analysed in GIS domain using in-house developed programme (BIOCAP and SPLAM) for the analysis of ecological parameters such as patchiness, porosity, interspersion, juxtaposition, biotic interferences, etc. (Forman and Godron 1986; Turner 1987; Turner and Garner 1990) to establish the forest fragmentation regimes and disturbance regimes. The field data were analysed for forest type-wise species richness, total importance value, representativeness, endemism, importance value index, similarity, etc. The above parameters were again analysed with terrain complexity to establish biodiversity richness regimes of the entire NW Himalaya. Details of ecological concepts, methodology for mapping, databases and landscape analysis can be referred in Roy et al. (2002a, b, c, d) and Roy et al. (2011).

The subtropical belt (~900–1800 m) has Binsar WLS and Mussoorie WLS (seven patches) in UK; and Chail WLS, Majathal WLS and Bandli WLS in Himachal Pradesh, none in Jammu and Kashmir. NW Himalaya landscape indicates high degree of land-use change to agriculture over the years, and appears very patchy or fragmented on satellite data. This region also has least number of protected areas and thus has maximum gap in PA network. This zone needs attention to identify PA from the remnants of forest patches. PAs in temperate, subalpine and alpine zones in Uttarakhand are Askot Musk Deer WLS, Nanda Devi NP, Valley of Flowers NP, Kedarnath WLS, Gangotri NP, and Govind Pashu Vihar WLS and in Himachal Pradesh are Sangla Valley WLS, Talra WLS, Daranghati WLS, Lipa Asrang WLS, Rupi Bhaba WLS, Great Himalayan NP, Pin Valley NP, Kanawar WLS, Kais WLS, Nargu WLS, Dhauladhar WLS, Kugti WLS, Tundah WLS, Chandra Taal WLS, Kibber WLS, Sechu Tuan Nala WLS, Kalatop Khajjiar WLS and in J&K Gamgul Siabehi WLS, Hemis NP, Overa Aru NP, Baltal Thajwas WLS, Dachigam NP, Hirpora WLS, Gulmarg WLS, Limber WLS, Lachipora WLS, Changthang WLS and Karakoram WLS and thus fairly good PA network. Areas above subtropical zone are reasonably represented in the PA network. The remnants of the subtropical vegetation will provide much-needed corridors between lower (tropical) and upper (temperate, subalpine) habitats.

9.13 Biodiversity Characterization and Management Gap Analyses

9.13.1 Uttarakhand and Uttar Pradesh Shivalik

The area comprises of hilly part of erstwhile Uttar Pradesh, now Uttarakhand, and the Shivalik hills of Saharanpur and Bijnor districts administratively are still under Uttar Pradesh. The altitude ranges from about 250 m in foot hills to 7816 m in the highest peak of Nanda Devi. The region has more than 50 peaks with >6000 m height. Forest cover is 24.24% of the geographical area. Maximum area is covered by snow (23.65%). Temperate and subalpine broadleaved forests occupy nearly 18% of the geographical area followed by pine forest (10.58%) and moist deciduous forest (6.59%). Sal forest covers about 1.378%. Agriculture is about 18.24%. There are 15 hydel projects in Uttarakhand (India-WRIS WebGIS 2014).

Table 9.1 Number and analysis of ground sample points for phytosociological characters

Forest types[a]	No. of sample	No. of families	No. of genera	Trees	Shrubs	Herbs	Total	Shannon-Wiener index	Average basal area (m^2/ha)
DD	30	28	50	20	20	9	49	0.79	8.50
Sal	42	44	102	10	56	47	113	4.83	17.90
MD	30	43	84	36	15	34	85	4.30	15.24
Pine	173	65	166	22	35	155	212	5.05	42.01
TBL	277	113	343	87	105	307	499	7.04	31.68
TC	79	61	134	24	45	126	195	5.56	29.78
AS	9	33	73	5	17	72	94	4.96	5.85
AP	38	35	73	–	12	94	106	4.98	0.00

Source: Roy et al. (2002c)

[a]*DD* dry deciduous forest, *Sal* Sal forest, *MD* moist deciduous forest, *Pine* pine forest, *TBL* temperate broadleaved forest, *TC* temperate coniferous forest, *AS* alpine scrub and *AP* alpine pastures

9.13.1.1 Biodiversity

A total of 678 sample plots were analysed for quantification of species richness and phytosociological analysis. The maximum number of species were recorded from temperate and subalpine broadleaved forest, whereas a minimum number of species were recorded from dry deciduous forest. A maximum number of genera and family were also recorded from temperate and subalpine broadleaved forest. The value of Shannon-Weiner index for the entire vegetation (cumulative for all layers) ranged between 0.79 (dry deciduous forest) and 7.04 (temperate and subalpine broadleaved forest) (Table 9.1) (Roy et al. 2002c).

Richness of trees species was the highest in temperate and subalpine broadleaved forests (3.510) followed by moist deciduous forest (2.94) and Sal forest (2.03). Sapling richness was the highest in moist deciduous forest (3.66) followed by dry deciduous forest (3.13) and temperate and subalpine broadleaved forest (2.57). Seedling diversity was the highest in moist deciduous forest (2.73) followed by temperate and subalpine broadleaved forest (2.57) and dry deciduous forest (2.50). The presence of seedling in alpine scrub and alpine pasture is nil which is a matter of concern and needs further investigations. Shrub diversity was the highest in temperate and subalpine broadleaved forest (5.13) followed by temperate and subalpine conifer (3.93) and moist deciduous forest (3.64). The presence of shrub diversity in alpine scrub and alpine pasture is a good sign and needs further study to understand the role of climate change. Temperate and subalpine broadleaved forests have the highest diversity (6.88) of herbaceous plants followed by temperate and subalpine broadleaved (5.48) and alpine scrub (4.95). The trend of species richness indicates that moisture and temperature are playing important roles. A total of 132 plants were found endemic to the region (Table 9.2) (Roy et al. 2002c).

Analysis of the total importance value (TIV) indicates that alpine scrubs (8.89) are the richest in terms of useful plants followed by alpine pasture (3.57) and

Table 9.2 Shannon-Wiener index for different storeys of vegetation in Uttarakhand and UP hills

Forest types	Trees	Saplings	Seedlings	Shrubs	Herbs
Dry deciduous forest	0.42	3.13	2.50	2.04	1.89
Moist deciduous forest	2.94	3.66	2.73	3.64	4.11
Sal forest	2.03	2.08	0.75	3.33	4.40
Pine forest	0.88	0.45	2.00	3.53	4.88
Temperate and subalpine conifer	2.02	2.61	2.29	3.93	5.48
Temperate and subalpine broadleaved	3.51	2.99	2.57	5.13	6.88
Alpine scrub	1.77	0.00	0.00	2.99	4.95
Alpine pasture	0.00	0.00	0.00	2.09	4.79

Source: Roy et al. (2002c)

temperate and subalpine conifer forest (3.35). Tropical forests of moist and dry deciduous including Sal forest have least TIV. A total of 316 plants were found to have medicinal value. Temperate, subalpine and alpine regions of Himalaya are well known for medicinal plants, mainly used in Ayurvedic and Tibetan system of medicines.

9.13.1.2 Landscape Analyses

Forest fragmentation analyses indicate that in general more than half of the forest area is intact, which is a healthy sign from biodiversity conservation point of view. Intact forest cover area varies from 27.40% in alpine pasture to 84.36 in temperate and subalpine conifer forests. About 33.22% of alpine scrub is low fragmented followed by pine forest (30.66%) and alpine pastures (29.79%). The highest area under medium level of fragmentation was observed in pine forest followed by alpine pastures (38.40%) and Sal forest (21.34%). Forested landscape of this region has very low area under the high level of fragmentation (Table 9.3) (Roy et al. 2002c).

Area under very high disturbance regime is very less which is a good indication and needs to be maintained. About 19.85% of the area is under undisturbed category and 49.85% under indication of disturbance category. Forest type-wise disturbance analysis indicates that the area under very high disturbance is very less. Alpine pasture has 42.99% of the area as undisturbed followed by dry deciduous forest (mainly in Shivalik) and moist deciduous forest (21.09%). Alpine scrub has about 63.12% of the area under indication of disturbance followed by moist deciduous (52.01%) and 51.46% in dry deciduous forest. Temperate coniferous forests have the highest moderately disturbed forest (41.81%) followed by alpine scrub (33.74%), pine forest (30.78%) and Sal forest (28.78%) (Table 9.4) (Roy et al. 2002c).

9.13.1.3 Biological Richness

The forests are very rich in biodiversity particularly in Uttarakhand where rainfall is more as compared to the other states in NW Himalaya. About 10% area of all forest

Table 9.3 Vegetation-wise distribution of fragmentation in Uttarakhand and UP hills

Forest	Fragmentation							
	Intact		Low		Medium		High	
Type	Area (km^2)	Area (%)	Area (km^2)	Area (%)	Area (km^2)	Area (%)	Area (km^2)	Area (%)
Dry deciduous	880.61	76.08	106.78	9.23	150.09	12.97	20.07	1.73
Sal	440.07	58.02	148.15	19.53	161.86	21.34	8.43	1.11
Moist deciduous	2324.55	64.06	558.56	15.39	700.35	19.30	45.14	1.24
Pine	1666.83	28.63	1784.99	30.66	2236.08	38.40	134.54	2.31
Temperate coniferous	2024.03	84.36	269.32	11.23	104.21	4.34	1.69	0.07
Temperate broadleaved	5277.48	53.08	2559.44	25.74	2044.81	20.57	60.17	0.61
Alpine scrub	468.54	26.13	595.62	33.22	697.80	38.92	30.83	1.72
Alpine pastures	302.42	27.40	328.77	29.79	409.30	37.08	63.24	5.73

Source: Roy et al. (2002c)

Table 9.4 Vegetation-wise distribution of disturbance classes in Uttaranchal and UP hills

Forest	Disturbance classes							
	Undisturbed		Indication		Moderate		High	
Types[a]	Area (km^2)	Area (%)	Area (km^2)	Area (%)	Area (km^2)	Area (%)	Area (km^2)	Area (%)
DD	366.34	31.65	595.69	51.46	192.47	16.63	3.06	0.26
Sal	127.56	16.82	402.47	53.06	218.31	28.78	10.17	1.34
MD	797.90	21.99	1887.05	52.01	866.59	23.88	77.04	2.12
Pine	1173.44	20.15	2841.43	48.80	1792.16	30.78	15.42	0.26
TC	363.98	15.17	1025.15	42.73	1003.18	41.81	6.94	0.29
TBL	1810.45	18.21	4977.48	50.07	3153.97	31.72	0.00	0.00
AS	50.24	2.80	1131.66	63.12	604.94	33.74	5.96	0.33
AP	474.54	42.99	485.01	43.94	141.38	12.81	2.81	0.25

Source: Roy et al. (2002c)

[a]*DD* dry deciduous forest, *Sal* Sal forest, *MD* moist deciduous forest, *Pine* pine forest, *TBL* temperate broadleaved forest, *TC* temperate coniferous forest, *AS* alpine scrub, and *AP* alpine pastures

types is under high biological richness and about 40.00% under medium richness. Nearly half of the area comes under low and very low biological richness. In pine forest about 90% area comes under very low biological richness. About 98.2% of alpine scrub and 91.7% of the alpine pastures are under high BR, which is very unique. About 93.9% of the temperate broadleaved forest is under moderate BR followed by moist deciduous forest (26%). Temperate conifer has about 84.5% area under low BR followed by dry deciduous forest (83%) and Sal forest (81.9%). Pine forest has the highest area (99.7%) under low BR followed by Sal forest (16.8%). Both of these are more or less mono-species forests (Table 9.5) (Roy et al. 2002c).

Table 9.5 Vegetation-wise distribution of biological richness classes in Uttarakhand and UP hills

Forest	Biological richness classes							
	Very low		Low		Medium		High	
Type[a]	Area (km²)	Area (%)	Area (km²)	Area (%)	Area (km²)	Area (%)	Area (km²)	Area (%)
DD	2.31	0.20	960.22	82.95	195.03	16.85	0.00	0.00
Sal	127.66	16.83	620.67	81.83	10.18	1.34	0.00	0.00
MD	0.37	0.01	2684.65	73.99	943.57	26.00	0.00	0.00
Pine	5805.05	99.70	15.36	0.26	1.98	0.03	0.05	0.00
TC	363.98	15.17	2028.34	84.54	6.94	0.29	0.00	0.00
TBL	0.44	0.00	543.83	5.47	9330.74	93.85	66.89	0.67
AS	4.94	0.28	0.00	0.00	27.69	1.54	1760.17	98.18
AP	92.14	8.35	0.00	0.00	0.00	0.00	1011.60	91.65

Source: Roy et al. (2002c)

[a]*DD* dry deciduous forest, *Sal* Sal forest, *MD* moist deciduous forest, *Pine* pine forest, *TBL* temperate broadleaved forest, *TC* temperate coniferous forest, *AS* alpine scrub and *AP* alpine pastures

9.13.1.4 Discussion and Recommendations

The Doon Valley of Uttarakhand and south-east part of Himachal Pradesh are the western limits of humid climate. Humidity decreases towards further west. Sal forests provide continuity to habitats in the lower region with Nepal forest in the east and NW Himalaya in the west. These forests support westernmost limit of Asian elephants who migrate from Rajaji National Park to Dudhwa National Park travelling long distances. With moderate to high floristic diversity and floral biological richness, the Sal forests of this region are well known for faunal diversity with umbrella species like tiger and other keystone species like elephant, leopard and several deer species.

9.13.1.5 Conservation Status and Gaps

The tropical region is well protected and provides good connectivity except near Hardwar and Rishikesh region where man-elephant conflict is very high. However, there are a lot of opportunities to extend and protect more forest areas. Protecting forests of Shivalik range of the Doon Valley and westward from Rajaji NP forest up to Kalesar NP will be a very good move. Because landscape-wise forests have the same characteristics and biological richness, therefore there is no reason to keep good forest area out of PA network. It will also facilitate the movement of animals up to Kalesar NP and beyond, which is important to provide new home range to increasing the population of tigers and elephants. The forest between Rajaji NP and Sonanadi-Corbett-Nandhaur PA needs to be regulated to maintain enough corridor width. The PA network between subtropical and subtemperate zone does not exist. In this vast landscape, only two wildlife sanctuaries (Binsar and Mussoorie) exist.

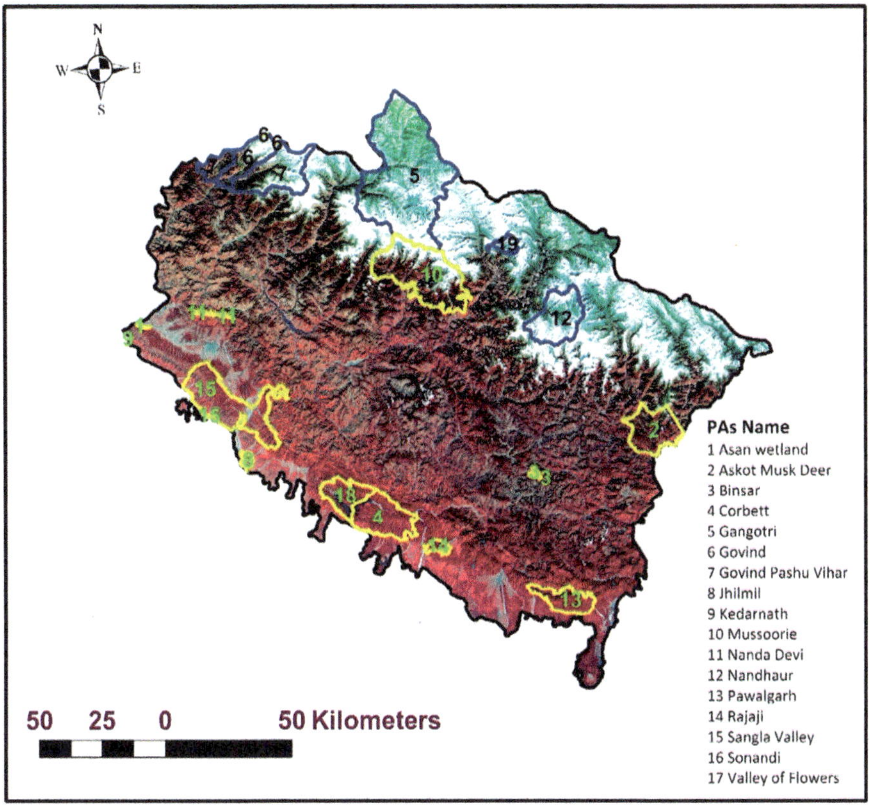

Fig. 9.1 Protected areas in Uttarakhand and Uttar Pradesh Shivaliks overlaid on False Colour Composite (RGB:3,2,1) of IRS AWiFS data (19 October, 2012)

Three possibilities exist to improve the PA network: (a) Rajaji NP-Mussoorie-Aglar watershed through Makhdet-Banari-Mugarsanti Range-Sigar Gaon-Molda Beet-Yamunotri and above, (b) Corbett National Park-Jogimani-Kundoli-Chatoli-Karnprayag-Bangthal-Sarpani-Chamoli-Joshimath-Badrinath-Kedarnath and (c) Samanora Range-Bhim Tal-Naina Range-Bhowali Range-Almora-Ranikhet-Binsar-Bageshwar-Munsyari to temperate zones. These areas still have good forest cover to provide corridor for animals. The forest landscape of Dwarahat Range and Dungar-Karnprayag-Rudraprayag-Kaploi will be good to protect and to provide another corridor. Himachal Pradesh has better representation of subtropical ecosystems. Temperate to alpine zone is very well represented. However, areas beyond Chakrata and Makhdet need protection to provide connectivity with Govind Pashu Vihar WLS. Temperate and subalpine vegetation are the richest regions in terms of species and economic importance and endemism in Uttarakhand; therefore, this region around Munsyari-Joshimath-Chamoli-Gopeshwar Valley of Flowers deserves attention to protect the indigenous germplasm. Nanda Devi NP needs redefining of the boundary based on scientific data to cover subalpine and temperate forests and should be extended on southern and western sides. We need to pay attention to very narrow corridor around Haldwani and above (Fig. 9.1).

9.13.2 Himachal Pradesh

The forest cover of the state is 13,880 km^2 which constitutes about 24.93% of the total geographical area (Roy et al. 2002b) (25.78% FSI 2003). Difference of 0.8% was observed with respect to FSI classification scheme. With 5 national parks, 32 wildlife sanctuaries and 3 conservation reserves, about 14% of the geographical area is under PA network suggesting that some more area can be brought under PA network for improved connectivity. The major vegetation cover types are mixed conifer (3226.72 km^2) followed by deodar (2193.6 km^2), dry deciduous (2153.25 km^2), chir pine (2005.5 km^2), moist deciduous (1573.61 km^2), oak (879.38 km^2), *Betula/Rhododendron* (455.08 km^2) and temperate broadleaved (408.82 km^2). HP has the highest ecosystem diversity. Mixed coniferous forest showed highest Shannon-Wiener diversity (9.15) followed by deodar (8.64), temperate broadleaved (8.52), temperate scrub (8.25) and dry deciduous (8.13) and then followed by moist deciduous, alpine scrub, oak, alpine meadows, chir pine, *Betula/Rhododendron* and temperate grassland in decreasing order. Interestingly the taxonomic diversity at family and genus level is the highest in temperate mixed conifer (88 and 235) and broadleaved forest (86 and 225) closely followed by mixed moist deciduous (80 and 221), temperate scrub (75 and 200) and mixed oak forest (75 and 199) militating common understanding that tropical forests are more diverse (Table 9.6) (Roy et al. 2002b).

9.13.2.1 Biodiversity

A total of 337 species (202 plus 125 species possessing both medicinal as well as economic properties) were recorded. Some of the medicinally important plants are *Podophyllum hexandrum*, *Taxus contorta Griff. (T. wallichiana)*, *Valeriana jatamansi*, *Dactylorhiza hatagirea*, *Aconitum* spp., *Picrorhiza kurroa*, *Boerhavia diffusa*, *Hedyotis diffusa*, *Ichnocarpus frutescens*, *Justicia adhatoda*, *Cordia dichotoma*, *Azadirachta indica*, *Aegle marmelos*, *Bacopa monnieri*, *Terminalia bellirica*, *Terminlaia chebula*, etc. The highest TIV (1135) is for moist deciduous forest followed by temperate broadleaved forest (1045), mixed conifer (1023) and mixed oak forest (991). Temperate forests of oak, scrub, deodar, blue pine, and chilgoza and tropical forests of Sal and mixed dry deciduous have high to moderate TIV. Forest with high TIV needs special attention and policy framework for sustainable resource utilization. Mean basal area is the highest in conifer forest of deodar (30.66 m^2/ha) and mixed conifer (26.67 m^2/ha) forest suggesting these forests have very high carbon stock and also high carbon sequestration potential. Therefore, conservation of these forests in Himalaya is crucial (Table 9.7) (Roy et al. 2002b).

Ecologically each species is important, but environmentally each has different level of significance. It is interesting to note that out of the 30 endemic species encountered during sampling, 25 are herbaceous species, 3 tree and 2 shrub species.

Table 9.6 Biodiversity status of the vegetation of Himachal Pradesh

				Number of species in each habitat			
Vegetation types	No. of plots	Families	Genera	Trees	Shrubs	Herbs	Total
Mixed conifer	52	88	235	48	84	308	440
Temperate broad Leaved	36	86	225	50	58	285	393
Moist deciduous	36	80	221	58	86	185	329
Temperate scrub	62	75	200	7	79	282	370
Oak	38	75	199	33	66	199	298
Deodar	31	66	144	42	57	188	287
Blue pine	19	63	137	16	46	151	213
Dry deciduous	25	61	155	27	48	97	172
Betula/Rhododendron	11	54	104	18	24	112	154
Scrub	17	53	94	–	60	96	156
Alpine meadows	145	43	98	0	2	143	145
Alpine scrub	204	41	114	0	46	175	221
Hippophae	19	35	68	–	14	81	95
Riverine	9	30	73	11	15	85	111
Temperate grassland	59	30	98	–	–	170	170
Juniper	9	28	55	6	12	50	68
Sal	6	24	58	9	22	41	72
Chir pine	39	23	31	35	63	132	230
Chilgoza	8	17	34	8	8	46	62
Ephedra	10	9	10	–	6	6	12

Source: Roy et al. (2002b)

Seven endangered species such as trees *Corylus jacquemontii* Decne. and *Taxus wallichiana* Zucc. and herbaceous *Habenaria edgeworthii* Hook. f., *Malaxis muscifera* (Lindl.) Kunze, *Podophyllum hexandrum* Royle, *Saussurea costus* (Falc.) Lipsch. and *Selaginella adunca* A.Br. ex Hieron were recorded. Species like *Dactylorhiza hatagirea* (D. Don.) Soo, *Dioscorea deltoidea* Wall. ex Kunth. and *Dioscorea melanophyma* Prain and Burkill are threatened, and *Allium stracheyi* Baker, *Picrorhiza kurroa* Royle ex Benth. and *Podophyllum hexandrum* Royle are vulnerable to over-exploitation and habitat loss.

9.13.2.2 Landscape Analyses

The forests of Himachal Pradesh show relatively higher degree of fragmentation. Fragmentation under high and very high category is about 23.11 per cent. Fragmentation map vs vegetation cover type indicated mixed coniferous, deodar and moist deciduous forests, and Alpine meadows have high level of fragmentation. Fragmentation vs species diversity vs forest types indicates that mixed coniferous, deodar and moist deciduous forests have 9.15 (highest), 8.64 (second highest) and 7.84 (sixth

Table 9.7 Forest type-wise Shannon-Wiener Index (H′), TIV and basal area (BA)

Vegetation types	H′	TIV	BA (m^2/ha)
Moist deciduous	7.84	1155	7.29
Temperate broad leaf	8.52	1045	25.59
Mixed conifer	9.15	1023	26.76
Oak	6.78	991	17.20
Temperate scrub	8.25	956	0.00
Dry deciduous	8.13	955	21.31
Deodar	8.64	775	30.66
Scrub	2.56	743	0.00
Blue pine	5.02	637	19.89
Betula/*Rhododendron*	5.34	574	23.92
Chilgoza	0.65	521	23.27
Sal	3.07	421	19.63
Hippophae	2.72	310	2.56
Temperate grassland	5.32	287	0.00
Alpine scrub	7.55	202	0.00
Chir pine	6.35	198	17.84
Juniper	1.73	184	9.21
Riverine	3.17	147	26.71
Alpine meadows	6.41	124	0.00
Ephedra	0.19	85	0.00

Source: Roy et al. (2002b)

highest) species diversity, suggesting a case of intermediate disturbance level theory. Since these forests account for very high biodiversity, there is a need to curve the further fragmentation of the landscape (Table 9.8) (Roy et al. 2002b).

Nearly 45% of the area is high to very highly disturbed, and about 7% is medium to least disturbed. Most of the forest cover types falling between tropical and temperate zone in valleys indicate higher degree of disturbance because of the concentration of the human activities. Of these chir pine forest indicates the highest area in different levels followed by blue pine, mixed conifer and dry deciduous. Tropical region scrub and moist deciduous and temperate region grassland, deodar, dry deciduous, alpine meadows, juniper, temperate broadleaf showed moderate to least disturbed (Table 9.9), which is following the general trend of disturbance in NW Himalaya. However, alpine meadows and *Juniperus* scrub are not contiguous by nature (Roy et al. 2002b).

9.13.2.3 Biological Richness

The forests of Himachal Pradesh are most varied of all the states in NW Himalaya and therefore, have the highest ecosystem diversity because of transitional location. The forests in the south-east are hot/humid moist deciduous Sal and in the north-west are dry cold desert. The Spiti valley in the north-west part is the eastern limit of cold

Table 9.8 Status of fragmentation of the vegetation of Himachal Pradesh

Vegetation type	Low	Medium	High	Very high
Moist deciduous	10.8	17.12	3.95	1.89
Dry deciduous	9.21	14.94	5.48	0.14
Deodar	1.61	0.71	8.75	23.1
Chir pine	6.33	3.49	10.72	2.38
Mixed conifer	4.33	1.3	9.75	39.84
Oak	3.04	12.2	3.79	2.57
Temperate broadleaf	0.19	0.21	0.79	6.66
Betula/Rhododendron	0.33	1.04	2.37	1.74
Juniper	2.24	2.17	0.51	0.86
Chilgoza	0.06	0.22	0.47	0.05
Sal	1.52	6.04	0.18	0.02
Blue pine	0.07	0.53	0.46	0.22
Riverine	0.01	0.05	0	0
Scrub	21.49	12.7	1.52	1.08
Temperate scrub	1.21	0.9	0.47	3.32
Alpine scrub	2.4	9.76	13.59	5.87
Temperate grassland	17.55	14.45	0.83	1.57
Alpine meadows	8.98	1.11	35.95	8.41
Ephedra	0.22	0.41	0.04	0.1
Hippophae	0.66	0.48	0.27	0.03

Source: Roy et al. (2002b)

desert, thus fostering myriad of habitats, ecosystems and landscapes. About 30% of forest area is covered in high to very high biological richness and about 23% in medium to low richness. It is heartening to note very high species-rich ($H' > 7$) forests of mixed conifer, deodar, alpine scrub and temperate broadleaf have large areas under high and very biological richness area. These are less disturbed. In medium species richness category ($H' = 5–7$), alpine meadows occupy the large area under high and very high BR. A number of PAs represented by these are three. However, plateau east of Kibber in Spiti valley represents very unique landscape of alpine scrub but is not part of PA network. Large-scale extraction of local fodder/ fuelwood plants such as *Polygonum* is done in the area (Table 9.10) (Roy et al. 2002b).

9.13.2.4 Discussions and Recommendations

Himachal Pradesh state is one of the fast developing economies, as is evident by the number of trucks one can see all over the state throughout the years ferrying various products. HP has 33 hydel projects causing submergence of riverine biodiversity and damage to local flora and fauna due to debris dumping, settlement, road network, etc. Anthropogenic disturbance is one of the key factors for habitat loss, and the forest

Table 9.9 Vegetation type-wise status of disturbance of the region

Vegetation type	Low	Medium	High	Very high
Chir pine	13.01	12.51	12.55	16.74
Blue pine	0.09	0.36	0.4	15.35
Riverine	0.02	0.06	0.01	15
Mixed conifer	15.12	16.66	9.76	11.83
Dry deciduous	1.05	2.54	6.74	9.97
Temperate grassland	8.91	8.59	6.78	8.23
Scrub	4.34	6.43	27.39	8.14
Deodar	10.29	11.34	6.63	8.09
Alpine scrub	3.67	3.76	8.13	7.21
Moist deciduous	16.19	12.52	5.3	3.76
Oak	11.38	8.1	2.65	2.15
Juniper	0.16	0.91	1.27	1.38
Betula/Rhododendron	3.37	3.48	1.52	1.15
Sal	0.24	0.4	1.29	0.99
Temperate broadleaf	4.55	4.36	1.34	0.76
Temperate scrub	4.01	3.09	1.07	0.63
Chilgoza	0	0.09	0.26	0.33
Alpine meadows	2.91	3.62	0.74	0.31
Hippophae	0.64	0.1	0.53	0.22
Ephedra	0.06	0	0.17	0.03

Source: Roy et al. (2002b)

patches characterize biodiversity depending on the existing environmental conditions and the associations. Being hilly state the surface area is very high, and therefore, impact of the developmental activities is less pertinent. People and government machinery are enterprising. And one can see developmental activities which could have been avoided, e.g. construction of new hotels and resorts almost within the part of PAs and construction of guest house on the only ridge connecting forest east and west in Kalatop Khajjiar WLS causing man-wildlife conflict (Chandrashekhar et al. 2004). The development activities in the state need to consider corridors and crucial forest patches for movement of biodiversity.

9.13.2.5 Conservation Gaps

One of the most characteristic alpine ecosystems occurs at the plateau along the north ridge of Spiti valley and south Kibber NP. This most unique alpine scrub ecosystem is characterized by very high biological richness and low disturbance and least fragmented and represented by species like *Ephedra*, wild *Cicer*, *Polygonum*, wild oats, etc. It is a large patch nearly intact and low to moderate biotic pressure. Of course *Polygonum* is harvested by locals for fodder. Subtemperate, temperate and subalpine forests are very well represented. Subtropical vegetation in eastern ranges

Table 9.10 Status of biological richness of the vegetation of Himachal Pradesh

Vegetation type	Low	Medium	High	Very high
Mixed conifer	4.33	1.3	9.75	39.84
Deodar	1.61	0.71	8.75	23.1
Alpine meadows	8.98	1.11	35.95	8.41
Temperate broad leaf	0.19	0.21	0.79	6.66
Alpine scrub	2.4	9.76	13.59	5.87
Temperate scrub	1.21	0.9	0.47	3.32
Oak	3.04	12.2	3.79	2.57
Chir pine	6.33	3.49	10.72	2.38
Moist deciduous	10.8	17.12	3.95	1.89
Betula/Rhododendron	0.33	1.04	2.37	1.74
Temperate grassland	17.55	14.45	0.83	1.57
Scrub	21.49	12.7	1.52	1.08
Juniper	2.24	2.17	0.51	0.86
Blue pine	0.07	0.53	0.46	0.22
Dry deciduous	9.21	14.94	5.48	0.14
Ephedra	0.22	0.41	0.04	0.1
Chilgoza	0.06	0.22	0.47	0.05
Hippophae	0.66	0.48	0.27	0.03
Sal	1.52	6.04	0.18	0.02
Riverine	0.01	0.05	0	0

Source: Roy et al. (2002b)

is moderately protected. Excellent ecosystems of the entire mountainous range between Khajjiar and Dhauladhar WLSs will be worth protecting. The range is least fragmented and has high biological richness and provides important corridor with subtropical and tropical biodiversity. Vast area under tropical and lower subtropical belt is highly fragmented and disturbed, and however, still has isolated patches of intact forest of different dimensions. These patches in Kangra Valley can be brought under some kind of protection.

Provision to connect forests of Dalhousie and Dhar Banjaut in HP and Bani and Kandhawara in J&K, respectively, requires immediate attention. This corridor is essential for biodiversity movement and gene exchange. Himachal has the eastern end of the cold desert and is therefore very unique biologically for medicinal plants, wild relatives of crop plants with high potential for bioprospecting. *Hippophae, Cicer, Barley*, several legumes, etc. are important plant groups in the cold desert. There is no measure to conserve *Hippophae* scrub, which is unique and has immense potential for economic gains for local and state – a potential bioprospecting species. It occurs in floodplains of rivers. The forest area adjacent to Govind Pashu Vihar in Uttarakhand and Chanshal-Kotigad range also needs attention for connectivity. The forested landscape suggested for extension and redefining are the PA boundaries of Shikari Devi, Churdhar, Talra and Lippa Asrang. More area towards north can be brought under Kais WLS. Part of pine forest Nahan-Solan-Swarghat needs some kind of protection and to maintain all important connectivity (Fig. 9.2).

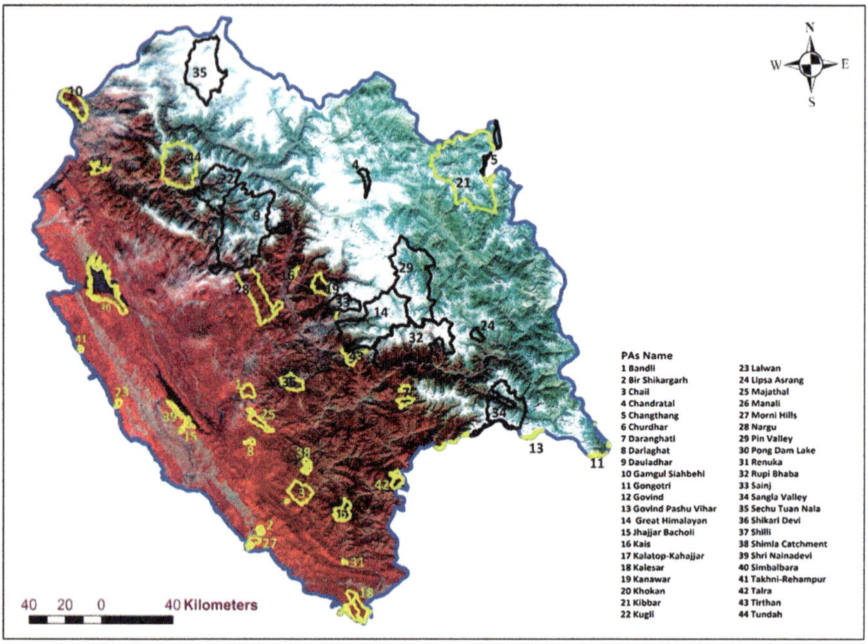

Fig. 9.2 Protected areas in Himachal Pradesh, Haryana and Punjab Shivalik overlaid on false colour composite (RGB:3,2,1) of IRS AWiFS data (19 October 2012)

9.13.3 Haryana Shivalik

The state of Haryana has agriculture-dominated landscape, and forest cover is only 3.43% which includes 1.37% of the areas under plantations of poplar, *Prosopis juliflora*, *Eucalyptus*, mixed plantation, scrub and *Prosopis* scrub. About 91.47% of the area is under agriculture and 4.24% is under settlement. Northern region of the state has Shivalik hills. The forests in Yamunanagar forest division are northern tropical dry deciduous type. *Shorea robusta* forests as well as northern tropical dry deciduous and Shivalik *Shorea robusta* forest in the Morni hills belong to the outer ranges of the Himalaya. The state is poor in biological resources and forest wealth; however, because of its location covering Shivalik and Aravalli hills and semiarid zone adjoining Rajasthan, it supports its own type of biodiversity. Shivalik hills have been and are well known to be the source of large number of medicinally and economically important plants. There are two hydel projects (India-WRIS WebGIS, 2014) (Roy et al. 2002d).

9.13.3.1 Biodiversity

The flora of the state is not much diverse in terms of taxa, habit and growth forms. However, the flora in Kalesar and Yamunanagar districts is considerably diverse

Table 9.11 Phytosociological data of Haryana

Forest type	Trees	Shrubs	Herbs	Climbers	Total
DDF	69	37	122	31	259
Thorn forest	13	16	77	17	123
Sal forest	29	17	54	18	118
Riverine forest	19	20	50	16	105
Khair forest	16	11	68	9	104
Scrub	23	15	41	5	84
Eucalyptus plantation	10	15	54		79
Chir pine forest	7	14	47	4	72
Prosopis scrub	17	9	29	12	67
Prosopis plantation	15	8	38		61
Riverine grassland	2	9	49		60
Mixed plantation	6	4	21		31

Source: Roy et al. (2002d)

having a number of medicinally and economically plant species. Kalesar wildlife sanctuary is having a dense vegetation of the Sal forest with a rich floral and faunal diversity. A total of 332 species under tree, herbs, shrubs and climbers have been recorded. Out of them a maximum of 259 species are reported from dry deciduous forest followed by thorn forest (123), Sal forest (118), riverine forest (105) and Khair forest (104), respectively. The rest of the area possesses less than 100 species under different forest categories. The species diversity is the highest, i.e. more than 3.5 in dry deciduous forest, and the minimum is less than 0.5 in riverine grasslands. The Sal forest which looks like single-species-dominated vegetation has a species diversity of more than 2. No endemic/threatened species were recorded during phytosociological sampling in Haryana (Table 9.11) (Roy et al. 2002d).

9.13.3.2 Landscape Analysis

It is observed that 75.63% of the Sal forest is still very intact showing low degree of fragmentation, whereas 60.41% of the pine forest shows medium fragmentation. It is also observed that 48.14% of the Khair forest is under medium to high degree of fragmentation and more than 20% of the chir pine is under high fragmentation category. The area statistics under various vegetation types indicates high disturbance in Sal forests (87.5%) followed by 67.1%, 65.8% and 56.4% for dry deciduous forest, Khair forest and thorn forest, respectively. The medium disturbed areas are high (61.9%) in riverine forests followed by 55.5% and 38.6% in pine forest and thorn forest, respectively. However, the *Eucalyptus* and mixed plantation areas are having more than 55% area under medium disturbance categories. The low disturbed areas are found high in teak plantation (94.4%), followed by pine forest (42.3%), riverine forest (22.3%) and *Prosopis juliflora* plantation (15.7%).

The overall area distribution of BR shows maximum at medium level (80.6%), followed by high (13.4%), low (3.8%) and very high (2.1%). Of the different forest types, dry deciduous forest represents the highest area (14.18%) under very high biological richness followed by pine forest (5.93%). The rest of the forest types are having less than 1% area under very high biological richness. The high biological richness (59.93%) was observed in pine forest followed by dry deciduous forest (34.85%), Sal forest (13.58%) and Khair forest (13.50%) (Roy et al. 2002d). Maximum area under medium biological richness was obtained in almost all vegetation types and the plantation areas. The study also indicates that high to very high biologically rich areas are occurring in Morni hills areas. Although the Sal forest areas near Kalesar are supporting remarkable number of medicinal plant, but due to very nearness to roads, these areas have gone into medium biological richness category except for the very few small patches under high to very high biological richness category where microclimatic conditions are favourable. Although the conservation measures are taken by the forest department and the Kálesar is already declared a wildlife sanctuary, the local people are still having a lot of interference for the collection of medicinal plants, because of its easy access. However, the dry deciduous forests are under high invasion by *Lantana camara* which prohibits the growth of indigenous species.

The disturbance regime map show that Sal forest, dry deciduous forest, Khair forest and thorn forest are under high degree of disturbance, and the Sal forest still has patches supporting very high biodiversity. Therefore, it is most important to have conservation measures to safeguard these areas from further degradation and encroachment. It is also observed that the forests of Haryana are under tremendous pressure from the surrounding population and the land grabbers are active to flatten the Shivalik for developing colonies. Conservation measures are to be taken in the periphery of Shivalik, and the forest fire management practices need to be strengthened for protecting the natural forests.

9.13.3.3 Discussion and Recommendation

Recent reports have reported the movement of elephant and tiger in Kalesar WLS, i.e. west to river Yamuna, and that is a good sign as increase population is trying to find and establish their own home range. Shivalik nearer to Panchkula is an example where the *Prosopis juliflora* scrub and the thorn forest are dominating almost everywhere. The most vital area is the Morni hills supporting chir pine forest, dry deciduous forest and also the Khair forest. Due to the development of industrial town Baddi and the opening of the roads to Chandigarh, most of the Khair forest areas are under low to medium density of forest. *Lantana* invasion is very high in open degraded areas. *Prosopis juliflora* has been planted by forest department of Haryana for soil conservation. The Shivalik Sal forest and dry deciduous forest provide important link between ecosystem on east and west of river Yamuna. Attempts to do afforestation in forest blank is recommended to avoid further spread of *Lantana* and *Parthenium* (congress grass).

9.13.3.4 Conservation Gaps

The state of Haryana has very less forested area as a part of Shivalik ranges. The forested area between Morni hills and Kala Amb bordering Himachal Pradesh can be considered for some kind to protection (Fig. 9.2).

9.13.4 Punjab and Chandigarh Shivalik

The majority of the forests are privately owned, and natural landscape of the region is seriously vulnerable to further degradation due to developmental activities such as resorts, amusement parks, hotels, industries, clear felling of trees, etc. The resulting landscape mosaic is a blend of natural and man-managed patches that vary in size, shape and arrangement. It has seriously affected the growth of indigenous plants and allowed impenetrable *Lantana camara* thickets. Forest covers about 14.65% (1404.06 km^2) of the total geographical area of the region. Non-forest area about 85.35 per cent. Among the forests, dry deciduous forest is widely distributed in the region from Chandigarh to Pathankot and covers an area of 775.85 km^2. Dry deciduous scrub which mainly constitutes of *Lantana* scrub is the next dominant forest type in the region followed by fragmented patches of moist deciduous forest found in Dhar, Pathankot, Dholna, Talwara, Nangal and Noorpur. The coniferous forest, covering an area of 6.51 km^2, is localized on the higher ridges or side slopes in the northern part of the Pathankot district. Punjab Shivalik has only two PAs (Takhni-Rehampur WLS and Lalwan Community Reserve). There are six hydel projects (India-WRIS WebGIS 2015).

9.13.4.1 Biodiversity

The dry deciduous forest shows highest diversity (3.54) with 363 species followed by moist deciduous forests (3.10) with 161 species and dry deciduous scrub (2.27) having 77 species, and coniferous forest shows least diversity (1.62) with 58 species. Familial biodiversity is higher in most deciduous forest (36 families) followed by dry deciduous (31) and deciduous scrub (25). It is important to note that deciduous scrub has good biodiversity, and since root stock is already existing, a little protection can help to rejuvenate it to high-quality forest ecosystem. Two rare species, viz. *Delphinium denudatum* Wall. ex HK.f. and Th. and *Peristylus constrictus* (Lindl.) Lindl., were found. In the Shivalik hills of Punjab state, 240 economically important plants were recorded. Total importance value was observed maximum for dry deciduous type (TIV of 10.53) and least for coniferous type (TIV of 8.01). Some of the economically important species are *Achyranthes aspera, Acacia catechu, Azadirachta indica, Justicia adhatoda, Ageratum conyzoides (alien invasive), Cannabis sativa, Moringa oleifera*, etc. Some of the medicinally important plants are

Justicia adhatoda, Aegle marmelos, Azadirachta indica, Cordia dichotoma, Bacopa monnieri, Terminalia bellirica, Terminalia chebula, etc. (Table 9.12) (Roy et al. 2002d).

9.13.4.2 Landscape Analyses

Low fragmented area in Punjab Shivalik is pointing to the fact that most of the terrain is unsuitable for other land uses. Fragmentation map overlaid over the vegetation types shows low fragmentation in the forest cover. The degree of fragmentation was observed to be relatively higher in dry deciduous scrub (40.20%) followed by dry deciduous forest (5.69%). Moist deciduous forest showed least degree of fragmentation. Species richness in highly fragmented dry deciduous scrub is on the lower side (2.27), suggesting impact of biotic pressure (Table 9.13) (Roy et al. 2002d).

Majority of the forest area is under medium to very low level of disturbance in spite of high settlement interspersed within forest, since area under medium to very high fragmentation is less indicating that good connectivity exists in forests. This may also be attributed to dissected nature of terrain. About 36.87% area of deciduous scrub is highly fragmented followed by deciduous forest (16.79%). Deciduous forest

Table 9.12 Biodiversity status in the Shivalik hills of Punjab state

Forest type	No. of families	No. of species			Total no. of species	Total importance value
		Trees	Shrubs	Herbs		
Moist deciduous	36	42	24	95	161	10.46
Dry deciduous	31	141	75	147	363	10.53
Deciduous scrub	25	7	12	58	77	9.75
Pine	11	7	15	36	58	8.01

Source: Roy et al. (2002d)

Table 9.13 Fragmentation status of the vegetation of Shivalik hills of Punjab state

Vegetation type	Low	Medium	High	Very high
Moist deciduous	226.74	41.74	7.58	0.68
	(81.93)	(15.08)	(2.74)	(0.25)
Dry deciduous	604.11	125.14	43.73	3.36
	(77.82)	(16.12)	(5.63)	(0.43)
Deciduous scrub	153.88	37.31	15.54	139
	(44.51)	(10.79)	(4.49)	(40.20)
Pine	4.98	1.29	0.23	0.04
	(76.15)	(19.72)	(0.00)	(0.61)

Source: Roy et al. (2002d)

showed higher degree of disturbance (50.77% and 16.79% in high and very high categories, respectively). Pine forest showed least disturbance (11.04% and 8.71% in high and very high categories, respectively) followed by moist deciduous forest (17.98% and 14.27% in high and very high categories, respectively) (Table 9.14) (Roy et al. 2002d).

Moist deciduous forest showed high degree of biological richness (55.09% and 12.86% in high and very high categories, respectively) followed by dry deciduous forest (17. 92% in high and 16.19% in very high categories). Deciduous scrub showed least richness (13.96% in high and 1.61% in very high categories) as compared to pine forest (Table 9.15) (Roy et al. 2002d).

9.13.4.3 Discussion and Recommendations

The state of Punjab has agriculturally dominant landscape. Natural or secondary forests exist mainly in and around Shivalik hills. There is shortage of land for the expansion of industries in the plains, and therefore, pressure is on these hills. Some areas have been flattened to do either cultivation or set up industries. Even though the numbers of PA in Punjab Shivalik are only two, the more important is that these provide continuity of the habitat. Large areas are covered by scrub and need

Table 9.14 Degree of disturbance of the various vegetation cover types

Vegetation type	Low	Medium	High	Very high
Moist deciduous forest	40.88	146.49	49.73	39.46
	(14.78)	(52.97)	(17.98)	(14.27)
Dry deciduous forest	57.58	193.93	393.63	130.15
	(7.43)	(25.01)	(50.77)	(16.79)
Dry deciduous scrub	16.91	39.13	163.91	128.47
	(4.85)	(11.23)	(47.04)	(36.87)
Pine	0.91	4.25	0.71	0.56
	(14.15)	(66.01)	(11.04)	(8.71)

Source: Roy et al. (2002d)

Table 9.15 Status of biological richness

Vegetation type	Low	Medium	High	Very high
Moist deciduous forest	46.54	42.19	152.53	35.59
	(16.81)	(15.24)	(55.09)	(12.86)
Dry deciduous forest	77.34	434.42	139.00	125.00
	(9.97)	(56.0)	(17.92)	(16.1)
Dry deciduous scrub	235.17	55.95	48.13	5.56
	(68.20)	(16.23)	(13.96)	(1.61)
Pine	3.25	1.6	1.03	0.28
	(52.76)	(25.97)	(16.72)	(4.55)

Source: Roy et al. (2002d)

protection to avoid further degradation. The region has a large number of medicinal plants. Biologically rich areas are limited to certain regions and fragmented pockets. The highest biological richness was found in moist deciduous forests followed by dry deciduous forest.

9.13.4.4 Conservation Gaps

The patches of deciduous scrub dominated by *Lantana* are impenetrable in most of the areas, though alien-invasive species provides protection to hilly soil from erosion which is mainly sandy and also provides shelter to wildlife; however, it retards the growth of indigenous plants, thus affecting overall biodiversity scenario in the region. These biodiversity 'refugia' in the unique and fragile ecosystem of Shivalik need to be protected on priority for future prosperity. Resin trapping in pine forests needs review for their sustainability and ecosystem protection. Moist deciduous forests near Thara Uparla, Dasuya Forest Division and adjacent areas need attention for a fit case to be part of PA network. Based on the findings of biodiversity characterization in Punjab Shivalik, the area has been declared an Important Bird Area (Roy et al. 2002d; Chandrasekhar et al. 2003) (Fig. 9.2).

9.13.5 Jammu and Kashmir

Jammu and Kashmir (J&K) occupies a crucial position in ecological and geographical context being at the confluence of Central and South Asian region providing corridor for the flow of flora and fauna over time and space. The characteristic climate and terrain pattern of the state renders it as unique for phytogeography, ecology, harbouring natural resources, tourism destinations, physiography, etc. The geographical area of the entire state extends over 222,236 km^2. The total vegetation area is about 10.34% of the total geographical area of the state. Biogeographically the state is categorized into four biomes, viz. tundra, alpine, temperate and subtropical (Rogers and Panwar, 1988), and some parts fall into tropical region also. Glaciers like Siachen and Baltoro moving from Hindukush and Karakoram ranges drain in to Central Asia. Brackish water lakes like Pangong Tso, Tso Kar and Tso Moriri owe their origin either to glaciation (Trans-Himalayan), whereas Wular, Dal, Sheshnag, Mansar and Surinsar prevail in non-glaciated regions of lower and middle terrain regions. The climate of Jammu region is tropical, while it is semi-arctic in Leh and Ladakh region and temperate in Srinagar region.

Phytogeographically J&K represents one of the most diverse ecosystem diversities. The flora have evolved through various stages during the geomorphological evolution of this region and elements colonized at different times by humid tropical Malayan forms, tropical African forms, temperate and alpine north Asiatic-European forms, sclerophyllous Mediterranean forms, temperate East Asian forms and semi-arid Central Asian forms. The severe environmental conditions have further acted upon these geographical forms leading to the extinction of species, breaking up of distribution ranges or induction of genetic variation with or without speciation. The Shivalik ranges in the south have typical morphology, and tropical vegetation is of semiarid type. The forests of subtropical and temperate zones have typical species such as chir pine, deodar, blue and fir and spruce associated with temperate broadleaved system. Further north alpine desert vegetation of Ladakh is almost a treeless expanse. The plants are generally found growing along moist rivers or moist rock crevices due to the scarcity of precipitation. The Kashmir valley abounds in lakes and swampy lagoons with distinctive hydrophytic formations. Assessments of biodiversity till date have addressed the taxonomic information (Dar and Sundarapandian 2016) or forest crown density. The information is certainly valuable in terms of addressing the species per se. Other major studies attempted are patterns of wildlife habitats, overview on cold deserts or overall geographic scale degradation (Blaise and Dawa 2004, Aswal and Merhotra 1999, Murty 2001). There are 19 hydel projects in J&K (India-WRIS WebGIS 2015).

9.13.5.1 Biodiversity

The alpine vegetation covers about 20.4% of the area. The highest number of species (434) was recorded in dry alpine pasture followed by moist alpine pasture with 398 species and moist alpine scrub with 374 species. Western mixed coniferous forests, which are the major community in the valley, showed 268 herbaceous species followed by open scrub (160) and degraded forest (147). Fir, chir pine and moist alpine pasture types showed subsequent herb counts (108, 58 and 86, respectively). The highest number of tree species were recorded in dry deciduous forests (21), western mixed forests (16) and degraded forests (19) followed by fir (13). It is interesting to note that degraded forest and scrub have high number of tree and herb species. However, it may be because of disturbance possibly leaving systems harbouring more elements than stabilized moderately diverse, temperate forests, etc. Analysis of alpine vegetation showed that dry alpine pasture has maximum number of herbs (295) followed by moist alpine pastures (260). Maximum shrubs were seen in moist alpine pastures (31) followed by moist alpine scrub (29) and dry alpine pastures (24). Trend indicates that dry alpine pastures are the richest habitats followed very closely by moist category of the same. Scrubs contain substantially less number of species.

The temperate scrub showed a diversity (3.00) matching that of mixed coniferous forests, which may be due to the fact that scrub contains large number of sparse trees. The fir forest and mixed pine forests show values of H′ as 1.67 and 2.42, respectively (Roy et al. 2002a). Trends in alpine region for herbaceous and arborescent communities showed no interference by human cultivation practices in the former, whereas the latter witness microclimatic influences brought out due to substantial planting of woody elements by population in the valley. *Cirsium-Artemisia* community showed maximum diversity (7.69), whereas *Rosa-Hippophae-Mentha* community was the most diverse with H′ index of 7.37. About 469 species across the region are economically important and 307 medicinally important. The dry deciduous forests have 21 tree species. Dry alpine pasture has 337 herb, shrub and tree (mostly planted) species followed by moist alpine pastures. On the average, every vegetation type studied in alpine region showed at least 140 herbaceous and 16 shrub species, which points to the fact that these systems are biologically very rich in terms of life forms and relatively far undisturbed.

Fourteen species listed under IUCN category out of which 4 species (*Podophyllum hexandrum, Ephedra gerardiana, Ulmus wallichiana* and *Dioscorea deltoidea*) are under threatened category, 2 endangered (*Nepeta eriostachys* and *Saussurea costus*) and 10 are rare species (e.g., *Artemisia gmelinii, Chenopodium tibeticum, Primula minutissima*) which were collected during field surveys. Only four endemic species (*Carex borii, Carex stenophylla, Christolea crassifolia, Cirsium falconeri*) were found, which might be due to the relatively lesser abundance in this region as compared to continentally isolated biogeographies.

9.13.5.2 Landscape Analysis

Due to climatic and physiographical setup, the alpine vegetation indicates high fragmentation, which is not the case and is because of uniform application of parameters NW Himalaya. These regions are sparsely inhabited and biotic pressure is low. It is interesting to note that phenological types of forests are degraded but least fragmented with 76% in the low fragmentation category followed by the gregarious formations 68% and moist gregarious formations 62%. The phenological type and the degraded formations having 2% and 4% of area are under the high fragmentation category. This indicates that the remnant forests in J&K still have undisturbed core forest with little or no anthropogenic disturbances. Degraded formations are highly disturbed (45%) and are followed by 39%, 32% and 27% for gregarious, alpine and phenological type, respectively. The medium disturbed areas are almost the same in all the four types, i.e. phenological types, gregarious, degraded and alpine formations with 46%, 46%, 42% and 47%, respectively (Roy et al. 2002a).

The overall distribution of BR indicates maximum area (40%) under very high BR is in alpine vegetation followed by 18% in phenological formations and about 8% in gregarious formations. Maximum area under high BR of about 71% is in phenological types followed by gregarious formations (37%) and 31% in alpine vegetation. Maximum area (69%) under low BR is in degraded formation followed by alpine vegetation (16%) and 9% in gregarious formations. The high BR areas are dominated by western mixed coniferous forests and pine/deodar forests. The very high biodiversity zones were found in between the alpine and temperate regions. The high biological richness in these regions is a result of climatic conditions and conservation measures in the region. High biological richness zone was found to be present over 27.3% of total vegetation cover mapped using remote sensing data. As far as trends in fragmentation were observed, 51.9% area of the entire mapped vegetation was under low fragmentation, whereas 10.1% of the geographic area is experiencing high fragmentation. Disturbance index showed that 34.5% of the area was under high disturbance category, whereas only 19.6% area shows low degree of disturbance. In respect of vegetation types, degraded vegetation showed 45% under high disturbance regimes (Roy et al. 2002a).

9.13.5.3 Discussion and Recommendation

The presence of westernmost limit of Sal forest up to Ravi river (not Sutlej as reported by Brandis 1874) beyond Katra has great significance, particularly in view of the favourite habitat for plants and animals like tiger, elephant, etc. J&K like Himachal has very high number of biomes and provides all the important corridor with Central Asian biodiversity with that of Indian subcontinent and is known as migratory routes during past glaciations and also possible route during global warming. It was observed that major area of biological richness coincides with Zanskar Range, which is the transition between warmer west and cold aridity of the east. Zanskar range is the least disturbed and biologically richest. The very high BR band running parallel to Zanskar mountains might be possible due to the abundance of alpine pastures prevalent as well as the contribution of terrain complexity.

9.13.5.4 Conservation Gaps

Changthang WLS provides good connectivity with Kibber WLS in Himachal Pradesh in east and Karakorum WLS in the north and Hemis NP in the west, all representing arctic and Tundra conditions. The connectivity between Gamgul Siabehi WLS in Himachal and Doda and Banihal in J&K requires urgent consideration. The area around Trikuta WLS can be further increased on the northern and eastern parts. Hirpora-Gulmarg-Lachipora-Limber have very optimal connectivity. Further west there are opportunities to develop PA till international boundary with temperate to alpine ecosystems. The hill ranges between Pahalgam-Sonmarg-Dawar-Dras have vast area without any PA network. This transition between subtropical and temperate with alpine ecosystem will be worthy of conservation (Fig. 9.3).

9.14 Capacity Building for Biodiversity Management

The issues such as loss of habitats and biodiversity, unique ecosystems, threat of extinction or local extinction, etc. have generated a lot of awareness and interest among the conservationists, researchers and government systems in the last four decades. A vast majority of the reports are compiled based on the existing inventory and very less information with new and original data. A large number of projects are

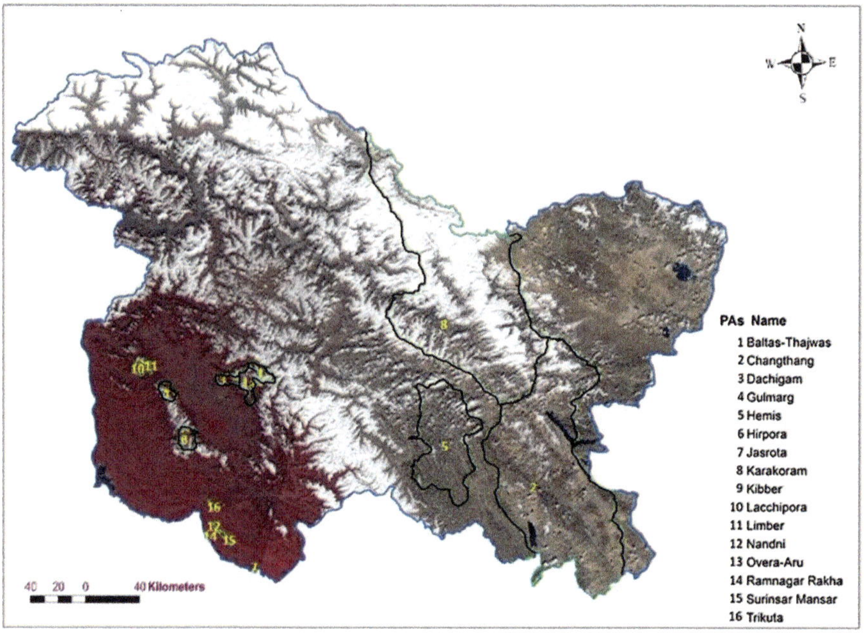

Fig. 9.3 Protected areas in Himachal Pradesh, Haryana and Punjab Shivalik overlaid on false colour composite (RGB:3,2,1) of IRS AWiFS data (13 October 2014 and 15, June, 2014)

funded by the government for 'biodiversity conservation' through government's own system, NGOs and other stakeholders. Of late the use of geospatial technology in biodiversity studies has increased, but the kind of awareness expected at administrative and managerial levels among the 'biodiversity managers' is either lacking or is very superficial. On the other side, the trained manpower does not have enough opportunities to contribute in biodiversity management. There is strong need of teaming taxonomist (botanist and zoologists), ecologists, geospatial experts and statisticians. ISRO has been bestowed with the responsibility of capacity building in natural resources management using geospatial technology. Of late there is also cessation of interest in taxonomy and fresh field data collection because of issues such as working conditions, policies, etc.

9.15 Conclusions

The NW Himalaya occupies a very crucial position in terms of link between biodiversity of Central Asia through Pamir Mountains and South Asia through Hindu Kush and Karakorum mountains to further east as evidenced by the presence of common/close relatives of floral and faunal elements. The region holds very important plant germplasm known for economic and medicinal uses and the future bioprospecting and biotechnological interventions and, therefore, needs to be conserved *in situ*, if possible *ex situ* as well. Since representative ecosystems in subtropical zone are poorly represented, it is recommended to conserve all the 'representative ecosystems' in the NW Himalaya to fill the gaps in management and conservation of biodiversity. The fragmented forested landscape in tropical and subtropical zones needs urgent attention to identify corridors from the 'refugia'. Earth observation data provide bird's eye view of the landscape giving opportunity to study the area, terrain and land use/land cover for visualizing and identifying potential areas for biodiversity conservation and gaps in management. Modelling approaches (graph, game, circuit theories, etc.) are available in GIS domain to identify critical forest patches for connectivity as well as for conservation. In order to understand and quantify the impact of potential climate change and its impact on the biodiversity, there is an urgent need to have long-term biodiversity monitoring system/programme. It is suggested that permanent sample sites need to be identified using remote sensing data, global positioning system (GPS) and homogeneity maps representing different microclimatic variability in zone. The sites sampled for floral diversity during 1998–2002 and 2007–2010 in biodiversity characterization study need to be revisited updating the information and noting the changes and its potential causes (Roy et al. 2002a, b, c, d and Roy et al. 2011). As discussed above, more areas can be brought under the purview of PA network. Lastly, it is recommended that concerted efforts and involvement of government and university systems are required to build capacity to use basic science data in conjunction with EO, GIS, GPS, etc.

Acknowledgements The state-wise information provided here is based on the reports of the project on 'Biodiversity Characterization at Landscape level using Remote Sensing and Geographic Information' carried out jointly by the Department of Space and Department of Biotechnology; and all the teams and contributors are gratefully acknowledged.

References

Agnihotri P, Husain T, Shirke PA, Sidhu, OP, Singh H, Dixit V, Khuroo AA, Amla DV, Nautiyal CS (2017) Climate change-driven shifts in elevation and eco-physiological traits of Himalayan plants during the past century. Current science 112(3): 595–601

Ahmad K, Sathyakumar S, Qureshi Q (2009) Conservation status of the last surviving wild population of hangul or Kashmir Deer *Cervus elaphus* in Kashmir. Journal of Bombay Natural History Society 016(3): 245–255

Anonymous (1991) Useful Plants of India. Publication & Information Directorate, Council of Scientific and Industrial Research, New Delhi

Arisdason W, Lakshminarasimhan P (2016) Status of Plant Diversity in India: An Overview. Status of Plant Diversity in India. Botanical Survey of India, Howrah. 1–9. http://www.bsienvis.nic.in/Database/Status_of_Plant_Diversity_in_India_17566.aspx) accessed on 23 September 2017)

Arora RK (1994) The Indian Gene Centre: Diversity in Crop Plants and their Wild relatives: In: Rana RS, Saxena RK, Tyagi RK, Saxena S, Mitter V (eds) Ex-situ Conservation of Plant Genetic Resources. National Bureau of Plant Genetic Resources New Delhi :29–37.

Arora RK, Nayar R (1991) Wild relatives of crop plant in India. NBPGR Sci. Monograph 7: 90

Aswal BS, Mehrotra BN (1999) Flora of Lahaul-Spiti : A cold desert in North West Himalaya. Bishen Singh Mahendra Pal Singh, Dehradun pp i–iii, 1–761.

Blaise HD, Dawa S (eds) (2004) Biodiversity of Ladakh: Strategy and Action Plan. Sampark and Ladakh Ecological Development Group pp 243

Brandis D (1874) Forest Flora of North-West and Central India. A Handbook of the indigenous Trees and Shrubs of these countries. Reprint Edition Bishen Singh Mahendra Pal Singh, Dehradun pp. 608

Burkill IH (1965) Chapters on the History of Botany in India, Botanical Survey of India, Kolkata

Census (2011) Census of India. http://www.census2011.co.in/census/state/uttarakhand.html). Accessed 23 Sept 2017

Champion HG, Seth SK (1968) A revised survey of the Forest Types of India. Government of India. Reprint Edition Natraj Publishers. Dehradun pp. i–xxvii, 1–404

Chandrashekhar MB, Singh S, Roy PS (2003) Geospatial modeling techniques for rapid assessment of Phytodiversity at landscape level in Western Himalayas, Himachal Pradesh. Current Science 84(5): 663–670

Chandrashkekhar MB, Singh S, Das NK, Roy PS (2004) Wildlife-Human conflict analysis in Kalatop Khajjiar Wildlife Sanctuary (H.P.): A geospatial approach. In Proceedings on Protected Habitats and biodiversity. Nature Conservators 8: 471–485. figs 7

Chaudhary P, Bawa KS (2011) Local perceptions of climate change validated by scientific evidence in the Himalayas. Biological Letters 7: 767–770

Chowdhary HJ (1999) Himachal Pradesh. In: Mudgal V, Hajra PK (eds). Floristic Diversity and Conservation Strategies in India, Botanical Survey of India, Kolkata vol. II: 845–903

CSIR-IHBT (2017) Patents: Council of Scientific and Industrial Research, Institute of Himalayan Bioresource Technology. http://www.ihbt.res.in/en/about-us/patents Accessed 19 Sept 2017

Dabla BA (2014) Migration Trends and Population Changes in Jammu and Kashmir. Kalpaz Publications. pp. 254

Dar JA, Sundarapandian S (2016) Patterns of plant diversity in seven temperate forest types of Western Himalaya, India. Journal of Asia-Pacific Biodiversity 9: 280–292

Duthie JF (1905) Flora of Upper Gangetic Plain and the Adjacent Siwalk and Sub-Himalayan Tracts. Vol. I–II. pp. 500

Dyer WTD (1872) Dipterocarpaceae. In: Hooker, J.D. (ed) Flora of British India. Reprint Edition Bishen Singh Mahendra Pal Singh, Dehradun and Periodical Experts, Delhi. vol III: pp 740

ENVIS (2017) ENVIS Centre on Wildlife & Protected Areas. Ministry of Environment, Forest and Climate Change. http://www.wiienvis.nic.in/Database/eri_8226.aspx Accessed 22 Sept 2017

Forman TTR, Godron M (1986) Landscape Ecology. Wiley and Sons, New York

FSI (2003) State of Forest Report. Forest Survey of India, Ministry of Environment, Forest and Climate Change, Dehradun

Hooker JD (ed.) (1872–1897) Flora of British India. Reprint Edition Bishen Singh Mahendra Pal Singh, Dehradun and Periodical Experts, Delhi. Vol. I–VII

Hunter Jr MJ (1995) Fundamental of Conservation Biology, Blackwell Science, NC USA

India-WRIS WebGIS (2014–15) Hydro Electric Projects in Himachal Pradesh. Water Resource Information System of India http://india-wris.nrsc.gov.in/wrpinfo/index.php?title=Hydro_Electric_Projects_in_Himachal_Pradesh Accessed 24 Sept 2017

IPCC (2007) Intergovernmental Panel on Climate Change, Climate Change 2007: The physical science basis Contribution of working group I to the fourth assessment report of the IPCC Cambridge University Press, Cambridge

Jaryan V, Uniyal SK, Gopichand, Singh RD, Lal B, Kumar A, Sharma V (2010) Role of traditional conservation practice: highlighting the importance of Shivbari sacred grove in biodiversity conservation. Environmentalist 30:101–110

Jhala YV, Qureshi Q, Gopal R, Sinha, PR (2011) Status of Tigers, Co-predators, and Prey in India, 2010. National Tiger Conservation Authority, Government of India, New Delhi and Wildlife Institute of India, Dehradun

Joshi NK, Tiwari SC (1990) Variation in woody species composition along an altitudinal gradient in a mountain flank of Garhwal Himalaya. Indian Journal of Forestry 13(4): 322–328
Joshi PK, Singh S, Agarwal S, Roy PS (2001a) Forest Cover Assessment in Western Himalayas, Himachal Pradesh using IRS 1C/1D WiFS data. Current Science 80(25): 941–947 figs. 4
Joshi PK, Singh S, Agarwal S, Roy PS (2001b) Land Cover Assessment in Jammu & Kashmir using Phenology as discriminant – a approach of wide swath satellite (IRS-WiFS). Current Science 81 (4): 392–399 figs. 4
Joshi PK, Yadav D, Singh S, Agarwal S, Roy PS (2002) Biome level characterization (BLC) of Western India a geospatial approach. Tropical Ecology 43(1): 213–228, f.1–3
Joshi PK, Roy PS, Singh S, Agrawal S, Yadav D (2006) Vegetation cover mapping in India using multi-temporal IRS Wide Field Sensor (WiFS) data. Remote Sensing of Environment 103(2): 190–202
Kachroo P (1954) Distribution of the Rebouliaceae in India. The Bryologist 56: 159–166
Kala CP (2010) Medicinal plants of the high altitude cold desert in India: Diversity, distribution and traditional uses. International Journal of Biodiversity Science & Management 2(1): 43–56
Kapur SK, Sarin YK (1990) Flora of Trikuta Hills, Bishen Singh Mahendra Pal Singh, Dehradun
Karthikeyan S (2000) A statistical analysis of Flowering plants of India. In: Singh NP, Singh DK, Hajra PK, Sharma BD (eds). Flora of India. Introductory volume Part II: 2001–217. New Delhi
Kumar S (1999) Haryana. In: Mudgal V, Hajra PK (eds). Floristic Diversity and Conservation Strategies in India, Botanical Survey of India, Kolkata vol. II: 807–844
Mamgain RP, Reddy DN (2015) Outmigration from Hill Region of Uttarakhand: Magnitude, Challenges and Policy Options, pp. 27. http://www.nird.org.in/nird_docs/srsc/srscrr261016-3.pdf Accessed 24 Sept 2017
Mathur V, Gopal R, Yadav S, Sinha P (2011) Management Effectiveness evaluation (MEE) of Tiger Reserves in India: Process and Outcomes. National Tiger Conservation Authority (NTCA), Government of India
Mathur VB, Sivakumar K, Onial M, Pande A, Singh Y, Kaur BJ, Ramesh C, Rosalind L, Bhattacharya I (2014a) In: Pande HK, Arora S (eds). India's Fifth National Report to the Convention on Biological Diversity. Ministry of Environment, Forest and Climate Change, Government of India, New Delhi pp. i–xxvi, 1–100
Mathur VB, Sivakumar K, Onial M, Ramesh C, Singh Y, Kaur BJ, Pande A (2014b). In: Pande HK, Arora S (eds). National Biodiversity Action Plan. Addendum 2014. Ministry of Environment, Forest and Climate Change, Government of India, New Delhi pp. 1–75
Murty SK (2001) Flora of cold deserts of Western Himalaya. Botanical Survey of India Calcutta
NBA (2004) The Biological Diversity Act, 2002 and Biological Diversity Rules, 2004. National Biodiversity Authority of India, Chennai pp 1–74
Nayar MP (1996) Hotspots of Endemic Plants of India, Nepal and Bhutan. Tropical Botanic Garden and Research Institute, Thiruvananthapuram
Pangti YPS, Joshi SC (eds) (1987) Western Himalaya. Gyanodaya Prakashan Nainital UP
Panwar P, Pal S, Tiwari AK (2016) Impact of *Lantana camara* Linn. Invasion on plant diversity, vegetation composition and soil properties in degraded soils of lower Himalayan Region, India. Indian Journal of Soil Conservation 44(2): 1770184
Parikh, K.S., Ravindranath, N.H., Murthy, I.K., Mehra, S., Kumar, S., James, E.J., Vivekanandan, E., and Mukhopadhyay, P. (2012). The economics of ecosystem and biodiversity-India: Initial assessment and scoping report, working document, TEEB-India
Ramakrishnan, P.S. (1998) Conserving the sacred for biodiversity: the conceptual framework. In: Ramakrishnan PS, Saxena KG, Chandrashekara UM (eds) Conserving the sacred for biodiversity management. Oxford and IBH Publishing Co, New Delhi
Rawat GS (2008) Special habitats and threatened plants of India. Wildlife and Protected Areas. In: Rawat GS (ed). Wildlife Institute of India. Envis Bulletin 11(1): 1–7
Rodgers WA, Panwar HS (1988) Planning a Wildlife Protected Area network in India vol. 1 and 2. Wildlife Institute of India, Department of Environment and Forests, Dehradun

Roy PS, Singh S, Hegde VS (2000) Biodiversity Characterization at Landscape level using satellite Remote Sensing and Geographic Information System. In: Roy PS, Singh S, Toxopeus AG (eds) Proceedings of Biodiversity & Environment. Remote Sensing and Geographic Information System Perspectives. Indian Institute of Remote Sensing, NRSSA, Dehradun. 18–47

Roy PS, Dutt CBS, Kant S, Gharai B, Pujar GS, Sharma N, Jhangir M (2002a) Biodiversity Characterization at Landscape level in Western Himalayas India using satellite Remote Sensing and Geographic Information System: Jammu & Kashmir. Indian Institute of Remote Sensing, National Remote Sensing Agency, Department of Space, Government of India, Dehradun. Pp. i–xxi, 1–234

Roy PS, Singh S, Chandrashekhar MB, Singh DK, Uniyal BP, Singh S, Hajra PK (2002b) Biodiversity Characterization at Landscape level in Western Himalayas India using satellite Remote Sensing and Geographic Information System: Himachal Pradesh. Indian Institute of Remote Sensing, National Remote Sensing Agency, Department of Space, Government of India, Dehradun. Pp. i–xxi, 1–234

Roy PS, Tiwari AK, Kumar S, Sharma N, Ghosh S, Mukherjee SK, Mathur VB, Saklani PL, Thapa R, Qureshi Q, Palni LMS, Sharma S, Hajra PK (2002c) Biodiversity Characterization at Landscape level in Western Himalayas India using satellite Remote Sensing and Geographic Information System: Uttarakhand (Uttaranchal). Indian Institute of Remote Sensing, National Remote Sensing Agency, Department of Space, Government of India, Dehradun. pp. i–xxi, 1–234

Roy PS, Singh S, Chandrashekhar MB, Joshi, PK, Singh DK, Uniyal BP, Singh S, Hajra PK, Jerath N, Prakash C (2002d) Biodiversity Characterization at Landscape level in Western Himalayas India using satellite Remote Sensing and Geographic Information System: Punjab Shivaliks. Indian Institute of Remote Sensing, National Remote Sensing Agency, Department of Space, Government of India, Dehradun. Pp. i–xxi, 1–234

Roy PS, Pujar GS, Peddi A, Yusuf AR, Jhangir A, Rashid H, Farooq M, Behera M, Chitale V, Romshoo S, Salroo I, Muslim M, Hussain A, Dar GH (2011) Biodiversity Characterization at Landscape level in Western Himalayas India using satellite Remote Sensing and Geographic Information System: Jammu & Kashmir. Indian Institute of Remote Sensing, Indian Space Research Organisation, Department of Space, Government of India, Dehradun. pp. i–xx, 1–135

Roy PS, Joshi PK, Jeganathan C, Yadav D, Agarwal S, S Singh (2004) Biome Level Characterization of Indian Vegetation using IRS WiFS data, ISRO-GBP Project Report, Indian Institute of Remote Sensing (NRSA), Dehra Dun, pp. 166

Roy PS, Singh S, Chandrashekhar MB (2006a) Biodiversity Characterization at Landscape level using satellite Remote Sensing and Geographic Information System. In Biodiversity in the Shivalik Ecosystem of Punjab. In: Jerath N, Puja, Chadha J (eds). Punjab State Council for Science & Technology, Chandigarh, pp. 21–64

Roy PS, Joshi PK, Yadav D, Agrawal S, Singh S, Jegannathan C (2006b) Biome mapping in India using multi-temporal satellite data and other inputs. Ecological Modeling 197(1–2): 148–158

Roy PS, Kushwaha SPS, Murthy MSR, Roy A, Kushwaha D, Reddy CS, Behera MD, Mathur VB, Padalia H, Saran S, Singh S, Jha CS, Porwal MC (2012) Biodiversity Characterization at Landscape level National Assessment. Indian Institute of Remote Sensing, Indian Space Research Organisation, Dehradun. pp. i–xx, 1–140

Roy PS, Murthy MSR, Roy A, Kushwaha SPS, Singh S, Jha CS, Behera MD, Joshi PK, Jagannathan C, Karnatak HC, Saran S, Reddy CS, Kushwaha D, Dutt CBS, Porwal MC, Sudhakar S, Srivastava VK (2013) Forest Fragmentation in India. Current Science 105(6): 774–750

Samant SS, Dhar U (1997) Diversity, endemism and economic potential of wild edible plants of Indian Himalaya. International Journal of Sustainable Development & World Ecology 4(3): 179–191

Saxena AK, Singh JS (1982) A Phytosociological analysis of woody species in forest communities of a part of Kumaon Himalayas. Vegetatio 50: 3–22

Sharma JR, Mudgal V, Hajra PK (1997) Floristic Diversity-Review, Scope and Perspectives. In: Mudgal V, Hajra PK (eds). Floristic Diversity and Conservation Strategies in India (Vol. 1 Cryptogams and Gymnosperms. Botanical Survey of India, Ministry of Environment, Forests and Climate Change, New Delhi

Sharma JR, Singh DK (2000) Status of Plant Diversity in India: An Overview. In: P.S. Roy, Sarnam Singh and A.G. Toxopeus. Proceedings of Biodiversity & Environment. Remote Sensing and Geographic Information System Perspectives. Indian Institute of Remote Sensing, NRSSA, Dehradun. 69–105

Shrestha UB, Gautam S, Bawa KS (2012) Widespread climate change in the Himalayas and associated changes in local ecosystems. PLoS ONE 7(5): 1–10 e36741. https://doi.org/10.1371/journal.pone.0036741 Accessed on 25 Sept 2017

Singh A, Lal M, Samant SS (2009) Diversity, indigenous uses and conservation prioritization of medicinal plants in Lahaul Valley, proposed Cold Desert Biosphere Reserve, India. International Journal of Biodiversity Science & Management 5(3): 132–154

Singh CP, Panigrahy S, Thapliyal A, Kimothi MM, Soni P, Parihar JS (2012) Monitoring the alpine tree-line shift in parts of the Indian Himalayas using remote sensing. Current science 102(4): 559–562

Singh, DK (1997a) Floristic Diversity (Angiosperms): An Overview. In: Dhar U (ed). Himalayan Biodiversity: Action Plan. GBPIHED Himavikas Publication No. 10: 33–42.

Singh DK (1997b) Liverworts In: Mudgal V, Hajra PK (eds) Floristic Diversity and Conservation Strategies in India Cryptogams and Gymnosperms. Botanical Survey of India, Ministry of Environment, Forests and Climate Change, New Delhi. vol. I: 235–300.

Singh DK, Hajra PK (1996) Floristic Diversity. In: Gujral GS, Sharma V (eds). Changing perspectives of biodiversity status in the Himalaya. British Council Division, British High Commission, New Delhi. pp.23–38

Singh DK, Uniyal BP, Mathur R (1999) Jammu & Kashmir. In Mudgal V, Hajra PK (eds). Floristic Diversity and Conservation Strategies in India, Botanical Survey of India, Kolkata vol. II: 905–974

Singh DK, Sharma JR, Pande HC, Gupta KK, Singh S, Das K, Singh SK, Kumar S (2006) Flora of Punjab Shivaliks. In: Jerath N, Puja, Chadha J (eds). Biodiversity in the Shivalik Ecosystem of Punjab. Punjab State Council for Science & Technology, Chandigarh, pp. 65–450

Singh DK, Uniyal BP (2002) Flora of Jammu & Kashmir. In: Singh NP, Singh DK, Uniyal BP (eds). Botanical Survey of India vol 1. I–v, 1–900

Singh JS, Singh SP (1992) Forests of Himalaya. Gyanodaya Prakashan, India, pp. 294

Singh JS, Rawat YS, Chaturvedi OP (1984) Replacement of oak Forest with Pine in the Himalayas affects the nitrogen cycle. Nature 311: 54–56

Singh KP, Sinha GP (1997) Lichens In: Mudgal V, Hajra PK (eds) Floristic Diversity and Conservation Strategies in India Cryptogams and Gymnosperms. Botanical Survey of India, Ministry of Environment, Forests and Climate Change, New Delhi. vol. I: 195–234

Singh KP, Mudgal V (1997) Gymnosperms. In: Mudgal V, Hajra PK (eds) Floristic Diversity and Conservation Strategies in India. Botanical Survey of India, Dehradun pp. 443–472

Singh N, Murugan MP, Bhoyar M, Angchok D, Srivastava RB (2011) Vegetables Scenario in Cold Desert Ladakh. Defence Institute of High Altitude Research (DIHAR) Defence Research & Development Organisation. Extension Bulletin 9: 1–7.

Singh P (1999) Punjab. In: Mudgal V, Hajra PK (eds). Floristic Diversity and Conservation Strategies in India, Botanical Survey of India, Kolkata vol. III: 1363–1382

Singh P, Dash SS (2014) Plant Discoveries 2013 – New Genera, Species and New records. Botanical Survey of India, Kolkata

Singh S, Chandrashekar K, Singh S, Singh DK, Uniyal BP (2003a). Three New Distributional Records for Himachal Pradesh Journal of Hill Research 16(1): 49–50

Singh S, Chandrashekhar K, Roy PS, Singh S, Singh DK, Uniyal BP, Chandrashekhar MB (2003b) Additions to the Flora of Lahaul-Spiti District, Himachal Pradesh. Annals Forestry 11(1): 59–62. pl.1

Singh, S. and G. Panigrahi (2005a). Ferns and Fern-Allies of Arunachal Pradesh. M/s Bishen Singh Mahendra Pal Singh, Dehardun, India, vol. I: i–xx, 1–426.

Singh, S. and G. Panigrahi (2005b). Ferns and Fern-Allies of Arunachal Pradesh Vol. II. M/s Bishen Singh Mahendra Pal Singh, Dehardun, India, vol. II: i–xviii, 427–881

Singh S, Roy PS, Chandrashekhar MB, Singh DK, Singh S, Uniyal BP, Joshi PK (2004) Phytodiversity analysis: a geospatial approach. Bulletin Botanical Survey of India (Seminar Volume) 46(1–4): 19–33

Singh SP, Singh JS (1985) Man and Environment: the Central Himalayan Case. Biological Memoirs 11(1): 47–59

Singh SP (2007) Himalayan Forest Ecosystem Services: Incorporating in National Accounting. Central Himalayan Environment Association, Nainital, India

Thakur AK, Singh G, Singh S, Rawat GS (2011) Impact of Pastoral Practices on Forest Cover and Regeneration in the Outer Fringes of Kedarnath Wildlife Sanctuary, Western Himalaya. Journal of Indian Society of Remote Sensing 39(1): 127–134

Tiwari AK (1998) Mapping and Monitoring of Biodiversity using Remote Sensing: A model study for Central Himalaya. Global Change Studies. Scientific Results from ISRO/GBP, ISRO/GBP Publication, Antarisksh Bhavan, Bengalure

Turner MG (1987) Landscape Heterogeneity and Disturbance. Springer-Verlag, New York

Turner MG, Gardner RH (eds) (1990) Quantitative Methods of Landscape Ecology. The Analysis and Interpretation of Landscape Heterogeneity. Ecological Studies Series, Springer-Verlag, New York

Uniyal BP (2002) Dipterocarpaceae. In: Singh NP, Singh DK (eds) Flora of Jammu & Kashmir Singh and B.P. Uniyal, Botanical Survey of India, Kolkata vol. 1:710

Uniyal BP, Khanna KK, Balodi B (1999) Uttar Pradesh. In: Mudgal V, Hajra PK (eds). Floristic Diversity and Conservation Strategies in India, Botanical Survey of India, Kolkata vol. III: 1529–1574

Verma M (2000) Economic valuation of forests of Himachal Pradesh. Himachal Pradesh Forestry Sector Review Report: International Institute of Environment and Development (IIED), London UK

Vohra JN, Aziz N (1997) Mosses. In: Mudgal V, Hajra PK (eds), Floristic Diversity and Conservation Strategies in India. pp. 301–374. BSI, Dehradun

Wilcox BA, Murphy DD (1985) Conservation Strategy: The Effects of fragmentation on extinction. American Naturalist 125: 879–887

Chapter 10
Himalayan Spatial Biodiversity Information System

Harish Karnatak and Arijit Roy

10.1 Introduction

Spatial distribution of environmental resource and its management issues are determined by complex processes and relationships. It involves several interrelating elements with many attributes and a dynamic behavior that required advanced spatial analytical capabilities in the GIS software. The technological solutions required to analyze the system include spatially distributed simulation and optimization models, interactive information system, decision support tools, and expert systems based on geospatial technologies. The primary paradigm of a GIS is the map, an inherently static concept of limited attributes. While modern GIS extends the scope of what can be done within this paradigm toward digital cartography considerably, elaborate applications can be built within existing GIS systems and powerful and flexible tool that involves spatial elements can be developed for different environmental applications. The Eastern Himalayan region is known as one of the global biodiversity hotspots. It includes several Global 200 eco-regions, two Endemic Bird Areas, and several centers for plant diversity. The high biological diversity of the Himalaya is mainly due to the multiple biogeographic origins. The climate variability as a result of being associated with the huge, complex, and steep terrain also gives the Himalayan region a plethora of habitats for the occurrence of the biodiversity hotspot in the region. Apart from being a storehouse of natural resources, the Himalaya is also

H. Karnatak (✉)
Geoweb Services, IT & Distance Learning, Indian Institute of Remote Sensing (IIRS), Indian Space Research Organisation (ISRO), Department of Space, Government of India, Dehradun, India
e-mail: harish@iirs.gov.in

A. Roy
Forestry and Ecology Department, Indian Institute of Remote Sensing (IIRS), Indian Space Research Organisation (ISRO), Department of Space, Government of India, Dehradun, India

R. R. Navalgund et al. (eds.), *Remote Sensing of Northwest Himalayan Ecosystems*,
https://doi.org/10.1007/978-981-13-2128-3_10

prone to innumerable natural and anthropogenically induced disasters. This is evident by the recurrent calamities like Kedarnath tragedy, which results in huge loss of life and property.

In this scenario there is a need to generate database in the region for understanding the structure and functioning of the fragile ecosystem. As part of the National Action Plan on Climate Change for sustaining the Himalayan Ecosystem, there is a need to collate the database on the Himalaya especially the geo-database which is already available as well as the database which is under preparation. This will enable researchers and planners to identify the areas for prioritization and conservation in the ever-changing scenarios. One of the most important aspects in the Himalaya is to understand when and where and how much in terms of natural resources and also infrastructure.

The information on distribution of biological diversity from gene to landscape level is very important for preparing an effective biodiversity conservation plan and taking up policy actions.

India which is known for its traditional knowledge on conservation of biological recourses is one of the mega biodiversity regions of the world. Authentic baseline data on biodiversity of India was the need of the hour due to large landscape of the country with unique floral and faunal diversity, vastness, endemism, and rarity of the many plant species. A Web-enabled information service delivers distributed but available information in one sharable framework (Karnatak et al. 2007). The Web-enabled information system is required to monitor, analyze, and plan action-oriented programs for conserving and preserving our biological wealth. The advanced information and communication technologies and its integration with GIS are offering to utilize these technologies for biodiversity conservation and prioritizations (Karnatak et al. 2007).

The vegetation-type map using multi-temporal satellite remote sensing data and model-driven Disturbance Index and Biological Richness map of India is generated under national-level project on Biodiversity Characterization at landscape level. This was jointly executed by the Department of Biotechnology and Department of Space, Government of India, during 1998–2012. The Western Himalayan region being a biodiversity hotspot was taken as priority area during this study. This major outcome of the project is primary and secondary geospatial data and repository of field sample plots as ground truth data on plant species information (Roy et al. 2015). The spatial data generated during the study is available as central data repository using enterprise-level geo-enabled relational database management system (RDBMS) for multiuser access through Web GIS application. The raster and vector data is made available in public domain as geospatial Web services using Service Oriented Architecture (SOA) and OGC Web Service specifications. The Web processing-based modeling tools are very effective for biological data and information services (Anthony et al. 2013). The Biodiversity Information System provides online geoprocessing services based on raster data models (Roy et al. 2012). The WPS-based online geoprocessing provides flexible and interoperable model for online raster and vector data analysis in multiuser environment (Cepicky and Becchi 2007). The system is accessible in the Internet domain using the URL http://bis.iirs.gov.in. This chapter presents a detailed status of BIS with various tools and functionalities available in the portal to study the biodiversity status of the country.

10.2 Biodiversity Characterizations at Landscape Level Using RS&GIS

The national-level Biodiversity Characterization has been taken up jointly by the Department of Space (DOS) and Department of Biotechnology (DBT) during 1998–2012. This project has generated geospatial data and information on (1) satellite-based primary information on vegetation type; (2) model-driven landscape indices on Disturbance, Fragmentation, and Biological Richness; and (3) geo-tagged field samples plots data on plant species information. The entire national level data is organized in a central data repository and made available as Web GIS-based geoportal for public access. The following geospatial data on biodiversity of India in the form of spatial data products are available in BIS:

- Vegetation-type map
- Fragmentation Index map
- Disturbance Index map
- Biological Richness map
  - Phytosociological database for 16,000+ sample plots for entire India

The study provides information of high to low disturbance and biological richness areas suggesting future management strategies and formulating action plans in 1:50,000 scale. This data repository is first of its kind to generate baseline database and information services for India which is one of the critical information required to study impact of climate change in the Indian subcontinent.

The spatial landscape model developed under this study provides a systematic graphical user interface (GUI) as well as an implementation of the spatial landscape algorithms such as fragmentation, patchiness, porosity, interspersion, juxtaposition, human disturbance (Euclidean distance), population density, terrain complexity, species richness, ecosystem uniqueness, and biodiversity values. The model generates the Biological Richness map of a study area using spatial landscape parameters.

10.3 Indian Biodiversity Information System (BIS)

The Web GIS-based geoportal is developed by using central data repository of national-level project on Biodiversity Characterization at landscape level in India. The geoportal (http://bis.iirs.gov.in) is developed using open system architecture for its wider dissemination and reach to the targeted user (s). The geospatial data is made available to its user (s) as OGC Web services using Service Oriented Architecture (SOA) for interoperable Geoweb services. Various GIS tools and utilities are available directly in the portal for geo-visualization and user-defined query generation. The Web Map Service (WMS)-, Web Feature Service (WFS)-, and Web Coverage Services (WCS)-based systems allow interactive geo-visualization, online query, and map outputs generation based on user requests and responses (Dubois et al. 2013). BIS is available as an interactive Web GIS application using typical mash-up architecture. The mash-up-based architectures are user centric and have great scope of generating user-defined data and information services for variety of applications (Karnatak et al. 2011).

The BIS geoportal provides online Geoweb services for plant biodiversity data sets of India including biodiversity hotspot in Western Himalaya. It has two major components, i.e., biodiversity spatial viewer and data download. These components are developed using open-source Application Programming Interface (API), i.e., OpenLayers for client side programming and PHP for server side programming. The important features available in the portal are:

- Design and development of central geo-data repository and Geoweb services using open system architecture.
- Central data repository of India's plant biodiversity information services.
- Data sharing and dissemination services – data download and Geoweb services based on user-defined area of interest (AOI) (simple online map drawing or shapefile upload).
- Online geoprocessing engine for raster-based data analysis.
- Responsive Web GIS application for geo-visualization, overlay of GIS layers using local and remote servers, navigation tools, spatial filters, measurement tools, layer swiping tool, etc.
- The search utility for plant species and their spatial distribution is available.
- The Geoweb services as OGC-compliant service are also available for interoperable GIS solution at user end.
- The geospatial data is downloaded by more than 800 users from different parts of the globe.
- At present the new version Spatial Landscape Model (SPLAM) is being developed in R platform. It is planned to develop the model as online modeling platform in open system architecture.

10.4 Technological Implementation of BIS

SOA-based solutions are presented for various services data sharing and dissemination. The Web GIS services for geospatial data using Web service standards developed by OGC are developed for central geo-data repository. The spatial queries and analysis for vector data is based on Structured Query Language (SQL) using PostGIS and Web Feature Service (WFS)-Transaction (T) operations

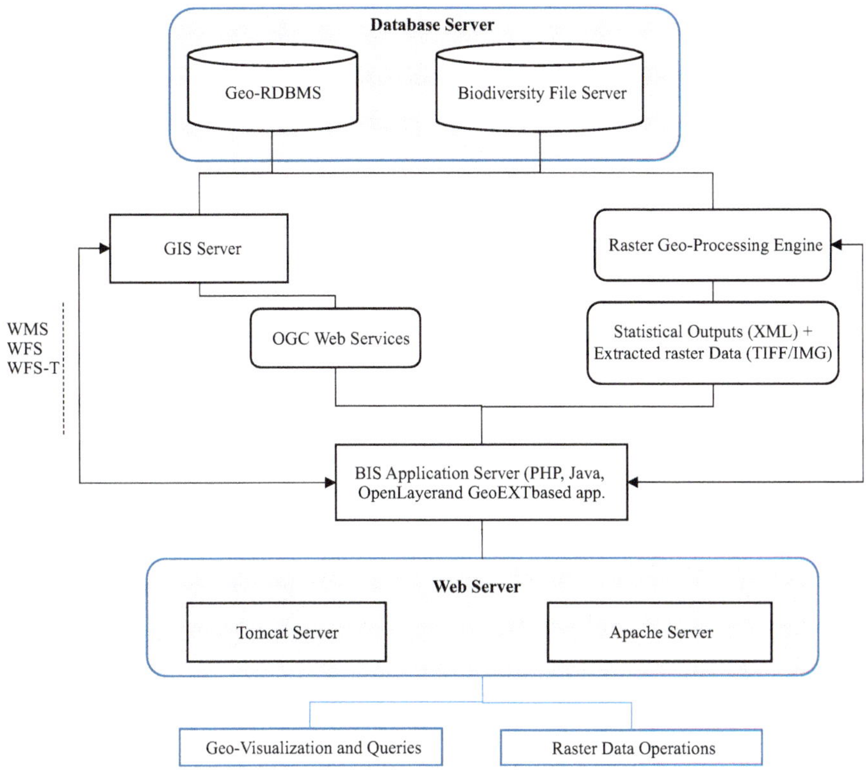

Fig. 10.1 Methodology – National Biodiversity Information System

(Karnatak 2014). The vector data is stored and managed in PostgreSQL and published as WMS and WFS. The development of online geoprocessing system for raster data in Web GIS environment is one of the important components of BIS. The technically established approach for the design and development of BIS is shown in Figs. 10.1 and 10.2.

The technological implementation of this portal is based on three basic principles, i.e., user inputs (well-known text (WKT) or shapefile, GIS operations (GDAL/OGR), and information presentation (XML and HTML). The major steps involve as follows.

10.4.1 Geoweb Application

The Web application is developed using open-source software development environment such as PHP, OpenLayers, and GeoExt APIs. The Web application has two major components for GIS operations, i.e., biodiversity spatial viewer and data download facility. The plant species data sets are also available for overlay on

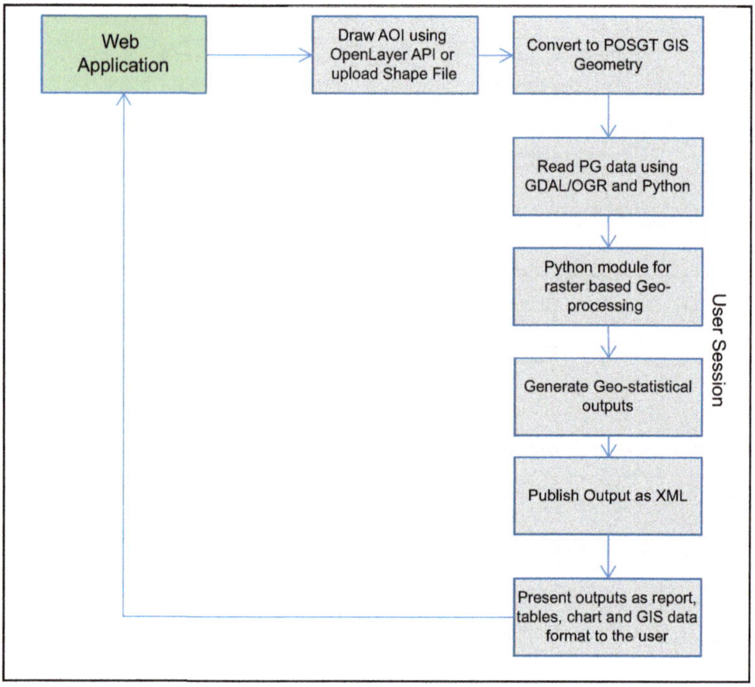

Fig. 10.2 Information flow

different map layers with online attributes queries and analysis. The user can filter the attribute(s) data with their location based on different characteristics such as medicinal, economic, and ecological importance, endemic and endangered plants, etc. The biodiversity spatial data viewer also provides the tool for searching a plant species with its scientific name, family, or local name and shows its spatial distribution on map.

10.4.2 Data Downloads and Online Analysis

This module is developed under data download section of BIS geoportal. The user can define his/her study area either by drawing on map or by uploading study area as shapefile (Karnatak et al. 2014). The system creates a user session for each request in the server under various user-defined data analysis tasks that are performed. The user-defined inputs are converted as PostGIS geometry and stored as a row in table of PostgreSQL database. This PostGIS geometry is used to generate raster data analysis outputs using Python programming language.

In BIS, the geoprocessing engine is based on GDAL/OGR library which is implemented in Python programming language. The geoprocessing engine allows online spatial analysis for raster data in multiuser environment.

The data downloads and analysis section of BIS geoportal provides a very interactive online tool for raster operations such as clip-zip-ship for online data delivery, calculation of area statistics, multilayer GIS operations for generation of user-defined spatial outputs, and diagrammatic representation of geostatistical analysis. In a background process, the software module creates XML files as output for spatial and nonspatial queries and analysis performed by a user. The spatial outputs are also presented in HTML format for better presentation (Karnatak et al. 2014). The original geospatial data sets for selected AOI are also available for download using HTTP protocol.

10.5 Outcomes of Biodiversity Information System: Specific to Himalaya

The spatial and nonspatial data for the Western Himalayan region in India is also available for user access. In BIS geoportal various tools and utilities are available to analyze biodiversity data for effective conservation planning (Roy et al. 2012). The area-specific data visualization and query system are available as interactive web application under biodiversity spatial viewer (Figs. 10.3 and 10.4).

In biodiversity spatial data viewer, the GIS layers from BIS data repository and distributed Geoweb service from any remote data servers can be accessed. For example, the WMS services of ISRO Bhuvan Geoportal are integrated with biodiversity spatial layers available in BIS repository (Fig. 10.4). The GIS layers are

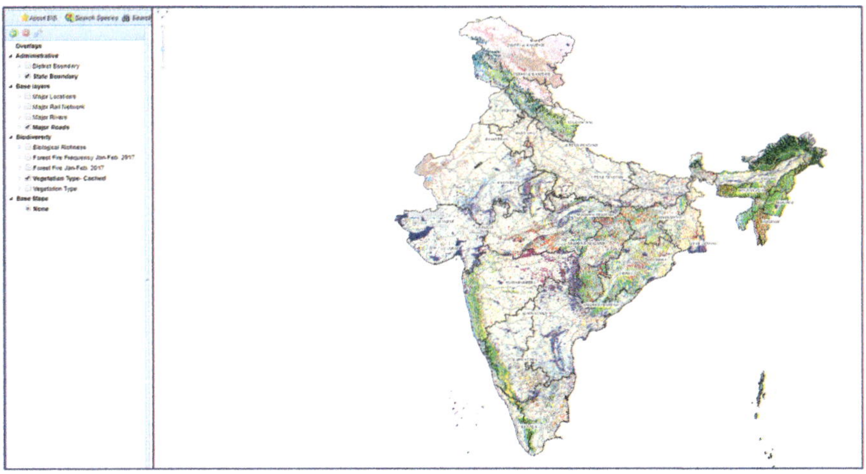

Fig. 10.3 Biodiversity spatial viewer – National Biodiversity Information System. (Source: http://bis.iirs.gov.in)

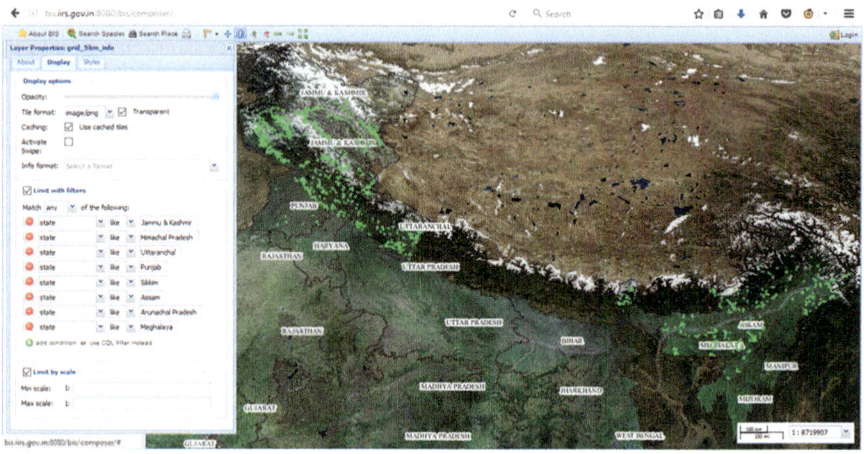

Fig. 10.4 Extraction of species data for Himalayan region in BIS

organized in layer tree (left panel) with a tool to add more layers from remote data repositories. The species attribute data of sample plots are linked with Web Feature Services of vector data for effective query builder in Web browser environment. This query builder allows generation of user-defined queries and spatial filters on vector data (Fig. 10.4).

One of the interesting tools available in BIS is "Layer Swiping tool." This tool provides an effective geo-visualization of spatial data sets especially for understanding the temporal changes in the data sets and comparing the two layers for establishing their relationship. The execution for mash-up-based architecture for geo-information services offers unique capability in a Web application to access data and information services from remote and local servers. This also provides an interoperable platform for online GIS environment. The vegetation-type map of the Himalayan region is hosted in BIS geoportal along with other landscape index maps as outcome of national biodiversity project (Fig. 10.5). The vegetation-type map generated for the Himalayan region using Biodiversity Information System is presented by Roy et al. (2013).

The Biological Richness (BR) of the region is one of the important information available under BIS geoportal. The BR map available for the Himalayan region is shown in Figs. 10.6 and 10.7 (P S Roy et al. 2013).

All the above information and data sets hosted in BIS geoportal are available for free download with various analytical tools. One of the unique features of this data download utility is that it will also analyze the spatial data with respect to area statistics. Once the user is satisfied with the data, then it will deliver the data by using the operation named as clip-zip-ship(s) (Fig. 10.8). The clip-zip-ship is the operation which will be performed online in real-time mode.

In BIS geoportal, the two modes are available to defining area of interest (AOI) by the user(s). In first method, the user can draw AOI on map and submit it for further

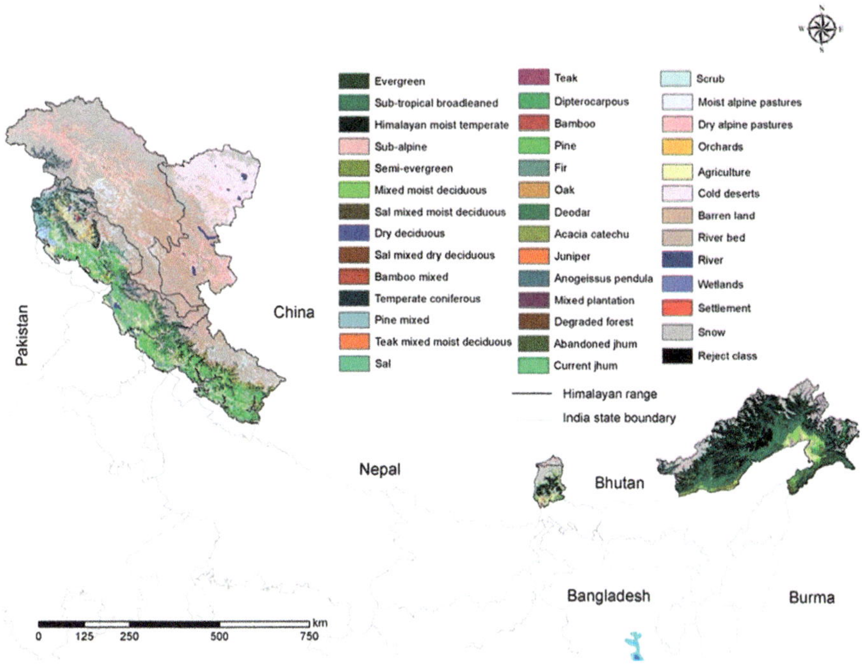

Fig. 10.5 Vegetation-type map of Himalayan region under BIS

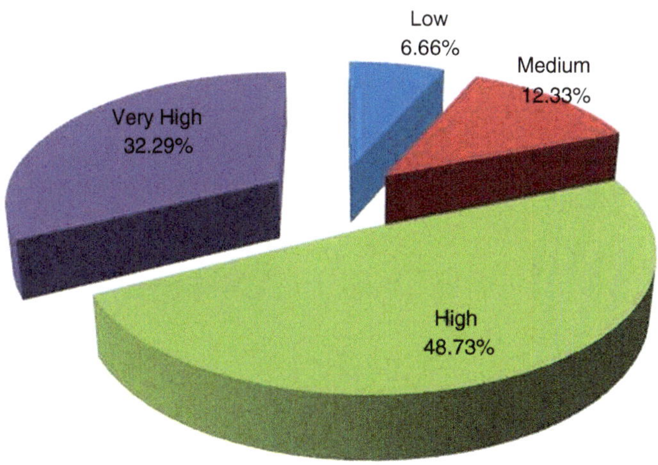

Fig. 10.6 Status of Biological Richness for Himalayan region under BIS portal

geoprocessing. In second method the shapefile of AOI can be uploaded in GCS and WG84 SRS. In background process, the AOI or shapefile are converted as PostGIS geometry value immediately after its successful upload into the server.

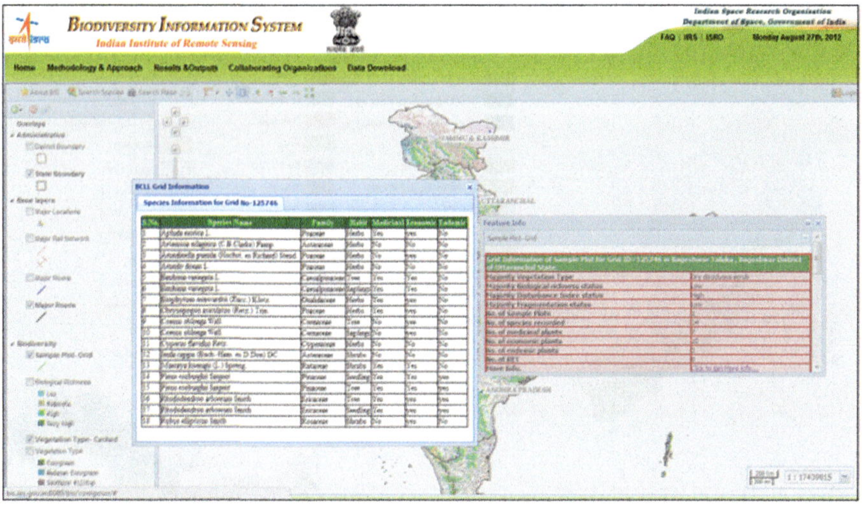

Fig. 10.7 Species queries and geo-tagging

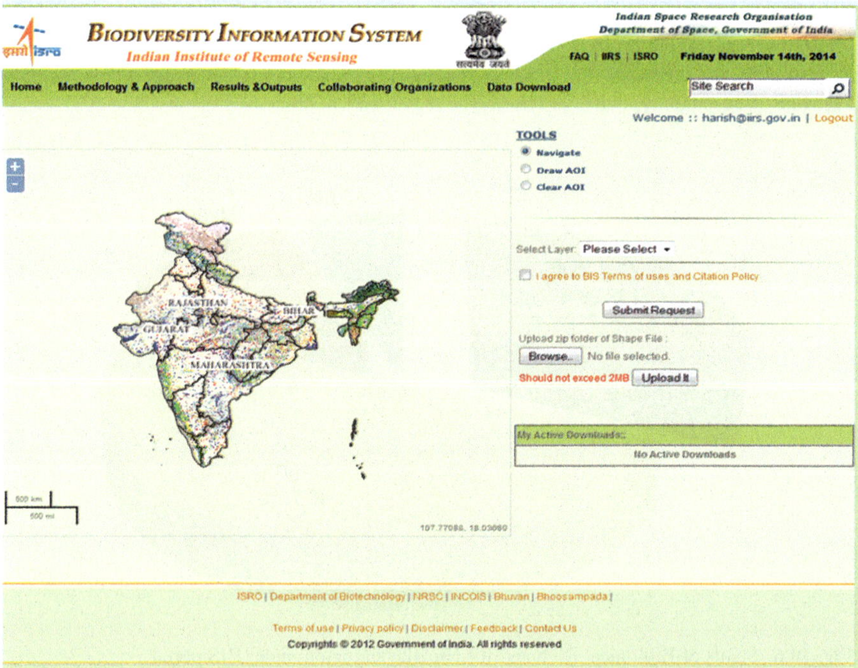

Fig. 10.8 Data download and analysis

Fig. 10.9 Home page of National Biodiversity Information System – http://bis.iirs.gov.in

The spatial data repository of BIS geoportal contains raster and vector data where each raster layer is of approximately 20 GB. The GIS operations such as clip-zip-ship, statistical analysis, multilayer GIS operations, chart preparations, etc. are performed on the fly based on user-defined AOI (Figs. 10.9 and 10.10).

The software development and data dissemination approach adopted in BIS geoportal for raster-based geoprocessing in Web GIS environment is important for many similar studies where online raster operations are required. The vector data analysis is made simpler by utilizing Structured Query Language (SQL). The performance of raster operations in Web GIS environment is a critical and important aspect for success of the system (Rautenbach et al. 2013). The performance enhancement for quick data delivery to the users is achieved by splitting computer processes into multiple parallel processes by adopting parallel computing for online geospatial analysis. The spatial and nonspatial outputs are generated as XML and presented as HTML.

10.6 Conclusions

The BIS geoportal (http://bis.iirs.gov.in) provides an online platform for sharing and dissemination of spatial biodiversity information services for the Himalayas. The central data repository of the system is based on outputs generated under DOS-DBT national project on Biodiversity Characterization at landscape level using RS&GIS. The data sharing and dissemination services are enabled through data download section which provides data download and online spatial data analysis utility for raster data. BIS geoportal is a unique centralized repository to access geospatial data-related studies on biodiversity conservation and prioritization. This information system provides an online GIS platform in Web browser environment for various

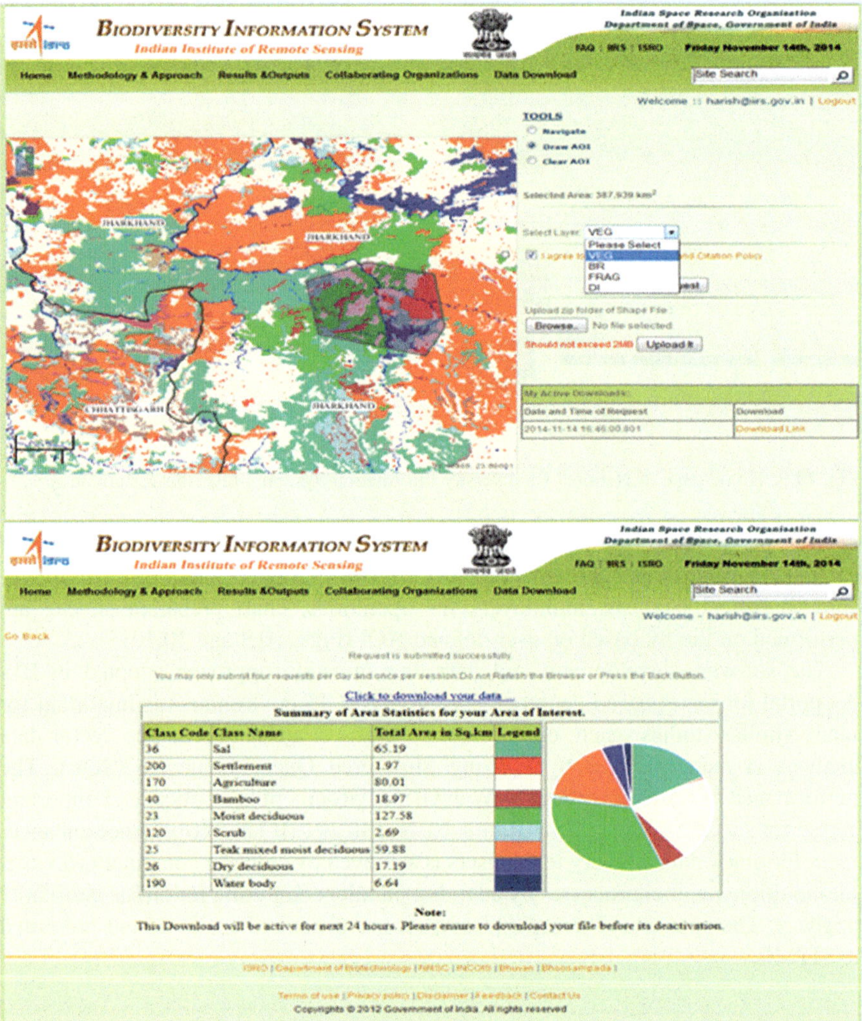

Fig. 10.10 Raster analysis and data download facility

geospatial data analysis and information services. The data is also available free of cost to its user(s) for analysis and use to generate scientific information. This data fulfils the requirements of an efficient and quality data on biological diversity at species, community, ecosystem, and landscape levels for identification of vulnerable ecosystem and species under risk. The spatially linked species database across the Himalaya along with spatial ecological data, risk species, and habitats under potential species loss risk can be identified. BIS is one of the important data repositories for prioritization of ecosystem conservation in the Himalayan region. In the coming decades, this archived data repository and information services with extensive documentation of biodiversity and its associated knowledge base will help in conservation and sustainable use of the biological resources for the benefit of society.

References

Anthony, M., Castronova, Jonathan, L., Goodall, Mostafa, M. and Elag. (2013). Model as web services using the Open Geospatial Consortium (OGC) Web Processing Service (WPS) standard. Environmental Modelling & Software, Vol. 41, pp.72–83.

Cepicky, J. and Becchi, L. (2007). Geospatial processing via Internet on remote servers– PyWPS. OSGeo Journal, Vol. 1, pp. 39–42. http://osgeo.org/journal. (Accessed 17.01.14).

Dubois, G., Schulz, M., Skoien, J., Bastin, L. and Peedell, S. (2013). eHabitat, a multi-purpose Web Processing Services for ecological modeling. Environmental Modelling & Software, Vol. 41, pp.123–133.

Harish C. Karnatak, Reedhi Shukla, Vinod Sharma, YVS Murthy and V Bhanumurthy (2011), "Spatial mashups technology and real time data integration in geo-web application using open source GIS- A case study for disaster management", Geocarto International © Taylor & Francis, Volume 27, Issue 6, pp-499–514, DOI: https://doi.org/10.1080/10106049.2011.650651.

Karnatak Harish Chandra, Sameer Saran, Karamjit Bhatia and P.S. Roy, (2007) "Multicriteria Decision Analysis in web GIS environment", Geoinformatica (2007) 11: pp 407–429: DOI https://doi.org/10.1007/s10707-006-0014-8.

Karnatak, H., Pandey, K., Oberai, K., Roy, A., Joshi, D., Singh, H., Raju, P. L. N., and Krishna Murthy, Y. V. N.: Geospatial data sharing, online spatial analysis and processing of Indian Biodiversity data in Internet GIS domain – A case study for raster based online geo-processing, Int. Arch. Photogramm. Remote Sens. Spatial Inf. Sci., XL-8, 1133–1137, doi:https://doi.org/10.5194/isprsarchives-XL-8-1133-2014, 2014.

Roy P.S., Roy, A and Karnatak H.C. (2013), "Himalayan Biodiversity Conservation: a Challenge in Climate Change Scenario", International Conference on Climate Change and Himalaya, 28–31st October, 2013, at NISCAIR, New Delhi.

Roy P.S., Roy, A and Karnatak H.C. (2012) "Contemporary tools for identification, assessment and monitoring biodiversity", Tropical Ecology 53(3):261–272, 2012 © International Society of Tropical Ecology, ISSN 0564-3295.

Rautenbach, V., Coetzee, S. and Iwaniak, A. (2013). Orchestrating OGC web services to produce thematic maps in a spatial information infrastructure. Computers, Environment and Urban Systems, Vol. 37, pp.107–120.

Roy P.S. et. All (2015), "New vegetation type map of India prepared using satellite remote sensing: Comparison with global vegetation maps and utilities", International Journal of Applied Earth Observation and Geo-information, vol. 39, pp-142–159, Publisher ELSEVIER.

Roy PS, Karnatak Harish, Kushwaha SPS, Roy A, Saran S (2012), " India's plant diversity database at landscape level on geospatial platform: prospects and utility in today's changing climate", Current Science, Vol 102, Issue 8, pp 1136–1142.

Chapter 11
Indian Bioresource Information Network (IBIN)

Sameer Saran, Hitendra Padalia, K. N. Ganeshaiah, Kapil Oberai, Priyanka Singh, A. K. Jha, K. Shiva Reddy, Prabhakar Alok Verma, Sanjay Uniyal, and A. Senthil Kumar

11.1 Background

The bioresource or biological resource includes all components of biological diversity with actual or potential value for humanity and the sustainability of the living systems. It is generally understood as the biotic component of ecosystems that includes organisms, parts thereof, populations, genetic resources and any other element that are of tangential and non-tangential benefit. Indian sub-continent supports varied bioclimatic regions and biodiversity. Despite several attempts from both the national and international survey organizations, we still have incomplete information about several taxonomic groups (e.g. plants, insects and other lower life forms) on their conservation status. This is because we have not been able to generate the comprehensive data on their distribution and population status. Even

S. Saran (✉) · K. Oberai · P. Singh · A. K. Jha · K. Shiva Reddy · P. A. Verma
Geoinformatics Department, Indian Institute of Remote Sensing (IIRS), Indian Space Research Organisation (ISRO), Department of Space, Government of India, Dehradun, India
e-mail: sameer@iirs.gov.in

H. Padalia
Forestry and Ecology Department, Indian Institute of Remote Sensing (IIRS), Indian Space Research Organisation (ISRO), Department of Space, Government of India, Dehradun, India

K. N. Ganeshaiah
University of Agricultural Sciences, Bangalore, India

S. Uniyal
CSIR-Institute of Himalayan Bioresource Technology, (Council of Scientific & Industrial Research), Palampur, India

A. Senthil Kumar
Indian Institute of Remote Sensing, Indian Space Research Organisation, Dehradun, Uttarakhand, India

R. R. Navalgund et al. (eds.), *Remote Sensing of Northwest Himalayan Ecosystems*,
https://doi.org/10.1007/978-981-13-2128-3_11

more, we hardly have reliable information on the levels of harvesting of these species from the natural habitats, their economic value and ecological significance.

In order to prospect and conserve the vast Indian bioresources, several organizations generate data sets in the country. Unfortunately all such data is highly scattered, not easily accessible with very little potential to add value to each other. Hence it was realized that the available information from different thematic specialities from different regions need to be networked in such a manner that various data sets can be seamlessly integrated and made available for any potential end user.

Globally, a number of online biodiversity database have been developed (Guralnick 2007; Kattge et al. 2011; Jetz et al. 2012) on the amphibians (http://www.amphibiaweb.org/), birds (http://www.birdlife.org/datazone/home), insects (http://www.brc.ac.uk/dbif/homepage.aspx), microbes (http://mbgd.genome.ad.jp/), plants (International Plant Name Index; http://www.ipni.org/), fishes (http://www.worldfishcenter.org/content/strategies-sustainability-online-open-access-biodiversity-databases) and bioresources in general (www.ibin.gov.in) (Gupta et al. 2016). For instance, some of the bioresource databases that provide online delivery of biodiversity information are Global Biodiversity Information Facility (http://www.gbif.org/), Australian Biological Resources Study (http://www.environment.gov.au/biodiversity/abrs/online-resources/index.html), Biodiversity Database Suriname (http://www.ethnobiobase.act-suriname.org/), Threatened Island Biodiversity Database (http://tib.islandconservation.org/), National Biodiversity Network's Gateway (https://data.nbn.org.uk/), National Biodiversity Data Centre (http://www.biodiversityireland.ie/) and Biodiversity of BC (http://ibis.geog.ubc.ca/biodiversity/BiodiversityDatabases.html).

In India, during the 1990s, DBT launched two independent programmes, viz. Digital Database of Indian Bio-resources and Biodiversity Characterization at Landscape Level (BCLL), using remote sensing and GIS. The *Jeeva Sampada* (Digital Database of Indian Bio-resources programme), launched under National Bioresource Development Board (NBDB) of DBT, began as an attempt to bring together the secondary data sets on the Indian bioresources. *Jeev Manchitra* is a unique effort to map the biodiversity at spatial level and was made by Indian Institute of Remote Sensing (IIRS) and National Remote Sensing Centre (NRSC), Department of Space, Government of India, in which spatial database on the vegetation type, plant species distribution patterns and landscape ecological analysis of the biodiversity-rich areas of the country was generated. The efforts were compiled in the form of *Jeev Manchitra* which provides a wide range of remote sensing and GIS-derived spatial layers and also nonspatial data for biodiversity analysis.

11.2 Genesis of IBIN

It is in this context that the *Indian Bioresource Information Network* (*IBIN*) was conceived by the Department of Biotechnology (DBT), Government of India, as a single digital window that brings together all the available databases and information on the bioresources and biodiversity of the country into one platform (Natesh 2006).

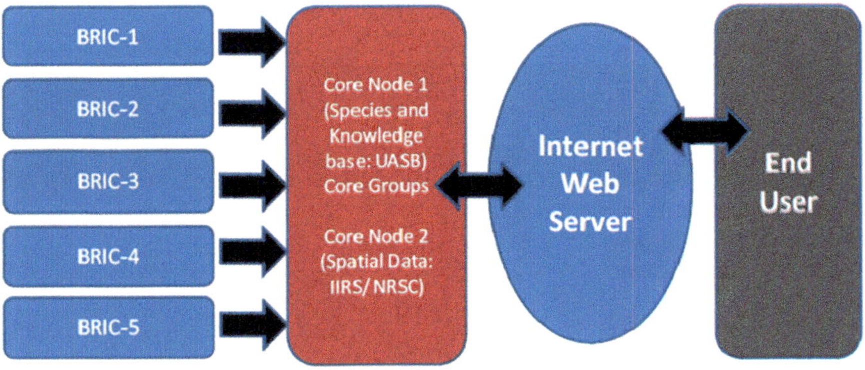

Fig. 11.1 IBIN conceptual level architecture

In view of the above, the Department of Biotechnology started a national level programme under the NBDB, Government of India, in 2006 for developing a website for providing data in different domains of bioresources and biodiversity, spatial distribution of vegetation, landscape features and natural resources. IBIN was planned to be uniquely developed as a single window service system for data on India's bioresources – plant, animal, marine, spatial distribution and microbial resources (Roy et al. 2012). The programme began as a collaborative effort between Indian Institute of Remote Sensing (IIRS), Department of Space and University of Agricultural Sciences as core nodes. Its main goal was to network and encourage an open-ended, coevolutionary growth of available digital databases related to bioresources of the country and to perform value addition to the databases by integration. The Indian Bioresource Information Network (IBIN) was planned to serve appropriate information on India's bioresources to the stakeholders engaged in bioprospecting, marketing, protecting biopiracy and the conservation of bioresources (Saran et al. 2012). As a result a single web platform (www.ibin.co.in) was developed and launched (Fig. 11.1).

Later various agencies/dataholders as partners of IBIN were brought together, and the website was developed to accommodate the data from other organizations. There are five institutions across the country (Ashoka Trust for Research in Ecology and the Environment (ATREE), Bangalore; Foundation for Revitalisation of Local Health Traditions (FRLHT), Bangalore; University of Calcutta, Kolkata (CU); Institute of Himalayan Bioresource Technology (IHBT), Palampur; and North-Eastern Hill University (NEHU), Shillong) that contribute databases to IBIN. These were recognized as Bio-resource Information Centres (BRICs), while the existing units are to be called as core nodes.

IBIN website was converted into IBIN portal (www.ibin.gov.in) (Fig. 11.2) with integration of five federally distributed BRICs (Bio-resource Information Centres) working on different aspects of biodiversity and bioresources of the country. The portal was launched on October 12, 2012 during the Conference of the Parties (COP 11) meeting at Hyderabad.

Fig. 11.2 Webpage of IBIN portal (www.ibin.gov.in)

11.3 IBIN Data: Standards

IBIN data is structured using data standards developed exclusively for the data integration and interoperability.

11.3.1 Species Data Standards

The data standards are defined as a set of concepts with basic attributes for integrating and retrieving information on living organisms for a wide range of users (Duval 2001). The standards for the species are being developed using a combination of concepts of globally reviewed metadata standards such as Darwin Core, Plinian Core, EOL Species Profile Model, Access to Biological Collection Data and Ecological Metadata Language (Table 11.1).

These set of concepts are classified into various groups containing different data elements. These groups are described as follows:

1. *Base elements*: This provides the basic information for the identification of the record. Its metadata type is NormalString and contains sub-data elements.

Table 11.1 List of metadata standards

Standards	Organization	Purpose
Darwin Core (DwC)	Taxonomic Databases Working Group (TDWG)	The Darwin Core is a metadata specification for information about the geographic occurrence of species and the existence of specimens in collections (Darwin Core 2016)
Ecological Metadata Language (EML)	Ecological Society of America	EML is a metadata description particularly developed for the ecology discipline (EML – Ecological Metadata Language/Digital Curation Centre 2016)
Plinian Core	Taxonomic Databases Working Group (TDWG)	Plinian Core is a standard oriented to share species level information (Plinian Core/Documentation 2016)
Species Profile Model (SPM)	Taxonomic Databases Working Group (TDWG) and Encyclopedia of Life (EOL)	SPM is intended to be a specification of data concepts and structure meant to support the integration of data on biology, ecology, evolution, behaviour, etc. of the species (tdwg/wiki-archive 2016)
Access to Biological Collection Data (ABCD)	Taxonomic Databases Working Group (TDWG)	It is meant to support the exchange of data about specimens and observations (tdwg/abcd 2016)

2. *Nomenclature and classification*: This gives the standard information for every known taxon. Its data type is standard information type and contains various sub-elements.
3. *Description*: This provides brief description, presented in a simple technical language, to distinguish the species from other close or similar ones. Its metadata type is text description type and contains various sub-elements.
4. *Natural history*: It is a natural history type, and it contains various sub-data elements.
5. *Molecular characteristics*: This gives information about the chemical structures and biological processes at the molecular level: DNA and protein sequences, protein structures and expression profiles of gene protein domains, families of genes, mutations, polymorphisms and involvement in disease. Its data type is like a placeholder for connecting with standards developed by specialists.
6. *Habitat location distribution*: This provides the general description of the sites where the species is found and geographical distribution of the species. It is habitat and distribution type.
7. *Demography and conservation*: This gives the information concerning the demographic aspects of the species like territory, population biology, threat status, direct threats, legislation and ancillary data and interventions undertaken designed to conserve species.

8. *Uses management*: This describes the ways in which the species are utilized by people and the actions directed at conserving or restoring species, and it is a complex type.
9. *Data set details*: This class explains the details of data set in the categorized form.
10. *Patent details*: These give the information about the granted patents.
11. *Specimen details*: This gives detail of an individual, item or part representative of a class or whole species for scientific study or display.
12. *Information listing*: If any data was obtained from the literature, a reference must be made, just as it is done in a regular scientific publication; similarly the referenced literature is placed under this category in the same format.
13. *Habitat description*: A description of the physical and biotic environment at the time and place of a collecting event. There are various elements defined in IBIN data standards to store multimedia. These are the following:

    (a) *Live image*: Displays images of the taxon in its natural habitat
    (b) *Herbarium specimen image*: Displays images of type specimen of taxon
    (c) *Paintings sketches*: Displays images of paintings or sketches related to the taxon
    (d) *Line diagrams*: Displays images of line diagrams related to the taxon
    (e) *Sonograms*: Displays a graph representing a sound, showing the distribution of energy associated with the taxon
    (f) *Karyograms*: Displays diagram or photograph of the chromosomes of a cell associated with the taxon
    (g) *Sound clippings*: Stores audio files associated with the species
    (h) *Videos*: Stores video files associated with the species

11.3.2 Spatial Data Standards

The web GIS portals provide a centralized and uniform medium to access the dispersed and varied resources and data services. Most of the web GIS-based portals available on the Internet have been developed for a particular theme and are directed to specific users. A single GIS service is insufficient to meet the needs of all users. Therefore, common international standards published by Open Geospatial Consortium (OGC) (http://www.opengeospatial.org/) for GIS data and services are being adopted. OGC has defined GIS services to build distributed systems based on the principles of service-oriented architectures (SOA). These systems unify distributed services through a message-oriented architecture. Web service standards are a common implementation of SOA ideals. The most popular services for the spatial data dissemination are OGC WMS (web mapping service) and OGC WFS (web feature service) (Nogueras-Iso et al. 2005). OGC WMS service maps are projected images, while OGC WFS services are features in GML (Geography Markup Language) format which are editable and can be spatially analysed.

11.4 IBIN Bioresource Database on Northwest Himalaya

Northwest Himalaya is the youngest mountain chain in the world involving orography-induced elevation regimes and monsoon-influenced climate. It is a global biodiversity hotspot with unique taxonomic hierarchies and high endemism in floral elements. Three distinct ecoregions, viz. the lesser Himalaya, the mid-Himalaya and the greater Himalaya, harbour a range of floral and faunal species. Climate, altitude and aspect primarily govern the vegetation distribution. The lower altitudes are dominated by subtropical forests and grasslands, while mid-elevation shows dominance of dry deciduous and conifers. The higher elevations are occupied by temperate conifers, alpine scrubs and pastures. The resource value spans from timber to non-timber category through wilderness ecotourism to medicinal aromatic-food-industrial gene pools. The region received large-scale clearing of forests for agriculture in the past. In the recent decades, chronic disturbances like illicit felling, grazing and forest fire are causing degradation of forests and loss of biodiversity. The landscape is characterized by commercial agriculture, horticulture, hydropower projects, tourism and mining activities.

11.4.1 Species Database

This module (Fig. 11.3) enables the potential user to view the list of species from the IBIN repository. Using this module the end user can search about the species in three ways: search based on the kingdom of species, namely, plants, animals, flora and fauna; search by species name based on alphabetical order, i.e. species starting by name 'A', 'B'...'Z'; search by entering the scientific name and/or common name of a particular specie; and text-based search by entering keywords. The search results can be filtered according to richness, title, public and BRICs (Bio-resource Information Centres that provide data to IBIN). This searching function results in the detail of species like its image, base elements, medicinal use, habitat, location, distribution, references, medicinal parts, etc.

11.4.2 Species Database from IIRS to UAS

The Indian Institute of Remote Sensing and University of Agriculture Sciences (UAS) have contributed nonspatial and spatial database on various groups of plant and animals for the Northwest Himalaya. The database includes primary and

Fig. 11.3 Species search module

Table 11.2 Statewise plant resource database for Northwest Himalaya by IIRS

NW region	Tree	Herbs	Shrubs	Saplings	Seedlings	Medicinal use	Total
Himachal Pradesh	18	160	29	3	NA	Yes: 80	
						No:132	210
Jammu and Kashmir	35	NA	9	NA	NA	No:42	
						Yes:2	44
Uttaranchal	42	40	24	9	6	No:54	121
						Yes:67	

Table 11.3 Nonspatial data on plants, animals and microbes compiled from secondary data sets by UAS

Plants	Birds	Butterflies	Domestic animals	Lichens	Silkworms	Microbes	Pests
7036	1667	536	15	490	18	7277	808

secondary data. Table 11.2 depicts the plant resource database for three states of NW Himalaya including information on plants of medicinal use contributed through IIRS sample plots. In addition, the database comprises data on 17,847 species of plants, animals and microbes compiled from secondary data sets (Table 11.3). The distribution data for animals and plants from **44,524** locations is collected from secondary data sets (Table 11.4).

Table 11.4 Spatial data (point locations) of plants, animals and microbes compiled from secondary data sets by UAS

Plants	Birds	Butterflies	Domestic animals	Lichens	Silkworms	Microbes	Pests
25892	2051	1255	43	490	22	13949	822

11.4.3 *Species Database from IHBT*

The Institute of Himalayan Bioresource Technology, Palampur, under CSIR is one of dedicated BRIC (Bio-resource Information Centre) of IBIN on floral resources of Northwest Himalaya. The webpage of the IHBT-BRIC is accessible at http://www.ihbt.res.in (Fig. 11.4).

Fig. 11.4 IHBT-BRIC webpage

IHBT-BRIC houses taxonomic database for over 1500 floral species from NW Himalaya with information species location and its characteristics. The database is a rich repository of herbaceous plant species including species of medicinal and economic importance. Apart from herb, shrub, tree and climber data archive, 106 families of plants are represented in the database with highest number of species belonging to Asteraceae family. In addition to the native plant species, the database has information on 497 alien or exotic plant species with information on origin, introduction and their status. The database has been validated using available floras and also by the Integrated Taxonomic Information System (http://www.itis.gov/) and The Plant List (http://www.theplantlist.org/).

11.4.4 Chemical Composition Database

This module provides the list of chemicals from the IBIN repository (Fig. 11.5). The chemical details include gene and chromosome information which provides a basic idea of the status, affinities and relationship of taxa which serve as prerequisites for undertaking any breeding programme. The list of chemicals is arranged in an alphabetical order so that the end user can view the particular chemical. Moreover the searching functionality is also provided to enable potential user to search by entering the chemical name and text (keywords).

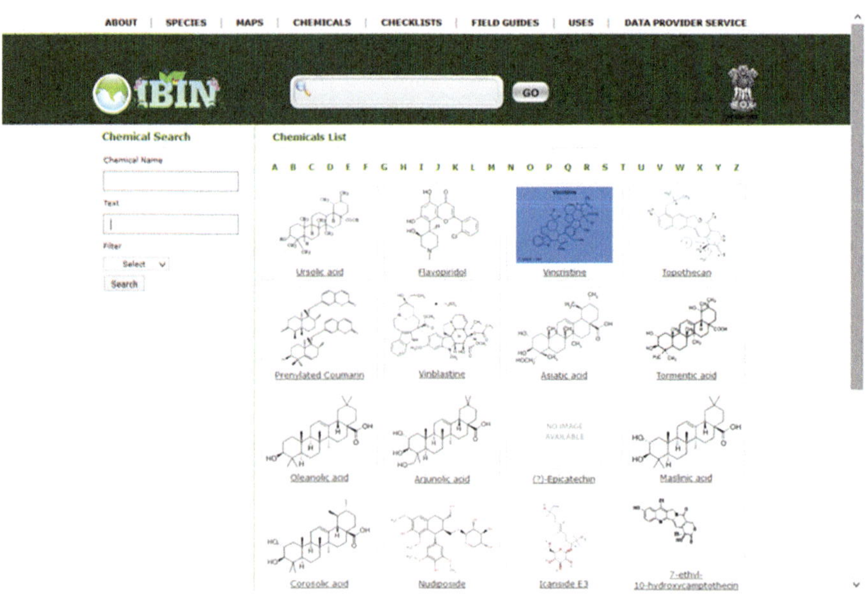

Fig. 11.5 Chemical list module

11.4.5 Field Guides

This module displays the list of identification kits from the IBIN database (Fig. 11.6). The identification kits are categorized into Pathanga Suchya, Phyllanthus identification and Rattans identification. These modules offer various taxonomic keys to the identification of species.

11.4.6 Spatial Data

The spatial database on vegetation cover types, disturbance regimes and biological richness is from DOS-DBT Biodiversity Characterisation at Landscape Level Project (2002–2004). The digital data on vegetation-type distribution is used as primary input to understand the landscape characteristics; analyse the disturbance pattern; integrate the ground-based phytosociological data,

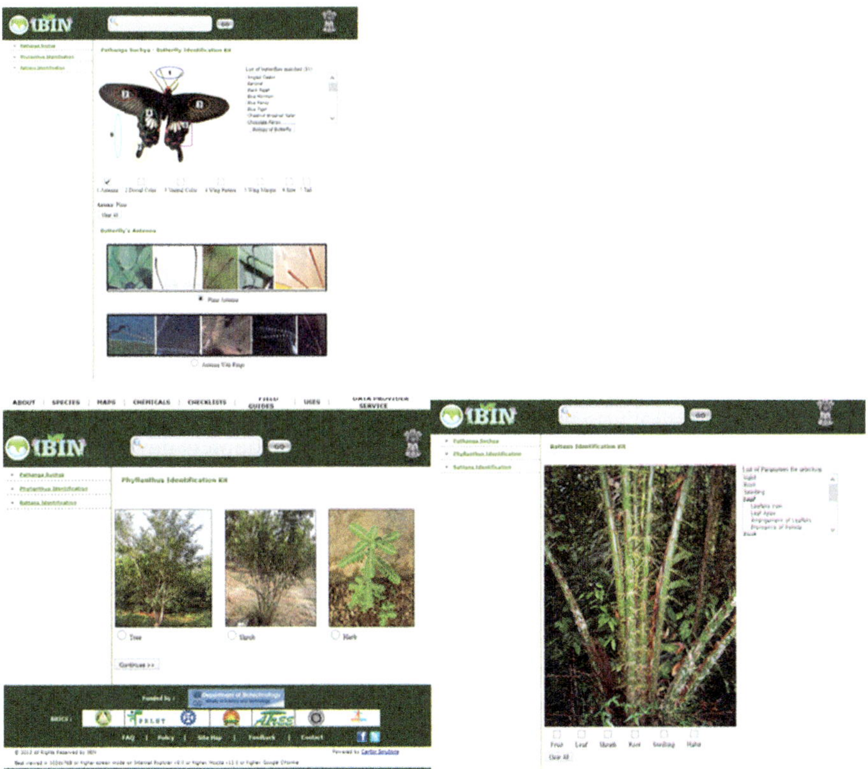

Fig. 11.6 Field guide module (*Pathanga suchya*, *Phyllanthus* spp. *and Rattans*)

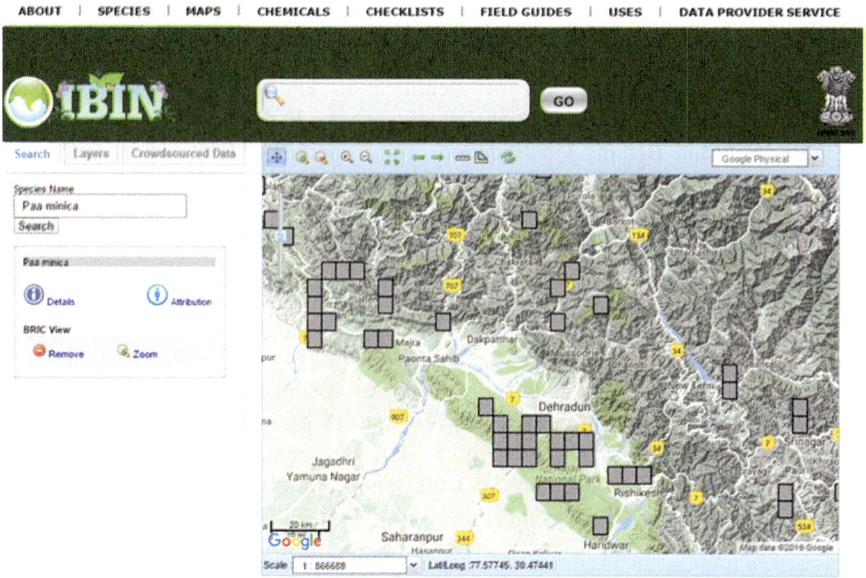

Fig. 11.7 Species grid locations

ecosystem uniqueness, economic importance and terrain variance; and generate a spatial biological richness map indicating the areas of high to low biological richness zones. The spatial data is linked to the species database of Botanical Survey of India-Red Data Book and field sample data, laid down in the different strata of vegetation. The field samples of key ecological characters have been used for geospatial extrapolation. The species database has been linked with above spatial details. The IBIN may facilitate the identification of gap areas, species-habitat relationship and biodiversity conservation planning by setting priority areas. These databases in conjunction with detailed site-specific field inventories support the identification of areas for bioprospecting (Fig. 11.7).

This application shows the geographical distribution of the searched species in the form of grids (5×5 KM) on map panel. The application is developed using AJAX (Asynchronous JavaScript and XML) technology wherein the user just needs to enter few characters identifying the species; automatically the species matching the entered text will be populated in the drop-down list which the user can select.

In the map panel, one can use Google Maps and OGC Web Map Services (WMS) like Bhuvan raster map, vegetation type, fragmentation type, disturbance index, biological richness map of India, etc. as the base map for visualization of grids. Attribution-related details are also available through this module.

11.5 Future of IBIN

The existing databases built during the past 15 years, both from the primary and secondary data sets, always need to be updated vertically and expanded horizontally to include new themes and also to fill the gaps. IBIN data sets have the potential of being very useful in shaping and aiding the national efforts in conservation, utilization and prospecting of the bioresources of the country. However the utility of the database becomes more effective if the conservation managers, the research units working on bioprospecting and the pedagogical units begin to use the information therein in the most effective way.

- IBIN database would be enriched through addition of new BRICs (species and spatial data) of different themes and domain on bioresources.
- IBIN with huge amount of data related to the bioresources can serve as a platform for decision-making using decision support system such as prioritization of conservation areas, assessment of habitat condition of RET and endemic species. Future efforts would demonstrate the use of IBIN data sets in utilization and bioprospecting of bioresources.
- A new framework has been conceptualized and proposed for integrating the IBIN services with ISRO's Bhuvan Geoportal. The continuous data flow and online availability of services on the web for all the web clients are the primary goal of IBIN. In order to ensure full access to all the spatial and species services, there will be a provision of IBIN data repository in the portal which shall consist of all the data from core nodes and BRICs and provide access to these services through IBIN interface.
- Crowdsourcing to gather data and also prepare mobile apps for IBIN such that it becomes more widely used. In crowdsourcing the public create and contribute georeferenced data which contain both spatial and nonspatial attributes of that location. The VGI/crowdsourcing can help in timely data integration at a very low cost, but such databases are compromised for data quality.

References

Duval, E. (2001). Metadata standards: What, who & why. Journal of Universal Computer Science, 7 (7), 591–601.

Gupta, A., Uniyal, S. K., Meenakshi, A. K., Kumar, A., & Singh, R. D. (2016). Designing and developing a Bioresource Information Centre for Floral Resources of Himachal Pradesh, Western Himalaya. *Current Science*, 111(5), 808–814.

Guralnick, R. P., Towards a collaborative, global infrastructure for biodiversity assessment. Ecol. Lett., 2007, 10, 663–672.

Darwin Core. (2016). Rs.tdwg.org. Retrieved 1 December 2016, from http://rs.tdwg.org/dwc tdwg/ wiki-archive. (2016). GitHub. Retrieved 1 December 2016, from http://wiki.tdwg.org/twiki/bin/ view/SPM/WebHome

EML – Ecological Metadata Language | Digital Curation Centre. (2016). Dcc.ac.uk. Retrieved 1 December 2016, from http://www.dcc.ac.uk/resources/metadata-standards/eml-ecological-metadata-language. tdwg/abcd. (2016). GitHub. Retrieved 1 December 2016, from http://www.tdwg.org/standards/115/

PlinianCore/Documentation. (2016). GitHub. Retrieved 1 December 2016, from https://github.com/PlinianCore/Documentation/wiki

Kattge, J., Diaz, S., Lavorel, S., Prentice, I. C., Leadley, P., Bönisch, G., ... & Cornelissen, J. H. C. (2011). TRY–a global database of plant traits. Global change biology, 17(9), 2905–2935.

Jetz, W. J., McPherson, J. M. and Guralnick, R. P., Integrating biodiversity distribution knowledge: toward a global map of life. Trends Ecol. Evol., 2012, 27, 151–159.

Natesh, S. (2006). Digitized inventory of biological resources of India. *Current Science* 91(7): 860.

Nogueras-Iso, J., Zarazaga-Soria, F. J., Béjar, R., Álvarez, P. J., & Muro-Medrano, P. R. (2005). OGC Catalog Services: a key element for the development of Spatial Data Infrastructures. Computers & Geosciences, 31(2), 199–209.

Roy, P., Karnataka, H., & Saran, S. (2012). Geospatial data processing in distributed computing environment. CSI Communications.

Saran, S., Kushwaha, S., Ganeshaiah, K., Roy, P., Krishna Murthy, Y. (2012). Indian Bioresource Information Network (IBIN): A distributed national bioresource portal. Indian Society of Geomatics, 18(3), 6.

Chapter 12
Western Himalayan Forests in Climate Change Scenario

Arijit Roy and Pooja Rathore

12.1 Introduction

The Himalayan mountain range is amongst the largest, newest and highest mountain chains on earth that form over 2400-km-long arch from east to west direction in the north of South Asia. Himalaya is home to over hundred million people with a small population inhabiting very high altitudes. The range prompts orographic precipitation and impacts weather of the region including the South Asian monsoon, acts as a storage of water in the form of snow and ice and goes about as a wellspring of vast rivers/waterways, for example, Ganges, Indus and Brahmaputra, thus making it the 'water tower' for millions of people of the Indo-Gangetic plains. In addition to this, they act as major stores of valued biodiversity resources due to their unique location and physiographic features and are a centre of age-old human cultural diversity. However, the Himalayan mountains are particularly vulnerable to climate change and variability due to their young and fragile nature coupled with sharp gradients; with the increase in population pressure, natural and socioeconomic systems in these mountain regions are at threat, especially with reference to rapid globalization. The quick change in the biological community, driven by both natural and anthropogenic determinants, represents a remarkable danger not exclusively to the source of revenue of the native people, biota and art but also to the people living in the downstream that are dependent on these natural resources and ultimately to the global environment.

Himalayan mountains are amongst the most biodiversity-rich ecosystems in the world. The complex and dynamic Himalayas with their varying climates and topographies exhibit diverse vegetation that provides a variety of ecosystem

---

A. Roy (✉) · P. Rathore
Forestry and Ecology Department, Indian Institute of Remote Sensing (IIRS), Indian Space Research Organisation (ISRO), Department of Space, Government of India, Dehradun, India
e-mail: arijitroy@iirs.gov.in

R. R. Navalgund et al. (eds.), *Remote Sensing of Northwest Himalayan Ecosystems*,
https://doi.org/10.1007/978-981-13-2128-3_12

services. The forest types of Himalaya range from moist deciduous forests at the base to the temperate broadleaf forests and coniferous forests; at a higher elevation and beyond the treeline lie alpine meadows with harbouring rare plants and unique biodiversity. Biodiversity in Himalayan mountains is significant because of its biogeographic location, habitat heterogeneity and pronounced endemism (Myers 1988). Alpine plant diversity recorded in the Himalayan region is greater than the global average (Korner 1999). These regions have fundamentally higher plant variety and abundance at the elevation range of 4200 and 4500 m as compared to any other mountain ranges at similar heights (Grytnes and Vetaas 2002).

Currently, the Himalayan ecosystems are affected by various anthropogenic pressures such as land use and land cover (LULC) changes, deforestation, overexploitation of natural resources and global warming-driven climate change (Pandit and Grumbine 2012). Climate change amongst these might prove to be a grave threat to Himalayan biodiversity. It is now a globally acknowledged fact backed by many scientific reports which proves that the average mean temperature at higher elevations, especially in Trans-Himalaya and Tibetan Plateau, is increasing at a faster rate than the low-lying areas (Liu and Hou 1998; Shrestha et al. 1999; Liu and Chen 2000; Nogues-Bravo et al. 2007; Solomon 2007). Global warming is causing effects on the climate of the greater Himalayan region (Cruz et al. 2007). The most broadly detailed impact is the quick retreat of Himalayan glaciers (Solomon 2007). Progressing environmental change over the following decades will probably have extra negative effects over these mountains, including significant falling effects on groundwater recharge, biodiversity, river flows, normal perils and natural hazards particularly biological system organization, structure and capacity as well as on human occupations (Nijssen et al. 2001; Parmesan 2006). Several studies anticipate that warming will cause a decline in biodiversity of the distinct alpine habitats in the Greater Himalaya, including tundra and rangelands (Klein et al. 2004). Endemic species, more importantly, are predominantly vulnerable to climate change due to their limited geographic range. There may be a reduction in the size of the niches for endemic species along elevation gradients. The climatic and geographical obstructions of the Himalaya are likely to hinder the movement of species along the latitudinal gradient under global warming (Liu et al. 2004).

Regardless of their remoteness and unavailability, the Himalayas have not been saved from human-incited biodiversity loss. Natural resources of the Himalaya provides to both, the people and wildlife of the region. Yet as increasing human populations are using the resources faster than the Himalaya can replenish them. Rapid economic and population growth in the region (population densities have increased by over 20% in the past decade) is causing competition for space between species which will likely result in shrinking ecosystems. Climate change gradually impacts nature at all scales consistently, while land use-based changes influence the forests more rapidly at constrained scales (Albert et al. 2008; Motta et al. 2006). A lack of habitat connectivity and hence disrupted dispersal will further result in the decline of the species in the Himalayan landscapes undergoing major human modifications. Landscape permeability and species interactions (Roland 1993) are significantly influenced by configuration of both habitat (Robertson and Radford 2009) and networks (Aben et al. 2012).

In spite of the increasing interest in climate change-oriented research in other parts of the world and the risk of global extinctions being an important concern, integrated ecological and land use-cover studies have yet not addressed the importance of the Greater Himalaya (Xu et al. 2008) in terms of biodiversity conservation. The region particularly lack prominent scientific reporting on the concerned issue of species' response to the changing global and regional climate (Pandit 2009) which prompted IPCC to call it as a 'white spot' due to data insufficiency on natural ecosystems as there are still many critical areas of research that need to be raised on global platform. The fundamental challenge of modern conservation biology includes an understanding of the biodiversity fragmentations, changes in species distributions and species richness in different geospatial contexts and assessment and quantification of species loss. This challenge has become more pertinent in the wake of ongoing deterioration of native ecosystems, declining populations and growing loss of biodiversity.

12.2 Climate Change in the Himalaya

The greater Himalayan region is strongly inflicted by global warming. Bhutiyani et al. (2007) observed that the Indian Western Himalayan region has witnessed a significant increase in the temperature at a rate of 1.6 °C during the last century, as compared to the rest of India, i.e. 0.5 °C (Kothawale and Kumar 2005), and the global average, i.e. 0.76 °C. An increase of 0.98 °C in annual maximum temperatures has been observed over the Western Himalaya (Dash and Hunt 2007).

A study by Christensen et al. (2007) has stated that expected warming on the higher-altitude Himalayan regions and Tibetan Plateau and is expected to be around 3.8 °C in the next 100 years. The consistent warming over the entire Himalayan region including Tibetan Plateau and the Indian subcontinent as reported by IPCC simulations has also been validated by the most extensively used RCM Indian: Providing REgional Climates for Impacts Studies (PRECIS) (Kumar et al. 2007). The results show that in most parts of the Himalayas, the increase in temperature in winter is consistently higher than other seasons, including the north-west Indian, Nepalese Himalaya and Chinese (Bhutiyani et al. 2007). The Trans-Himalaya of Nepal showed winter warming at the rate of 0.9 °C/decade as compared to 0.9 °C/decade in Tmax during 1971–2000, whereas for the high mountains, it is recorded at 1.2 °C/decade with respect to 0.6 °C/decade increase in Tmax during 1971–2000 (Shrestha and Devkota 2010). This seasonal trend (greatest warming in winter, smallest in summer) has also been agreed upon by several other noted researchers in the Tibetan Plateau (Liu and Chen 2000; Du et al. 2004; You et al. 2008).

One of the most prominent impacts of global warming that has been widely reported and discussed is the rapid retreat of glaciers (Solomon 2007). The ongoing trend of climate change which is expected to be continued over the succeeding decades will further aggregate the adverse impacts across these mountains and have severe implications on groundwater recharge, river flows and ecosystem composition, structure and function which will result in undesirable impacts on human livelihood (Nijssen et al. 2001; Parmesan 2006). The increase in temperature will also affect various phonological processes in species such as the timing of leaf flush and flowering, activities of plant pollinators and plant reproduction in monsoonal Asia (Corlett and Lafrankie 1998). The flowering of most alpine species has already been reported to be strongly affected by the change in the speed of snowmelt (Kudo 1991).

Climate change and variability can also have major impacts on the socio-economy of the region in terms of food and water security in the Himalayan region, under the lack of well-equipped and inadequate storage systems (natural or man-made). About 60% of the agricultural land in the Himalayan region is rain-fed and therefore making it vulnerable to changes in rainfall timing and frequency.

12.3 Western Himalayan Forests

The Western Himalaya physiographically extends from the Shiwaliks that constitutes its foothills in the south, up to Tibetan Plateau on the north (Trans-Himalaya). The Karakoram Mountains form its northernmost range of mountains that distributes up to Pakistan and China. Zanskar ranges lie in the south of the Karakoram Range. Pir Panjal forms the parallel range to the Zanskar ranges. The states of Jammu and Kashmir constitute most of the areas of these three mountain ranges that lie adjacent to each other in the north-western part of India. The region is known for having some of the highest mountains on earth. Major geological fault lines separated the three major geographical units, the Himadri (Greater Himalaya), Himachal (Lesser Himalaya) and the Shiwaliks (Outer Himalaya), extending almost uninterrupted throughout its length and expanding relatively continuous all through its length. Older streams like the Indus, Ganga, Yamuna, Sutlej, Kali, Kosi and Brahmaputra have their origins in these ranges that slice through sharp gorges to reach Great Plains and form the most productive aggregation parts of the country.

The region is world-renowned for its biological, hydrogeological and cultural values. The ecoclimatic zones vary from tropical (<500 m a.s.l) to high alpine (>6000 m a.s.l), with majority of the area dominated by temperate broadleaf, deciduous or mixed forests, temperate coniferous forests, tropical and subtropical rainforests, alpine moist and dry scrub, meadows and desert steppe (Pei 1995; Guangwei 2002). Concerning general physiognomic and floristic designs at upper timberline, the characteristic vegetation of the subalpine belt in the North-West and West Himalaya essentially comprises coniferous *Picea smithiana*, *Pinus wallichiana*, *Abies pindrow* and *Betula utilis* forests and *Juniperus* sp. (Schickhoff 2005). These species also make the timberline on south-facing slopes all through the mountain chain. *Juniperus recurva* especially is more prominent as the treeline species in Eastern Himalaya. The normal elevation constraints for *Betula utilis* and

Juniperus sp. have been accounted for to be around 3700–3800 m in North-West Himalaya. However, sometimes it is also reported to be further up to 4200 m (Rau 1974).

According to the work done by Roy et al. 2013, there are around 35 vegetation-type classes in the Western Himalaya, and almost all the three states have very high forest and natural cover. Since the Western Himalaya is part of the Himalayan ecosystem which has been identified as 1 of the 36 biodiversity hotspots of the world, these regions are also under tremendous threat. Uttarakhand has around 65%, Himachal Pradesh has 66.52% and Jammu and Kashmir have 19.95% of the total geographic area under forest cover. The forests of this region are highly fragmented due to human settlement in these regions resulting in landscape with forest and agriculture mosaic.

12.4 Impact of Climate Change on the Himalayan Forests

Himalaya being one of the highly fragile ecosystems of the world is highly prone to various anthropogenic as well as a natural perturbation.

12.4.1 Treeline Shifts

The establishment of the treeline and its spatiotemporal changes largely influences the biodiversity patterns, carbon cycle and the overall LULC changes in mountain ecosystems. The species distribution patterns and especially treeline on mountain ecosystems develop as a function of climatic regulations, orography and species adaptation capabilities. As temperature diminishes with elevation, moderately short-distance upward migration is required by the species for their population constancy. Although this migration is feasible for the warmer climatic and natural zones beneath the mountain crests (Penuelas and Boada 2003). Mountain ridges provide extensive deterrents to dispersal for some species, which have a tendency to compel movements to incline upward movement (Pounds et al. 2006). This further limits a species' geographical range (mountaintops provide less space than their bases). This is relied upon to decrease genetic diversity within species and build the dangers of extinction due to additional stresses (Gottfried et al. 1999), a hypothesis affirmed by recent investigations based on genetic analysis indicating past climate changes driven gene drift effects (Bonin et al. 2006).

A reshuffling of species on elevation gradients might be an outcome of individualistic species reactions that are interceded by shifting life spans and survival rates. These, thus, are the consequence of a high level of evolutionary specialization to severe mountain climatic conditions, and sometimes, they incorporate impacts prompted by invading alien species. A study focusing on the genetic evidence for *Fagus sylvatica* proposes an in situ climate change-based adaptive reaction that populations may show in some capacity (Jump et al. 2006). However, continuous distributional changes (Penuelas and Boada 2003) demonstrate that this reaction won't really enable this species to hold on all through its range.

For a long time, it was thought that carbon balance party controls the upper treelines, which refers to the transition zone between subalpine forests and alpine meadows (Stevens and Fox 1991). The hypothesis was challenged by Korner (2003). Around the world, seasonal mean air temperatures of 6 °C (Zha et al. 2005) seem to characterize the treelines. However, due to grazing, or anthropogenic disturbances, wind or fire in many mountains, the upper treeline is pushed below its potential climatic position. The trend is also evident in the Himalaya, as well, where much of the environment has been transformed due to deforestation leading to fragmented ecosystems. In past few decades, although temperature control may be a crucial contributing factor of geographical range, the inability of tree species to migrate with the pace of changing temperature zones also poses a major challenge.

The average timberline in the Himalaya is at an elevation of ca. 3500 m. Even the slightest change in atmosphere that upsets the harmony amongst the vegetation and atmosphere will provoke imperative changes in the demography of these species. Any change in climate, which perturbs the vegetation-climate equilibrium, will lead to significant changes in the demographic patterns of these species. The recent climate change in the Western Himalaya has formed a gateway for the introduction of invasive alien species, which were considered as a second-worst threat of existence of native biodiversity after habitat loss. A global warming scenario will lead to the upward shift in the treeline and will lead to changes in the orography of the habitat. This may lead to changes in the ecosystem properties and loss of important associated flora and fauna, which may be a keystone species in the community and also an invasion of alien species.

12.4.2 Species Loss

The Himalayan region is one of the most noted and distinguished global biodiversity hotspots (Myers et al. 2000). It comprises as high as 200 global ecoregions (Olson and Dinerstein 1998), several centres for plant diversity (WWF 1995) and 2 endemic bird areas (Stattersfield et al. 1998). The multiple biogeographic origins contribute to the exceptional biological diversity of the Himalaya. Its location area at the crossroads of two mainland plates places it in an ecotone characterized by widely varied vegetation of both the Asian mainland shells and the Indian plate having the Gondwana origin. Further, the Palearctic realms also contribute to the plant species in the higher elevations like fir (*Abies*) and spruce (*Picea*) accompanying many charismatic faunal species like snow leopard (*Uncia uncia*), wolf (*Canis lupus*) and brown bear (*Ursus arctos*). Further the climate variability as a result of associated with the vast, steep and complex topography also gives the Himalayan region a plethora of habitats for the occurrence of the biodiversity hotspot in the region. A brief description of the Himalayan flora and fauna can be observed from Table 12.1.

Table 12.1 Endemic species distribution of Himalayan flora and fauna

Taxonomic group	Species	Endemic species	Percent endemism
Plants	10,000	3160	31.6
Mammals	300	12	4.0
Birds	977	15	1.5
Reptiles	176	48	27.3
Amphibians	105	42	40.0
Freshwater fish	269	33	12.3

Source: http://www.biodiversityhotspots.org/xp/hotspots/himalaya/

Various factors such as birth, growth, death and dispersal rates of an individual comprising a population influence the distribution and abundance of a species or a group of species. These essential rates are, thusly, impacted by ecological elements, including local climate, other than resource availability, survivorship and fecundity (Hansen and Rotella 1999). These elements, when accumulated across populations, change in key rates showing local extinction and colonization events, which are the system by which species range changes. Being the world's most astounding mountain chain, the Himalaya is described by a complex geologic structure, snow-topped pinnacles, substantial valley ice sheets, profound stream gorges and rich vegetation. A perplexing interaction of climatic and landforms, patterns of resource use and economic conditions have prompted resource degradation and related ecological outcomes in the Himalayan biological system. (Jodha 2001).

The Himalayan ranges are geologically relatively young and still in the formative stages, and formation of new habitats and corridors for evolution and migration of species is still in progress. Paleoecological-based studies recommend that most extraordinary changes in the Himalaya occurred amid the Pliocene when the Dipterocarpus and Anisoptera forests completely vanished from the Western Himalaya and the wet forests were changed into moist or dry types. As recent as ca. 5000 years back, the forests of North Western Himalaya encountered a quick decay of oaks, figs and laurels and the progressive prevalence of *Pinus*, *Cedrus*, spruce and graminoids. In this way, the current floral and faunal collections in the region are a result of consistent climatic interactions. It is being hypnotized that climate change amid mid-Holocene prompted the quick colonization of higher elevations by present-day vegetation along with an expansion in human population.

12.4.3 Upward Shift in Species Range Areas

Temperature is an important determinant of timberline position and distribution (Shi and Ning 2013). Hence, the expected regional warming in Western Himalaya which is three times the rate in plain areas may induce species range expansions towards higher elevation for maintaining their populations (Danby and Hik 2007; Kharuk et al. 2010; Moiseev et al. 2010). Some field studies also suggest that warming has resulted in geographical range shifts in endemic plant species in Himalaya. A study based on field experiments by Telwala et al. (2013) using historical data (1849–1950) and the recent data (2007–2010) on endemic species' elevation ranges and temperature suggest that warming has contributed to shifts in the geographical range of 87% of the 124 endemic plant species studied in the Himalayan region. Another study by Zomer et al. (2014) using modelling approach indicated that upward shift in mean elevation of ecoregions (371 m) and bioclimatic zones (357 m) is one of the potential impacts of global warming. However, this upward movement may cause a decrease in biodiversity (Walker et al. 2006) due to increased competition for the resources and land availability. Many studies demonstrate strong general connections between climatic parameters and treeline position and repeated climatically induced treeline fluctuations which demonstrate the significance of treeline biological system as a pointer of climatic change.

There is sufficient confirmation on treeline shift in the Himalaya over the previous decades. Panigrahy et al. (2010) used satellite images to map the treeline over nearly 20 years. In another study, the imagery obtained for a biosphere reserve—Nanda Devi Biosphere Reserve—in the central Indian Himalaya from 2004 revealed an increase in the vegetation cover at the areas which were glaciated in 1986. However, with varying rates of shifts and sensitivity towards climate, alpine plant species have been observed to shift to higher altitudes. Various studies capture the significant changes in Himalayan ecosystems since 1960 (Sushma et al. 2010; Panigrahy et al. 2010). Changes in snow precipitation have been identified to be more related to treeline dynamics than to global warming (Negi et al. 2012). The treeline has already shifted 388 $\pm$ 80 m upwards in the Uttarakhand Himalaya between 1970 and 2006 as indicated in a remote-sensing study by Singh et al. (2012). Glacier recession and an advance in the treeline have also been confirmed by another study using repeat photography and supplementary measurements in the Eastern Himalaya (north-west Yunnan) (Baker and Moseley 2007).

A study conducted at Forestry and Ecology Department, Indian Institute of Remote Sensing (IIRS), on the western Himalayan subalpine and temperate species shows impacts of climate change-induced global warming on the future distribution area of these species under climate change. *Abies pindrow*, *Betula utilis*, *Juniperus recurva*, *Picea smithiana*, *Quercus semecarpifolia*, *Taxus wallichiana*, and *Rhododendron arboreum* were included in the study. For the study, simulation results projected under the IPCC-AR5 Representative Concentration Pathway (RCP) 4.5 and 6.0 for 2050 from Hadley Global Environment Model (HadGEM2) and Community Climate System Model (CCSM) models were used. The results (Figs. 12.1, 12.2, 12.3, and 12.4) show a mounting shift towards higher elevations with the anticipated rise in global temperature, encroaching grasslands prevailing in the upper Himalayan region for all the species. However, as opposed to *Abies pindrow*, *Betula utilis*, *Juniperus recurva*, *Picea smithiana*, *Rhododendron arboreum*, *Quercus semecarpifolia* and *Taxus wallichiana* are expected to lose their current range area under climate change scenario (RCP 6.0, HadGEM2). *Quercus semecarpifolia* has emerged as the most vulnerable tree species to the climate change amongst all the species modelled.

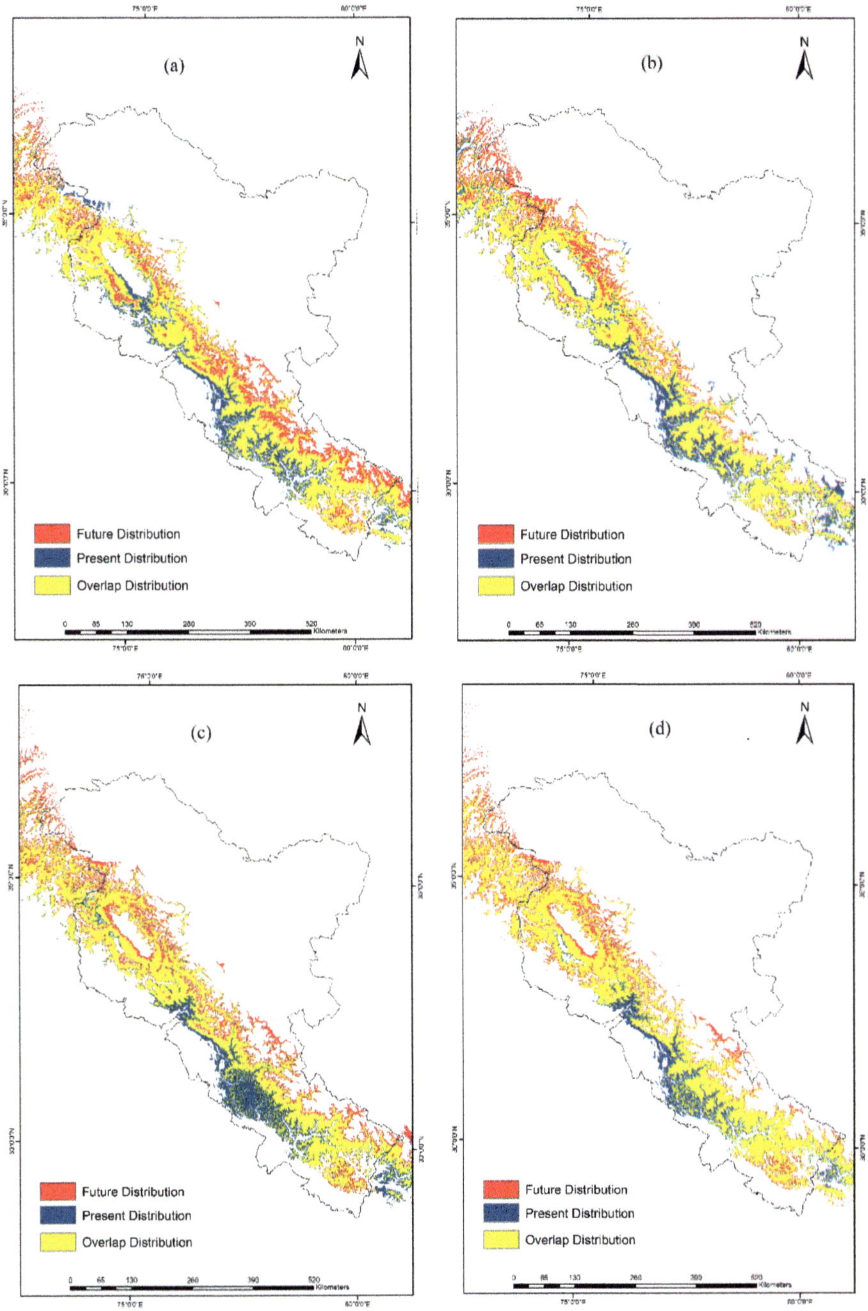

Fig. 12.1 Probable range shift in *Abies pindrow* under different scenarios of climate change: (**a**) CCSM (RCP4.5); (**b**) HadGEM2 (RCP 4.5); (**c**) CCSM (RCP 6.0); (**d**) HadGEM2 (RCP 6.0)

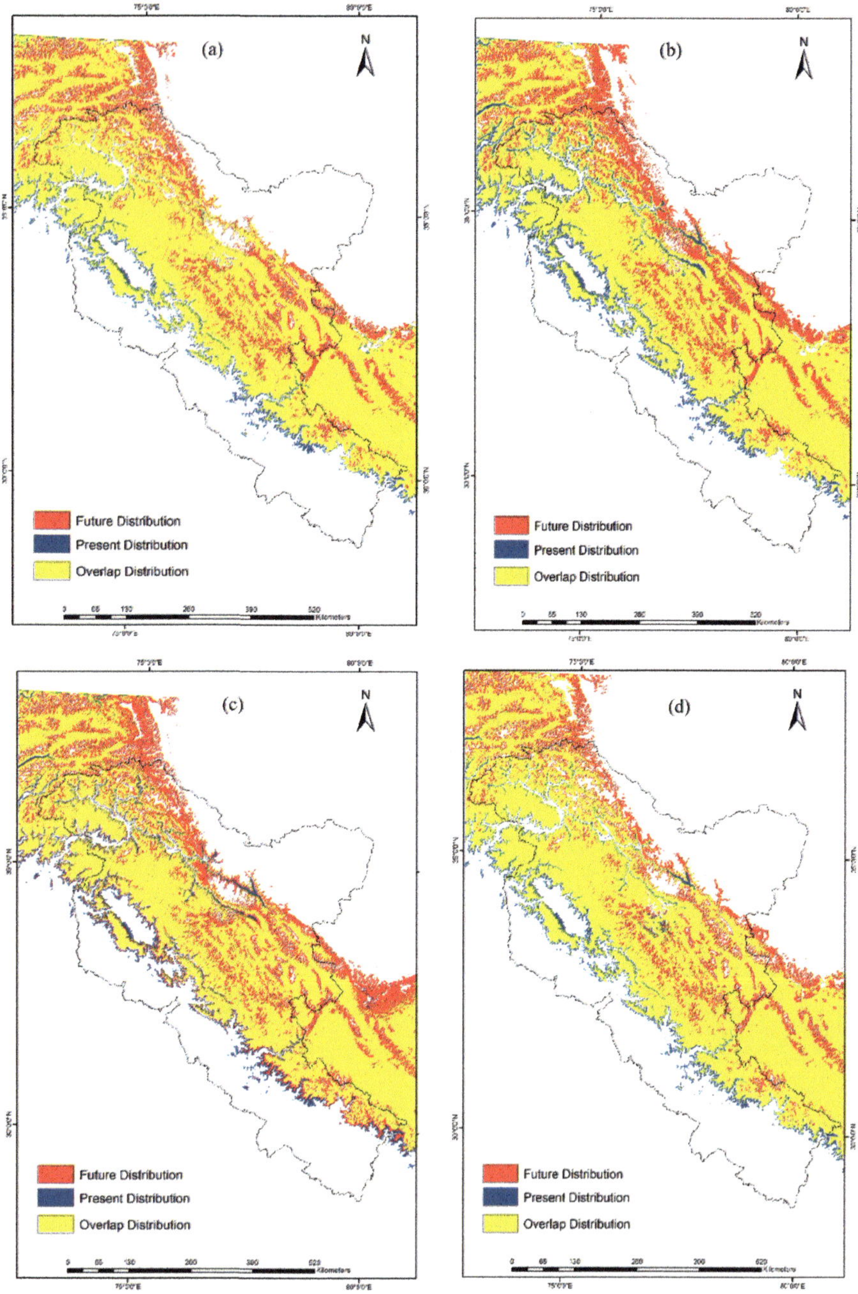

Fig. 12.2 Probable range shift in *Betula utilis* under different scenarios of climate change: (**a**) CCSM (RCP4.5); (**b**) HadGEM2 (RCP 4.5); (**c**) CCSM (RCP 6.0); (**d**) HadGEM2 (RCP 6.0)

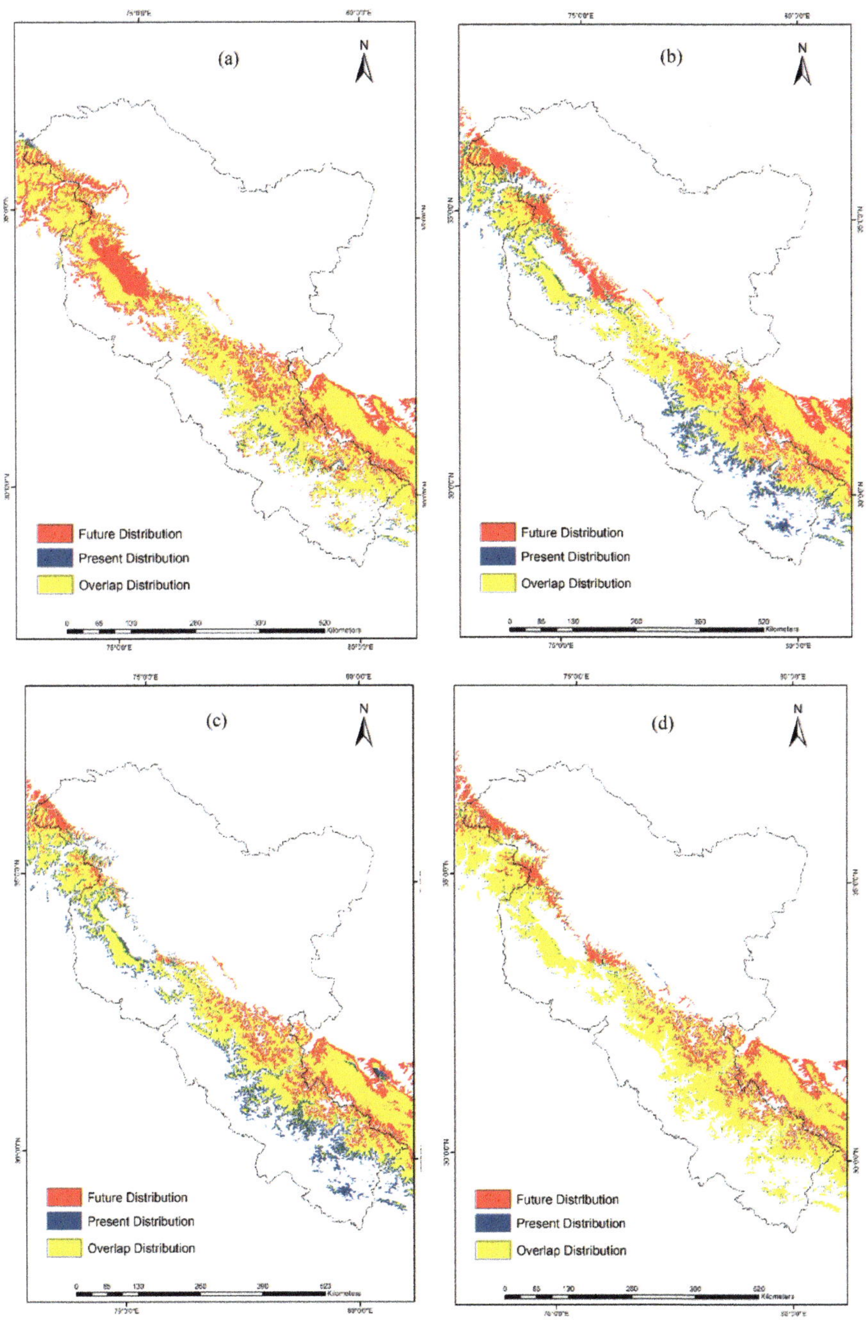

Fig. 12.3 Probable range shift in *Juniperus recurva* under different scenarios of climate change: (**a**) CCSM (RCP4.5); (**b**) HadGEM2 (RCP 4.5); (**c**) CCSM (RCP 6.0); (**d**) HadGEM2 (RCP 6.0)

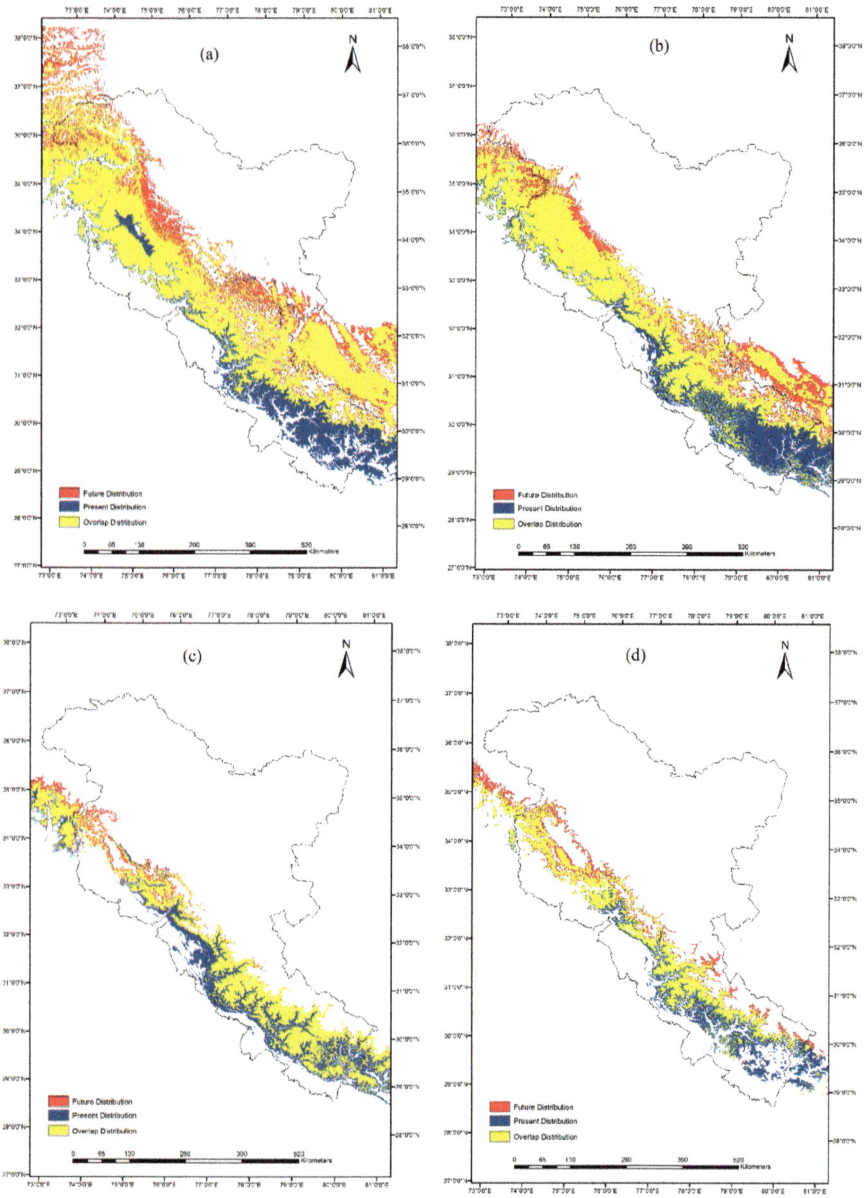

Fig. 12.4 Probable range shift in (**a**) *Picea smithiana*, (**b**) *Rhododendron arboreum*, (**c**) *Quercus semecarpifolia* and (**d**) *Taxus wallichiana* under RCP 6.0 (HadGEM) scenario of climate change

12.4.4 Role of Invasive Species in Biodiversity Loss

The alpine regions of the Himalaya are facing a critical threat in the form of loss of the endemic species. It is of significance that the potential hazard zones and vulnerable ecosystems are distinguished for the enactment of appropriate conservation actions. Some reports account invasive species as a key driver behind biodiversity loss that may soon outperform the damage caused by habitat degradation. Biological invasion is considered as a noteworthy part of the human-caused worldwide ecological change. (Reddy 2008). It is an ideal opportunity to focus on the biological effect of invasive aliens both at the species and at the environment levels. Millennium Ecosystem Assessment (2005) reported biological invasion of alien species as the worst threat to indigenous biodiversity and as an indicator of global change. Invasive alien species mainly *Ageratina adenophora*, *Ageratum conyzoides*, *Ageratum houstonianum*, *Chromolaena odorata*, *Galinsoga parviflora*, *Gnaphalium coarctatum*, *Lantana camara*, *Mikania micrantha*, *Mimosa pigra*, *Parthenium hysterophorus*, *Ulex europaeus*, and *Urena lobata* are posing threat to indigenous species and ecosystems. Besides, weedy species with a wide ecological tolerance have an advantage over native species (Reddy et al. 2008).

12.4.4.1 Forest Cover Loss

The impacts of various demands and pressures on the forest ecosystems have led to the gradual degradation of the forests in India. Presently, all the forests in India are in some stage of degradation. It is important that the degradation of the forest ecosystems are monitored and critical damage to the forest controlled so that we do not lose the remnant natural ecosystems available to us. Some of the indicators of the forest ecosystem damage can be very accurately monitored through remote sensing and GIS.

A remote-sensing-based analysis of the forest cover in the Western Himalaya during 1985–2005 indicates significant degradation in the forest cover in this region. During the time period 1985–1995, major changes were seen for classes such as built-up land, barren land, scrubland, fallow land and snow and ice. The area under built-up land, fallow land, scrubland and barren land increases by 8.2%, 6.2%, 9.3% and 9.9%, respectively, whereas the area under snow and ice decreased by 5.1%.

For 1995–2005, a noticeable increase of 21.4% was found for built-up land. Another increase was found for cropland which increased by 8%. The area under plantation, evergreen broadleaf forest and wasteland decreased by 6.9%, 5.1% and 5.5%, respectively (Fig. 12.5).

For 1985–2005, the area under built-up land, cropland, fallow land, scrubland and grassland increased by 31.3%, 7.2%, 9.2%, 6.5% and 2.6%. The decrease in areas was observed for plantation, evergreen needle-leaf forest and evergreen broadleaf forest by 5.8%, 2.0% and 7.7%, respectively. During the period, 22,928.9 ha cropland, 678 ha evergreen needle-leaf forest area and 1263.8 ha evergreen

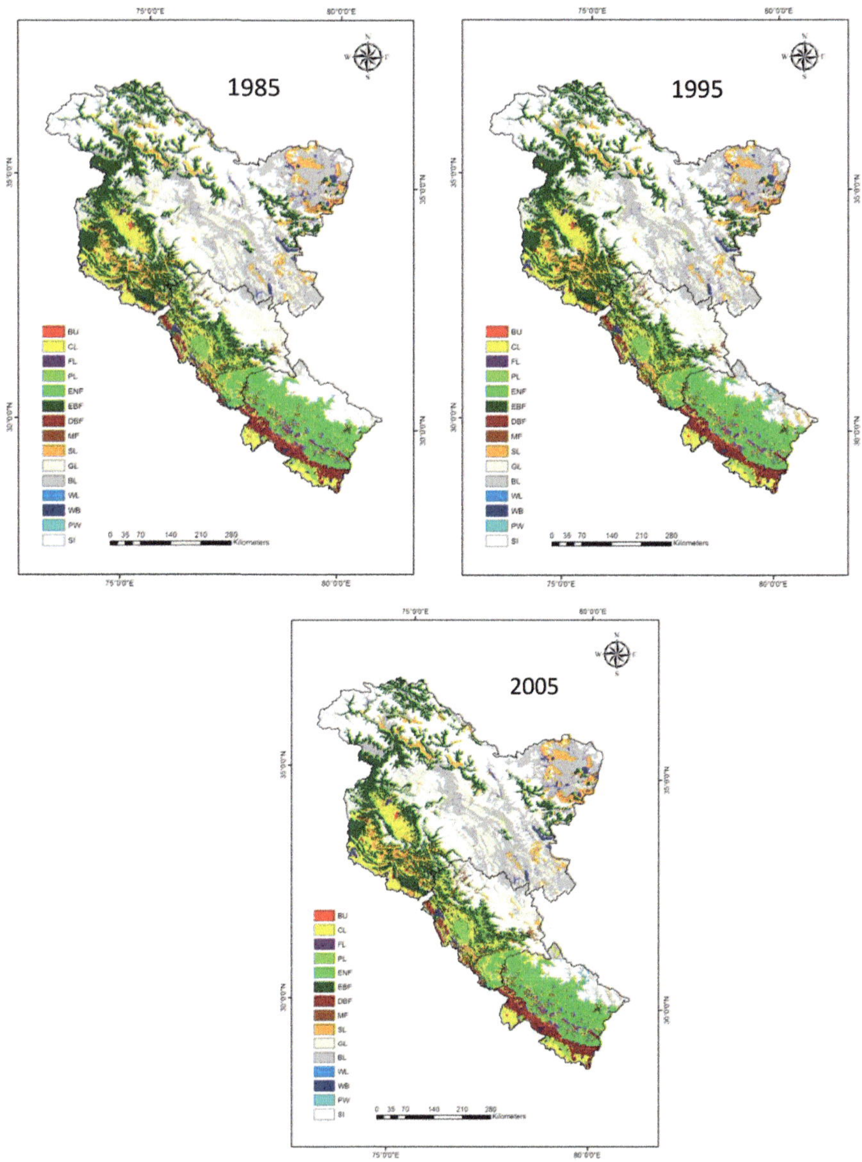

Fig. 12.5 Forest cover and land use in Western Himalaya during 1985–2005

broadleaf forest area were soaked into built-up land. Another significant conversion includes forests getting converted to cropland. 87,136.3 ha forest land was converted to cropland during the two decades. 10,077.2 ha forest land was converted to fallow land. 80,936.7 ha forest area was changed into scrubland. The total forest area decreases by 489,597 ha, i.e. 5% (Fig. 12.6).

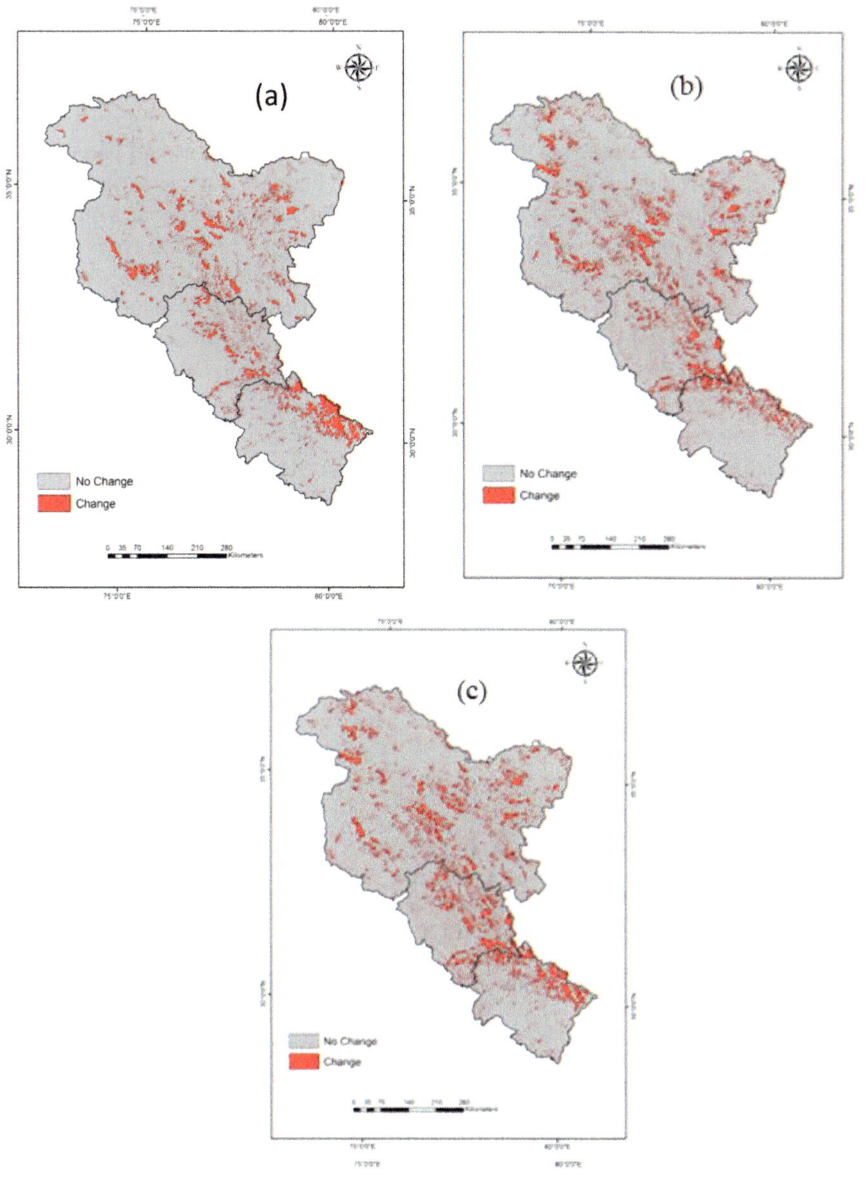

Fig. 12.6 The changes in the natural areas in Western Himalaya. (**a**) Changes in between 1985 and 1995; (**b**) changes in between 1995 and 2005; (**c**) cumulative changes in the natural areas in Western Himalaya during 1985–2005

12.5 Conclusions

The study describes the climate change impacts on the Himalayas, in particular on western Himalayan forests. The warming trends over the whole Himalayan region have been quantified based on the review of several temperatures and precipitation patterns in the region. However, the trends of temperature vary across subregions and across seasons. Nevertheless, the research on climate change impacts on western Himalayan flora and fauna are notably lacking. Different vegetation types on these mountainous forests including alpine, subalpine, moist temperate and dry temperate forests are vulnerable to climate change initiated with an Earth-wide temperature boost. The unique forest ecosystems at high elevations in north-western Indian states are susceptible to projected global and regional climate change. The growth patterns in the West and central Himalaya especially are sensitive to temperature and humidity conditions. The outcomes of the study propose a high inclination to react to expanding temperatures and a huge future treeline progress towards higher elevations. Treeline shifts are of great natural significance because of conceivable consequences for territorial biodiversity and biological-ecological integrity. A broad upward infringement of subalpine forests would perhaps uproot regionally unique alpine species, whereas the temperate species may incur a substantial loss in their population in the absence of physiological adaptions with climate change.

References

Aben J, Adriaensen F, Thijs KW, Pellikka P, Siljander M, Lens L, Matthysen E (2012) Effects of matrix composition and configuration on forest bird movements in a fragmented Afromontane biodiversity hot spot. Anim Conserv 15:658–668

Albert C.H., Thuiller W., Lavorel S., Davies I.D., and Garbolino E., 2008. Land-use change and sub alpine tree dynamics: colonization of Larix decidua in French sub alpine grasslands. J. Appl. Ecol. 45: 659–669.

Assessment, M. E. (2005). Ecosystems and human well-being. Washington, DC.at http://www.gisinetwork.org

Baker, B. B., & Moseley, R. K. (2007). Advancing treeline and retreating glaciers: implications for conservation in Yunnan, PR China. Arctic, Antarctic, and Alpine Research, 39(2), 200–209.

Bhutiyani, M.R., Kale, V.S. and Pawar, N.J. (2007). Long-term trends in maximum, minimum and mean annual air temperatures across the Northwestern Himalaya during the twentieth century. Climatic Change, 2007, 85, 159–177.

Bonin, A., Taberlet, P., Miaud, C., & Pompanon, F. (2006). Explorative genome scan to detect candidate loci for adaptation along a gradient of altitude in the common frog (Rana temporaria). Molecular Biology and Evolution, 23(4), 773–783.

Christensen J.H., B. Hewitson, A. Busuioc, A. Chen, X. Gao, I. Held, R. Jones, R.K. Kolli, W-T. Kown, R. Laprise, V. Magana Rueda, L. Mearns, C.G. Menendez, J. Raisanen, A. Rinke, A. Sarr, and P. Whetton (2007). Regional climate projections. In: Solomon S., D. Qin, M. Manning, Z. Chen, M. Marquis et al. (Eds.) Climate Change 2007: The Physical Science Basis. Contribution of Working Group I to the Fourth Assessment Report of the Intergovernmental Panel on Climate Change. Cambridge University Press, Cambridge, UK and New York. pp. 996.

Corlett, R., and Lafrankie, J. (1998). Potential impacts of climate change on tropical Asian forests through an influence on phenology. Cli- mate Change 39:439–453.

Cruz, R., et al. (2007). Asia. Pages 469–506 in M. Parry, et al. editors. Climate change 2007: impacts, adaptation and vulnerability. Contri- bution of Working Group II to the Fourth assessment report of the Intergovernmental Panel on Climate Change. Cambridge University Press, Cambridge, United Kingdom.

Danby, R.K. and Hik, D.S. (2007). Variability, contingency and rapid change in recent subarctic alpine treeline dynamics, J. Ecol., 95,352–363.

Dash, S. K., and J. C. R. Hunt. "Variability of climate change in India." Current Science (2007): 782–788.

Du, M., Kawashima, S., Yonemura, S., Zhang, X., & Chen, S. (2004). Mutual influence between human activities and climate change in the Tibetan Plateau during recent years. Global and Planetary Change, 41(3), 241–249.

Gottfried, M., Pauli, H., Reiter, K., & Grabherr, G. (1999). A fine-scaled predictive model for changes in species distribution patterns of high mountain plants induced by climate warming. Diversity and Distributions, 5(6), 241–251.

Grytnes, J.A., Vetaas, O.R. (2002). Species richness and altitude: A comparison between null models and interpolated plant species richness along the Himalayan altitudinal gradient, Nepal. Am Nat 159: 294–304.

Guangwei, C. (2002). Biodiversity in the Eastern Himalayas: conservation through dialogue. Summary reports of the workshops on biodiversity conservation in the Hindu Kush-Himalayan Ecoregion. Kathmandu: ICIMOD

Hansen, A., & Rotella, J. (1999). Abiotic factors. Maintaining biodiversity in forest ecosystems. Cambridge University Press, Cambridge, United Kingdom, 161–209.

Jodha, N. S. (2001). Life on the edge: sustaining agriculture and community resources in fragile environments. Oxford University Press.

Jump, Alistair S., et al. "Natural selection and climate change: temperature-linked spatial and temporal trends in gene frequency in Fagus sylvatica. "Molecular Ecology 15.11 (2006): 3469–3480.

Kharuk, V.I., Ranson, K.J., Im S.T., and Vdovin, A.S. (2010). Spatial distribution and temporal dynamics of high-elevation forest stands in southern Siberia, Global Ecol. Biogeogr., 19, 822–830

Klein, J. A., J. Harte, and X. Q. Zhao. 2004. Experimental warming causes large and rapid species loss, dampened by simulated grazing, on the Tibetan Plateau. Ecology Letters 7:1170–1179.

Korner C. (1999). Alpine Plant Life: Functional Plant Ecology of High Mountain Ecosystems. Berlin: Springer

Korner, C. (2003). Alpine plant life: functional plant ecology of high mountain ecosystems; with 47 tables. Springer Science & Business Media.

Kothawale, D.R. and Rupa Kumar, K. (2005). On the recent changes in surface temperature trends over India. Geophys. Res. Lett, 32, 18714

Kudo, G. (1991). Effects of snow-free period on the phenology of alpine plants inhabiting snow patches. Arctic Alpine Research 23:436–443.

Kumar, V., Singh, P., and Singh, V. (2007). Snow and glacier melt contribution in the Beas River at Pandoh Dam, Himachal Pradesh, India. Hydrological Sciences Journal 52: 376–388.

Liu, Q., Jiang X., Xie S.-P., and Liu, W. T. (2004). A gap in the IndoPacific warm pool over the South China Sea in boreal winter: Seasonal development and interannual variability. J. Geophys. Res., 109, C07012, doi:https://doi.org/10.1029/2003JC002179.

Liu, X., and Chen, B. (2000). Climatic warming in the Tibetan Plateau during recent decades. International Journal of Climatology 20:1729–1742.

Liu, X., and Hou, P. (1998). Relationship between the climatic warming over the Qinghai-Xizang Plateau and its surrounding areas in recent 30 years. Plateau Meteorology 17:245–249 (in Chinese).

Moiseev, P.A., Bartysh, A., and Nagimov, Z.Y. (2010). Climate changes and tree stand dynamics at the upper limit of their growth in the North Ural Mountains, Russ. J. Ecol., 41, 486–497
Motta R., Morales M., and Nola P. (2006). Human land-use, forest dynamics and tree growth at the treeline in the Western Italian Alps. Ann. For. Sci. 63: 739–747.
Myers N. (1988). Threatened biotas: "Hot spots" in tropical forests. The Environmentalist 8: 1–20
Myers, N., Mittermeier, R. A., Mittermeier, C. G., Da Fonseca, G. A., & Kent, J. (2000). Biodiversity hotspots for conservation priorities. Nature, 403(6772), 853.
Negi, G. C. S., Samal, P. K., Kuniyal, J. C., Kothyari, B. P., Sharma, R. K., & Dhyani, P. P. (2012). Impact of climate change on the western Himalayan mountain ecosystems: An overview. Tropical Ecology, 53(3), 345–356.
Nijssen, B., Donnell, O'. G. M., Hamlet, A., and Letternmaier, D.P. (2001). Hydrological sensitivity of global rivers to climate change. Climate Change 50:143–175.
Nogues-Bravo, D., Araujo, M.B., Errea, M.P., and Martinez-Rica, J.P. (2007). Exposure of global mountain systems to climate warming during the 21st century. Global Environmental Change 17:420–428.
Olson, D. M., & Dinerstein, E. (1998). The Global 200: a representation approach to conserving the Earth's most biologically valuable ecoregions.Conservation Biology, 12(3), 502–515
Pandit M.K., Grumbine R.E., (2012). Potential effects of ongoing and proposed hydropower development on terrestrial biological diversity in the Indian Himalaya. Conserv Biol 26: 1061–1071.
Pandit, M.K. (2009). Other factors at work in the melting Himalaya: follow-up to Xu, et al. ConservBiol 23: 1346–1347
Panigrahy, S., Anitha, D., Kimothi, M. M., & Singh, S. P. (2010). Timberline change detection using topographic map and satellite imagery. Tropical Ecology, 51(1), 87–91.
Parmesan, C. (2006). Ecological and evolutionary responses to recent climate change. Annual Review of Ecology, Evolution and Systematics 37:637–669.
Pei, S. (1995). Banking on biodiversity: report on the regional consultations on biodiversity assessment in the Hindu Kush Himalaya. Kathmandu: ICIMOD
Peñuelas, J., & Boada, M. (2003). A global change-induced biome shift in the Montseny mountains (NE Spain). Global change biology, 9(2), 131–140.
Pounds, J. A., Bustamante, M. R., Coloma, L. A., Consuegra, J. A., Fogden, M. P., Foster, P. N., ... & Ron, S. R. (2006). Widespread amphibian extinctions from epidemic disease driven by global warming. Nature, 439(7073), 161–167
Rau, M.A. (1974). Vegetation and phyto-geography of the Himalaya. In: Ecology and Biogeography in India. (ed.) M.S. Mani, The Hague
Reddy, C. S. (2008). Catalogue of invasive alien flora of India. Life Science Journal, 5(2), 84–89.
Reddy, C.S., Bagyanarayana, G., Reddy, K.N. and Raju, V.S. (2008). Invasive Alien Flora of India. NBII/USGS, Washington, DC. Available
Robertson, O. J. and Radford, J. Q. (2009). Gap-crossing decisions of forest birds in a fragmented landscape. Austral Ecology, 34: 435–446. doi:https://doi.org/10.1111/j.1442-9993.2009.01945.x
Roland, J. (1993). Large-scale forest fragmentation increases the duration of tent caterpillar outbreak. Oecologia, 93(1), 25–30.
Roy, P.S., Kushwaha, S.P.S., Roy, A., Karnataka, H., & Saran, S. (2013). Biodiversity characterization at landscape level using geospatial model.Anais XVI Simpósio Brasileiro de Sensoriamento Remoto–SBSR, Foz do Iguacu, PR, Brasil, 3321–3328.
Schickhoff, U. (2005) The upper timberline in the Himalayas, Hindu Kush and Karakorum: a review of geographical and ecological aspects. In Mountain Ecosystems (pp. 275–354). Springer Berlin Heidelberg
Shi, P, and Ning, W. (2013). "The Timberline Ecotone in the Himalayan Region: An Ecological Review." High-Altitude Rangelands and their Interfaces in the Hindu Kush Himalayas: 108.
Shrestha, A. B., & Devkota, L. P. (2010). Climate change in the Eastern Himalayas: observed trends and model projections. International Centre for Integrated Mountain Development (ICIMOD).

Shrestha, A.B., Wake, C.P., Mayewski, P.A., and Dibb, J.E. (1999) Maximum temperature trends in the Himalaya and its vicinity: an Analysis based on temperature records from Nepal for the Period 1971–94. Journal of Climate 12:2775–2787.

Shrestha, A. B., Wake, C. P., Mayewski, P. A., Dibb, J. E. (1999). Maximum temperature trends in the Himalaya and its vicinity: an analysis based on temperature records from Nepal for the period 1971–94. *Journal of climate*, *12*(9), 2775–2786.

Singh, CP; Panigrahy, S; Thapliyal, A; Kimothi, MM; Soni, P; Parihar, JS (2012) 'Monitoring the alpine treeline shift in parts of the Indian Himalayas using remote sensing'. Current Science 102: 559–562

Solomon, S. (Ed.). (2007). Climate change 2007-the physical science basis: Working group I contribution to the fourth assessment report of the IPCC(Vol. 4). Cambridge University Press.

Stattersfield, A. J., Crosby, M. J., Long, A. J., & Wege, D. C. (1998). Global directory of endemic bird areas. BirdLife International, Cambridge, United Kingdom.

Stevens, G. C., & Fox, J. F. (1991). The causes of treeline. Annual review of ecology and systematics, 22, 177–191

Sushma, P; Singh, CP; Kimothi, MM; Soni, P; Parihar, JS (2010) 'The upward migration of alpine vegetation as an indicator of climate change: observations from Indian Himalayan region using remote sensing data'. In Hegde, VS; Dadhwal, VK; Roy, PS; Parihar, JS (eds) Bulletin of the National Natural Resources Management System NNRMS (B) – 35

Telwala, Y., Brook B.W., Manish K., Pandit M.K. (2013). Climate-induced elevational range shifts and increase in plant species richness in a Himalayan biodiversity epicenter. PloS one, 8(2), e57103.

Walker M., Wahren C.H., Hollister, R.D. (2006). Plant community responses to experimental warming across the tundra biome. Proceedings of the National Academy of Sciences of the United States of America 103:1342–1346.

World Wildlife Fund (1995). Ecotourism: conservation tool or threat? Conserv. Iss. 2 (3), 1–10.

Xu, Z.X., Gong, T.L., and Li, J.Y. (2008). Decadal trend of climate in the Tibetan Plateau—regional temperature and precipitation. Hydrological Processes 22(16): 3056–3065.

You, Q., Kang, S., Aguilar, E., & Yan, Y. (2008). Changes in daily climate extremes in the eastern and central Tibetan Plateau during 1961–2005.Journal of Geophysical Research: Atmospheres, 113(D7).

Zha, Y., J. Gao and Zhang, Y. (2005): Grassland productivity in an alpine environment in response to climate change. Area, 37, 332–340.

Zomer, R. J., Trabucco, A., Wang, M., Lang, R., Chen, H., Metzger, M. J., ... & Xu, J. (2014). Environmental stratification to model climate change impacts on biodiversity and rubber production in Xishuangbanna, Yunnan, China. Biological Conservation, 170, 264–273.

Chapter 13
Remote Sensing-Based Forest Biomass Assessment in Northwest Himalayan Landscape

Subrata Nandy, Surajit Ghosh, S. P. S. Kushwaha, and A. Senthil Kumar

13.1 Introduction

Forests cover around one-third of the global land cover (4.03 billion hectares) (FAO 2010; Pan et al. 2013) and are among the richest ecosystems in terms of biological and genetic diversity (Köhl et al. 2015). Forests are considered as reservoirs of carbon, and it is stored as biomass (phytomass). The total amount of above- and belowground organic matter of both living and dead plant parts is called biomass (FAO 2005). Net primary productivity (NPP) is majorly accumulated as biomass. Around two-thirds (262.1 PgC) of the global terrestrial biomass is stored by the tropical forests (Pan et al. 2013; Negrón-Juárez et al. 2015). Therefore, forests act as one of the keystones of the global carbon cycle and play a vital role in designing the mitigation strategies for climate change and reducing the emission of greenhouse gases. Hence, forest biomass estimation is useful in quantifying the carbon stock, carbon emissions due to forest degradation and disturbances, carbon budget, productivity, forest planning and management and policy-making (Caputo 2009). Biomass monitoring in regular interval is utmost necessary for understanding the nature (source/sink) of the forest (Kushwaha et al. 2014). In addition, forests are vital sources of livelihood and economic development of any country (Köhl et al. 2015).

S. Nandy (✉) · S. Ghosh
Forestry and Ecology Department, Indian Institute of Remote Sensing (IIRS), Indian Space Research Organisation (ISRO), Department of Space, Government of India, Dehradun, India
e-mail: subrato.nandy@gmail.com

S. P. S. Kushwaha
Forest Research Institute, Dehradun, India

A. Senthil Kumar
Indian Institute of Remote Sensing, Indian Space Research Organisation, Dehradun, Uttarakhand, India

R. R. Navalgund et al. (eds.), *Remote Sensing of Northwest Himalayan Ecosystems*,
https://doi.org/10.1007/978-981-13-2128-3_13

Forest ecosystems offer numerous goods (timber, fodder, food, etc.) and ecological services (MEA 2005).

Growing stock and biomass can be estimated by harvest method, field inventory and integration of field inventory and remote sensing (RS) data (Kushwaha et al. 2014). Aboveground biomass (AGB) of forests can be easily measured at a broad scale, while belowground biomass (BGB) is still poorly known due to measurement limitations. The BGB is generally considered as a fraction of AGB. Conventionally, the AGB assessment and monitoring require exhaustive field inventory. It is laborious and inapplicable in inaccessible areas, thus making it practical only in relatively smaller areas. On the other hand, RS in conjunction with geographic information system (GIS) offers an efficient and economical means for AGB mapping, monitoring and modelling (Nelson et al. 2000; Lu 2005; Sales et al. 2007; Kushwaha et al. 2014; Lu et al. 2014; Manna et al. 2014; Heyojoo and Nandy 2014; Yadav and Nandy 2015; Watham et al. 2016). However, field-measured data is essential for RS-based AGB estimation for training and validation purposes. A variety of optical multispectral and hyperspectral images, active sensor RADAR (Radio Detection And Ranging) data, and LiDAR (Light Detection and Ranging) data (e.g. Landsat, WiFS, AWiFS, LISS-III, ASTER, LISS-IV, SPOT, QuickBird, IKONOS, WorldView, Cartosat, MODIS, AVHRR, Radarsat, RISAT, ALOS PALSAR, Hyperion, ICESat/GLAS, etc.) are now available for AGB studies.

The Indian Himalaya extends over seven states, viz. Jammu and Kashmir, Himachal Pradesh, Uttarakhand, Sikkim, Arunachal Pradesh, Assam and West Bengal. Jammu and Kashmir, Himachal Pradesh and Uttarakhand are part of Northwest Himalaya (NWH). Physiographically, the region is divided into Shivaliks or the outer Himalaya, lesser Himalaya, greater Himalaya and Trans-Himalaya, extending from the foothills of the south to north. The present status of forest in the three states of NWH is given in Table 13.1.

The forest degradation and deforestation are the major problems in the Himalayan region of India (Nandy et al. 2011). The rapid economic development, agricultural growth, overgrazing, increasing requirement for timber, firewood and fodder, excessive tapping of resin, and frequent forest fires have caused forest degradation in different parts of the Himalaya (Negi 1982; Somanathan 1991; Awasthi et al. 2003; Nandy et al. 2011), which affected the forest landscape significantly (Sharma et al. 1999). Regular monitoring of the forests of the mountainous regions is, thus, important to achieve the sustainable development goals.

Table 13.1 Status of forests of NWH states (FSI 2015)

Sl no.	State	Forests (%) of states' geographical area	Forests (%) of India's geographical area	Growing stock (million cum)	Carbon stock (million Mg)
1.	Himachal Pradesh	66.52	4.84	317.58	1,61,224
2.	Jammu and Kashmir	9.10	2.65	236.82	2,41,711
3.	Uttarakhand	71.05	4.97	440.72	2,85,689
	Total		12.46	995.12	6,88,624

13.2 Sampling Design and Field Data Collection for Biomass Calculation

Field data collection, an important part of RS-based AGB studies, involves (i) identifying appropriate variables derived from RS data by finding the relations between those variables and field-measured biomass (considered as reference data), (ii) developing models for biomass estimation by linking reference data and selected remotely sensed variables, (iii) evaluating the model estimates with another set of reference data which were not used for developing the model or comparing different model estimates (generally 70% of the data is randomly identified as training set, and the remaining 30% data are used for validation) and (iv) conducting error and uncertainty analysis for identifying the factors influencing the estimation accuracy .

Hence, collecting high-quality and representative stratum-wise field-measured biomass data is crucial for spatial biomass estimation. Generally, field biomass is calculated using destructive sampling, allometric models and volume to biomass conversion approaches (Lu 2006). Destructive method of biomass estimation involves cutting the sample trees and generating species-specific allometric models using diameter at breast height (*dbh*) of trees, tree height (*h*) and/or wood density/ specific gravity (*Sg*) (Overman et al. 1994; Chave et al. 2004). Biomass estimation using allometric equations is a common method. However, in many cases, allometric equations are not available. In these cases, species-specific volumetric equations are generally used for AGB calculation. The volume calculated from volumetric equations multiplied by *Sg* gives the AGB.

For field data collection related to RS-based forest biomass studies, stratified random sampling approach is generally used where forest type and forest canopy density are considered as a stratum (Yadav and Nandy 2015). Square plots of 31.62 m × 31.62 m, i.e. 0.1 ha sample size, can be laid down randomly in different strata. Samples are selected from each stratum based on probability proportional sampling. A pilot study is carried out to calculate the number of sampling units following Chacko (1965):

$$N = \frac{t^2 \times \mathrm{CV}^2}{(\mathrm{SE}\%)^2} \tag{13.1}$$

where N is the number of sample plots, t is the statistical value at 95% significance level, CV is the coefficient of variation and SE % is the standard error percentage.

Using Eq. 13.1, the total number of sample plots to be laid in different strata can be determined. Then, the sample plots are to be proportionally allocated to the different strata following Cochran (1963):

$$n_{\mathrm{h}} = \frac{N_{\mathrm{h}}}{N} \times n \tag{13.2}$$

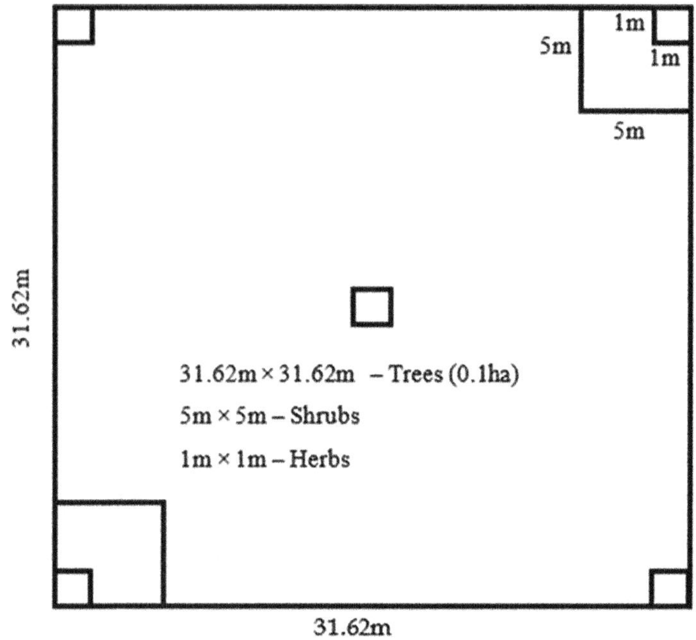

Fig. 13.1 Field inventory design for tree, shrub and herb sampling

where n_h is the number of sample in h stratum, $N_h$is the size of h stratum, N is the total population size and n is the total number of samples.

Figure 13.1 shows the field inventory design. At each sample plot of 0.1 ha, species name and girth at breast height (*gbh*) of all trees at 1.37 m above ground are noted down. Two sample plots of 5 m × 5 m for shrubs at opposite corners, five sample plots of 1 m × 1 m at four corners and one 1 m × 1 m at the centre of the plot for herbs and litters are also laid in each sample plot. The total number of individuals for each shrub species needs to be noted, and specimen samples should be collected. Herbs in each 1 m × 1 m plot may be harvested and weighted. Litter at each 1 m × 1 m plot is weighted on-site using electronic balance and noted, and only 100 g of litter can be collected as a representative sample. Representative samples of each shrub species, herbs and litters were kept in a hot air oven at 80 °C for drying till constant weight and finally total shrub, herb and litter biomass can be calculated for each plot.

The volume of each individual tree in a sample plot can be calculated using dbh (from gbh) value in the species-specific volumetric equations (FSI 1996). The AGB can be obtained by multiplying the tree volume with *Sg* of wood (FRI 2002) of the tree species and further multiplying by biomass expansion factor (BEF) (Haripriya 2000). BGB can be estimated using a root-shoot ratio. For example, the root-shoot ratio for sal (*Shorea robusta*) and other species are 0.30 and 0.26, respectively, in the NWH (Negi 1984). Total biomass of the sample plot is worked out by adding AGB,

BGB, shrub, herb and litter biomass. Further, carbon stock can be estimated by multiplying 0.47 with the total biomass (IPCC 2006).

13.3 Remote Sensing Data

RS data have many advantages, viz. synoptic view, large area coverage, multi-temporal, etc. Due to such benefits as well as accessibility to remote areas, RS technology has been used widely for AGB estimation using optical and radar RS data (Dobson et al. 1992; Muukkonen and Heiskanen 2007; Manna et al. 2014; Kushwaha et al. 2014; Heyojoo and Nandy 2014; Yadav and Nandy 2015; Mangla et al. 2016). Optical RS is used to retrieve forest variables based on reflectance and normalized indices (Zhang et al. 2014). Cloud cover limits the applications of the optical sensors. Passive optical sensors detect the top of the surface feature, and hence understory vegetation properties cannot be addressed properly (Joshi et al. 2016). Even the spectral saturation may occur for a high biomass region. Radar data has an advantage in this scenario. However, both optical and radar (L-band and shorter wavelengths) data cannot address high biomass accurately (Waring et al. 1995; Mitchard et al. 2012). Shugart et al. (2010) used L-band synthetic aperture radar data to assess forest biomass and found that the sensitivity to biomass saturated at about 100–150 Mg ha^{-1}. However, ESA's future BIOMASS mission with P-band will be capable of estimating high biomass more effectively (Le Toan et al. 2011). In case of high biomass areas, optical or radar sensors are not capable of rightly estimating the biomass. On other hand, LiDAR data can provide the forest structure and can estimate biomass as high as 1300 Mg ha^{-1} (Means et al. 1999; Mitchard et al. 2012).

13.3.1 RS Data-Derived Variables

The predictor variables for biomass prediction modelling can be derived from passive optical multispectral or hyperspectral images, active sensor RADAR data and LiDAR data (Table 13.2).

Vegetation indices (VIs) have been widely used for generating empirical relations of biomass estimates. Normalized difference vegetation index (NDVI) has been used extensively for forest productivity and biomass studies (Mather 1999; Foody et al. 2001; Li et al. 2007; Santin-Janin et al. 2009). Other VIs have also been widely used to estimate vegetation biomass, e.g. the enhanced vegetation index (EVI) (Mutanga and Skidmore 2004; Lu 2005; Anaya et al. 2009; Casady et al. 2013), soil-adjusted vegetation index (SAVI) (Ren et al. 2011; Yan et al. 2013), modified soil-adjusted vegetation index (MSAVI) (Boschetti et al. 2007; Ren et al. 2011; Yan et al. 2013), etc. Frequently used vegetation indices are listed in Table 13.3.

Table 13.2 Remotely sensed variables for biomass modelling

Sl no	Sensor type	Variables
1.	*Optical*	
	Optical sensors use visible, near-infrared and shortwave bands to acquire images of the Earth surface	*Spectral*
		Spectral resolution facilitates discrimination of different features based on their spectral response in each band. Spectral bands, band ratio or vegetation indices and transformed images like PCA, etc. are considered as spectral variables
		Textural
		Texture analysis techniques are used with various criteria of feature extraction: statistical methods (grey level co-occurrence matrix, semi-variogram analysis); filter techniques (energy filters, Gabor filters); or based on wavelet decomposition
2.	LiDAR	
	A LiDAR uses a laser (light amplification by stimulated emission of radiation) to transmit a light pulse and a receiver with sensitive detectors to measure the backscattered or reflected light. By recording the time between transmitted and reflected pulses and by using the speed of light, distance to the object is determined	*Space-borne LiDAR*
		Maximum canopy height, total return waveform energy, canopy return energy, matrices, etc. can be extracted at plot level from space-borne LiDAR full-waveform data
		Air-borne LiDAR
		Canopy height model and other biophysical features can be calculated at individual tree level from airborne LiDAR data
		Terrestrial LiDAR (TLS)
		Point cloud information can be extracted to estimate individual tree height, *dbh*, canopy projection area (CPA), etc.
3.	RADAR	
	An active radar sensor emits microwave radiation in a series of pulses from an antenna. Some of the energy is reflected back towards the sensor, and this backscattered microwave radiation is detected, measured and timed. By recording the range and magnitude of the energy reflected from the targets, a 2D image of the surface can be made	Backscattering coefficients
		SAR responses to vegetation structure. Different backscattering coefficients in different polarization characterize the condition of vegetation
		Textural
		Texture measurement is useful to identify various forest stand structure attributes and hence considered important for high-resolution SAR images

Table 13.3 List of satellite data-derived vegetation indices

Sl. no.	Indices	Formula	Description	References	
1.	Simple ratio (SR)	$\frac{NIR}{RED}$	This index is a ratio of NIR and Red band. NIR band has high reflectance, and the Red band has maximum absorption for vegetation. This index value may get saturated in case of dense vegetation	Birth and McVey (1968)	Broadband
2.	Moisture stress index (MSI)	$\frac{SWIR}{NIR}$	This index is sensitive to leaf water content. The absorption in the SWIR band increases with an increase in the leaf water content, whereas changing water content does not affect the absorption at NIR, hence used as a reference	Hunt and Rock (1989)	
3.	Land surface water index (LSWI)	$\frac{NIR-SWIR}{NIR+SWIR}$	LSWI is sensitive to liquid water in vegetation and its soil background, as there is absorption by liquid water in SWIR region	Xiao et al. (2002)	
4.	Normalized difference vegetation index (NDVI)	$\frac{NIR-RED}{NIR+RED}$	It is a measure of healthy and green vegetation. The use of the highest absorption and reflectance regions of chlorophyll make it robust. The values can get saturated in dense vegetation conditions	Rouse et al. (1974)	
5.	Wide range vegetation index (WDRVI)	$\frac{a*NIR-RED}{a*NIR+RED}$ a has a value of 0.1–0.2	The sensitivity of this index is greater than NDVI in moderate-to-high LAI conditions	Gitelson (2004)	
6.	Enhanced vegetation index (EVI)	$\frac{2.5*(NIR-RED)}{NIR+6*RED+7.5*BLUE+1}$	The blue reflectance region is used by this index to correct soil background signals and reduces atmospheric influences, including aerosol scattering	Huete et al. (2002)	
7.	EVI2	$\frac{2.5*(NIR-RED)}{NIR+2.4*RED+1}$	It is used for sensors without a blue band to produce EVI-like vegetation index	Jiang et al. (2008)	

(continued)

Table 13.3 (continued)

Sl. no.	Indices	Formula	Description	References
8.	Visible atmospherically resistant index (VARI)	$\frac{\text{GREEN}-\text{RED}}{\text{GREEN}+\text{RED}-\text{BLUE}}$	The fraction of vegetation is estimated in a scene with low sensitivity to atmospheric effects	Gitelson et al. (2002)
9.	Global vegetation moisture index (GVMI)	$\frac{(\text{NIR}+0.1)-(\text{SWIR}+0.02)}{(\text{NIR}+0.1)+(\text{SWIR}+0.02)}$	It is a good predictor of vegetation moisture measured as equivalent water thickness	Ceccato et al. (2002)
10.	Soil-adjusted vegetation index (SAVI)	$\frac{(1+\text{L})(\text{NIR}-\text{RED})}{(\text{L}+\text{NIR}+\text{RED})}$	Same as NDVI but the effects of soil pixels are suppressed. It is best suited for areas with relatively sparse vegetation	Huete (1988)
11.	Optimized soil-adjusted vegetation index (OSAVI)	$\frac{\text{NIR}-\text{RED}}{\text{NIR}+\text{RED}+0.16}$	This index is based on SAVI. It is best suited for areas with relatively sparse vegetation	Rondeaux et al. (1996)
12.	Renormalized difference vegetation index (RDVI)	$\frac{\text{NIR}-\text{RED}}{(\text{NIR}+\text{RED})^{0.5}}$	This index is used for highlighting healthy vegetation	Roujean and Breon (1995)
13.	Perpendicular vegetation index (PVI)	$\frac{\text{NIR}-\alpha*\text{RED}-\text{b}}{1+\alpha^2}$ $\text{NIR}_{\text{Soil}} = \alpha * \text{RED}_{\text{Soil}} + b$	It calculates the vertical distance between the vegetation spot on the NIR-Red scatterplot and the soil line. Since vegetation has higher near-infrared and lower red reflectance than the underlying soil, the vegetation spot will be on the top left corner of the scatterplot	Richardson and Wiegand (1977)
14.	Transformed soil-adjusted vegetation index (TSAVI)	$\frac{\alpha*(\text{NIR}-\alpha*\text{RED}-b)}{\text{RED}+\alpha*\text{NIR}-\alpha b+0.08(1+\alpha^2)}$ $\text{NIR}_{\text{Soil}} = \alpha * \text{RED}_{\text{Soil}} + b$	The improvement of this index over SAVI is to take the soil line slope (α) and intercept (b) into consideration instead of 1 and 0, respectively, as assumed in SAVI	Baret and Guyot (1991)
15.	Modified soil-adjusted vegetation index (MSAVI)	$(2 * \text{NIR} + 1 - ((2 * \text{NIR} + 1)^2 - 8(\text{NIR} - \text{RED})^{0.5})/2$	Like SAVI, this index also uses soil-adjustment factors. The difference is that it uses a self-adjustment L, while SAVI uses a manual adjustment L	Qi et al. (1994)
16.	Difference vegetation index (DVI)	$\text{NIR} - \text{RED}$	Soil and vegetation can be distinguished but does not consider the difference, caused by atmospheric effects or shadows, between reflectance and radiance	Tucker (1979)

17.	Modified chlorophyll absorption ratio index (MCARI)	$((\rho_{700} - \rho_{670}) - 0.2 * (\rho_{700} + \rho_{550}) * (\rho_{700}/\rho_{670})$	The relative abundance of chlorophyll is indicated by this index. It minimizes the combined effects of soil and non-photosynthetic surfaces	Daughtry et al. (2000)	
18.	MSAVI	$(2 * \rho_{800} + 1 - ((2 * \rho_{800} + 1)^2 - 8(\rho_{800} - \rho_{670})^{0.5})/2$	This is same as of the traditional broadband MSAVI	Qi et al. 1994	
19.	Triangular vegetation index (TVI)	$0.5 * (120(\rho_{682} - \rho_{553}) -200 * (\rho_{682} - \rho_{553}))$	This index is used for green LAI estimation. With an increase in canopy density, the sensitivity of this index to chlorophyll increases	Broge and Leblanc (2001)	
20.	Red edge normalized difference vegetation index (RENDVI)	$\frac{\rho_{750} - \rho_{705}}{\rho_{750} + \rho_{705}}$	It is a modification of the traditional broadband NDVI. However, this differs from NDVI by using bands along the red edge. It exploits on the sensitivity of vegetation red edge to small variations in canopy foliage	Gitelson and Merzlyak (1994)	
21.	Modified red edge normalized difference vegetation index (MRENDVI)	$\frac{\rho_{750} - \rho_{705}}{\rho_{750} + \rho_{705} - 2 * \rho_{445}}$	This index corrects for leaf specular reflection. It takes advantage of the sensitivity of the vegetation red edge to small changes in canopy foliage	Datt (1999)	
22.	Radar vegetation index (RVI)	$\frac{8 * \sigma_{hv}^0}{\sigma_{hh}^0 + \sigma_{hh}^0 + 2 * \sigma_{hv}^0}$	This index is used to characterize vegetation scattering. It may not be applicable if there is little to no volume scattering present	Kim and van Zyl (2009)	Radar

13.4 Techniques and Models Involved in RS-Based Forest Biomass Estimation

To establish the relation between RS data and AGB, empirical and process-based models are used. In empirical modelling, the predictor variables derived from RS data are empirically linked to field-measured AGB (Gasparri et al. 2010; Liang et al. 2012; Manna et al. 2014; Yadav and Nandy 2015). The empirical approaches vary from simple linear regression (SLR) (Sarker and Nichol 2011; Kushwaha et al. 2014), multiple linear regression (MLR) (Lu 2005), geostatistical techniques (Yadav and Nandy 2015; Watham et al. 2016) to complex machine learning algorithms (Powell et al. 2010; Dhanda et al. 2017) or non-parametric methods. Statistical methods are classified on the basis of what we know about the population we are studying. The parametric statistical test is one that makes assumptions about the parameters (defining properties) of the population distribution(s) of the data, while a non-parametric test does not have that kind of assumptions. Parametric methods are those for which the population is approximately normal or can be approximated using a normal distribution after invoking the central limit theorem. The non-parametric methods are statistical techniques for which we do not have to make any assumption of normality for the population. Indeed, the methods do not have any dependence on the population of interest. Biomass and productivity are directly related and limited by alike ecological factors (Knapp and Smith 2001). Process-based model such as dynamic global vegetation models (DGVMs) integrates both biophysical and ecophysiological processes to model present-day carbon production and subsequent carbon storage, past reconstructions and future scenarios addressing climate change feedback (Cramer et al. 2001; Bonan et al. 2003; Joos et al. 2004; Keeling and Phillips 2007).

13.4.1 Simple Linear Regression (SLR) and Multiple Linear Regression (MLR)

In SLR or direct radiometric relation (DRR), a regression relationship is established between the independent variables, e.g. spectral bands, band ratios, vegetation indices, etc., and the dependent variable, i.e. field-measured biomass (Viana et al. 2012). MLR is used when more than one independent variable is applied to estimate biomass. This is basically many to one mapping technique. MLR is one of the most frequently used techniques for developing models to estimate biomass (Lu 2005). The regression models, both SLR and MLR, assume that spectral responses are correlated with the biomass and there exists a very limited correlation among the independent variables (Lu et al. 2004).

Yadav and Nandy (2015) mapped aboveground woody biomass (AGWB) using field inventory data and IRS P6 LISS-III (Linear Imaging Self-Scanning Sensor) imagery in part of Shivalik Himalaya. They did not find any significant relationships

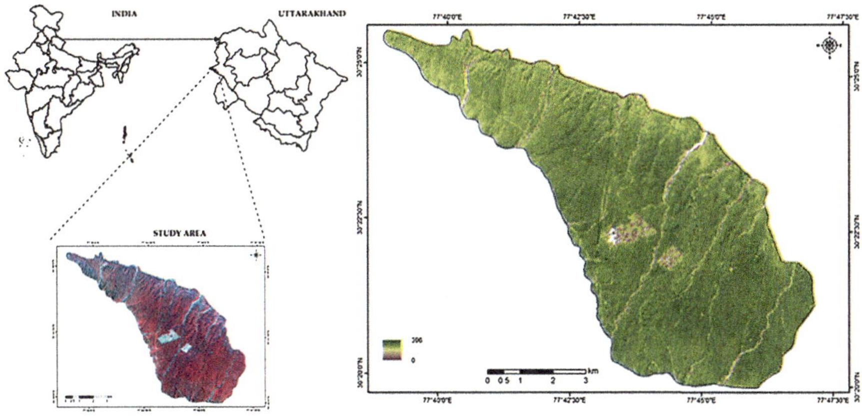

Fig. 13.2 Predicted biomass (Mgha^{-1}) using SLR

between individual spectral bands and AGWB as the values of spectral bands were saturated in high biomass area. This is well established that passive optical RS-based biomass models perform better in low biomass region (Anaya et al. 2009; Kushwaha et al. 2014; Manna et al. 2014). However, a positive relation was found between NDVI and biomass, though it was very low. A biomass map (Fig. 13.2) was prepared using NDVI with root mean square error (RMSE) of 67.17 Mgha^{-1} and $R^2 = 0.15$.

Resourcesat-2 LISS-III dataset of different seasons was used to estimate AGB, instead of using single-dated image. This study was also done in part of Shivalik Himalaya, where AGB is high. Seasonal NDVIs derived from LISS-III imagery were used to nullify the saturation problem of NDVI. Stepwise MLR was applied to predict AGB. It was observed that R^2 increased from 0.2 to 0.46 and RMSE reduced from 132 to 108 Mg ha^{-1} when seasonal NDVI images were used instead of using single-season image.

13.4.2 Non-parametric Methods

13.4.2.1 k-Nearest Neighbour (k-NN) and Co-Kriging (CoK)

The regression-based biomass models assume that the biomass, the dependent variable, is linearly correlated with the independent spectral variables and there exists a limited correlation among the spectral variables (Lu et al. 2012). However, this method fails in many cases as spectral variables are often found to be linearly correlated (Lu 2005) which may have nonlinear relationships with the biomass (Li 2010). To overcome these limitations, non-parametric approaches can be used. k-nearest neighbour (k-NN) is one of the most effective non-parametric methods for mapping forest biomass and other attributes using RS data (Chirici et al. 2012). The k-NN method is extensively used for volume and AGB estimation (Franco-Lopez et al. 2001; Holmström and Fransson 2003; Labrecque et al. 2006; Lu 2006; Chirici et al. 2008; Yadav and Nandy 2015).

Co-kriging (CoK) allows predicting the dependent variable using multiple variables based on their inter-variability and spatial structure (Carr et al. 1985). In this method, a cross-variogram is used to quantify the spatial autocovariance between the primary and secondary variable (Webster and Oliver 2001). It works perfectly when the primary variable is less densely sampled (Eldeiry and Garcia 2009). CoK was used for estimating forest biomass by many studies (Sales et al. 2007; Dwyer 2011).

Yadav and Nandy (2015) mapped AGWB using k-NN and CoK in Timli forest range of Shivalik Himalaya. In the k-NN approach, the estimate of each location is calculated through the weighted mean of k spectrally nearest neighbours by inverse weighted distance (Lu et al. 2012). k-NN using Mahalanobis distance showed the best result (RMSE = 42.25 Mgha^{-1}) (Fig. 13.3) followed by fuzzy distance (RMSE = 44.23 Mgha^{-1}) and Euclidean distance (RMSE = 45.13 Mgha^{-1}), respectively, whereas RMSE was found to be 52.2 Mgha^{-1} using CoK technique. The study highlighted the integration of field-measured parameters, RS and non-parametric methods (such as k-NN and CoK), for forest biomass mapping, especially in high biomass regions.

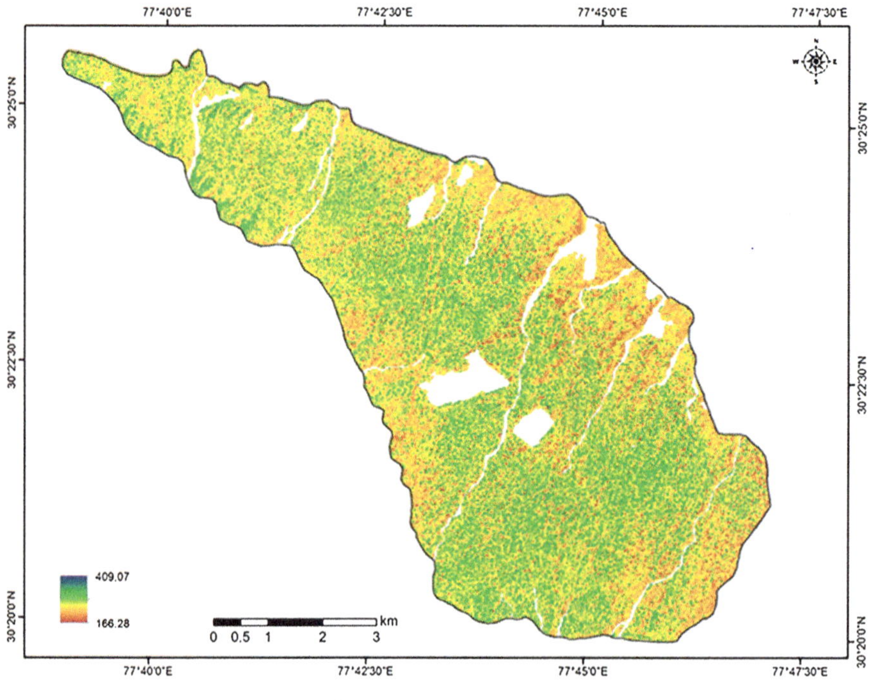

Fig. 13.3 Predicted biomass (Mgha^{-1}) using *k*-NN

13.4.2.2 Random Forest (RF), Support Vector Machine (SVM) and Artificial Neural Network (ANN)

The non-parametric methods, such as RF, SVM and ANN, can be used for variable optimization obtained from RS data. RF algorithm randomly and iteratively samples the data to generate a large group (or forest) for classification through classification and regression tree (CART) technique. CART uses many decision trees and is more robust as well as accurate than single tree-based classification method (Breiman 2001). SVM is based on statistical learning (Vapnik 2006). It emphasizes the boundary between classes rather than the mean and variances of classes. ANN has been inspired by the neuronal architecture of the brain, and it simulates the thinking process of human being, whose brain uses interconnected neurons to process incoming information (Haykin 1994). Multilayer perceptron (MLP) neural network is the most widely used neural network model (Mather and Tso 2009). It has a layered architecture consisting of input, hidden and output layers.

Nandy et al. (2017) assessed forest biomass by incorporating RS data with field-measured biomass using ANN in Barkot forest, Uttarakhand. The feedforward MLP was used in the study because of its desirable computational and approximation capabilities (Cybenko 1989). LISS-III (Resourcesat-1) data of April 24, 2013 was used in this study. Numerous spectral and textural variables were retrieved from the RS data. ANN was used for identifying the contribution of these spectral and textural variables, extracted from LISS-III data, to field-measured forest biomass. The top ten variables were selected based on their ranking to generate an MLR model for predicting the biomass. The predicted biomass well-accorded ($R^2 = 0.74$) with field biomass. On validation, the model yielded $R^2 = 0.70$ and RMSE = 93.41 Mg ha^{-1}. Overall, the study revealed the capability and usefulness of LISS-III data for estimating forest biomass. The study also highlighted the utility of ANN technique for optimizing the independent variables and predicting the AGB with a minimum number of spectral and textural variables.

One more study was carried out in Jhajra forest of Uttarakhand Shivaliks. Resourcesat-2 LISS-III satellite data was used in this study. ANN and *k*-NN were applied separately to various RS-based spectral and textural variables for optimizing the number of highly correlated variables, and, then, MLR analysis was performed to generate biomass map. In *k*-NN, the Euclidean, Mahalanobis and fuzzy distance were used for this purpose. *k*-NN with Mahalanobis distance (RMSE = 51.07 Mg ha^{-1}) was the best method followed by the fuzzy distance and Euclidean distance with RMSE of 62.84 $Mgha^{-1}$ and 74.87 Mg ha^{-1}, respectively. In case of ANN, RMSE was found to be 81.23 Mg ha^{-1}.

Three non-parametric methods, viz. ANN, RF and SVM, were compared for AGB estimation in Barkot forest using seasonal NDVI imagery. Among all these methods, RF regression was found to be the best. The evaluation was made based on R^2 and RMSE. R^2 values of 0.76, 0.85 and 0.89 were found for ANN, SVM and RF, respectively, whereas RMSEs of ANN, SVM and RF were found to be 72, 64 and 55 $Mgha^{-1}$, respectively.

13.4.3 Object-Based Image Analysis

Very-high-resolution satellite (VHRS) imagery like WorldView-2 (WV-2), GeoEye, IKONOS and QuickBird can be used for the precise estimation of carbon stocks at individual tree level (Baral 2011; Maharjan 2012; Karna et al. 2015). Object-based image analysis (OBIA) technique is used for extracting individual tree crowns/canopy projection area (CPA) from the VHRS imagery (Jing et al. 2012). For estimating carbon of a tree, dbh is a key parameter. Among the forest inventory parameters, dbh is highly related to AGB. But the passive optical RS data can record the canopy reflectance, not the tree dbh. However, the CPA and dbh of a tree are related (Shimano 1997). Hence, by developing a relationship between CPA and dbh, the carbon stock of forest can be assessed.

One study was carried out to quantify and map the aboveground carbon (AGC) stock of sal forests (Fig. 13.4) of Shivalik Himalaya using VHRS imagery and field

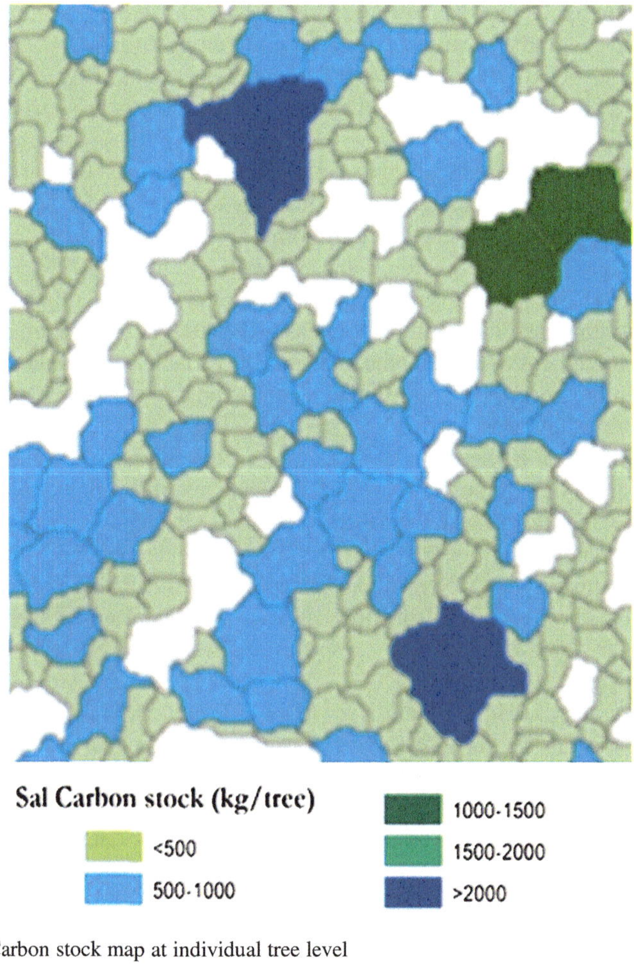

Fig. 13.4 Carbon stock map at individual tree level

data. OBIA technique was used for image segmentation (accuracy – 72.12%) and classification (accuracy – 84.82%). It facilitated to delineate individual tree crowns and CPA calculation. The study showed that a strong relationship exists between *dbh* and CPA of trees and CPA and tree carbon. The study highlighted the utility of VHRS imagery for quantifying and mapping individual tree carbon stock. The study also suggested that a significant relationship exists between CPA and biomass/ carbon of sal trees.

Another study has endeavoured to assess and map AGC stock using Resourcesat-2 LISS-IV and Cartosat-1 satellite data coupled with field-based inputs in Timli forest of Uttarakhand Shivaliks. Both multispectral and panchromatic images were pan-sharpened to get spectrally and spatially high-quality image. Multi-resolution segmentation followed by watershed transformation and morphology was carried out to split the cluster of trees. The segmented image was classified as forest, non-forest and shadow area. Total volume and AGB were estimated using species-specific volumetric equations. The biomass was converted into carbon stock using a conversion factor of 0.47. The relation between CPA and carbon was developed using 134 trees identified in the field and CPA-derived image. Multi-resolution segmentation gave 67% accuracy, while the overall classification accuracy was 94.5%. Average carbon in the forest was 122.18 MgC ha^{-1}. The total carbon stock of the study area was 1103480.34 MgC.

13.4.4 Model Transferability

Generally, biomass is estimated at particular study site using data acquired for that particular site as characteristics of the data differ when surface characteristics change. This is called 'one place one time' approach. There is also another approach where the relation/model established in one site can be transferred to another site. Therefore, the predictive relation can be applied to the sites (test site) which were developed for a site (train site) with similar characteristics. This approach is called model transferability. The accuracy of the predictive model is evaluated at the test site. If any predictive model is successfully transferred to a similar site, it is expected that the model will work at other similar sites with a similar level of accuracy.

A study was conducted to map AGB using a non-parametric method by integrating RS data and forest inventory data in Barkot-Rishikesh forest (training site), Uttarakhand, India, and transfer the model to assess the AGB in Timli forest range (testing site), Uttarakhand, India. Resourcesat-2 LISS-III dataset of different seasons was used to estimate AGB, instead of using the single-dated image. Seasonal NDVIs derived from LISS-III imagery were used, as single-dated NDVI imagery leads to saturation problems. The AGB was estimated with RMSE of 55 Mg ha^{-1} and 32.67 Mg ha^{-1} in the training and testing sites, respectively. Hence, the study demonstrated that the model transfer is possible for large-area AGB mapping by wise selection of RS-based variables and models.

13.4.5 Data Integration Approach

Integration of spatial data from multiple sources is important for accurate estimation of biomass. Data from multiple sensors have been used in the recent past for forest biomass estimation. Combined with better methods of integration, the multi-sensor approach can overcome the limitations of the single sensor data (Koch 2010). Information about the forest type, crown cover and forest structure is important for getting the reliable forest biomass estimates. No single RS data can fulfil all the requirements as it is limited by different acquisition technique, time frame, weather and other biophysical conditions. The use of passive optical, RADAR and LiDAR data can improve the estimation of biomass owing to their complementary nature.

13.4.5.1 High-Resolution Optical and TLS

Carbon stock of individual trees was estimated using WV-2 and TLS data in Barkot-Rishikesh forest of Dehradun Forest Division of Uttarakhand, India. TLS data acquisition was done using a Terrestrial Laser Scanner-RIEGL VZ-400, which operates at 1550 nm wavelength. The TLS data acquisition was done in the field and further processed to extract various inventory parameters such as *dbh*, height and CPA from the point cloud data (Fig. 13.5). The WV-2 satellite image was segmented (Fig. 13.6a) and classified using OBIA technique. A model was developed between TLS-derived CPA and field-measured carbon and was implemented on the classified image to get the carbon stock at individual tree level.

Multi-resolution segmentation followed by watershed transformation and morphology segmented (Fig. 13.6b) the objects which were classified depending upon the parameters NDVI and mean of the near-infrared band. The objects were classified into five classes: dry riverbed, sal, teak, other vegetation and shadow (Fig. 13.6c). An overall classification accuracy of 87.12% was achieved. Hence,

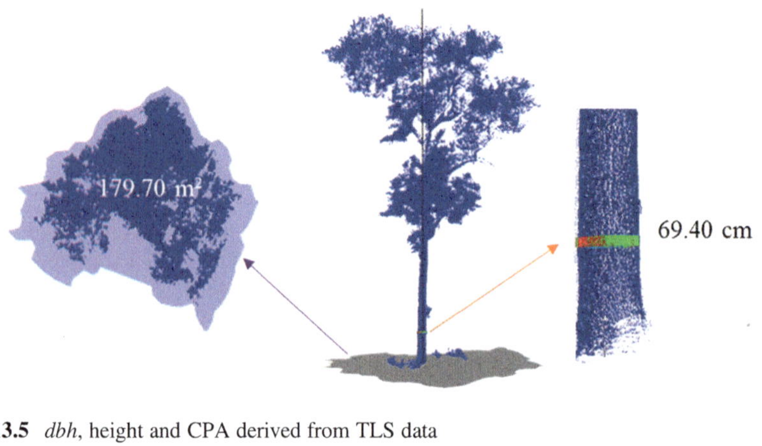

Fig. 13.5 *dbh*, height and CPA derived from TLS data

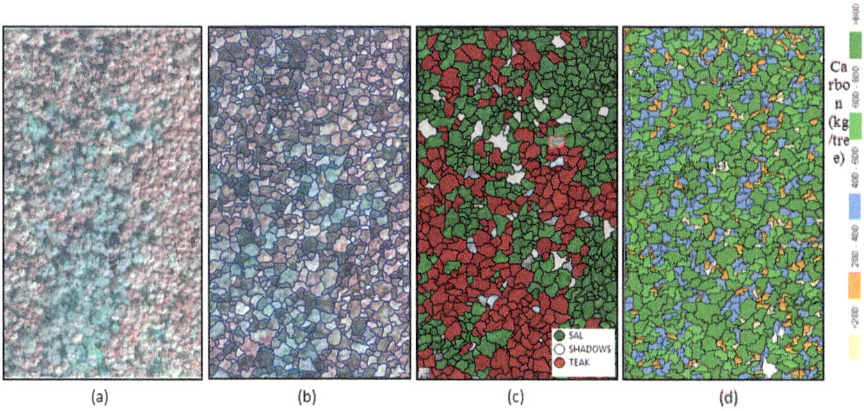

Fig. 13.6 (**a**) WV-2 image, (**b**) segmented image, (**c**) classified image, (**d**) carbon stock map

CPA-carbon model for both the species was developed independently. It was observed that TLS-derived CPA and carbon of teak and sal maintained a nonlinear relationship with R^2 of 0.84 and 0.81, respectively. The carbon stock of individual trees was mapped (Fig. 13.6d) using the relationship of CPA and carbon. Mean carbon stock for individual sal and teak trees were found to be 434.75 and 182.83 kg C, respectively. The results were further validated and found that predicted and observed carbon was highly correlated (sal: $R^2 = 0.93$, RMSE = 14.30 kg C/tree; teak: $R^2 = 0.92$, RMSE = 10.13 kg C/tree). The study highlighted the integration of VHRS imagery and TLS data for AGC estimation at individual tree level.

13.4.5.2 Radar and TLS

AGB was estimated using point cloud data derived from TLS and RISAT-1 fully polarimetric synthetic aperture radar (SAR) data in Timli forest (dominated by sal) of Shivalik Himalaya, Uttarakhand (Mangla et al. 2016). Fully polarimetric SAR data has the advantage to retrieve information of different components of forest structure, and LiDAR system has the ability to measure structure. Random sampling with sample plot size 31.62 m × 31.62 m, i.e. 0.1 ha, was carried out for TLS and field inventory. Surface scattering, double-bounce scattering, volume scattering, helix scattering and wire scattering were derived from RISAT-1 through polarimetric decomposition. Stem diameter and tree height were calculated from TLS point cloud data. All the variables obtained from RADAR and LiDAR data were used as independent variables in a machine learning-based RF regression model for forest AGB estimation. The predicted forest AGB showed reliable accuracy (RMSE = 27.68 Mg ha^{-1} and $R^2 = 0.63$). Further, a sensitivity analysis was performed, and it showed that the model was very much sensitive to stem diameter and volume scattering values, and interestingly these variables were derived from different sensor types.

13.4.5.3 Moderate-Resolution Optical and TLS

There are several variables to indirectly estimate the biomass and carbon of a forest. Forest canopy density (FCD) is one of these variables which facilitate the rapid estimation of biomass and carbon. One such study was carried out to assess forest canopy density by using FCD Mapper and to develop a model between FCD classes, spectral indices and biomass based on TLS data in Barkot-Rishikesh forest of Uttarakhand, India. FCD classes were generated from the Landsat-8 OLI data which was validated with ground measurement. The point cloud data were collected from each FCD class using TLS. The multi-scan point cloud data were first registered, and then each individual tree was extracted. The diameter and height of each tree were estimated from TLS point cloud data which were validated using the field data. The volume, biomass and carbon were estimated from the validated *dbh* and height. The FCD measured from FCD Mapper and ground gave a positive correlation of 0.91. The correlation between TLS and field-measured dbh is 0.99 with RMSE of 2.34 cm, and the correlation between TLS and range finder's height is 0.97 with RMSE of 2.01 m. The model between biomass and FCD classes is exponential in nature, and the model among biomass, indices and FCD classes is linear in nature. The average biomass and carbon estimated from the developed linear model are 374 Mg ha^{-1} and 176 Mg ha^{-1}, respectively.

13.4.5.4 High-Resolution Optical and Space-Borne LiDAR

AGB was estimated at ICESat/GLAS footprints by integrating data from passive optical and space-borne LiDAR sensors using three machine learning algorithms, viz. RF, SVM and ANN with MLP (Dhanda et al. 2017). The forest height was predicted with an RMSE of 1.35 m. Six most important variables derived from LiDAR and optical data could explain 78.7% (adjusted) variation in the observed AGB with an RMSE of 13.9 Mg ha^{-1}. SVM regression algorithm could explain 88.7% variation in AGB with an RMSE of 13.6 Mg ha^{-1}, while RF regression algorithm explained 83.5% variation in AGB with an RMSE of 20.57 Mg ha^{-1}. ANN with MLP algorithm explained 83.7% variation of biomass with an RMSE value of 13.29 Mg ha^{-1}. The study conclusively established that multi-sensor integration approach is better than single-sensor approach in AGB estimation.

13.5 Challenges and Gaps in Forest Biomass Estimation in NWH

The forest biomass estimation process is limited by various factors, for example, sampling type and time, and spatial resolution of RS data. Multi-temporal RS datasets are used for biomass dynamics study; however, the selection of suitable dates of imagery is crucial to understand biomass changes. The incongruence between RS data acquisition and field inventory dates can affect the accuracy of the estimated biomass. The availability of RS time series data (both spatial resolution and temporal intensity) is also an important factor for modelling field measurements. Additionally, the spatial variability of biomass among various forest types as well as topography controls the accuracy of biomass. Light availability on the surface is influenced by the topographic variation. Due to this, reflectance from the mountainous area is adversely affected (Deng et al. 2007; Veraverbeke et al. 2010; Wang et al. 2012). To reduce the topographic effect, various methods have been used which are based on solar incidence angle (Dubayah and Rich 1995; Valeriano et al. 2016). Similarly, rugged topography may cause a change in backscattering values in SAR data (Soenen et al. 2010; Liu et al. 2008; Folkesson et al. 2009; Attarchi and Gloaguen 2014). On the same way, slope correction due to topographic variation is recommended for LiDAR data (Xing et al. 2010; Dhanda et al. 2017).

Effect of topography such as elevation, slope and aspect on biomass distribution is needed to be studied considering the topographical variation of the NWH. It is expected that not only topography but also climatic variables, e.g. precipitation, temperature, etc., can play a crucial role in biomass distribution in this area. There are numerous parametric and non-parametric methods that have been evolved, but there is not any single algorithm which can be applied for all forest types for biomass modelling. Therefore, selection of an ideal algorithm for modelling the biomass is very poorly understood (Lu et al. 2014). It is also important to calculate the uncertainties of the measurements and the factors which involve in it. Furthermore, substantial effort is required to reduce the uncertainties in biomass estimation.

Considering data point of view, passive optical sensor data were mostly used for biomass estimation in the NWH. Instead of horizontal vegetation profile, such as vegetation canopy cover, there is also immense scope of using passive optical data for extracting vegetation structure. The stereo-viewing capability of the RS satellites (e.g. ALOS/PRISM, Terra ASTER and SPOT) can provide a vertical structure of vegetation. Appropriate integration of vegetation structure derived from the stereo-viewing satellite and spectral response and textures from non-stereo-viewing optical sensors in biomass estimation models may be a different way for refining the accuracy of biomass estimation, but there has not been much research on it as required (Lu et al. 2014).

Radar datasets are another source for estimating forest biomass, especially in cloud cover scenario when optical sensors are not capable to provide proper land information. Due to penetration capability of radar pulses inside the vegetation, radar sensors can capture the vertical forest structure which can make radar sensor more accurate for biomass estimation than optical sensors. However, speckle problem and incapability to differentiate vegetation types can affect the accuracy of biomass. Further investigation is required to effectively use InSAR data for extraction of forest canopy height for improving accuracy in biomass estimation. The NISAR mission, a new proposed joint mission of ISRO and NASA, might be helpful in assessing the status of forest biomass in NWH. Saturation in both optical and radar data always affects the accuracy especially in forests with complex stand structures. Therefore, further research is looked for on data integration techniques to achieve the desired accuracy level.

LiDAR data seems to be the most promising, compared to both optical and radar data, for biomass estimation. LiDAR has the capability to provide the forest structure information which in turn relates well to the high biomass values. The integration of airborne LiDAR and passive optical satellite imagery is another promising approach for large-area biomass mapping. However, at present, the airborne LiDAR data is not available for NWH region. The physical methods or radiative transfer models (RTM) are complex and need expert knowledge. Still now there are no studies found which used RTM for AGB estimation with the help of RS data. Therefore, there is a huge scope for study in this aspect.

Multi-sensor data have been widely used in the recent past for forest biomass estimation. Multi-sensor data in combination with improved methods of integration can solve some of the limitations of single data (Leal et al. 1997; Hyde et al. 2006; Koch 2010; Guo et al. 2010; Kellndorfer et al. 2004; Swatantran et al. 2011; Montesano et al. 2013) and can improve the biomass estimates (Koch 2010). There is a requirement to generate biomass or carbon stock map of NWH, specifically using multi-sensor approach. Also, the biomass or carbon stock may relate with gross primary productivity measured at flux tower sites (Watham et al. 2017; Ahongshangbam et al. 2016) to get a broader view of productivity status of this area.

13.6 Conclusions

RS applications in natural resources inventory and condition assessment are growing. Several factors are contributing to this development including (i) an increasing number of satellite and sensors with improvements in terms of spectral, spatial, radiometric and temporal resolution and the acquisition techniques, (ii) efforts of integrating data from multiple data sources using their individual advantages, (iii) methods for handling large data (spatially and/or temporally) with advanced data mining/machine learning/big data analysis techniques, (iv) development of new spectral indices related to different environmental/climatic variables, etc.

Forest biomass maps are important for forest management and planning, carbon accounting, carbon dynamics and forest productivity modelling. Therefore, a reliable method of forest biomass mapping and monitoring has to be devised to address all these issues effectively. RS data has been used for biomass estimation on various spatiotemporal scales. The RS applications provide a reasonable AGB estimates when compared to labour-intensive, economically expensive and time-consuming traditional techniques. Briefly, biomass estimation using geospatial technology includes few steps: field survey, field data collection, biomass calculation at the plot level, remotely sensed data selection, suitable variable extraction from RS data, appropriate algorithm selection, biomass prediction and error evaluation of the estimation. Local topography and biophysical conditions also significantly influence

AGB estimation performance. Forests play a crucial role in the carbon cycle, and hence, mapping and monitoring of forest carbon stock through RS can act as a vital indicator of climate change. The present chapter has reviewed the various RS applications in AGB assessment, emphasizing the limitations and the prospects linked to these techniques, particularly in case of NWH. Further research is required on the application of RS data for estimating the AGB, especially for the whole NWH region in order to meet the Kyoto Protocol objectives.

Acknowledgements The authors wish to acknowledge the Forest Department, Government of Uttarakhand, India, and field staff of Barkot Flux Research Site for their field support. The authors are thankful to NSIDC for providing the ICESat/GLAS data.

References

Ahongshangbam J, Patel NR, Kushwaha SPS, Watham T, Dadhwal VK (2016) Estimating Gross Primary Production of a Forest Plantation Area Using Eddy Covariance Data and Satellite Imagery. J Ind Soc Remote Sens 44(6): 895–904.

Anaya JA, Chuvieco E, Palacios-Orueta A (2009) Aboveground biomass assessment in Colombia: A remote sensing approach. For Ecol Manag 257:1237–1246

Attarchi S, Gloaguen R (2014) Improving the estimation of above ground biomass using dual polarimetric PALSAR and ETM+ data in the Hyrcanian mountain forest (Iran). Rem Sens 6 (5):3693–3715

Awasthi A, Uniyal SK, Rawat GS, Rajvanshi A (2003) Forest resource availability and its use by the migratory villages of Uttarkashi, Garhwal Himalaya (India). For Ecol Manag 174: 13–24

Baral S (2011) Mapping Carbon Stock using High Resolution Satellite Images in Sub-tropical Forest of Nepal, Dessertation, Faculty of Geo-Information and Earth Observation (ITC), University of Twente, Enschede, The Netherlands

Baret F, Guyot G (1991) Potentials and limits of vegetation indices for LAI and APAR assessment. Remote Sens Environ 35(2–3):161–173

Birth GS, McVey GR (1968) Measuring the color of growing turf with a reflectance spectrophotometer. Agron J 60(6):640–643

Bonan GB, Levis S, Sitch S, Vertenstein M, Oleson KW (2003) A dynamic global vegetation model for use with climate models: concepts and description of simulated vegetation dynamics. Glob Change Biol 9(11):1543–1566

Boschetti M, Bocchi S, Brivio PA (2007) Assessment of pasture production in the Italian Alps using spectrometric and remote sensing information. Agric Ecosyst Environ 118:267–272

Breiman L (2001) Random forests. Mach Learn 45(1):5–32

Broge NH, Leblanc E (2001) Comparing prediction power and stability of broadband and hyperspectral vegetation indices for estimation of green leaf area index and canopy chlorophyll density. Remote Sens Environ 76(2):156–172

Caputo J (2009) Sustainable forest biomass: promoting renewable energy and forest stewardship. Policy paper, Environmental and Energy Study Institute

Carr JR, Myers DE, Glass CE (1985) Cokriging—a computer program. Comput Geosci 11(2): 111–127

Casady G, van Leeuwen W, Reed B (2013) Estimating winter annual biomass in the Sonoran and Mojave deserts with satellite- and ground-based observations. Remote Sens 5:909–926

Ceccato P, Flasse S, Gregoire JM (2002) Designing a spectral index to estimate vegetation water content from remote sensing data: Part 2. Validation and applications. Remote Sens Environ 82 (2):198–207

Chacko VJ (1965) A manual on sampling techniques for forest surveys. New Delhi: Manager of Publications, Government of India

Chave J, Condit R, Aguilar S, Hernandez A, Lao S, Perez R. 2004. Error Propagation and Scaling for Tropical Forest Biomass Estimates. Philos Trans Royal Soc B: Biol Sci 359: 409–420

Chirici G, Barbati A, Corona P, Marchetti M, Travaglini D, Maselli F, Bertini, R. 2008. Non-parametric and parametric methods using satellite images for estimating growing stock volume in Alpine and Mediterranean forest ecosystems. Remote Sens Environ 112 (5):2686–2700

Chirici G, Corona P, Marchetti M, Mastronardi A, Maselli F, Bottai L, Travaglini D (2012) K-NN FOREST: a software for the non-parametric prediction and mapping of environmental variables by the k nearest neighbors algorithm. Eur J Remote Sens 45:433–442

Cochran, W. G. 1963. Sampling techniques. John Wiley and Sons Inc, New York

Cramer W, Bondeau A, Woodward FI, Prentice IC, Betts RA, Brovkin V, Cox PM, Fisher V, Foley JA, Friend AD, Kucharik C (2001) Global response of terrestrial ecosystem structure and function to CO2 and climate change: results from six dynamic global vegetation models. Global Change Biol 7(4):357–373

Cybenko G (1989) Approximation by superpositions of a sigmoidal function. Math Control Signal Syst 2(4): 303–314

Datt B (1999) Remote sensing of water content in Eucalyptus leaves. Aust J Bot 47(6): 909–923

Daughtry CST, Walthall CL, Kim MS, De Colstoun EB, McMurtrey JE (2000) Estimating corn leaf chlorophyll concentration from leaf and canopy reflectance. Remote Sens Environ 74(2): 229–239

Deng Y, Chen X, Chuvieco E, Warner T, Wilson JP (2007) Multi-scale linkages between topographic attributes and vegetation indices in a mountainous landscape. Remote Sens Environ 111:122–134

Dhanda P, Nandy S, Kushwaha SPS., Ghosh S, Murthy YVNK, Dadhwal VK (2017) Optimizing spaceborne LiDAR and very high resolution optical sensor parameters for biomass estimation at ICESat/GLAS footprint level using regression algorithms. Prog Phys Geog 41(3): 247–267

Dobson MC, Ulaby FT, LeToan T, Beaudoin A, Kasischke ES, Christensen N (1992) Dependence of radar backscatter on coniferous forest biomass. IEEE Trans Geosci Remote Sens 30(2), 412–415

Dubayah R, Rich PM (1995) Topographic solar radiation models for GIS. Int J Geogr Inf Syst 9 (4):405–419

Dwyer PC (2011) A spatial estimation of herbaceous biomass using remote sensing in southern African savannas. M. Sc. thesis, University of the Witwatersrand, Johannesburg

Eldeiry A, Garcia LA (2009) Comparison of regression kriging and cokriging techniques to estimate soil salinity using Landsat images. Civil and Environmental Engineering Department, Colorado State University, Fort Collins, CO 80523–1372, Hydrology Day, pp. 27–38

FAO (2005) Global Forest Resources Assessment Update 2005, Terms and Definitions (Final Version) (p. 33). Rome: Forest Resources Assessment Program, Working Paper 83, Forest Resources Development Service, Forest Resources Division, FAO

FAO (2010) Global forest resources assessment 2010. Rome, Italy

Folkesson K, Smith-Jonforsen G, Ulander LM (2009) Model-based compensation of topographic effects for improved stem-volume retrieval from CARABAS-II VHF-band SAR images. IEEE Trans Geosci Rem Sens 47:1045–1055

Foody GM, Cutler ME, Mcmorrow J, Pelz D, Tangki H, Boyd DS, Douglas I (2001) Mapping the biomass of Bornean tropical rain forest from remotely sensed data. Global Ecol Biogeogr 10 (4):379–386

Franco-Lopez H, Ek AR, Bauer ME (2001) Estimating and mapping of forest stand density, volume, and cover type using the k-nearest neighbors method. Remote Sens Environ 77:251–274

FRI (2002) Indian woods: their identification, properties and uses, (Revised edition). Dehradun: Forest Research Institute, Indian Council of Forestry Research and Education, Ministry of Environment and Forests, Government of India, I-VI

FSI (1996) Volume equations for forests of India, Nepal and Bhutan. Dehradun: Forest Survey of India, Ministry of Environment and Forests, Government of India
FSI (2015) India State of Forest Report: Forest Survey of India, Ministry of Environment, Forest and Climate Change, Government of India
Gasparri NI, Parmuchi MG, Bono J, Karszenbaum H, Montenegro CL (2010) Assessing multi-temporal Landsat 7 ETM+ images for estimating above-ground biomass in subtropical dry forests of Argentina. J Arid Environ 74:1262–1270
Gitelson A, Merzlyak MN (1994) Spectral reflectance changes associated with autumn senescence of Aesculus hippocastanum L. and Acer platanoides L. leaves. Spectral features and relation to chlorophyll estimation. J Plant Physiol 143(3):286–292
Gitelson AA (2004) Wide dynamic range vegetation index for remote quantification of biophysical characteristics of vegetation. J Plant Physiol 161(2):165–173
Gitelson AA, Kaufman YJ, Stark R, Rundquist D (2002) Novel algorithms for remote estimation of vegetation fraction. Remote Sens Environ 80(1):76–87
Guo Z, Chi H, Sun G (2010) Estimating forest aboveground biomass using HJ-1 Satellite CCD and ICESat GLAS waveform data. Science China Earth Sci 53(1):16–25
Haripriya GS (2000) Estimates of biomass in Indian forests. Biomass Bioenerg 19(4):245–258
Haykin S (1994) Neural Networks: A Comprehensive Foundation. Prentice Hall PTR Upper Saddle River, New Jersey, USA
Heyojoo BP, Nandy S (2014) Estimation of above-ground phytomass and carbon in tree resources outside the forest (TROF): A geo-spatial approach. Banko Janakari 24(1):34–40
Holmström H, Fransson JES (2003) Combining remotely sensed optical and radar data in kNN estimation of forest variables. For Sci 49(3):409–418
Huete A, Didan K, Miura T, Rodriguez EP, Gao X, Ferreira LG (2002) Overview of the radiometric and biophysical performance of the MODIS vegetation indices. Remote Sens Environ 83 (1):195–213
Huete AR (1988) A soil-adjusted vegetation index (SAVI). Remote Sens Environ 25(3):295–309
Hunt ER, Rock BN (1989) Detection of changes in leaf water content using near-and middle-infrared reflectances. Remote Sens Environ 30(1):43–54
Hyde P, Dubayah R, Walker W, Blair JB, Hofton M, Hunsaker C (2006) Mapping forest structure for wildlife habitat analysis using multi-sensor (LiDAR, SAR/InSAR, ETM+, Quickbird) synergy. Remote Sens Environ 102(1–2):63–73
IPCC (2006) IPCC guidelines for national greenhouse gas inventories, Prepared by the National Greenhouse Gas Inventories Programme, Eggleston HS, Buendia L, Miwa K, Ngara T, Tanabe K (eds) Published: IGES, Japan
Jiang G, Zhao D, Zhang G (2008) Seismic evidence for a metastable olivine wedge in the subducting Pacific slab under Japan Sea. Earth Planet Sci Lett 270(3):300–307
Jing L, Hu B, Noland T, Li J (2012) An individual tree crown delineation method based on multi-scale segmentation of imagery. ISPRS J. Photogramm. Remote Sens 70:88–98
Joos F, Gerber S, Prentice IC, Otto Bliesner BL, Valdes PJ (2004) Transient simulations of Holocene atmospheric carbon dioxide and terrestrial carbon since the Last Glacial Maximum. Global Biogeochem Cy 18(2)
Joshi N, Baumann M, Ehammer A, Fensholt R, Grogan K, Hostert P, Jepsen MR, Kuemmerle T, Meyfroidt P, Mitchard ET, Reiche J (2016) A review of the application of optical and radar remote sensing data fusion to land use mapping and monitoring. Remote Sens 8(1):70
Karna YK, Hussin YA, Gilani H, Bronsveld MC, Murthy MSR, Qamer FM, Karky BS, Bhattarai T, Aigong X, Baniya CB (2015) Integration of WorldView-2 and airborne LiDAR data for tree species level carbon stock mapping in Kayar Khola watershed, Nepal. Int J Appl Earth Obs Geoinform 38:280–291
Keeling HC, Phillips OL (2007) The global relationship between forest productivity and biomass. Global Ecol Biogeogr 16(5):618–631
Kellndorfer J, W Walker, L Pierce, C Dobson, JA Fites, C Hunsaker, J Vona, M Clutter (2004) Vegetation Height Estimation from Shuttle Radar Topography Mission and National Elevation Datasets. Remote Sens Environ 93 (3):339–358

Kim Y, van Zyl JJ (2009) A time-series approach to estimate soil moisture using polarimetric radar data. IEEE T Geosci Remote Sens 47(8):2519–2527

Knapp AK, Smith MD (2001) Variation among biomes in temporal dynamics of aboveground primary production. Sci 291(5503):481–484

Koch B (2010) Status and Future of Laser Scanning, Synthetic Aperture Radar and Hyperspectral Remote Sensing Data for Forest Biomass Assessment. ISPRS J Photogramm Remote Sens 65 (6):581–590

Köhl M, Lasco R, Cifuentes M, Jonsson Ö, Korhonen KT, Mundhenk P, de Jesus Navar J, Stinson G (2015) Changes in forest production, biomass and carbon: Results from the 2015 UN FAO Global Forest Resource Assessment. For Ecol Manag 352:21–34

Kushwaha SPS, Nandy S, Gupta M (2014) Growing stock and woody biomass assessment in Asola-Bhatti Wildlife Sanctuary, Delhi, India. Environ Monitor Assess 186(9):5911–5920

Labrecque S, Fournier RA, Luther JE, Piercey D (2006) A comparison of four methods to map biomass from LandsatTM and inventory data in western Newfoundland. For Ecol Manag 226:129–144

Le Toan T, Quegan S, Davidson MW, Balzter H, Paillou P, Papathanassiou K, Plummer S, Rocca F, Saatchi S, Shugart H, Ulander L (2011) The BIOMASS mission: Mapping global forest biomass to better understand the terrestrial carbon cycle. Remote Sens Environ 115(11): 2850–2860

Leal RR, Butler P, Lane P, Payne PA (1997) Data fusion and artificial neural networks for biomass estimation. IEE Proceedings-Science, Measurement and Technology 144(2): 69–72

Li D (2010) Remotely Sensed Images and GIS Data Fusion for Automatic Change Detection. Int J Image Data Fusion 1(1): 99–108

Li X, Gar-On Yeh A, Wang S, Liu K, Liu X, Qian J, Chen X (2007) Regression and analytical models for estimating mangrove wetland biomass in South China using Radarsat images. Int J Remote Sens 28(24):5567–5582

Liang S, Li X, Wang J (2012) Advanced Remote Sensing: Terrestrial Information Extraction and Applications. Academic Press, Oxford

Liu W, Song C, Schroeder TA, Cohen WB (2008) Predicting forest successional stages using multi-temporal Landsat imagery with forest inventory and analysis data. Int J Remote Sens 29: 3855–3872

Lu D (2005) Aboveground biomass estimation using Landsat TM data in the Brazilian Amazon. Int J Rem Sens 26:2509–2525

Lu D (2006) The potential and Challenge of Remote Sensing-based Biomass Estimation. Int J Remote Sens 27 (7):1297–1328

Lu D, Chen Q, Wang G, Liu L, Li G, Moran E (2014) A survey of remote sensing-based aboveground biomass estimation methods in forest ecosystems. Int J Digit Earth 9(1):63–105

Lu D, Mausel P, Brond'ızio E, Moran E (2004) Relationships between forest stand parameters and Landsat TM spectral responses in the Brazilian Amazon Basin. For Ecol Manag 198 (1–3):149–167

Lu D, Q Chen, G Wang, E Moran, M Batistella, M Zhang, G VaglioLaurin, D Saah. (2012) Aboveground Forest Biomass Estimation with Landsat and LiDAR Data and Uncertainty Analysis of the Estimates. Int J For Res 2012:436537

Maharjan S (2012) Estimation and mapping above ground woody carbon stocks using lidar data and digital camera imagery in the hilly forests of Gorkha, Nepal. Dessertation, Faculty of Geo-Information and Earth Observation (ITC), University of Twente, Enschede, The Netherlands

Mangla R, Kumar S, Nandy S (2016) Random forest regression modelling for forest aboveground biomass estimation using RISAT-1 PolSAR and terrestrial LiDAR data. In SPIE Asia-Pacific Remote Sensing (pp. 98790Q–98790Q); doi:https://doi.org/10.1117/12.2227380.

Manna S, Nandy S, Chanda A, Akhand A, Hazra S, Dadhwal VK (2014) Estimating aboveground biomass in Avicennia marina plantation in Indian Sundarbans using high-resolution satellite data. J Appl Remote Sens 8(1):083638

Mather P, Tso B (2009) Classification methods for remotely sensed data. CRC Press, New York

Mather PM (1999) Computer processing of remotely-sensed images. John Wiley & Sons, England

Means JE, Acker SA, Harding DJ, Blair JB, Lefsky MA, Cohen WB, Harmon ME, McKee WA (1999) Use of large-footprint scanning airborne lidar to estimate forest stand characteristics in the western cascades of Oregon. Remote Sens Environ 67(3):298–308

Millennium Ecosystem Assessment (2005) Ecosystems and Human Well-Being: biodiversity synthesis. World Resources Institute, Washington, DC

Mitchard ET, Saatchi SS, White L, Abernethy K, Jeffery KJ, Lewis SL, Collins M, Lefsky MA, Leal ME, Woodhouse IH, Meir P (2012) Mapping tropical forest biomass with radar and spaceborne LiDAR in Lopé National Park Gabon: overcoming problems of high biomass and persistent cloud. Biogeosci 9:179–191

Montesano PM, BD Cook, G Sun, M Simard, RF Nelson, KJ Ranson, Z Zhang, S Luthcke (2013) Achieving Accuracy Requirements for Forest Biomass Mapping: A Spaceborne Data Fusion Method for Estimating Forest Biomass and LiDAR Sampling Error. Remote Sens Environ 130:153–170

Mutanga O, Skidmore AK (2004) Hyperspectral band depth analysis for a better estimation of grass biomass (Cenchrus ciliaris) measured under controlled laboratory conditions. Int J Appl Earth Obs Geoinf 5:87–96

Muukkonen P, Heiskanen J (2007) Biomass estimation over a large area based on standwise forest inventory data and ASTER and MODIS satellite data: a possibility to verify carbon inventories. Remote Sens Environ 107(4):617–624

Nandy S, Kushwaha, SPS, Dadhwal VK (2011) Forest degradation assessment in the upper catchment of the river Tons using remote sensing and GIS. Ecolo Indic 11:509–513

Nandy S, Singh RP, Ghosh S, Watham T, Kushwaha SPS, Senthil Kumar A, Dadhwal VK (2017) Neural Network-based Modelling for Forest Biomass Assessment. Carbon Manag 8(4):305–317

Negi JDS (1984) Biological productivity and cycling of nutrients in managed and man-made ecosystems; Ph.D. Thesis, Garhwal University, Srinagar, India

Negi SS (1982) Environmental Problems in the Himalaya. Bishen Singh Mahendra Pal Singh, Dehradun, pp 188

Negrón-Juárez RI, Koven CD, Riley WJ, Knox RG, Chambers JQ (2015) Observed allocations of productivity and biomass, and turnover times in tropical forests are not accurately represented in CMIP5 Earth system models. Environ Res Lett 10(6):064017

Nelson RF, Kimes DS, Salas WA, Routhier M (2000) Secondary forest age and tropical forest biomass estimation using Thematic Mapper imagery. Biogeosci 50:419–431

Overman JPM, HJL Witte, JG Saldarriaga (1994) Evaluation of Regression Models for Above-ground Biomass Determination in Amazon Rainforest. J Trop Ecol 10 (02):207–218

Pan Y, Birdsey RA, Phillips OL, Jackson, RB (2013) The structure, distribution, and biomass of the world's forests. Annu Rev Ecol Evol Syst 44:593–622

Powell SL, WB Cohen, SP Healey, RE Kennedy, GG Moisen, KB Pierce, JL Ohmann (2010) Quantification of Live Aboveground Forest Biomass Dynamics with Landsat Time-series and Field Inventory Data: A Comparison of Empirical Modeling Approaches. Remote Sens Environ 114 (5):1053–1068

Qi J, Chehbouni A, Huete AR, Kerr YH, Sorooshian S (1994) A modified soil adjusted vegetation index. Remote Sens Environ 48(2):119–126

Ren HR, Zhou GS, Zhang XS (2011) Estimation of green aboveground biomass of desert steppe in Inner Mongolia based on red-edge reflectance curve area method. Biosyst Eng 109:385–395

Richardson AJ, Wiegand CL (1977) Distinguishing vegetation from soil background information. Photogramm Eng Remote Sens 43(12):1541–1552

Rondeaux G, Steven M, Baret F (1996) Optimization of soil-adjusted vegetation indices. Remote Sens Environ 55(2):95–107

Roujean JL, Breon FM (1995) Estimating PAR absorbed by vegetation from bidirectional reflectance measurements. Remote Sens Environ 51(3):375–384

Rouse Jr J, Haas RH, Schell JA, Deering DW (1974) Monitoring vegetation systems in the Great Plains with ERTS. https://ntrs.nasa.gov/archive/nasa/casi.ntrs.nasa.gov/19740022614.pdf (last accessed 22 July 2017)

Sales MH, Souza Jr CM, Kyriakidis PC, Roberts DA, Vidal E (2007) Improving spatial distribution estimation of forest biomass with geostatistics: a case study for rondônia, Brazil. Ecol Model 205:221–230

Santin-Janin H, Garel M, Chapuis JL, Pontier D (2009) Assessing the performance of NDVI as a proxy for plant biomass using non-linear models: a case study on the Kerguelen archipelago. Pol Biol 32(6):861–871

Sarker LR, Nichol JE (2011) Improved forest biomass estimates using ALOS AVNIR-2 texture indices. Remote Sens Environ 115: 968–977

Sharma A, Prasad R, Saksena S, Joshi V (1999) Micro-level sustainable biomass system development in central Himalayas: stress computation and biomass planning. Sust Dev 7 (3):132–139

Shimano K (1997) Analysis of the relationship between DBH and crown projection area using a new model. J For Res 2(4): 237–242

Shugart HH, Saatchi S, Hall FG (2010) Importance of structure and its measurement in quantifying function of forest ecosystems. J Geophys Res 115 (G2): G00E13

Soenen SA, Peddle DR, Hall RJ, Coburn CA, Hall FG (2010) Estimating aboveground forest biomass from canopy reflectance model inversion in mountainous terrain. Remote Sens Environ 114:1325–1337

Somanathan E (1991) Deforestation, property rights, and incentives in central Himalaya. Econ Pol Wkly 26:37–46

Swatantran A, Dubayah R, Roberts D, Hofton M, Blair JB (2011) Mapping biomass and stress in the Sierra Nevada using lidar and hyperspectral data fusion. Remote Sens Environ 115(11): 2917–2930

Tucker CJ (1979) Red and photographic infrared linear combinations for monitoring vegetation. Remote Sens Environ 8(2):127–150

Valeriano MDM, Sanches IDA, Formaggio AR (2016) Topographic effect on spectral vegetation indices from landsat tm data: is topographic correction necessary? B Cienc Geod 22(1):95–107

Vapnik V (2006) Estimation of Dependences Based on Empirical Data. Springer Science & Business Media

Veraverbeke S, Verstraeten WW, Lhermitte S, Goossens R (2010) Illumination effects on the differenced Normalized Burn Ratio's optimality for assessing fire severity. Int J Appl Earth Obs 2:60–70

Viana HJ, Lopes AD, Cohenc WB (2012) Estimation of crown biomass of Pinus pinaster stands and shrubland above-ground biomass using forest inventory data, remotely sensed imagery and spatial prediction models. Ecol Model 226:22–35

Wang Y, Hou X, Wang M, Wang M, Wu L, Ying L, Feng Y (2012) Topographic controls on vegetation index in a hilly landscape: a case study in the Jiaodong Peninsula, eastern China. Environ Earth Sci 70:625–634

Waring RH, Way J, Hunt ER, Morrissey L, Ranson KJ, Weishampel JF, Oren R, Franklin SE (1995) Imaging radar for ecosystem studies. BioSci 45:715–723

Watham T, Kushwaha SPS, Nandy S, Patel NR, Ghosh S (2016) Forest carbon stock assessment at Barkot Flux tower Site (BFS) using field inventory, Landsat-8 OLI data and geostatistical techniques. Int J Multidisc Res Dev 3 (5):111–119

Watham T, Patel NR, Kushwaha SPS, Dadhwal VK, Kumar AS (2017) Evaluation of remote-sensing-based models of gross primary productivity over Indian sal forest using flux tower and MODIS satellite data. Int J Remote sens 38(18): 5069–5090

Webster R, Oliver MA. (2001) Geostatistics for environmental scientists. New York: Wiley.

Xiao X, Boles S, Frolking S, Salas W, Moore Iii B, Li C, He L, Zhao R (2002) Observation of flooding and rice transplanting of paddy rice fields at the site to landscape scales in China using VEGETATION sensor data. Int J Remote Sens 23(15):3009–3022

Xing Y, de Gier A, Zhang J,Wang L (2010) An improved method for estimating forest canopy height using ICESat-GLAS full waveform data over sloping terrain: A case study in Changbai mountains, China. Int J Appl Earth Obs 12(5):385–392

Yadav BKV, Nandy S (2015) Mapping aboveground woody biomass using forest inventory, remote sensing and geostatistical techniques. Environ Monitor Assess 187(5):1–12

Yan F, Wu B, Wang YJ (2013) Estimating aboveground biomass in Mu Us Sandy Land using Landsat spectral derived vegetation indices over the past 30 years. J Arid Land 5:521–530

Zhang G, Ganguly S, Nemani RR, White MA, Milesi C, Hashimoto H, Wang W, Saatchi S, Yu Y, Myneni RB (2014) Estimation of forest aboveground biomass in California using canopy height and leaf area index estimated from satellite data. Remote Sens Environ 151:44–56

Chapter 14
CO_2 Flux Tower and Remote Sensing: Tools for Monitoring Carbon Exchange over Ecosystem Scale in Northwest Himalaya

N. R. Patel, Hitendra Padalia, S. P. S. Kushwaha, Subrata Nandy, Taibanganba Watham, Joyson Ahongshangbam, Rakesh Kumar, V. K. Dadhwal, and A. Senthil Kumar

14.1 Introduction

Carbon accounts for nearly half of the total dry mass of all living things (Schlesinger 1991). Forests are the major reservoir of terrestrial carbon on the Earth and play vital role in balancing the steadily rising concentration of carbon dioxide in the atmosphere owing to fossil fuel and biomass burning (IPCC 2005). A forest is called the sink or source of carbon dioxide depending on net removal or release of carbon dioxide into the atmosphere. India supports a vast mosaic of forest ecosystems and contributes significantly to its carbon dynamics (Chhabra and Dadhwal 2004). Accurate quantification of carbon fluxes of forest ecosystems at local, regional, and global scales is necessary for understanding the feedback mechanism between the terrestrial biosphere and the atmosphere. Deep insight into the role of forests in

N. R. Patel (✉)
Agriculture and Soils Department, Indian Institute of Remote Sensing (IIRS), Indian Space Research Organisation (ISRO), Department of Space, Government of India, Dehradun, India
e-mail: nrpatel@iirs.gov.in

H. Padalia · S. Nandy · T. Watham · J. Ahongshangbam · R. Kumar
Forestry and Ecology Department, Indian Institute of Remote Sensing (IIRS), Indian Space Research Organisation (ISRO), Department of Space, Government of India, Dehradun, India

S. P. S. Kushwaha
Forest Research Institute, Dehradun, India

V. K. Dadhwal
Indian Institute of Space Science and Technology, Department of Space, Government of India, Thiruvananthapuram, India

A. Senthil Kumar
Indian Institute of Remote Sensing, Indian Space Research Organisation, Dehradun, Uttarakhand, India

R. R. Navalgund et al. (eds.), *Remote Sensing of Northwest Himalayan Ecosystems*,
https://doi.org/10.1007/978-981-13-2128-3_14

the regional carbon cycle is critical for taking policy-oriented decisions on forest-based initiatives to mitigate global warming.

The conventional means of accounting carbon fluxes of a forest ecosystem over several years include measuring temporal changes in biomass (Clark et al. 2001) and underlying soil carbon pool (Amundson et al. 1998). Forest inventory of biomass change yields estimates of annual net primary productivity (NPP) and depends upon allometric relations to scale incremental changes in diameter at breast height (dbh) with NPP at plot and landscape scales (Barford et al. 2001). Field inventories, however, are practically difficult to operate and unable to generate an accurate and consistent estimate of carbon fluxes at daily, monthly, and yearly time scales.

Eddy covariance (EC) technique has been developed as an alternative method to assess net ecosystem exchange (NEE) of CO_2 (Running et al. 1999; Canadell et al. 2000). It directly measures NEE across the canopy-atmosphere interface. It computes CO_2 exchange rate across the interface between the atmosphere and vegetation by measuring the covariance between fluctuations in vertical wind velocity and the CO_2 mixing ratio (Baldocchi et al. 1988). The measured fluxes represent an average exchange rate from an area upwind from the flux tower. The area sampled with EC technique is called flux footprint, and it may range from a few meters to kilometers (Schmid 1994), and it offers ecosystem-level carbon dioxide exchanges across a range of time scales, varying from seconds to hours to years (Wofsy et al. 1993; Baldocchi et al. 2001).

A flux tower site is a micrometeorological tower that uses eddy covariance methods. Globally more than 680 flux towers have been installed till April 2014

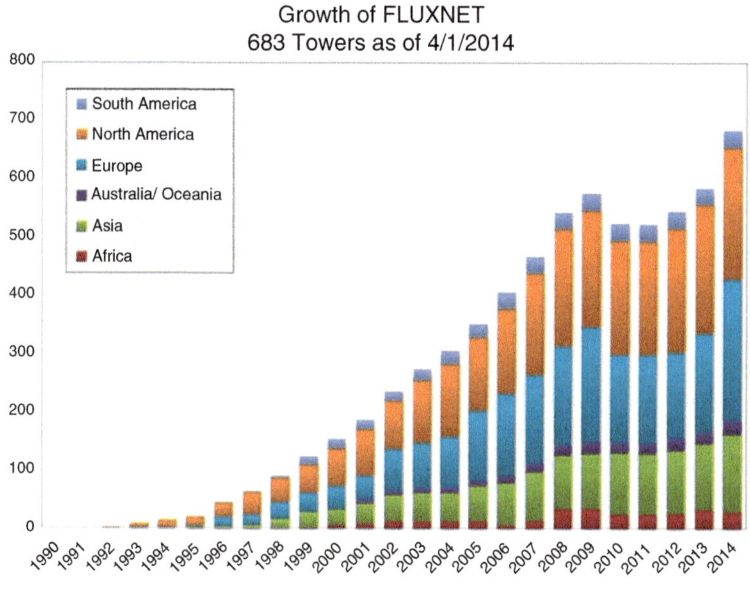

Fig. 14.1 Distribution of EC flux towers worldwide

(Fig. 14.1) by different nations in various ecosystems including forests. Under FLUXNET, a global network is formed by the combination of regional networks like AMERIFLUX, CARBOEURO FLUX, ASIAFLUX, OZFLUX, and other nonnetwork sites. The majority of existing flux towers are located in the northern hemisphere midlatitude locations with poor representation of tropical forests. The National Carbon Project (NCP) under Geosphere-Biosphere Programme (GBP) of Indian Space Research Organisation (ISRO) focuses on spatiotemporal patterns of carbon dioxide and its source-sink relations over India. The EC towers being operated by ISRO over different forest ecosystems are (i) moist sal forest, Barkot (Uttarakhand); (ii) dry deciduous forest, Betul (Madhya Pradesh); (iii) mixed forest plantation, Haldwani (Uttarakhand); and (iv) Sunderban mangrove (West Bengal) in India (Dadhwal et al. 2010; Jha et al. 2013).

Satellite remote sensing offers the unequivocal potential for synoptic monitoring of vegetation functioning with wider coverage and near real-time observations (Matson and Ustin 1991). It has proven its usefulness in monitoring inter-annual and intra-annual activity of vegetation like phenology, NPP, etc. (Myneni et al. 1998). Several satellite data-driven modeling approaches have been suggested to estimate GPP or NPP at large spatial scales in India (Nayak et al. 2010, 2013) and overseas (Ruimy et al. 1994; Running et al. 1999). Various remote sensing-driven modeling approaches, viz., temperature-greenness, light-use efficiency (LUE), and regional-scale CASA (Carnegie-Ames-Stanford Approach) models, have been evaluated to estimate GPP over terrestrial ecosystems in India (Nayak et al. 2010; Patel et al. 2010, 2011). Though remote sensing-driven LUE method has been widely used globally, the performance of such models mainly depends on the parameterization of canopy parameters such as maximum LUE (ε^*) and the temperature optima (T_{opt}) of gross photosynthesis. EC method is being used to derive these ecosystem parameters as input to LUE models. In a nutshell, remote sensing technique plays a key role in optimization of ecosystem model parameters, validation of its outputs, and up-scaling of CO_2 flux across spatial or temporal scales.

14.2 Approach for Carbon Flux Monitoring and Modeling

14.2.1 Principle of Eddy Covariance System

In an EC system, airflow can be considered as a horizontal flow of many rotating eddies, and each rotating eddy has three-dimensional components, including vertical movement of air, and it constitutes the turbulent motions of rising and descending moving air that transport gases (e.g., CO_2 (Fig. 14.2)). The EC technique samples such turbulent motions to compute the net difference of scalar moving between the atmosphere-vegetation interfaces.

The common principle of flux measurement is to quantify (i) how many molecules of any gas are moving upward and downward over the time and (ii) how fast these molecules are traveling. To compute the fluxes of heat, water vapor, and CO_2,

Fig. 14.2 Principle of operation of an eddy covariance system. (Modified from Wolf 2010)

an equation $F = \overline{\rho_a W'S'}$ is used, where ρ_a represents air density and W' and S' represent fluctuation in vertical wind speed and mixing ratio of air. The overbars on each term denote time averaging (usually, 30 min), and primes ($'$) represent instantaneous deviations from this average value for every time unit like 0.1 or 0.05 s. The mean vertical flux density of CO_2 is obtained as 30-min covariance between mean product of air density, vertical wind speed, and mixing ratio of the CO_2 (Burba and Anderson 2010). The instantaneous vertical mass flux density is statistically analyzed using Reynolds' rules of averaging (Reynolds 1895). A positive value of covariance represents a net transfer of CO_2 into the atmosphere, and negative value signifies the net transfer of CO_2 toward vegetation and soil from the atmosphere.

14.2.2 Processing of Eddy Flux Data

The EC technique is based on high-speed measurements of wind speed, direction, and a scalar of interest. This raw data often consists of noises, i.e., spikes, dropouts, constant value, etc., which may be due to inaccurate adjustment of the transducers of ultrasonic anemometers, low voltage, water contamination of sensors, rain, snowflakes, etc. During measurements, random errors may also be introduced due to an electronic system, turbulence transport (Hollinger and Richardson 2005), and instrument limits such as acquisition frequency, sensor separation, etc.

During the night, low-turbulence environments may generate a decoupling between the soil surface and the canopy top. In these situations, the phenomenon advection is a significant term in the flux balance inflicting errors in flux estimation (Massman and Lee 2002) of most of the flux tower sites (Finnigan et al. 2006). The portion of flux measurements inflicted with nighttime advection problem can be filtered out by applying some correction procedures like "u-star correction" (Papale 2006). Friction velocity (u^*) is used to differentiate low and well-mixed periods, usually known as "u-star correction," and is one of the commonly used methods to deal with the nighttime advection problem (Papale 2006).

Calculation of flux from raw data involves the following steps:

1. Transforming the sensor output signal into meteorological variables
2. Quality testing to flag and/or remove spikes present in the raw data
3. Spectral correction, WPL, and other sensor-specific corrections
4. Correction of nighttime flux
5. Filling up the gaps

The EC data can be analyzed using software like EddyPro (www.licor.com/eddypro) and TK3 (Mauder and Foken 2011). These softwares take care of the necessary steps such as de-spiking, rotation of coordinates, correction of time delay, de-trending, applying frequency response, WPL, and other corrections as well as quality control for accurate flux calculation (Burba 2013; Fratini and Mauder 2014).

Despite all these advancements in the EC technique, data gaps are unavoidable which may be caused by power failure, damage of instruments by strong wind and wild animals, lightning, incorrect system calibration, low turbulence in the night, etc. Data gaps pose difficulty in (i) annual estimation of NEE, latent heat, and the sensible heat; (ii) biased relationships between NEE, LE, and H with climatic variables; and (iii) low-quality data for validation of models (Hui et al. 2004). To overcome the problem of data gaps, various gap-filling techniques, viz., mean diurnal variation (Falge et al. 2001), look-up table (Falge et al. 2001), neural network (Papale and Valentini 2003), marginal distribution sampling (Reichstein et al. 2005), and nonlinear regressions (Noormets et al. 2007), have been developed. Moffat et al. (2007) reviewed 15 gap-filling techniques and found overall performance of nonlinear regressions, look-up table, and marginal distribution sampling to be effective.

14.2.3 Partitioning of NEE into GPP and Re

EC measurement system gives the net carbon exchange from ecosystem, i.e., NEE as output. The daytime NEE given by EC is a resultant of both photosynthesis (GPP) and ecosystem respiration (R*e*), but nighttime NEE comprises of ecosystem respiration only. Partitioning of daytime NEE into GPP and R*e* can improve our understanding of carbon sequestration or release processes (Falge et al. 2002). To estimate daytime R*e*, it is assumed that the temperature response of daytime R*e* resembles with that of nighttime R*e*. Based on this assumption, the exponential relationship between nighttime NEE and the nighttime temperature (T_a) is established as follows (Falge et al. 2002):

$$NEE_{night} = a\ exp^{bT_a}$$

where NEE_{night} is nighttime ecosystem respiration and T_a is nighttime mean air temperature. The constants, *a* and *b,* are determined by nonlinear optimization. The temperature sensitivity coefficient of respiration (Q_{10}) is calculated as:

$$Q_{10} = \exp(10*b)$$

Ecosystem respiration (R*e*) (day time) is obtained by applying the nighttime CO_2 exchange-temperature relationship to daytime temperature (Zhou et al. 2009).

Finally, GPP is calculated as the residual of NEE and R*e* as:

$$GPP = -NEE + Re$$

CO_2 flux from the atmosphere to the surface is denoted as negative, but the GPP and R*e* are always positive.

14.2.4 Remote Sensing-Based Up-Scaling and Validation of Modeled CO_2 Fluxes

Various ecosystem process models along with remote sensing-derived parameters are used to extrapolate flux measurements over large areas. Light-use efficiency (LUE) model is one of the most commonly used ecosystem process models. The biophysical parameters derived from remote sensing and meteorological observations are used as inputs to this model (Fig. 14.3). Spectral indices such as NDVI (normalized difference vegetation index), LSWI (land surface wetness index), etc. are used to derive the model parameters and its comparison with field observations. The realized light-use efficiency of vegetation is calibrated by incorporating temperature and water scalar.

14.3 Eddy Covariance Studies in Northwestern Himalaya

This section provides research highlights of eddy flux studies carried out in a mixed forest plantation and moist deciduous sal forest in Uttarakhand state as part of the National Carbon Project (NCP) of ISRO's Geosphere-Biosphere Programme (IGBP).

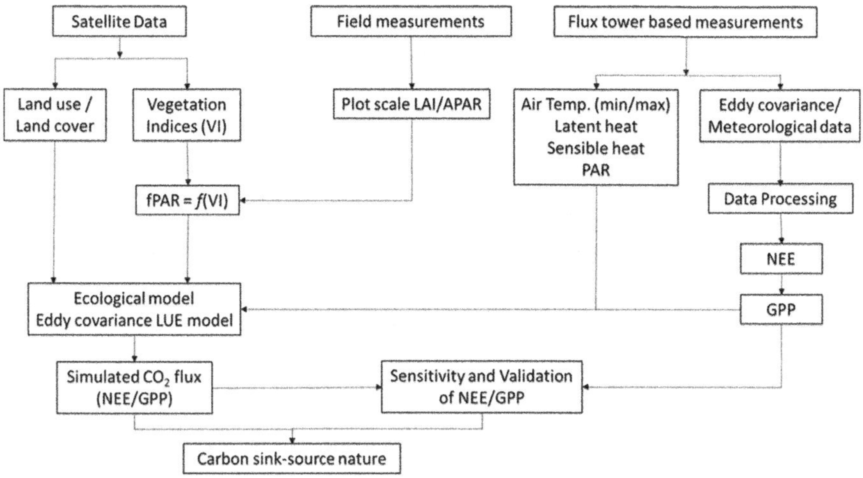

Fig. 14.3 Approach for estimating carbon fluxes using flux measurements and remote sensing. (Source: Watham 2016)

14.3.1 Mixed Forest Plantation Site

14.3.1.1 Site Characteristics and Instrumentation

The Uttarakhand forest department has raised a mixed forest plantation in Terai Central Forest Division (29°8′57.55″N and 79°25′15.97″E) in Uttarakhand in 2004 comprising species such as *Holoptelea integrifolia*, *Dalbergia sissoo*, *Acacia catechu*, and *Albizia procera* (Fig. 14.4). The topography of the area is nearly flat. Annual temperature ranges between 5 °C and 40 °C, and annual mean rainfall is 1500 mm. This site was chosen for establishing EC tower in 2009 with the aim to assess carbon sequestration potential of native forest plantation species. The flux tower was established with the support of Tuscia University, Italy, and Uttarakhand Forest Department.

At this site, eddy covariance sensors, viz., the open path IRGASON CO_2/H_2O analyzer and the sonic anemometer, are installed at 17 m height. These sensors measure 3D wind speed, air temperature, carbon dioxide, and water vapor at 10 Hz sampling frequency. Micrometeorological sensors, such as net radiometer to measure short-wave and long-wave radiation (incoming and outgoing) at 15 m height; hygrometer to measure air temperature and humidity at 5, 10, and 15 m heights; and anemometer to measure wind speed and direction at 5, 10, and 15 m heights, are installed. Soil moisture and temperature are being measured at 30, 60, and 100 cm depth levels. Barometric pressure and precipitation are also recorded at 15 m height. An automated data logger is being used to archive flux data and micrometeorological data.

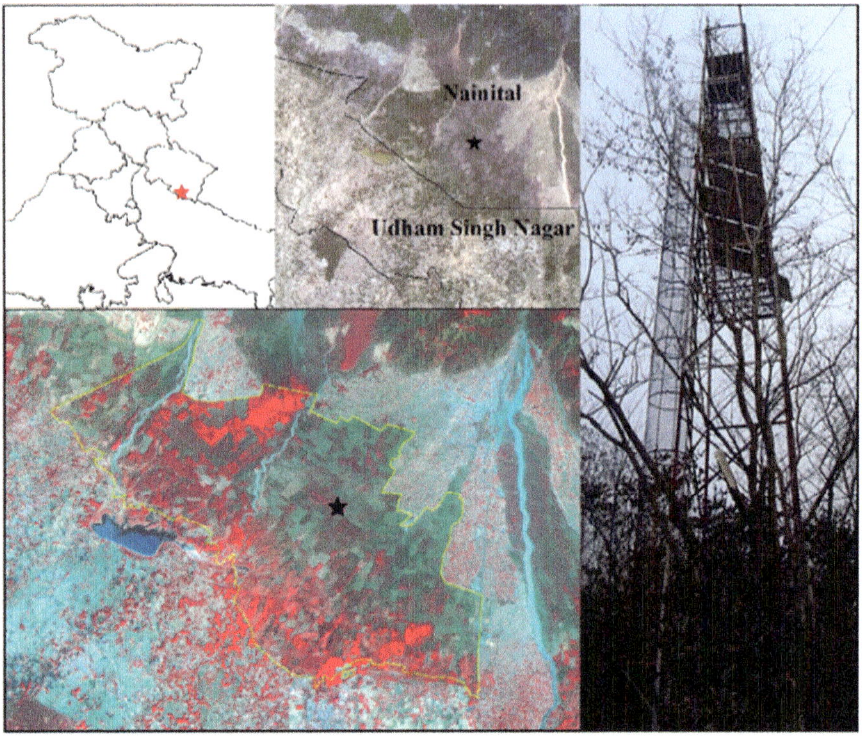

Fig. 14.4 Location of mixed forest plantation site of Haldwani

14.3.1.2 Research Highlights

This section presents the eddy covariance and remote sensing-based assessment of carbon flux carried out for mixed forest plantation site for the period 2013–2014. The raw flux data was averaged at 30 min interval before processing in EddyPro software package for estimation of flux, followed by gap-filling and intra- and inter-annual trend analysis. The analysis revealed that deciduous mixed forest plantation is a net sink of carbon (Fig. 14.5). The plantation sequestrated 1124 g C m^{-2} $year^{-1}$ during 2013–2014. The higher rate of carbon sequestration was observed during monsoon and post-monsoon season; however, significantly lower carbon uptake was observed in winter due to leaf shedding. It is noteworthy that the plantation acted as net source of carbon during January and February. GPP, which is a measure of gross productivity, also followed trends similar to NEE and reached to peak value of 10.7–11.7 g C m^{-2} day^{-1} during monsoon. The lower GPP of winter season is mainly due to leaf fall in the study area.

The daily GPP for the entire plantation area was simulated following LUE model given by Monteith (Monteith 1972), using satellite imagery of October 2013 to March 2014 and site-specific data on PAR (photosynthetically active radiation),

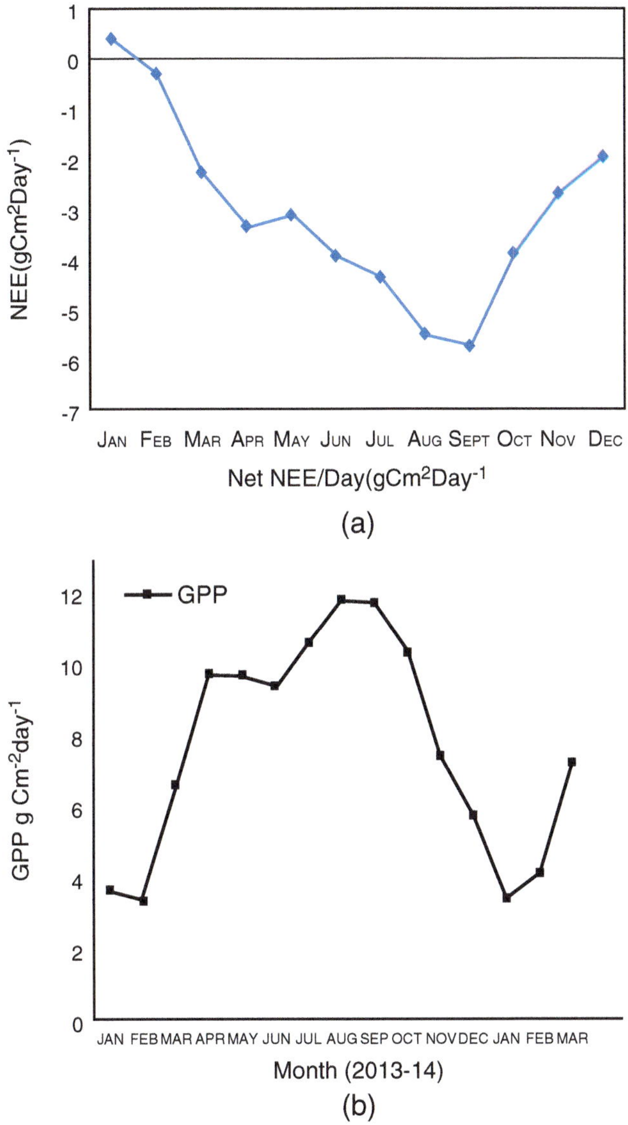

Fig. 14.5 (**a**) NEE (daily mean) and (**b**) GPP (diurnal mean) budget for 2013–2014. (Source: Watham et al. 2014)

*f*PAR (fraction of absorbed PAR), temperature, and water stress. It was observed that seasonal dynamics of predicted GPP follows an identical trend in all plantations. Mixed plantation, however, had higher GPP with respect to teak, poplar, and eucalyptus plantations. The modeled GPP across the study area varied between 2.84 and 14.22 g C m^{-2} day^{-1} (Fig. 14.6). The temporal dynamics of GPP_{pred}

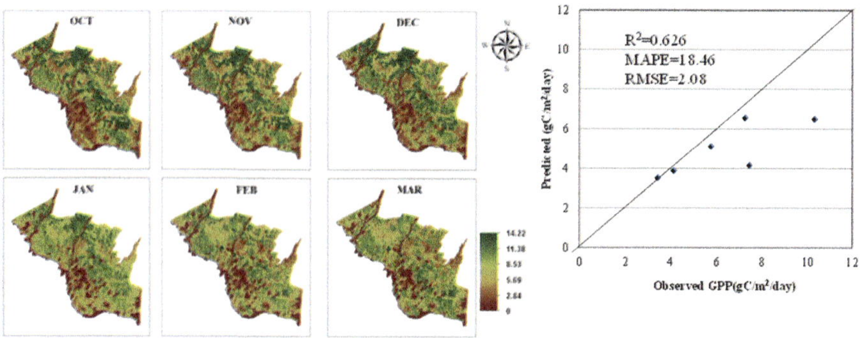

(a) (b)

Fig. 14.6 (**a**) Modeled GPP (g C m^{-2} day^{-1}) and (**b**) relationship between predicted and observed GPP. (Source: Ahongshangbam et al. 2016)

agreed well ($R^2 = 0.62$) with the GPP_{obs} at flux tower site. The predicted GPP was found to be −30% to −3% lower than observed GPP. The underestimation of GPP_{pred} might be because of inconsistencies in realized LUE value and NDVI. The noise or errors in the satellite reflectance values may lead to the incorrect value of indices. These errors will further propagate and contribute to the values of *f*PAR and water surface index and finally affect the prediction of GPP. The modeled GPP explained about 70% of the observed variations of daily GPP. The NPP estimated from observed GPP showed a significant net sink nature of all the plantation types of the forest division. It was also observed that the seasonal dynamics of GPP was predominantly controlled by PAR and temperature. The study showed the effectiveness of using EC and remote sensing techniques in addressing the net carbon budget at ecosystem scale.

14.3.2 Moist Deciduous Sal Forest Site

14.3.2.1 Site Characteristics and Instrumentation

The moist deciduous sal (*Shorea robusta*) is a dominant forest biome of Western Himalayan foothills. The primary objective of establishing a flux tower in the sal forest was to know about the carbon sequestration potential of a climatic climax forest in the region. The flux tower was established in Barkot forest range, Dehradun (Uttarakhand) (30°6′44.391″N and 78°12′43.06″ E), in 2011 (Fig. 14.7). The site has relatively flat topography (slope < 5°). The annual temperature ranges between 4.50 °C and 41.16 °C and receives 2073.3 mm of annual rainfall. The average height of forest canopy is 32 m.

At this site, the flux tower is equipped with a high-speed sonic anemometer and open path IRGASON CO_2/H_2O analyzer installed at 46 m height. It measures 3D

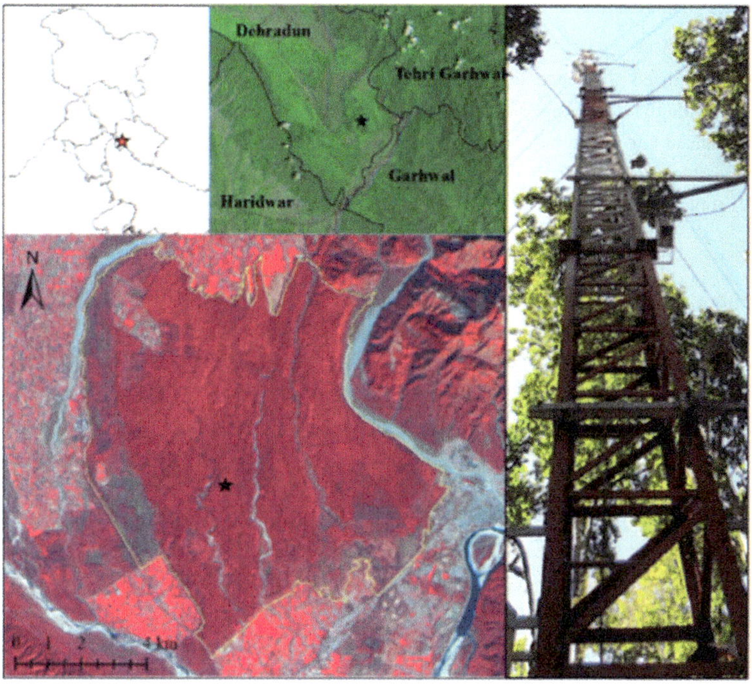

Fig. 14.7 Location of moist deciduous sal forest site

wind speed, air temperature, carbon dioxide, and water vapor at 10 Hz sampling frequency. Meteorological parameters such as short-wave and long-wave radiation (incoming and outgoing) at 40 m height; air temperature and humidity at 2, 4, 8, 16, 32, and 48 m heights; wind speed and direction at 5, 10, and 15 m heights; barometric pressure at 2 m height; and precipitation at 50 m height are usually recorded at 30 min intervals. In addition, soil temperature, soil moisture, and soil flux are also recorded at different depth levels.

14.3.2.2 Research Highlights

The analysis of eddy covariance data from moist deciduous sal forest site indicates that the sal forests are the net sink of carbon. However, mature sal forest was less efficient in sequestrating carbon (NEE = 500 g C m^{-2} $year^{-1}$) that mixed forest plantation of Haldwani. The value of NEE was found to be varying with seasons. Higher daily absorption of carbon was observed in September to March compared to the dry season (April–June) and monsoon season (July–August). A higher value of NEE during September to March was due to ideal climatic conditions, viz., clear sky condition clubbed with high LAI, and adequate water supply for photosynthesis was found in this period. The highest rate of carbon storage by the forest was found in

Table 14.1 The relationship between satellite data-derived vegetation indices and GPP

Vegetation indices	*a*	*b*	R^2	*p*-value	*N*
NDVI (normalized difference vegetation index)	3.03	6.82	0.12	0.047443	32
WDRVI (wide dynamic range vegetation index)	7.47	3.64	0.14	0.032034	32
EVI (enhanced vegetation index)	−0.60	23.89	0.62	9.89E-08	32
VARI (visible atmospherically resistant index)	5.62	12.44	0.35	0.000385	32
OSAVI (optimized soil-adjusted vegetation index)	−1.92	14.90	0.48	1.17E-05	32

November, but a higher rate of ecosystem respiration was observed during monsoon season. A gradual increase followed by the decrease in vapor pressure deficit (VPD) during the period of study was observed. Maximum GPP was observed when the value of VPD is about 20 kPa. The GPP was also found to be increasing with the increase in air temperature (T_a) and touched its peak when the temperature was around 30 °C. But the GPP was found to be negatively affected by the further increase in air temperature.

The GPP of climax sal forest of Barkot forest range was assesed using vegetation indices derived from 8-day interval MODIS (Moderate Resolution Imaging Spectroradiometer). Regression equations were developed for time-series vegetation indices and flux data for flux tower footprint area, and developed equation (with highest R^2 value) was used to predict GPP of entire forest range. Enhanced vegetation index (EVI) was most promising ($R^2 = 0.62, p < 0.05$) in capturing the spatial and temporal variability of GPP (Table 14.1). EVI act as a proxy for leaf area index and fPAR which in turn governs the GPP. EVI was, therefore, used to model the spatiotemporal variability of GPP in the study area (Fig. 14.8). The modeled GPP was found to be relatively higher in August to October but lowest in December to February.

14.4 Conclusions

Existing flux towers in Northwest Himalaya have provided valuable information on carbon source and sink status of a climax sal forest and a young mixed forest plantation. It revealed that both mature sal forest and young plantation are net sink of carbon; however, young mixed plantation (associate species of sal) sequestrated twice higher amount of carbon than that of mature sal forests. The assessment of source/sink status of the NWH requires expanded network of flux towers covering dominant plant functional types of the region. More number of flux towers will aid in effective calibration and validation of the remote sensing-based LUE models. The uncertainties in flux measurements and model assimilation methodologies need to be better understood. To refine estimate of GPP at ecosystem scale, attempts are being made to incorporate vegetation indices derived from new satellite sensors (e.g., Landsat 8 OLI, Sentinel-2 etc.). Long-term pursuance of eddy flux studies would

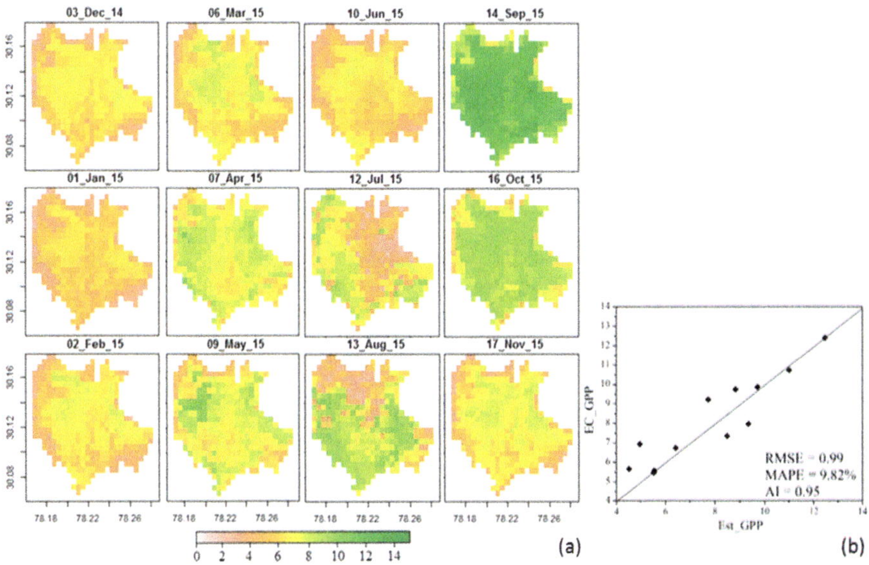

Fig. 14.8 (**a**) Modeled daily GPP (g C m^{-2} day^{-1}) and (**b**) relationship between estimated and observed daily GPP of Barkot flux site

provide novel insights into carbon and water exchange from forest canopies and response of forests to changing climate.

Acknowledgment Authors are thankful to ISRO-GBP for supporting eddy covariance research in NWH. Uttarakhand Forest Department is duly acknowledged for providing desired infrastructure for setting and maintenance of eddy flux towers.

References

Ahongshangbam J, Patel NR, Kushwaha, SPS, Watham T, Dadhwal VK (2016) Estimating gross primary production of a forest plantation area using eddy covariance data and satellite imagery. J Indian Soc Remote Sens 44:895

Amundson R, Stern L, Raisden T, Wang Y (1998) The isotopic composition of soil and soil-respired CO_2. Geoderma 82: 83–114

Baldocchi DD, Falge E, Gu L, Olson R, Hollinger D, Running S, Anthoni P, Bernhofer C, Davis K, Evans R, Fuentes J, Goldstein A, Katul G, Law B, Lee X, Malhi Y, Meyers T, Munger, W, Oechel W, Paw KT, Pilegaard K, Schmid HP, Valentini R, Verma S, Vesala T, Wilson K, Wofsy S (2001) FLUXNET: A New Tool to Study the Temporal and Spatial Variability of Ecosystem–Scale Carbon Dioxide Water Vapor and Energy Flux Densities. Bull Am Meteorol Soc 82:2415–2434

Baldocchi DD, Hicks BB, Meyers TP (1988) Measuring biosphere-atmosphere exchanges of biologically related gases with micrometeorological methods. Ecology 69:1331–340.

Barford CC, Wofsy SC, Goulden ML, Munger JW, Pyle EH, Urbanski SP, Hutyra L, Saleska S R, Fitzjarrald D, Moore K (2001) Factors controlling long- and short-term sequestration of atmospheric CO2 in a mid-latitude forest. Science 294:1688–1691.

Burba G (2013) Eddy Covariance Method for Scientific Industrial Agricultural and Regulatory Applications: A Field Book on Measuring Ecosystem Gas Exchange and Areal Emission Rates, LI-COR Biosciences, Lincoln NE, USA

Burba G, Anderson D (2010) A brief practical guide to eddy covariance flux measurements: principles and workflow examples for scientific and industrial applications. Li-Cor Biosciences, Lincoln NE, USA

Canadell JG, Mooney HA, Baldocchi DD, Berry JA, Ehleringer B, Field CB, Gower ST, Hollinger DY, Hunt JE, Jackson RB, Running SW, Shaver GR, Steffen W, Trumbore S E, Valentini R, Bond BY (2000) Carbon metabolism of the terrestrial biosphere: a multi-technique approach for improved understanding. Ecosystems 3:115–130

Chhabra A, Dadhwal VK (2004) Estimating terrestrial net primary productivity over India using satellite data. Curr Sci 86(2): 269–271

Clark DA, Brown S, Kicklighter DW, Chambers JQ, Thomlinson JR, Ni J (2001) Measuring net primary production in forests: Concepts and field methods. Ecol Appl 11: 356–370

Dadhwal VK, Kushwaha SPS, Patel NR, Yogesh Kant (2010) Carbon flux monitoring in India. ENVIS Forestry Bulletin 9(2):46–49

Falge E, Baldocchi DD, Olson R, Anthoni P, Aubinet M, Bernhofer C, Burba G, Ceulemans R, Clement R, Dolman H et al. (2001) Gap filling strategies for defensible annual sums of net ecosystem exchange. Agric For Meteorol 107:43–69

Falge E, Baldocchi DD, Tenhunen J, Aubinet M, et al. (2002) Seasonality of ecosystem respiration and gross primary production as derived from FLUXNET measurements. Agric For Meteorol 113:53–74

Finnigan J, Aubinet M, Katul G, Leuning R, Schimel D (2006) Report of a Specialist Workshop on "Flux Measurements in Difficult Conditions". Bull Am Meteorol Soc. Boulder, Colorado

Fratini G, Mauder M (2014) Towards a consistent eddy-covariance processing: an intercomparison of EddyPro and TK3. Atmos Meas Tech 7:2273–2281

Hollinger DY, Richardson AD (2005) Uncertainty in eddy covariance measurements and its applications to physiological models. Tree Physiol 25:873–885

Hui DF, Wan SQ, Su B, Katul G, Monson R, Luo YQ (2004) Gap-filling missing data in eddy covariance measurements using multiple imputation (multiple imputation) for annual estimations. Agric Forest Meteorol 121:93–111

IPCC, IPCC (2005) Special Report on Carbon Dioxide Capture and Storage, eds E Calvo and E Jochem, Cambridge Univ Press, Cambridge

Jha CS, Kiran Chand T, Suraj RR, Raghavendra KV, Kushwaha SPS, Patel NR, Dadhwal VK (2013) Establishment of Eddy-Flux Network in India for NEE Monitoring. Asia Flux news letter 35:21–25

Massman WJ, Lee X (2002) Eddy covariance flux corrections and uncertainties in long-term studies of carbon and energy exchanges. Agric For Meteorol 113:121–144.

Matson PA, Ustin SL (1991) Special Feature: The Future of Remote Sensing in Ecological Studies. Ecology 72 (6):1917

Mauder M, Foken T (2011) Documentation and Instruction Manual of the Eddy-Covariance Software Package TK3. Universität Bayreuth, Germany

Moffat AM, Papale D, Reichstein M *et al.* (2007) A comprehensive comparison of gap filling techniques for eddy covariance net carbon fluxes. Agric For Meteorol 147:209–232

Monteith J (1972) Solar radiation and productivity in tropical ecosystems. J Appl Ecology 9 (3):747–766

Myneni RB, Tucker CJ, Asrar G, Keeling CD (1998) Interannual variations in satellite-sensed vegetation index data from 1981 to 1991. J Geophys Res 103: 6145–6160

Nayak RB, Patel NR, Dadhwal VK (2010) Estimation and analysis of terrestrial net primary productivity over India by using remote sensing driven CASA model. Environ Moni Assess 170 (1–4):195–213

Nayak RB, Patel NR, Dadhwal VK (2013) Inter-annual variability of terrestrial net primary productivity over India. Int J Climatol 33:132–142

Noormets A, Chen J, Crow TR (2007) Age-dependent changes in ecosystem carbon fluxes in managed forests in northern Wisconsin, USA. Ecosystems 10:187–203

Papale D (2006) Towards a standardized processing of Net Ecosystem Exchange measured with eddy covariance technique: algorithms and uncertainty estimation. Biogeosciences 3:571–583

Papale D, Valentini R (2003) A new assessment of European forests carbon exchanges by eddy fluxes and artificial neural network spatialization. Global Change Biol 9:525–535

Patel NR, Dadhwal VK, Saha SK (2011) Measurement and scaling of carbon dioxide (CO_2) exchange in wheat using flux-tower and remote sensing. J Indian Soc Remote Sens 39 (3): 383–391

Patel NR, Dadhwal VK, Saha, SK, Garg A, and Sharma N. (2010) Evaluation of MODIS data Potential to infer water stress for wheat NPP estimation. J Int Soc Tro Eco 51 (1): 93–105

Reichstein M, Falge E, Baldocchi DD *et al.* (2005) On the separation of net ecosystem exchange into assimilation and ecosystem respiration: review and improved algorithm. Global Change Biol 11(9)1424–1439

Reynolds O (1895) On the dynamical theory of incompressible viscous fluids and the determination of the criterion. Philos Trans R Soc Lond 186: 123–164

Ruimy A, Saugier B, Dedieu G (1994) Methodology for the estimation of terrestrial net primary production from remotely sensed data. J Geophys Res 97:18515–18521

Running SW, Baldocchi DD, Turner DP, Gower ST, Bakwin PS, Hibbard KA (1999) A global terrestrial monitoring network integrating tower fluxes, flask sampling, ecosystem modeling and EOS satellite data. Remote Sen Environ 70:108–127

Schlesinger WH (1991) Biogeochemistry an Analysis of Global Change. Academic Press, New York.

Schmid HP (1994) Source areas for scalars and scalar flux. Bound-Layer Meteor 67:293–318

Watham T (2016) Monitoring of carbon exchange over moist sal forest using eddy covariance data and remote sensing driven LUE model. Ph.D thesis, Forest Research Institute, Dehradun.

Watham T, Kushwaha SPS, Patel NR, Dadhwal VK (2014) Monitoring of carbon dioxide and water vapour exchange over a young mixed forest plantation using eddy covariance technique. Curr Sci 107(5):858–867

Wofsy SC, Goulden ML, Munger JW, Fan S-M, Bakwin PS, Daube BC, Bassow SL Bazzaz FA (1993) Net exchange of CO2 in a mid-latitude forest. Science 260:1314–1317

Wolf, S. (2010). *Carbon dioxide and water vapour fluxes of tropical pasture and afforestation.* ETH Zurich.

Zhou L, Zhou GS, Jia QY (2009) Annual cycle of CO_2 exchange over a reed (*Phragmites australis*) wetland in northeast China. Aquat Bot 91:91–98

Chapter 15
Carbon Monoxide Plume over Northwestern Himalaya: A Remote Sensing and Modeling Approach

S. Srivastava, I. Nandi, Y. Yarragunta, and A. Senthil Kumar

15.1 Introduction

Forest fire has a considerable impact on the atmospheric abundance of trace gases and aerosols. Large forest fire emits significant amount of carbon dioxide, carbon monoxide, nitric oxide, nitrogen dioxide, methane, and non-methane hydrocarbons (Crutzen and Andreae 1990; Andreae and Merlet 2001; Duncan et al. 2003). Most of these gases are greenhouse gases and/or pollutants. Several research articles are published on the enhanced mixing ratios of these gases over Africa, America, Australia, and Southeast Asia due to fire emissions (Takegawa et al. 2003; Kondo et al. 2004; Pfister et al. 2008). Favorable winds can transport these emitted pollutants to great distances, far away from their local sources. Pollutants, transported over long distances, can contribute to the background air at the downwind region. Thus, biomass burning may have several impacts on local and global air qualities.

Carbon monoxide (CO) is one of the most important trace constituents emitted by biomass burning as it plays a vital role in atmospheric photochemistry (Savage et al. 2001). CO is a criteria pollutant having severe implications on human health. It is one of the main precursors of criteria pollutant ozone and affects the ambient air quality (Crutzen and Zimmermann 1991). This gas indirectly contributes to global warming and climate change by enhancing the concentration of major greenhouse

S. Srivastava (✉) · I. Nandi · Y. Yarragunta
Marine & Atmospheric Sciences Department, Indian Institute of Remote Sensing (IIRS), Indian Space Research Organisation (ISRO), Department of Space, Government of India, Dehradun, India
e-mail: shuchita@iirs.gov.in

A. Senthil Kumar
Indian Institute of Remote Sensing, Indian Space Research Organisation, Dehradun, Uttarakhand, India

R. R. Navalgund et al. (eds.), *Remote Sensing of Northwest Himalayan Ecosystems*,
https://doi.org/10.1007/978-981-13-2128-3_15

gases like methane and ozone (Daniel and Solomon 1998). Widespread forest fires can enhance CO emission by 4–6 times higher than normal days over one location (Vander Werf et al. 2006).

Intense forest fire near southern slope of the Himalayas has very important climatic implications. The high level of pollution over this region during pre-monsoon season can potentially alter the strength of the South Asian monsoon (Lau et al. 2006). Black carbon emitted by forest fire near the Himalaya may deposit onto the snow cover. Yasunari et al. (2010) showed that black carbon deposited on snow cover could reduce the snow albedo, thereby triggering the melting process of Himalayan glaciers. In addition to carbon dioxide, carbon monoxide emitted due to fire activities may get trapped over this region due to Himalayan peaks in the north and east directions. The accumulation of this gas may trigger the production of other greenhouse gases and can contribute to glacier melting process. Despite these important consequences, study on air pollutants like carbon monoxide emitted from biomass burning near the Himalayan region is limited. Kumar et al. (2011) investigated the influence of northern Indian biomass burning on the air quality of central Himalayas. However, their study has included complete biomass burning (forest fire as well as agriculture residue burning) over a bigger Indian region which covered the entire North India. In the present research work, we have investigated carbon monoxide emitted from short duration and intense fire episode that occurred over a region very close to Northwestern Himalayan glaciers. In this case study, attempt has been made to quantify the contribution of different emission sources on carbon monoxide.

An intense forest fire occurred over Uttarakhand during dry and warm phase of El Niño in April 2016 (Panmao et al. 2016). El Niño-Southern oscillation 2016 might be responsible for very dry atmospheric conditions over Uttarakhand which could have triggered biomass burning of high intensity. It caused the spread of wildfire across the dense pine forest of Uttarakhand and created 2166 km^2 burned area (Jha et al. 2016). This event added enormous amount of carbon monoxide in the pristine environment of Uttarakhand. WRF-Chem model is used to quantify the contribution of different emission sources toward CO enhancement during this fire episode.

15.2 Study Region

Uttarakhand is a mountainous state on the southern slope of Northwestern Himalaya. Figure 15.1 shows its geographical location in India. Northern and northeastern parts of Uttarakhand are mostly snow-covered Himalayan peaks. In the west and south, it is surrounded by Himachal Pradesh, Haryana, and Uttar Pradesh states of India. It has a total geographical area of 53,483 km^2, of which 24,240 km^2 is covered by forests. Chir pine forest, conifer forest, and oak forest are major forest types over this state (Singh et al. 2016). Fire over this region mostly occurs in pre-monsoon season (Semwal and Mehta 1996). Other pollution sources over this state are almost

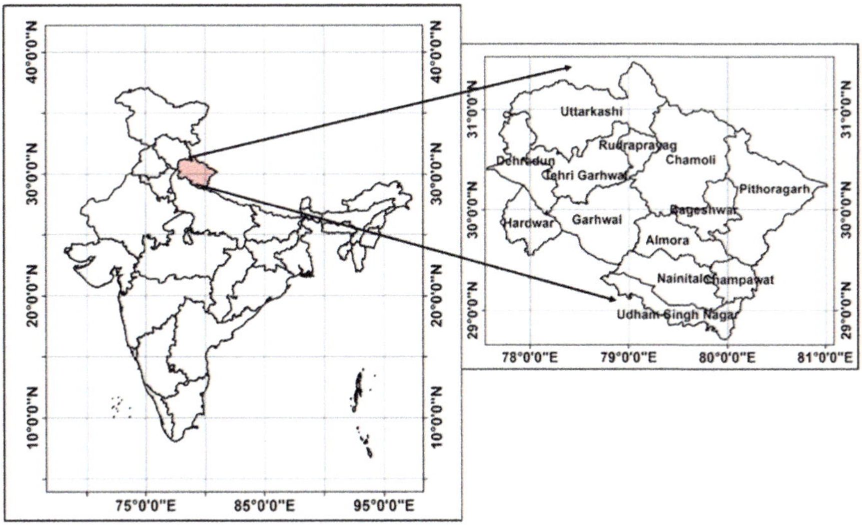

Fig. 15.1 Geographical location of Uttarakhand

negligible due to very few small-scale industries and relatively lower population density with respect to other surrounding states. Systematic surface measurement of carbon monoxide is available over Nainital (29.37°N, 79.45°E, 1958 m amsl), a high altitude site in southeast Uttarakhand (Sarangi et al. 2014). The regional background level of carbon monoxide is found to be 142 ± 47 ppbv over this region.

15.3 Datasets and WRF-Chem Model

15.3.1 Atmospheric Infrared Sounder (AIRS)

Atmospheric infrared sounder is a grating spectrometer onboard the Aqua spacecraft which works in cross track scanning mode. This instrument measures infrared radiation in the spectral range of 3.7 to 16 μm at 2378 spectral channels and provides information about the atmospheric abundance of several trace gases. AIRS retrieves tropospheric distribution of carbon monoxide (CO) by sensing infrared radiation in the spectral range of 4.58–4.50 μm (Chahine et al. 2006). Systematic comparison of AIRS CO profiles with in situ profiles indicates 15% accuracy in AIRS CO retrieval.

15.3.2 Carbon Monoxide Analyzer

A surface analyzer (HORIBA, APMA 370) is used for the continuous measurements of CO at every 5-min interval over Dehradun. This analyzer works on the principle of cross modulation nondispersive infrared absorption method. The EM signal passes through ambient air parcel aspirated by the analyzer. CO mixing ratio is calculated in the air parcel according to Beer-Lambert law. The minimum detection limit of this instrument is 50 ppbv.

15.3.3 Moderate Resolution Imaging Spectroradiometer (MODIS)

MODIS is an imaging sensor onboard two NASA spacecrafts: Terra and Aqua. These satellites orbit in near-polar sun-synchronous orbit at 705 km and cross equator at local solar time of 10:30 and 13:30, respectively. MODIS records visible and thermal infrared radiation in the range of 0.47–2.1 μm at 36 spectral channels and provides information on various land, atmospheric and oceanic parameters. In the present work, Aqua/Terra MODIS fire counts have been utilized.

15.3.4 WRF-Chem Model

WRF-Chem is a regional numerical weather prediction model coupled with chemistry. WRF-Chem version 3.8.1 has been used in this study to simulate the chemical weather over northwest Indian region. The gridded GFS (Global Forecast System) data was used as the input to the WRF Preprocessing System (WPS). The meteorological output from WPS was fed as input for the WRF-Chem model along with the Model of Emissions of Gases and Aerosols from Nature (MEGAN) biogenic emissions, the Fire Inventory from the National Center for Atmospheric Research (FINN) biomass burning emissions, and the Emission Database for Global Atmospheric Research (EDGAR)-Hemispheric Transport of Air Pollution (HTAP) anthropogenic emissions. Initial and lateral boundary conditions for simulation of meteorology have been taken from the National Centers for Environmental Prediction (NCEP) final analysis fields (GFS-FNL) (available at 6 hourly intervals with a resolution of $1^\circ \times 1^\circ$ globally).

Chemical boundary conditions have been taken from Model for Ozone and Related chemical Tracers version 4 (MOZART-4) (Emmons et al. 2010). Different schemes used for the parameterization of atmospheric processes in the WRF-Chem configuration are given in Table 15.1. The model output accounts a range of constituents present in the atmosphere that are also important for climate, air quality, and surface solar radiation.

Table 15.1 Major physics and chemistry schemes used in this study

Atmospheric process	Scheme used
Cloud microphysics	Lin et al. scheme for microphysics (Lin et al. 1983)
Long-wave radiation	Rapid radiative transfer model (Iacono et al. 2008)
Short-wave radiation	Rapid radiative transfer model (Iacono et al. 2008)
Surface layer	Monin-Obukhov (Janjic 1996)
Land surface model	Noah land surface model (Chen and Dudhia 2001)
Planetary boundary layer	Yonsei University scheme (Hong et al. 2006)
Cumulus	Grell-Freitas ensemble scheme (Grell and Freitas 2013)
Photolysis	Fast-J photolysis (Wild et al. 2000)
Dry deposition	Wesely (Wesely 1989)

15.4 Results

15.4.1 Carbon Monoxide in Fire Plume

Figure 15.2 shows the spatial distribution of CO mixing ratio obtained from AIRS over Uttarakhand on April 30, 2016, at 925 hPa (0.7 km), 850 hPa (1.5 km), and 700 hPa (3 km). Fire locations active on this date are also shown in this figure (Jha et al. 2016). This figure shows very high CO mixing ratio extended over the entire Uttarakhand. This is to be noted that CO mixing ratio is particularly higher over the region of intense biomass burning. This confirms that CO plume has risen due to widespread fire activities over Uttarakhand.

According to spatial distribution of forest fire, a fire-affected region is classified as the rectangular area bound by latitudes 29–31°N and longitudes 78–81°E. Figure 15.3 shows the time series variation of daily averaged CO over this region at 925 hPa, 850 hPa, and 700 hPa during April 18 to May 4, 2016. This time series variation clearly shows a systematic enhancement in CO mixing ratio from April 23 onward at all three pressure levels. CO showed peak mixing ratio on April 30 and declining tendency afterward. Fire-affected CO mixing ratio is found to be higher by 60–125 ppbv than the CO mixing ratio before the fire episode. Interestingly, concentration of CO is higher at 700 hPa and 850 hPa with respect to 925 hPa. Uplifting of carbon monoxide plume from biomass burning region to the higher altitude has been reported in several studies (Ding et al. 2009, 2015; Srivastava and Sheel 2013). This may be associated with complex topography, local meteorology, and forest cover.

15.4.2 WRF-Chem Simulation of Fire Event

WRF-Chem model has been used to quantify CO contribution coming from different sources in 2016. For this purpose, model simulation has been made for one calm year 2015 and one fire-active year 2016. To depict the significant difference in fire

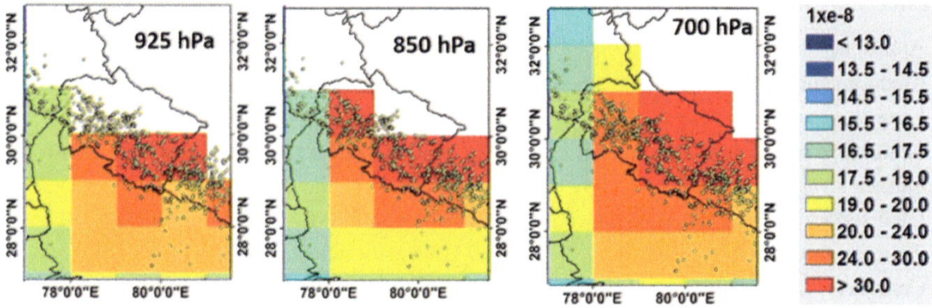

Fig. 15.2 Spatial distribution of CO mixing ratio over Uttarakhand on April 30, 2016

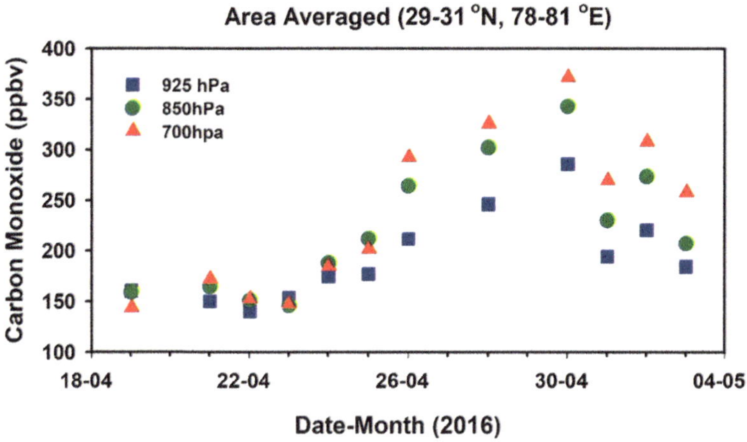

Fig. 15.3 Daily averaged CO distribution over fire-affected region during April 18 to May 4, 2016

activities between two consecutive years, accumulated MODIS fire hot spots (MCD14ML) from April 20 to May 05 for these years are shown in Fig. 15.4. The number of fire-active locations is less than 20 over Uttarakhand during 2015 whereas more than 300 during 2016. Thus, these two years are ideal for the model simulation.

The model simulation has been made from April 15 to May 5 for both 2015 and 2016. For this study, the model domain was centered at 78°E and 23°N covering approximately the entire South Asian region at a horizontal resolution of 30 km, with 115 × 140 grid cells. Simulation has been made at 37 pressure levels ranging from surface to 50 hPa. We introduced three different tracers of carbon monoxide to track carbon monoxide contribution from different emission source types. These tracers were artificial tracers added as separate species within the simulation. Just like the total simulated carbon monoxide, these tracers were also able to provide information on chemistry, transport, and loss activities (Pfister et al. 2011; Kumar et al. 2013). Those identified carbon monoxides account all the possible sources of near-surface total carbon monoxide within the WRF-Chem model. We have introduced

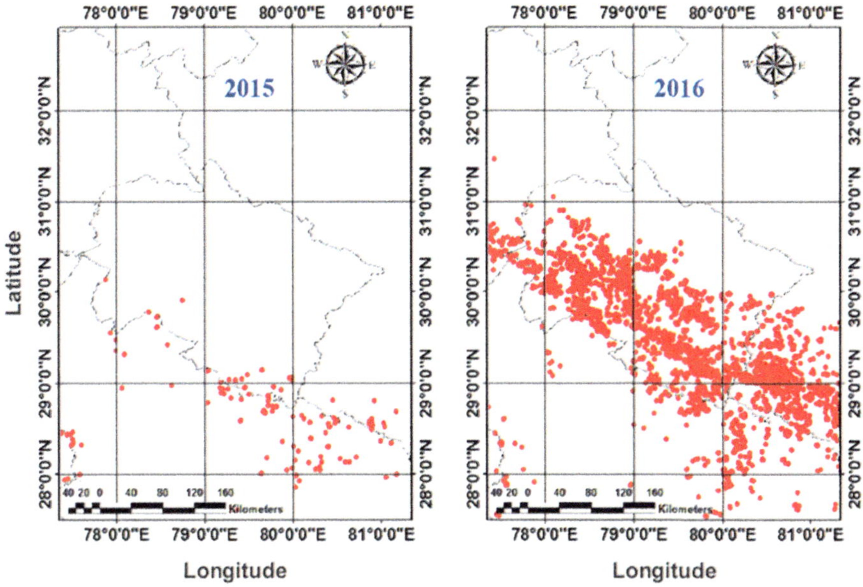

Fig. 15.4 Fire locations over study region during April 20 to May 5, 2016

"CO_ANTHO," "CO_BIOG," and "CO_BIOM" tracers to quantitatively identify the contribution of carbon monoxide coming from anthropogenic sources, biogenic sources, and biomass burning sources, respectively (Kumar et al. 2013). In order to avoid the effects due to model spin-up, the evaluation of the model results was restricted to the period from 20 April to 5 May for both 2015 and 2016.

15.4.3 Validation of Model

To validate the model, model simulation has been compared with CO in situ observations at Dehradun (30.31°N, 78.03°E), a valley in Northwestern Himalayas. This comparison has been made for the fire duration. Figure 15.5 shows the distribution of CO observations over Dehradun and corresponding CO simulation made by WRF-Chem model at 900 hPa and 850 hPa. The model is able to reproduce the buildup tendency of CO during the fire event but significantly underestimates the magnitude of this gas. This is to be noted that fire emission inventory FINN used in WRF chemical simulation is developed on the basis of MODIS fire counts. Padalia et al. (2016) reported that MODIS significantly underestimated the fire count with respect to VIIRS (Visible Infrared Imaging Radiometer Suite) sensor onboard SNPP satellite which has better horizontal resolution. Thus, fire emissions estimated in FINN emission inventory might be relatively lower than actual CO emissions. This might be a cause of model underestimation with respect to real-time observations.

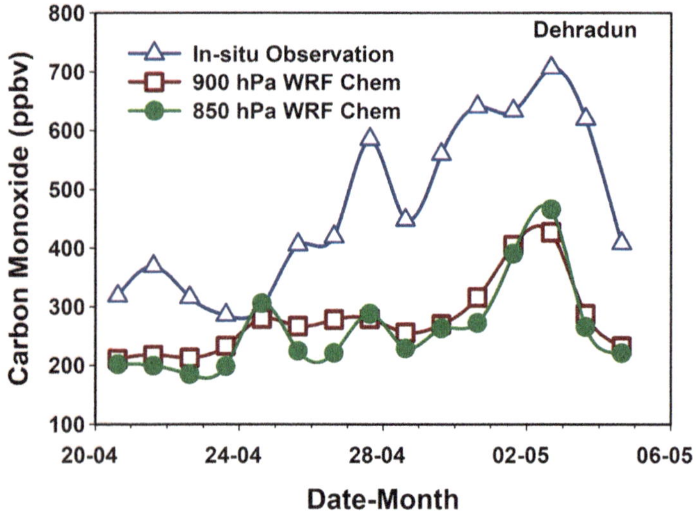

Fig. 15.5 In situ observation of CO over Dehradun during fire event and corresponding CO simulation made by WRF-Chem model at 900 hPa and 850 hPa

15.4.4 Quantification of CO Contribution from Different Sources

Figure 15.6 shows spatial variation of anthropogenic, biogenic, and biomass CO tracers over the study region. Large amount of anthropogenic CO is evident over Delhi and Western Uttar Pradesh which are part of highly populated Indo-Gangetic Belt. This figure shows that there is negligible CO difference between 2015 and 2016. Over fire-affected Uttarakhand region defined in section 15.4.1, anthropogenic CO is found to be 144 $\pm$ 22 ppbv and 91 $\pm$ 24 ppbv for 2015 and 2016, respectively. Similarly, biogenic CO is found to be 2.0 $\pm$ 0.3 and 1.9 $\pm$ 0.6 ppbv for these two consecutive years. Biomass CO is significantly lower in 2015 with respect to 2016 as expected. Biomass CO contribution in total CO was 4.3 $\pm$ 2.9 ppbv and 79.9 $\pm$ 56.6 ppbv in 2015 and 2016, respectively. This highlights the tremendous amount of CO contributed from forest fire in 2016.

15.5 Conclusions

An intense forest fire occurred over Uttarakhand during April 24 to May 2, 2016, which created several thousand km^2 burned area over western and southwestern Uttarakhand. This event injected enormous amount of CO in the clean environment of Uttarakhand. Criteria pollutant CO showed significant enhancement in the lower troposphere of Uttarakhand. These features were registered both in satellite-borne and in situ observations. CO during fire-affected period was found to be

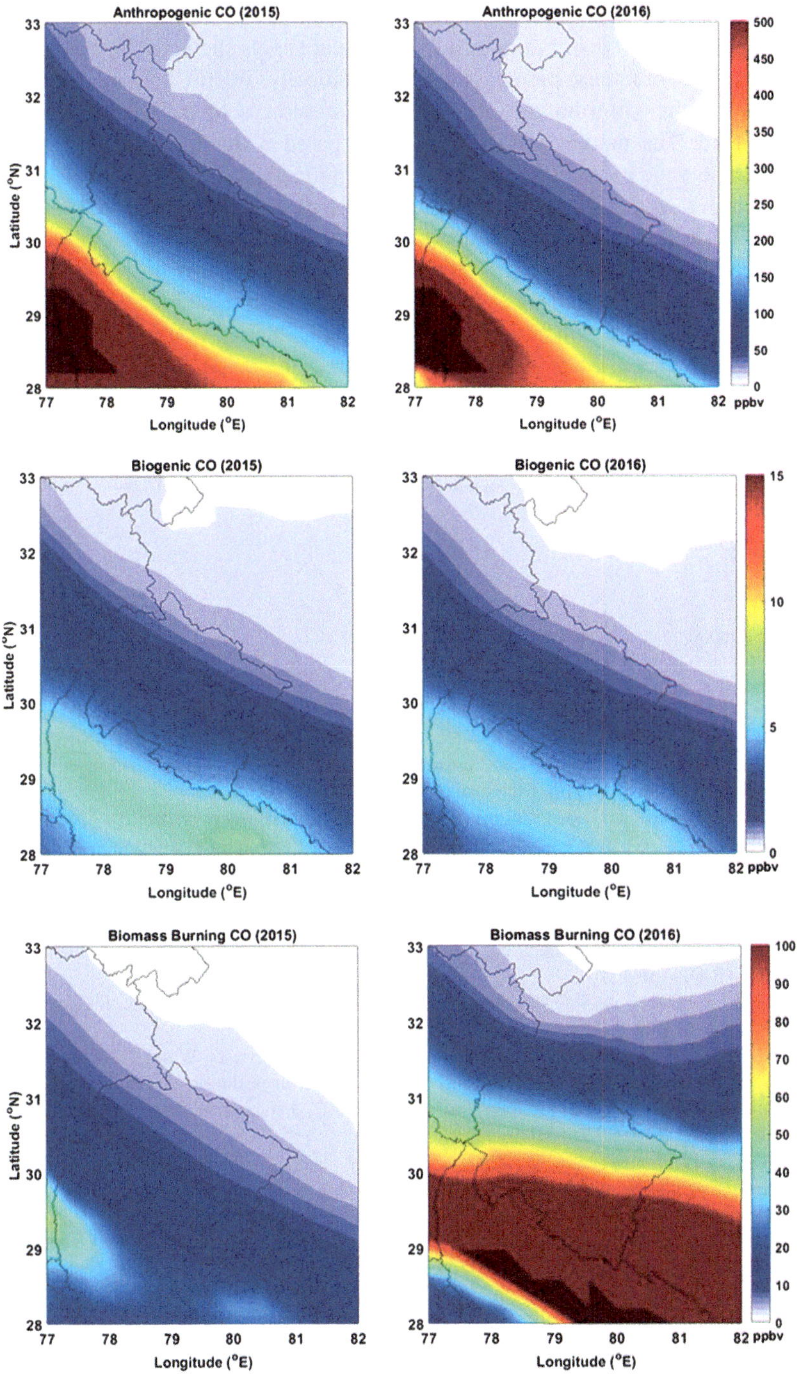

Fig. 15.6 Spatial variation of anthropogenic, biogenic, and biomass CO tracers over the study region. These tracers have been averaged for a period of April 20 to May 5, 2016

60–125 ppbv higher with respect to CO before the fire event at 925 hPa, 850 hPa, and 700 hPa. Spatial distribution of CO and fire hot spots showed a high level of CO over the region of intense biomass burning specifically. WRF-Chem model was used to quantify the contribution of different emission sources in total CO over Uttarakhand. The model simulation was compared with in situ observation over Dehradun to validate the model. The model significantly underestimated CO during this fire event. Model simulation was made for two consecutive years 2015 and 2016 to identify the contribution of different emission sources during a calm year and a fire-affected year. Anthropogenic sources showed relatively lower contribution during fire-affected year. There was no significant change in biogenic CO contribution during these two consecutive years. However, biomass burning CO showed a tremendous increase from 4.3 $\pm$ 2.9 ppbv in 2015 to 79.9 $\pm$ 56.6 ppbv in 2016.

Acknowledgments We are thankful to AIRS and MODIS teams for making their satellite data freely available for scientific research. CO in-situ observation over Dehradun is funded by ISRO. SS is grateful to Head MASD, Dean Academics and Director IIRS for their encouragement and support.

References

Andreae, M. O., and Merlet, P., (2001), Emission of trace gases and aerosols from biomass burning, Global Biogeochem. Cycles, 15, 955–966, doi:https://doi.org/10.1029/2000GB001382.

Chahine, M. T., et al. (2006), AIRS: Improving weather forecasting and providing new data on greenhouse gases, Bull. Am. Meteorol. Soc., 87, 911–926.

Chen, F., and Dudhia, J. (2001), Coupling an Advanced Land Surface–Hydrology Model with the Penn State–NCAR MM5 Modeling System. Part I: Model Implementation and Sensitivity. Monthly Weather Review, 129(4), 569–585.

Crutzen, P. J., and Andreae, M. O., (1990), Biomass burning in the tropics: Impact on atmospheric chemistry and biogeochemical cycles, Science, 250, 1669–1678.

Crutzen, P.J. and Zimmermann, P.H. (1991), The changing photochemistry of the troposphere. Tellus 43AB, 136-151.

Daniel, J.S. and Solomon, S. (1998), On the climate forcing of carbon monoxide. J. Geophys. Res. 103, 13249–13260.

Ding, A.,Wang, T., Xue, L., Gao, J., Stohl, A., Lei, H., Jin, D., Ren, Y., Wang, X., Wei, X., Qi, Y., Liu, J., and Zhang, X. (2009), Transport of north China air pollution by midlatitude cyclones: case study of aircraft measurements in summer 2007, J. Geophys. Res., 114, D08304, doi: https://doi.org/10.1029/2008JD011023.

Ding K., Liu, J., Ding, A., Liu, Q., Zhao, T. L., Shi, J., Han, Y., Wang, H., and Jiang, F. (2015), Uplifting of carbon monoxide from biomass burning and anthropogenic sources to the free troposphere in East Asia, Atmos. Chem. Phys., 15, 2843–2866.

Duncan, B. N., Bey, I., Chin, M., Mickley, L. J., Fairlie, T. D., Martin, R. V, and Matsueda, H. (2003), Indonesian wildfires of 1997: Impact on tropospheric chemistry, J. Geophys. Res., 108(D15), 4458, doi:https://doi.org/10.1029/2002JD003195.

Emmons, L. K., Walters, S., Hess, P. G., Lamarque, J. F., Pfister, G. G., Fillmore, D., Kloster, S. (2010), Description and evaluation of the Model for Ozone and Related chemical Tracers, version 4 (MOZART-4), Geoscientific Model Development, 3, 43–67. https://doi.org/10.5194/gmd-3-43-2010.

Grell, G. A., and Freitas, S. R. (2013), A scale and aerosol aware stochastic convective parameterization for weather and air quality modeling. Atmospheric Chemistry and Physics Discuss, 13, 23845–23893, doi:http://doi.org/10.5194/acpd-13-23845-2013.

Hong, S.Y., Noh, Y., and Dudhia, J. (2006), A New Vertical Diffusion Package with an Explicit Treatment of Entrainment Processes, Monthly Weather Review, 134(9), 2318–2341.

Iacono, M. J., Delamere, J. S., Mlawer, E. J., Shephard, M. W., Clough, S. A., and Collins, W. D. (2008), Radiative forcing by long-lived greenhouse gases: Calculations with the AER radiative transfer models, J. Geophys. Res., 113, D13103, doi:http://doi.org/10.1029/2008JD009944.

Janjic, Z. I. (1996), The surface layer in the NCEP Eta Model, Eleventh Conference on Numerical Weather Prediction, Norfolk, VA, 19–23 August, Amer. Meteor. Soc., Boston, MA, 354–355.

Jha C. S. et al. (2016), Monitoring of forest fires from space – ISRO's initiative for near real-time monitoring of the recent forest fires in Uttarakhand, India, Current Science, 110, 2057–2060.

Kumar, R., Naja, M., Satheesh, S. K., Ojha, N., Joshi, H., Sarangi, T., Pant, P., Dumka, U. C., Hegde, P., and Venkataramani, S. (2011), Influences of the springtime northern Indian biomass burning over the central Himalayas, J. Geophys. Res., 116, D19302, doi:https://doi.org/10.1029/2010JD015509.

Kumar, R., Naja, M., Pfister, G. G., Barth, M. C., and Brasseur, G. P. (2013), Source attribution of carbon monoxide in India and surrounding regions during wintertime. J. Geophys. Res., 118(4), 1981–1995.

Kondo, Y., Morino, Y., Takegawa, N., Koike, M., Kita, K., Miyazaki, Y., Sachse, G. W., Vay, S. A., Avery, M. A., Flocke, F., Weinheimer, A. J., Eisele, F. L., Zondlo, M. A., Weber, R. J., Singh, H. B., Chen, G., Crawford, J., Blake, D. R., Fuelberg, H. E., Clarke, A. D., Talbot, R. W., Sandholm, S. T., Browell, E. V., Streets, D. G., and Liley, B (2004), Impacts of biomass burning in Southeast Asia on ozone and reactive nitrogen over the western Pacific in spring, J. Geophys. Res., 109(D15).

Lau, K. M., Kim, M. K., and Kim, K. M. (2006), Asian summer monsoon anomalies induced by aerosol direct forcing: The role of the Tibetan Plateau, Clim. Dyn., 26, 855–864, doi:https://doi.org/10.1007/s00382-006-0114-z.

Lin, Y. L., Farley, R. D., and Orville, H. D. (1983), Bulk Parameterization of the Snow Field in a Cloud Model, Journal of Climate and Applied Meteorology, 22(6), 1065–1092.

Panmao, Z., Rong, Y., Yanjun, G., et al. (2016), The strong El Niño of 2015/16 and its dominant impacts on global and China's climate, J. Meteor. Res., 30(3), 283–297, doi:https://doi.org/10.1007/s13-351-016-6101-3.

Padalia H., Singh S. and Kumar, A. S. (2016), A comparison of VIIRS and MODIS active fire products for Uttarakhand Episodic forest fire 2016, IIRS CONTACT, 18, 12–13. (https://www.iirs.gov.in/iirs/sites/default/files/pdf/newsletter/Contact_December_2016.pdf)

Pfister, G. G., Wiedinmyer, C., and Emmons, L. K. (2008), Impact of the fall 2007 California wildfires on surface ozone: Integrating local observations with global model simulations, Geophys. Res. Lett., 35, L19814, doi:https://doi.org/10.1029/2008GL034747.

Pfister, G. G., Avise, J., Wiedinmyer, C., Edwards, D. P., Emmons, L. K., Diskin, G. D., Wisthaler, A. (2011), CO source contribution analysis for California during ARCTAS-CARB. Atmospheric Chemistry and Physics, 11(15), 7515–7532. https://doi.org/10.5194/acp-11-7515-2011.

Sarangi, T., Naja, M., Ojha, N., Kumar, R., Lal, S., Venkataramani, S., Kumar, A., Sagar, R., and Chandola, H. C. (2014), First simultaneous measurements of ozone, CO, and NOy at a high-altitude regional representative site in the central Himalayas, J. Geophys. Res. Atmos., 119,1592–1611, doi:https://doi.org/10.1002/2013JD020631.

Savage, N.H., Harrison, R.M., Monks, P.S., and Salisbury, G. (2001), Steady-state modeling of hydroxyl radical concentrations at Mace Head during the EASE' 97 campaign, May 1997, Atmos. Environ, 35, 515–524.

Semwal, R. L. and Mehta, J. P. (1996), Ecology of forest fires in chirpine (Pinus roxburghii. Sarg.) forests of Garhwal Himalaya. Curr. Sci., 70, 426–427.

Singh, R. P., Gumber, S., Tewari, P., and Singh, S. P. (2016), Nature of forest fires in Uttarakhand: frequency, size and seasonal patterns in relation to pre-monsoonal environment, Current Science, 111, 398–403.

Srivastava, S., and Sheel, V. (2013), Study of tropospheric CO and O3 enhancement episode over Indonesia during Autumn 2006 using the Model for Ozone and Related chemical Tracers (MOZART-4), Atmospheric Environment, 67, 53–62. https://doi.org/10.1016/j.atmosenv.2012.09.067.

Takegawa, N., et al. (2003), Photochemical production of O3 in biomass burning plumes in the boundary layer over northern Australia, Geophys. Res. Lett., 30(10), 1500, doi:https://doi.org/10.1029/2003GL017017.

Vander Werf, G. R., Randerson, J. T., Giglio, L., Collatz, G. J., Kasibhatla, P. S., and Arellano, A. F. (2006), Interannual variability of global biomass burning emissions from 1997 to 2004, Atmos. Chem. Phys. Discuss., 6(2), 3175–3226, doi:https://doi.org/10.5194/acpd-6-3175-2006.

Wesely, M. L. (1989), Parameterization of surface resistances to gaseous dry deposition in regional-scale numerical models, Atmospheric Environment, 23(6), 1293–1304. doi:https://doi.org/10.1016/0004-6981(89)90153-4.

Wild, O., Zhu, X., and Prather, M. J. (2000), Accurate Simulation of In- and Below-Cloud Photolysis in Tropospheric Chemical Models, Atmospheric Chemistry, 37, 245–282.

Yasunari, T. J., et al. (2010), Estimated impact of black carbon deposition during pre-monsoon season from Nepal Climate Observatory—Pyramid data and snow albedo changes over Himalayan glaciers, Atmos. Chem. Phys., 10, 6603–6615, doi:https://doi.org/10.5194/acp-10-6603-2010.

Chapter 16
Wildlife Habitat Evaluation in Mountainous Landscapes

Subrata Nandy, S. P. S. Kushwaha, and Ritika Srinet

16.1 Introduction

A habitat is defined as the place where a plant or animal naturally lives (Darwin 1859). The habitat of an organism is an area with a combination of resources and biotic and abiotic factors that promote the occupancy of individuals of a species and allows them to survive and reproduce. The concepts of habitat, habitat preferences and habitat association of species are vital parameters in wildlife ecology and management, as the rate of survival and reproduction of organisms depends on the quality of habitat. To study the habitat occupied by an organism it is required to understand its evolutionary history, ecological requirements and interactions, the climatic history of the area, and even the history of the movements of landmasses. The identification of the constituents of the habitat of a species is the basis on which the wildlife management activities depend.

Habitat loss poses the greatest threat to species. Increasing anthropogenic pressure has led to the depletion of natural environments. The natural habitats are continuously shrinking due to the unprecedented growth of human population, commercial exploitation of natural resources and industrial development. Encroachment into natural habitats for various purposes including agriculture, grazing, and poaching has been the major threat to wildlife. Designing the studies that identify habitat condition requires considerable thought about the needs and perpetual abilities of the species under investigation, the spatial and temporal scales at which the study is to be conducted, and the methods for measuring environmental features (Morrison et al. 2006). Mapping and monitoring of wildlife habitat have become the

S. Nandy (✉) · R. Srinet
Forestry and Ecology Department, Indian Institute of Remote Sensing (IIRS), Indian Space Research Organisation (ISRO), Department of Space, Government of India, Dehradun, India
e-mail: subrato.nandy@gmail.com

S. P. S. Kushwaha
Forest Research Institute, Dehradun, India

R. R. Navalgund et al. (eds.), *Remote Sensing of Northwest Himalayan Ecosystems*,
https://doi.org/10.1007/978-981-13-2128-3_16

vital components for interpretation, analysis, and prediction of species distribution. Habitat evaluation is considered as the first and crucial step for planning effective wildlife conservation and management activities. For a particular species, the habitat evaluation for suitability requires information on a host of parameters including the biotic and abiotic factors.

The field-based approaches generate measurements with high accuracy, but it is generally impractical for studies beyond local scales as it is labour and time intensive (Aplin 2005). Remote sensing (RS) is an effective technology for mapping and monitoring the wildlife habitats at broader spatial extents. Geospatial technology, including RS, geographic information system (GIS) and global navigation satellite system (GNSS) with habitat suitability index (HSI) model, can provide an efficient, quick, cost-effective and accurate method of habitat evaluation (Schamberger and Krohn 1982). Traditionally, the ground survey-based approaches have been used to evaluate wildlife habitats. Geospatial technology can supplement the tedious ground survey methods for mapping and monitoring the habitats (Kushwaha and Roy 2002). RS data and techniques observe the target or area of interest ranging from local to global (Kerr and Ostrovsky 2003). Satellite data, utilized to analyse the trends in forest cover, can provide direct estimates of land use/land cover change and habitat loss, thus increasing the scope of applied ecological and wildlife studies to detect changes in species distributions. Geospatial technology is increasingly being used as a vital tool in wildlife studies including habitat evaluation, suitability analysis and wildlife corridor monitoring. Habitat evaluation serves as a tool to predict the habitat suitability of a given species. For habitat mapping, field knowledge for habitat preferences of the target species is combined with RS data, biophysical, geophysical and meteorological data (Leyequien et al. 2007). HSI modelling provides a probability that the habitat is suitable for the target species and that species can occur in similar habitats.

16.2 Variables Used for Wildlife Habitat Suitability Modelling

For wildlife habitat modelling, various input variables are taken into consideration. These inputs vary with species as well as with the study area. The source of these input variables is either field data or remote sensing data (Table 16.1). The dependent variable is the present locations of the species. GPS is used to record the geographic coordinates of the presence locations, i.e. the direct sightings as well as the indirect evidences of the species. The independent variables are the vegetation type/land use, forest canopy density, elevation, slope, aspect, road, settlement, water body/water holes, landscape matrices, vegetation indices, prey density (for carnivores), climate data, etc. However, the choice of the variables is completely dependent on the species as well as the area. Hence, one has to study the target species first before starting the habitat suitability modelling.

Table 16.1 Input variables for wildlife habitat suitability modelling

Variables		Source
Dependent variable	Direct and indirect evidences of wildlife	Field data
Independent variables	Vegetation type/land use	Remote sensing (RS) data
	Forest canopy density	-do-
	Elevation	-do-
	Slope	-do-
	Aspect	-do-
	Road	High-resolution RS data
	Settlement	-do-
	Water body	RS data
	Water holes	High-resolution RS data
	Landscape/patch/class matrices	Computed using FRAGSTATS (vegetation type/land use map – RS data)
	Vegetation indices	RS data
	Prey density	Field data (for carnivores)
	Climate data	WorldClim

Modified from Roy and Nandy (2016)

16.3 Techniques and Models Involved in RS-Based Wildlife Habitat Suitability Modelling in Northwest Himalaya

The northwest Himalayan ecosystem is fragmented both by the topography and by human disturbance. The increased anthropogenic pressure and development schemes have led to acceleration in forest fragmentation and loss of forest cover, which makes the wildlife of the region more vulnerable. The expert opinion is considered as the primary source of information for wildlife habitat evaluations which is framed based on habitat suitability index (HSI). HSI is an additive, multiplicative or logical equation with coefficients representing the relative value of the habitat requirement parameters. Based on the best available expert knowledge or published literature, the coefficients are scaled between 0 and 1. HSI equations can be used to generate maps of ranked habitat by integrating GIS with the data representing the spatial distribution of model inputs. In habitat suitability modelling, a higher value of the index in a particular location indicates a greater chance of occurrence of that species.

Started with habitat ranking, based on the knowledge of the expert and from published literature, HSI models now have started using geostatistics, logistic regression, refined logistic regression, multi-criteria analysis (MCA), and various data integration techniques to determine the species occurrence (Brown et al. 2000) and provide an effective method of evaluating habitat quality.

16.3.1 Overlay Analysis Technique

Overlay analysis technique is used to apply a common set of values to varied inputs for an integrated analysis. This requires an analysis of different factors on which the habitat preferences of a species can depend. Kushwaha et al. (2000) evaluated the habitat suitability of goral (*Naemorhedus goral*) in Chilla Sanctuary of Rajaji National Park in Uttarakhand using this technique. The habitat preferences for goral include open forests with intermittent grasslands, steep slopes and availability of water. Using IRS 1B LISS-II satellite imagery, topographic maps and ground observations, the information on forest cover, water sources, slope, settlements and the road network in and around the Sanctuary was synthesized. The study highlighted that about 14% Sanctuary area is highly to moderately suitable for goral. According to the results, an extra 5% area can be available for goral habitat if *Gujjars* (tribals living inside Sanctuary) can be settled outside.

16.3.2 Multi-criteria Analysis

MCA can be used to address complex problems involving multi-perspectives, various data and information and objectives (Wang et al. 2010). This approach can be used for finding the best or satisfactory choice for making decisions in solving complex problems. It is mainly concerned with how to integrate the information from various criteria to form a single index of evaluation. It provides a logical and well-structured process to identify and prioritize the different factors. For wildlife habitat characterization, the MCA can prove helpful as it considers various criteria essential for suitable habitat preferences of a species. It can also be used for quick assessment of potential habitat of wildlife.

Nandy et al. (2012) used MCA to assess the potential habitat of swamp deer (*Cervus duvauceli duvauceli* Cuvier) in Jhilmil Jheel Conservation Reserve, Uttarakhand, India. Swamp deer, classified as vulnerable, feeds mainly on grasses and aquatic plants. Jhilmil Jheel Conservation Reserve is considered as one of the remaining habitats of swamp deer. The study emphasized a situation of limited scope for conservation and management of swamp deer in the Indo-Gangetic plains. As per the findings, 6.08% (225 km^2) of the study area was falling under the highly suitable to suitable category, whereas 60.23% (135.52 km^2) area was found to be moderately suitable (Fig. 16.1). It was disappointing from a conservation point of view to find that only 14 km^2 area was fit for the population of 134 individuals of this flagship species. It was suggested that the agricultural land (2.47 km^2) nearby the Conservation Reserve can be included under the reserve management which can provide a surplus area to meet the fodder requirements.

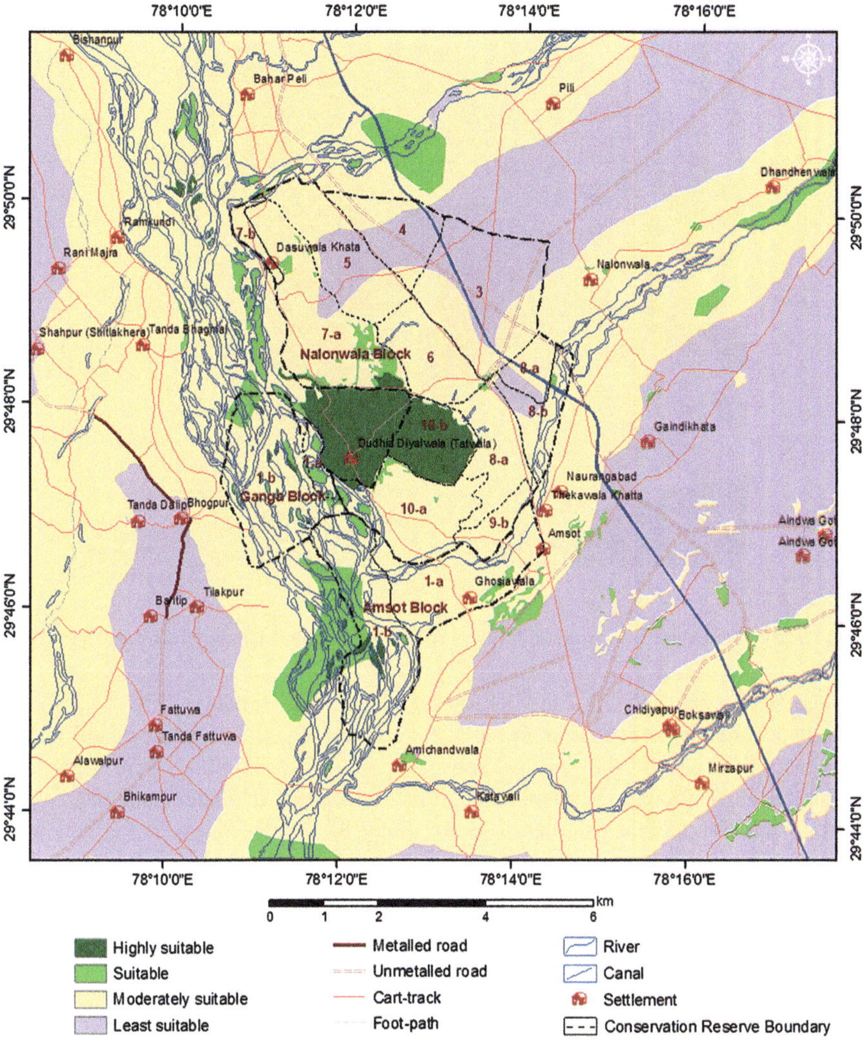

Fig. 16.1 Potential habitat map of swamp deer in Jhilmil Jheel Conservation Reserve (Nandy et al. 2012)

16.3.3 Geostatistical Modelling

Geostatistics is used for mapping and modelling the spatial variability of any variable. For wildlife habitat evaluation, field data are statistically analysed to know the habitat use pattern of the target species. Then, the distribution of the species is derived using multiple regression. Kushwaha et al. (2004) used a synergistic approach, using field survey, geostatistical analysis and geospatial tools, for evaluation of habitat of sambar (*Cervus unicolor* var. *niger*) and muntjak (*Muntiacus*

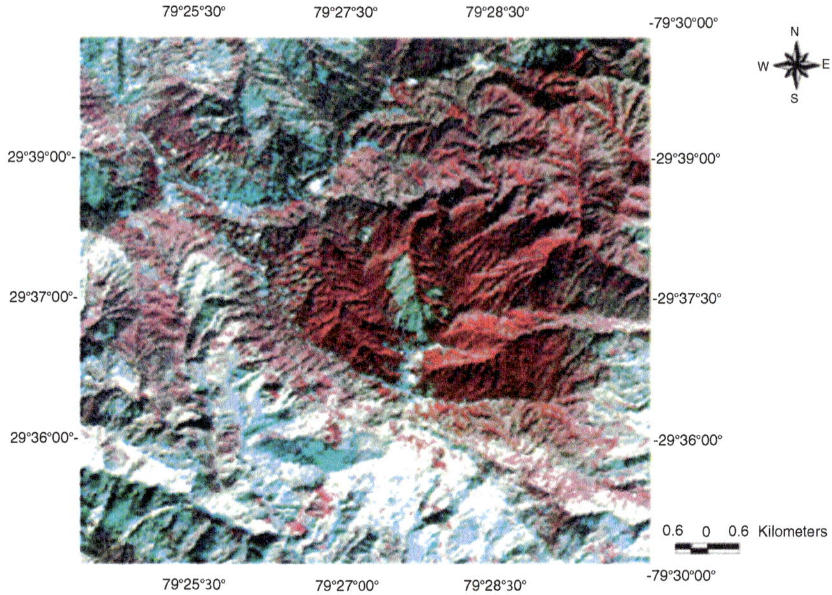

Fig. 16.2 False colour composite of Ranikhet forests, Kumaon Himalaya (Kushwaha et al. 2004)

muntjak var. *vaginalis*) in Ranikhet forests of Kumaon Himalaya (Fig. 16.2). They attempted the geostatistical analysis at three levels. Firstly, to find out the micro-habitat preference of the species, the cases of animal sighting were taken and subjected to principal component analysis. Secondly, to understand the factors responsible for habitat use, cases where either of the species were sighted were clustered into three categories, viz. pure sambar, pure muntjak and mixed; then successively it was subjected to discriminant function analysis. Finally, a binomial multiple logistic regression was run to derive the distribution of the respective species. For regression, the locations of animal sighting (as Boolean values) were considered as the dependent variable and six physical habitat variables, viz., forest density, drainage, elevation, slope, aspect and the human settlements, as the independent variables. It was found that both sambar and muntjak had a preference for oak forests. Out of the total area, the highly suitable area for sambar (Fig. 16.3) was found to be only 4.07%, and for muntjak (Fig. 16.4), it was 2.37%. About 20% and 10% area in case of sambar and muntjak, respectively, was found to be moderately to less suitable. The habitat overlap between the species was only 0.35% (29.71 ha). The sensitivity of the model in this study was 87.8% for sambar and 97.62% for muntjak.

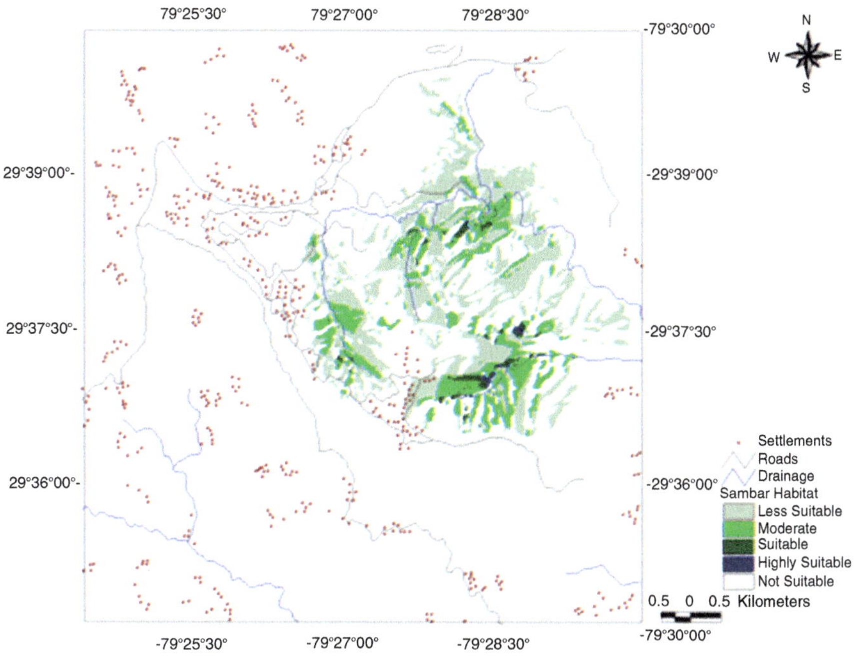

Fig. 16.3 Habitat suitability for sambar (Kushwaha et al. 2004)

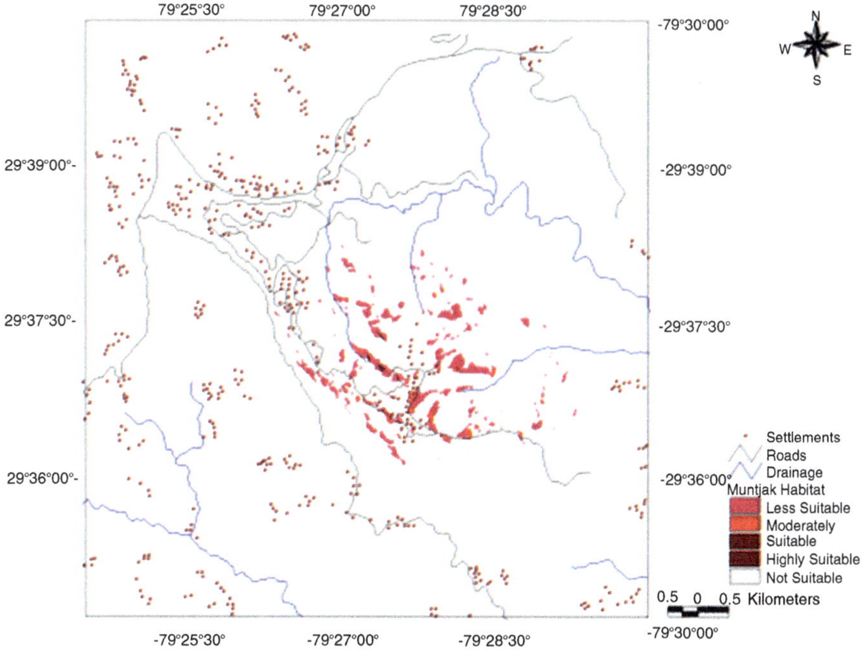

Fig. 16.4 Habitat suitability for muntjak (Kushwaha et al. 2004)

16.3.4 Logistic Regression Modelling

Logistic regression is one of the statistical methods used to evaluate data where there are one or more independent variables that regulate the dependent variable, which is dichotomous. Logistic regression modelling is appropriate for habitat suitability studies which employ presence–absence data of the target species. Himalayan musk deer (*Moschus chrysogaster*), the most primitive and smallest Himalayan ungulate, is endangered owing to poaching for its musk and habitat loss. A study was carried out to identify the potential habitat of Himalayan musk deer in Kedarnath Wildlife Sanctuary in Uttarakhand, India, using binomial multiple logistic regression (BMLR) model. The results of the study suggested that 7.39% (72.04 km^2) of the Sanctuary area is highly suitable for Himalayan musk deer. The study also showed that about 71.59% of highly suitable habitat was found in subalpine forest. The study highlighted that vegetation type/land use happens to be the most important variable governing the habitat suitability for Himalayan musk deer in the study area. The BMLR model predicted the potential habitat of Himalayan musk deer with reasonably high accuracy (ROC = 0.85). Such studies can be immensely useful in protected area management through the judicious allocation of available resources for protection and conservation of endangered species.

16.3.5 Refined Logistic Regression Modelling

To improve the logistic regression models, Singh and Kushwaha (2011) proposed a simulation-based approach, which is extensively used for habitat suitability modelling. This modelling strategy was tested using field survey conducted for habitat suitability evaluation for muntjak (*Muntiacus muntjak*) and goral (*Naemorhedus goral*) in the central Himalaya, India. The habitat suitability models were developed with refined data using the coefficients resulting from generalized linear mixed models (GLMMs). Simulated results matched the expectations in terms of model behaviour and published habitat associations of the target species. The proposed technique can be used for rapid assessment of wildlife habitat.

16.3.6 Ensemble Modelling

Species distribution models (SDMs) are increasingly used in order to define the ecological niche of a species. These models statistically relate the species occurrence data with environmental predictor variables to predict the distribution of a species in a geographical location. These models are also used to generalize the prediction outside the locations for which data exist. SDMs were used to predict the distribution of tigers in relation to prey availability in Corbett Tiger Reserve (CTR) in

Uttarakhand, India. In this study, biomod2 package of R including ten different SDM techniques was employed. The modelling was also performed within an ensemble forecast framework by using seven ensemble modelling techniques. The effects of anthropogenic stress (settlements and roads) on the distribution of tigers were assessed in the study area. Ensemble models, particularly ensemble by the weighted mean of probabilities (AUC = 0.96), predicted the distribution of tigers in the study area accurately and outperformed all individual models. All individual models except surface range envelope (SRE) performed fairly well in predicting the distribution of tigers in the study area. Amongst individual models random forest (RF), generalized linear models (GLM) and artificial neural network (ANN) consistently performed well across all prey and predator species.

16.4 Wildlife Corridors

Wildlife corridor is a linear landscape element which connects two or more wildlife habitat patches (Soule and Gilpin 1991). From today's perspective, with an increase in the instances of habitat fragmentation, corridors play a crucial role in facilitating the movement and dispersal of animals amongst habitat islands and reduce the impact of habitat isolations. This decreases the pressure on smaller habitats, facilitates habitat restoration and also decreases the extinction probabilities. Understanding the importance of wildlife corridors in India, it was proposed to connect a network of protected areas by corridors as a conservation strategy, which resulted in the identification of several wildlife corridors across the country (Rodgers and Panwar 1988). Some of the important wildlife corridors in India are Aryankavu Pass in Tamil Nadu and Kerala, the Chilla–Motichur corridor and the Rajaji–Corbett corridor in Uttarakhand, the Kallar–Jaccanari corridor in Tamil Nadu and the Siju–Rewak corridor in Meghalaya (Johnsingh and Williams 1999). The increasing land, infrastructure and energy requirements have put wildlife corridors under threat. The growth of linear infrastructure network, i.e. roads, railway lines and power lines, increasing demand of land for agriculture and human habitation has led to depletion of corridors.

Geospatial technology is considered as an effective approach for assessment and monitoring of wildlife corridors. Nandy et al. (2007) assessed the status of Chilla–Motichur wildlife corridor of Rajaji National Park, Uttarakhand, India, using temporal satellite imagery of 1972, 1990 and 2005 (Fig. 16.5). The study highlighted the changes occurred in the Chilla–Motichur corridor during 1972 to 2005. The corridor area was mapped into different classes using on-screen visual interpretation technique. The change maps, generated using change matrix analysis, indicated that considerable corridor loss occurred between 1972 and 2005 (17.56 km^2) (Fig. 16.6). The loss was more prominent during 1972 to 1990 (11.18 km^2) than during 1990 to 2005 (6.38 km^2). The loss of corridor was due to roads, rail, hydropower canal projects, expansion of townships in nearby areas, the resettlement of people evacuated from the Tehri dam site and establishment of the army cantonment. This study highlighted the conservation concerns related to the depletion of the corridor and the potential of RS and GIS in the assessment and monitoring of the sensitive areas.

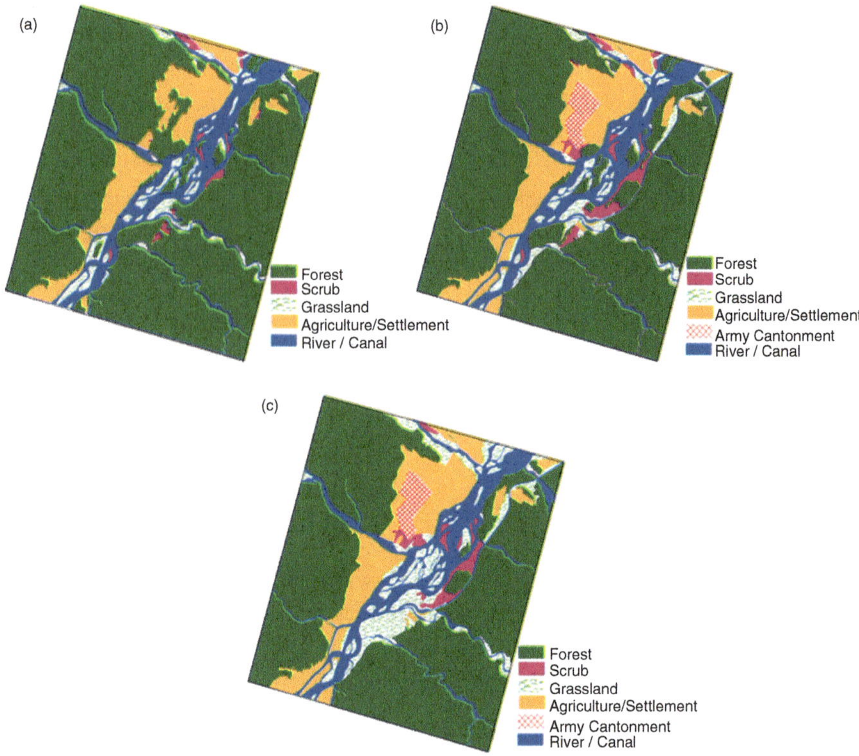

Fig. 16.5 Land use/land cover in Chilla–Motichur corridor in (**a**) 1972, (**b**) 1990 and (**c**) 2005 (Nandy et al. 2007)

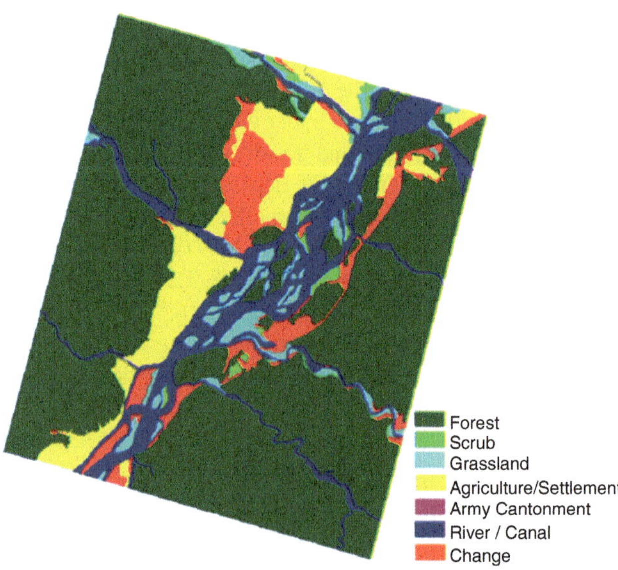

Fig. 16.6 Change in land use/land cover in Chilla–Motichur corridor between 1972 and 2005 (Nandy et al. 2007)

16.5 Conclusions

The habitat distribution maps and models serve as important tools for wildlife habitat conservation and management. The recent techniques such as machine learning including RF and ANN are increasingly being used with GIS to enhance the ecological knowledge to develop these models. In comparison to the traditional statistical methods, machine learning algorithms are capable of addressing the nonlinear relationships and can provide logical models which can be inspected to get the detailed insight (DĽeroski 2009). Different modelling approaches such as GARP, MaxEnt and BioMapper have also shown their potential in wildlife research in the recent years (Baldwin 2009). For any single sensor, to deliver information on all aspects of vegetation composition and structure important for wildlife habitat characterization is difficult. In addition to optical RS data, LiDAR data and multi-sensor (LiDAR, SAR/InSAR, ETM+, Quickbird) data in synergy (Hyde et al. 2006) are also used for wildlife habitat analysis. Repetitive RS data from various sensors can characterize the vertical and horizontal distribution of habitat, thus increasing our understanding of drivers of habitat selection and species distributions at multi-scales (Vogeler and Cohen 2016). Over the past decades, geospatial technology has already proved its potential in wildlife habitat analysis. With the advent of newer and higher-resolution RS data in combination with new modelling and data integration techniques, the future studies will continue to expand the species and geographic range of habitat modelling.

References

Aplin P (2005) Remote sensing: ecology. Prog Phys Geog 29:104–113.

Baldwin RA (2009) Use of Maximum Entropy Modeling in wildlife research. Entropy 11:854–866.

Brown M, Lewis HG, Gunn SR (2000) Linear spectral mixture models and support vector machines for remote sensing. IEEE Trans Geosci Remote Sens 38:2346–2360

Darwin C (ed) (1859) On the origin of the species by natural selection or the preservation of favoured races in the struggle for life. John Murray, Albemarle Street, London

DĽeroski S (2009) Machine learning applications in habitat suitability modeling. In: Haupt SE, Pasini A, Marzban C (eds) Artificial Intelligence methods in the environmental sciences, 397-411.

Hyde P, Dubayah R, Walker W, Blair JB, Hofton M, Hunsaker C (2006) Mapping forest structure for wildlife habitat analysis using multi-sensor (LiDAR, SAR/InSAR, ETM+, Quickbird) synergy. Remote Sens Env 102:63–73.

Johnsingh, AJT, Williams, AC (1999) Elephant corridors in India: Lessons for other elephant range countries. Oryx 33(3):210–214.

Kerr JT and Ostrovsky M (2003) From space to species: ecological applications for remote sensing. Trends Ecol Evol 18:299–305.

Kushwaha SPS and Roy PS (2002) Geospatial technology for wildlife habitat evaluation. Trop Ecol 43:137–150.

Kushwaha SPS, Khan A, Habib B, Quadri A, Singh A (2004) Evaluation of sambar and muntjak habitats using geostatistical modelling. Curr Sci 86(10):1390–1400.

Kushwaha SPS, Munkhtuya S, Roy PS (2000) Geospatial modelling for goral habitat evaluation. J Ind Soc Remote Sens 28(4):293.

Leyequien E, Verrelst J, Slot M, Schaepman-Strub G, Heitkönig IM, Skidmore A (2007) Capturing the fugitive: Applying remote sensing to terrestrial animal distribution and diversity. Int J Appl Earth Observ Geoinform 9(1):1–20

Morrison ML, Marot BG, Mannan RW (2006) Wildlife-habitat relationships: Concepts and application (Third edition), Washington DC: Island Press 128

Nandy S, Kushwaha SPS, Gaur P (2012) Identification of swamp deer (*Cervus duvauceli duvauceli* Cuvier) potential habitat in Jhilmil Jheel Conservation Reserve, Uttarakhand, India using multi-criteria analysis. Env Manage 49:902–914

Nandy S, Kushwaha SPS, Mukhopadhyay S (2007) Monitoring the Chilla–Motichur wildlife corridor using geospatial tools. J Nat Conserv 15:237–244

Rodgers WA, Panwar HS (1988) Planning a wildlife protected area network in India, Vols. I and II. Dehradun: Wildlife Institute of India

Roy PS, Nandy S (2016) Remote Sensing and Geographic Information System in Wildlife Habitat Analysis. In: Balakrishnan M (ed) Wildlife Ecology and Conservation. Scientific Publishers, Jodhpur

Schamberger M, Krohn WB (1982) Status of the habitat evaluation procedures. Trans. North American Wildlife Nat Resour 47:154–164

Singh A, Kushwaha SPS (2011) Refining logistic regression models for wildlife habitat suitability modeling - a case study with muntjak and goral in the Central Himalayas, India. Ecol Model 222 (8):1354–1366

Soule ME, Gilpin ME (1991) The theory of wildlife corridor capability. In Saunders DA, Hobbs RJ (eds) Nature conservation 2: The role of corridors. Surrey Beatty and Sons, Chipping Norton, Australia

Vogeler JC, Cohen WB (2016) A review of the role of active remote sensing and data fusion for characterizing forest in wildlife habitat models. Revista de Teledetección 45:1–14

Wang JJ, Jing YY, Zhang CF (2010) Review on multi-criteria decision analysis aid in sustainable energy decision-making. Renew Sustain Energ Rev 13:2263–2278

Part V
Agriculture

Summary

Mountain agriculture is a socially and culturally unique system, but is also a regionally important economic sector. Mountain areas of the developing countries face alarming increase in population pressure, degradation of the environment and production resource base. Mountain environment especially in Northwest Himalaya (NWH) offers tremendous challenges to government planners in their attempts to institute rational, efficient programs for agricultural resource development planning. By and large before the advent of geospatial technology, the development interventions had been undertaken without consideration to mountain conditions (e.g. resource and fragility) and their imperatives. Development strategies were simply extension of approaches and practices meant for plain areas and henceforth, development efforts were predominantly lacking in mountain perspective of NWH.

Modern tools such as satellite remote sensing and GIS have been providing newer dimensions to effectively monitor and manage natural resources. It has been well conceived that remote sensing and GIS have a great role to play in resource characterization, zonation, soil-water conservation and crop planning. Agro-ecological zoning and land suitability assessment are the best examples of key geospatial applications being developed for the niche based agricultural planning in North western Himalayan regions. Usage of the modern geospatial tools in crop diversification activities with high-value horticultural crops (fruit trees and vegetables) also gained momentum during last few years. Although geospatial technology offers tremendous capability towards resource inventory and land use planning, its role in modelling per se of agricultural system is still not adequately explored. In this context, Indian Institute of Remote Sensing has taken up many studies related to mountain agriculture. In this chapter, while one study highlights the use of geospatial approach in modeling soil erosion processes in predicting soil erosion and nutrient loss in hilly and mountainous landscape, the other study pertains to impact assessment of climate change on mountain agriculture.

Chapter 17
Geospatial Approach in Modeling Soil Erosion Processes in Predicting Soil Erosion and Nutrient Loss in Hilly and Mountainous Landscape

Suresh Kumar

17.1 Introduction

Soil erosion due to water is one of the most important land degradation processes and considered as major land degradation type in the world (UNEP 1994; Jain et al. 2010). The entire Himalayan region is facing serious problem of land degradation due to soil erosion. Deforestation and inappropriate land utilization coupled with steep sloping terrain, fragile, and young soil with erosive rainfall pattern have accelerated soil erosion in the Himalayan landscape. It reduces soil fertility by removing top soil layer and large amount of soil nutrients along with sediments (Oldeman 1994; Bai et al. 2008). It results in reduction in soil quality that adversely affects the suitability of soils for various agricultural crops and vegetation types. Harmonized statistics by ICAR (NBSS&LUP) and ISRO (NRSC) (Trivedi 2010) reported that the total area under degraded and wastelands in India is 120.72 M ha (arable land and open forest) and out of it nearly 73.27 M ha land is affected by water erosion. The average annual soil erosion rate is estimated to be 16.4 ton ha^{-1} $year^{-1}$, resulting in an annual total soil loss of 5.3 billion tons throughout the country in India (Dhruvanarayana and Ram Babu 1983; Pandey et al. 2008). Nearly 29% of total eroded soil is permanently lost to the sea, while 61% is simply translocated from one place to another, and the remaining 10% is deposited in reservoirs.

The Indian Himalayas region covers an area of 53.7 Mha that constitutes 16.4% of India's total geographic area of 329 Mha. The Himalayan landscapes are under constant threat of soil erosion and landslides and getting enhanced by continuous degradation or depletion of forest cover and nonscientific agricultural practices prevalent in the region. Garde and Kothyari (1986) reported soil erosion rate of

S. Kumar (✉)
Agriculture and Soils Department, Indian Institute of Remote Sensing (IIRS), Indian Space Research Organisation (ISRO), Department of Space, Government of India, Dehradun, India
e-mail: suresh_kumar@iirs.gov.in

R. R. Navalgund et al. (eds.), *Remote Sensing of Northwest Himalayan Ecosystems*,
https://doi.org/10.1007/978-981-13-2128-3_17

20–25 t ha^{-1} year^{-1} in the Northern Himalayan region in Nepal. Mandal and Sharda (2011) suggested soil loss tolerance (*T*-value) value from 2.5 to 12.5 t ha^{-1} year^{-1} depending upon soil resistibility to erosion and soil depth in India. The soil loss tolerance or is defined as the upper threshold soil erosion rate that can be permitted without degrading long term productivity of soil. They observed that nearly 59% of land within the hilly region in India has soil erosion rate higher than soil tolerance limit. Therefore, there is high need to adopt appropriate conservation measures in the Himalayan region in various land use/land cover types. Considering the magnitude of the problems caused by water erosion, Government of India (GOI) is executing soil and water conservation planning through watershed management programs in various states of the country. It is implementing Integrated Watershed Management Programme (IWMP) project to address soil and water resource conservation and improving crop productivity in the watershed.

Understanding the soil erosion processes in the Himalayan landscape is necessitated to establish interaction and association of various environmental factors governing soil loss in various land cover types. The characterization and quantification of soil loss is required in order to devise effective control mechanisms and suggesting appropriate land management practices. The estimation of soil erosion is more difficult as it is controlled by interplay of several factors such as climate, land use/land cover, soil, topography, and anthropogenic activities. The quantification of soil erosion rate is one of the most complex and challenging in natural resources and environmental planning. Monitoring and quantification of soil erosion rates requires installations of various runoff and erosion gauging stations, which is rather expensive and time-consuming and often unaffordable. Several erosion models ranging from empirical to physical process based are developed over the years to simulate soil erosion on daily, weekly, seasonal, and annual basis. These models are used as predictive tools to estimate soil erosion. However, these models differ greatly in terms of complexity, processes, and data requirement. These models can provide a quantitative and consistent estimation of soil erosion rate in various land cover types as well as at watershed scale. Erosion models provide quantitative estimates of soil erosion rates and other quantitative parameters that are often used by land use planners and decision-makers in preparing suitable soil and water conservation measures for protecting soils from further erosion in the watershed. In recent years, the availability of high-resolution remote sensing data has facilitated in providing spatial distribution of land use/land cover, soil types, and terrain information more accurately. Availability of digital elevation model (DEM) at various resolutions has revolutionized retrieving terrain parameters needed in soil erosion estimation. Today, various commercial and open-source GIS software are available and serve as an important tool in integrating and analyzing these spatial data layers in predicting spatial distributed erosion risk area for better conservation planning in the watershed. Soil erosion can also be monitored with the integration of ancillary data and remotely sensed data in a GIS environment.

17.2 Soil Erosion Processes

Soil erosion processes include detachment, transport, and deposition of soil particles over the landscape. Water erosion is caused by rainfall (rainfall erosion) or by runoff water (runoff erosion). During rainfall, raindrops directly hits exposed soil surfaces and detach soil particles that transported by surface runoff water and get deposited into pores of soils resulting sealing of soil surface. Therefore, it inhibits water infiltration rate and results in higher surface runoff (overland flow) generation initiating surface soil erosion. Surface runoff get accumulated in depression and got channelized initiating formation of rills and gullies in the landscape (Poesen et al. 2003). The formation of these rills and gullies is affected by several factors such as topography (slope), terrain complexity, soil type, woody debris, exposed rock surface, and coarse material (surface stones and gravels). Runoff water flows on surface cause inter-rill (sheet) erosion and when it flows in depression as concentrated flow generates rill and gully erosion. Inter-rill erosion process is rainfall dominated, whereas rill erosion is mostly defined by runoff. Gully erosion is, in general, similar to rill erosion.

Soil is being detached by surface runoff water and transported to elsewhere from the point of detachment. The surface runoff water, i.e., overland flow, is generated by rainstorms, and its quantity depends on rainfall intensity and duration of rainfall conditioned by land cover, soil types, and terrain variation. In addition to it, the runoff generation depends upon the physical properties (infiltration capacity, hydraulic conductivity, and bulk density and soil texture) of the soils, vegetation cover, and terrain type and catchment size. Soil infiltration capacity varies spatially with the soil properties (texture, structure, organic matter content, antecedent soil moisture, etc.). In hilly and mountainous landscape, soil characteristics are largely altered by presence of rock fragments, rock cover, micro-topography, and vegetation cover. It primarily influences soil hydrological behaviors governing surface runoff generation in the landscape.

Soils in the Himalayan region are shallow to moderately deep and coarse in texture having low water-holding capacity. Besides this, topography (slope and aspect) largely influence the runoff generation and triggering soil erosion. Steepness of the terrain enhances the runoff water velocity and thus increases kinetic energy of runoff water that facilitates more soil detachment and promoting soil erosion. Aspect affects temperature and in situ soil moisture that determine vegetation types and its growth in large extent. Southern aspect (sun facing) is warmer and subject to marked fluctuation of soil moisture, whereas northern aspect is cooler and subject to higher and stable soil moisture during the year. Therefore, it results poor vegetation growth in southern aspect in contrast to northern aspect. Northern aspect had good forest cover, whereas southern aspects are predominantly under agriculture. Southern aspect witnesses higher soil erosion rate due to poor vegetation growth and soil development in contrast to northern aspect. Therefore, soil surfaces in southern aspect had high percentage of rock fragments. The concentration of rock fragments on the soil surface is due to the selective removal of fine material by runoff.

Hilly and mountainous landscape of Himalaya is characterized by steep to very steep sloping hillslopes and witnesses extreme rainfall in rainy season that triggers severity of soil erosion. Surface runoff water with gravitational force in association with anthropogenic activities (livestock grazing and tillage operations) enhances soil erosion processes in such environment during the rainy season. Runoff and erosion processes in these landscapes are largely controlled by topography and land use management practices. Limited soil depth, coarse texture, and poor soil development in these landscapes lead to low water-holding capacity and weak soil structure development that induces soil erosion. Understanding soil erosion and runoff generation requires in employing soil erosion modeling as well as implementing soil and water conservation measures for appropriate land use planning. Selection of suitable watershed simulation models requires comprehensive knowledge of hydrological process operating in the hilly and mountainous landscape.

17.3 Modeling Watershed Hydrology

The watershed refers to the natural hydrologic unit. It is defined as the "catchment area" from which surface runoff water (overland flow) drains downslope to the lowest point. In the hilly and mountainous regions, watershed is characterized by a unique blend of climate, topography, geology, soils, vegetation cover, and anthropogenic activities. Surface runoff generation and soil erosion in the watershed are largely governed by the watershed hydrology. Hydrological behaviors of any watershed are determined by its topography, land use, vegetation cover, and soil types. Watershed is comprised of several hillslopes, and these hillslopes are characterized by hillslope positions (landscape positions) of hilltop, upper back-slope, lower back-slope, and toe slope. Soil surface conditions (land use, vegetation cover, soil type, topography, rock fragments, surface roughness, etc.) vary at each landscape position, and these have an important influence on infiltration rate, runoff generation, and erosion parameters (Auzet et al. 1995). Soil development varies with the landscape position that is closely associated with the topography. Soils at each landscape position responds differently to runoff and erosion processes (Brunner et al. 2004).

Runoff and sediment productions from a hillslope segment are highly variable (Huang et al. 2002). At the hillslope scale, surface conditions vary with the topographic positions and result in different hydrologic regimes, runoff, and erosion. The soil water retention capacity is strongly influenced by rainfall patterns, soil infiltration rate, and soil water-holding capacity. These properties play a major role in predicting surface runoff rates and its spatial pattern (Singh and Woolhiser 2002). Spatial distribution of soil hydrological properties has a significant impact of surface runoff generation at watershed scale. It influences the amount and distribution of infiltration and the routing of surface runoff (overland flow) in the watershed. The relationship between soil hydrological properties with soil and terrain types in various surface conditions helps in quantification of average hydraulic parameters and erosion properties at the field scale. It will help in upscaling soil erosion estimation from field scale to watershed scale as well as region and basin (Le Bissonnais et al. 1998).

A watershed model simulates hydrologic processes resulting in soil erosion and nutrient loss in a more comprehensive approach as compared to the models which primarily account individual process or combination of processes at relatively field-scale. They help to understand dynamic and complex interactions between climate and land surface controlling soil hydrology. Watershed models simulate hydrological processes of the flow of water, sediment, soil nutrients, and soil organic matter within watersheds. It also quantifies the impact of anthropogenic activities on hydrological processes. Watershed models have emerged as an important scientific research and management tool in order to understand soil erosion processes and sediment loss in at watershed/catchment scale. These models are being utilized to quantify the impacts of soil and water conservation measures as well as prevailing traditional management practices with respect to surface runoff (overland flow), soil erosion, and sediment and nutrient loss within the watershed and at the outlet of the watershed.

Satellite remote sensing data and integrated use of geographic information systems (GIS) and Global Positioning System (GPS) technologies offer a unique potential in providing spatial information such as land use/land cover, soil hydrological, terrain, and watershed characteristics parameters. Geostatistics in association with GIS serves as an advanced method in quantifying the spatial pattern of soil properties and environmental variables. Remote sensing data are providing spatial information over large and inaccessible areas and helping in developing spatially distributed models for watersheds. These models require large quantities of spatial data of topography, land use, land cover, vegetation, soil attributes, and climatic data which can be stored, retrieved, managed, and manipulated with the use of GIS. Advances of GIS have grown beyond simple data management, storage, and mapping. They are being used to integrate various mathematical and computer-generated models with spatial data within the GIS. These simulation models are emerging useful tools for analysis of hydrological processes and evaluating various watershed management scenarios (He 2003).

17.4 Soil Erosion Models

Soil erosion models mathematically describe the erosion processes of detachment, transport, and deposition of soil particles on the land surface. It is based on an understanding of the physical laws and processes in generating surface runoff and their detachment capacity and sedimentation over the landscape. Erosion models are expression of mathematical relationship between soil erosion factors and soil erosion processes. Soil erosion factors include climate, topography, soil, and land use/land cover and management practices. Models help in improving understanding of interaction between soil erosion factors and processes and impact of various land use/land cover and management practices on soil erosion rates. It helps in simulating

Table 17.1 Most commonly used soil erosion models

S. no.	Model types	Model	Spatial scale	Temporal scale	Author
1.	Empirical	USLE	Plot/hillslope	Event/annual	Wischmeier and Smith (1965, 1978)
		MUSLE	Hillslope/ catchment	Annual	Williams (1975)
		RUSLE	Hillslope/ watershed	Annual	Renard et al. (1997)
2.	Conceptual	MMF	Hillslope/ watershed	Annual	Morgan et al. (1984)
		SWAT	Watershed/ basin	Continuous	Arnold et al. (1998)
		AGNPS	Small watershed	Event-based	Young et al. (1987)
3.	Physical based	WEPP	Hillslope/ watershed	Distributed, event-based, continuous	Nearing et al. (1989)
		KINEROS	Hillslope/ small watershed	Event-based	Smith et al.(1995)
		ANSWERS	Small watershed	Distributed, event-based	Beasley et al. (1980)

impact of cropping systems and various soil conservation measures on soil erosion rate and to suggest appropriate management practices in the landscape. Several erosion models have been developed in last three decades. These models can be grouped into three categories of empirical, conceptual, and physically based models (Table 17.1).

Empirical Models These models are primarily developed by establishing relationship between factors of soil erosion and soil erosion rates for a particular landscape/ region. They are statistical in nature and site specific. They are most commonly used as it is simple in application and require very less parameters. They are based on merely relationship of observations and do not provide any detail of physical process of erosion.

Conceptual Models They are also known as semiempirical model**.** These models better described some of the erosion factors and its relation with the soil erosion. They are based on spatially lumped forms of water and sediment continuity equations. They consider spatially averaged parameters and compute soil loss and sediment yield at watershed or catchment level. These models commonly used to simulate land use change and management practices on performance of watershed with similar spatially and temporally distributed input data.

Physical Process-Based Models Process-based models are advanced models that describe basic erosion processes including impact of raindrop, detachment of soil particles by rainfall and runoff, and transport and deposition of soil particles by runoff water (overland flow) on the landscape. These models take into account fundamental processes of hydrology, vegetation and plant growth, soil erosion, and sedimentation. They emerged with the development of computers as they need to simulate numerous mathematical expression using differential equations. These are mainly used as research tool to analyze impact of various scenarios of climate, land use/land cover, cropping system, and management practices on natural landscape in watershed or catchment scale. They are used by researcher, planner, and conservationists to estimate and validate soil erosion, soil nutrients, and sediment yield at various time scales ranging from rain event, daily, monthly, seasonal, and on annual basis.

These models in brief are discussed in the following sections:

(i) Universal Soil Loss Equation (USLE)

It is one of the most widely used erosion model used to predict soil erosion at plot level/field scale under a variety of crop management systems. It is an empirical model developed from the analysis of more than 10,000 plot-years of runoff and soil loss data from small plots (Wischmeier and Smith 1965, 1978). The USLE model can be written as

$$A = R\ KL\ S\ CP$$

where A is the average annual soil loss (tons. ha^{-1} .year^{-1}), R is the rainfall erosivity index, K is the soil erodibility factor, L is the slope length factor, S is the slope steepness factor, C is the vegetation cover factor, and P is the management practice factor in various land uses/land covers.

Erosion map generated using empirical model provide qualitative analysis which help to identify area and ranking of the degree of intensity and probability of occurrence of erosion risk. USLE model (Wischmeier and Smith 1978) is being widely used in soil conservation planning over the past 30 years. In these years, it has been revised to improve mathematical computation of erosion factors to extend its applications in various conditions. The model was modified as Modified Universal Soil Loss Equation (MUSLE) to compute soil loss and sediment loss at watershed level as well as revised as Revised Universal Soil Loss Equation (RUSLE) (Renard et al. 1997) to compute soil erosion in complex terrain and various land use/land cover conditions.

Modified Universal Soil Loss Equation (MUSLE)

USLE model was modified to compute sediment loss at watershed/catchment scale. The rainfall erosivity (R) factor was replaced by a runoff rate factor in the USLE model, while other factors kept unchanged. Williams (1975) reported that the sediment delivery ratio is not necessary if the rainfall energy factor in the USLE is replaced by a runoff rate factor. The R factor is calculated by runoff volume and peak runoff rate of the catchment/watershed. This model is known as the Modified Universal Soil Loss Equation (MUSLE). The MUSLE equation can be written as:

$$Sye = Xe\ K\ L\ S\ C\ P$$

where, *SYe* is the rainfall event sediment yield (metric tons)

$$Xe = 11.8(Qe\ qp)^{0.56}$$

where, *Qe* is the surface runoff amount (mm ha^{-1}) and *qp* is the peak runoff rate (m^3 s^{-1}) obtained during the rain event, and K, L, S, C, P as defined for the USLE model.

(ii) Revised Universal Soil Loss Equation (RUSLE)

RUSLE model is upgraded version of USLE model. It retained the basic structure of USLE model, but algorithms used to calculate individual erosion factors have been changed significantly (Renard et al. 1997). It has improved the computation of rainfall erosivity and soil erodibility factors depending on seasons, revised slope steepness, and slope length and new method to calculate the crop cover and management factors. RUSLE model estimates inter-rill and rill erosion but not estimate gully or river bank erosion. It is currently the most widely used model to compute long-term annual soil erosion loss in various landscapes all over the world. It has been improved for applications to different land cover conditions such as croplands, rangelands, and forest lands to estimate soil erosion and soil erosion risk assessment and to guide in preparing soil conservation plan (Millward and Mersey 1999). The RUSLE model is described as (A) as follows:

$$A = R*K*LS*C*P$$

where *A* (t $ha^{-1}y^{-1}$) is the annual average soil loss per year, *R* (mt ha-cm^{-1}) is the rainfall erosivity factor, *K* (t $ha^{-1}R^{-1}$) is the soil erodibility factor, *LS* (dimensionless) is the topographic factor, *C* (dimensionless) is the land cover factor, and *P* (dimensionless) is the soil conservation or prevention practice factor.

RUSLE-3D RSUSLE-3D has all other factor similar to RUSLE except topographic factor (LS factor). The slope length was replaced by the upslope contributing area per unit in RUSLE-3D (Mitasova et al. 1996). Slope length (l) was approximated by accounting number of grid cell contributing in flow accumulation at particular grid cell. Replacing slope length with overland flow accumulation provided opportunity to integrate RUSLE in GIS environment using digital elevation model (DEM). It has merit to reflect the impact of concentrated flow on increased erosion in the landscape. The RUSLE-3D model calculates potential average soil loss (*A*) as follows:

$$A(r) = R*K*LS(r)*C*P$$

where *A*(*r*) (t $ha^{-1}y^{-1}$) is the average soil loss per year of a grid cell, i.e., at a point *r* (geographic location of grid cell).

(iii) Morgan, Morgan, and Finney (MMF) Model

Morgan et al. (1984) proposed MMF model to predict annual soil erosion rate at field scale to hillslope and watershed scale. It provides better description of impact of rainfall on detachment and transportation of soil particles by runoff over the landscape. The model considers the soil erosion in two phases, i.e., the water phase and the sediment phase. The water phase determines rainfall energy and generation of runoff volume, whereas sediment phase accounts rate of detachment of soil particles by rainfall and runoff and the transporting capacity of runoff water. The MMF model was revised to improve the description of erosion processes and to guide users to define input parameters for its ease to applications (Morgan 2001). The revised model incorporated detachment of soil particles by runoff which was ignored earlier.

(iv) Soil and Water Assessment Tool (SWAT) Model

SWAT model was jointly developed by the United States Department of Agriculture (USDA) and Agricultural Service and Agricultural Experiment Station, Texas, USA. It is a physically based, spatially distributed, continuous time model designed to stimulate water, sediment, nutrient, and pesticide transport at watershed/ catchment scale on a daily time scale (Setegn et al. 2008). The model calculates the water balance and considers hydrologic balance of the watershed. It computes hydrologic components such as runoff, streamflow, and evapotranspiration (Arnold et al. 1998). It calculates the transpiration and evaporation components separately. SWAT uses a modified version of the SCS-CN method for predicting runoff. Peak runoff is calculated based on Modified Rational Formula. It calculates the components of the landscape's water balance over a daily time step. The data required for SWAT model are terrain, soil, land cover, and daily weather data. The model follows primary equation of the water balance which is represented as:

SWAT-VSA Model SWAT-VSA (Soil and Water Assessment Tool-Variable Source Area) is a reconceptualization of SWAT model by integrating terrain index to simulate the spatial distribution of saturation-excess runoff within the watershed. It uses SCS-CN equation to estimate surface runoff volume where CN values of various land uses/land covers are modified by accounting soil wetness index (SWI) classes. The soil wetness index is computed for watershed using digital elevation model (DEM). The significant difference in the SWAT and SWAT-VSA lies in the redefining of the hydrological response units (HRUs) based on soil wetness index (SWI). In Swat model, HRUs are defined by land use and soil types, whereas in SWAT-VSA, soil wetness index (SWI) is used in combination with land use to define the HRUs. SWAT model primarily assumes infiltration excess concept for runoff generation, whereas the SWAT-VSA consider saturation excess. In SWAT-VSA, the CN equation was reinterpreted in terms of a saturation-excess runoff generation process (Schneiderman et al. 2007).

(v) The Agricultural Nonpoint Source (AGNPS) Pollution
The agricultural nonpoint source (AGNPS) pollutant model is a continuous watershed model used to predict surface runoff, sediment and nutrient, and pesticide load from agricultural watershed. It is an advanced model that uses physical parameters of the watershed in simulation for un-gauged watershed in GIS environment. The

watershed is divided into cells or grids, and the model simulates at each cell and then computes at the outlet of the watershed. It considers each cell as a separate hydrologic unit. It uses SCS Curve Number method to estimate surface runoff from rainfall event. The Revised Universal Soil Loss Equation (RUSLE) is used to estimate the daily sheet and rill erosion. A basic mass conservation equilibrium is used to estimate nutrients generation and loading for rainfall event. Later on, an improved version of AGNPS was developed as AnnAGNPS which support continuous simulation with latest data manipulation technology and physical characteristics of watershed in GIS (Bingner and Theurer 2005). It is used as tool to evaluate nonpoint source pollution from agricultural area.

(v) EUROSEM

The EUROpean Soil Erosion Model, EUROSEM, is continuous simulation model used to predict rill and inter-rill soil erosion from hillslope and watershed (Morgan et al. 1998). The model computes surface runoff, soil erosion, and sediment loss at watershed. It requires comprehensive data on daily weather, soil hydrologic parameters, watershed characteristics, and plant growth parameters. The model was designed as an event-based model, since it was thought that erosion was dominated by only a few events per year.

(vi) Areal Nonpoint Source Watershed Environment Response Simulation (ANSWERS)

The ANSWERS (Areal Nonpoint Source Watershed Response Simulation) is a continuous simulation model used to compute surface runoff and soil erosion (Beasley et al. 1980). It divides watershed into small and independent unit. It describes runoff processes by empirical method of SCS-CN method, whereas soil erosion and sediment transport processes by physics-based continuity equations. The model explicitly deals with the effect of rainfall intensity and spatial variability of infiltration capacity of soil and terrain conditions. The model differs with AGNPS model as it considers more physically based approach for erosion and transport modeling. Application of the model is limited as it requires large spatial and temporal input data.

(vii) The Kinematic Runoff and Erosion Model (KINEROS)

It is a distributed, event-based, deterministic, and physically based model that is primarily useful for predicting surface runoff and erosion of small agricultural watersheds (Smith et al. 1995). The KINEROS model solves kinematic wave equations by a four-point implicit method. It uses equations to describe the sediment dynamics at any point along a surface flow path is a mass balance equation similar to that for kinematic water flow. The watershed is divided into homogeneous overland flow planes and channel segments. The dimensions of planes are chosen to completely cover the watershed, so rainfall on the channel is not considered directly. The model simulates water movement over these planes in a cascading fashion. Channel may receive water and sediment uniformly from either or both sides of the channel or from a plane at the upstream boundary. The spatial variation of rainfall, infiltration, runoff, and erosion parameters can be accommodated.

(viii) Water Erosion Prediction Project (WEPP)

Water Erosion Prediction Project (WEPP) model is the most advanced physically process-based continuous simulation model used to predict surface runoff and sediment loss from single hillslopes, agricultural field, and small-sized watersheds approximately less than 400 hectares (Nearing et al. 1994). It is available for both hillslopes and watershed versions. The watershed version developed for field areas characterized by ephemeral gullies. WEPP model is comprised of sub-models of weather generation, Green and Ampt infiltration, surface runoff, erosion mechanics, plant growth, residue management, tillage effects on the soil, and soil consolidation. It describes soil detachment, transport, and deposition processes using steady-state sediment continuity equation representing rill and inter-rill processes. The WEPP model broadly describes erosional processes, hydrological processes, plant growth and residue processes, water use processes, hydraulic processes, and soil processes. The spatial and temporal variability of topography, soil parameters, hydrology, surface roughness, and land use conditions can be defined for hillslopes to compute surface runoff and erosion at event basis, daily, monthly, and yearly basis.

GeoWEPP

The Geospatial interface of WEPP model is known as GeoWEPP. The model was developed to delineate the watershed configuration of channel and representative hillslopes (Renschler 2003). The digital elevation model (DEM) was used for topographic analysis to automate delineation of drainage network and to define major channel in each representative watershed. The slopes of each hillslopes were computed. The version of GeoWEPP model delineates a single representative hillslope with a single soil and land use in each watershed. The GIS interface creates all necessary files and organizes databases and WEPP simulations. It allows assessment for small watersheds (<500 hectares).

(ix) APEX Model

The Agricultural Policy Environmental eXtender (APEX) is a distributed, continuous daily time step model used to simulate surface runoff and sediment and nutrient loss at watershed scale (Williams et al. 2008). It is an extension of Environmental Policy Integrated Climate (EPIC) (Williams 1990). APEX model is capable of simulating for long term. It divides the watershed into homogenous land area in terms of soil, topography, and land use/land cover for simulation. Its routing component simulates flow from one subarea to another through channels and flood plains to the watershed outlet and transports sediment, nutrients, and pesticides. It has major components of weather simulation, hydrology, erosion, sedimentation, nutrient cycling, crop growth, tillage practices, and plant environmental control. It provides routing mechanism between subareas involving surface runoff, sediment and nutrient transport, and deposition at the watershed outlet.

Hilly and mountainous landscape is geographically dynamic, and its behavior varies both spatially and temporarily. The Himalayan region is characterized by rugged terrain that develops spatial varying soil hydrological condition due to varying soil types and land use/land cover. Inaccessible and difficult terrain conditions in the landscape resulted in scarcity of data that has restricted the use of data-intensive process-based model for runoff and erosion studies. The Revised Universal Soil Loss Equation (RUSLE/RUSLE-3D) model is commonly used for spatial prediction of soil loss and erosion risk potential in the hilly and mountainous landscape. RUSLE model incorporates the influence of terrain convexity/concavity and improved empirical equations for the computation of topographic (LS) factor (Renard et al. 1997). The SCS-CN-based models are widely used for estimating surface runoff and sediment and nutrient loss at watershed/catchment scale in the hilly and mountainous landscape. Most commonly used models are SWAT, SWAT-VSA, APEX and WEPP models. These models are semi-distributed and consider watershed as lumped hydrological units. Himalayan landscapes are characterized by rugged terrain that develops variable source area (VSA) contributing in varying amount of runoff generation on the landscape. Simulation results showed that modified SWAT-VSA is most suitable for estimating surface runoff and runoff source area in the hilly watershed of Himalayan landscape having scarcity of hydrological data. There are very few studies have been conducted to calibrate and validate the models in the Himalayan region due to scarcity of runoff measurement at watershed scale. Applicability of these models in the landscape has been demonstrated through case studies as example in the chapter.

17.5 Modeling Sediment and Nutrient Loss at Watershed Scale

Understanding the hydrology of watershed and hydrological processes within the watershed are important to model sediment and nutrient loss at watershed scale. Hydrological models primarily compute surface runoff generation (overland flow) accounting soil, topography and vegetation cover distributed in the watershed. These models are coupled with erosion and sediment transport equations that provide estimation of sediment and nutrient loss at watershed scale. These models provide scientific framework for watershed analysis of surface water generation, soil erosion, and water quality of the watershed/catchment.

Watershed models are considered as a cost-effective scientific tool for the assessment of soil nutrient load and nonpoint source pollution (Shrestha et al. 2006). Nitrogen (N), phosphorus (P), and potassium (K) are the major nutrients for crops production, each having its own characteristic against transport by runoff. N and P are more mobile compared with that of K, so those nutrients are easy to transport by water through runoff process. Since N, P, and K are adsorbed on the soil particles especially clay particle. Thus these nutrients are removed along with the sediment during soil erosion.

There are several models that are being used to simulate runoff, sediment and nutrient loss, and management effects with alternative land uses at watershed scale: SWAT, WEPP, EPIC, AGNPS, APEX, etc. Newly developed models are coming with GIS interfaced; GIS has emerged as a powerful tool in handling spatial datasets, so models are interfaced with GIS. Several GIS software provide hydrological tool that helps in extracting various physical terrain parameters of the watershed. They process DEM for stream/drainage network delineation and to automatic delineation of watershed. GIS provide efficient platform for spatial integration of and analysis of various thematic layers such as land use/land cover, physiography, soils, elevation, slope, and hydrography data for the watershed.

Merritt et al. (2003) provided a detail review of erosion and sediment transport models. These models differ in terms of complexity and in representing hydrological and erosion processes. Selection of appropriate model will depend on characteristics of the watershed, availability of input data, and output requirement of the users. Modeling requires calibration of models. Hydrological models need to calibrate with the observed runoff at watershed scale. The performance and accuracy of model need to be tested and also necessitates identifying sensitive parameters of the model. Uncertainty in the output of hydrological models incorporates additional uncertainty in the sediment and nutrient prediction. Integration of GIS with hydrological models serves as tool in understanding soil erosion and sediment and nutrient transport mechanism in the watershed and quantifying these variables to suggest suitable soil conservation measures to watershed managers (Xu et al. 2009).

17.6 Geospatial Approach in Modeling Soil Erosion

The rapid development of geospatial technologies (RS, GIS, GPS, and computer software) in recent years has made available new tools and capabilities to resource planner and users for inventory and mapping of natural resources at various scales to meet the need of various user groups for soil and water conservation planning. The growth of open-source software has had a particular impact on the potential to analyze geospatial data. RS data available in variety of spatial, spectral, and temporal resolutions are being widely used in characterizing watershed and assessing soil erosion and sediment and nutrient loss in the watershed. Assessing soil erosion risk or potential area and effect of soil conservation measures on soil erosion rates in large geographical area requires information on landscape components such as vegetation, soils, and terrain information. Geospatial technologies enabled resource planners to generate these natural resource maps of large areas at various scales and allowed updating these maps efficiently. The combined use of remote sensing, GIS, and erosion models has been shown to be an effective geospatial approach for estimating the severity and spatial distribution of soil erosion potential area. Remote sensing data and GIS provide reliable estimate of the soil erosion risk assessment (Kumar and Kushwaha 2013).

Erosion models consider erosion factors of climate (rainfall), soil, topography, vegetation cover, and land management practices in assessing soil erosion. They serve as tool to resource planners and conservationist in making watershed development plan. Reliable estimation of soil erosion estimation requires precise and accurate quantification of model input parameters. These model input parameters can be obtained directly by field measurement or derived from existing literature sources. Remote sensing and GIS have provided unique opportunity to quantifying these model input parameters. The spatial assessment of soil erosion can be improved by calculation of soil erosion factors at required scale in various region using GIS. GIS also facilitates in extracting watershed model input parameters required in hydrological models to automate the computation for larger area. Most of these models need information related with soil type, land use/land cover, landform, climate, and topography to estimate soil loss. Remote sensing data are used to generate various model input parameters are discussed in brief in the Table 17.2.

Spatial information of land use/land cover, soil, and terrain information are needed for geospatial modeling of soil erosion and hydrological modeling of watershed. Remote sensing and geographic information system (GIS) are powerful tools to derive accurate and timely information on the spatial distribution of land use/land cover and their change over large areas. Land use/land cover map can be prepared by visual interpretation or digital classification methods. The digital classification is much more cost-effective and quicker than the visual method. Digital classification uses supervised and unsupervised classification methods to generate land use/land cover maps using remote sensing data. In supervised classification, ground truth data as training sets are defined for the area, whereas different spectral classes are identified in unsupervised classification that are later defined as land use classes. The digital techniques require less time and field work, and it classifies each pixel of the area. Land use/land cover map is used to derive crop cover and management practice factors for erosion modeling.

Soil erodibility is one of the most important factors used in spatial soil erosion risk assessment. Soil information derived from soil map is used to generate soil erodibility factor map. The standard false color composite (FCC) and satellite imagery are used to interpret soil landscape based on visual interpretation considering landforms, topography, and dominant land use/land cover for preparing soil resource inventory at semi-detail and reconnaissance scales. However, with the availability of high-resolution PAN (5.8 m) and LISS III (23 m) data from IRS-1C/1D satellites and multispectral high-resolution (5.8 m) data from Resourcesat-1 and Resourcesat-2 satellites, it is now possible to utilize these data for large-scale soil mapping. Remote sensing methods facilitate mapping inaccessible areas by reducing the need for extensive time-consuming and costly field survey. Spatial soil erodibility factor map for any hilly terrain areas can be generated using DEM and limited soil sampling (Kumar and Gupta 2016).

Topography plays an important role in soil erosion prediction. High-resolution stereo satellite data and microwave SAR (Synthetic Aperture Radar) data have been

Table 17.2 Remote sensing and GIS applications in deriving input parameters for soil erosion models

S. No	Soil erosion factors	Factor elements	Thematic maps	Remote sensing/ ancillary data	Methods of analysis
1.	Vegetation cover (C) factor	Crop types, cropping pattern, natural/perennial land cover types	Land use/land cover map	Multi-date, remote sensing data at appropriate sensor/ scale	Visual analysis, digital supervised/ unsupervised classification, spectral indices
2.	Land management practices (P) factor	Field management practices – bunding, terracing, contouring, field size, etc.	Detail land use/land cover types showing management practices	High spatial resolution remote sensing data	Visual analysis/ based on farmers interview
3.	Soil erodibility (K) factor	Soil texture (sand, silt, and clay percentage), organic matter, soil structure, soil permeability	Soil map	Remote sensing data at appropriate scale	Visual analysis for physiographic-soil analysis
4.	Topographic (L S) factor	Elevation, slope, drainage network	Slope map, drainage map	Digital elevation model (DEM) – CartoDEM, ASTER, SRTM	Terrain analysis based on digital elevation model (DEM) using GIS/image analysis software
				Survey of India (SOI) toposheet, stereo satellite data, microwave satellite data	
5.	Climate (R) factor	Rainfall, rain intensity, rainy days	Location-based rain erosivity	Weather station	Point interpolation (Thiessen map) using GIS

used to generate digital elevation models (DEMs) to study topography of earth surface. Today, several global digital elevation models (DEMs) such as SRTM and ASTER DEM are freely available to download from the respective websites (http://earthexplorer.usgs.gov; http://glovis.usgs.gov; http://ws.csiss.gmu.edu/DEMExplorer). National Remote Sensing Centre (NRSC), India, generated CartoDEM from stereo Cartosat satellite data which can be downloaded from the website http://www.bhuvan.nrsc.gov.in. The DEMs are used to derive terrain parameters such as slope, aspect, drainage network, flow accumulation, and flow direction of the watershed. DEM of appropriate resolutions are required to extract terrain parameters for assessing soil erosion at various scale. Coarse-resolution DEM underestimates the terrain parameters and results in erroneous assessment of topographic factors. Hilly and mountainous landscapes having complex topography required high-resolution DEM to represent reliable terrain geometry. It will improve computation of flow accumulation and upslope contribution area for estimating topographic factors of slope length (L) and slope steepness (S) factors.

17.7 Geospatial Modeling of Soil Erosion in Hilly and Mountainous Landscape: Some Experiences in Northwest Himalaya

An experimental watershed representing Himalayan landscape covering an area of 57 ha was instrumented with measurement of weather parameters and surface runoff at watershed scale. The watershed is located near to Langha village, Dehradun, Uttarakhand state, India. It is micro-watershed of Sitlarao watershed covering an area of 52 km^2. The watershed had elevation range of 960–1480 m and characterized moderate sloping to steep sloping hills. The gauging civil structure is constructed at second order stream for measuring surface runoff and sediment yield of the watershed. Digital stage level recorder is placed in the gauging structure to record runoff at 15 min interval during rainy season. There is one automatic weather station established at the water divide line of the watershed and one digital rain gauge near to the watershed outlet (Fig. 17.1a) to record daily weather data (Fig. 17.1b). The watershed comprises of forest cover (30%), scrub (18%), and agricultural land (51%). Major crops grown are paddy and maize in kharif (summer) season and wheat in winter crop in rabi (winter) season. Soils are deep to moderately deep and loamy and sandy loam in crop land and loamy skeletal in forest land. Scrub land characterized as skeletal to fragmental in texture.

Fig. 17.1 (**a**) Surface runoff gauging structure, (**b**) Automated Weather Station (AWS), Dehradun, India

Several erosion and hydrological models such as RUSLE, MMF, SWAT, APEX, AGNPS, and WEPP were used to understand hydrological processes governing surface runoff generation, soil erosion, and soil nutrient loss in the watershed. These models were calibrated and tested in the watershed. Some of the results obtained from these models are summarized as below.

17.8 Modeling Soil Erosion Using RUSLE Model

The study attempted to assess soil erosion risk in a watershed of Himalayan region employing RUSLE model integrated with GIS (Fig. 17.2). Watershed selected for the study falls in Dehradun district (Uttarakhand state) which is located between latitude 30°25′ N to 30°30′ N and longitude 77°45′ E to 78°0′ E, covering an area of 805 ha. The runoff water from watershed flows into the main river Sitlarao. IRS LISS III remote sensing data was used to prepare land use/land cover and soil map to generate vegetation cover and soil erodebility factors, respectively. Soil erodebility (K) factor was computed based on soil properties using soil map prepared at scale of 1:25,000. CartoDEM (30 m) was used to derive slope length (L0 and steepness (S) factors using flow accumulation method. The topographic factor (LS) was estimated to be highest in hills; therefore LS factor appears more dominant for soil erosion because the erosion rate was predicted more in steep to very steep slope.

The slope study with erosion revealed that the gentle and moderate slope had lowest predicted soil erosion rate of 18.02 tha^{-1} $year^{-1}$ and 43.55tha^{-1} $year^{-1}$,

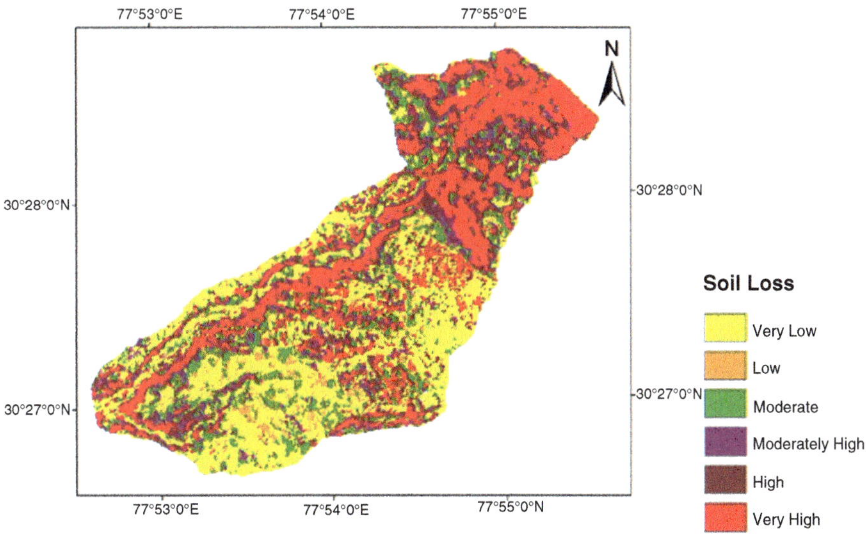

Fig. 17.2 Soil erosion risk predicted using RUSLE model

respectively, whereas, highest soil erosion rate (80 t ha^{-1} $year^{-1}$) was predicted under steep to very steep slope. The vegetation cover and management practices played a major role for controlling the soil erosion. Annual soil erosion rates predicted under different land uses/land covers revealed that the highest erosion rate (51.98 t ha^{-1} $year^{-1}$) was under scrub land followed by dense (48.75 t ha^{-1} $year^{-1}$) and open forest (17.64 t ha^{-1} $year^{-1}$), whereas agriculture land was predicted with the erosion rate of 14 t ha^{-1} $year^{-1}$.

17.8.1 Modeling Erosion Processes on Hillslope: WEPP – A Hillslope Model

The WEPP model is available in hillslope and watershed versions. It is basically designed for hillslopes. Each hillslope is divided into overland flow elements (OFEs) representing hillslope elements. A hillslope must define at least one overland flow element. Soil properties, land use/land cover, management practice, and slope steepness are defined for each OFE, i.e., hillslope elements such as hilltop, backslope, and toe slopes. Watershed version of the model divides the watershed into hillslopes to define model input parameters. WEPP model runs simulation for each hillslope in the watershed and then compute final simulation by merging the output and simulates channels. An experiment was conducted to observe surface runoff generation at various hillslope positions of a hillslope. WEPP – hillslope version model – was run to understand runoff generation and soil erosion on various hillslope units.

It was found that surface runoff generated on all rain events at various hillslope positions (Fig. 17.3). Surface ponding of rainwater was observed at the hilltop. These soils get early saturation conditions due to shallow depth of soil and poor soil hydraulic characteristics. It may also be attributed to decrease in permeability with soil depth due to paralithic, i.e., weathered, material below the surface layer. The higher amount of runoff in upslope may have resulted due to steep slope, large amount of rock fragments, and higher bulk density of the soil. These conditions facilitated to unsaturated runoff generation (Hortonian overland flow) as indicated by high amount of surface runoff observed at all rain events. Huang et al. (2002) observed high runoff generation at portion of high slope gradient and decreasing downslope at a much higher intensity. Soils of higher mid-slope also have better water retention capacity than the upslope. It showed more saturation of soil at lower mid-slope than higher mid-slope. But the surface runoff was observed lesser than lower mid-slope. It indicated that most of rainwater infiltrated at higher mid-slope and lead to saturation of soils of lower mid-slope and generation of unsaturated surface runoff (Hortonian overland flow). Lower mid-slope had shown more amount of surface runoff generation then higher mid-slope at most of the rain events. Lower hillslope characterized as terraced paddy fields. Water exfiltrating from riser of terraced field were observed just after rains. It indicates high infiltration of rainwater

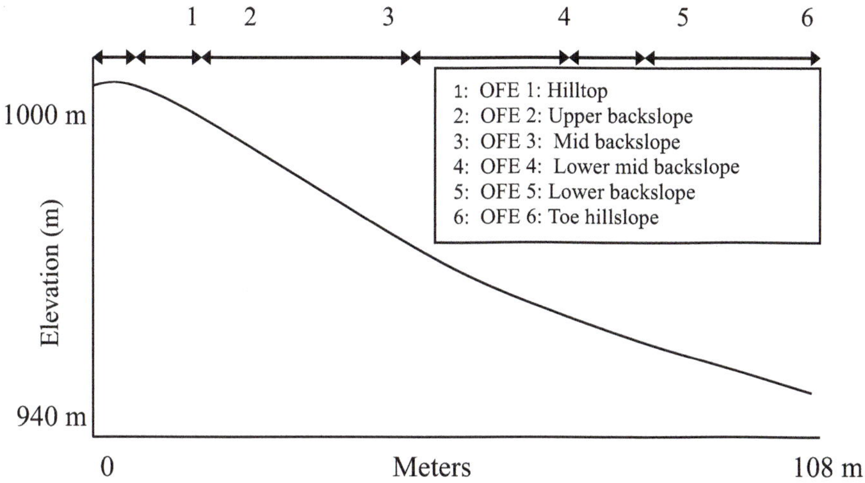

Fig. 17.3 Hillslope units (OFEs) identified on a representative hillslope in the watershed

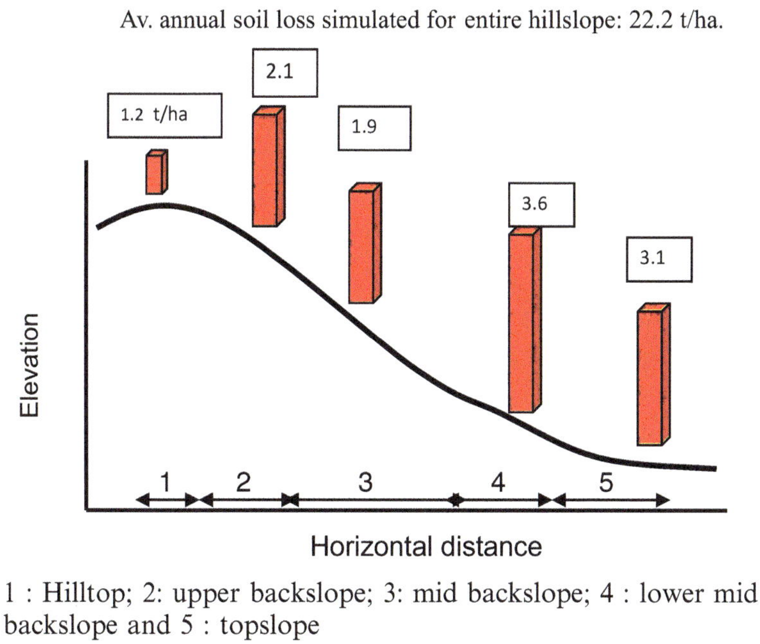

Fig. 17.4 Soil erosion rate at various hillslope positions based on WEPP model

and consequently saturation of soil and the subsurface flow of water at lower slope (Kumar and Sterk 2005).

WEPP model was run to predict soil erosion at various hillslope positions (Fig. 17.4). Among the hillslope positions, lower hillslope was predicted to have

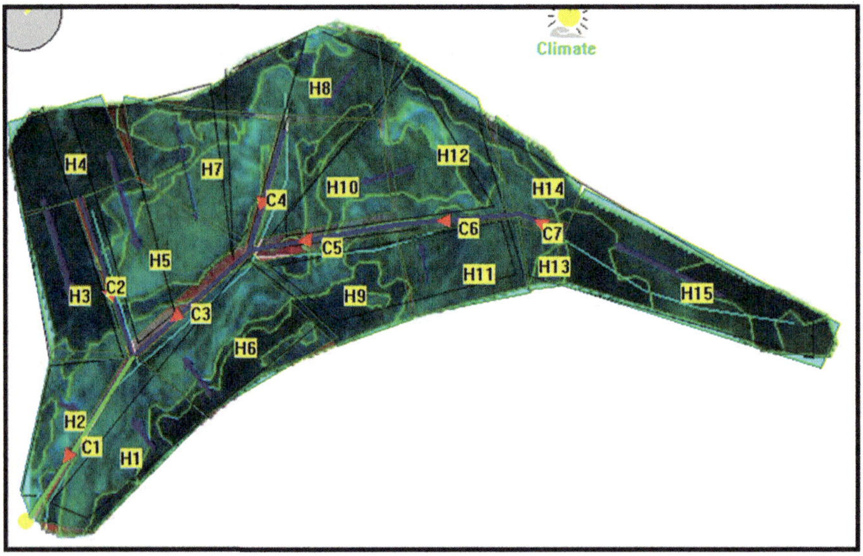

Fig. 17.5 Various hillslopes defined in the watershed shown on the Std. FCC of LISS IV satellite

the highest soil erosion (3.6 t/ha) followed by toe slope, upper back-slope, and mid back-slope. The lowest soil erosion (1.2 t/ha) was predicted at hilltop position. The lower mid-slope position was most critical among hillslopes positions in the watershed.

17.8.2 Modeling of Surface Runoff and Erosion: WEPP Watershed-Scale Model

Watershed version of WEPP model was run to simulate soil erosion and surface runoff for experimental watershed located in Dehradun, India (Kumar et al. 2006). The watershed was defined by 15 numbers of hillslopes with 7 numbers of channels (Fig. 17.5). Soil types, land cover, and slope parameters were defined for each hillslope. The model was run for data of year 2003. The surface runoff at watershed level was calibrated and validated to simulate soil erosion in the watershed. Study revealed that the WEPP model well simulated for low to medium rain intensity (< 50 mm hr.$^{-}$1) rainfall where as it fails to simulate for rain events of higher rainfall intensity (> 50 mm hr.$^{-1}$). The average annual soil erosion was predicted of 45 tones ha^{-1} year^{-1} in the watershed. It also predicted surface runoff generation and erosion of each hillslope of the watershed. It provides detailed information regarding quantity of surface runoff generation and soil erosion rates of each hillslope in the watershed for each rain event as well as daily, monthly, and on yearly basis. The study provided critical information to identify critical hillslope for suggesting suitable soil conservation plans. WEPP model is used as research tool to understand hydrological processes responsible for soil erosion and nutrient loss in the watershed.

17.8.3 Modeling Sediment Loss and Soil Nutrient Loss: APEX Watershed-Scale Model

The study was undertaken to study surface runoff, soil erosion, and nutrient loss at watershed scale using physical/process-based (APEX) model. The model used spatial and nonspatial database. Spatial dataset was prepared using digital terrain model (DTM) derived from Cartosat data, land use/land cover, and soil map prepared at large scale using IRS LISS IV data. Further, nonspatial dataset includes fertilizer application, management operations, and crop database including maximum LAI, seeding rate, plant population, etc. Weather data from AWS was used to prepare weather files including minimum and maximum temperature and precipitation data file. Inputs required by the model have been collected from study area with comprehensive field visits. The sensitivity analysis was carried out to evaluate surface runoff and sediment loss response with changes in model input of hydrologic parameters. Further the model was calibrated and validated for daily runoff and sediment and nutrients loading at watershed outlet. Calibration was done for low to medium and high rainfall events.

Model was calibrated for surface runoff, sediment loss, and nutrient loss to optimize the input given to the model to predict the sediment loss, erosion, and nutrient loss. The calibration was done by changing the sensitive parameters. Analysis showed that SCS CN number was found most sensitive to runoff, followed by saturated hydraulic conductivity, available water-holding capacity, CN retention parameter, and C factor, whereas erosion control practice (P) factor was found to be most sensitive, followed by C factor, sediment routing coefficient, average upland slope, and soil erodibility (K) factor for the sediment and nutrient loss. APEX model was calibrated for the sub-watershed, and it predicted quite well for the surface runoff, sediment loss, and various nutrients of total carbon, total nitrogen, and available phosphorus. Surface runoff was predicted quite well for low and medium rainfall; however it was over predicted for high rainfall events. Soil erosion predicted in the watershed shown in Fig. 17.6. The hydrological assessment of this model will facilitate future modeling applications using APEX to the Himalayan watersheds for watershed analysis including water quality management, impacts of alternative land management practices, etc.

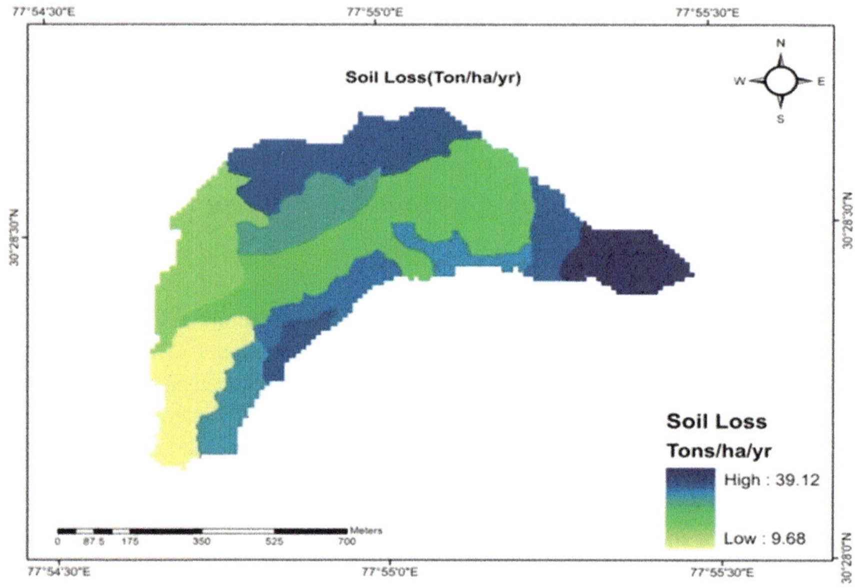

Fig. 17.6 Soil erosion rate predicted using APEX model

17.9 Up Scaling of Measurement from Small Watershed to the Region

Measurement of surface runoff (overland flow) and sediment loss at watershed scale necessitated to quantify the rate of soil erosion and nutrient loss in the hilly and mountainous landscape. Measurement alone provides empirical evidences that are difficult to extrapolate in time and space (Stroosnijder 2005) Erosion processes at different scales range from point, plot, landscape, hillslope, and watershed scale. The point ($1m^2$) scale is used to study inter-rill (splash) erosion, whereas plot scale (< $100m^2$) for rill erosion, hillslope (< $500m^2$) for sediment deposition, and landscape (<1 ha) for channel erosion. A small watershed (<50 ha) is used to study spatial interaction effect on hydrological processes as wholesome to quantify soil erosion and sediment and nutrient loss within the watershed and at watershed scale. Watershed scale provides integrated effect of spatial heterogeneity controlled by topography, land use/land cover, soil, and management practices on soil erosion and nutrient loss. Erosion measurement techniques are the scientific erosion research that is highly accurate and provide clues of causes and effect of soil erosion. Erosion modeling required to understand hydrological behavior and erosion processes as well as to identify sensitive parameters controlling soil erosion and runoff generation in the watershed. Soil erosion and watershed simulation model varies with distinct processes and governed by several soil hydrological parameters. Understanding of erosion and runoff generation processes operating at the landscape and in the watershed is required for the selection of appropriate model for the landscape.

Watershed-scale model has been established as a powerful tool in understanding hydrological processes controlling runoff generation and sediment and nutrient loss at watershed/catchment level. Its effectiveness depends on both calibration and performance during the validation. The calibrated model should be tested against an independent set of measured data. There are uncertainties related to how the simulation processes are represented in the model. The uncertainties related to the simulation processes may arise from the assumptions used in the model. The uncertainties exist at all levels, and it is not possible to determine which uncertainties cause the deviation between simulated and measured variables. The calibrated parameters established after validation for the selected watershed model of Himalayan landscape can be used to apply for other watershed similar in their terrain and environmental condition. The entire Northwest Himalayan regions are characterized in three distinct regions, lower (Siwalik), middle, and higher Himalayan region, owing to its distinct climatic condition, topography, soil, land use, and vegetation cover. There is a need to establish calibrated and validated model parameters of the representative watersheds located in these three distinct regions to quantify soil erosion, sediment and nutrient loss for the Indian Himalayan region.

17.10 Conclusions

Understanding the processes of soil erosion, its causes, and impacts in a watershed is needed to suggest suitable soil and water conservation measures as well as appropriate land management practices. Various hillslope positions in the hilly and mountainous landscape attributed to varying soil hydrological condition leading to varying surface runoff generation and soil erosion in the watershed. Remote sensing data and GIS provide reliable estimate of the soil erosion risk assessment. Remote sensing and GIS allow realistic assessment of the potential soil erosion risk with validation. Integrating soil erosion models with remote sensing (RS) and GIS emerged indispensable tool in assessing soil erosion potential of the landscape and monitoring and modeling the effect of land use on soil erosion potential.

Large areas of hilly and mountainous landscape are inaccessible. Therefore, RS and GIS are found to be the most vital tools in deriving model input parameters of land use/land cover, soil types, and terrain information. Topography plays a dominant role in spatial distribution of soil properties and hydrological processes and thus controlling runoff generation processes in the hilly terrain. Watershed model computes runoff and soil erosion based on hillslope elements which are efficiently characterized with DEM using GIS. In the recent years, digital elevation model (DEM) terrain data demonstrated potential application in predicting soil hydrological parameters and their spatial distribution, and it had helped in improving runoff estimation in the hilly landscape. Availability of DEMs facilitated in great extent to characterize complex slope of the hilly landscape. Progress in availability of freely available DEMs (digital elevation models) has offered new opportunities for the scientific community for detailed representations of surface terrain features and its integration in runoff modeling with geographical information systems.

The results from hydrological and erosion models simulation of processes have been the primary information for land use planning in the watershed, especially to better understanding how the hydrological cycle, soils, and vegetation interact between themselves. One of the most significant state-of-art studies involves erosion models to simulate climate change impacts on the water erosion and hydrology of watershed in the future. GIS and remote sense tools, searching to obtain a distributed model in both space and time, have supported the models, aiming to reduce the uncertainties and providing basis for simulation of different scenarios in the watersheds.

References

Arnold JG, Srinivasan R, Muttiah RS, Williams JR (1998) Large-area hydrologic modeling and assessment: Part I Model development. J. American Water Resour. Assoc., 34(1): 73–89.

Auzet AV, Boiffin J, Ludwig B (1995) Concentrated flow erosion in cultivated catchments: influence of soil surface state. Earth Surface Processes and Landforms 20(8): 759–767.

Bai ZG, Dent DL, Olsson L, Schaepman ME (2008) Proxy global assessment of land degradation. Soil Use and Management 24(3): 223–234.

Beasley DB, Huggins LF, Monke A (1980) ANSWERS: A model for watershed planning. Transactions of ASAE 23(4): 938–944.

Bingner, RL, Theurer, FD (2005). AnnAGNPS technical processes documentation, Version 3.2,

Brunner AC, Park SJ, Ruecker GR, Dikau R, Vlek PLG (2004) Catenary soil development influencing erosion susceptibility along a hillslope in Uganda. Catena 58(1): 1–22.

Dhruvanarayana, V.V., Ram Babu (1983) Estimation of soil erosion in India. Journal of Irrigation and Drainage Engineering, 109(4): 419–434.

Garde RJ, Kothyari UC (1986) Erosion in Indian catchments, 3rd Int. Symposium on River Sedimentation, USA.

He C (2003). Integration of geographic information systems and simulation model for watershed management. Environmental Modelling & Software, 18(8–9): 809–813.

Huang C, Gascuel-Odoux C, Cros-Cayot S (2002) Hillslope topographic and hydrologic effects on overland flow and erosion. Catena 46(2): 177–188.

Jain MK, Mishra SK, Shah RB (2010) Estimation of sediment yield and areas vulnerable to soil erosion and deposition in a Himalayan watershed using GIS. Current Science 98(2): 213–221.

Kumar S, Gupta S (2016) Geospatial approach in mapping soil erodibility using CartoDEM–A case study in hilly watershed of Lower Himalayan Range. Journal of Earth System Science 125(7): 1463–1472.

Kumar S, Kushwaha SPS (2013) Modelling soil erosion risk based on RUSLE-3D using GIS in a Shivalik sub-watershed. Journal of Earth System Science 122(2): 389–398.

Kumar S, Sterk G (2005) Process based modeling in understanding erosion processes and soil erosion assessment at hillslope scale in the lesser Himalayas, India. In: Proceedings of the International Conference on Hydrological Perspectives for Sustainable Development, Roorkee, India, 23–25 February 2005 (pp. 420–427).

Kumar S, Sterk, G and Dadhwal, V.K. (2006). Process based modeling for simulating surface

Le Bissonnais Y, Benkhadra H, Chaplot V, Fox D, King D, Daroussin J (1998) Crusting, runoff and sheet erosion on silty loamy soils at various scales and upscaling from m^2 to small catchments. Soil and Tillage Research 46(1–2): 69–80.

Mandal D, Sharda VN (2011) Assessment of permissible soil loss in India employing a quantitative bio-physical model. Current Science 100(3): 383–390.

Merritt WS, Letcher RA, Jakeman AJ (2003) A review of erosion and sediment transport models. Environmental Modelling & Software 18(8): 761–799.

Millward AA, Mersey JE (1999) Adapting the RUSLE to model soil erosion potential in a mountainous tropical watershed. Catena 38(2): 109–129.

Mitasova H, Hofierka J, Zlocha M, Iverson LR (1996) Modelling topographic potential for erosion and deposition using GIS. International Journal of Geographical Information Systems 10(5): 629–641.

Morgan RPC, Morgan DDV, Finney HJ (1984) A predictive model for the assessment of soil erosion risk. Journal of Agricultural Engineering Research 30: 245–253.

Morgan, RPC (2001). A simple approach to soil loss prediction: a revised Morgan–Morgan–Finney model. *Catena*, *44*(4), 305–322.

Morgan, RPC, Quinton, JN, Smith, RE, Govers, G, Poesen, JWA, Auerswald, K, Chisci, G, Torri, D and Styczen, ME. 1998. The European Soil Erosion Model (EUROSEM): a dynamic approach for predicting sediment transport from fields and small catchments. Earth surface processes and landforms, 23(6): 527–544.

Nearing MA, Foster GR, Lane LJ, Finkner SC (1989) A process-based soil erosion model for USDA-Water Erosion Prediction Project technology. Transactions of ASAE 32(5): 1587–1593.

Nearing MA, Lane LJ, Lopes VL (1994) Modeling soil erosion. In: R. Lal, (Ed.), Soil Erosion Research Methods, Soil and Water Conservation Society and St. Lucie Press, Delray Beach, FL, pp. 127–156.

Oldeman LR (1994) The global extent of soil degradation. Soil Resilience and Sustainable Land Use, 9.

Pandey A, Chowdary VM, Mal BC, Billib M (2008) Runoff and sediment yield modeling from a small agricultural watershed in India using the WEPP model. Journal of Hydrology 348(3): 305–319.

Poesen J, Nachtergaele J, Verstraeten G, Valentin C (2003) Gully erosion and environmental change: importance and research needs. Catena 50(2): 91–133.

Renard KG, Foster GR, Weesies GA, McCool DK, Yoder DC (1997) Predicting soil erosion by water: a guide to conservation planning with the Revised Universal Soil Loss Equation (RUSLE) (Vol. 703). Washington, DC: US Government Printing Office.

Renschler, CS (2003) Designing geo-spatial interfaces to scale process models: The GeoWEPP approach. Hydrological Processes, 17(5): 1005–1017.

Schneiderman EM, Steenhuis TS, Thongs DJ, Easton ZM, Zion MS, Neal AL, Mendoza GF, Todd Walter M (2007) Incorporating variable source area hydrology into a curve-number-based watershed model. Hydrological Processes 21(25): 3420–3430.

Setegn S, Srinivasan GR, Dargahi, B (2008) Hydrological modelling in the Lake Tana Basin, Ethiopia using SWAT model. The Open Hydrol. Journal, 2: 49–62.

Shrestha S, Babel MS, Gupta AD, Kazama F (2006) Evaluation of annualized agricultural nonpoint source model for a watershed in the Siwalik Hills of Nepal. Environmental Modelling & Software 21(7): 961–975.

Singh VP, Woolhiser DA (2002) Mathematical modeling of watershed hydrology. Journal of Hydrologic Engineering 7(4): 270–292.

Smith RE, Goodrich DC, Woolhiser DA, Unkrich CL (1995) KINEROS: A Kinematic Runoff and Erosion Model. In: V.P. Singh (Ed.), Computer Models of Watershed Hydrology. Water Resources Publications, Colorado, pp. 697–732.

Stroosnijder L (2005) Measurement of erosion: Is it possible? Catena 64(2): 162–173.

Trivedi TP (2010) Degraded and Wastelands of India Status and Spatial Distribution, Directorate of Information and Publications of Agriculture, Indian Council of Agricultural Research, Krishi Anusandhan Bhavan I, Pusa, New Delhi.

UNEP (1994) Land Degradation in South Asia: Its Severity, Causes and Effects upon the People. USDA-ARS

Williams JR (1975) Sediment-Yield Prediction with Universal Equation using Runoff Energy Factor. In: Present and Prospective Technology for Predicting Sediment Yield And Sources ARS, S-40, USDA, WA.

Williams JR (1990) The erosion-productivity impact calculator (EPIC) model: a case history. Philosophical Transactions of the Royal Society of London B: Biological Sciences 329 (1255): 421–428.

Williams JR, Izaurralde RC, Steglich EM (2008) Agricultural policy/environmental eXtender model: Theoretical documentation version 0604 (draft). BREC report # 2008–17. Texas AgriLIFE Research, Texas A&M University, Temple, TX.

Wischmeier WH, Smith DD (1965) Predicting Rainfall Erosion Losses from Cropland East of the Rocky Mountains. In: Agriculture Handbook 282. US Government Print Office, Washington, DC.

Wischmeier WH, Smith DD (1978) Predicting Rainfall Erosion Losses-A Guide to Conservation Planning. In: Agriculture Handbook 537. US Government Print Office, Washington, DC.

Xu YQ, Shao XM, Peng J (2009) Assessment of soil erosion using RUSLE and GIS: a case study of the Maotiao River watershed, Guizhou Province, China. Environmental Geology 56: 1643–1652.

Young RA, Onstad CA, Bosch DD, Anderson WP (1987) AGNPS, Agricultural Non-Point Source Pollution Model: A Watershed Analysis Tool. Conservation Res. Report, 35. US Dept. Of Agriculture, Agricultural Research Service, Morris, MN, USA.

Chapter 18
Geospatial Technology for Climate Change Impact Assessment of Mountain Agriculture

N. R. Patel, A. Akarsh, A. Ponraj, and Jyoti Singh

18.1 Introduction

Northwestern Himalaya comprises states of Himachal Pradesh, Uttarakhand, and Jammu and Kashmir (J&K). It is a mountain region with varied agroecological situations such as subtropical to temperate climate, undulating topography, and erodible but fertile soils. The uniqueness of geography in the Himalayan region supports diversified farming systems (e.g., terrace cultivation, poultry, horticulture, and livestock raising). In western Himalaya, wheat is the principal crop (36.4% of area) followed by rice (30.7%) and maize (26.3%). Valley bottoms and river basins with assured irrigation water are used for growing rice and wheat as summer and winter crops, respectively. Maize is cultivated on upland slope terraces under rain-fed conditions. However, the area under these food grains sharply declines or remains constant due to the lack of pests and disease-resistant varieties and has caused a shift toward cultivation of fruit and trees. In addition, this Himalayan region has subtropical to temperate climate favorable for growing a wide range of fruits, vegetables, and other cash crops. Small areas with their own microclimatic conditions can provide suitable sites for growing these crops. It experiences sharp variation in climatic variables over very short distances. The vulnerability of the region to human induced disturbance, and extreme climate is well recognized. The Himalayan ecosystem experiences noticeable effects of climate change in the form of rising temperatures, altered precipitation patterns, reduction of rainy days, increase in frequency of droughts, and other biotic influences.

The mountain region has witnessed above-average warming in the twentieth century (IPCC 2007). For example, Himalaya particularly at higher altitudes

N. R. Patel (✉) · A. Akarsh · A. Ponraj · J. Singh
Agriculture and Soils Department, Indian Institute of Remote Sensing (IIRS), Indian Space Research Organisation (ISRO), Department of Space, Government of India, Dehradun, India
e-mail: nrpatel@iirs.gov.in

R. R. Navalgund et al. (eds.), *Remote Sensing of Northwest Himalayan Ecosystems*,
https://doi.org/10.1007/978-981-13-2128-3_18

experiencing warming phenomenon almost three times the global average (Xu et al. 2009; Shrestha et al. 2012). Some other studies also showed much higher warming in the Himalaya than the global level warming during the last hundred years (Du et al. 2004). Historical trend of climate provides ominous sign of climate change in western Himalaya. The mean temperature over a period of 1951–2010 showed a remarkably increasing trend for most seasons, viz., winters (0.02 °C/year), monsoons (0.03 °C/year), and post-monsoon (0.02 °C/year), but not significantly increasing for summers (0.01 °C/year) (Rathore et al. 2013). The consequences of climate change are quite visible in the western Himalayan regions, and they are posing serious implications for the country's food security and economy. Adverse effects of climate change linked to western Himalaya includes reduced water availability, decline in productivity of food grains and apples, shortening of *rabi* season, shift of temperate fruit belt, more extreme weather events (drought/flood), and increased incidence of pests/diseases outbreaks (Negi et al. 2012).

Model-based projections of climate change impacts indicate near certainty that global crop production will decline owing to climate change. Based on a meta-analysis of 1700 model simulations, the most recent IPCC assessment demonstrated that, despite uncertainties, on average, global mean crop yields of rice, maize, and wheat are projected to decrease between 3 and 10% per degree of warming above historical levels. Possible effects of global warming scenarios may be positive or negative depending upon the magnitude of atmospheric CO_2 and temperature changes (Aggarwal 2003). As of now, the literature report reveals that research focus on the causes and impact of climate change-related various sectors (e.g., forests, water, agricultural resources, etc.) is meagre and still in infancy stage in the Himalayan mountains. Agroecosystems of mountain are highly prone to deterioration by various forces of degradation such as water erosion, landslides, and frequent occurrence of extreme events. The natural fragility of these ecosystems makes them highly susceptible to small changes in temperature and water availability. Crop growth and development processes are highly sensitive to changes in temperature and water availability, and as a result the effect of climate change on agriculture in western Himalaya has become reality. In the past few decades, crop models have emerged as promising tool to assess climate change impact on crop yields at station level in mountain region (Rana et al. 2016). These crop models have the ability to simulate crop growth development and crop yield at experimental field scale by means of input variables (e.g., daily meteorological data, soil characteristics, crop parameters, agronomic management options). Crop model projections from these experiments may be inaccurate in diverse mountain region with large degree of heterogeneity in topography, soils, and crop management practices. It is well conceived that modern geospatial tools have great role to play in effective monitoring and management of fragile mountain ecosystem. Many successful applications of these technologies have been made to map and monitor natural resource base and subsequently characterization of agro-environments to improve sustainability of mountain agriculture (Patel 2009).

18.2 Geospatial Technology: Key Roles in Climate Impact Assessment

Climate varies greatly over space particularly in mountain landscape due to influence of altitude and physiographic setting. This climatic variation is further aggravated in recent times as a result of global climate change. The issue of climate change and its consequences on natural resources (agriculture, forest, and water) and livelihood security of Himalayan region assume crucial importance for stakeholders and policy makers operating at local and regional scales. To better address the issue, the geographical analysis of climatic changes and associated impacts in spatially explicit manner has been gaining importance worldwide. This growing trend of projecting climate change (CC) impact on agriculture and forestry sector in particular has been paralleled by the rapid development of satellite imaging, computer-based GIS, and open-source data/tools. Firstly, the geographical information system (GIS) is extensively used to store, analyze, and display both spatial and nonspatial dimensions, and thus it provides spatial dimension to solve problems related to agriculture in mountain ecosystem. Secondly, the land use/land cover, soil information, and digital elevation model (DEM) from satellite data are nowadays easily available at fine-resolution grid from public geo-data portals, e.g., Bhuvan in India (http://www.bhuvan.nrsc.gov.in). Such spatially explicit and geographically representative database helps to incorporate the spatial heterogeneity of landscape units and to drive agroecosystem models on grid basis. The *Fourth Assessment Report* of the Intergovernmental Panel on Climate Change (IPCC) has heavily emphasized the use of crop models within GIS environment for regional climate impact assessments (Easterling et al. 2007). The combined use of GIS and crop models would offer capability to simulate spatiotemporal plant processes, crop growth, and finally yield. It is foreseen that GIS and crop model interface will find more extensive use in climate change impact assessment and agronomic decision-making. Thirdly, the output from global circulation models (GCMs) on various climate scenarios is nowadays easily exported to GIS data formats, and it helps to simulate crop performance over space by using process-based crop models in both current and future climate scenarios (Lobell et al. 2008; Neelin et al. 2006). In addition, an increase in computer processing power has made it possible to undertake agricultural system modeling at a regional scale or global scale. Salient features of integrated technologies, viz., GIS, remote sensing, and GCM-based climate change scenarios, have greatly improved scientific capability to address issue of crop model projections and optimized crop planning in changing climate.

A number of approaches ranging from simple meteorological indices to sophisticated crop models have been employed for agroclimatic risk assessment over the past few decades. Approaches being used to analyze climate change impact on agriculture include (1) agroclimatic indices (e.g., growing degree days/chilling units within geographic information systems (GIS)), (2) multivariate regression models and yield functions, (3) crop simulation models (e.g., Decision Support System for Agrotechnology Transfer (DSSAT) family models), and (4) large area climate-coupled crop model (e.g., general large area model, GLAM). The capability of geospatial tools and models to support climate change impact studies is summarized.

18.2.1 Geographical Information System

A GIS is a computer-based database management system which offers spatial solutions to many problems of agriculture including agroclimatic risk assessment and agronomic decision-making. It has major functionality which includes ability to input, store, manipulate, analyze, and display spatially geo-referenced and point data. It supports integration of spatial and nonspatial data obtained from various sources with other decision support tools such as crop models. In the context of mountain landscape, the GIS has seen widespread use in mapping of land suitability, watershed characterization, and agroecological zonation (Patel et al. 2005; Kumar et al. 2013). The capability of GIS to generate spatial maps of terrain, soils, and other environmental variables makes it a versatile tool to address agricultural problems which are spatially referenced. (e.g., Benson 1996). With availability of continuous environmental variables in gridded format from various GCM models (e.g., HadCM3, CSIRO MK3), it has become easier to implement agroecosystem models on cell-by-cell basis.

18.2.2 Process-Based Crop Models for Climate Impact Assessment

Process-based models are the simplified functions or inbuilt program that simulates behavior of plant as a function of environmental and management conditions (i.e., climate, soils, and management). These dynamic models have emerged as versatile tools for climate impact assessments. These models are designed as decision support tools for agronomic management, especially for optimizing appropriate agronomic management practices, crop yield forecasting, and climate change impact assessment in agriculture. The family of models from ICASA/IBSNAT has been used widely for evaluating climate impacts in agriculture at different levels ranging from individual sites to wide geographic areas (Rosenzweig and Iglesias 1998). These models are structured as DSSAT (Decision Support System for Agrotechnology Transfer) family of models with graphical user interface that allows input and output analysis. Other generic models, viz., WOFOST (WOrld FOod STudies), EPIC (Erosion Impact Productivity Calculator), and WTGROW, were also established based on simplified crop growth functions in order to study crop responses to varying climate, soil, and management (Supit et al. 1994; Sharpley and Williams 1990). EPIC (Erosion Productivity Impact Calculator) is a daily time-step, field-scale model, and it has been used in some climatic impact assessments. Crop simulation models when integrated with GIS showed immense capability to assess the impact of climate change in agricultural systems. These models are modern tools and publicly available for assisting in assessing vulnerability and adaptation to climate change.

18.2.3 GIS-Based Agroclimatic Indices Generation

Agroclimatic indices are in general used as proxies to depict broad picture of climate change assessment and quantify regional or national crop production. These indices are built either from single or a combination of climate parameters (e.g., temperature, precipitation, potential evaporation, etc.) and have significant influence on soil moisture conditions, suitability of crops, drought vulnerability, and crop yields. On global scale, agroecological zone (AEZ)-based methodology and databases for assessment of climate change impact developed jointly by the International Institute for Applied Systems Analysis (IIASA) and Food and Agricultural Organization (FAO) of the United Nations (Fischer et al. 2002). Major agroclimatic indices which are mapped quite easily within GIS environment are accumulated temperature, length of growing period, moisture index, chilling unit accumulation, etc. (Piao et al. 2010; Tian et al. 2014). Many studies reported the use of agroclimatic indices such as accumulated temperature over the growing season to predict crop yields (Holden 2001), advancement or delay in fulfillment of chilling unit requirement for early or late flowing of apple, and shortening of length of growing period in rain-fed conditions due to drought (Tian et al. 2014).

18.3 GIS-Based Crop Modeling and Climate Change Impact Studies

Integration of crop models and GIS has shown enormous potential to address large number of applications related to agricultural decision-making and environmental issues (Hartkamp et al. 1999). As of now several crop models including DSSAT family models, EPIC, and WOFOST are linked to GIS and extensively used in operational applications (Subash and Ram Mohan 2012). Further, coupling crop growth models and satellite data into GIS enables researchers to characterize agro-environment, predict crop yield, and support problem-solving, climatic risk assessment, and decision-making. In India, a conceptual approach of crop growth monitoring system (CGMS) for wheat developed using GIS and WTGROWS simulation model for Haryana, India (Sehgal et al. 2002). A mechanistic WOFOST model has been successfully applied for regional level studies (de Wit et al. 2005; de Wit and van Diepen 2007), including in-season wheat yield prediction in Punjab (India) with assimilation of remote sensing input (Tripathy et al. 2013).

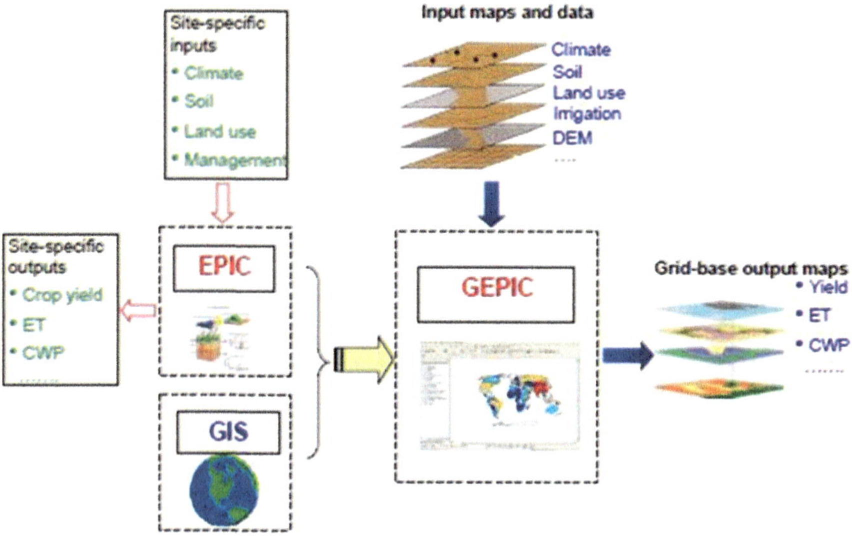

Fig. 18.1 Framework and spatial databases of GEPIC model. (Source: Liu et al. 2007)

18.3.1 Crop Model Framework and GIS Integration

The EPIC crop model is designed especially for simulating biophysical and biogeochemical processes which cover growth, productivity, soil erosion, and carbon sequestration rates and water balance at experimental scale. The EPIC model is parameterized for several food crops and extensively used to predict yields of crops (e.g., wheat, corn, rice, soybean, sunflower, etc.) at several sites across the world (Williams et al. 1989). The GEPIC is a GIS-based interface of a biophysical EPIC model (Environmental Policy Integrated Climate) and aimed at simulation of spatiotemporal dynamics of biophysical and hydrological processes occurring within agricultural system (Liu et al. 2007). GEPIC is developed based on simple coupling that relies on exchanges of input and output files between GIS and simulation models (Huang and Jiang 2002). The EPIC model is designed to simulate crop-related processes for specific sites with site-specific inputs of weather, soils, and management information. However, GEPIC has versatile capability for assessing agricultural systems at regional, continental, and global scale. The framework of spatial database within the GEPIC model is expressed in Fig. 18.1. In the GEPIC model, ArcGIS is mainly used for inputing, editing, and visualizing of spatial input drivers and simulated output maps. As part of application framework, ArcGIS has inbuilt VBA programming which provides capability to design the interface of GEPIC with subroutine of input data access, text output data generation, and output map creation. In GEPIC setup, the area or administrative region is divided into several grids of uniform size and model treats each grid cell as a site. It simulates the crop-related processes for each predefined grid cell with spatially distributed inputs. The inputs are provided to the model in terms of GIS raster maps as well as text files. Necessary maps include land use maps, elevation and slope maps, irrigation maps, fertilizer maps, climate code maps, and soil code maps. The land use maps provide information on crop distribution (code 0 indicates the absence of a specific crop, while 1 and 2 indicate existence of the crop under rain-fed and irrigated conditions, respectively). The elevation and slope maps show the average elevation and slope in each grid cell. The irrigation and fertilizer maps show the annual maximum irrigation depth and fertilizer application rate. The climate and soil code maps indicate the code numbers of the climate and soil files in each grid cell. These code numbers correspond to the text files of climate and soil data.

Climate files contain daily weather data (e.g., daily precipitation, daily minimum and maximum temperatures) and monthly weather statistics. Soil files contain several soil parameters (e.g., soil depth, percent sand and silt, pH, organic carbon content, etc.). Annual irrigation and fertilizer inputs are provided in irrigation and fertilizer maps. Model parameters were selected from the literature (Doraiswamy et al. 2003; Ren et al. 2010). The outputs of the GEPIC model are raster GIS maps representing the spatial distribution of output variables such as crop yield and evapotranspiration.

18.3.2 Climate Change Impact on Productivity of Food Grains in Dehradun Valley

Crop models are the most widely accepted tools to investigate effects of climate change on crop productivity. Henceforth, we tested the applicability of GIS-based GEPIC model with spatially explicit geo-databases on soils, weather, terrain, and crop land use information. This pilot study was attempted to investigate magnitude of changes in crop productivity of major crops (rice, wheat, and maize) in an agricultural landscape of a mountain ecosystem in Dehradun Valley, Uttarakhand. The selected study area, namely, Doon Valley in Uttarakhand, is a perfect example for a mountainous agroecosystem, due to its vulnerability toward extreme climatic conditions. GEPIC model was first parameterized for local conditions based on field observations of leaf area index and crop yield of three crops at selected nine sites. Sensitivity analysis of key model parameters was analyzed, and later best parameters have been fixed to run model for over regional scale. The site-specific calibration and validation result shows the capability of the model to predict crop productivity based on local conditions. The calibration and validation of the model are performed for rice and wheat crop in the study area with an RMSE of 0.38 t ha^{-1} and modeling efficiency of 0.78 for rice. LAI correction strategy is performed for wheat crop calibration, which in turn led to RMSE of 0.24 t ha^{-1} and a modeling efficiency of 0.88.

The applicability of the model to predict impact of climate change on crop productivity was investigated in the study area on grid basis. The monthly climate

Table 18.1 Spatial and nonspatial input databases of GEPIC model

Sl No.	Input data	Parameters	Source
1	Climate (current and future scenario)	Tmin, Tmax, precipitation	World Clime (1 × 1 km)
			Baseline (1960–90)
			HadCM3- A2a & B2a Scenario data for 2020, 2050, 2080
2	Soil	Depth, texture, OC, CaCO3, pH, EC,CEC, bulk density, coarse fragment	1:50000 scale (NR-Census Report, Doon Valley)
			Field measurement and laboratory analysis
			1:250,000-NBSS-LUP soil map +68 soil pedon databases (Himachal Pradesh)
3	Management	Date of planting and harvesting, fertilizer use, tillage practices	Field observations and queries to farmers
4	Topography	Elevation, slope, geolocation	DEM (SRTM 90 m)
5	Crop	Crop area and distribution, LAI	LULC map

data (maximum temperature, minimum temperature, precipitation, and number of rainy days) with a spatial resolution of 30 arc seconds (~1 km) obtained from WorldClim global climate data archive for baseline period (1950–2000) and future three time frames, the 2020s (2010–2039), 2050s (2040–2069), and 2080s (2070–2099), were used for climate change impact assessment. The soil physiographic units are integrated with major soil series, and codes are assigned to each series. This consists of layer-wise information such as soil pH, EC, bulk density, sand, silt, clay, soil organic carbon, rock fragment, and $CaCO_3$.The digital elevation model (DEM) of SRTM (Shuttle Radar Topography Mission) data with a spatial resolution of 90 m was acquired from USGS earth explorer archive. Elevation and slope map for the study area were prepared from this dataset. All these datasets such as LULC, soil, climate, elevation, and slope maps were extracted based on 1 km × 1 km grid cells and used as input to EPIC model (Table 18.1). Due to large variability in topography, climate, and soils in the study area, simulations are performed with a spatial resolution of 1 km × 1 km by incorporating same resolution datasets under baseline conditions and future scenarios.

The results on simulation revealed distinct spatial variability in modeled yield of crops during baseline and changes during three future time frames of climate change scenarios (Fig. 18.2). It could be inferred that rice yield will likely decline under A2a scenario without CO_2 fertilization with magnitude of yield reduction ranging from -1.03% in the 2020s to -19.22% in the 2080s.

However under A2a scenario with CO_2 fertilization, the rice crop will witness an improvement in yield to the tune of 5% till 2050, but later in the 2080s, there will be a slight reduction in yield by 2.4%. Under B2a scenario both in the 2020s and 2050s, there is possibility of much higher reduction in rice yield by 3% because the rise in atmospheric CO_2 concentration in the atmosphere may not be sufficient to

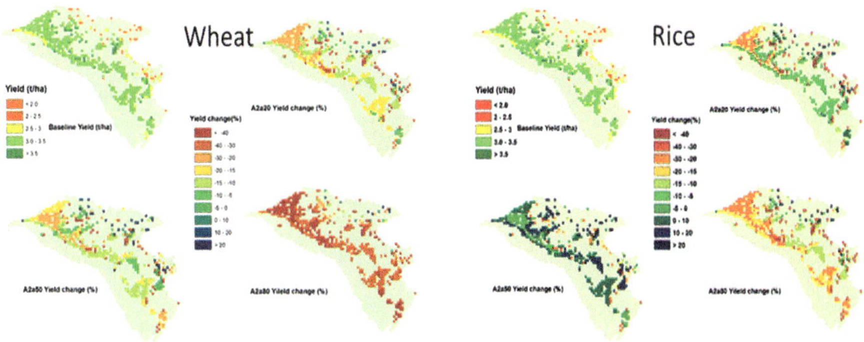

Fig. 18.2 Changes in productivity of rice and wheat under various scenarios without CO_2 fertilization

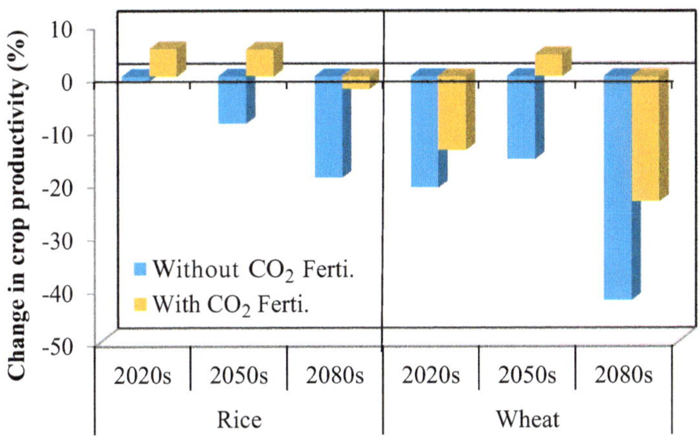

Fig. 18.3 Percent changes (negative/positive) of crop productivity in Doon Valley

compensate the loss due to increase in temperature. However, in the 2080s, there could be a slight improvement in yield of about 1.3%. To assess the vulnerability, wheat crop simulations were performed assuming unirrigated conditions. Under A2a scenario without CO_2 fertilization, the results showed a large reduction in wheat yield with magnitude of 42% in the 2080s. The vulnerability of wheat crop to climate change is less if we consider CO_2 fertilization, as evident from improvement in yield of about 4% in 2050 due to increase in rainfall (+20%). But in the 2020s and 2080s, there could be a reduction in productivity of about 13% and 23%, respectively. The results showed that even increasing CO_2 would not be able to offset losses incurred due to extreme rise in temperature in the 2080s. Decline in crop yields (%) in the 2080s without CO_2 fertilization for maize, rice, and wheat was found to be in the order of 14, 19, and 42%, respectively, in Doon Valley (Fig. 18.3).

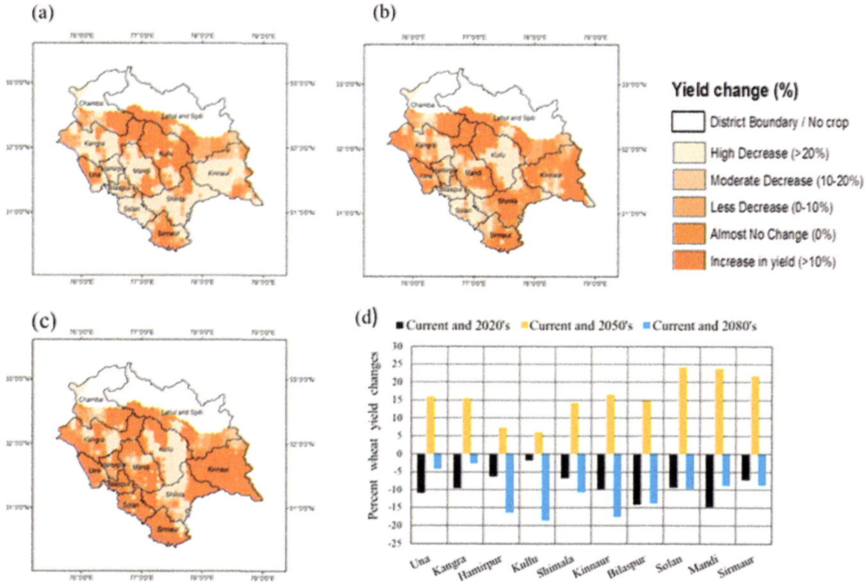

Fig. 18.4 Spatially pattern of climate change impact on productivity of wheat in different scenarios: (**a**) 2020s, (**b**) 2050s, (**c**) 2080s, and (**d**) corresponding district level changes

18.3.3 Climate Change Impact on Wheat Productivity in Himachal Pradesh

Impact of climate change on productivity of major food crops is well understood at district level, but studies on climate change impact on crop's productivity over high-resolution grids are limited. In this study, a high-resolution gridded climate data (statistically downscaled HadGM2 GCM) are used from WorldClim. We used GIS-based crop model, namely, GEPIC (GIS-based Environmental Policy Integrated Climate model), to simulate the productivity of wheat during baseline (1960–1990) and future climate periods (2020s, 2050s, and 2080s) under A2a scenario of IPCC. The spatial data in study area is divided into homogeneous grids. The grid is taken as the basic simulation unit, and each grid has a complete set of input data (meteorological, soil, cultivar, and management). The spatial monthly climate data of 30 arc seconds (~1 km) for the baseline period and future projected scenarios were obtained from WorldClim global climate data archive (Hijmans et al. 2005). The GEPIC model runs on 1 km grid with monthly weather inputs comprising maximum temperature, minimum temperature, solar radiation, and rainfall parameters. Similarly static terrain and management databases (soil, elevation, slope, land use, irrigation, and fertilizer) were created in grid format and used for GEPIC simulation.

Simulated wheat yields in Himachal Pradesh were found to be declined in the 2020s and 2080s as evident from negative changes in the 2020s and 2080s (Fig. 18.4). In addition, it showed that wheat yield could be increased in 2050s due to positive impact of increased atmospheric CO2 and favorable rainfall conditions. Decline of wheat yields in 2020s and 2080s is 9 and 11%, respectively. However, wheat yields in HP as a whole will be increased by 15% in 2050s as compared to baseline period.

18.4 Climate Change Impact on Land Suitability of Crops in Himachal Pradesh

Multi-criteria analysis was done for identifying suitability of land for maize and wheat crop in baseline (1960–1990) and future scenarios (2020s, 2050s, and 2080s). Extensive literature exists on the combination of analytical hierarchal process (AHP) and GIS in various disciplines (Malczewski 2006). Ying et al. (2007) combined AHP with GIS to provide an effective means for studies of regional eco-environmental evaluation. AHP (Saaty 1990) is a multi-criteria decision-making method based on pairwise comparisons for elements in a hierarchy. It decomposes problems in a hierarchical structure and explicitly incorporates the decision-maker's expertise/experience in AHP evaluation. The approach included creation of a spatial database and its integration in GIS environment by developing a suitable rating and ranking scheme for the generation of spatially explicit information on land suitability for major crops in Himachal Pradesh, India (Fig. 18.5).

The factors identified for land suitability analysis were climate (temperature and precipitation), soil (soil texture, soil depth, soil pH), and topography (slope). The spatial databases on climatic parameters (monthly precipitation and maximum/minimum temperatures) at 1 × 1 km grid resolution during baseline period (1960–1990) and HadCM3-A2a future climate scenarios during the 2020s (2010–2039), 2050s (2040–2069), and 2080s (2070–2099) were obtained from world climate data archive (Hijmans et al. 2005). The spatial information on classes of soil texture, depth, pH, and slope were derived assigning attributes to digital soil map (1:250000 scale) obtained from National Bureau of Soil Survey and Land Use Planning (NBSS-LUP).

18.4.1 Impact of CC on Agroclimatic Suitability of Maize

In the context of precipitation factor for maize suitability, the highly suitable areas are situated in the lower parts of Himachal Pradesh in baseline period 1960–1990, but in the future scenarios, the highly suitable areas are shifting toward the upper parts of the state. Similarly, areas (Lahaul and Spiti districts) unsuitable due to inadequate rainfall in baseline period are becoming marginally to moderately suitable for maize production because of increase in amount of precipitation. In the 2020s, highly suitable areas are becoming moderately suitable areas due to

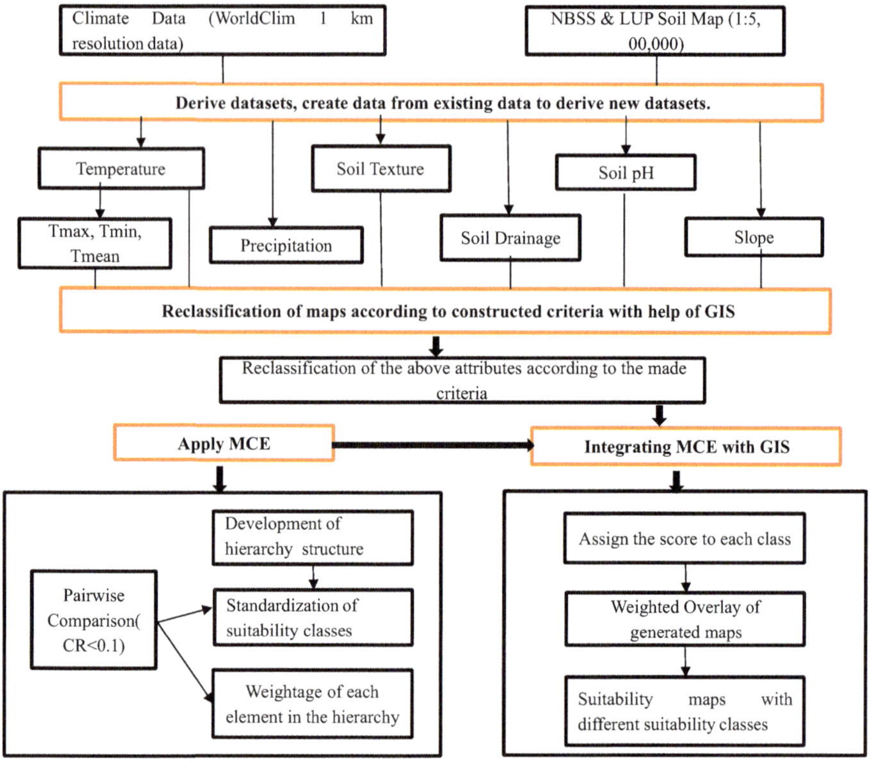

Fig. 18.5 Methodology for agroclimatic suitability using multi-criteria approach

precipitation constraint. With respect to temperature rising in future scenarios, the highly suitable areas of maize are decreasing because of increase in temperature in lower parts of Himachal Pradesh, whereas the upper areas of Himachal Pradesh which are not suitable (cooler temperature limiting maize growing) for maize production will become suitable because of increase in temperature in future scenarios. The overall suitability based on precipitation, temperature, and terrain (soil/topography) information reveals that suitable areas are increasing from baseline period to 2080s (Fig. 18.6). Hence, we can say land suitable for cultivating maize will increase, but their production potential will decrease in future.

18.4.2 Impact of CC on Agroclimatic Suitability of Wheat

Criteria related to precipitation in Himachal Pradesh were considered as non-limiting criteria since wheat is irrigated crop unlike maize. However, suitable area is increasing from current climate to projected climate up to the 2050s, but later in the 2080s, it

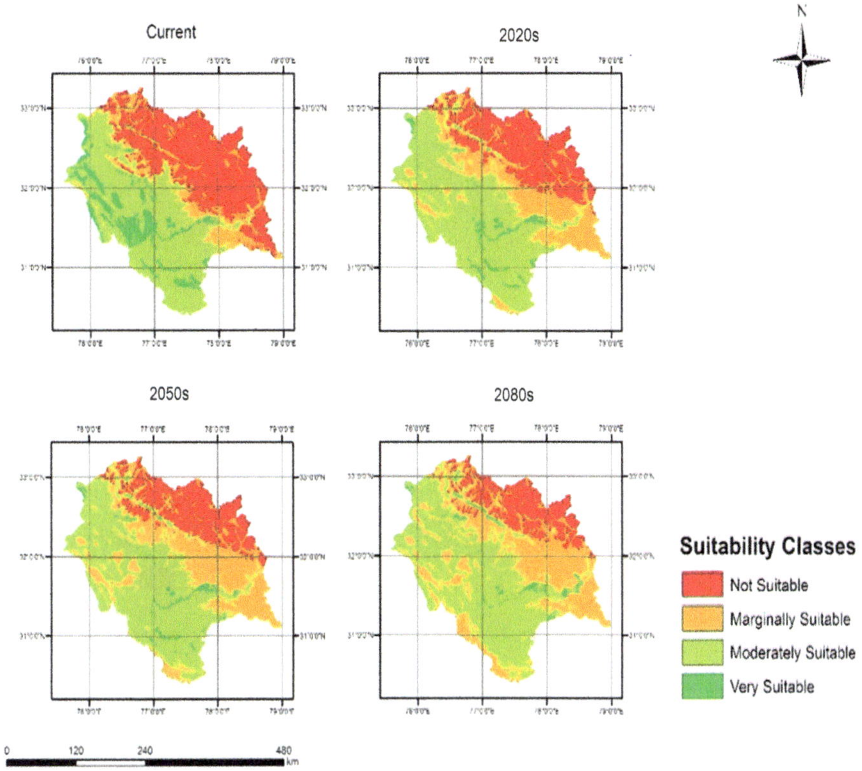

Fig. 18.6 Agroclimatic suitability of maize in baseline and future CC scenarios

decreases drastically. In addition to this, it appeared that marginally suitable areas of wheat showed shift to a class of moderately suitable areas in the 2080s.

According to WorldClim data, temperature is increasing gradually from current scenario to projected climate (2020, 2050, and 2080) but still not up to the extent that a large area can become suitable for cultivation of winter wheat. Agroclimatic suitability analysis showed that the highly suitable areas that lie in northwest part of the state are areas having high temperature, precipitation, soil, and topographic suitability (Fig. 18.7). Northern and eastern parts of Himachal Pradesh are not suitable according to precipitation, temperature, soil, and topography.

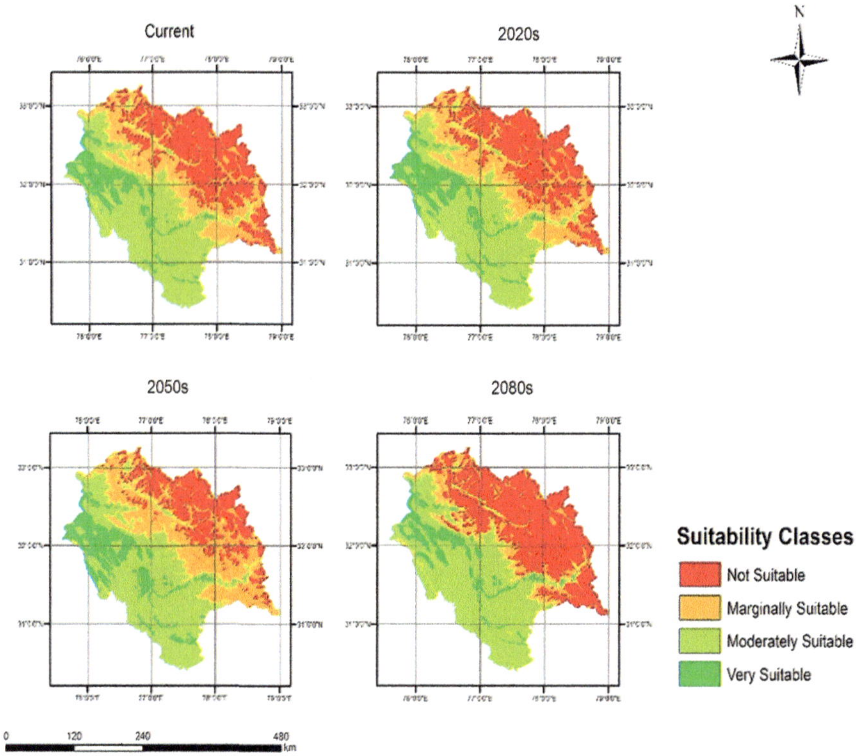

Fig. 18.7 Agroclimatic suitability of maize in baseline and future CC scenarios

18.5 Impact of CC on Apple Growing Areas in Himachal Pradesh

Horticulture crops play a significant role in the food and nutritional security of the country. These crops are gaining importance in hilly and undulating terrains of NW Himalaya. Apple is the most important temperate fruit of the Northwestern Himalayan region in India. It is predominantly grown in Jammu and Kashmir, Himachal Pradesh ,and some parts of Uttarakhand, accounting for about 90% of the total production. Temperate fruit/orchards in India are most vulnerable targets of climate change, as temperature is one major factor that affects the fruit yield in these (Awasthi et al. 2001). Adverse impact of climate change in the western Himalayas on apple-producing areas and productivity is being reported in recent decades (Vedwan and Rhoades 2001; Chaudhary and Bawa 2011; Rana et al. 2009). Kullu Valley, the apple belt of the state, has gone under sharp decline of apple productivity, and apple orchards had been replaced by vegetables and fruit varieties (with lower chilling requirements) in several areas.

Areas previously unsuitable because of very low temperatures are becoming suitable for apple cultivation like Lahaul and Spiti. Apple cultivation is collapsing in lower altitudes and expanding toward upper altitudes (Rana et al. 2009). Effort is made here to study apple cultivation areas and shift in apple zones under changing climate and time variability (36 years). Thus, various chilling zones, suitable classes for apple cultivation, and shift in apple belt are spatially mapped in the entire Himachal Pradesh.

18.5.1 GIS-Based Chill Unit Model

Apple crop require 1000–1500 h of chilling units below 7 °C during winter to break the dormancy period. During this dormant period, plants cannot resume growth until the chilling requirement is met. Point-based chill unit models have been widely to calculate effective chilling units (ECU) for the weather station or given point for assessing apple suitability. However, addressing the shift in apple suitability over a large area, a spatial calculation and analysis of ECU is required. The use of geospatial technique and simulation model was used successfully earlier in modeling ecological niche of apple orchards in Jammu and Kashmir and in Himachal Pradesh (Panigrahi et al. 2014). They have reported that IRS-LISS III is better suited to map apple orchards and investigated the changes in distribution of apple areas under climate change scenarios by using GARP model and GIS. The GIS technology was employed to map land suitability of apple in Uttarakhand by integrating thematic information (e.g., average temperature, Precipitation Distribution Index, chilling hours, and soil depth) using the weighted overlay method (Dhami et al. 2012). In this study, a simple GIS-based chill unit model was developed to calculate ECU (effective chill unit) on grid basis over Himachal Pradesh. This model consists of two broad steps: (1) synthesis of hourly temperature from daily maximum and minimum temperature and (2) ECU calculation. Inputs for the model are daily maximum temperature and daily minimum temperature. The daily data on maximum and minimum temperature for a period of the 1972–2013 were used as an input to the UTAH chill unit model (Richardson et al. 1974). According to the UTAH model, most effective temperature range which contributes in dormancy release is between 1.5 and 12.4 °C. However, the effectiveness of chilling hour is different for each range of temperature and is weighted according to criteria of UTAH model. Effective chill unit for the whole season is calculated by adding by adding C_U for each hour from October to March of next year.

The suitability assessment for apple is done on the basis of the chilling requirements which clearly reflect the importance of winter temperatures (Singh and Patel 2017). The threshold for the range of ECU indicating different suitability classes was derived from various literatures and expert advice of horticulture scientists from Himachal Pradesh. These suitability classes are assigned according to chilling zones; areas having ECU between 800–1000 are mapped as highly suitable. Parts of Kullu, Chamba, Kangra, Lahaul and Spiti, and Kinnaur districts come under very suitable

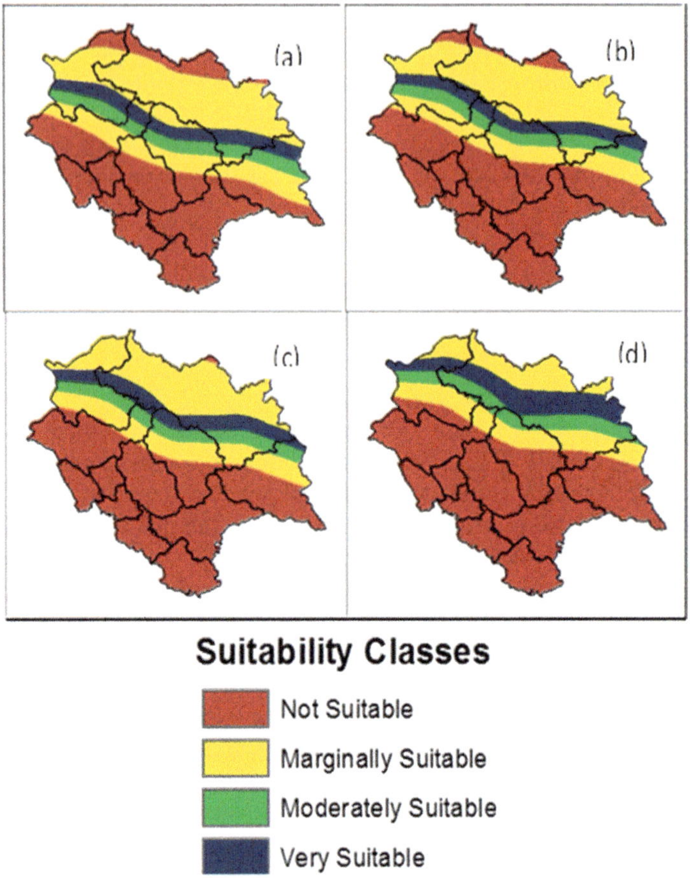

Fig. 18.8 Changes in apple suitability in Himachal Pradesh over climatic periods (**a**) 1978–86, (**b**) 1987–1995, (**c**) 1996–2004, and (**d**) 2005–2013

zone in the first decade, but in the fourth decade (2005–2013), the Kullu and Kangra districts are no more in very suitable zone. Moderately suitable class contains two chilling zones, one with lower ECU (600–800) then very suitable zone and the other one with higher ECU (1000–1500). Marginally suitable classes contains single chilling zone (300–600) which is below very suitable zone as well as moderately suitable zone. This area has shifted upward causing lower areas to become unsuitable in the progressing decades (Fig. 18.8). Unsuitable zone with ECU < 300 increased continuously from the first to fourth decade; in the first decade, it consists of only lower portion of Kullu district, but by coming up to the fourth decade, this region occupied more than half of the area of Kullu district. Sustaining productivity of apples in low-lying areas of Himachal Pradesh, farmers can be advised to opt for the plantation of low chill cultivars (Gala, Spur, Vance, etc.) in orchards which are less sensitive to climatic changes (Rana et al. 2016). As second option of adaptation strategy, farmers choose shifting to cultivation of other fruit crops (e.g., pears, pomegranate, etc.) and commercial vegetables which fetched higher market prices.

18.6 Conclusions

This chapter presents an overview of geospatial technology and its role in assessing climate change impacts on agriculture in the mountain ecosystem which is more or less constrained in terms of climate observing network and resources availability. Major emphasis is laid on exploiting complementary power of geographical information system (GIS) and crop simulation models to strengthen climate change research and applications for decision-making in mountain agriculture. Few successful applications of geospatial tools (GIS and crop model) for predicting crop yield responses to climate change and possible shift in agroclimatic suitability of some dominant crops of Northwestern Himalaya are discussed. Integrated use of climate scenario from global circulation model (GCM), crop models, and geospatial data (e.g., topography, soil, and land cover) within GIS could improve spatial representation of climate-induced projection of crop productivity and land suitability changes in future climate scenarios. Crop model projection in Doon Valley showed negative changes in productivity of major food grain crops in extreme climate change scenarios. Crop yields of major food grain crops, viz., rice, wheat, and maize, appeared to decline in the 2080s under CO2 fertilization to the tune of 14, 20, and 40%, respectively, in Doon Valley. However, the GEPIC model simulation for Himachal Pradesh showed wheat yield decline in the 2020s and 2080s, but wheat yield was likely to be increased in the 2050s compared to baseline (1960–1990) period. Multi-criteria evaluation within GIS in Himachal Pradesh depicts that overall suitable areas for summer maize and winter wheat increasing trend in future climate. Chill unit accumulation has decreased significantly from 1997 to 2013, and this decreasing trend continued to affect apple cultivation. In general, suitable areas for apple cultivation are shifting toward high altitude in Himachal Pradesh due to changing climate.

References

Aggarwal PK (2003) Impact of climate change on Indian agriculture. *J. Plant Biology* 30(2): 189–198.

Awasthi RP, Verma HS, Sharma RD, Bhardwaj SP and Bhardwaj SV (2001) Causes of low productivity in apple orchards and suggested remedial measures. In: K.K. Jindal and D.R. Gautam (Ed.), Productivity of temperate fruits, Dr YSParmar University of Horticulture and Forestry, Solan, HP. pp. 1-8

Benson T (1996) The use of Geographic Information Systems in agricultural research in Malawi. SoilFertNet Network Methods Working Paper 1. International Maize and Wheat Improvement Center (CIMMYT), Mexico, DF.
Chaudhary P, Bawa KS.(2011) Local perceptions of climate change validated by scientific evidence in the Himalayas. Biology Letters, rsbl20110269
de Wit AJW, Boogaard HL, van Diepen CA (2005) Spatial resolution of precipitation and radiation: the effect on regional crop yield forecasts. Agricultural and Forest Meteorology, 135: 156–168
de Wit AJW and van Diepen CA (2007) Crop model data assimilation with the Ensemble Kalman filter for improving regional crop yield forecasts. Agricultural and Forest Meteorology, 146: 38–56
Dhami J, Roy S, Nain AS and Panwar R (2012)Suitability analysis of apple and pear using remote sensing and GIS in Uttarakhand. Journal of Agrometeorology, 14:464-474.
Doraiswamy PC, Moulin S, Cook PW,Stern A (2003) Crop yield assessment from remote sensing. Photogramm. Eng. Remote Sens. 69: 665–674.
Du MY, Kawashima S, Yonemura S, Zhang XZ and Chen SB(2004)Mutual influence between human activities and climate change in the tibetan plateau during recent years. Global and Planetary Change, 41: 241-249
Easterling WE, Aggarwal PK, Batima P, Brander KM, Erda L, Howden SM, Kirilenko A, Morton J, Soussana JF, Schmidhuber J,Tubiello FN (2007) Food, fibre and forest products. In: Parry ML, Canziani OF, Palutikof JP, van der Linden PJ, Hanson CE (eds) Climate Change 2007: Impacts, Adaptation and Vulnerability. Contribution of Working Group II to the Fourth Assessment Report of the Intergovernmental Panel on Climate Change. Cambridge University Press, Cambridge, UK, pp. 273–313.
Fischer G, van Velthuizen H, Shah M and Nachtergaele F (2002) Global Agro-Ecological Assessment for Agriculture in the 21st Century : Methodology and Results. IIASA: Laxenburg, Austria.
Hartkamp AD, White JW and Hoogenboom G (1999) Interfacing Geographic Information Systems with agronomic modeling: a review. Agronomy Journal, 91: 761–772.
Hijmans RJ, Cameron SE, Parra JL, Jones PG, Jarvis A (2005) Very high resolution interpolated climate surfaces for global land areas. International Journal of Climatology 25 :1965–1978
Huang B and Jiang B (2002). AVTOP: a full integration of TOPMODEL into GIS *Environmental Modelling & Software*, 17(3): 261-268.
IPCC(2007) Intergovernmental Panel on Climate Change (IPCC) Fourth Assessment Report: Climate Change 2007. IPCC, Geneva. Available at: http://www.ipcc.ch/
Kumar S, Patel NR, Sarkar A, Dadhwal VK (2013) Geo-spatial approach of agro-climatic suitability of soybean in Rainfed agroecosystem. J. Indian Soc. of Remote Sensing 41:609-618.
Liu J, Williams JR, Zehnder AJ, Yang H (2007) GEPIC–modelling wheat yield and crop water productivity with high resolution on a global scale. Agric. Syst. 94: 478–493
Lobell DB, Burke MB, Tebaldi C, Mastrandrea MD, Falcon WP, Naylor RL (2008) Prioritizing climate change adaptation needs for food security in 2030. Science 319: 607–610
Malczewski J (2006) GIS based multicriteria decision analysis: a survey of the literature. Int J GeogrInfSci 20(7):703–726
Neelin JD, Munnich SU, Meyerson MH, Holloway JE (2006) Tropical drying trends in global warming models and observations. Proceedings of the National Academy of Sciences, 103(16) : 6110–6115.
Negi GCS, Samal PK, Kuniyal JC, Kothyari BP, Sharma RK and Dhyani PP (2012) Impact of climate change on the western Himalayan mountain ecosystems: An overview. Trop. Ecol., 53 (3):345-356.
Patel NR, Endang P, Kumar S, PandeLM (2005) Agro-ecological zoning using remote sensing and GIS – A case study in part of Kumaon region. In: Sustainable agriculture development, (Eds) B. Bandopadyay, KV Sundaram, M. Moni and M. Zha (Eds), Northen Book Depo, New Delhi. pp. 265-280.

Patel NR (2009) Geo-information for Sustainable Development of Agriculture – Some Examples of North India. In: Geoinformatics for Natural Resource Management. Joshi et al (Eds), NOVA Science Publishers pp: 291–304.
Panigrahi S, Singh CP and Parihar JS (2014)Geospatial Planning Towards Climate Change Impact on The Apple (Malus domestica) Orchards in India. eJournal of Applied Forest Ecology, 2(1):1-8.
Piao S, Ciais P, Huang Y, Shen Z, Peng S, Li J, Zhou L, Liu H, Ma Y and Ding Y (2010) The impacts of climate change on water resources and agriculture in China.Nature, 467: 43–51.
Rana RS, Bhagat RM, Kalia V, Lal H (2009) Impact of climate change on shift of apple belt in Himachal Pradesh. In Proceedings of the workshop on Impact of Climate Change on Agriculture, SAC, Ahmedabad, pp. 17–18
Rana RS, Kalia V, Pathania RR, Singh KK (2016) Managing impacts of extreme weather event on crop productivity of mountain agriculture. Mausam 67(1): 223-232
Rathore LS, Attri SD and Jaswal AK (2013. State level climate change trends in India. Meteorological Monograph No. ESSO/IMD/EMRC/02/2013
Ren J, Yu F, Chen Z, Qin J(2010) Regional yield estimation of summer maize based on assimilation of remotely sensed LAI into EPIC model, in: Geoscience and Remote Sensing (IITA-GRS), 2010 Second IITA International Conference on. IEEE, pp. 361–365.
Richardson EA, Seeley SD, Walker DR (1974) A model for estimating the completion of rest for Redhaven and Elberta peach trees.HortScience 9 (4): 331–332
Rosenzweig C, Iglesias A (1998) The use of crop models for international climate change impact assessment. In Understanding Options for Agricultural Production, G.Y. Tsuji, G. Hoogenboom, and P.K. Thornton (eds.). Kluwer Academic Publishers, Dordrecht, The Netherlands, pp. 267-292.
Saaty TL (1990) How to make a decision: the analytic hierarchy process. European journal of operational research 48(1): 9-26
Sehgal VK, Rajak DR, Chaudari KN and Dadhwal VK (2002) Improved regional yield prediction by crop growth monitoring system using remote sensing derived crop phenology. In Proceeding of ISPRS Technical Commission VII Symposium on Resource and Environmental Monitoring held on 3–6 December, 2002 at Hyderabad.
Sharpley, A.N. and Williams, J. R. (1990).EPIC – Erosion/Productivity Impact Calculator. 1. Model Documentation. U.S. Department of Agriculture Technical Bulletin No. 1768.
Singh J and Patel NR (2017) Assessment of agro-climatic suitability of apple orchards in Himachal Pradesh under changing climate. Journal of Agrometeorology, 19(2): 110-113.
Subash N and Ram Mohan HS (2012) Evaluation of the impact of climatic trends and variability in rice–wheat system productivity using Cropping System Model DSSAT over the Indo-Gangetic Plains of India.Agricultural and Forest Meteorology, 164:15
Supit I, Hooijer AA, van Diepen CA (eds.). 1994. System description of the WOFOST 6.0 Crop Simulation Model Implemented in CGMS. Volume 1: Theory and Algorithms. EUR 15956, Office for Official Publications of the European Communities, Luxembourg.
Tian Z, Yang X, Sun L, Fischer G, Liang Z and Pan J (2014) Agroclimatic conditions in China under climate change scenarios projected from regional climate models. Int. J. Climatol. 34: 2988–3000.
Tripathy R, Chaudhari KN, Mukherjee J, Ray SS, Patel NK, Panigrahi S and Parihar JS (2013) Forecasting wheat yield in Punjab state of India by combining crop simulation model WOFOST and remotely sensed inputs, RemoteSensing Letters, 4 (1): 19-28.
Vedwan N, Rhoades RE (2001) Climate change in the Western Himalayas of India: a study of local perception and response. Climate Research 19(2) :109-117
Williams JR, Jones CA, Kiniry JR and Spanel DA (1989) The EPIC crop growth model. *Transaction of the ASAE*, 32: 497-511.
Xu J, Grumbine R E, Shrestha A, Eriksson M, Yang X, Wang Y and Wilkes A (2009) The melting Himalayas: Cascading effects of climate change on water, biodiversity, and livelihoods, Conservation Biology, 23: 520-53.

Ying X, Zeng GM, Chen GQ, Tang L, Wang KL, Huang DY (2007) Combining AHP with GIS in synthetic evaluation of eco-environment quality—a case study of Hunan Province, China. Ecol Model 209(24):97–109

Shrestha, U. B., Gautam, S., Bawa, K. S. (2012). Widespread climate change in the Himalayas and associated changes in local ecosystems. PLoS One, 7(5), e36741.

Holden, N. M. (2001). Modelling concepts. Agro-meteorological modelling–principles, data and applications. Dublin: Agmet, 1–22.

Part VI
Urban Environment

Summary

The Himalayan towns and cities have been stressed beyond sustainable limits due to population migration from rural areas, in search of better livelihood and basic amenities. The generic ills that plague Indian urban sector also applies to these Himalayan urban settlements in equal potency with added constraint of limited land and financial resources. As a consequence these cities are facing lack of adequate physical and social infrastructure resulting in untreated sewage, seasonal water shortages, heaps of trash, increased air pollution levels, etc. Since, urban areas are a complex system with myriad dimensions, it is imperative to adopt a holistic approach while addressing urban issues. Studies presented in this section attempt to address various dimensions of the urban system viz. population dynamics, unregulated urban growth, effects on urban environment, urban vulnerability to hazards, alternate energy sources and the exposure of urban population to various vector borne diseases. The studies have been carried out for the North-Western Himalaya (NWH) region consisting of three states viz. Uttarakhand, Himachal Pradesh and Jammu & Kashmir. The geospatial database required for the study was generated primarily using remote sensing datasets, while the analysis was carried out in GIS environment.

The first section describes the urban population dynamics in three NWH states and subsequently, the settlement pattern and growth dynamics of urban areas were analysed using geospatial techniques. The future urban growth scenarios were generated using an Artificial Neural Networks (ANN) based simulation model. The chapter also elaborates on ISRO's geoportal *Bhuvan*, National urban Information system (NUIS) and night-time light datasets application in urban studies.

The second section, discusses the characteristics of urban environment in terms of natural and built-up spaces. The application of remote sensing data, geographical information system and Global Navigation Satellite System (GNSS) in extraction of built-up typography, urban heritage and regeneration studies has been illustrated.

The exposure of urban population to various air pollutants emitted as a result of vehicular activities and the vulnerability of urban population to seismic hazards has also been elaborated. The section also discusses various sources of renewable energy, which can be utilized for meeting the spirally increasing power demand of urban area. The third section details with various Vector-borne diseases prevalent in Uttarakhand region along with parameters that propagate these diseases. The potential application of geospatial technology in developing a spatial information infrastructure, which provides the necessary data integration framework to combine information on human environment interactions from variety of sources has been emphasized.

The three sections, have thus attempted to address the various dimensions of urban environment using geospatial technology, which should provide valuable inputs to the urban planners for ensuring a sustainable urban environment.

Chapter 19
Understanding Urban Environment in Northwest Himalaya: Role of Geospatial Technology

Pramod Kumar, Asfa Siddiqui, Kshama Gupta, Sadhana Jain, B. D. Bharath, and Sandeep Maithani

19.1 Introduction

An urban area is a complex ensemble of interrelated and entangled socio-economic, spatial and environmental processes. The irreversible demographic changes being witnessed in an urban set-up pose threat to the rapidly depleting resources. Therefore, the study of urban footprints and the services with associated infrastructure is a subject that intrigues urban planners. Although planning has more to say about cities, it is equally imperative to explore all perspectives of a holistic regional development keeping in view the social, economic, cultural, environmental and governance issues. Interestingly, with the advent of smart city concept, the conceptualization of six key enablers of the mission, viz. smart governance, smart living, smart people, smart mobility, smart environment and smart economy, finds its complete implementation scope in spatial planning aided with information and communication technology (ICT) solutions. Various pillars of smart planning of an urban set-up ranging from infrastructure to resource management and smart energy to smart environment are achievable with the help of geospatial technologies.

Geospatial technology, i.e. a system involving remote sensing (RS), geographic information system (GIS) and global navigation satellite system (GNSS), plays a

P. Kumar (✉) · A. Siddiqui · K. Gupta · S. Maithani
Urban & Regional Studies Department, Indian Institute of Remote Sensing (IIRS), Indian Space Research Organisation (ISRO), Dehradun, India
e-mail: pramod@iirs.gov.in

S. Jain
Regional Remote Sensing Centre – Central, National Remote Sensing Centre, Indian Space Research Organisation (ISRO), Nagpur, Maharashtra, India

B. D. Bharath
National Remote Sensing Centre, Indian Space Research Organisation (ISRO), Hyderabad, India

R. R. Navalgund et al. (eds.), *Remote Sensing of Northwest Himalayan Ecosystems*,
https://doi.org/10.1007/978-981-13-2128-3_19

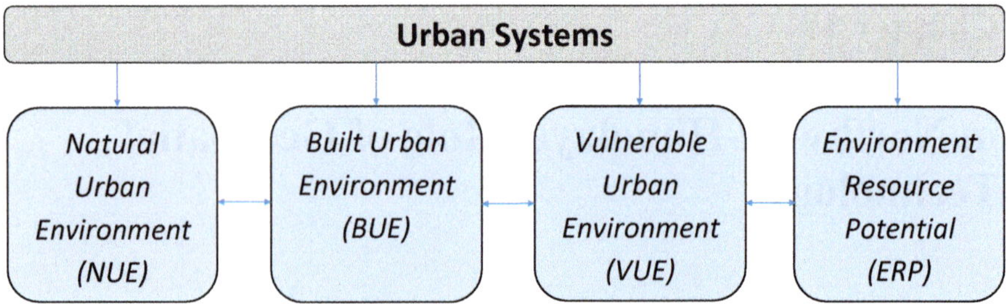

Fig. 19.1 Urban systems and its components

pivotal role in analysing the numerous facets of urban environment. It helps us in monitoring green spaces, understanding the climatological concepts within the region, pollution monitoring, identifying suitable sites for infrastructure, mapping and delineating infrastructure deficit, identifying potential zones for harnessing energy, hazardous site mapping, etc.

The dynamics of urban areas, especially the ones situated in the Himalayan terrain, require prompt responsiveness for their sustainable development. In such a scenario, understanding the facets of an urban purlieu is a challenging task that requires methodical solutions and urgent commitments. The fundamental aspects of urban planning related to the study and analysis of the landscape pattern prevailing in the NWH terrain have been dealt in detail in the subsequent parts of this chapter.

19.2 Urban Systems

Broadly, an urban system of a resourceful expanse, like the NWH, can be categorized into four major heads (Fig. 19.1), viz. natural urban environment (NUE), built urban environment (BUE), vulnerable urban environment (VUE) and environmental resource potential (ERP).

19.2.1 Natural Urban Environment (NUE)

As understood, the urban settlements are emergent and mostly not planned in India and in several other countries of the world. The natural urban environment (NUE) consists of challenges and problems associated with the natural features within the urban set-up. The Himalayan terrain acts as oxygen bank and carbon sink, for leisure and recreation, combating pollution and provide more intangible benefits that if not studied will intensify the environmental issues. NUE consists of the urban areas amid this serenity in the form of open spaces, green spaces, hydrological aspect, urban climatology and biodiversity, etc.

19.2.2 Built Urban Environment (BUE)

High-, medium- and low-density housing, slum areas, urban form and function, transport and transit-oriented development (TOD), social and physical infrastructure, waste (municipal, construction and industrial) management, risk assessment, urban heritage, etc. create the built urban environment (BUE). A healthy urban environment with the capability to contain its demographic set-up of the society can be achieved only by keeping the natural environment in harmony with it and can be achieved with the help of data-driven assessment planning.

19.2.3 Vulnerable Urban Environment (VUE)

Encroachment of built-up regime on the natural upbringing of an area poses a threat to the serenity around any settlement. The natural environment and the built environment are both exposed to threats. Moreover, the synchronization between the built and nature reduces the proneness to risk. Vulnerable urban environment (VUE) in terms of pollution (air, water and noise) and the risk towards natural/man-made hazards (flood, seismic, etc.) can be studied for an overall understanding.

19.2.4 Environmental Resource Potential (ERP)

There is an abundance of energy, resources, material and people in our urban ecosystems. The renewable energy potential in the form of wind, sun and water can be reaped in and around the urban set-up for their sustenance and growth. The beautiful landscapes of Himalayan terrain also offer tourism potential which could be optimally harnessed to boost economy of the countries in the vicinity. The integration of natural and built urban environment can be materialised with the advancement in geospatial technologies. These technologies can assume critical role by giving the datasets for inventory and evaluation of various aspects of urban purlieu.

19.3 Natural Urban Environment (NUE)

19.3.1 Quantifying Urban Landscapes-Green Space Assessment

Urban areas are characterised by heterogeneous landscapes and fragmentation is witnessed abundantly within the range of the Himalayas. These landscapes can be identified as areas having interacting patches, and their existence is supported by human activities (Turner et al. 2001; Paudel and Yuan 2012). The natural urban landscapes within the closed urban system in the form of parks and greenbelts are extremely helpful in counteracting the local climatic influences within an urban morphology. With increasing influx to the cities and restricted permissible city expansion limits, the phenomenon of de-densification within cities inexorably damages the sustainability rate of urban set-up. The delineation of landscapes into discrete patches can be useful in estimating landscape fragmentation, connectivity and human influences (Plexida et al. 2014). The thematic maps can be used for the purpose of precisely quantifying the urban natural environment (Herold et al. 2003).

The spatial measurements are measures to evaluate the structure and pattern of landscapes. It can be utilised in conjunction with remote sensing data for enhanced comprehension and portrayal of urban morphology. Spatial metrics conjointly act as a tool to create alternate conceptions of urban spatial structure to help the change modelling processes. The landscape pattern has been explored using percentage of landscape (PLAND), mean patch size (MPS) and number of patches (NP) measurements (Jain et al. 2013) using IRS-1C/1D LISS-III and Landsat TM images of Dehradun City. The Dehradun City is regarded as an ecologically sensitive and has seen rapid developments both spatially and economically, in the wake of turning into the interim capital of Uttarakhand state in India. Five land use/land cover (LULC) classes were classified using supervised classification technique. The amount of green spaces were evaluated inside seven buffer zones of 1 km each from city centre to assess green space pattern and changes due to anthropogenic influence. Table 19.1 shows the pattern through spatial metrics of most recent two decades based on fragmentation analysis. The compactness in pattern was observed amid the years 1986 and 2011, in both the classes, i.e. built-up and green spaces. The reduction in open spaces is also visible from PLAND and MPS values.

Consequently, seven buffers of 1 km each around the city centre (clock tower) were examined to check the urban environmental quality (Fig. 19.2). A drastic decrease in green spaces was witnessed from 34.62% to 7.78% during 1986–1998 and then further reduced to 2.81% in 2011 within the 1 km zone. Up to 38% change was observed within 4 km buffer zone from the city centre. Most of the green spaces are concentrated outside the city limit. It can also be observed that percentage of green spaces increases in buffer zones from 6 km onwards from the city centre. The

Table 19.1 Value of PLAND, NP and MPS metrics for different LULC classes

	PLAND			NP			MPS		
	1986	1998	2011	1986	1998	2011	1986	1998	2011
Built-up	13.32	19.57	36.93	2969	7468	2859	0.99	0.59	2.86
Forest	16.15	19.37	20.95	801	1850	1735	4.45	2.34	2.68
Green space	36.26	37.96	31.07	2458	4290	3145	3.25	1.98	2.19
Water/river	7.00	4.16	2.73	480	2510	731	3.22	0.37	0.83
Open space	26.81	18.58	8.16	1645	4868	2993	3.59	0.85	0.60

Source: Jain et al. (2013)

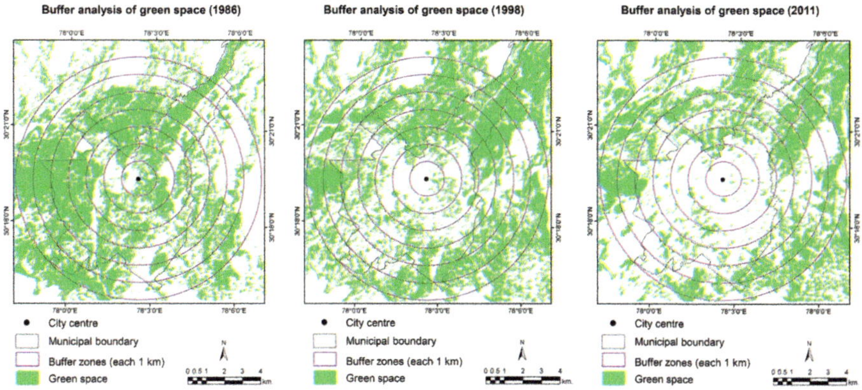

Fig. 19.2 Distribution of green spaces within seven buffer zones of 1 km from city centre for the years 1986, 1998 and 2011. (Source: http://technical.cloud- journals.com/index.php/IJARSG/article/download/Tech-162/pdf)

pattern of urban landscapes is mostly in the form of patches and has multiple nuclei. The concentric zone patterns of landscape cannot be seen in the study area. Most of the landscape can be seen concentrated in the government institutions. The built-up area has high PLAND and MPS values in 2011, which demonstrates the urban agglomeration amid most recent one decade. Hence, landscape planning is essential to curtail the urbanization effects on environment and also to preserve fertile agriculture land.

19.4 Built Urban Environment (BUE)

19.4.1 Land Suitability for Solid Waste Management

Most of the land within the urban premises are allocated to one or the other land use mentioned in master plans. Often, very less of these lands are allocated on the basis of any suitability analysis. Problem arises when the non-suitable sites convert themselves into hazard-prone areas or show reflections of their misuse. These discrepancies can represent major issues for the future and the intensity may only be explained rationally.

Solid waste management involves generation, collection, transport and disposal of solid waste in an environment-friendly manner while following the norms of economy, aesthetics, energy and conservation. Among various environmental issues existent in an urban area, urban solid waste disposal is considered as a serious environmental challenge (Rajan et al. 2014). Huge increment in solid waste is observed in India from 100 gm/person/day to 500 gm/person/day and also the disposal is generally unscientific (Akolkar 2005). In order to keep abreast of land-

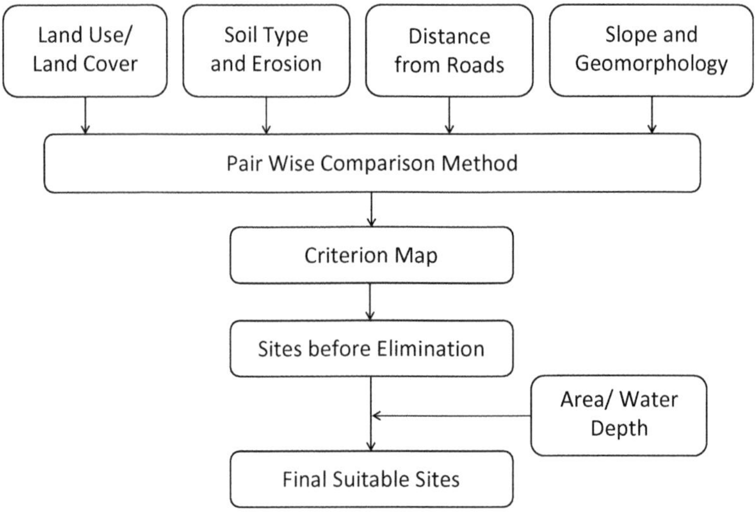

Fig. 19.3 Methodology for solid waste site selection

Table 19.2 Pairwise comparison matrix

	LULC	Slope	Roads	Soils	Erosion	Geology	RIW
LULC	1	3	4	7	5	7	0.434
Slope	3	1	3	5	3	5	0.236
Road	¼	1/3	1	5	3	5	0.156
Soils	1/7	1/5	1/5	1	1/5	3	0.045
Erosion	1/5	1/3	1/3	5	1	5	0.105
Geology	1/7	1/5	1/3	1/3	1/5	1	0.024

filling needs, new landfill sites have to be identified with the rapid exhaustion of older ones. A study has been carried out to identify landfill sites in Dehradun urban area (DUA) using analytical hierarchy process (AHP) based on site suitability modelling. The multidisciplinary datasets were integrated into composite information set following a model, which can answer a broader range of questions in spatial context (Fig. 19.3). On the basis of relative importance of different parameters, a set of weights were decided by carrying out a suitable data analysis and the weights obtained from data analysis were assigned to different parameters, which aided in designing of decision rules (Table 19.2).

The drainage system of the Dehradun City is driven by Bindal and Rispana rivers. The flow direction of these seasonal streams in the eastern part is north to south and in the western part is north to south-west. The dense forest patches exist in the north and along the outer limits of regulated area. The region is marked by high undulations and has mostly governed the physical development. Figure 19.4 shows various input maps used and the ranking of solid waste disposal sites based on AHP.

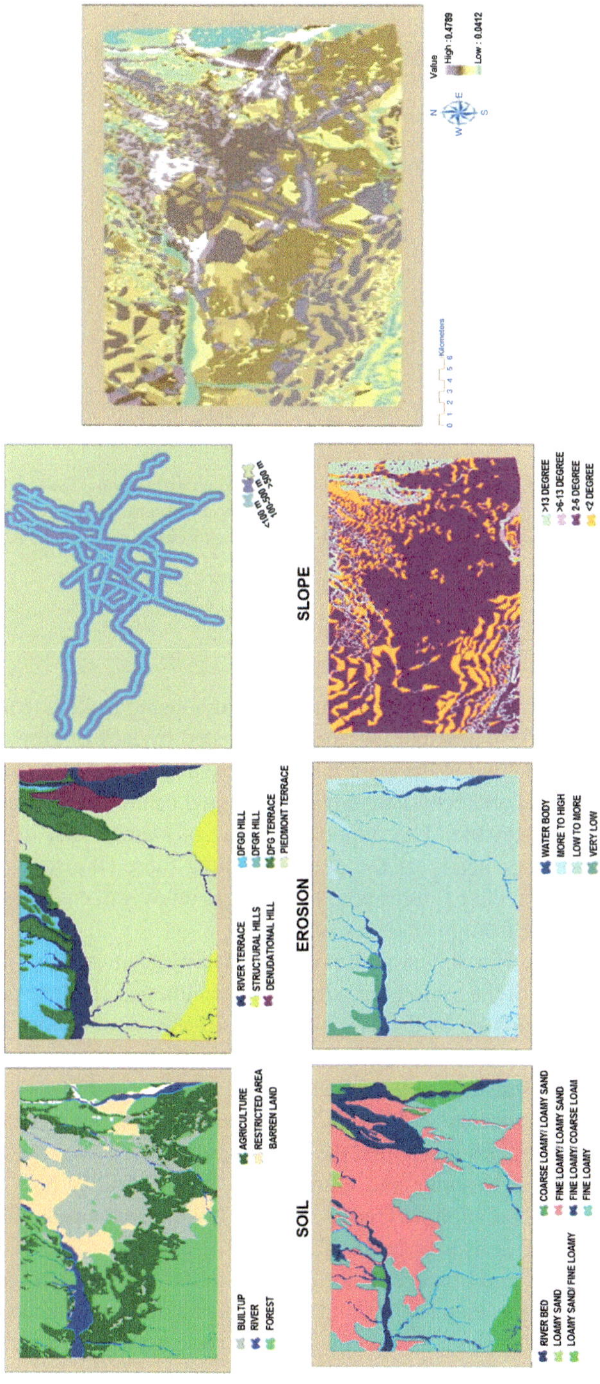

Fig. 19.4 Ranking of solid waste disposal sites based on AHP

Table 19.3 Slum population in Northwest Himalayan states of Jammu and Kashmir, Himachal Pradesh and Uttarakhand (2001 and 2011)

Location/number of towns			Type wise slum population			
State	Statutory towns	Slum reported towns	Total slum population	Notified slums	Recognised slums	Identified slums
2011						
J&K	86	40	6,62,062	1,62,909	1,36,649	3,62,504
Himachal Pradesh	56	22	61,312	60,201	0	1,111
Uttarakhand	74	31	4,87,741	1,85,832	52,278	2,49,631
2001						
J&K	69	12	268,513	–	–	–
Himachal Pradesh	–	–	–	–	–	–
Uttarakhand	74	6	195,470	–	–	–

Source: http://www.censusindia.gov.in/(S(uhiplu45xmcwkd552p4kbq30))/Census_Data_2001/Census_data_finder/A_Series/Total_Slum_population.htm, http://www.nbo.nic.in/Images/PDF/Slums_in_India_Compendium_English_Version.pdf and http://www.censusindia.gov.in/2011-Documents/Slum-26-09-13.pdf

19.4.2 Slums

A slum is a down area of a city characterized by substandard housing lacking tenure security (UN-HABITAT 2003). These are the areas dominated by lack of employment, illiteracy and poverty leading to decay in the urban environment. Slums have problems of geographic, social and economic environment and cannot be separated from the social system in urban life. The important aspect of slum is that it is a characteristic of place, not specifically of the people residing there (Weeks et al. 2007). The haphazard development of slums poses challenges on overall environs as well as quality of life, ranging from local to regional level.

The pattern of slums in the Himalayan region is governed by the topography of the area. Major urban centres of the region face similar environmental challenges as about 80% of slum developments are on the river bed in these areas. Poor quality and dilapidated house structures dominate the slums, largely due to the meagre resources available to its residents and thus ruin the aesthetic appeal of the city. Most of the settlements are illegal on encroached public land, and they also do not have the benefit of land tenure. The statistics reveal that the slum population has increased in both the states of Jammu and Kashmir and Uttarakhand from 2001 to 2011. In Himachal Pradesh, the slum population was reported in 2011, while in the year 2001, no slums were reported as per Census of India (Table 19.3).

Slum pockets are dynamic and observe major changes in the built environment over a period of time. With the increase in land prices of the surrounding areas, usually they sell the property and settle to other place. Also, *pucca* houses replace the *kutcha* houses, with the improvement in residents' economic conditions. The

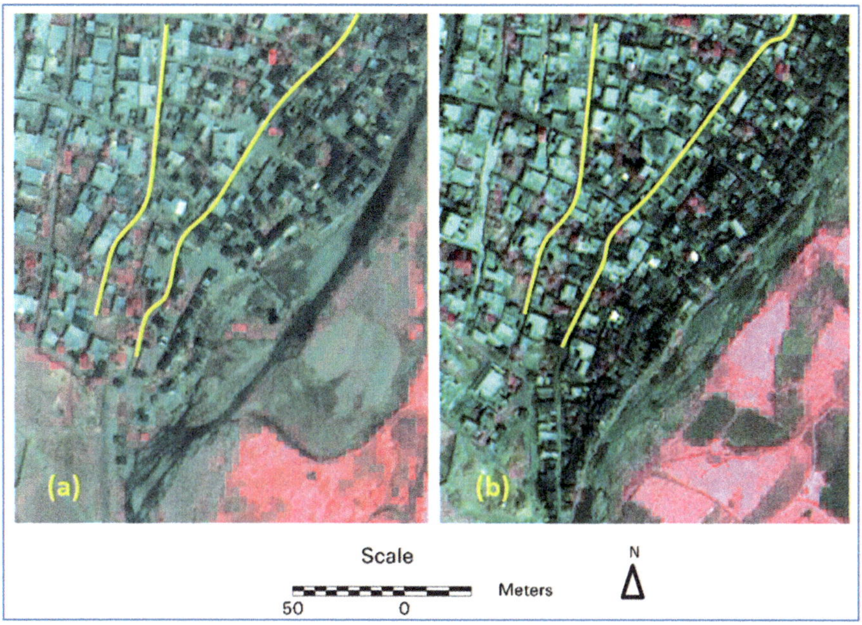

Fig. 19.5 Pan-sharpened MS image of (**a**) IKONOS image 2000 and (**b**) IKONOS image 2005 of Dehradun, highlighting the dynamic environment of slum areas

changes in socio-economic conditions of its residents can be traced out through reinforced cement concrete (RCC) structure and improved infrastructure. Most of these squatter settlements are near the existing high- and middle-income residential areas or small commercial areas where these people are employed in different activities. It is feasible to distinguish slums using satellite imageries based on these characteristics. The improved availability of high-resolution satellite data has facilitated superior extraction of ground objects, rather than regions of certain land cover classes only, and has turned out to be progressively vital for an assortment of remote sensing and GIS applications (Zhang and Maxwell 2006). In Dehradun City, about 80% of the informal settlements are situated along Rispana and Bindal Rao rivers and also along railway lines (Fig. 19.5). The rural to urban relocation is one of the major reasons behind the rapid development of slum areas in the city and its densification, etc. The condition of houses near the river is mostly dilapidated in comparison to the houses away from the river bed. These settlements are vulnerable to the flash floods due to the close proximity to river bed.

19.4.3 Urban Heritage: Preserving the Built

NWH states are rich in natural and cultural heritage. Its cultural heritage narrates a history of different time periods and includes temples, churches, rock inscriptions, excavated sites, water structures, cave and cemeteries. The preservation of cultural heritage is not only important for a country's identity but its sustainable management is necessary to boost country's economic progress and development. However, it is equally important to understand and create a balance between new development and preservation of cultural and historical aspects of urban environment (Kumar and Singh 2011). It is necessary to have a comprehensive and integrated approach for urban revitalisation and transformation for sustainable preservation of cultural heritage.

The advancements in geospatial technologies provide potential datasets and analysis platforms for analysing the existing scenario, detailed documentation and assessment of heritage structures, prioritization of heritage for taking up conservation measures, etc. The advanced sensors such as terrestrial laser scanner (TLS) provide ample opportunities for preparing the realistic 3D models of heritage areas (Alshawabkeh and Haala 2004; Campana and Remondino 2008), for assessment of structural deformations and for keeping a permanent record of historic monuments. GIS helps in the development of systematic database that facilitates data analysis. GIS also assists in scoring of the contributing factors that can have impact on the preservation of heritage buildings.

A study on urban heritage was taken up in Nainital town of Uttarakhand state in NWH. The heritage conservation in Nainital basically faces two fundamental problems: incomplete legal protection and redevelopment pressure. The legislation adopted for heritage at present provides inadequate and incomprehensive protection, while urban decay and hunger for usable land in urban keep pressure on redevelopment. Therefore, it is important to develop a holistic approach for heritage conservation planning in the city.

In this study, a GIS database for 40 listed heritage buildings by Indian National Trust for Art and Cultural Heritage (INTACH) was generated with the help of high-resolution satellite data, GPS and extensive field survey. The data consist of location, age, physical condition, height, construction material and ownership status as push factors and distance from lake, connectivity from highways and accessibility to infrastructures as pull factors for each building. An index system in GIS was adopted for the evaluation of these factors, and two indices (Ko 2008), namely, redevelopment potential index (RI) and urgency index (UI), were used in the present study. Each push and pull factor was assigned different ranking for RI and UI, e.g. good building condition was assigned highest ranking for redevelopment potential and lowest ranking for UI (Table 19.4).

To bring objectivity in the derived weights of each parameter, AHP-based evaluation of all parameters was carried out using Saaty's pairwise comparison method (Malczeweski 2006). For each group of parameters, different weights were computed for both the index. The final RI and UI maps are presented in Fig. 19.6. The urgency index value of St. John Wilderness Church was highest; hence it was selected for further investigation using TLS. This church was established in year 1844 by Daniel Wilson and is a fine example of Catholic Church architecture. The whole building is constructed of stone and is situated near High Court in Mallital area of Nainital. The 3D laser scanning system RIEGL 3D laser imaging scanners of

Table 19.4 Scores for assessment of redevelopment potential and urgency index

Push factor scoring			Pull factor scoring	
Building age scoring	*RI*	*UI*	*Market demand (distance from Nainital lake)*	*RI*
Less than 50 years old	0	0	Less than 50 m from lake	3
50–100 years old	1	1	50–150 m from lake	2
100–150 years old	2	2	150–250 m from lake	1
150–200 years old	3	3	More than 250 m from lake	0
Building condition scoring	*RI*	*UI*	*Connectivity from NH and SH*	*RI*
Normal/good condition	3	0	Less than 50 m from main road	3
Average condition	2	1	150 m away from main road	2
Bad condition	1	2	250 m away from main road	1
Abundant condition	0	3	More than 250 m from main road	0
Building construction material scoring	*RI*	*UI*	*Accessibility to infrastructure (bus stand, railways station)*	*RI*
Brick masonry construction	2	1	Less than 200 m buffer	3
Timber construction with wooden ornamentation	2	3	Between 400 m buffer	2
Stone walls, foundation with glass and wooden catholic windows	3	3	Between 600 m buffer	1
			More than 800 m buffer	0
Building height scoring w.r.t its surrounding buffer zone	*RI*	*UI*		
Ground	3	3		
G + 1	2	2		
G + 2	1	1		
G + 3	0	0		
Ownership status	**RI**	**UI**		
Private ownership	1	3		
Public ownership	3	0		

LMSZ series was used to capture exteriors of church building. Overall 15 scans were obtained to ascertain the high density of point cloud and full coverage of the building. Figure 19.7 shows the location of scans around the church building and Fig. 19.8 demonstrates resultant point cloud of church after the geo-registration.

The obtained point cloud was registered and textured with digital images obtained during scanning process. Some of the applications of resultant 3D model are structural assessment of building, deformation analysis, cross-sectional measurements, measured drawings (documentation of building), identification of cracks in foundation and exterior cladding, jamming of doors or windows, rubbing or refusing to close, roof damage, deteriorated mortar, materials conservation, identification of remedial measures, etc. The TLS and GIS combination is a promising solution to generate a high-quality photorealistic model and useful for documentation/preservation of complex heritage sites. It can provide a useful realistic 3D documentation which can be utilised for better planning and to boost tourism in cultural heritage area.

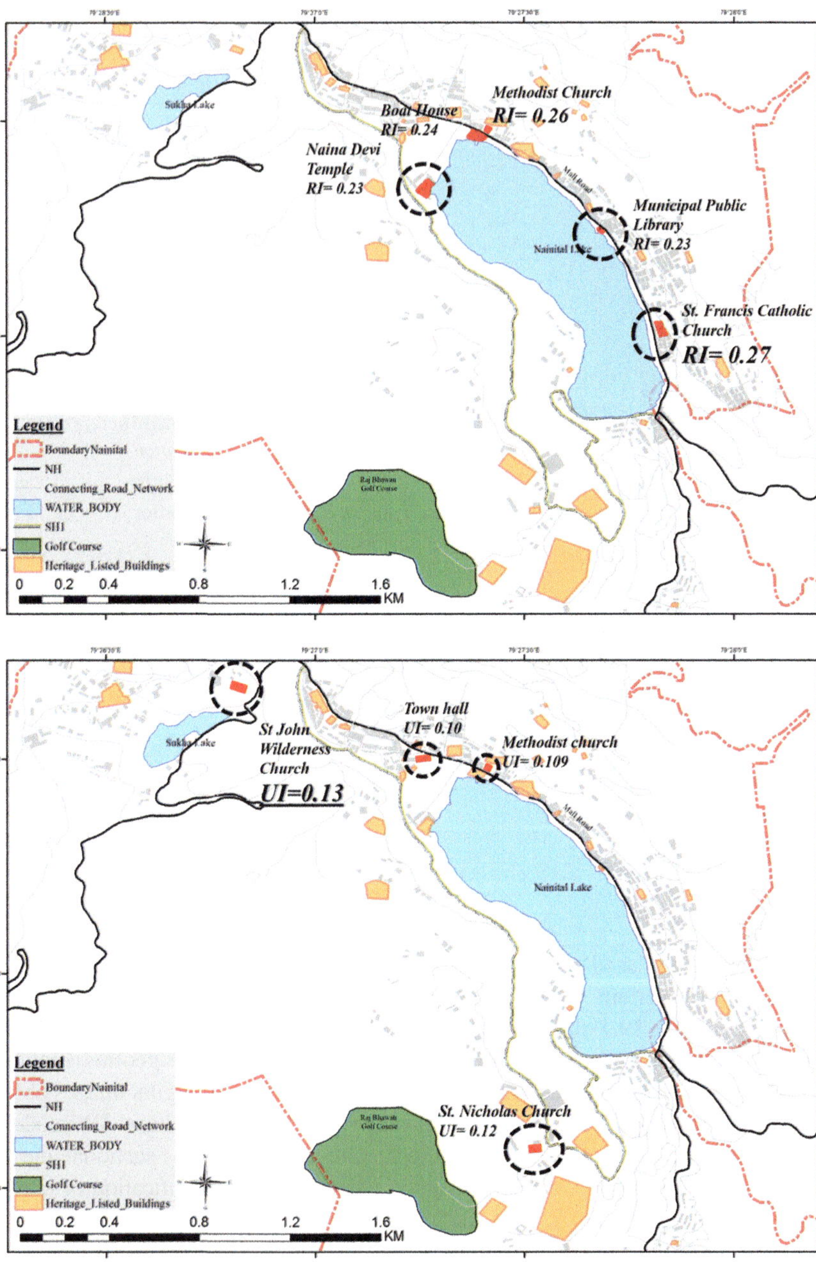

Fig. 19.6 Redevelopment potential index map and urgency index map

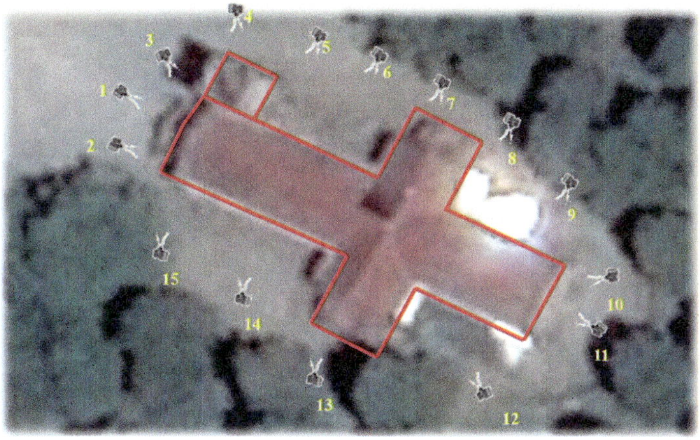

Fig. 19.7 Locations for 3D laser scanning

Fig. 19.8 Resultant point cloud of church after geo-registration

19.4.4 Regeneration of City Core

Growth and infrastructure development are two concurrent processes if not that the growth in terms of population and uses has adverse effect on existing infrastructure. Owing to negative development, the degeneration of the area begins to happen. As the activity increases manifold in the core areas of cities predominantly with commercial activities, it leads to faster degeneration in the areas. Hence, an intervention is required for regeneration of the city core areas. The urban regeneration is an integrated approach for redevelopment of any area going through the

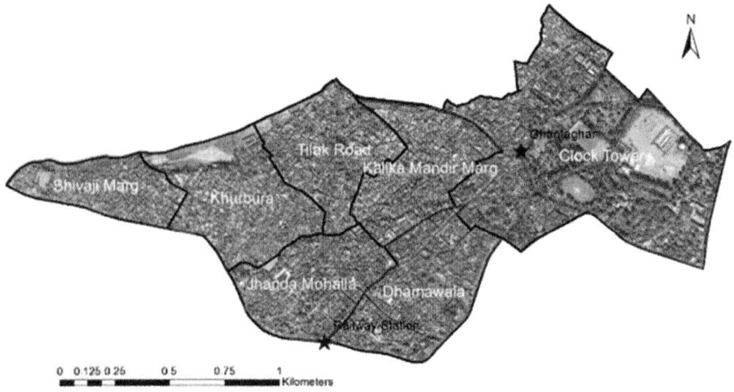

Fig. 19.9 Study area wards (Dehradun Core)

degeneration process and addresses all aspects comprehensively be it physical, economic, social, ecological or environmental concerns. Geospatial technologies assume a critical part in regeneration process as it gives up-to-date spatial information much required for mapping and analysis. Besides making available the temporal data, it also assists in monitoring of changes.

The Dehradun City core area (Fig. 19.9) comprising of seven commercial wards is densely populated and highly commercialized (Table 19.5). A marker-based approach was adopted to analyse the diverse aspects of study area. The major indicators were per capita green, percentage green, land use mix ratio, road density (geographical area and population wise), built-up density, land use ratio, service ratio index and parking accessibility ratio. Finally, a multi-criteria appraisal approach was associated for determining final regeneration index. The weights of various indices were drawn utilizing Saaty's method. The investigation of study area revealed that it has colossal deficiency of urban green spaces which are valuable for natural ventilation and controlling microclimate. The study also focused on the analysis of land use as well as the space use (Fig. 19.10). Space use is the use of space in third dimension, i.e. use of the built-up space at each floor. Space use maps are useful to monitor the growth, actual calculation and collection of taxes, to monitor the complex ownership issues, etc. In this study, the visualization of 3D space use maps was generated in commercial off-the-shelf (COTS) software which was till now mostly done through 2D cartographic mapping.

The space use investigation in the study area revealed that there is an abnormal state of commercial activity which has pulled in enormous vehicular traffic as well as pedestrian population and necessitates greater measure towards creating parking spaces. The whole core area is confronting traffic management issues. Some of the transitory measures, for example, time-bound entry of vehicle, have been adopted in the study area; however, it requires a coordinated traffic management plan for long-term arrangement. In this study, remote sensing-derived parameters have been utilised for index generation; however, other parameters related to age of building, floating population, economic characteristics, façade control, etc. can likewise be

Table 19.5 Study area wards with population

Ward No.	Ward name	Ward population
15	Clock tower	5341
18	Kalika Mandir Marg	5326
19	Tilak road	8687
20	Khurbura	7374
21	Shivaji Marg	10,237
23	Dhamawala	5295
24	Jhanda Mohalla	6727

Fig. 19.10 Space use map

incorporated based on information accessibility to derive the regeneration recommendations.

19.5 Vulnerable Urban Environment (VUE)

19.5.1 Traffic-Induced Air Pollution in an Urban Canopy

The vehicular count on the streets is expanding rapidly because of rising urban populace and in addition because of their rising purchase power. This in turn is increasing the levels of pollutants inside and encompassing the domain of urban agglomerations. As per World Health Organization (WHO), India is the ninth most polluted country of the world. The higher level of pollutants in the air like suspended particulate matter (SPM), respirable suspended particulate matter (RSPM), sulphur oxides (SO_x), nitrogen oxides (NO_x) and other pollutants like trace metals have adverse effects on human beings and environment and in majority is due to vehicular emissions (Caselles et al. 2002; Maitre et al. 2006; Kaushik et al. 2006; Sharma et al. 2006; Jayaraman 2007, Curtis et al. 2006). The composition of the air is drastically changing primarily due to particulate matter (Zanini et al. 2006).

The pollution data collected from 20 monitoring locations in Himachal Pradesh state in NWH along with trends of annual average for SO_2, NO_2 and RSPM are depicted in Fig. 19.11. It can be observed that the annual average values of SO_2 and NO_x are well below the permissible annual average. However, the value of RSPM is observed to be higher than the permissible limits. The cities in Uttarakhand state

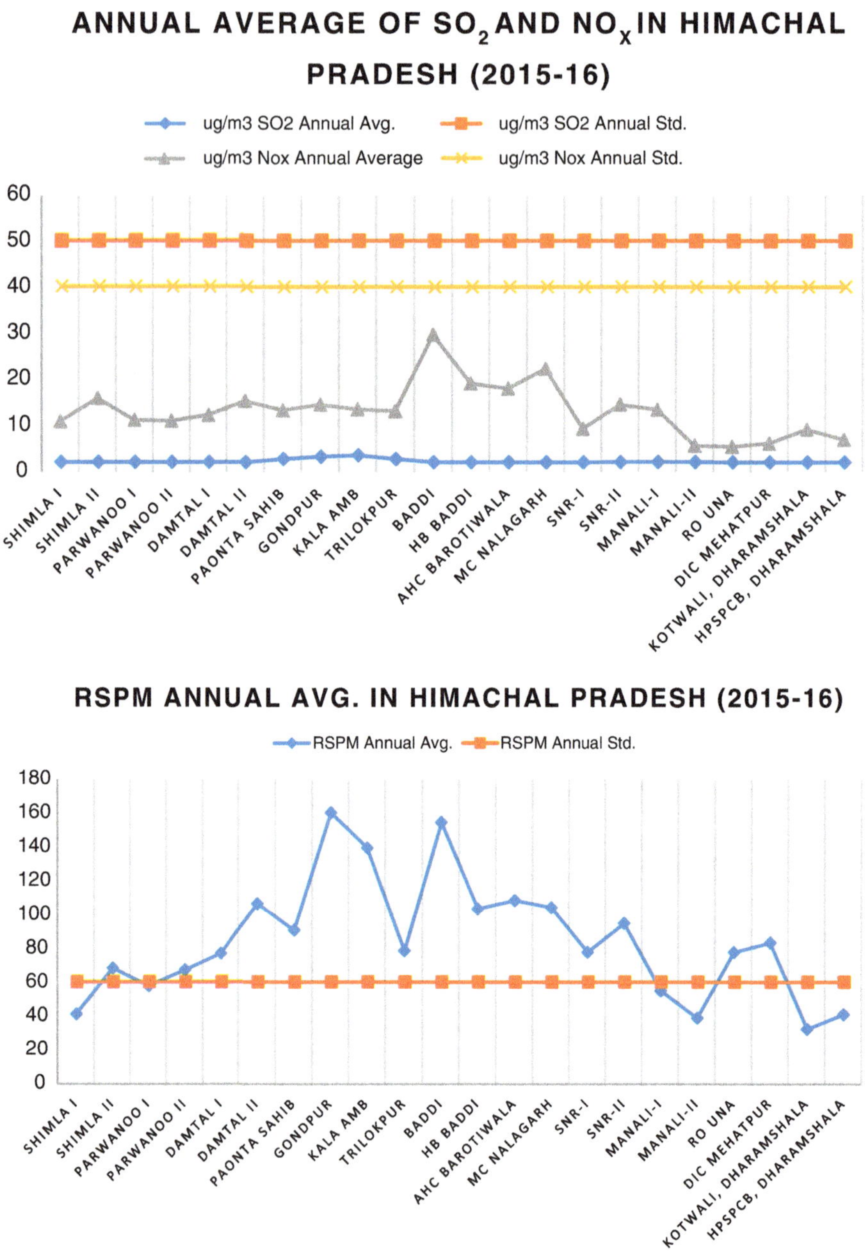

Fig. 19.11 Annual average of SO_2, NO_x and RSPM in Himachal Pradesh during 2012–2013. (Source: ENVIS Centre: Himachal Pradesh Status of Environment and Related Issues, MoEF, GoI)

experience higher pollution values for RSPM, SO_2 and NO_x in all the months of the year 2015 (Fig. 19.12). This clearly indicates that Uttarakhand needs immediate attention in terms of monitoring and planning for improving the air quality. Also, in

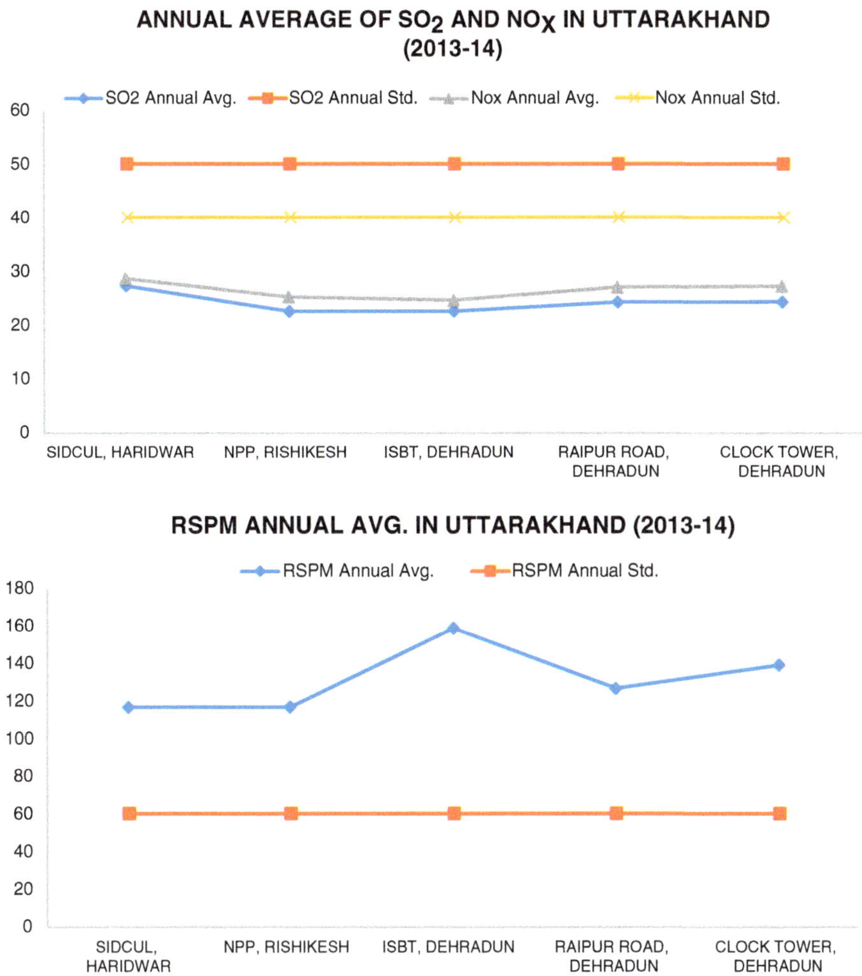

Fig. 19.12 Annual average of SO_2, NO_x and RSPM in Uttarakhand for the year 2015. (Source: ENVIS Centre: Uttarakhand, Status of Environment and Related Issues, MoEF, GoI)

Uttarakhand state, only four monitoring stations (three at Dehradun town and one at Rishikesh town) are insufficient to conclude the air pollution status of the state.

Dehradun City is taken as a case study example to analyse the status of pollution using ground-based measurements. Three air pollution monitoring stations (clock tower, ISBT and Rajpur Road) are installed in the city which seems insufficient to capture the status of pollution trend of the entire city. Therefore, the pollutants were measured at eight locations for 3 h in the morning for eight working days (excluding Saturday and Sunday) at major traffic junctions of the city during April 2016 using a handheld portable sensor (Graywolf Advanced Sense Pro). These junctions, viz. clock tower, Ballupur Chowk, ISBT Chowk, Saharanpur Chowk, Dilaram Chowk,

Table 19.6 Air quality standards for CO from NAAQS, 2009

Pollutant	Time-weighted average	Concentration in ambient air			Measurement method
Carbon monoxide (CO)		Industrial areas	Residential, rural and other areas	Sensitive areas	Nondispersive infrared spectroscopy
	8 h	5.0 mg/m^3	2.0 mg/m^3	1.0 mg/m^3	
	1 h	10.0 mg/m^3	4.0 mg/m^3	2.0 mg/m^3	

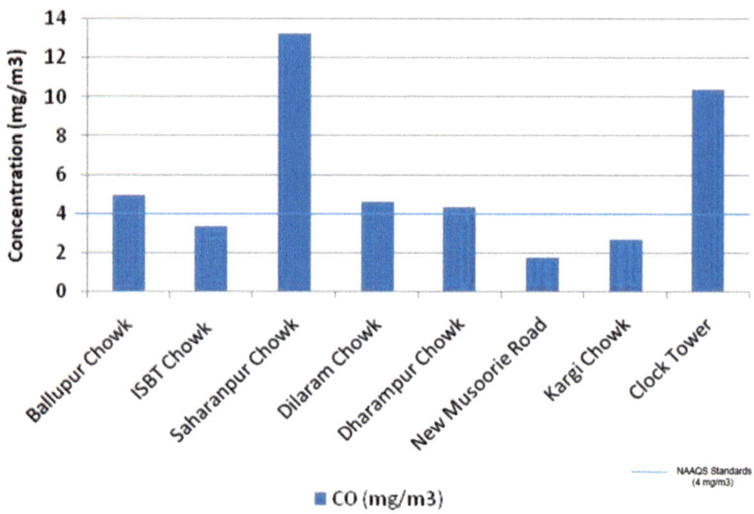

Fig. 19.13 CO (in mg/m^3) concentration at eight site locations and comparison with NAAQS, 2009

Dharampur Chowk, Kargi Chowk and New Mussoorie Road Chowk were selected based on traffic volume count, road hierarchy and adjacent land use. Additionally, traffic volume count, modal split survey and speed spot point survey were also carried out in conjunction with the pollutant measurement survey. The survey was conducted to analyse the effect of vehicular pollution on the junctions during peak hours, and hence, the morning peak interval from 0800 h to 1100 h was chosen for the same. The sampling interval was chosen to be 1 min for computational purpose and automatic logging in the logging meter of the instrument.

Carbon monoxide is an odourless and colourless gas which is responsible for smog due to burning of gasoline in vehicles and has serious health threats. The National Ambient Air Quality Standards (NAAQS) reveal that 8 h average permissible concentration of CO in industrial, residential and sensitive areas are 5.0 mg/m^3, 2.0 mg/m^3 and 1.0 mg/m^3, respectively (Table 19.6). A considerable difference is observed in the maximum and minimum values for the concentration of pollutants at various intersections (Fig. 19.13). The value of pollutant at New Mussoorie Road and ISBT Chowk exhibited less variation in pollutant concentration. The CO concentration was highest at Saharanpur Chowk (13.2 mg/m^3) and lowest at New Mussoorie road (1.71 mg/m^3).

Higher CO concentration was observed due to large number of auto rickshaws (17%). The total passenger car unit (PCU) at Saharanpur Chowk was observed to be 1233 PCUs. The CO concentration at Saharanpur Chowk exceeded the NAAQS by 230% (NAAQS standard is 4 mg/m^3). Other intersections, namely, clock tower, Ballupur Chowk, Dilaram Chowk and Dharampur Chowk, too exceeded the NAAQ Standards. Further interpolation in GIS environment was carried out using point observation data of seven stations and one station (clock tower) data was used for validation. The spline, local polynomial and inverse weighted average methods were used in the analysis. The spline interpolation resulted in least error of 1.82% for CO. The interpolated map depicted that central and western part of the city has poor to very poor air quality as compared with Air Quality Index (AQI). These studies are essential for understanding the relation between the vehicular traffic and emissions with pollution-level estimation through ground surveys and are helpful in understanding the environmental degradation in terms of pollution level.

19.5.2 Seismic Hazard

The Bureau of Indian Standards (BIS) based on studies and earthquake data investigated by Indian Meteorological Department (IMD) has grouped the country into four seismic zones from Zone V (the most active) to Zone II (the least active). Jammu and Kashmir, Himachal Pradesh and Uttarakhand fall under the category of Zone IV and Zone V (severe intensity zone). Hence, study of seismic hazards holds more and more importance in the NWH.

A study was carried out to implement a seismic risk assessment methodology, risk assessment tools for diagnosis of urban areas against seismic disasters (RADIUS) developed by the United Nations (http://www.unisdr.org/files/2752_RADIUSRiskAssessment.pdf) in GIS environment. RADIUS is a basic and simple tool useful for decision-makers and urban planners alike to comprehend the damage and human causalities in the event that an earthquake strikes a city. The study area, Dehradun City as the interim capital of Uttarakhand state, is also a major business as well as service centre. Various national level civil and defence institutions signify its importance in the country. The investigation using RADIUS went for preliminary estimation of earthquake damages to bring issues to light about earthquake risks.

The urban seismic hazard analysis requires earthquake scenario analysis. This incorporates various parameters for earthquake scenario like its epicentre, extent and at times its location. The peak ground acceleration (PGA) values are produced from this model. The PGA computation is done based on following three equations:

Joyner and Boore (1981):

$$\mathrm{PGA} = 10\wedge(0.249^*\mathrm{M} - \mathrm{Log}(\mathrm{D}) - 0.00255^*\mathrm{D} - 1.02); \mathrm{D} = (\mathrm{E}\wedge 2 + 7.3\wedge 2)\wedge 0.5 \tag{19.1}$$

Campbell (1981):

$$\mathrm{PGA} = 0.0185^*\mathrm{EXP}(1.28^*\mathrm{M}) * \mathrm{D}\wedge(-1.75); \mathrm{D} = \mathrm{E} + 0.147^*\mathrm{EXP}(0.732^*\mathrm{M}) \tag{19.2}$$

Fukushima and Tanaka (1990):

$$\begin{aligned} \mathrm{PGA} = \big(10\wedge(0.41^*\mathrm{M} - \mathrm{Log}\ \ 10(\mathrm{R} + 0.032^*10\wedge(0.41^*\mathrm{M}) \\ -0.0034^*\mathrm{R} + 1.30)/980 \end{aligned} \tag{19.3}$$

The PGA value depends upon the actual ground conditions which after modification yields the Net-PGA values. The damages are estimated from hazard map which is produced as modified Mercalli intensity (MMI) map in RADIUS and is resultant from earthquake scenario. The MMI map is produced on the basis of Trifunac and Brady (1975) equation:

$$\mathrm{MMI} = (1/0.3) * (\mathrm{Log10}(\mathrm{PGA}^*980) - 0.014) \tag{19.4}$$

The hazard map produced from above analysis and represented as MMI values shows the potential damage to various structures and life-saver facilities. The earthquake scenario created was the same as that of the Chamoli earthquake that happened on March 29, 1999. The following parameters were assumed in the analysis: Richter magnitude, 6.8; depth of earthquake, 21 km; and epicentral distance = 13 km. The attenuation function utilised in the methodology was proposed by Joyner and Boore (1981). A soil amplification function of 0.7 was assumed for the entire study area. The PGA values were changed over to MMI scale by utilising the Trifunac and Brady (1975) equation. The estimated MMI values were utilised to assess percentage damage for each building type while using the primary damage curve data based on Applied Technology Council (ATC) document (ATC 1992). Figure 19.14 demonstrates the ward wise total buildings damaged (in percentage). The RADIUS tool has ease of usage and it simulates earthquake hazard. With a wider part of India now being examined for seismic zonation and the cost-effective tool like RADIUS is effectively accessible, we have better odds of enhancing the state of crisis administration in our country.

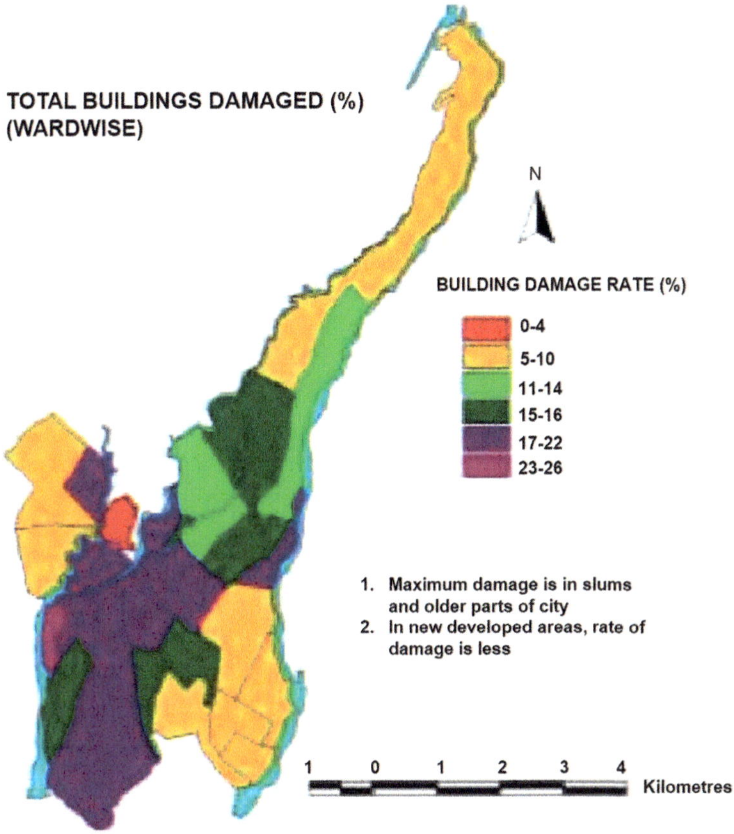

Fig. 19.14 Total buildings damaged (in percentage) ward wise

19.6 Environment Resource Potential (ERP)

19.6.1 Tourism Potential

Tourism has turned into a prevalent worldwide recreation activity. It assumes an imperative part in keeping up social harmony, recreation and aides in enhancing excellence in community skills. Tourism can be domestic or international and has significance towards country's balance of payments and economic growth. Today, tourism is considered as an important source of income for many countries. The annual tourist footfall in Uttaranchal was 11,619,457 for the period of April 2005–March 2006 (http://tourism.gov.in/sites/default/files/Other/07%20uttaranchal.pdf), in 2012–2013 for Himachal Pradesh was 12,329,907 (http://tourism.gov.in/sites/default/files/Other/Himachal%20Pradesh%20Final%20Report_%20new.pdf) and in Jammu and Kashmir was 5,268,246 in 2001 (http://tourism.gov.in/sites/default/files/jammu%20kashmir.pdf) contributing heavily to the states' GDP.

A study was carried out using remote sensing data along with ancillary information to assess the tourism carrying capacity of Nainital town in Uttarakhand state. The tourism carrying capacity (TCC) is characterised as the maximum number of people that may visit the tourist destination without causing destruction of the physical, economic and sociocultural environment and an unsuitable reduction in the quality of visitors' satisfaction (Alvin Chandy 2009). The study area, Nainital, is a popular tourist destination of Uttarakhand state. It is also known as the Lake District of Uttarakhand. The tourist's influx during the peak season is nearly six to nine times of the local population. Nainital is now being inflicted by the growing traffic, anomalous growth and big turnout of tourists in the recent years, and it is leading to inadequate accommodation and infrastructure facilities for tourists, which could be detrimental to the growth of tourism industry in the area. Since, Nainital's entire economy is dependent on the tourism industry with diminutive production in town, it can have far-reaching effects on the city.

TCC has been assessed based on parameters such as number of domestic and foreign tourists arriving in peak season, residential population, municipal boundary area, average number of days of tourists' stay in city, number of days in peak season and density and normalizing density methods as defined in urban development plans formulation and implementation (UDPFI) guidelines. The methodology proposed can be used by planners for the timely upgradation of infrastructure and facilities in tourist areas. The carrying capacity of Nainital City was calculated as a difference between carrying capacity minus the existing load (Table 19.7). The appraisal of TCC depends on three major indicators: (a) physical and ecological indicators and flexible components (infrastructure systems like water supply, electricity, transportation, etc.), (b) socio-demographic indicators and (c) political-economic indicators. High-resolution satellite imagery, Cartosat DEM of 30 m resolution, ward map, land use map, topographical map and Nainital guide map have been used in the present study. Secondary data emphasizing on perception and opinion of tourists, persons engaged with tourism sector and experts were also considered. The network analysis was performed to make routing of tourist destinations such as heritage sites, lake sites and also adventure tour path around Nainital town.

Geospatial techniques are useful for mapping, visualization and analysis of places of tourist importance. Major tourist destinations of Nainital include High Court, Raj Bhawan, ARIES, Naini Lake, etc., and the peak season for tourists is during the period of May–June. It was observed that during the peak season, tourist influx is beyond the carrying capacity and causes problems to tourist infrastructure/facilities.

Table 19.7 Available carrying capacity in Nainital City

District	Tourist town	Destinations covered	Carrying capacity	Existing load	Available capacity
Nainital	Nainital	Naini Lake, Raj Bhawan, High Court, etc.	67,252	78,452	−11,199

Table 19.8 Renewable energy generation potential in NWH

Sl. No.	States/UTs	Wind power (MW)	Small hydropower (MW)	Bioenergy – biomass power (MW)	Bioenergy – bagasse cogeneration (MW)	Bioenergy – waste to energy (MW)	Total
1.	Himachal Pradesh	64	2398	142	NA	2	2606
2.	Jammu and Kashmir	5685	1431	43	NA	NA	7158
3.	Uttarakhand	534	1708	24	NA	5	2271
	Total	6283	5536	209	0	7	12,034

Source: https://data.gov.in/resources/state-wise-potential-various-renewable-energy-technologies/download

19.6.2 Energy Potential

Large quantity of rainfall, abundant solar insolation and huge wind capturing locations offer opportunities to tap vast energy resources available in India. Being close to Tropic of Cancer and the equator, it has huge solar potential. There is a requirement for countries to secure energy potential and production of their country alongside developing the resources (Warren et al. 2005). A significant part of most countries energy production originates from consuming non-renewable energy sources. Table 19.8 shows the renewable energy generation potential in NWH.

19.6.2.1 Solar Energy

The total irradiation that hits the ground surface and the energy generated due to this is not enough keeping in mind today's energy consumption requirement scenario. Taking advantage, radiation captured can also be utilized for meeting day-to-day needs (http://www.smartcities-infosystem.eu/renewable-energy-sources). Solar energy is one of the most promising sustainable energy sources due to its accessibility. India has a potential of about 750 GW of solar energy, according to the National Institute of Solar Energy (https://cleantechnica.com/2014/11/29/indias-solar-power-potential-estimated-750-gw/). Solar rooftop is an emerging renewable energy option for the cities to initiate smart city concept using smart energy.

Understanding the requirement for solar resource data, the National Centres for Environmental Information (NCEI), Department of Energy's National Renewable Energy Laboratory (NREL), the National Aeronautics and Space Administration, the Northeast Regional Climate Centre alongside various universities and companies teamed up in building the National Solar Radiation Database (NSRDB). It was updated in 2012 containing data from 1991 to 2010 for over 1500 stations in the United States. Similarly, in the Solar Energy Centre (SEC)-NREL collaborative project on 'solar resource assessment' under the Indo-US energy dialogue, solar

Table 19.9 Mean direct normal irradiance (DNI) at 10 km spatial resolution based on hourly estimates of radiation over 7 years (2002–2008) averaged for entire states in NWH

Mean DNI in KWh/m^2/day	Jammu and Kashmir	Himachal Pradesh	Uttarakhand
January	3.47	3.72	4.55
February	3.46	3.67	4.73
March	4.17	4.60	5.67
April	4.88	4.95	5.40
May	5.89	5.89	5.75
June	6.10	5.11	4.62
July	5.87	4.10	3.14
August	5.38	3.90	3.18
September	6.04	5.17	4.57
October	6.61	6.97	6.76
November	5.74	6.33	6.75
December	3.80	4.61	5.73
Annual	*5.13*	*4.92*	*5.07*

Source: MNRE, http://www.mnre.gov.in

maps were generated using satellite imagery-based measurements. These maps are available at 10 km spatial resolution for the entire India (Table 19.9). Initially, the solar maps containing direct normal irradiance (DNI) and global horizontal irradiance (GHI) were prepared from January 2002 to December 2008. The maps are later updated extending the data up to 2014 using weather satellite METEOSAT measurements (http://www.nrel.gov/international/ra_india.html).

In order to respond to the growing urbanization process and subsequent energy demand, methodological approaches which implement alternative urban models are required to support the indispensable change towards more energy efficient cities. Smart energy is one of the important pillars to make city smart and smarter in the days to come.

19.6.2.2 Hydropower

Hydroelectric power (HEP) systems produce electrical energy from the potential and kinetic energy of water and offer emission-free power in many parts of the world. HEPs are the clean energy alternatives without causing any air contamination and exhaustion of fossil fuels. Perennial inflow of water and suitable location offering adequate 'head' are the basic requirements to site a HEP scheme. These must be complemented with favourable geology, environmental and socio-economic characteristics. The small hydropower plants (SHP) generally produce less than 10 MW. Based on energy production capacity, the SHP are characterised as pico (<5 kilowatts), micro (5–100 kW), mini (100 kW-1 MW) and small (1–10 MW) (https://www.renewablesfirst.co.uk/hydropower/hydropower-learning-centre/what-is-the-difference-between-micro-mini-and-small-hydro/). As the urban settlements are generally sparsely placed in Himalayan region, SHPs provide viable solution to meet the energy demand. While pico level of hydropower plants could be operated in conjunction with water mills, relatively larger size plants could be installed along rivers/channels in vicinity to urban settlements.

The hydropower sites feasibility assessment demands estimation of flow characteristics and various terrain parameters. The geospatial technology has tremendous scope and advantages for the assessment of many of these parameters. Among the topographic parameters, the fundamental prerequisite in hydropower scheme is adequate 'head', which can be assessed using digital elevation model (DEM) prepared from high-resolution satellite stereo data. Such DEMs are also useful to analyse submergence region upstream, capacity assessment and delineating the catchment area of a hydropower site. The multispectral remote sensing data enables the preparation of LULC surrounding the hydropower locations to assess the environmental impacts of such schemes; geological characteristics such as lithological and structural information and geomorphological evaluation are useful in terms of nature and type of valley sections and landforms. Various catchment characteristics derived from satellite multispectral and stereo data are useful to assess discharge from the catchment at the identified hydropower diversion site. The decision criteria that can be followed for analysing site suitability for HEP schemes are given as follows: (a) valley segments with abrupt drop in their elevation levels in a very short distance and which are at sufficiently high drainage order (to provide necessary catchment area) are given preference in view of the probability to generate sufficient head (assessed with the help of satellite data, DEM and topographic profiles generated from DEM); (b) site should have sufficient discharge to sustain uninterrupted power generation; (c) site falls downstream of confluence of major tributaries and is away from meandering sections; (d) river sections forming loop with favourable bed gradient and a straight section immediate upstream are evaluated for diversion schemes (assessed using DEM and multispectral satellite data); and (e) the tail water level (TWL) of any upstream site are kept above the full reservoir level (FRL) of downstream sites (assessed using topographic profiles drawn from DEM).

The geospatial technology-driven hydropower sites selection and potential estimates have been carried in nine valleys of Alakananda river of Uttarakhand state. Cartosat-1 stereo data and GCPs were used to generate DEM for the study area. Resourcesat-1 LISS-4 images were used to derive geological and LULC characteristics. Based on the above decision criteria, 68 potential HEP sites have been identified. The topographic 'head' computed for these sites varies from 66 to 1242 m and power potential varies from 0.3 MW to 33.2 MW for 50% dependable year. The aggregate power potential was estimated as 411 MW for 50% dependable year. The potential sites were recommended for detailed studies. The present study is an endeavour where geospatial approach could expediently be used to identify potential hydropower sites at pre-feasibility level for untapped catchments in Himalayan terrain. Figure 19.15 shows HEP sites identified in Birahi Ganga river valley.

Fig. 19.15 Potential hydropower sites identified in Birahi Ganga valley

19.7 Conclusions

This chapter dealt with four major aspects of the urban settlements in NWH, viz. natural environment, built environment, vulnerable urban environment and environmental resource potential. The geospatial applications in each aspect and their subthemes were described and highlighted. The natural built environment needs to be protected and geospatial technologies provide tools for its inventory, mapping and assessment. Spatial metrics is one such tool for analysing green spaces in urban areas. The case study of Dehradun City presented in this chapter shows degradation in natural environment from the analysis. The urban settlements in Himalayas are responsible for loss of natural habitat as well as disrupting and polluting the water resources as well. The solid waste management is also an issue facing the urban settlements. The changes in consumer behaviour have led to increasing amount of solid waste, and the scientific disposal of this waste is a big challenge in front of planner's community. The unscientific and unmanaged solid waste disposal is responsible for polluting ground water and surface water resources in NWH. The surface water bodies or hydrological system in NWH is facing threats from unplanned, unorganized growth of slums in drainage flow areas. It does hamper not only the free flow of drainage channels but also generates potential areas for flood water flooding. A case study to analyse slum areas revealed that the slums in Dehradun urban area are highly vulnerable as they are placed in active flood zones of existing water bodies.

Within the built environment, the NWH states have a rich history of cultural heritage which can be seen in the form of temples, churches, monuments and so on. There is a need to distinguish the heritage buildings for protection and conservation. The advanced geospatial technologies provide ample tools for documentation and assessment of heritage buildings. It has great potential for 3D modelling, structural analysis, deformation analysis and many more applications. However, there is a need to popularize these technologies for their optimum utilization in this field.

NWH cities are highly vulnerable from seismic activity point of view as this region has high prevalence of active seismic activity and coincides with sensitive to very sensitive zones. RADIUS methodology provides a comprehensive method for the assessment of vulnerability of urban structures in case of seismic activity, and it could be a useful tool in the hand of planners for the preparation of action plans, mitigation and adaptation measures in the form of appropriate planning and construction guidelines. Air pollution is another aspect of vulnerable urban environment which is posing threat to urban liveability. The GIS-based interpolation methods are useful for assessing the spatial distribution of air pollution. However, development of new tools and methods are required to improving modelling in this area of research.

NWH cities also have ample resources to be harnessed for improving the economy and urban sustainability. The tourism potential of NWH cities have to be explored and managed in an optimum manner by taking care of the carrying capacity of Himalayan towns for their sustainable development. Similarly, harnessing the renewable energy towards sustainable development has to be addressed critically in this region. The region has ample potential and resources for solar and hydropower generation, but care should be taken to develop them in a sustainable manner.

References

Akolkar, A.B., 2005. Status of solid waste management in India, implementation status of municipal solid wastes, management and handling rules. Central Pollution Control Board, New Delhi.

Alshawabkeh, Y., and Haala, N., 2004. Laser scanning and photogrammetry: a hybrid approach for heritage documentation. In Third International Conference on Science & Technology in Archaeology & Conservation, The Hashimite University, Jordan.

ATC, 1992. A model methodology for assessment of seismic vulnerability and impact distribution of water supply systems. Report No: ATC – 25 – 1, Redwood City, CA.

Campana, S. and Remondino, F., 2008. Fast and detailed digital documentation of archaeological excavations and heritage artefacts. In Posluschny, A., K. Lambers and I. Herzog (eds.), Layers of Perception. Proceedings of the 35th International Conference on Computer Applications and Quantitative Methods in Archaeology (CAA), Berlin, Germany, April 2–6, 2007 (Kolloquien zur Vor- und Frühgeschichte, Vol. 10). Dr. Rudolf Habelt GmbH, Bonn, pp. 36–42, http://proceedings.caaconference.org/paper/18_campana_remondino_caa2007/.

Campbell, K.W., 1981. Near-source attenuation of peak horizontal acceleration. Bulletin of the Seismological Society of America, 71(6), pp.2039–2070.

Caselles, J., Colliga, C. and Zornoza, P., 2002. Evaluation of trace element pollution from vehicle emissions in petunia plants. Water, Air, & Soil Pollution, 136(1), pp.1–9, https://doi.org/10.1023/A:1015229714374.

Chandy, A., 2009. Carrying capacity analysis, identification of tourism circuits across India submitted to Ministry of Tourism, India.

Curtis, L., W. Rea, P. Smith-Willis, E. Fenyves and Y. Pan, 2006. Adverse health effects of outdoor air pollutants. Environ. Intern. 32, 815–830.

Fukushima, Y. and Tanaka, T., 1990. A new attenuation relation for peak horizontal acceleration of strong earthquake ground motion in Japan. Bulletin of the Seismological Society of America, 80 (4), pp.757–783.

Herold, M., Goldstein, N. and Clarke, K., 2003. The spatiotemporal form of urban growth: measurement, analysis and modelling. Remote Sensing of Env., 86, 286–302.

http://tourism.gov.in/sites/default/files/Other/07%20uttaranchal.pdf

http://tourism.gov.in/sites/default/files/Other/Himachal%20Pradesh%20Final%20Report_%20new.pdf

http://www.cea.nic.in/installed_capacity.html

http://www.censusindia.gov.in/(S(uhiplu45xmcwkd552p4kbq30))/Census_Data_2001/Census_data_finder/A_Series/Total_Slum_population.htm and

http://www.censusindia.gov.in/2011-Documents/Slum-26-09-13.pdf

http://www.mnre.gov.in

http://www.nrel.gov/international/ra_india.html

http://www.smartcities-infosystem.eu/renewable-energy-sources

http://www.unisdr.org/files/2752_RADIUSRiskAssessment.pdf

https://cleantechnica.com/2014/11/29/indias-solar-power-potential-estimated-750-gw/

https://data.gov.in/resources/state-wise-potential-various-renewable-energy-technologies/download

https://www.renewablesfirst.co.uk/hydropower/hydropower-learning-centre/what-is-the-difference-between-micro-mini-and-small-hydro/

Jain, S., Laphawan, S. and Singh, P.K., 2013. Tracing the Changes in the Pattern of Urban Landscape of Dehradun over Last Two Decades using RS and GIS. International Journal of Advanced Remote Sensing and GIS, 2(1), pp. 351.

Jayaraman, G., 2007. Air quality and respiratory health in Delhi. Environmental monitoring and assessment, 135(1), pp.313–325, DOI https://doi.org/10.1007/s10661-007-9651-0.

Joyner, W.B. and D.M. Boore, 1981. Peak horizontal acceleration and velocity from strong motion records including records from the 1979 Imperial Valley, California, earthquake. Bull. Seism. Soc. Am., 71, pp.2011–2038.

Kaushik, C.P., Ravindra, K., Yadav, K., Mehta, S. and Haritash, A.K., 2006. Assessment of ambient air quality in urban centres of Haryana (India) in relation to different anthropogenic activities and health risks. Environmental Monitoring and Assessment, 122(1), pp.27–40.

Ko, H., 2008. Using GIS in assessing the redevelopment potentials of heritage buildings in Wan Chai, Hong Kong. (Thesis). University of Hong Kong, Pokfulam, Hong Kong SAR, https://doi.org/10.5353/th_b4154959.

Kumar Prabhat and Singh Ashish K., 2011. Preservation of heritage buildings and monuments, NBMCW February 2008.

Maitre, A., Bonneterre, V., Huillard, L., Sabatier, P. and de Gaudemaris, R., 2006. Impact of urban atmospheric pollution on coronary disease. European Heart Journal, 27(19), pp.2275–2284.

Malczewski, J., 2006. GIS-based multicriteria decision analysis: a survey of the literature. International Journal of Geographical Information Science, 20, 703–726.

Paudel, S. and Yuan, F., 2012. Assessing landscape changes and dynamics using patch analysis and GIS modelling. International Journal of Applied Earth Observation and Geoinformation, 16, pp.66–76.

Plexida, S.G., Sfougaris, A.I., Ispikoudis, I.P. and Papanastasis, V.P., 2014. Selecting landscape metrics as indicators of spatial heterogeneity—A comparison among Greek landscapes. International Journal of Applied Earth Observation and Geoinformation, 26, pp.26–35.

Rajan, S.S., Yeshodha, L. and Babu, S.S., 2014. RS and GIS Based Site Suitability Analysis for Solid Waste Disposal in Hosur Municipality, Krishnagiri District, Int. Journal of Innovative Research in Science, Engineering and Technology, 3(3).

Sharma, K., Singh, R., Barman, S.C., Mishra, D., Kumar, R., Negi, M.P.S., Mandal, S.K., Kisku, G. C., Khan, A.H., Kidwai, M.M. and Bhargava, S.K., 2006. Comparison of trace metals

concentration in PM10 of different locations of Lucknow City, India. Bulletin of Environmental Contamination and Toxicology, 77(3), pp.419–426.
Trifunac, M.D. and Brady, A.G., 1975. On the correlation of seismic intensity scales with the peaks of recorded strong ground motion. Bulletin of the Seismological Society of America, 65(1), pp.139–162.
Turner, M.G., Gardner, R.H. and O'neill, R.V., 2001. Landscape ecology in theory and practice (Vol. 401). New York: Springer.
UN-HABITAT, 2003. The Challenge of Slums: Global Report on Human Settlements. https://www.un.org/ruleoflaw/files/Challenge%20of%20Slums.pdf.
Warren, C.R., Lumsden, C., O'Dowd, S. and Birnie, R.V., 2005. Green on green: public perceptions of wind power in Scotland and Ireland. Journal of Environmental Planning and Management, 48(6), pp.853–875.
Weeks, J.R., Hill, A., Stow, D., Getis, A. and Fugate, D., 2007. Can we spot a neighborhood from the air? Defining neighborhood structure in Accra, Ghana. GeoJournal, 69(1–2), pp.9–22.
Zanini, G., Berico, M., Monforti, F., Vitali, L., Zambonelli, S., Chiavarini, S., Georgiadis, T. and Nardino, M., 2006. Concentration measurement in a road tunnel as a method to assess "real-world" vehicles exhaust emissions. Atmospheric Environment, 40(7), pp.1242–1254.
Zhang, Y. and Maxwell, T., 2006. A fuzzy logic approach to supervised segmentation for object-oriented classification. In ASPRS 2006 Annual Conference Reno, Nevada May (pp. 1–5).

Chapter 20
Urban Settlement Pattern and Growth Dynamics in Northwest Himalaya

Sandeep Maithani, Kshama Gupta, Asfa Siddiqui, Arifa Begum, Aniruddha Deshmukh, and Pramod Kumar

20.1 Introduction

The urban settlements in Northwest Himalaya (NWH) are experiencing a rapid pace of urbanization. However, this development is rather skewed in nature. The hilly regions of NWH are experiencing a negligible growth in their population, whereas the population pressure is rising in urban settlements at foothills. One of the primary reasons is the lack of economic opportunities and infrastructural facilities. The economic base of these hilly urban settlements, which is mainly based on tourism, was further eroded due to the recent series of natural disasters. A direct outcome has been reduction in the tax base of urban local bodies, due to which they are now struggling to sustain their existing infrastructure leave apart its augmentation.

In sharp contrast, the urban settlements at foothills are failing to cater the needs of the burgeoning migratory population in terms of adequate social and physical infrastructure. This population migration to existing urban areas could not be accommodated and has subsequently spilled onto the contiguous-rich agriculture lands in a rather unplanned manner, thus affecting the arable lands and hydrological and ecological cycles. It is therefore imperative to study the urban settlement patterns and their growth dynamics in the NWH, not only to sustain the unregulated growth of foothill urban settlements but also to provide impetus to the stagnated urbanization process of hill towns.

The spatio-temporal dynamic studies require time-series geospatial data, which is often found lacking in Indian conditions. Even if time-series data is available, inconsistencies in datasets often induce errors. Remote sensing can address these

S. Maithani (✉) · K. Gupta · A. Siddiqui · A. Begum · A. Deshmukh · P. Kumar
Urban & Regional Studies Department, Indian Institute of Remote Sensing (IIRS), Indian Space Research Organisation (ISRO), Department of Space, Goverment of India, Dehradun, India
e-mail: maithani@iirs.gov.in

R. R. Navalgund et al. (eds.), *Remote Sensing of Northwest Himalayan Ecosystems*,
https://doi.org/10.1007/978-981-13-2128-3_20

problems, as it provides accurate and consistent time-series geospatial data and the environment for analysing these spatial datasets. The present chapter highlights some of the applications of geospatial technology in studying the settlement pattern and growth dynamics of urban areas. These studies will provide a rational basis to the urban planners for taking future planning decisions in order to ensure a sustainable urban development.

20.2 Definition of Urban Settlements

The definition of urban areas as per Census of India is as follows (http://censusindia.gov.in/2011-prov-results):

(a) *Statutory Towns*: Towns which have urban local bodies and are notified under law by the respective state/union territory government.
(b) *Census Towns*: Towns which have a population of more than 5000 persons and a population density of more than 400 persons/hectare with more than 75% of male working population employed in nonagricultural activities.
(c) *Urban Agglomerations (UAs)*: A continuous urban spread consisting of one or more towns and their adjoining outgrowth(s). An urban agglomeration must consist of at least a statutory town, and its total population should not be less than 20,000 as per the 2001 census.
(d) *Outgrowths (OGs)*: It is the area around a city or town, consisting of well-recognized places with urban character, normally lying beyond the urban limits (e.g. railway colony, university campus, port area, etc.).

The Census of India classifies towns based on their population size as per sixfold classification scheme: (i) Class I, 1,00,000 and above; (ii) Class II, 50,000–99,999; (iii) Class III, 20,000–49,999; (iv) Class IV, 10,000–19,999; (v) Class V, 5000–9999; and (vi) Class VI, less than 5000 persons. The classification of towns in the NWH (i.e. Uttarakhand, Himachal Pradesh and Jammu and Kashmir states) as per the above-mentioned sixfold classification, for the census years 2001 and 2011,

Table 20.1 Classification of towns in NWH states during census years 2001 and 2011

	Uttarakhand		Himachal Pradesh		Jammu and Kashmir	
Class of town	Census 2001	Census 2011	Census 2001	Census 2011	Census 2001	Census 2011
Class I	4	6	1	1	2	3
Class II	3	6	0	0	5	4
Class III	14	19	6	7	4	15
Class IV	13	33	7	7	19	30
Class V	27	37	16	15	18	46
Class VI	15	15	26	29	21	24
Total	*76*	*116*	*56*	*59*	*69*	*122*

is depicted in Table 20.1. It can be observed from Table 20.1 that the town strength in the states of Uttarakhand and Jammu and Kashmir (J&K) has grown by 53% and 77%, respectively, whereas the increase in number of towns is marginal in the state of Himachal Pradesh (H.P.).

20.3 Pattern of Human Settlements

Remote sensing images provide an updated and synoptic view of settlements across landscape pattern which helps to analyse the settlement patterns. Geographic information system (GIS)-based approaches have been used to analyse spatial pattern of settlements such as nearest neighbour (Linard et al. 2012; Tian et al. 2012; Zhang et al. 2014), spatial autocorrelation, Ripley's K function (Uria-Diez et al. 2013), high/low clustering method, etc. Nearest neighbour (NN) distance technique is widely used due to its simplicity and ease of implementation (Kint et al. 2004; Yang and Lee 2007). It investigates the arrangement and distance of the sample points by calculating the average NN distance. This index depicts if the settlement points are spatially clustered or randomly located within the study area. If the average NN ratio is less than 1, then the pattern exhibits clustering. If the average NN ratio is greater than 1, then the trend is towards dispersion.

The urban settlement pattern in NWH was analysed for the three constituent states, i.e. H.P., J&K and Uttarakhand, based on NN analysis. The list of urban settlements based on 2001 census was downloaded from the Census of India website, and their location was identified with the help of high-resolution images. The generated vector data was then used as an input in GIS environment for calculating the NN ratio (refer to Table 20.2).The analysis showed that H.P. state had a random pattern of urban settlements, whereas J&K and UK states exhibited clustered pattern. All the three states have a hilly terrain where location of urban settlements is largely determined by the topography, accessibility to water sources and availability of agricultural land. However, H.P. differed from both other states in its settlement pattern, one of the reasons being the good level of infrastructure facilities across the state due to which the tendency of settlements to cluster was reduced.

Table 20.2 NN ratio of towns in NWH states

State	No. of points	Observed mean distance	Expected mean distance	Nearest neighbour ratio	Z-score	Pattern
H.P.	54	14,282.50	13,640.53	1.047	0.66	Random
J&K	51	23,321.54	23,171.74	0.606	−2.088	Clustered
UK	62	11,264.17	14,174.96	0.794	−3.093	Clustered

20.4 Demographic Analysis

For analysing the urban population dynamics of H.P., J&K and Uttarakhand state, the district-wise urban population was analysed for the respective states, on a decadal basis (i.e. 2001–2011), using the census data. The statewise results are as follows:

20.4.1 Himachal Pradesh

The state of Himachal Pradesh (H.P.) is constituted by 12 districts, with its capital city located in the Shimla district. The Shimla district is the most urbanized with a level of urbanization of 23% and 25% during the census years 2001 and 2011, respectively. Solan and Sirmaur had an urbanization rate of 18% and 10% in 2001, respectively, while it was 18% and 11%, respectively, during the year 2011. Kangra and Bilaspur remain the least urbanized districts of the state after Lahaul, Spiti and Kinnaur districts. The average level of urbanization in the state was 10% in 2001 which increased to 11% in 2011. Figure 20.1 shows the levels of urbanizations in H.P. during census years 2001 and 2011.

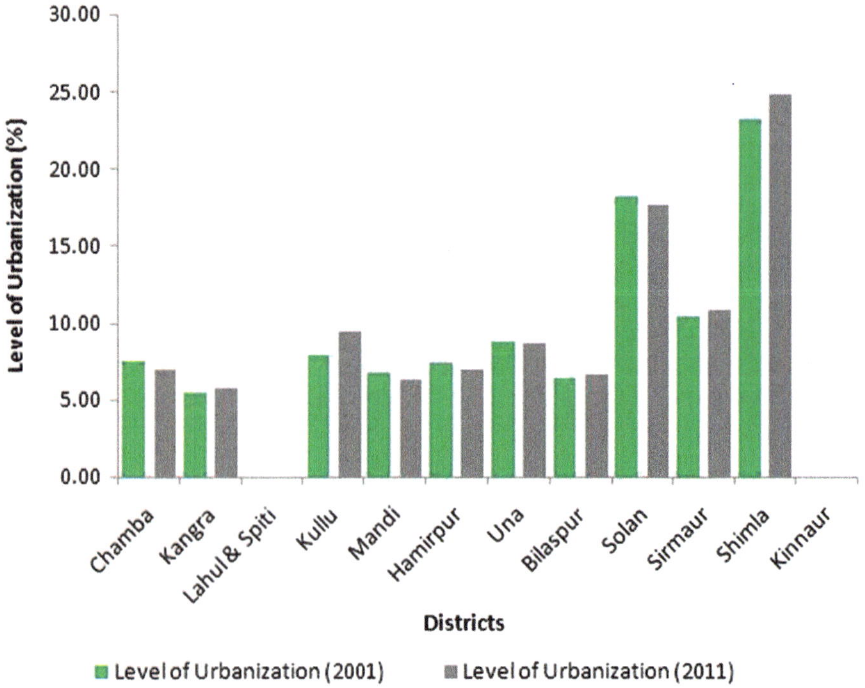

Fig. 20.1 Urbanization levels in H.P. during 2001 and 2011

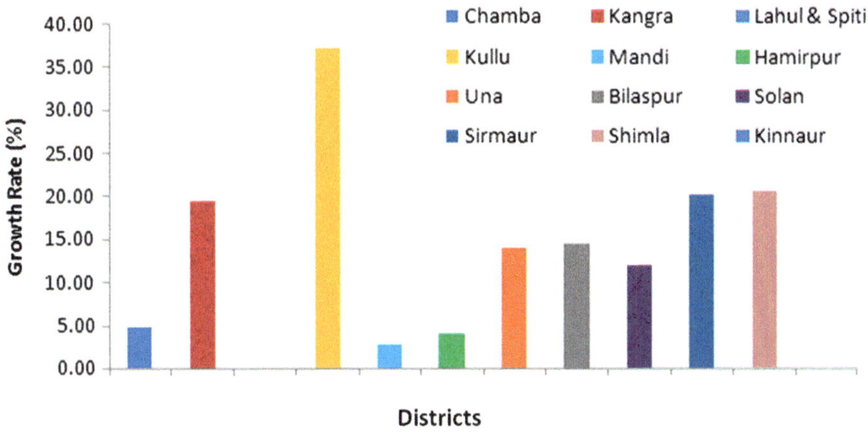

Fig. 20.2 Urban population growth rate in H.P. during 2001–2011

In terms of urban population growth, the districts of Kullu had the highest growth rate; Shimla and Sirmaur stood second and third, respectively, during the census years 2001–2011 (Fig. 20.2). Lahaul, Spiti and Kinnaur districts had negligible urban population growth rates. The average urban population growth rate was 16%.

20.4.2 Jammu and Kashmir

The Srinagar and Jammu districts in the Jammu and Kashmir state had the highest percentage of urbanization, i.e. 79% and 44%, respectively, during the census year 2001. This has gone up to 99% and 50%, respectively, during the census year 2011. Leh stood at third with 24% in 2001 which rose to 43% in 2011. Figures 20.3 and 20.4 show the levels of urbanization during the census years 2001 and 2011, respectively.

The state average level of urbanization during census year 2001 was 25%, while in 2011 census it increased to 27%. Jammu and Kashmir state constituted of fourteen (14) districts during 2001 census; however post-2001 census, eight new districts were carved out (viz. Ganderbal, Kulgam, Shopian, Bandipora, Samba, Reasi, Kishtwar and Ramban); thus the total number of districts in the state increased to twenty-two (22). Henceforth, district-wise comparative analysis of the urban growth for census years 2001 and 2011 was not plausible.

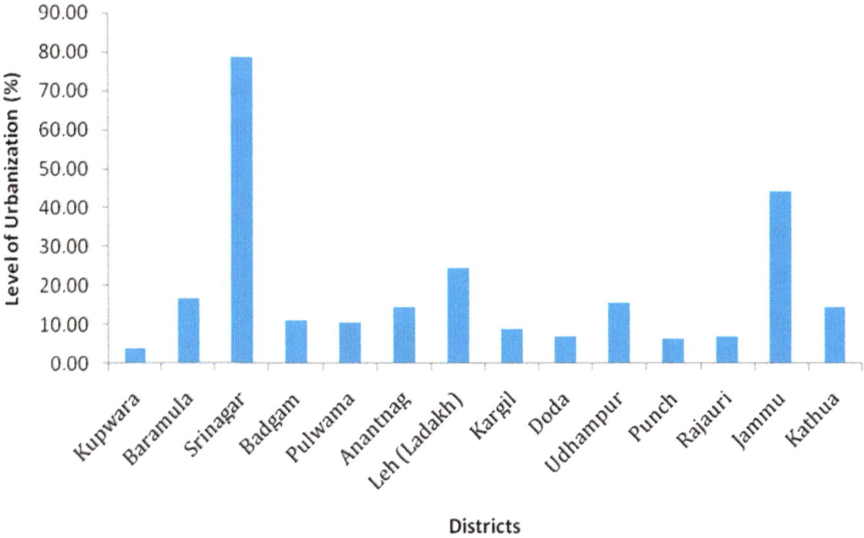

Fig. 20.3 Urbanization levels in Jammu and Kashmir for 2001

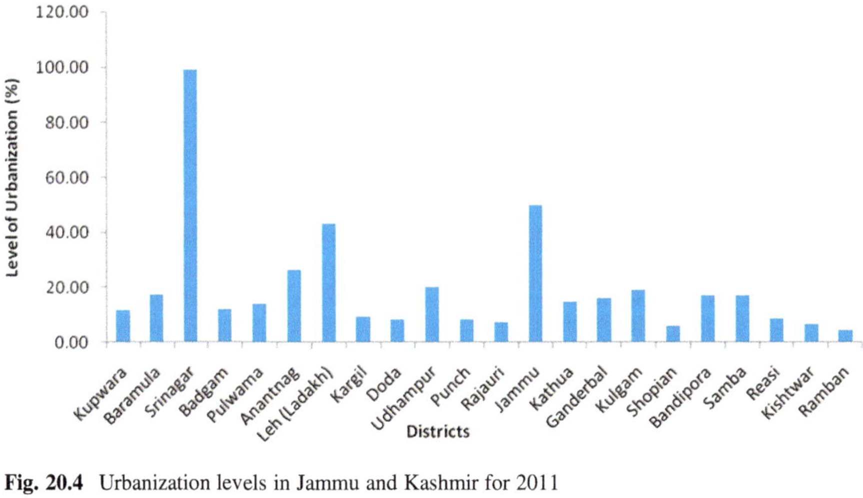

Fig. 20.4 Urbanization levels in Jammu and Kashmir for 2011

20.4.3 Uttarakhand

During census year 2001, the four highly urbanized districts of Dehradun (31%), Nainital (12%), Haridwar (18%) and Udham Singh Nagar (20%) collectively accounted for 82% of state urban population. The rest of nine districts accommodated only 18% of total urban population. The districts of Almora (3%), Uttarkashi (1%), Rudraprayag (0.4%) and Bageshwar (0.1%) had the least share of urban population.

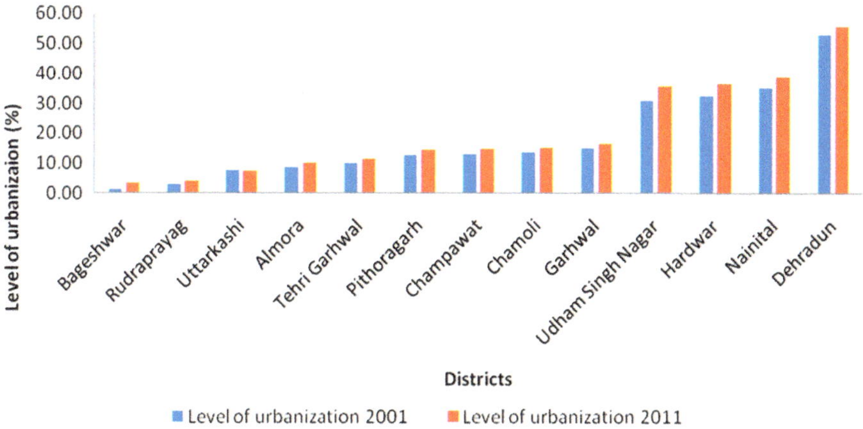

Fig. 20.5 Urbanization levels for 2001 and 2011 in Uttarakhand state

The analysis of disaggregated percentage figures of level of urbanization at the district level within Uttarakhand indicates high level of disparity. The data obtained from the past two censuses, i.e. 2001 and 2011, substantiate this fact. During 2001, the districts of Dehradun (53%), Haridwar (33%), Nainital (39%) and Udham Singh Nagar (36%) reported a level of urbanization above the Uttarakhand (26%) and also the national average (28%). On the other hand, out of the total 13 districts, 9 districts reported level of urbanization below 20%. The lowest among them were Almora (9%), Uttarkashi (8%), Rudraprayag (3%) and Bageshwar (1.20%). Thus, the coefficient of variation of the level of urbanization across districts worked out to be as high as 79.63, reflecting a high degree of dispersion.

The 2011 census also reveals a similar pattern. The highly urbanized districts are Dehradun (56%), Haridwar (37%), Nainital (35%) and Udham Singh Nagar (31%) (Fig. 20.5), which have a level of urbanization higher than the Uttarakhand state (30%) and the national average (31%). Similar to 2001 census, 9 out of 13 districts reported level of urbanization below 20%. The lowest among them were Almora (10%), Uttarkashi (7%), Rudraprayag (4%) and Bageshwar (3%). The urbanization process in the state has thus maintained the high degree of disparity across the districts with the coefficient of variation being 76. Thus, the highly urbanized districts have been able to maintain their relative positions in the two census, while the less urbanized districts are stagnant in terms of their positions in the urban hierarchy.

The pattern of percentage of urban population across the districts has not changed even in 2011. The developed districts have been able to maintain their relative positions, with Dehradun (31%), Nainital (12%), Haridwar (23%) and Udham Singh Nagar (19%) collectively accounting for 85% of the total state urban population. Haridwar district has recorded an increase in share of urban population, i.e. from 18% to 23%, Garhwal district from 2% to 4% and Bageshwar district from 0.1% to 0.3% during census years 2001 and 2011. The less urbanized districts

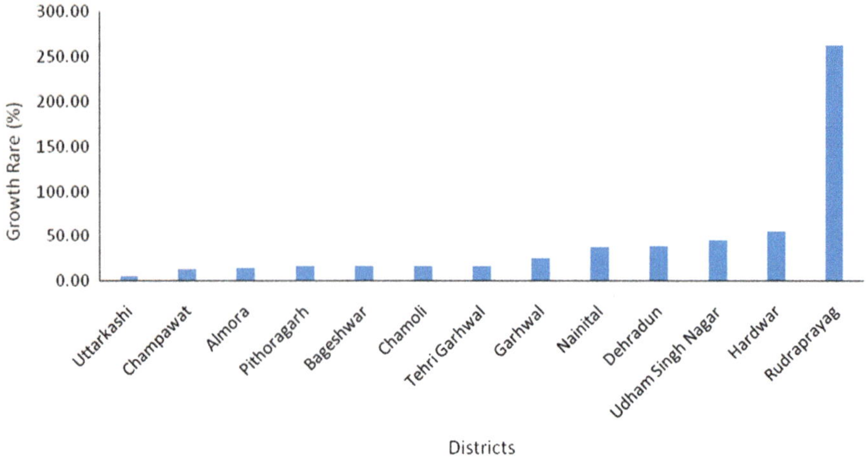

Fig. 20.6 Urban population growth rate during 2001–2011 in Uttarakhand state

of Almora (2%), Uttarkashi (0.8%), Rudraprayag (0.3%) and Bageshwar (0.3%) are stagnant in terms of their share of total state urban population. Pithoragarh has recorded a decrease in percentage of urban population, from 4% to 2% and Uttarkashi from 1% to 0.8% during 2001–2011. The disparity in percentage of urban population has continued during the 2001–2011 decade with the coefficient of variation being 123 and 128 in 2001 and 2011, respectively. Rudraprayag town had an urban population growth rate of 263%, followed by Haridwar (55%); surprisingly Dehradun which has the highest level of urbanization and percentage of urban population had a population growth rate of 39%. Bageshwar had a population growth rate of 16%, while it ranked last in both the previous analyses. Champawat had a population growth rate of 14%, while Uttarkashi had the lowest growth rate of 6% (Fig. 20.6).

20.5 Urban Growth: A Critique

Globally, urban areas are experiencing a rapid growth and proportion of global urban population which is estimated to reach 69% by 2050 (United Nations Populations division 2011). In India also, the share of urban population has risen from 18% in 1961 to 31% in 2011 (Census of India 2011). It is imperative to study urban growth in spatio-temporal scale as this understanding helps in urban planning. Remote sensing is an important tool for analysing the spatio-temporal growth patterns due to repetitive coverage, synoptic view and GIS tools available for spatial data analysis. The temporal remote sensing data have been used to analyse growth rate using landscape metrics such as Shannon's entropy, largest patch index, patch

Table 20.3 Change in built-up area in major towns of Himachal Pradesh during 2005–2015

		Built-up area (km^2)		
Sl. no.	Name of town	2005–2006	2014–2015	Change
1.	Baddi Barotiwala Nalagarh	27.25	29.30	2.05
2.	Dharamshala	4.12	5.18	1.06
3.	Kasauli	1.44	1.90	0.46
4.	Mandi	2.81	2.98	0.17
5.	Paonta Sahib	4.17	6.33	2.16
6.	Parwanoo	3.11	3.23	0.12
7.	Shimla	7.75	10.01	2.26
8.	Solan	5.52	6.08	0.56

Table 20.4 Change in built-up area in major towns of Jammu and Kashmir during 2005–2015

		Built-up area (km^2)		
Sl. no.	Name of town	2005–2006	2014–2015	Change
1.	Anantnag	2.40	6.26	3.87
2.	Baramulla	2.70	7.05	4.34
3.	Jammu	104.68	105.88	1.19
4.	Kathua	15.04	15.09	0.05
5.	Srinagar	48.07	99.24	51.17
6.	Udhampur	6.61	6.66	0.05

density, etc. (Pathan and Jothimani 1985; Kong and Nakagoshi 2006; Vinoth Kumar et al. 2007; Taubenbock et al. 2008).

The towns and cities in three states of NWH are experiencing sprawl, especially in valleys and foothill regions, primarily due to population influx and growth of industries, institutes, etc. For analysing the growth of built-up area in the three states, land use/land cover (LULC) maps for years 2005–2006 and 2014–2015 were downloaded from the Bhuvan website. These LULC maps were generated using IRS-P6 AWiFS under the 'Natural Resources Census (NR-Census)' programme launched by the Department of Space (DOS), Government of India (GoI). The urban growth characteristics for the major towns in three states based on built-up area analysis have been shown in Tables 20.3, 20.4 and 20.5. The contiguous expanse of built-up class was segregated around major towns and cities to understand the urban growth. The decadal urban growth in major towns of H.P. state is mostly insignificant, whereas many towns and cities in J&K and Uttarakhand states have witnessed significant urban growth.

20.5.1 Urban Growth Analysis in Doon Valley

Dehradun is a Class I city and the biggest urban agglomeration in Uttarakhand state. It is one of the major business and service centres besides being the administrative

Table 20.5 Change in built-up area in major towns of Uttarakhand during 2005–2015

		Built-up area (km^2)		
Sl. no.	Name of town	2005–2006	2014–2015	Change
1.	Almora	0.79	4.08	3.29
2.	Dehradun	19.22	81.63	62.41
3.	Haldwani and Kathgodam	6.88	30.41	23.53
4.	Haridwar	20.83	39.44	18.61
5.	Jaspur	1.27	2.20	0.93
6.	Kashipur	5.47	11.90	6.42
7.	Kotdwar	1.18	6.47	5.29
8.	Nainital	1.34	3.20	1.86
9.	New Tehri	1.03	1.97	0.94
10.	Pithoragarh	1.28	6.86	5.57
11.	Rishikesh	3.21	5.28	2.07
12.	Roorkee	10.99	22.52	11.53
13.	Rudrapur	7.46	24.91	17.44
14.	Srinagar	2.05	7.06	5.01

capital of state. The city has experienced a high population growth rate which has almost doubled during the last decade, i.e. from 447,808 in 2001 to 706,124 in 2011 (Census of India, 2011). Its scenic beauty and prominent location in hill state have led to unorganized, haphazard growth, with the rising tide of urbanization (Rawat, 2002).

Indian Remote Sensing Satellite datasets (LISS-II, 36.5 m, and LISS-III, 23.5 m) and Landsat TM (79 m) datasets, at a temporal resolution of approximately 5 years (1987, 1992, 1998, 2003, 2008 and 2013), were used to carry out the present study. After geo-referencing and co-registration of the datasets, LULC maps having six classes, namely, urban built-up, forest, agriculture, waterbody, scrubland and tea garden, were generated using maximum likelihood classifier (MLC) (Fig. 20.7). The urban growth map was analysed for the above-mentioned five time periods using an eight-directional cardinal scheme (i.e. north, north-east, east, south-east, south, south-west, west and north-west) with 1 km incremental buffer from growth centroid of Dehradun.

It was observed that Dehradun urban area has tripled during the period 1987–2013 (Table 20.6). The maximum urban growth rate of 49% was during the period 2003–2008 followed by 32% during the period 1998–2003 and 25% in 2008–2013. This rapid growth could be explained due to the fact that in year 2000, Dehradun became an intermittent capital of newly carved Uttarakhand state. This change in the city status and government's policies attracted a large amount of institutional and commercial activities in Dehradun. As a result, Dehradun has developed as a regional service centre for the entire Garhwal region and pulled a large amount of hill population. This resulting unprecedented growth of urban area has led to chaos, traffic congestion, overcrowding and mass encroachments on the drainage system of the city. A directional analysis of urban growth of Dehradun

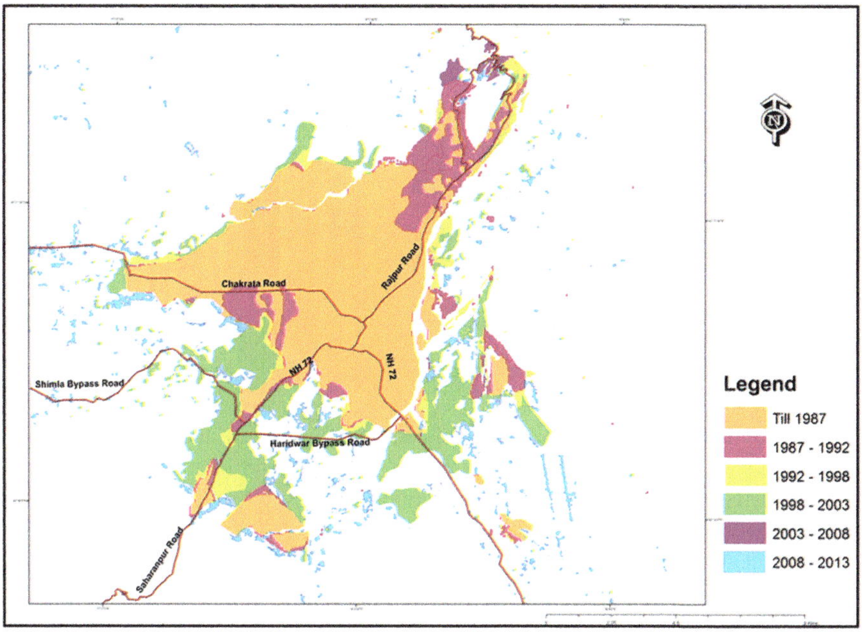

Fig. 20.7 Growth profile of Dehradun urban area during time period 1987–2013

Table 20.6 Urban growth of Dehradun urban area during different time periods (1987–2013)

Year	Area (km^2)	% change
Till 1987	56.2	–
1987–1992	69.1	23.0
1992–1998	74.6	7.9
1998–2003	98.3	31.7
2003–2008	146.5	48.9
2008–2013	183.4	25.2

urban area revealed that major urban growth has taken place in north, west and south-west directions of Dehradun city (Fig. 20.8).

It can be inferred from Fig. 20.8 that urban growth in NWN direction was limited because of its topographical locations. In NNE direction, the growth could be visible due to the development taking place along Rajpur road till Mussoorie. Growth continued in this direction due to the establishments of the educational institutes during 2003–2013. In NEE and ESE directions, post 2003 maximum developments took place along the major arterial roads. From 2003 to 2013, the city displayed tremendous growth and its radius increased from 9 to 15 km distance from the city centre. Post-2003 urban expansion was visible towards south-east and south-eastern (SES, SSW and SWW) parts of the city and is due to a shift of interstate bus terminals (ISBT) and development of residential and educational activities. The analysis of urban growth pattern with respect to direction and distance from city

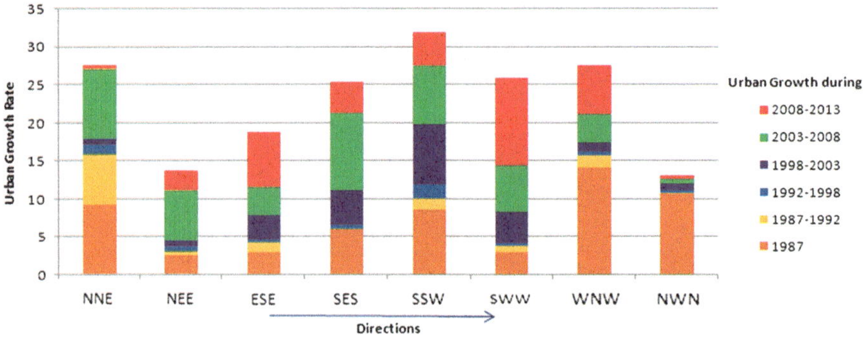

Fig. 20.8 Directional analysis of Dehradun urban growth

centre provides critical inputs for understanding the causes and consequences of growth in particular area. The analysis can be further used to understand the LULC dynamics in the area, further modelling the growth scenarios and predicting the future growth patterns.

20.5.2 Monitoring Urban Growth Using Night-Time Data

The present study attempts to study the applicability of human settlement index (HSI) for monitoring urban growth in the Indo-Gangetic plain (during 2001–2007 period). The HSI was derived using night-time Operational Linescan System (OLS) and the MODIS-NDVI datasets.

OLS sensor operates in visible near-infrared (0.4–1.10 μm) and thermal infrared (10.5–13.4 μm) channels. The sensor has spatial and radiometric resolution of 2.7 km and 6 bits, respectively, and provides a daily coverage of the earth. On a cloudless night, even the faint radiation sources on the earth surface can be detected by the sensor (Sutton et al. 1997, 2001). The National Geophysical Data Centre (NOAA-NGDC) has generated annual stable light composites at 1 km resolution, in which all ephemeral events have been filtered.

20.6 Human Settlement Index (HSI)

NDVI is inversely correlated with the impervious surfaces (e.g. built-up areas). Lu et al. (2008) used this negative relationship for generating human settlement index (HSI):

$$\mathrm{HSI} = [(1 - \mathrm{NDVI}_{max}) + \mathrm{OLS}_{nor}]/[(1 - \mathrm{OLS}_{nor}) + \mathrm{NDVI}_{max} + (\mathrm{OLS}_{nor}{}^{*}\mathrm{NDVI}_{max})] \tag{20.1}$$

where

$$\mathrm{OLS}_{nor} = (\mathrm{OLS} - \mathrm{OLS}_{min})/(\mathrm{OLS}_{max} - \mathrm{OLS}_{min})$$

OLS_{min} and OLS_{max} are minimum and maximum values in the night-time OLS image.

The relationship between the HSI, OLS and NDVI was examined using regression analysis. The regression between HSI and OLS data gave a value of 0.77 and 0.6 for years 2001 and 2007, respectively, while the regression between HSI and NDVI gave a value of −0.79 and − 0.89 for years 2001 and 2007, respectively. It can be inferred from these results that HSI is positively correlated with OLS data, while it shows a negative correlation with NDVI values. A map depicting the growth of HSI values in the study area was generated for the period 2001–2007. The map was classified into three categories, viz. A, low change in HSI; B, medium change in HSI; and C, high change in HSI. Results showed that category A had 46% of the values, while categories B and C constituted 48% and 6% of the HSI change values.

20.6.1 HSI-Based Urban Growth Analysis

The HSI growth map was used for analysing the growth patterns in different states of Indo-Gangetic plain. In states of Delhi and West Bengal, primarily less dense growth was taking place (category A), while Chandigarh had the least percentage of category A. The maximum percentage of dense growth (category C) took place in Bihar followed by Uttar Pradesh, with the least percentage of category C located in Punjab (Fig. 20.9).

20.7 Urban Growth Modelling

Urban areas are expanding in an haphazard manner on the contiguous croplands and forests, leading to their irreversible loss. Hence, the urban planners are now looking for tools and techniques to model and predict the urban growth patterns. These models also provide a rational basis for taking future planning decisions.

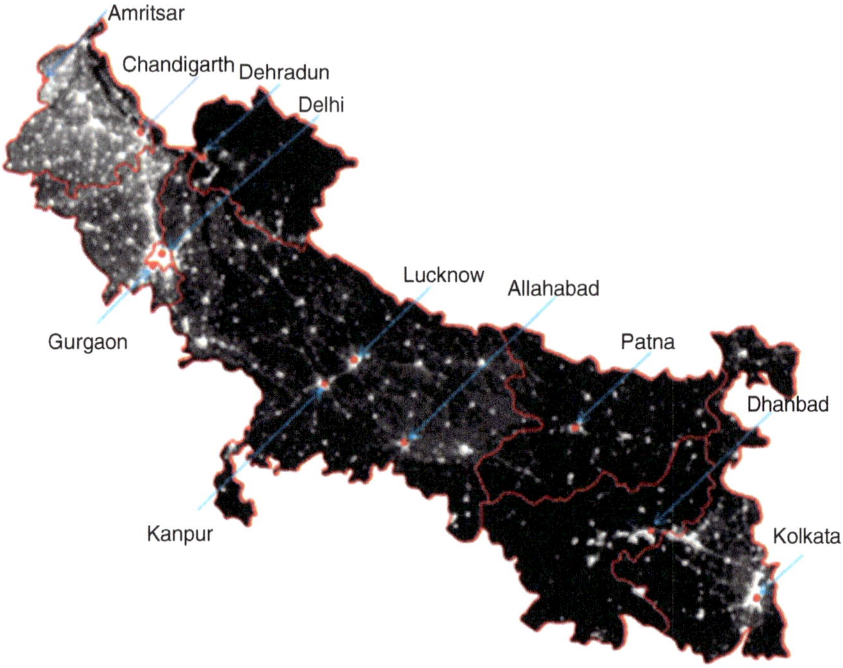

Fig. 20.9 Night-time OLS stable lights yearly composite for the year 2001

20.7.1 LULC Dynamics in Uttarakhand

The study aims to analyse the land cover (LC) transformation in the Uttarakhand state and extrapolate the LC trajectory using a cellular automata (CA)-based spatial predictive model. The CA approach was used as it is simple to understand and implement, can capture the non-linear and complex behaviour of urban systems, has affinity with remote sensing data and is dynamic in nature (Torrens and O'Sullivan 2001). CA is composed of five elements, namely,

(a) lattice, (b) cell states, (c) discrete time steps, (d) transition rules and (e) cell neighbourhood.

A Markovian chain-based CA model was used for simulation of land cover dynamics. Since Markovian process is a probabilistic method with no spatial component, hence CA was utilized in the model for spatial allocation of future land cover changes (Eastman 2001, Yeh and Li 2001, Pontius and Chen 2006). Land cover maps of Uttarakhand state for the years 2006, 2010 and 2015 were downloaded from the Bhuvan site. These maps were reclassified into nine classes for analysis purpose, i.e. built up, agriculture, plantation, forest, swamp, grassland, wasteland, waterbodies and snow (Fig. 20.10). Map overlay analysis was done in a GIS environment for analysing the LC during the periods 2006–2010 (T_1) and

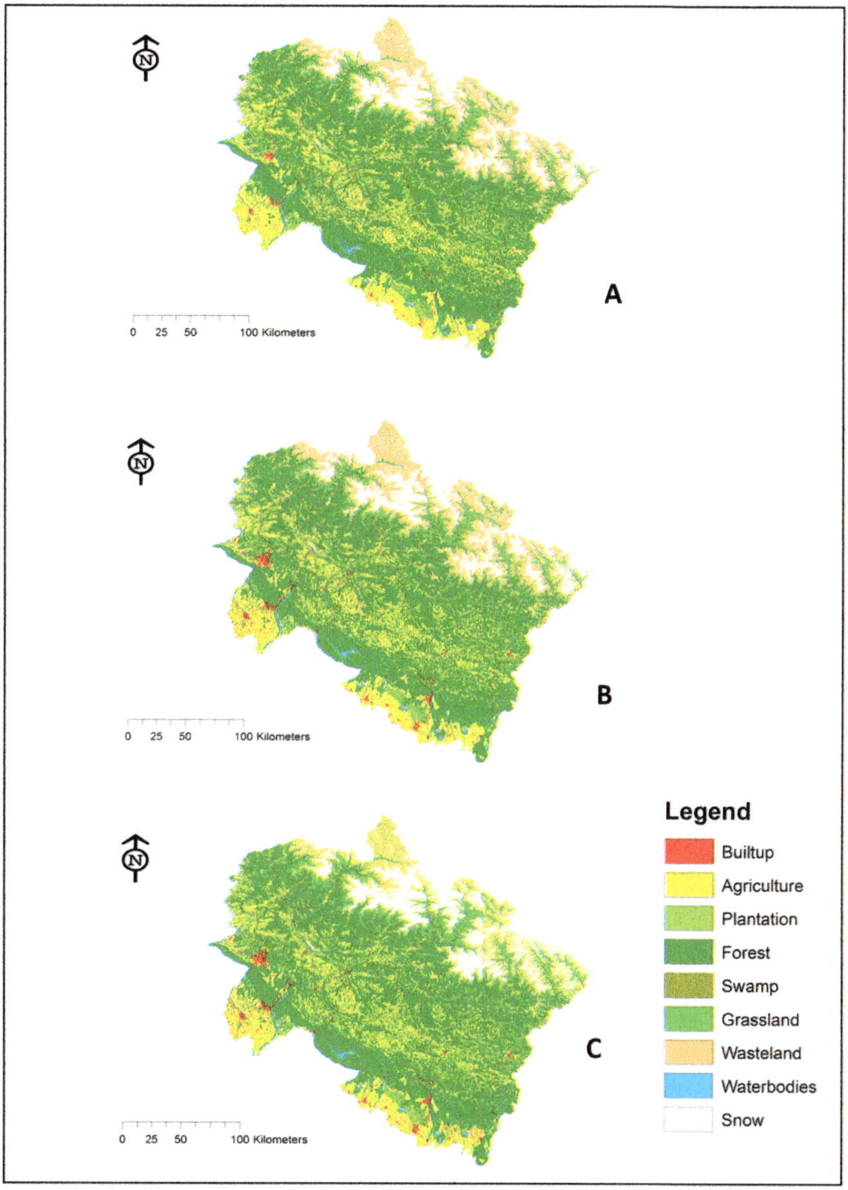

Fig. 20.10 LULC map of Uttarakhand for 2005–2006 (**a**), 2009–2010 (**b**) and 2010–2015(**c**)

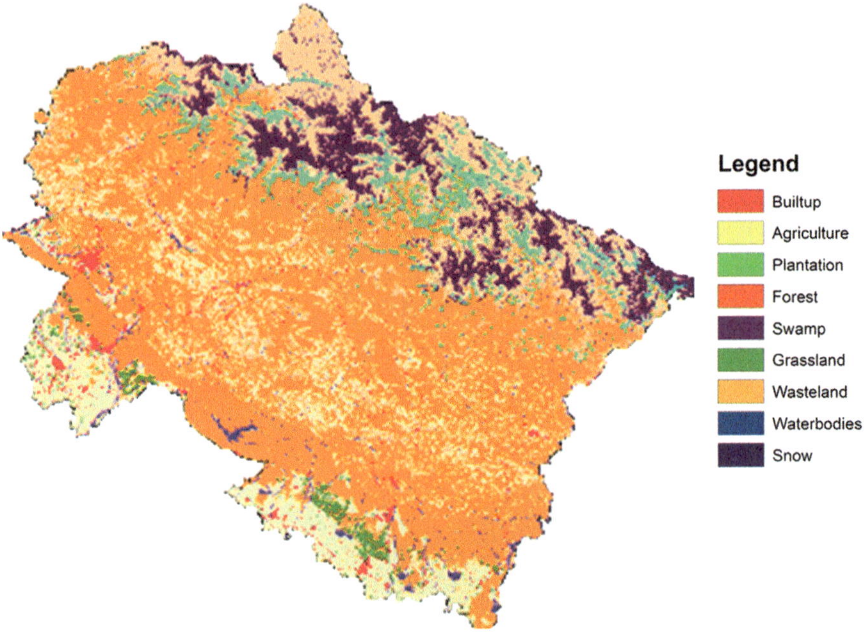

Fig. 20.11 Simulated LULC map of Uttarakhand 2020

2010–2015 (T_2). Markov transitional probability matrix was calculated based on the LC map of period T_1. The CA-Markov was calibrated for period T_1 by changing the cell neighbourhood size and model runs. Model validation was done for period T_2, and subsequently the model was executed to simulate the land cover for year 2020 (Fig. 20.11).

20.7.2 Urban Growth Modelling in Doon Valley, India

The urban growth pattern analysis has been carried out in the Doon valley, with a geographical extent of 29°57′ to 30°35′ N latitude and 77°35′ to 78°20′E longitude. The valley in the recent decades has been experiencing rapid urbanization which has severely affected various hydrological and ecological cycles. In order to ensure a sustainable form of development, it is imperative for urban planners to have a vision of future development scenarios as a response to various policy measures. An artificial neural network (ANN)-based model has been used to simulate future growth scenarios. The model is basically data driven and reduces subjectivity in the urban modelling process.

IRS-1C/1D LISS-III data of years 2001, 2005 and 2009 was used for generation of land cover maps and transport network layer (depicting major and secondary

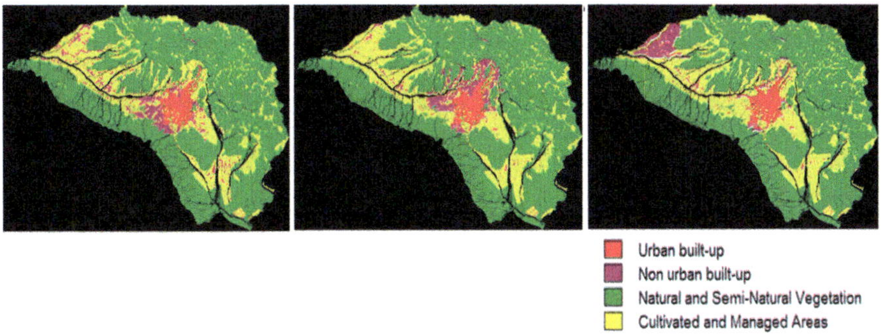

Fig. 20.12 Urban growth pattern under various scenarios for Doon valley

roads). The slope and aspect maps were generated using ASTER DEM, while secondary data was collected from the local city authorities.

The land cover transformation during the periods 2001–2005 and 2005–2009 was analysed. It was found that mainly the cropland was transiting to build up, while in some areas, encroachment of agricultural land into forests was also taking place. These land cover transitions were modelled as a function of some causative factors.

The model results were evaluated based on spatial autocorrelation indicators like Moran's index which accounted for similarity of patterns between the simulated and reference map. For evaluating the model results, land cover map of year 2005 was taken as the reference map. The future urban growth was modelled based on different policy scenarios (Fig. 20.12):

(a) *Scenario A*: This scenario simply extrapolates the existing urban growth trend in the future.
(b) *Scenario B*: This policy scenario predicts a dense urban growth in the study area. This would result in saving of infrastructure cost and make the process of urban growth sustainable in nature.
(c) *Scenario C*: This scenario predicts a hierarchy of urban centres, instead of a concentrated growth. While Dehradun will be the primate city, the other towns would serve as service centre for the surrounding hinterland.

Scenario A predicted high-density growth in existing built-up areas by 2021, while the periurban growth would be concentrated mainly in the south-west and east directions onto the contiguous croplands. In contrast, scenario B predicted most of the future urban growth (dense and compact) in the southern part of the city. Scenario C predicted a hierarchy of settlements, with Dehradun being the mother city and second level of towns in the hierarchy located in the west side of study area mainly along the major transport corridors.

20.8 Bhuvan NUIS and Master Plan Preparation

The 74th Constitution Amendment Act of 1992 bestows the municipal authorities with the power to prepare spatial plans for development of urban areas. In continuation of these policies, MoUD initiated the National Urban Information System (NUIS) as a national mission in the year 2006 for the generation of thematic database for urban areas at scales of 1:10,000, 1:2000 and 1:1000 with the help of remote sensing and GPR techniques. Under NUIS, 12 thematic layers were generated at 1:10,000 scale using Cartosat-1 (PAN) and IRS-P6 LISS-IV MX sensor data; besides four incorporated layers of administrative, forest, city/town boundaries and settlement village locations/names were also generated. The local urban authorities can utilize this data for various urban planning activities, viz. master plan generation and suitability analysis (http://www.nrsc.gov.in/NUIS).

Initially, 152 towns across the country were selected for implementing the scheme. The project has been planned and executed jointly by three organizations, namely, the Town and Country Planning Organization, Survey of India and National Remote Sensing Centre. The database generated under the scheme is available at Bhuvan geoportal (www.bhuvan.nrsc.gov.in).

The data available at Bhuvan portal can be used by the respective city authorities for master plan preparation. The proposed master plan can be uploaded at the Bhuvan portal for public review. Based on the suggestions received from public or various organizations, the master plan can be amended. This process can be carried out iteratively till all the suggestions are incorporated in the master plan. Thus, use of the portal makes planning process more public participatory.

20.9 Conclusions

The study of population dynamics in the three states of NWH exhibited a highly skewed urban population distribution. In Uttarakhand state, during the census year 2001, the four highly urbanized districts of Dehradun (31%), Nainital (12%), Haridwar (18%) and Udham Singh Nagar (20%) collectively accounted for 82% of the total state urban population. These highly urbanized districts were able to maintain their relative positions during 2011 census, while the less urbanized districts remained stagnant in terms of their positions in the urban hierarchy. In state of H.P., Shimla was the primate city during census years 2001 and 2011, while in the state of J&K, the districts of Jammu and Srinagar had the maximum share of urban population. These findings further augment the findings that urbanization process in the NWH tends to be more of monocentric, with a few cities dominating the urban landscape while the rest of the towns continue to stagnate. This necessitates an urgent augmentation of the infrastructure level in these small and medium towns, in order to ensure a sustainable hierarchy of urban system. Thus, it can be inferred from the above discussion that the urban system in NWH states is composed of two subsystems. The first subsystem consists of foothill towns which have a strong economic base and are facing a heavy influx of migratory population. The second subsystem consists of urban centres located in the hills; these towns are tourist, summer stations or administrative centres (i.e. district headquarters) and do not have the necessary economic and infrastructure base to support the process of urbanization and are thus experiencing a heavy outflux of population.

In order to sustain the process of urbanization in NWH, it is mandatory to address the difficulties of the two urban systems, for which one of the prerequisite is to have an updated database for taking future decisions. Geomatics provides a solution to the need for developing an accurate, updated spatial database for urban planning. These spatial databases can be used to study the pattern of urbanization and subsequently develop a spatial predictive model of future urban growth.

The geospatial techniques allow to quantify the growth morphology of the urban centres using NN and other advanced techniques. The use of new datasets like night-time lights in conjunction with other remote sensing datasets also helps in monitoring the urban growth process at a regional level in a rapid and efficient manner. In order to provide a rational basis for urban planners on which they can base their future decisions, spatial predictive models like cellular automata models can also be utilized for future land use/land cover simulation. The subjectivity in CA-based urban modelling can be reduced by using artificial neural networks or other artificial intelligence (AI) techniques. In these models the remote sensing provides the empirical data, GIS provides the site attributes, and AI techniques help in estimating the development potential of a site for future urbanization process. These coupled models provide an understanding of the urban growth process and also provide a realization of the different urban future forms.

The initiative taken by the Ministry of Urban Development under NUIS scheme for providing large-scale urban thematic databases would provide important inputs to the master plan generation process, which can be easily carried out on the *Bhuvan* platform. As illustrated in the above-mentioned case studies, monitoring of spatio-temporal dynamics of urban areas and its future simulation can be done on a routine basis for the NWH cities in a GIS environment using inputs from remote sensing. The large-scale databases developed under the NUIS scheme will expedite the master plan generation of NWH cities, in the absence of which the hill cities are growing in a haphazard manner and are becoming increasingly vulnerable to various hazards as exemplified by the Kedarnath disaster in Uttarakhand and flooding of the capital city of Srinagar in Jammu and Kashmir state.

References

Bhuvan, Indian Geo-portal of ISRO. Retrieved from http://www.bhuvan.nrsc.gov.in. Accessed on April, 2016.

Census of India (2011) Provisional Population Totals Urban Agglomerations and Cities, Data highlights retrieved from: http://censusindia.gov.in/2011-prov-results/paper2/data_files/India2/1.%20Data%20Highlight.pdf, Last accessed on August 25, 2016.

Eastman, J. R. (2001). Guide to GIS and Image Processing Idrisi 3.2 release 2. Worcester: Clark Labs.

Kint V., Robert D.W., Noel L., (2004) Evaluation of sampling methods for the estimation of structural indices in forest stands. Ecological Modelling, 180:461–476.

Kong, F. and Nakagoshi, N (2006) Spatial-temporal gradient analysis of urban green spaces in Jinan, China, Landscape and Urban Planning 78: 147–164

Linard C., Gilbert M., Snow R.W., Noor A.M., Tatem A.J., (2012) Population distribution, settlement patterns and accessibility across Africa in 2010. PLoS One. 2012: 7(2)

Lu, D., Tian, H., Zhou, G. and Ge, H., (2008) Regional mapping of human settlements in southeastern China with multisensory remotely sensed data. Remote Sensing of Environment 112: 3668–3679

National Remote Sensing Centre, National Urban Information System (NUIS) Retrieved on http://www.nrsc.gov.in/NUIS. Accessed on April, 2016.

Pathan, S.K. and Jothimani, P., (1985) Mapping and monitoring of urban sprawl using Landsat MSS data: case studies of three major cities (Rajkot, Bhavangar, and Jamnagar) of Saurashtra region, Gujarat, ISRS symposium on Remote Sensing for Planning and Environmental aspects of Urban and Rural settlements, December 27–28, 1985, Visakhapatnam, India, 45–52.

Pontius, G. R. and Chen, H. (2006). GEOMOD Modelling: Land use and cover change modelling tutorial. Retrieved from http://www2.clarku.edu/~rpontius/pontius_chen_2006_idrisi.pdf. Last accessed on August 2016.

Rawat R (2002) Governance and Citizen Intervention in Urban Planning Processes of Dehradun, India at **url** http://prayaga.org/documents/paper-upp.pdf, Last Accessed July 25, 2013

Sutton, P., Roberts, D., Elvidge, C.D. and Baugh, K., (2001) Census from Heaven: an estimate of the global human population using night-time satellite imagery. International Journal of Remote Sensing 22(16): 3061–3076

Sutton, P., Roberts, D., Elvidge, C.D. and Meij, H., (1997) A comparison of nighttime satellite imagery and population density for the continental United States. Photogrammetric Engineering and Remote Sensing 63: 1303–1313

Taubenbock H., Wegmann M., Berger C., Breunig M, Roth A. and Mehl H., (2008) Spatio-temporal analysis of Indian Mega cities, The international archives of the Photogrammetry, remote Sensing and Spatial Information Sciences, 38(B2), 75–82.

Tian G., Qiao Z., Zhang Y., (2012) The investigation of relationship between rural settlement density, size, spatial distribution and its geophysical parameters of China using Landsat TM images. Ecol. Model.; 231:25–36.

Torrens, P. M., and O'Sullivan, D. (2001) Cellular automata and urban simulation: where do we go from here? Environment and Planning B: Planning and Design, 28, 163–168.

United Nations Populations division, (2011) Urban Population, Development and environment, Department of Economics and Social affairs, http://www.un.org/esa/population/publications/2011UrbanPopDevEnv_Chart/urban_wallchart_2011-web-smaller.pdf, Last Accessed July 25, 2013.

Uria-Diez J, Ibanez R, Mateu J (2013) Importance of habitat heterogeneity and biotic processes in the spatial distribution of a riparian herb (Carex remota L.): a point process approach. Stoch Environ Res Risk A 27:59–76. doi:https://doi.org/10.1007/s00477-012-0569-x

Kumar, J. A. V., Pathan, S. K., & Bhanderi, R. J. (2007). Spatio-temporal analysis for monitoring urban growth–a case study of Indore city. Journal of the Indian Society of Remote Sensing, 35 (1), 11–20.

Yang T.Y., Lee J.C., (2007) Bayesian nearest-neighbor analysis via record value Statistics and nonhomogeneous spatial Poisson processes. Comput. Stat. Data Anal.; 51: 4438–4449.

Yeh, A.G.O., Li, X., (2001) Measurement and monitoring of urban sprawl in a rapidly growing region using entropy. Photogr. Eng. Remote Sens., Vol. 67, No. 1, pp. 83–90. www.asprs.org/publications/pers/2001journal/january/2001_jan _83–90.pdf (accessed 11 November 2008).

Zhang, Z., Xiao, R., Shortridge, A., Wu, J. (2014). Spatial point pattern analysis of human settlements and geographical associations in eastern coastal China—A case study. International journal of environmental research and public health, 11(3), 2818–2833.

Chapter 21
A Reappraisal on Factors for Vector-Borne Diseases (VBDs) in Uttarakhand, India

Ritwik Mondal, R. K. Jauhari, N. Pemola Devi, Sameer Saran, and A. Senthil Kumar

21.1 Introduction

Horizontal transmission is the most propagated parameter for numerous infectious diseases of humankind emphasizing direct enrolment of either vertebrates or invertebrates which involves complex social and demographic factors. Geographical distribution of many vector-borne diseases along with higher altitudes and latitudes is unambiguously related with climate change (Dhara et al. 2013). Due to the high rate of mortality in most of the diseases, there is a need for eco-epidemiological approach for individual disease with emphasis over climatic determinants on vulnerable areas of micro-niche of the vectors. According to the United Nations Framework Convention on Climate Change (UNFCCC), change in climate variability attributes directly or indirectly to human activity causing proportionately alteration of the composition of global atmosphere over comparable time phase (IPCC 2001). In the case of VBDs basically due to increase in upper threshold limit of temperature, the parasites/vectors cope either by curtailing transmission or accelerating the rate of metamorphosis, whereas increase in lower threshold limit of temperature may cause acceleration of the transmission including invasion of new periphery (TIME 2007; Dhiman et al. 2010).

R. Mondal · R. K. Jauhari
Department of Zoology, D.A.V. (P.G.) College, Dehradun, Uttarakhand, India

N. Pemola Devi
Department of Zoology, D.B.S. (P.G.) College, Dehradun, Uttarakhand, India

S. Saran (✉)
Indian Institute of Remote Sensing (IIRS), Indian Space Research Organisation (ISRO), Department of Space, Government of India, Dehradun, India
e-mail: sameer@iirs.gov.in

A. Senthil Kumar
Indian Institute of Remote Sensing, Indian Space Research Organisation, Dehradun, Uttarakhand, India

R. R. Navalgund et al. (eds.), *Remote Sensing of Northwest Himalayan Ecosystems*,
https://doi.org/10.1007/978-981-13-2128-3_21

21.2 Climatic Scenario of Uttarakhand

Uttarakhand is poised on high vulnerability of the climate change being designated as one of the most densely populated Himalayan states of India. The geographical altitude designates the climate perpetually from frozen zone (above 4800 m) to warm temperature (900–1800 m) with underlying zonation, viz. glacier zone (4000–4800 m), alpine zone (3000–4000 m), cold zone (2400–3000 m) and cool temperate zone (1800–2400 m). Glacier melting and unspontaneous downpour are affecting directly the availability of water resources for effective agricultural practices which traditionally depend on the abundance of water. Climate change is being evidenced by the shiftiest of monsoon rains from July/August to August/September and winter precipitation from December/January to January/February, leading to serious environmental, economic and social impacts on the Himalayas, where livelihood depends on the use of the natural resources particularly in the rural areas along with the flourishment of the vector-borne diseases (USCST 2010). In Uttarakhand mainly four seasons are prevalent, viz. summer (March–May), monsoon (June–August), post-monsoon (September–November) and winter (December–February).

The worth-mentioning seasons are monsoon and post-monsoon because breeding habitats and/or onset of immature stages is directly proportionate to the intensity and amount of precipitation. Heavy downpour may either directly increase the breeding sites of the vectors like mosquitoes in large water bodies, viz. lakes and rivers, or indirectly eliminate the immature stages by flushing them off from the breeding sites. The amount, intensity and duration of rainfall proportionately affect the population of mosquitoes and other such vectors (Russell et al. 1963). Relative humidity (RH) attains higher values in post-precipitation phase which directly affects the longevity, survivability and fecundity of mosquitoes, thus contour in transmission of disease (Molineaux and Gramiccia 1980).

21.3 Parameters for Spreading VBDs

Insects are poikilothermic creatures, thus directly depending on the climatic peripheral conditions for their encroachment. The infective parasitic conditions were fully intermingled with the life cycle of vectors, and henceforth their life cycle and developmental features are also affected by climatic conditions. Host, pathogen and transmitting agent as vectors consist of the three pillars of the epidemiological triangle of VBDs which directly depend on their integrity with environment and meteorological parameters (Dhiman et al. 2010). Distribution pattern, endemicity degree and epidemicity intensity of a particular vector-borne disease in a specified area are directly correlated with climatic parameters. The regions where year-round climatic parameters (temperature, relative humidity and precipitation) strikingly remain in optimum conditions are more prone for disease transmission. The

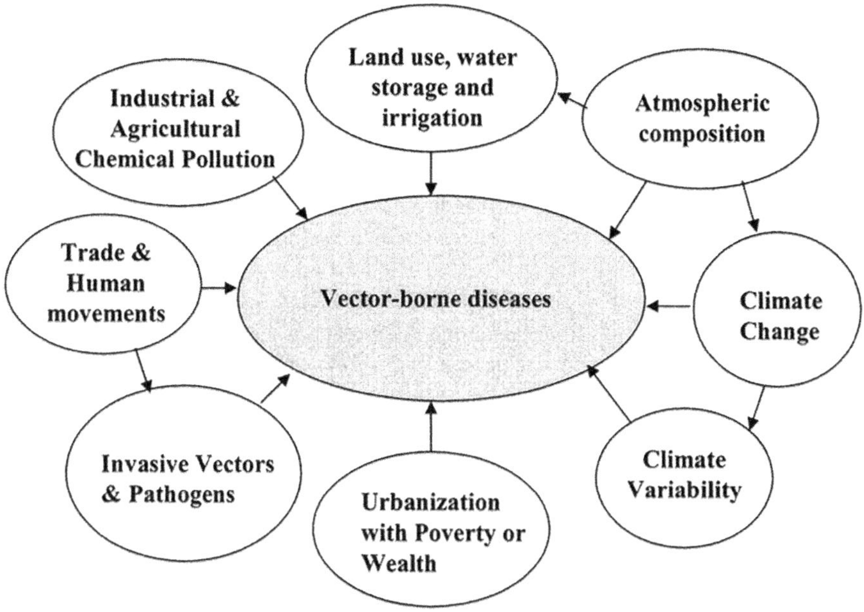

Fig. 21.1 Schematic display showing various parameters supporting the vector-borne diseases

epidemiology of VBDs is complex and involves many factors. However, distribution portfolio of any vector species may not imply the mandate phenomena of the disease transmitted by it. The various developmental features like construction of water reservoirs, intense agricultural practices and phenomenal measures of the health infrastructure along with improvement of education, economic and behavioural conditions that aided improvement of mankind also indirectly propagate the transmission of VBDs. Thus the grass root level information about disease transmission suggests not only the involvement causative and transmitting agents but also accounts demographic and social aspects. The overall impact of both anthropogenic and the natural factors affecting the reappraisal of VBDs is depicted in Fig. 21.1.

21.4 Potential Threat Aspects of VBDs Based on Spatial and Temporal Spectra

Remarkable anthropological activities like improper urbanization and devastating approach of deforestation cause adverse effect on climatic scenarios directly challenging public health for zoonosis. The urge of mankind for socioeconomic development leads them to encroachment of forest, wasteland, etc., redeemed changes in ecology causing altered epidemiological scenario of diseases (Dhiman 2014). The constant changes of the ecology cause altered niche selection for wild animals which may lead to more enormous human-vector conflict, thus posing enormous risk to human health. The immediate effect of deforestation may cause the elevation of the temperature in local scenario which directly propagates the vectors specially increasing their inertia of resting time. Open stagnant pools which occurred either due to deforestation or vegetation clearance are preferred by malaria vectors and hence induce transmission phenomenon (IPCC 2001). In addition meteorological parameters like downpour frequency, elevation in mean optimum threshold temperature and RH are supposed to be the main concrete factors for ensuing transmission (Ansari and Razdan 1998; Sharma et al. 2005). Hence the relationship of vector surveillance distribution pattern along with abiotic meteorological parameters is very complex. Vectors that are found endemic in certain areas can increase the affection of suppressing population immunity in greater proportions, resulting in epidemics in later years (Gage et al. 2008).

21.4.1 Dengue

The role of climatic factors playing the pivotal role in the biomics of *Aedes* mosquitoes along with the transmission of dengue is well established (Fay 1964; Moore et al. 1978 and Focks et al. 1995). Minimum threshold temperature plays a vital role in dengue viruses' survival in *Aedes* mosquitoes in as low as 11.9 °C (IPCC 2001) though multiplication of the virus gets stalked when temperature is lower than 18 °C (Watts et al. 1987). However, temperature rises beyond 42 °C create inimical for immature survival of *Aedes* species (Rueda et al. 1990). In the northern sub-Himalayan region, it was found that transmission potential of the dengue transmission may increase two to five times with an increase of 4 °C in mean temperature (Dhiman et al. 2010). The reason behind the epidemicity may be due to the introduction of new serotypes in the new areas which may be a result of and lack of surveillance of vector due to improper way of urbanization (Gubler 1997, 1998; Gupta et al. 2005) that may include improper and/or scarcity of water leading to water storage practices (Sharma 1998). The overall features about the urbanization issues project in the rise of temperature which will probably extend the dengue incidence in the northern region of India. In India since 1950, almost more than 50 outbreaks have been reported which is directly linked with the efflux of temperature and precipitation (Gage et al. 2008).

In Uttarakhand, the first two cases of dengue were reported in Haldwani town, situated in the foothills of the Kumaon hills in the western Himalayas, in 1996 (Shukla and Sharma 1999). Thereafter eight cases of dengue with one death were reported from Uttarakhand in 2003 (unofficial report from CMO office). As per the records of health department, during 2006–2012 the numbers of dengue cases in Uttarakhand were 12, 67, 140, 76, 4140, 454 and 538. However, during 2010 a major outbreak of dengue was reported in this state in which eight deaths occurred (unofficial report from CMO office). The health department report ascertained the outbreak was reported for the first time in August 2009 in Kumaon region of

Uttarakhand. Some earlier works on related vectors and disease clinical studies were done by Pemola and Jauhari (2012), Singh et al. (2010, 2013), Dhar et al. (2013), Sarangi and Padhi (2014) and Sarkar et al. (2015).

21.4.2 Malaria

Malaria is considered as one of the most serious and complex health problems causing annually almost 1 million deaths among 400–500 million cases globally (IPCC 2001). Global resurgence of malaria occurred due to many reasons which include resistance against drugs and insecticides and human demographic features along with deterioration in public health scenario of malaria. The abiotic changes that happened due to anthropogenic features like irrelevant land uses and deforestation programme causing drastic changes in the meteorological parameters also affect malaria transmission. Dhara et al. (2013) reported that the factors affecting the dispersion of the mosquito densities are indirectly helping the development of *Plasmodium* parasites and thus expanding the malaria distribution to higher latitudes and altitudes. It was postulated by Kiszewski et al. (2004) that by 2050s, the malaria vectors will get shifted more towards temperate zones of northern India. The scenario of the malaria incidence is in surge perspectives in Uttarakhand causing more number of cases in successive years as evidenced by earlier work (Pemola and Jauhari 2006a, b, 2013; Shukla et al. 2007; Srivastava et al. 2011; Pemola et al. 2013 and annual report of NVBDCP 2014–2015). The epidemicity of malaria may be due to the involvement of multifactorial nature which cannot be predicted by simple disease model. Henceforth research on malaria transmission must get intermingled with disease surveillance tools combining trend analysis for multiple sites to account for local factors (Dhara et al. 2013).

21.4.3 Filariasis

Filariasis is emerging as a threat in India, because of its cosmopolitan endemic distribution in both urban and rural areas. *Culex quinquefasciatus* that acts as the vector of the causative parasite *Wuchereria bancrofti* prefers to breed in polluted water but can also breed in clear water. Due to the wide distribution of the *Culex* species for filariasis, it is supposed to be much more independent of the climatic variables than that of malaria and dengue (Dhiman et al. 2010). However, NVBDCP report 2014–2015 suggests that filariasis is still in nascent stage for Uttarakhand state, and this may be due to the absence of dormant condition of the *Culex* mosquito along with their comparatively long process of metamorphosis and henceforth less adaptability with the temperate climate (TIME 2007).

21.4.4 Leishmaniasis

Leishmaniasis is caused by the bite of *Phlebotomus* sand flies categorized mainly under three forms as visceral, cutaneous and mucocutaneous based on the transmission (Park 2014; Singh 2014). Sand fly mainly prefers to breed in temperature ranging from 7 °C to 37 °C and RH above 70% in India (Napier 1926) preferring alluvial soil which contain high subsoil water table, and henceforth it can be said that leishmaniasis is a climate-sensitive disease. Visceral leishmaniasis (VL), profoundly known by the name kala-azar, is the most severe form of leishmaniasis, mainly confined to plains, but recently regular reports of the cases have been reported from sub-Himalayan region of India as because of the presence of various species of *Phlebotomus* (*P. argentipes*, *P. longiductus* and *P. major*) and strains of obligate intracellular parasite *Leishmania* donovani complex (*L. donovani* and *L. infantum*) in Himalayan and sub-Himalayan regions (Raina et al. 2009). Both Garhwal and Kumaon regions of Uttarakhand were evoked with kala-azar in various frequencies and intensities (Singh et al. 1999; Verma et al. 2007; Kumar et al. 2013; Chandra et al. 2013). The increasing scenario of kala-azar in the higher altitudes reflects the contribution of anthropogenic behaviour (deforestation, colonization, urbanization, etc.) in attributing climatic parameters suitable for transmission. Various initiatives were taken by the government of India to eradicate kala-azar, but on the contrary, it happens to re-emerge as a major health issue (Singh 2014). In recent past various cases of lymphadenopathy were reported from nonendemic sub-Himalayan regions of Uttarakhand with unusual characteristic (Verma et al. 2007; Raina et al. 2009). According to Chufal et al. (2016) in nonendemic region where clinical suspicion is low, bone marrow findings even though it is clinically reported as negative for LD bodies can be a strong indicator for VL.

21.4.5 Chikungunya

Chikungunya was first reported from Calcutta (now Kolkata) and its endimicity continued until 1973. Due to the absence of standardized surveillance system for chikungunya, it happened to attain a position of epidemic since its re-emergence from 2005 with a report of more than one million cases (Mavalankar et al. 2007; Epstein Paul 2007); the inclination of the diseases spreads spatially towards northern parts of India (Dhiman et al. 2010). CHIK virus, transmitted mainly by *Aedes aegypti*, is the main causative agent for chikungunya fever. Two particular cycles of transmission prevailed in CHIKV designated as man-mosquito-man (urban cycle) and animal-mosquito-man (sylvatic cycle). Among the prevailing two transmitting cycles, in India urban cycle of CHIKV is mainly found, whereas in Africa mainly the report is on sylvatic cycle. Recently the report of *Ae. albopictus* role in periurban transmission evoked the parameters of sylvatic cycle transmission in India warranting in-depth investigations (Dhiman 2014). In Uttarakhand, chikungunya is still waving its firm paws fixation as negligible cases are reported (unpublished data of CMO Dehradun).

21.4.6 Japanese Encephalitis (JE)

Culex tritaeniorhynchus and *Culex vishnui* are reported as main vectors for the transmission of viral disease Japanese encephalitis. The epidemiology involves wild birds as reservoir host, pig as amplifier host and man as the dead-end host. JE is found to propagate in the rice growing areas as vector species prefer to breed in rice fields (Sunish and Reuben 2001; Tyagi et al. 2013). Symptoms and transmission of JE consisting of the typical biotope in the endemic areas are well defined; however, the propagation and occurrence of acute encephalitis syndrome are quite complicated due to lacunae of knowledge about aetiology and parameters of risk involved. The information about the effect of climatic parameters on JE is negligible, and so it demands necessitating for analysis. However significant number of cases of Japanese encephalitis were registered from the foothills of the Himalaya especially from Udham Singh Nagar and Dehradun of Garhwal region, and from some parts of Haridwar district signifying the fact that the vector gets associated with the changing environmental patterns (NVBDCP report 2015).

21.4.7 Scrub Typhus

The zoonotic disease under this category causes infection like rickettsia (bacteria), and small rodents found in newly cleared forest/scrub grassy areas may act as pivotal role for the transmission. In general nymph and adult mite do not prefer feeding on human, and transmission to man is purely an accidental phenomenon through larval stage. The transmission is mainly reported as transovarial in nature. The most possible way of the larval mite infestation to human occurs when ecological equilibrium is disturbed by their intrusion mainly through forest deforestation (Tsai and Yeh 2013). Patchy-type distribution is often found in mites which indirectly supports the propagation of larva. The gradual rise of the local climate may result in preventing the mites in digging deep into the soil for egg laying (Vaz and Gupta 2006). In Uttarakhand, the rugged mountains of the Garhwal region received almost twice the annual rainfall, which may cause the soil to remain loose along with inductive industrial development which may probably result in conducive vector breeding (Ahmad et al. 2010).

21.4.8 *Cercarial Dermatitis*

Cercarial dermatitis having cosmopolitan distribution is categorized as parasitic disease, encountered in both fresh and saline water, mainly designated for affecting the skin (Hoeffler 1977). Mollusca acts as host, and pathogenicity is likely to happen when molluscs share common water resources with warm-blooded organism like human beings causing severe dermatitis that may last for weeks. Itching rashes caused by cercarial dermatitis are quite synonymous with insect bites. It causes skin dermatitis to the people mainly engaged in agriculture and pisciculture and other such water-oriented work and henceforth popularly also known as swimmer's itch or paddy field dermatitis. Jauhari and Pemola (2014) reported the hazardous effect in rural India, with relative evidences of swimmer's itch occurring in Doon Valley where villagers are largely dependent on water ponds.

The overall views of the major vector-borne diseases pathogenicity and other characteristics can be enumerated in Table 21.1.

21.5 Environmental Monitoring and Surveillance: A Future Aspect

Due to lacunae of routine surveillance for most of the above-said zoonotic diseases, there is an urge for periodic survey of arbovirus infections via systemic serological surveys in vulnerable areas. Monitoring on the early warning signs of the spread of infectious diseases will have both health and economic benefits (Dhara et al. 2013). There is a substantial requirement of new initiatives of research on climate-related health association for prediction of future scenarios. Propagation of risk indicators like surveillance of mosquito as pathogen load in extreme meteorological parameters is also supposed to be a mandatory step in environmental monitoring. The meteorological parameters along with clinical data can be propagated along with geographic information system and spatial analysis which can be further developed to conduct the vulnerability assessments prioritizing public health infrastructure (Jerrett et al. 2010). Demographic data correlated along with GIS data (land use and land cover) can provide information on the sensitivity and adaptive capacity of the VBDs. The spatial information providing the data integration network in combination with distribution of resource for human-environment interaction could be an effective control measure for various vector-borne diseases including malaria (Ageep et al. 2009; Haque et al. 2010; Reid et al. 2010; Tonnang et al. 2010). The assessments of vector vulnerability conducted in aspects of ecological and social periphery in both spatially and temporally integrated social and environmental data can provide in an assembled way the future risk map scenario.

Table 21.1 List of major vector-borne diseases reported from Uttarakhand

Sl. no.	Vector-borne diseases	Causative agent	Common name	Major disease manifestation	Reported vector	Host	Preferred habitat
1.	Dengue	Dengue virus (DENV)	Dengue fever (DF)	Sudden high fever, vomiting, headache, nausea and bleeding gums. In extreme condition, it leads to DSS and DHS	*Aedes aegypti* and *Ae. albopictus*	Human	Standing clear water with rainfall with steady interval
2.	Malaria	*Plasmodium* sp.	Fowl air	Clinical symptoms like fever, fatigue vomiting and headaches. In extreme conditions yellow skin, seizures, coma or even death may occur	*Anopheles* sp.	Human Another host – monkey (Africa)	Stagnant water and humid environment
3.	Filariasis	*Wuchereria bancrofti*, *Brugia malayi* and *Brugia timori*	Elephantiasis	Inflammation in legs, arms, vulva, breasts and scrotum causing oedema in the lymphatic system	*Culex quinquefasciatus*	Human Reservoirs – domesticated mammalian pets	Moist with affinity towards maximum temperature
4.	Leishmaniasis	*Leishmania* sp.	Kala-azar	Fever, damage to the spleen and liver and anaemia	*Phlebotomus* sp.	Rodent and dogs Accidental – human	Agricultural field with rodent burrow
5.	Chikungunya	Chikungunya virus	–	Sudden onset with high fever, joint pain and rash. Other symptoms may occur, including headache, fatigue, digestive complaints and conjunctivitis	*Aedes aegypti* and *Ae. albopictus*	Human	Standing clear water

(continued)

Table 21.1 (continued)

Sl. no.	Vector-borne diseases	Causative agent	Common name	Major disease manifestation	Reported vector	Host	Preferred habitat
6.	Japanese encephalitis	Japanese encephalitis virus (JEV)	Formerly known as Japanese B encephalitis	Onset causes severe rigours. Symptoms like fever, headache and malaise. In severe acute encephalitic cases, cachexia, hemiparesis and convulsions are common with body temperature raised between 38 and 41 °C (100.4–105.8 °F) along with neck rigidity. Mental retardation may occur	*Culex tritaeniorhynchus* and *Cx. vishnui*	Human Reservoirs – pigs and wild birds	Close proximity to human and pigs (amplifying host) enhance the epidemiology of the disease
7.	Scrub typhus	*Orientia* (formerly *Rickettsia*) *tsutsugamushi*	Bush typhus	Fever featured with headache, muscle pain, mainly cough and gastrointestinal symptoms are common. In more virulent strains, haemorrhage and intravascular coagulation occur along with morbilliform rash, scar, splenomegaly and lymphadenopathies. Abnormal liver function may occur in the late phase of illness	*Leptotrombidium deliense*	Normal host – chigger mites on rodents Accidental host – human	Terrain between woods and clearings
8.	Cercarial dermatitis	Cercariae of certain species of schistosomes	Swimmer's itch	Affected part with swollen skin and reddish spot. Cercarial dermatitis occurs on the exposed skin outside of close-fitting garments	*Indoplanorbis exustus* and *Melanoides tuberculata*	Normal host – water fowl Accidental host –humans	Stagnant shallow water bodies

References

Ageep TB, Cox J, M'oawia MH, Knols BG, Benedict MQ, Malcolm CA, Babiker A and El Sayed BB (2009): Spatial and temporal distribution of the malaria mosquito *Anopheles arabiensis* in northern Sudan: influence of environmental factors and implications for vector control. Malar. J., 8:123.

Ahmad S, Srivastava S, Verma SK, Puri P and Shirazi N (2010): Scrub Typhus in Uttarakhand, India: a common rickettsial disease in an uncommon geographical region. Trop. Doc., 40: 188–190.

Ansari MA and Razdan RK (1998): Seasonal prevalence of *Aedes aegypti* in five localities of Delhi, India. Dengue Bull 22:28–32.

Chandra H, Chandra S and Kaushik RM (2013): Visceral leishmaniasis with associated common, uncommon, and atypical morphological features on bone marrow aspirate cytology in nonendemic region. J. Trop. Med., 861032.

Chufal SS, Pant P, Chachra U, Singh P, Thapliyal N and Rawat V (2016): Role of haematological changes in predicting occurrence of Leishmaniasis-a study on Kumaoun region of Uttarakhand. J. Clinical and Diag. Res., 10(5) EC 39-EC43.

Dhar M, Shirazi N, Ahmad S, Srivastava S, Biswas D and Bhat NK (2013): De novo experience of a single outbreak of dengue infection at a tertiary referred centre of Uttarakhand, North India. J. Ind. Academy of Clin. Med., 14(3–4): 225–229.

Dhara VR, Schramm PJ and Luber G (2013): Climate change and infectious diseases in India: Implications for health care providers. Ind. J. Med. Res., 138: 847–852.

Dhiman RC (2014): Emerging vector-borne zoonoses: eco-epidemiology and public health implications in India. Frontiers of Pub. Hlth., 168 (2): 1–6.

Dhiman RC, Pahwa S, Dhillon GPS and Dash AP (2010): Climate change and threat of vector borne diseases in India: are we prepared? Parasitol. Res., 106: 763–773.

Epstein Paul R (2007): Chikungunya fever resurgence and global warming. Am. J. Trop. Med. Hyg., 76: 403–404.

Fay RW (1964): The Biology and bionomics of *Aedes aegypti* in the laboratory. Mosq. News 24:300–308.

Focks DA, Daniels E, Haile DG and Keesling JE (1995): A simulation model of the epidemiology of urban dengue fever: literature analysis, model development, preliminary validation, and samples of simulation results. Am. J. Trop. Med. Hyg., 53:489–506.

Gage KL, Burkot T, Eisen RJ and Hayes N (2008): Climate and vector borne diseases. Am. J. Prev. Med., 35: 436–50.

Gubler DJ (1997): Dengue hemorrhagic fever: its history and resurgence as a global health problem. In: Gubler DJ, Kuno G (eds) Dengue hemorrhagic fever. CAB International, New York, 1–2.

Gubler DJ (1998): Dengue and dengue hemorrhagic fever. Clin. Microbiol. Rev. 11:480–496.

Gupta E, Dar L, Narang P, Srivastava VK and Broor S (2005): Serodiagnosis of dengue during an outbreak at a tertiary care hospital in Delhi. Indian J Med Res 121:36–38.

Haque U, Magalhães RJ, Reid HL, Clements AC, Ahmed SM, Islam A, Yamamoto T, Haque R and Glass GE (2010): Spatial prediction of malaria prevalence in an endemic area of Bangladesh. Malar. J., 9:120.

Hoeffler DF (1977): "Swimmers itch" (cercarial dermatitis). Cutis, 19 (4): 461–467.

Intergovernmental Panel on Climate Change (IPCC). Climate change (2001): The scientific basis. Contribution of Working Group I to the Third Assessment Report of the Intergovernmental Panel on Climate Change. Houghton JT, Ding Y, Griggs DJ, Noguer M, van der Linden PJ, Dai X, Maskell K, Johnson CA, editors. Cambridge University Press, Cambridge, United Kingdom. 1–881.

Jauhari RK and Pemola Devi N (2014): Occurrence of a snail borne disease, cercarial dermatitis (Swimmer itch) in Doon valley (Uttarakhand), India. Iranian J. Pub. Hlth., 43 (2): 162–167.

Jerrett M, Gale S and Kontgis C. (2010): Spatial modelling in environmental and public health. Int. J. Environ. Res. Pub. Hlth., 7:1302–1329.

Kiszewski A, Mellinger A, Spielman A, Malaney P, Sachs E and Sachs J (2004): A global index representing the stability of malaria transmission. Am. J. Trop. Med. Hyg., 70: 486–498.

Kumar A, Rawat V, Thapliyal N and Saxena SR (2013): Kala-azar-A case series from nonendemic area, Uttarakhand. Ann. Trop. Med. Public Hlth., 6:355–357.

Mavalankar D, Shastri P and Raman P (2007): Chikungunya epidemic in India: a major public-health disaster. Lancet Infect. Dis., 7: 306–307.

Molineaux L and Gramiccia G (1980): The Garki project: research on the epidemiology and control of malaria in the Sudan Savanna of West Africa. WHO.

Moore CG, Cline BL, Ruiz-Tiben E, Lee D, Romney-Joseph H and Rivera-Correa E (1978): *Aedes aegypti* in Puerto Rico: environmental determinants of larval abundance and relation to dengue virus transmission. Am. J. Trop. Med. Hyg., 27:1225–1231.

Napier LE (1926): An epidemiological consideration of the transmission of Kala—azar in India. In Reports of the Kala–azar commission, India, Report No 1 (1924–25). Indian Medical Research Memoirs, Memoir No–4, November, 4–6(102): 219–265.

National Vector Borne Disease Control Program (2015): Annual report 2014–15. Govt. of India, NVBDCP, Delhi. 1–38.

Park K (2014): Leishmaniasis. In Textbook of Preventive and Social Medicine. 22edn, Jabalpur, India, Banarasi Das Bhanot Publishers, 267–268.

Pemola Devi N and Jauhari RK (2006a): A relationship between *Anopheles fluviatilis* and *An. stephensi* (Diptera: Culicidae) catches and the prevalence of malaria cases at Kalsi area in Dehradun district (Uttaranchal). Indian J. Med. Res. 123:151–158.

Pemola Devi N and Jauhari RK (2006b): Climatic variables and malaria incidence in Dehradun, Uttaranchal. J. Vector Borne Dis., 43:21–28.

Pemola Devi N and Jauhari RK (2012): Records of potential dengue vector mosquitoes in Doon Valley, Uttarakhand, India. Flora and Fauna, 18(2): 229–232.

Pemola Devi N and Jauhari RK (2013): Meteorological variables and malaria cases based on 12 years data analysis in Dehradun (Uttarakhand) India. European J. Expt. Biol., 3(1): 28–37.

Pemola Devi N, Jauhari RK and Bahuguna JP (2013): Mosquitoes and mosquito borne diseases in Dehradun district, Uttarakhand. J. Env. & Bio-Sciences, 27(2):177–182.

Raina S, Mahesh DM, Kaul R, Satinder KS, Gupta D, Sharma A and Thakur S (2009): A new focus of visceral leishmaniasis in the Himalayas, India. J. Vector Borne. Dis., 46:303–306.

Reid H, Haque U, Clements AC, Tatem AJ, Vallely A, Ahmed SM, Islam A and Haque R (2010): Mapping malaria risk in Bangladesh using Bayesian geostatistical models. Am. J. Trop. Med. Hyg., 83:861–867.

Rueda LM, Patel KJ, Axtell RC and Stinner RE (1990): Temperature dependent development and survival rates of *Culex quinquefasciatus* and *Aedes aegypti* (Diptera: Culicidae). J. Med. Entomol., 27:892–898.

Russell PF, West LS, Manwell RD and MacDonald G (1963): Practical malariology. Oxford University Press, London.

Sarangi MK and Padhi S (2014): Dengue and its phytotherapy: a review. Int. J. Pharmaceutical Phytopharmacological Res., 4(1): 37–46.

Sarkar M, Kumar K, Sharma AK and Gupta AK (2015): Ento epidemiological characterization of dengue in Uttarakhand (India): J. Mosq. Res., 5(17): 1–10.

Sharma RS, Joshi PL, Tiwari KN, Katyal R and Gill KS (2005): Outbreak of dengue in national capital territory of Delhi, India during 2003. Vector Ecol., 30:337–338.

Sharma SK (1998): Entomological investigations of DF/DHF outbreak in rural areas of Hisar District, Haryana. India. Dengue Bull., 22:36–41.

Shukla RP and Sharma SN (1999): *Aedes aegypti* survey of West Himalayan foothill town of Haldwani, district Nainital, India. Dengue Bull., 23: 113–114.

Shukla RP, Sharma SN and Dhiman RC (2007): Seasonal prevalence of malaria vectors and its relationship with malaria transmission in three physiographic zones in Uttaranchal state, India. J. Vect. Borne Dis., 44:75–77.

Singh R, Singh SP and Ahmad N (2013): A study on clinical and laboratory profile of dengue fever in tertiary care centre of Uttarakhand. Int. J. Res. Med. Sci., 2(1): 160–163.

Singh RK, Dhiman RC, Dua VK and Joshi BC (2010): Entomological investigations during an outbreak of dengue fever in Lalkuan town, Nanital district of Uttarakhand, India. J. Vect. Borne Dis., 47: 189–192.

Singh S (2014): Changing trends in the epidemiology, clinical presentation and diagnosis of Leishmania-HIV co-infection in India. Int. J. Infet. Dis., 29:103–112.

Singh S, Biswas A, Wig N, Aggarwal P, Sood R and Wali JP (1999): A new focus of visceral leishmaniasis in sub-Himalayan (Kumaon) region of northern India. J. Commun. Dis., 31:73–77.

Srivastava S, Ahmad S, Shirazi N, Verma SK and Puri P (2011): Retrospective analysis of *vivax* malaria patients presenting to tertiary referral centre of Uttarakhand. Acta Tropica. 117:82–85.

Sunish IP and Reuben R (2001): Factors influencing the abundance of Japanese encephalitis vectors in rice fields in India-I Abiotic. Med. Vet. Entomol., 15:381–392.

TIME. (2007). Vagobond Virus. Dengue fever is spreading and some think climate change is to blame. TIME magazine 2007.

Tonnang HEZ, Kangalawe RYM and Yanda PZ (2010): Predicting and mapping malaria under climate change scenarios: the potential redistribution of malaria vectors in Africa. Malar. J., 9:111.

Tsai PJ and Yeh HC (2013): Scrub typhus islands in the Taiwan area and the association between scrub typhus disease and forest land use and farmer population density: geographically weighted regression. BMC Infect. Dis., 13:191.

Tyagi RK, Das MK, Singh SS and Sharma YD (2013): Discordance in drug resistance associated mutation patterns in marker genes of *Plasmodium falciparum* and *Plasmodium knowlesi* during co-infections. J. Antimicrob. Chemother., 68(5):1081–1088.

Uttarakhand State Council for Science and Technology (2010): Report of brain storming session on impact of climate change on Himalayan live hood. USCST; UCOST. 1–12.

Vaz LS and Gupta NK (2006): Outbreak of scrub typhus in Jammu – a report. MJAFI. 62:342–343.

Verma SK, Ahmad S, Shirazi N, Kusum A, Kaushik RM and Barthwal SP (2007): Sodium stibogluconate-sensitive visceral leishmaniasis in the non-endemic hilly region of Uttarakhand, India. Trans. R. Soc. Trop. Med. Hyg., 101:730–732.

Watts DM, Burke DS, Harrison BA, Whitmire RE and Nisalak A (1987): Effect of temperature on the vector efficiency of *Aedes aegypti* for dengue 2 virus. Am. J. Trop. Med. Hyg. 36:143–152.

Part VII
Geospatial Data, Web Services and Analysis Tools

Summary

Systematic investigations pertaining to various resources/themes of any region requires reliable data in spatial format at different scales and at different time intervals. Northwest Himalaya is no exception. In addition to thematic data, one also requires high resolution elevation information. It is also likely that all data required is not available from a single source. Methods of accessing data/products from different sources as well as the need to have appropriate tools to integrate them and to carry out spatial analysis in an open environment are necessary. In this context, earth observation data requirements for NWH, mountainous region with steep slopes, snow covered peaks, biodiversity rich forests, prone to many disasters of both geologic and hydrometerological origin have been evaluated vis-à-vis current availability and gaps. The first article discusses these aspects and also the radiometric and geometric corrections specifically required for such terrains. The second article describes the importance of web services, geo-portals, open source GIS services, concepts of standards and interoperability. It also demonstrates use of available data from open web services, data servers for a case study of flood warning, flood inundation and generating a flood inundation map with administrative maps to be overlaid for parts of Jammu and Kashmir. It also provides an extensive list of all open source web sites. How to interpolate observations for regions in a spatial format and tools developed are discussed in the third article.

Chapter 22
Geospatial Data for the Himalayan Region: Requirements, Availability, and Challenges

S. Agrawal, S. Raghavendra, Shashi Kumar, and Hina Pande

22.1 Introduction

Availability of appropriate geospatial data over the Himalayan and adjacent Tibetan Plateau region has emerged as one of the key data requirements to understand the fragile landscape dynamics, changing climate and its implications, and assessment of natural resources of the region. In the Himalayas, there are three principal agents of change: plate tectonics, climate change, and human interaction; all three work in a very intricate manner to modulate all natural processes and features. Data requirements can be as diverse as the processes and features of the Himalayas. Spatial resolution requirements of RS data vary from tens of centimeters to tens of meters and a few minutes to years in terms of temporal resolution. Availability of data was limited to systematic aerial photography by the mapping agencies carried out in the second half of the last century and resultant topographical maps on 1:50,000 scale earlier. Satellite coverages are available from the early 1970s. In spite of enormous progress in satellite imaging, large tracts of Himalaya remain unexplored in terms of data availability at a high spatial resolution. Therefore, in spite of wide spread concern, development needs, and environment and security issues, the Himalayas have emerged as "data-scarce" region of world. In this context, it is very pertinent to evaluate existing datasets, and then a systematic attempt can be made on data acquisition strategy involving near-ground (UAVs (unmanned aerial vehicles))

S. Agrawal (✉)
Geospatial Technology and Outreach Programme Group (GT&OP Group), Indian Institute of Remote Sensing (IIRS), Indian Space Research Organisation (ISRO), Department of Space, Govt. of India, Dehradun, India
e-mail: shefali_a@iirs.gov.in

S. Raghavendra · S. Kumar · H. Pande
Photogrammetry & Remote Sensing Department, Indian Institute of Remote Sensing (IIRS), Indian Space Research Organisation (ISRO), Department of Space, Govt. of India, Dehradun, India

R. R. Navalgund et al. (eds.), *Remote Sensing of Northwest Himalayan Ecosystems*,
https://doi.org/10.1007/978-981-13-2128-3_22

platforms to aerial and spacecraft platforms. The spatial data acquisition strategy should be driven by the most immediate concerns of the region, i.e., disaster monitoring and mitigation, natural resource management including cryosphere status vis-à-vis climate change impact assessment, infrastructure development, and crustal deformation.

Topography and relief are the most critical components of any mountain system to understand surface processes and features associated with it. Relief in the form of digital elevation models (DEMs) is used for visual and mathematical analysis of topography, landforms, as well as modeling of surface processes. It provides easy extraction of commonly used terrain parameters such as slope, aspect, curvature, and slope length with finer details. Satellite images have proven their capabilities in generation of DEM, and with the advent of high-resolution satellite data ranging from 10 m from SPOT-1 to 2.5 m in case of Cartosat-1, high-resolution DEMs even up to submeter-level accuracy have been attempted. Other remote sensing techniques used to collect topographic information include interferometric synthetic aperture radar (InSAR), radar altimetry, and light detection and ranging (LiDAR). They can provide height accuracy in the range of submeter to centimeters. However, in the Himalayan region, it is observed that there are challenges in obtaining a high-quality DEM mainly due to steep topography and shadows in case of optical remote sensing and layover and foreshortening in case of microwave remote sensing. As a result, high-quality DEM with a spatial resolution less than 10 m and with height accuracy of 3–4 m or higher is not available as a regular product. The need for this was felt during execution of several projects under Uttarakhand Disaster Recovery Project and all-weather road development project of Government of India. Other projects related to mining and excavation also require higher height accuracy (<1 m), for which alternate techniques like UAVs are being explored by the state governments. Overall the requirement is to have a DEM with 5–10 m spatial resolution and height accuracy of 3–4 m or less than a meter in case of mining and infrastructure-related projects. The temporal resolution also varies from once in many years to every year in case of landslide hazard assessment, river erosion, aggradation, and flood zoning-related studies. Mining may require data even at higher frequency due to periodic volume estimation requirements. After the Nepal earthquake in 2015, every point on ground (in the affected region) has moved up/down or laterally by around a meter or so, thus disturbing all positions of ground control points. Therefore, after every major earthquake, it is imperative to map the entire region with accuracy that can meet the highest standards of topographical survey as well as developmental needs.

Next to topography is the surface cover, which is crucial to understand many processes active in the Himalayan region. In addition to having data with high spatial resolution, frequent repetitivity, and stereo capability, it is important to have data with high radiometric resolution, since the Himalayan surface has not only forest shadow regions with low albedo but also snow-covered regions with very high albedo. In view of the cloud cover present over the mountain regions very often, it is necessary to have SAR observations at appropriate frequencies and resolutions. Although space technology has progressed well in recent times, still there are many challenges due to limitations inherent to technology and terrain. In the present context, all the three segments (ground, near ground, and space) have developed, and clearly the winning strategy lies in synergy between the three with clear advantage of each segment in terms of technology and deployability in extreme topography. Therefore, in this chapter, all known issues related to data acquisition and processing are discussed, and possible solutions are provided to address multitude of applications with a focus on synergistic use.

22.2 Earth Observation Satellites

A number of Earth observation platforms are used for observation and dynamic monitoring of the Earth's land, atmosphere, and oceans. EO satellites (Indian & Global) provide necessary data in different spatial, spectral, and temporal resolutions covering different parts of the electromagnetic (EM) spectrum (visible, infrared, thermal infrared, and microwave) to cater to different user requirements. Currently, there are 12 Indian remote sensing satellites for Earth observation (Resourcesat-1, 1A, 2, Cartosat-1, Cartosat-2 series, Oceansat-2, Megha-Tropique, SARAL, RISAT, etc.). A summary list of all international EO satellites operating in VNIR, thermal, and microwave (SAR) is provided in Tables 22.1, 22.2, and 22.3. Even with the availability of such a large number of satellites, certain gaps exist. Table 22.4 summarizes requirements and availability vis-à-vis gaps. Some of the challenges in data processing and the work carried out at IIRS in this direction are described in the next few sections.

22.3 Geospatial Data Challenges

The planimetric accuracy of satellite images is extremely variable, due to geometric distortions introduced by topography, Earth curvature, sensor errors, and satellite viewing angle, and therefore, requires corrections with best possible accuracy. The overall geolocation accuracy can vary significantly in the case of hilly/mountain areas or when images are acquired with large off-nadir angles. Therefore, getting precise terrain coordinates for large-scale mapping using very high-resolution images is a challenging task even today with all technological advancements.

Secondly, satellite imagery in hilly terrain is highly affected by variations in the topographic parameters such as altitude, slope, and aspect which affect the brightness value of the satellite images having different spectral and spatial resolutions. Differential illumination results in significant variation in the spectral characteristics of similar other land cover features. Sun-facing or illuminated slopes have high spectral radiance or reflectance than the opposite and other slopes. Differential illumination effects in satellite imagery conceal the information on the north-facing or leeward slopes in Himalayan context.

Table 22.1 Some of the Earth observation systems and their characteristics[a]

Satellite	Sensors	Spectral range	Spatial resolution (m)	Swath (km)
Landsat 1 and 2 (1972, 1975)	MSS 4	VNIR	79	185
	RBV	VNIR	30	
Landsat 3 (1978)	MSS 5	VNIR+thermal	79	185
	RBV	PAN	30	
Landsat 4 and 5 (1982, 1984)	MSS 5	VNIR		185
	TM	VNIR,SWIR, thermal	30, 120 (T)	
Landsat 7	ETM	VNIR,SWIR, thermal, PAN	30, 60, 15	185
Landsat 8	OLI	VNIR, SWIR, PAN, thermal (2)	30, 15, 100	185
IRS-1A, 1B, and P2 (1988, 1991, 1993)	LISS-1	VNIR	72.5	148
	LISS-2	VNIR	36.25	74
IRS-1C, 1D (1995, 1998)	WiFS	VNIR (2)	188	770
	LISS-3	VNIR, SWIR (4)	23	140
	PAN		5.8	70
Resourcesat-1, 2, and 2A (2003, 2011, 2016)	AWiFS	VNIR, SWIR (4)	56	740
	LISS-3	VNIR, SWIR (4)	23	140
	LISS-4	VNIR (3)	5.8	70
SPOT 1, 2, and 3 (1986, 1990, 1993)	PAN	PAN	10	117
	Multispectral	VNIR	20	117
SPOT 4 and 5 (1998, 2002)	PAN	PAN	2.5	60
	Multispectral	VNIR	10–20	60
SPOT 6 and 7 (2012, 2014)	PAN	PAN	1.5	60
	Multispectral	VNIR	6	60
MOS-1A and 1B (1987, 1990)	MESSR	VNIR, TIR		100
JERS-1 (1992)	Optical	VNIR, SWIR	18 × 24	75
Sentinel-2 (2015)	Multispectral	VNIR, SWIR	10, 20 (red edge)	290
	IMAGER	(0.443–2190 mm)	60 (atmo. corr)	
Terra (1999)	ASTER	VNIR (3), SWIR (6), TIR (5)	15, 30, 90	60

[a]Adapted from Navalgund and Rajani, Current Science special section on Geospatial techniques in archaeology (under publication)
VNIR visible and near infrared, *SWIR* shortwave infrared, *TIR* thermal infrared

In applications which require temporal/multisensor data integration, precise radiometric and atmospheric correction or normalization between these temporal/multisensor images is essential. Some applications require geometrically preprocessed images in near real time irrespective of platforms and sensors. Hence the need arises for rectifying the data for the geometric and radiometric distortions

Table 22.2 High spatial resolution satellites[a]

Satellite	Sensor	Spatial resolution (m)	Swath (km)
IKONOS (1999)	PAN	0.82	11.3
	MS (multispectral)	3.2	
QuickBird (2001)	PAN	0.61	16.8
	Multispectral	2.44	
WorldView-1 (2007)	PAN	0.50	17.2
WorldView-2 (2009)	PAN	0.46	16.4
	MS	1.85	
WorldView-3 and 4 (2014, 2016)	PAN	0.31	13.1
	MS	1.24	
	SWIR	3.70	
Cartosat-1 (2005)	PAN (stereo)	2.5	30
Cartosat-2 (2007)	PAN	0.83	~10
GeoEye-1 (2008)	PAN	0.46	
Pleiades-1A/1B (2011, 2012)	PAN	0.7	20
	MS	2.8	
KOMPSAT-3A (2015)	PAN	0.7	16.8
	MS	2.8	

[a]Adapted from Navalgund and Rajani, Current Science special section on Geospatial techniques in archaeology (under publication)

Table 22.3 Synthetic Aperture Radar (SAR) missions[a]

Mission	Frequency (GHz)	Nominal resolution (m)	Swath
Seasat (1978)	L (1.275)	25	100
SIR-A (1981)	L (1.275)	40	50
SIR-B (1984)	L (1.275)	15–50	20–50
SIR-C (1994 and 1995)	X (9.6)	10–200	15–90
	C (5.3)		
	L (1.25)		
ERS-1/2 (1991)	C (5.3)	26 × 28	100
JERS-1 (1992)	L (1.275)	18 × 18	75
Radarsat-1 and 2 (1995, 2002)	C (5.3)	9 × 9 to 100 × 100	45/510
Envisat-ASAR (2002)	C (5.3)	30	56
ALOS-1 and 2 (2006, 2014)	L (1.257)	3	25
RISAT-1 (2012)	C (5.3)	3 × 3 to 5 × 50	/250
Sentinel-1A and 1B (2014, 2016)	C (5.405)	5 × 5 to 25 × 100	80–400

[a]Adapted from Navalgund and Rajani, Current Science special section on Geospatial techniques in archaeology (under publication)

Table 22.4 EO data requirements, availability, and gaps for the Himalayas

Sl. no.	Requirements	Availability	Gaps
1.	Multispectral (VNIR)	Landsat, Resourcesat, Sentinel, etc.	Nil
	~10 m, ≥5 days, > 10 bit radiometry	(Refer Table 22.1)	
2.	Panchromatic/ multispectral data	Cartosat-2, WorldView, QuickBird, Pleiades, etc.	Lack of systematic coverage
	~1 m, daily coverage/ ~5 days	(Refer to Table 22.2)	Data not available for all areas at the required time
		Planet Labs satellite constella tion: global coverage at 3–5 m spatial resolution	
3.	Stereo coverage and DEM	Cartosat-1, ALOS-PRISM, ASTER, SRTM (10–30 m DEM)	High-resolution (~1 m) stereo data
	~5–10 m	Z earth (proposed from CNES)	LiDAR-based stereo for specific areas
	~1 m and better		
4.	Microwave SAR: L,C,X band data	RISAT (C-band), Sentinel (C-band), ALOS PALSAR (L band), Radarsat-1 and 2, etc.	Constellation of SAR satellites for disaster monitoring
	~ 10 m, ≥10 days	(Refer to Table 22.3)	Interferometric data for surface deformation, cryosphere: NISAR L and S band SAR may fill this gap (2020)
	~10 m, daily		
	Interferometric SAR		
5.	Thermal data	Landsat, ASTER, etc. Refer to Table 22.1	Multiband high spatial, temporal, and radiometric resolution data for forest fire monitoring
	~100 m, ~10 days in 10–14 μm		
	~100 m, daily in SWIR		
6.	Hyperspectral data	Hyperion, Chris-Proba	Hyperspectral data currently unavailable on a continuity basis from EO missions
	~20 m, 10 days, ~3–5 nm bandwidth	Airborne AVIRIS data available over specific sites	Aircraft with HS sensor should be available ready to fly
7.	Spaceborne LiDAR data	ICESat, CALIPSO, Slicer, etc. (spaceborne missions)	Restrictive data dissemination policy regarding aerial/UAV data
	Aerial data for high-resolution DEM		Dense LiDAR footprints at fine spatial resolution from EO platforms
	UAV		Continuity of space missions
	Airborne and terrestrial LiDAR		

induced not only due to sensor and platform geometry but also due to relief variations. This is addressed by employing two prominent techniques known as ortho-rectification that removes the relief distortions and topographic corrections or topographic normalization that compensates for the differential solar illumination due to rugged mountainous terrain (Garg et al. 2008, Hahn et al. 2003, Negi et al. 2009, Berthier et al. 2014). However, these two set of corrections cannot be applied in a straightforward manner due to complexities involved in data acquisition and terrain geometry.

22.4 Geometric Distortions of Optical Images

The geometric distortions vary with respect to the platform (airborne/satellite), sensor, resolution (V/IR or SAR; low to high resolution), view angle (its position relative to objects), and topography. The sources of distortion can be categorized into two broad classes: acquisition system related (platform, imaging sensor, and other instruments, such as gyroscope, INS, etc.) and medium and object related (atmosphere and the Earth).

In order to integrate data from different sources, each dataset must be geometrically corrected for above distortions so that these can be registered, compared, and combined at pixel level. Hence, the data must be ortho-corrected or terrain-corrected where all distortions due to relief are rectified. The concept of terrain-geocoded images was initially introduced in Canada in defining value-added products for developmental needs (Guertin and Shaw 1981). Subsequently these have gained popularity all over the world due to their easy integration with GIS database for mapping, monitoring, and modeling purposes.

Conventional procedure relies on the availability of an accurate map of the area, and if the position and attitude of the satellite platform with respect to the Earth are not known, then the only alternative is to use ground control points using GPS (Mather and Koch 2010). Nowadays all optical satellites utilize linear scanners to acquire images, and in such cases, every line has its own exterior orientation parameters (position and attitude) which provide improved accuracy, but still not sufficient for many applications requiring precise geometry. This can be resolved by using ground control points (GCPs) in addition to sensor parameters. High-resolution satellite images are equipped with rational polynomial coefficients (RPCs) (Tao and Hu 2001), which are currently used for geometric corrections; however, they have their own limitations.

Many automated and fast rectification systems have also been developed that can correct, rectify, and register the images quickly and efficiently. Xie et al. (2003) developed an automatic image registration algorithm based on fast Fourier transform (FFT). Keith and Wang (2005) developed a spectral matching method for improving image-to-image co-registration accuracy.

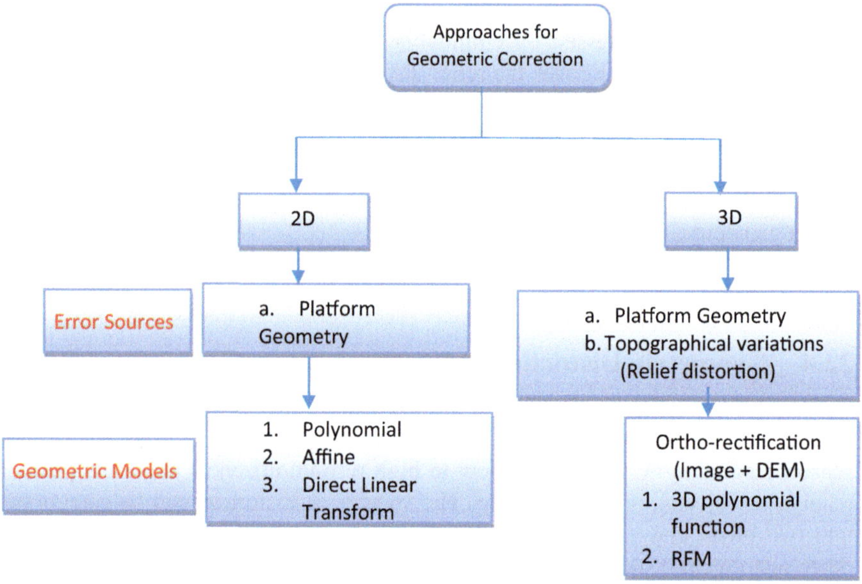

Fig. 22.1 Approaches for geometric correction

22.4.1 Geometric Correction of Optical Images

The topographical variations on the Earth and the tilt of the satellite or aerial sensor affect the distance of features on raw images. For topographically diverse landscape, the distortion is inherent to the imaging and is directly proportional to relief variations. Image data acquired by airborne and satellites have terrain distortions induced due to systematic sensor and platform attitude and altitude variations and sensor look angle. Terrain displacements can be in the order of hundreds of meters and, therefore, need to be rectified to preserve geometric integrity of spatial objects. Mathematical functions like 2D/3D empirical models (polynomial or rational function) or 2D/3D physical sensor-based models are used to correct the images with geometric distortions (Fig. 22.1). The 2D/3D empirical models are used when the parameters of the acquisition systems are not known. These empirical models use ground control points and mathematical functions like 2D or 3D polynomials.

The 2D/3D physical sensor-based methods mathematically model platform distortions (position, velocity, and attitude), the sensor parameters (viewing angles and panoramic effect), and topography. Two-dimensional models do not use elevation information, whereas 3D models take it into account to create precise ortho-rectified images. Hence, in order to accurately remove the relief distortions, a digital elevation model (DEM) is required. Figure 22.2 shows the raw image of Cartosat-1 satellite having spatial resolution of 2.5 m, DEM generated at 10 m spatial resolution using Cartosat-1 stereo data, and the ortho-rectified output. The overall planimetric accuracy of ortho-product was 2 m.

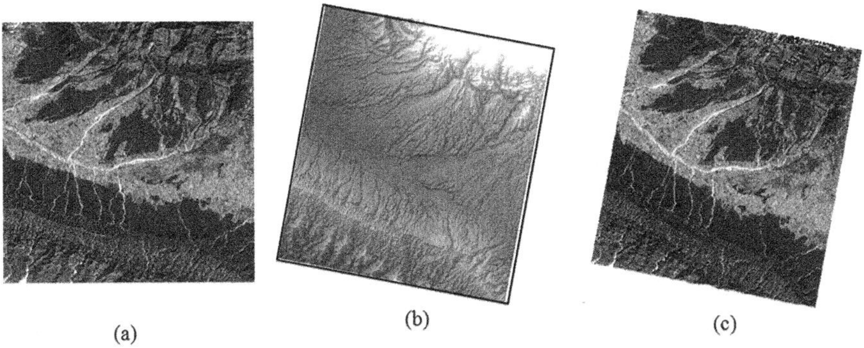

Fig. 22.2 (**a**) Raw Cartosat-1 image, (**b**) CartoDEM area, (**c**) ortho-rectified Cartosat image

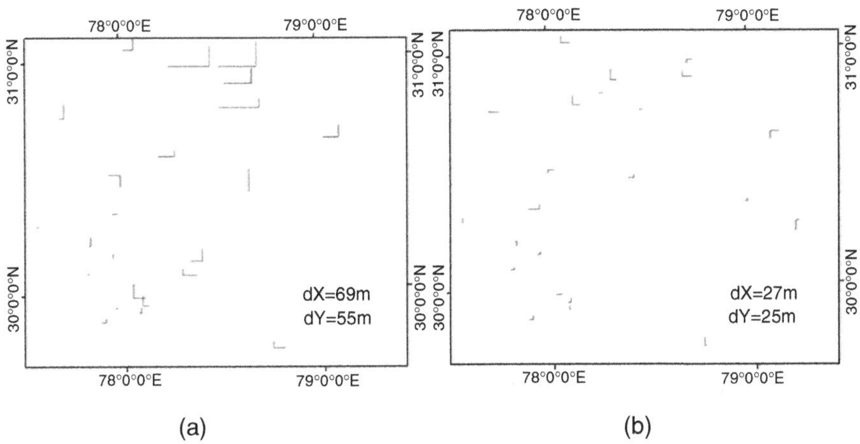

Fig. 22.3 (**a**) Relative errors in simple rectification. (**b**) Relative errors in ortho-rectification (Garg et al. 2008). dX and dY have been plotted as 100 times their actual length for graphical representation

Garg et al. (2008) investigated the accuracy of multidate registration using simple polynomial rectification and ortho-rectification technique over a mixed plain and hilly terrain in northern India. The results highlighted that ortho-rectified product had better sub-pixel registration accuracy (half pixel) in comparison to simple rectification (2–3 pixels). The relative errors as obtained are shown in form of deviations in X and Y direction in Fig. 22.3a, b, respectively.

22.4.2 *Radiometric Distortions of Optical Images*

Radiometric distortions in satellite images are primarily caused due to variation/degradation in sensor sensitivity with time, topographic effects, sun angle variation, and atmospheric influence. In optical data, the atmospheric particles cause selective scattering, absorption, and emission attenuating the signal recorded by the sensor. The radiation is attenuated twice as the radiation travels through the atmospheric twice, once from the Sun to the Earth and then from the Earth to the sensor. Thermal images measure the emitted radiation; hence the effect of atmospheric influence is considered only once. SAR images are least affected by atmospheric parameters. Surface topography influences the radiometric measurement in all wavelength ranges: optical, thermal, and SAR. This influence becomes significant in high-altitude areas of Himalaya. Also in temporal images, the target may have different radiometric response due to various factors like different imaging seasons or dates, solar altitudes and azimuth, variation in sensor illumination and observation angles, meteorological conditions, variation in sensor calibration with time, atmospheric effects, topographic effects (i.e., slope aspect), and actual changes in the target reflectance.

22.4.3 Radiometric Normalization of Optical Images

The aim of radiometric corrections/normalization is to remove or reduce radiometric distortions due to the factors mentioned above. Two different approaches are envisaged: relative and absolute normalization. Absolute radiometric correction converts the digital numbers used to represent the radiances measured by the satellites to at-sensor radiance (W m^{-2} sr^{-1} μm^{-1}) and then through atmospheric correction to surface reflectance. Absolute correction techniques require in situ measurements. Relative radiometric corrections normalize the variation between images and produce a radiometrically normalized data. The choice of one of the two approaches depends on the particular application and availability of input parameters/information. Absolute corrections are required when quantitative estimates are to be made from remotely sensed images. Relative radiometric corrections can be applied for studies using multi-temporal datasets for change detection, etc. Absolute radiometric correction follows a two-step process. First the sensor calibration parameters, gain and offset, are used to convert the recorded DN values into radiance (Eq. 22.1) and then apply the appropriate correction model to derive the surface reflectance (Eq. 22.2) or land surface temperature from the radiance values.

$$L_{\mathrm{sat}} = \mathrm{DN}^{*}\mathrm{Gain} + \mathrm{Offset}, \tag{22.1}$$

where L_{sat} is spectral radiance detected by a satellite sensor, DN is the digital number of the sensor measurement, and gain and offset are sensor-specific calibration parameters.

$$\rho_{\mathrm{surface}} = \left(L_{\mathrm{sat}} - L_{\mathrm{path}}\right)\pi/\mathrm{E}.\tau, \tag{22.2}$$

where $\rho_{surface}$ is the ground surface reflectance of the target, L_{path} is the path radiance, E is the irradiance on the ground target, and τ is the transmission of the atmosphere (Lillesand and Kiefer 1994). Radiative transfer models (RTM) are used to predict the atmospheric radiative properties and model the radiance at the sensor by simulating the interactions (absorption, emission, and scattering) between radiation and the medium (atmosphere). These models require information on the satellite geometry, illumination and viewing angles, and atmospheric conditions (visibility, aerosol type, water vapor, etc.) at the time of data acquisition. Several atmospheric correction models such as ATCOR (Richter 1996 and Richter and Schläpfer 2002), ATREM (Gao and Davis 1997; Gao et al. 2000), ENVI/FLAASH (Research Systems, Inc.), ACORN (Analytical Imaging and Geophysics, LLC), and HATCH (Qu et al. 2000) and Second Simulation of Satellite Signal in the Solar Spectrum (6S) (Vermote et al. 1997) are available for converting the DN values into surface reflectance.

Several techniques for relative radiometric correction of optical and thermal images exist (Yuan and Elvidge 1996); some common approaches include histogram matching (Chavez and MacKinnon 1994), normalization using pseudo-invariant features (Schott et al. 1988, Vries et al. 2004), linear regression, dark object subtraction, Gaussian method, flat-field correction, internal average relative reflectance correction, and empirical line correction. Relative correction techniques require no atmospheric or viewing geometry parameters; it utilizes image statistics to compute parameters for correction.

22.4.4 Radiometric Normalization Due to Topographic Effect

The influence of topography on satellite imagery causes a problem of differential illumination due to steep and varying slopes in rugged Himalayan terrain. This results in variation of spectral response of the same land cover present in shadowed regions having lower reflectance as compared to the sunny side. Therefore, topographic corrections are essential for qualitative and quantitative analysis. One of the essential requirements for correcting for topographic effects is a high-resolution digital elevation model (DEM). Although worldwide high-resolution DEM is not available at a scale comparable to image data (submeter resolution), it is available at medium resolution from NASA Shuttle Radar Topographic Mission (Rabus et al. 2003), ASTER Global Digital Elevation Map (Slater et al. 2011), and CartoDEM for the Indian region. There are several approaches to correct for the topographic errors; these can be broadly classified as Lambertian (reflectance is independent on observation angle) and non-Lambertian methods. Lambertian method is a simple photometric function that corrects for differences in illumination due to different sun positions. There are several variants of such approach which are known as cosine and cosine-C (Civco 1989), cosine-T, C-correction, and smooth-C (Dozier and Marks 1987). Non-Lambertian approaches model the diffused irradiance by using a constant, which depends on the wavelength and reflection characteristics that are estimated separately for each land cover class. These non-Lambertian methods can further be divided into empirical (Teillet et al. 1982), semiempirical (Teillet et al. 1982), and physically based models (Sun-canopy-sensor (SCS), Gu and Gillespie 1998). Six different topographic correction methods are listed in Table 22.5.

Figures 22.4 and 22.5 highlight the result of applying topographic correction using cosine and Minnaert correction over snow-covered and hilly areas of Himalaya.

Table 22.5 Topographic correction methods

Topographic correction models	Formula	References
Cosine correction	$L_m = L\left(\frac{\cos\theta_s}{\cos\theta_i}\right)$	Teillet et al. (1982)
Cosine correction (two stage)	$L_m = L + L\left(\frac{1-\cos\theta_i}{\cos\theta_i}\right)$	Civco (1989)
C-correction	$L_m = L\left(\frac{\cos\theta_s + c_\lambda}{\cos\theta_i + c_\lambda}\right)$	Teillet et al. (1982)
Minnaert correction	$L_m = L\left(\frac{\cos\theta_s}{\cos\theta_i}\right)^k$	Colby (1991) and Smith et al. (1980)
Minnaert SCS (Sun-canopy-sensor) correction	$L_m = L\left(\frac{\cos S\ \cos\theta_s}{\cos\theta_i}\right)$	Gu and Gillespie (1998)
SCS + C-correction	$L_m = L\left(\frac{\cos S\ \cos\theta_s + c_\lambda}{\cos\theta_i + c_\lambda}\right)$	Soenen et al. (2005)

where:
$\cos\theta_i = \cos\theta_s \cos S + \sin\theta_s \cos(\Omega - A)$
L_m = Normalized radiance in horizontal terrain (per pixel)
L = Uncorrected radiance of a pixel in rugged terrain
θ_s = Solar zenith angle
θ_i = Incidence angle/illumination
$\theta_{\bar{i}}$ = Mean illumination angle
c_λ = Semiempirical moderator
k = Minnaert constant
S=Slope
A = Aspect
Ω = Solar azimuth angle

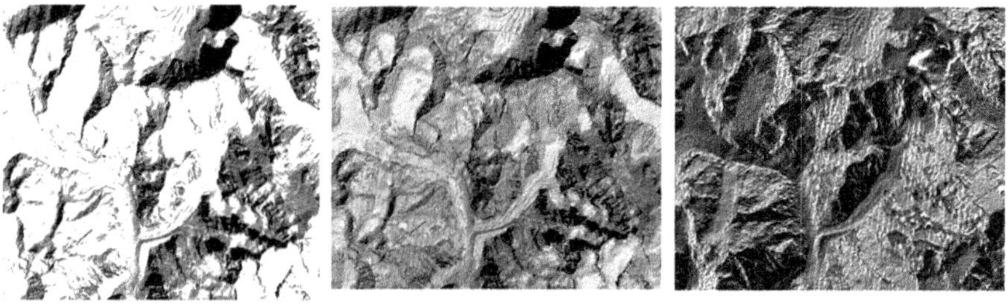

(a) (b) (c)

Fig. 22.4 (**a**) Landsat 7 original IR band March 2000 (snow area), (**b**) Minnaert correction, (**c**) cosine correction

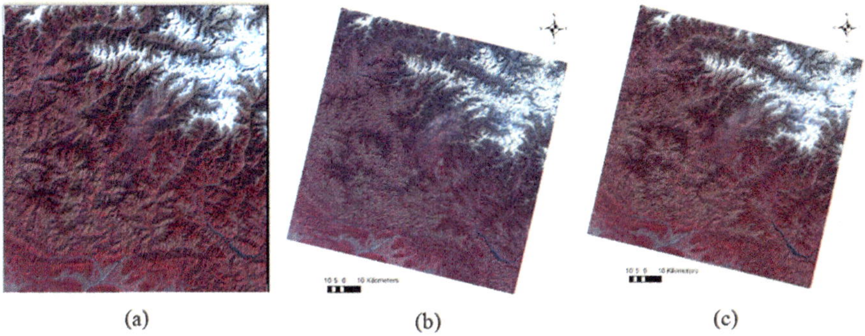

Fig. 22.5 (**a**) Ortho-rectified LISS-3 image, 16 November 2008, (**b**) cosine correction, (**c**) slope correction. (Source: Deepika Baru, M.Tech. Report 2010–11)

22.5 Synthetic Aperture Radar Interferometry

Synthetic aperture radar (SAR) interferometry is a well-established technique in satellite remote sensing for derivation of information primarily related to surface cover, elevation, and deformation (change in elevation). The main advantage is that data acquisition is independent of daylight and to a great extent atmospheric conditions. SAR interferometry involves two or more SAR data acquisitions collected from the same geometric configuration of the sensor over the same area with a distance/separation between platform positions to create a small phase difference between the two acquisitions. The difference in phase is sensitive to height and displacement information of the Earth surface. Generation of interferogram for elevation/height retrieval of terrain involves the steps of accurate image co-registration of InSAR pairs at sub-pixel level, interferogram generation, flattening, and phase unwrapping. Very limited studies have been carried over Himalaya to show the potential of SAR interferometry for height and displacement measurements (Bhattacharya and Mukherjee 2017; Mangla and Kumar 2014). Surface deformation and subsurface slip had been measured for Western Himalaya, and coseismic uplift of 60 mm was reported by Satyabala and Bilham (Satyabala and Bilham 2006). Inter-seismic uplift on a transect crossing the whole Himalaya measured from interferometric SAR data revealed uplift velocity of 7 mm/year (Grandin et al. 2012). However, such information is required to be derived from different structural domains of Himalaya, and surficial deformation pattern is to be correlated with crustal deformation pattern to decipher on landform evolution across Himalaya. Besides Himalayan region is one of the most seismically active regions; hence, it requires high temporal observation for crustal modeling. Although attempts have been made to improve acquisition frequency by combining ascending and descending mode data for interferogram generation, fully reliable operational strategy is yet to evolve and demonstrate.

22.5.1 Geometric Correction of SAR Images

Synthetic aperture radar sensors have proven their potential over other remote sensing techniques in information retrieval due to cloud penetration and night vision capabilities. The longer wavelength of microwave region of electromagnetic (EM) spectrum provides the cloud penetration capability, and the active sensor of SAR has the property to acquire the data in nighttime. SAR sensor transmits EM pulses toward the Earth surface to capture the information of any object within SAR resolution cell. The interaction of incident EM pulses with the object within resolution cell creates a scattering pattern, and the incident EM pulse is distributed in all the possible direction. The SAR sensor receives an amount of backscattered energy to generate an image of the illuminated target on the Earth surface. The strength of backscattered signal depends on the structure, shape, size, orientation, and dielectric properties of the illuminated target. Object parameters are not the only parameters, which affect the strength of backscatter properties (van der Sanden 1997 and Sullivan 2004). Geometry of SAR system also plays an important role in backscattering phenomena. Parameters of SAR sensors, which may affect the backscatter properties, are mainly wavelength used in data acquisition and angle of incidence. SAR sensor moves in the direction of flight or azimuth direction and transmits EM pulses in range direction. Due to these two directions, each resolution cell of SAR data keeps two spatial resolutions, and these resolutions are named as azimuth and range resolution. The length of resolution cell in azimuth direction is known as azimuth resolution, and this resolution is directly depending on the length of the antenna size of the radar system (Jensen 2006). To place a very large physical aperture in space to create a high azimuth resolution is very difficult and challenging task. The problem of large aperture was resolved by creating a virtually synthetic aperture antenna in space with the help of motion of the sensor in forward direction. Another resolution is range resolution, which is defined as the length of resolution cell in range direction. Range resolution depends on the pulse width, and this resolution may be increased by increasing the radar bandwidth. Due to the slant range direction, all the objects, in near range of SAR geometry, appear as a compressed object in the image. The compressed appearance of the features in their shape and size in the SAR imagery is known as slant range ambiguity. The minimization of slant range ambiguity is generally considered as first step in SAR SLC image processing. This ambiguity is removed by slant to ground range conversion (Luckman 1998). Though topographic variation and side-looking property of SAR sensor cause distortion in SAR imagery (Bayanudin and Jatmiko 2016), it can be addressed to a great extent. The geometric rectification process reconstructs the precise SAR geometry for each resolution cell to represent its corresponding location on the Earth. A number of researches have been carried out to develop methods for precise and accurate terrain geocoding of SAR data (Shimada 2010). However, the problems of layover, foreshortening, and shadow still remain an issue in high-altitude terrain, albeit the solutions envisaged by combining ascending and descending node data as well as acquisitions with different incident angles.

22.5.2 Radiometric Correction of SAR Images

Synthetic aperture radar (SAR) sensors have been widely used for information extraction of man-made and natural features of the Earth surface. Several researches have been carried to show the potential of SAR data for vegetation height extraction (Kumar et al. 2017a, b, c), aboveground biomass of the forest area (Kumar et al. 2012, 2017d; Mangla et al. 2016; Tomar et al. 2016), oil spill detection (Kumar et al. 2016a), land subsidence studies (Gonnuru and Kumar 2018), feature identification (Kumar et al. 2016b; M. N. S., R. and Kumar 2016; Joshi and Kumar 2017), and several other applications (Agrawal et al. 2016; Prajapati et al. 2016; Awasthi et al. 2017; Joshi and Kumar 2017). The backscattered energy for any object on the Earth surface is recorded by transmitting the electromagnetic pulses toward the object. The transmitted electromagnetic pulse interacts with different objects within SAR resolution cell, as a result of interaction the incident EM pulse gets scattered in all the directions with some reflected back in the direction of the antenna, which is recorded and converted into digital data. The measured backscattered energy not only depends on the target parameters but also on system parameters like aperture length of the antenna, gain, system loss, etc., which contributes radiometric biasness in the backscatter data (El-Darymli et al. 2014). Radiometric calibration of SAR data is a normalization process to generate backscatter cross section of individual resolution cells to represent actual scattering from the data (Kumar 2009). Figure 22.6 shows a backscatter image generated from the HH polarization of Radasat-2 data over forest ranges of Doon Valley.

It can be easily seen from the image that different Earth surface features exhibit different scattering characteristics; calm smooth water body and plane surfaces have low backscatter, so these surface appear dark. Rough surface like vegetation, buildings of urban area, and barrage on the river have very high backscatter and thus appear as very rough surface on the image. SAR backscattering depends on the wavelength of transmitted wave, angle of incidence shape, size, structure, and electrical property of the targeted object. Out of these parameters, only angle of incidence of SAR system varies with topography of the Earth surface, and any wrong measurement in the angle of incidence will result in wrong interpretation of backscatter property of the feature. The angle of incidence will be exactly equal to the look angle of the SAR system for flat terrain, and this information can be easily retrieved from the metadata of the SAR data. If the topography is not flat, then local incidence angle of SAR data needs to be estimated, which can be retrieved with the help of external DEM.

Besides measurement of local incidence angle, the ambiguity in the SAR image appearing due to slant range geometry is also a major challenge in SAR data processing. The time taken between the transmission and receiving the signals

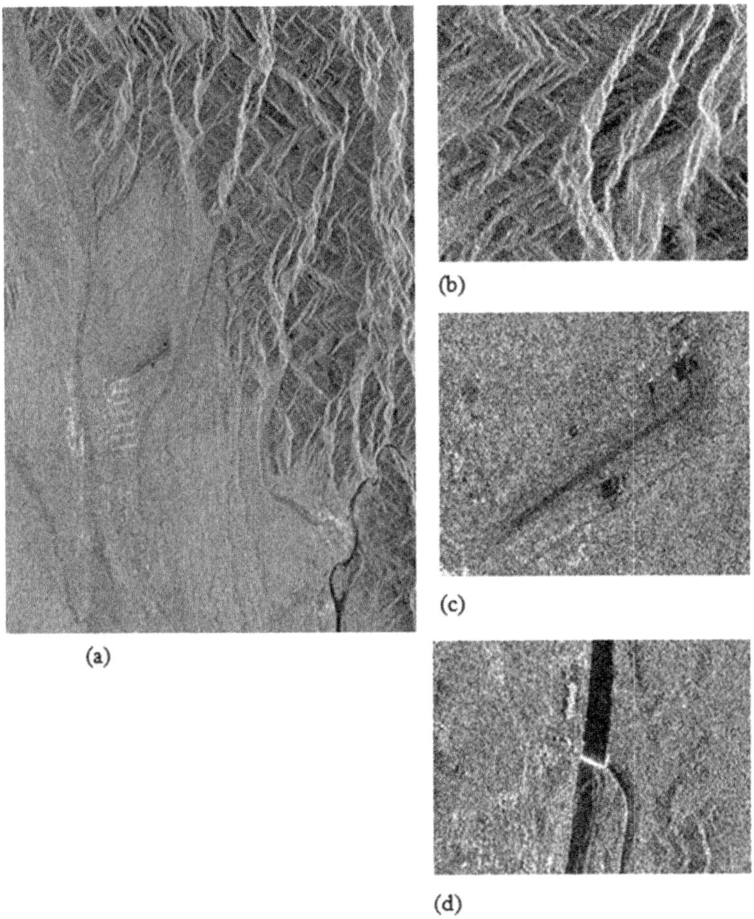

Fig. 22.6 Backscatter image of HH polarization Radarsat-2 data. (**a**) Full scene of the data. (**b**) Hilly terrain. (**c**) Dehradun Airport. (**d**) River Ganga

provides the total distance of the object from the SAR sensor in range direction or the direction in which SAR sensor transmits electromagnetic pulses.

Due to the slant range distortion, shape and size of the objects are not imaged in their actual geographic and structural position. The ambiguity was removed by slant to ground range conversion. Multi-looking and terrain correction was carried out with the help of external DEM and precise SAR geometry. Generally slant to ground range conversion and multi-looking steps are not required for geocoded terrain-corrected SAR data. Range Doppler terrain correction algorithm was implemented on the HH polarization of Radarsat-2 data to derive terrain-corrected geocoded backscatter SAR image as shown in Fig. 22.7.

From Fig. 22.7, it is clear that after terrain correction of the SAR data, actual ground information provided to all the features make it more visible in comparison to

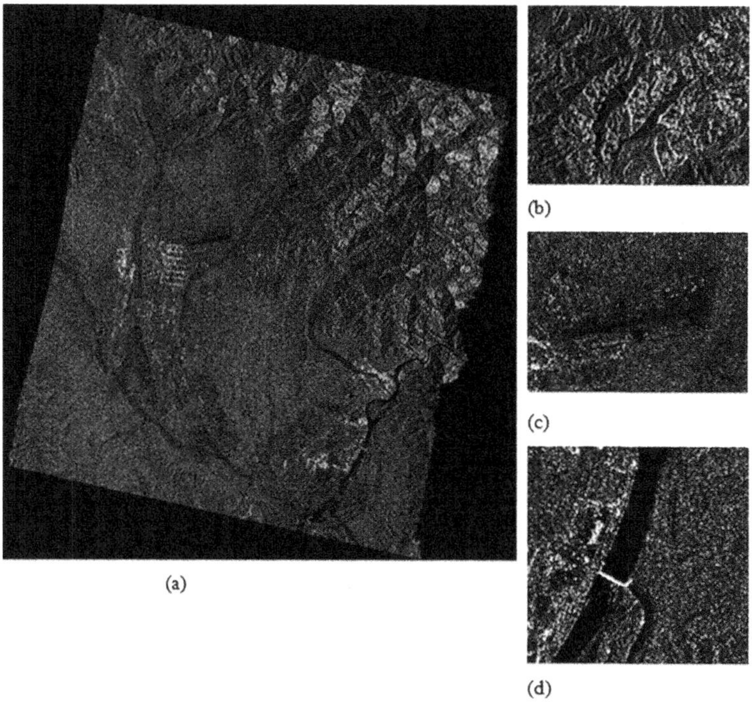

Fig. 22.7 Terrain-corrected backscatter image of HH polarization of Radarsat-2 (**a**) Full scene of the data. (**b**) Hilly terrain. (**c**) Dehradun Airport. (d) River Ganga

the SAR data without correction as shown in the first Fig. 22.6. The range Doppler terrain correction algorithm is the widely used technique to orthorectify the data by calculating local incidence angle with the help of external DEM and state vector information from the metadata of the sensor.

22.6 Sources of DEM

As described in the previous sections, geometric correction and atmospheric and radiometric corrections require digital elevation models (DEMs). DEMs/DSMs are also required for many other applications which are out of scope of present discussion. There are three major sources of DEM data: (1) ground survey including GNSS supported techniques, (2) existing topographic maps produced at long temporal interval, and (3) satellite/aerial remote sensing. A number of satellite systems exist that acquire imagery in stereo viewing mode, and the stereo data can be processed to derive elevation information (DEMs) using photogrammetric techniques. DEM can also be derived from SAR images using interferometric techniques and LiDAR (Jacobsen 2004).

22.6.1 Photogrammetric Approach

Extraction of digital elevation models (DEMs) from optical stereo images is through physical sensor or generalized sensor models. The physical sensor model takes into account the physical imaging process, and the parameters involved describe the sensor's position and orientation with respect to an object-space coordinate system. The physical models are rigorous, and are not sensitive to GCP distribution and hence yield high modeling accuracy. Physical sensor models are sensor dependent, i.e., different types of sensors need different models. They require the information about the imaging parameters such as orbit parameters, ephemeris data, relief displacement, Earth curvature, atmospheric refraction, lens distortion, etc., without which it would be difficult to develop a rigorous physical sensor model. The physical senor model was first generated for SPOT HRV sensor (Toutin 2004).

Generalized sensor model is independent of sensor platforms and sensors. In a generalized sensor model, the transformation between object and image space is represented as some general function without modeling the physical imaging process. The function used can be polynomials or rational functions. In general, generalized sensor models do not require knowledge of the sensor geometry; thus they can be adapted to different sensor types. Use of RFM (rational function model) to approximate the physical sensor models has become popular due to its unique characteristics of sensor independence and real-time calculation. Rational polynomial coefficients (RPCs) are determined from RFM by the space agencies and provided to the end user (Grodecki 2001, Grodecki and Dial 2003). After applying the RFM, biases or errors still exist which need to be improved by use of few GCPs and 2D polynomial transformations (Fraser and Hanley 2005).

The main drawback of DEM derived from optical data is that it generates DSMs and it cannot produce accurate DEMs due to shadows, occlusions, rugged terrain, and steep slopes (Giribabu et al. 2013a, Bahuguna et al. 2007, Martha et al. 2010). Giribabu et al. (2013b) used stereo triplet in reverse stereo mode from Cartosat-1 to generate DEM to overcome the above problem and observed that better DEM could be derived in occluded region as compared to simple stereo pair-based DEM.

The accuracy of the DEM depends on many factors such as the resolution of the satellite image, algorithm used for height estimation, and interpolation technique used. DEMs generated from satellite imagery (ASTER, SRTM, SPOT-5, IRS-1C, Cartosat-1, etc.) are being used for Himalayan region. Several researches have been carried out to evaluate the accuracy of DEM in hilly terrains (e.g., Yamane et al. 2008; Bahuguna et al. 2004). Gupta et al. (2014) assessed the performance of ASTER and SRTM DEMs in the Himalayan region and observed a RMSEz (altitudinal difference with respect to ground survey data) of 11 m and 11.3 m, respectively.

22.6.2 LiDAR

LiDAR (light detection and ranging) is an active remote sensing technique that uses electromagnetic radiation in the form of lasers and is produced by stimulating the emission of radiation by using external source of energy. This is a nonimaging technique which provides 3D terrain surface in a discrete point cloud form. In this process, the EMR has special properties as follows: high energy (highly directional-normal EMR spreads in all direction, but the laser is a focused beam EMR); monochromatic (the photons of laser have the same wavelength, and for topographical mapping purpose, lasers in infrared range are generally used); and coherent (i.e., the phase difference is either constant or zero).

The principle used in LiDAR is that it generates laser pulses, which are sent to the surface/target, measures the pulses reflected by target, and computes precise time of flight of laser pulse transmitted by the laser sensor. Using the time of flight, the distance between sensor and target can be computed using Eq. 22.3 for ranging.

$$R = C^* T \tag{22.3}$$

where R is range, i.e., distance between laser sensor and target; C is the velocity of light; and T is the time of flight of laser pulse.

22.6.2.1 Platforms for Laser Scanning

Like any other remote sensing sensor, the laser sensor can be mounted on satellite, aerial or terrestrial platform. Some of the spaceborne laser remote sensing sensors include LiDAR in-Space Technology Experiment (LITE), Shuttle Laser Altimeter (SLA), Geoscience Laser Altimeter System (GLAS), and CALIPSO. LITE and CALIPSO are mainly used for studying vertical structure of cloud, aerosol, and other atmospheric-related applications. List of spaceborne missions carrying laser profilers is given in Table 22.6.

ICESat mission carried a payload GLAS with a major objective to study changes in ice sheets and topography. GLAS sensors emitted laser pulses of 4 nanoseconds in two wavelengths 532 nm and 1064 nm which makes a footprint of 70 meters with a ground spacing of 170 m. GLAS sensor records the full waveforms of backscattered energy so that entire vertical structure of the objects within the footprint could be analyzed. There are many demonstrated applications of GLAS data in estimating the changes in ice sheets and canopy heights for biomass estimation, atmospheric studies, and hydrological studies. Noting the potential of ICESat, a second mission, ICESat-2, is scheduled to launch in 2017/2018.

ICESat-2 will carry a single payload "Advanced Topographic Laser Altimeter Systems" (ATLAS), which would emit laser pulses at higher rate (10,000 pulse/second) compared to GLAS. ATLAS would operate at 532 nm and can record the waveform by photon counting, so that the vertical structure can be mapped in a

Table 22.6 Spaceborne missions with laser payload

S. No.	Satellite/sensor	LASER wavelengths	Footprint on ground	Objectives
1.	LITE	355 nm,532 nm and 1064 nm	300 m	Atmospheric studies
2.	SLA	1064 nm	100 m with spacing at 700 m	Land and vegetation elevation studies
3.	ICESat	532 nm and 1064 nm,	70 m with spacing of 170 m	Atmospheric and topographical (vegetation height/land elevation)
4.	ICESat-2 (to be launched in 2017/18)	532 nm	50 m with spacing of 140 m	
5.	CALIPSO	532 nm and 1064 nm	70 m with spacing of 333 m	Atmospheric studies

precise and detailed manner. ATLAS data can be utilized for various topographic applications.

With the recent developments in laser scanner and positioning sensors, airborne laser scanning is using drones/unmanned aerial vehicles (UAVs) that have gained popularity. Manned missions are undertaken for larger areas, whereas unmanned mission using drones are found to be economical for smaller areas. GNSS observations are processed for differential correction using a GNSS base station data for positioning the trajectory of aircraft at an accuracy of few centimeters. All the sensors are operated simultaneously with time synchronization, and datasets are further processed to geo-locate all the footprints of laser with high positional accuracy.

In terrestrial mode of laser scanning, laser scanners are mounted on static tripod or on moving platform (car/motorbike/ships) and accordingly are termed as terrestrial laser scanners (TLS) or mobile mapping systems (MMS). The basic difference lies in the mechanism of scanning: TLSs are 3D scanner, whereas MMS are 2D scanners. MMS are normally accompanied by IMU and DMI sensors. List of few commercially available TLS is provided in Table 22.7.

TLS has wide range of application from industrial reverse engineering applications to natural resource management such as monitoring landslides, generating very high-resolution DEM, estimating precise forest inventory, etc. Figure 22.8b, c shows the DEM surface of Kaliasaur/Serogarh landslide location at 350.6 km on NH-58, which majorly has toe cutting, weak strata (friable, brecciated, and sheared quartzite), and fault. DEM was generated from co-registered point cloud acquired using TLS from two exposure stations.

Various profiles across and along the landslide can be extracted from DEM; Fig. 22.9 shows the horizontal and vertical sectional profiles of Kaliasaur landslide.

Since laser fired from TLS can have multiple returns and can also penetrate into canopy, it is immensely a useful tool for forest plot inventory data collection like

Table 22.7 List of commercial TLS

	I-Site 8810	IMAGER 5010	ILRIS-LR	Leica ScanStation C10	RIEGL VZ-400	Focus3D
Year of introduction	2012	2010	2010	2009	2008	2010
Total weight (kg)	14.8	9.8	21	13.8	9.6	5
Scan time per battery (h)	2.5	2.5	5	3.5	8	5
Min.-Max range (m)	2.5–2000	0.3–187	3–3000	0.1–300	1.5–600	0.6–120
Max. vertical*horizontal field of view (deg)	80*360	320*360	360*360	270*360	100*360	300*360

(a) (b) (c)

Fig. 22.8 (**a**) Ground photograph of Kaliasaur, (**b**) DEM of Kaliasaur landslide, (**c**) shaded relief of landslide

individual tree heights, stem diameter/volume, and crown diameter and volume and further can be used to estimate the biomass using the allometric equations defined by Forest Survey of India. A sample of TLS point cloud of a forest plot (20 m × 20 m) near Barkot forest, Dehradun, is shown in Fig. 22.10.

22.6.2.2 Multiple Returns/Full-Waveform LiDAR

As laser sensors record the backscattered energy against time, it is possible that within the footprint of laser, the backscatter may come from different elevations. Hence, detector of laser sensor may receive more than one pulse depending on the elevation differences within the footprint. As a result multiple returns are recorded for a single transmitted pulse. The return that reaches the sensor first is named as first return, subsequently second, third, etc. The number of returns that can be obtained over an area depends on the capability of sensor to detect and record the echo/pulse.

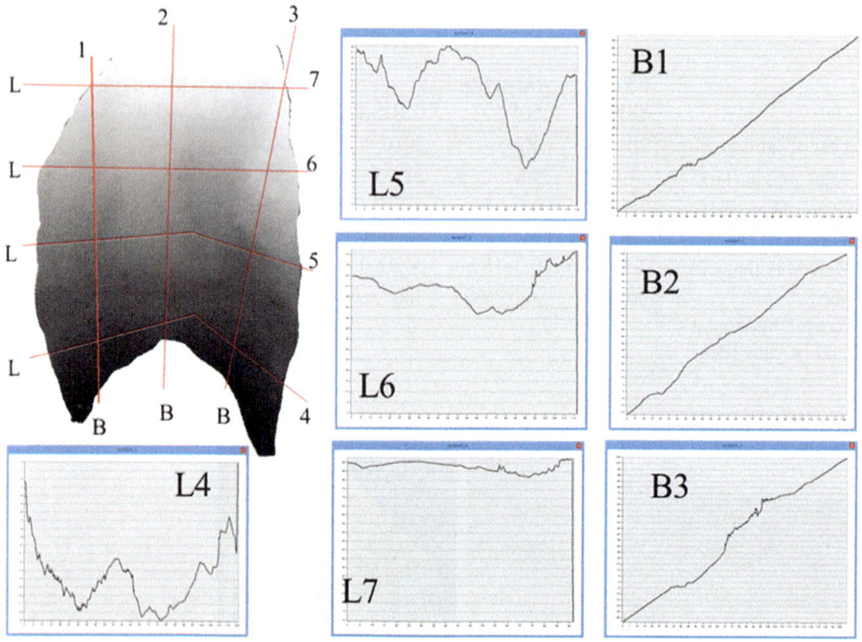

Fig. 22.9 Profiles of landslides derived from DEM generated from TLS

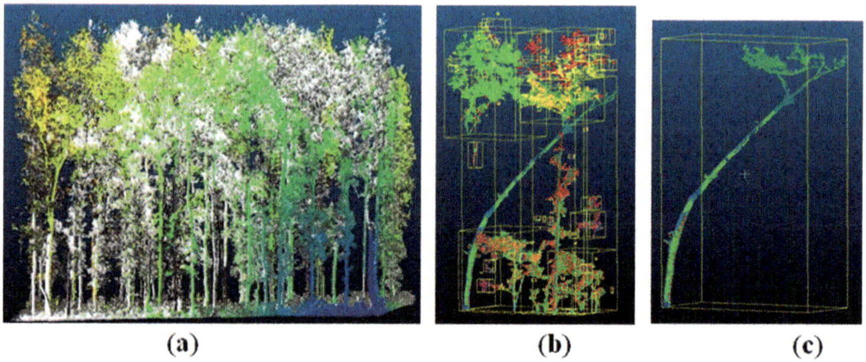

Fig. 22.10 (**a**) Front view of TLS data of forest plot, (**b**) point cloud of individual tree, (**c**) filtered point cloud of stem

Most importantly, the first returns are from top surfaces, and the last returns are from ground. Nowadays sensors are capable of recording the backscattered energy with a fine sampling rate which results in waveform rather than the discrete pulse. Compared to discrete laser systems, the full-waveform LiDAR records the entire reflected

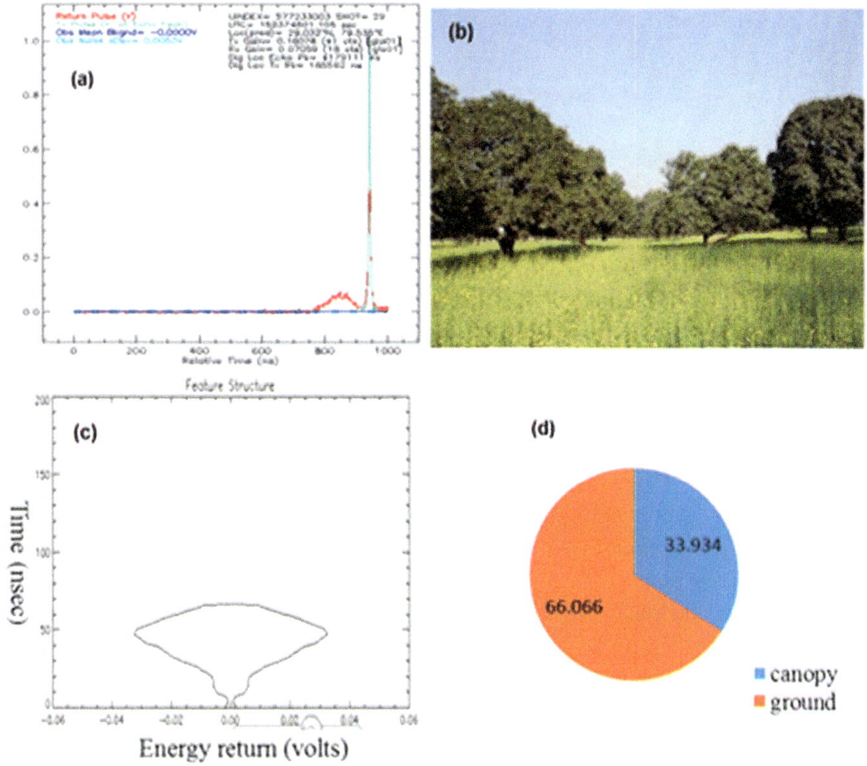

Fig. 22.11 (**a**) ICESat waveform over mango orchid, (**b**) ground photograph of mango orchard, (**c**) extracted canopy stand profile, (**d**) footprint LULC

signal as a function of time. This helps in extraction of more detailed information on the structure of the scanned surface. Full-waveform LiDAR data are useful in applications requiring detection and differentiation of vertical structures such as tree canopy structure, width of tree tops, etc. (Lefsky et al. 1999; Dubayah and Blair 2000). LiDAR waveforms over forest areas digitize the entire surface of canopy and represent the stand profile of all trees within the footprint. Denoising and decomposing the waveforms can yield precise stand canopy height estimation, and further by identifying the waveform signal from crown, canopy stand profile can be extracted. Figure 22.11 shows an ICESat GLAS (Geoscience Laser Altimeter Sensor) waveform and the canopy stand profile and heights extracted from the waveform.

(a) (b)

Fig. 22.12 (**a**) Dense point cloud of the landslide area collected from TLS, (**b**) digital elevation model. (Source: Hilal et al. 2015)

22.6.3 Close-Range Photogrammetry

Close-range photogrammetry makes use of accurate imaging techniques to analyze the three-dimensional shape of a wide range of man-made and natural objects. Documenting and monitoring the natural and cultural resources would form the basis for future decision-making especially in vulnerable areas such as hilly and mountainous terrain. In hilly regions rainfall-triggered mass movement is one of the most widespread Earth surface hazards. Mapping the spatial distribution of vulnerable built-up areas can help the decision-makers for proper planning and initiating preventive measures.

Hilal et al. (2015) effectively demonstrate that in case the variation in slope is not very large, the conventional Earth observation (EO) datasets with large to medium resolution are not suitable for capturing the minute elevation details of the study area. In such scenarios, close-range photogrammetry (CRP) and terrestrial laser scanning (TLS) enable very high-resolution data acquisition and hence become a useful tool for studying small landslide which are difficult to monitor using satellite imageries. The study is carried out in a rainfall-triggered debris flow site that occurred due to incessant rains in 2014, at Kathbangla in Dehradun, India (Fig. 22.12a).

The study (Hilal et al. 2015) illustrates an application of 3D GIS for estimating damage in vulnerable built-up areas on slopes susceptible to collapse. Since the study area is very small (2910 m^2) and the variation in slope is also not very large, hence the conventional Earth observation (EO) datasets with large to medium resolution are not suitable for capturing the minute elevation details of the study area. Terrestrial laser scanner (TLS) was used to generate a surface model (Fig. 22.12b) with a very high point density (RIEGl 2015).This surface model was taken as prime data source for simulation of debris flow. An open-source tool, Trivim (Raghavendra et al. 2015), which uses the concept of close-range digital terrestrial photogrammetry, was used to generate a 3D building scenarios (Fig. 22.13) using sequence of overlapping geotagged images. Corresponding database was attached to individual building. The RAMMS (rapid mass movement

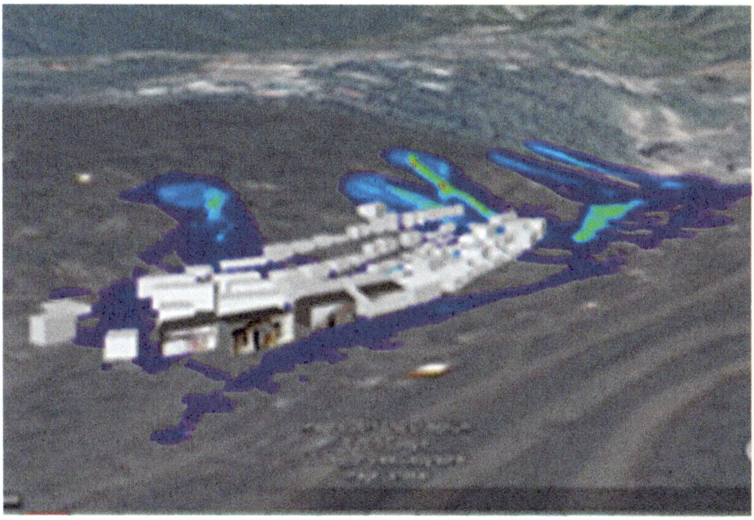

Fig. 22.13 3D simulation of affected area. (Source: Hilal et al. 2015)

simulation) numerical dynamic model developed by Swiss Federal Institute for Snow Avalanche Research was used to model the runout of the debris flow (Bartelt et al. 2013). Trivim results in aggregating all the data pertaining to a block and reconstructing photo-textured 3D models with submeter accuracy in planimetry as well as elevation that can be visualized on a geo-portal. Trivim generated 3D buildings with associated database was used in conjunction with the simulated debris flow for estimating the risk zones. The technique developed can be an effective method for identifying potential risk-prone built-up areas.

22.7 Way Forward

With the availability of multi-resolution, multifrequency, and multi-temporal image data from a suite of operational as well as experimental Earth observation satellites, research and operational applications have gained momentum. However, these are faced by challenges that are of technological, operational, and economic in nature. The technological challenges are mainly related to topography, geometry, atmosphere and sensor-related parameters. These can be addressed to a great extent through improved modeling and usage of ground data. Operational applications require data at very high temporal interval, which is at the moment limited by the inadequate number of optical, thermal, and SAR satellites. This is being addressed by some of the space agencies, but in the near future, it is not very clear how these issues can be completely resolved to address the data requirement of fragile ecosystem of Himalaya. There is a need to develop comprehensive integrated mountain

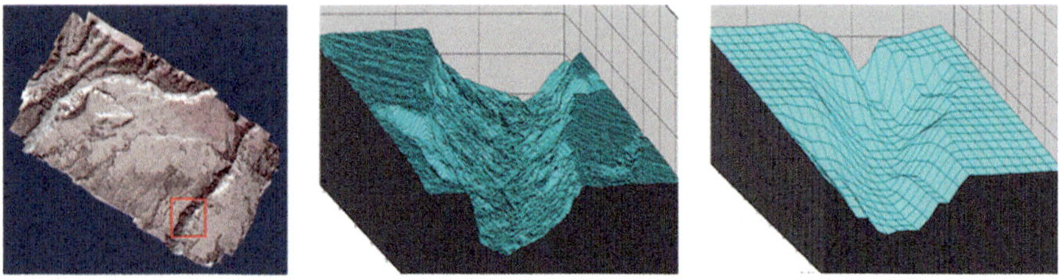

(a) (b) (c)

Fig. 22.14 (**a**) UAV photogrammetric DEM, (**b**) zoomed view of DEM from UAV, (**c**) zoomed view of DEM from Cartosat-1

monitoring system (IMMS) that can address needs of Himalayan population which could be replicated in other parts of the world. Timely availability of satellite data at a predetermined format is a critical requirement for many applications dealing with dynamic phenomena such as natural hazards and disaster management-related applications. Partially this can be addressed by acquiring data through aerial platform such as UAVs. Although UAVs are well suited for small-scale research and operational needs, their limited payload and short flight endurance still remain areas of weakness for their wide-scale implementation. Major advantages of UAVs lie in acquiring data from high-risk areas without endangering human life, by operating at low altitude and at flight profiles closer to targets which is otherwise impossible to acquire by manned aerial vehicles. UAV can provide detailed and accurate information of terrain especially in undulating mountainous regions like Northwestern Himalayas. Figure 22.14 shows the DEM (30 cm) generated from sequence of overlapping images acquired from UAV of Koti, Dehradun. Microrelief variations of the terrain are better captured with UAV DEM as compared to Cartosat-DEM (Fig. 22.14b, c).

Further, in cloudy and hazy weather conditions, the data acquisition with UAVs is still a possibility, albeit quality issues. Significant advantage is the ability for fast data acquisition and transmission of image, video, and orientation data in real time to the ground control station, although geometric and radiometric calibrations of UAV images remain a challenge, which has to be addressed through improved algorithm and processing chain. Overall, different remote sensing techniques have their own advantages and limitations which could be overcome in efficient manner to get a geospatial product with minimum error and ambiguity by integration of satellite and aerial data acquired by different techniques. Dataset collected from different sensors and their integration requires a meticulous strategy and very rigorous processing chain for radiometric/geometric corrections, normalization, and calibration to relate real-world phenomena to multiparametric satellite/aerial observation.

References

Agrawal, N., Kumar, S. and Tolpekin, V. (2016) 'Polarimetric SAR interferometry-based decomposition modelling for reliable scattering retrieval', Proc. SPIE, pp. 987708–987710.

Awasthi, S. et al. (2017) 'Pol-InSAR based snow depth retrieval using spaceborne TerraSAR-X data', in 2017 8th International Conference on Computing, Communication and Networking Technologies (ICCCNT). New Delhi: IEEE, pp. 1–7.

Bahuguna, I. M., Kulkarni, A. V, & Nayak, S. (2004). Technical Note: DEM from IRS-1C PAN stereo coverages over Himalayan glaciated region—accuracy and its utility. International Journal of Remote Sensing, 25(19), 4029–4041.

Bahuguna, I. M., Kulkarni, A. V, Nayak, S., Rathore, B. P., Negi, H. S., & Mathur, P. (2007). Himalayan glacier retreat using IRS 1C PAN stereo data. International Journal of Remote Sensing, 28(2), 437–442.

Bartelt, P., Bieler, C., Bühler, Y., Christen, M., Deubelbeiss, Y., Graf, C., McArdell, B.W., Salz, M., Schneider, M., 2013. RAMMS - rapid mass movements simulation, A numerical model for debris flows in research and practice, User Manual v1.7.0, Debris Flow, Manuscript update November 2017. WSL Institute for Snow and Avalanche Research SLF 1–116.

Bayanudin, A. A., & Jatmiko, R. H. (2016). Orthorectification of Sentinel-1 SAR (Synthetic Aperture Radar) Data in Some Parts of South-eastern Sulawesi Using Sentinel-1 Toolbox. IOP Conference Series: Earth and Environmental Science, 47(1), 12007.

Berthier E., Vincent C., Magnússon E., Gunnlaugsson Á. Þ., Pitte P., Le Meur E., Masiokas M., Ruiz L., Pálsson F., Belart J. M. C., and Wagnon, P. 2014 Glacier topography and elevation changes derived from Pléiades sub-meter stereo images. The Cryosphere, 8, 2275–2291, 2014.

Bhattacharya, A., & Mukherjee, K. (2017). Review on InSAR based displacement monitoring of Indian Himalayas: issues, challenges and possible advanced alternatives. Geocarto International, 32(3), 298–321.

Chavez, P.S., Jr., and D.J. MacKinnon. 1994. Automatic detection of vegetation changes in the Southwestern United States using remotely sensed images. Journal of Photogrammetric Engineering and Remote Sensing 60(5):571–583.

Civco, D. L. 1989. Topographic normalization of Landsat Thematic Mapper digital imagery", Photogramm. Eng. Remote Sens., vol. 55, pp. 1303–1309.

Colby, J.D., 1991. Topographic normalization in rugged terrain. Photogrammetric Engineering and Remote Sensing 57, 531–537.

Dozier, J.and D. Marks, 1987. Snow mapping and classification from Landsat Thematic Mapper data, Ann. Glaciol., 9, 97–103.

Dubayah, R., Blair, J., 2000. LiDAR Remote Sensing for Forestry Applications, Journal of Forestry 98 (6), pp. 44–46.

El-Darymli, K. et al. (2014) 'Understanding the significance of radiometric calibration for synthetic aperture radar imagery', in 2014 I.E. 27th Canadian Conference on Electrical and Computer Engineering (CCECE), pp. 1–6.

Fraser, C., and H. B. Hanley (2005). Bias compensated RPCs for sensor orientation of high-resolution satellite imagery. Photogrammetric Engineering and Remote sensing, 71(8): 909–915

Gao, B.C. & Davis, C. O., 1997. Development of a line-by-line-based atmosphere removal algorithm for airborne and spaceborne imaging spectrometers, in Imaging Spectrometry III, Eds. Michael R. Descour, Sylvia S. Shen, Proceedings of SPIE Vol. 3118, 132–141.

Gao, B.C., Montes, M. J., Ahmad, Z. and Davis, C. O., 2000. Atmospheric Correction Algorithm for Hyperspectral Remote Sensing of Ocean Color from Space, Applied Optics, 39(6): 887–896.

Garg, R.D, Agrawal Shefali, Dadhwal, V.K., 2008. Evaluation of approaches for AWiFS multi-date registration. International Journal of Applied Earth Observation and Geoinformation 10 (2008) 175–180

Giribabu, D., Kumar, P., Mathew, J., Sharma, K.P., Murthy, Y.V.N.K. (2013b). DEM generation using Cartosat-1 stereo data: issues and complexities in Himalayan terrain, European Journal of Remote Sensing 46, 431–443.

Giribabu, D., Srinivasa Rao, S., & Krishna Murthy, Y. V. N. (2013a). Improving Cartosat-1 DEM accuracy using synthetic stereo pair and triplet. ISPRS Journal of Photogrammetry and Remote Sensing, 77, 31–43.

Gonnuru, P. and Kumar, S. (2018) 'PsInSAR based land subsidence estimation of Burgan oil field using TerraSAR-X data', Remote Sensing Applications: Society and Environment, 9(January 2018), pp. 17–25.

Grandin, R., Doin, M.-P., Bollinger, L., Pinel-Puysségur, B., Ducret, G., Jolivet, R., & Sapkota, S. N. (2012). Long-term growth of the Himalaya inferred from interseismic InSAR measurement. Geology.

Grodecki, J. and Dial, G. 2003. Block adjustment of high-resolution satellite images described by rational polynomials, Photogramm. Eng. Remote Sens., vol. 69, pp. 59–68.

Grodecki, J., 2001. IKONOS stereo feature extraction-RPC approach. Proceedings of 2001 ASPRS Annual Convention (CD ROM), 23-27 April.

Gu, D.; Gillespie, A. 1998. Topographic Normalization of Landsat TM Images of Forest Based on Subpixel Sun–Canopy–Sensor Geometry. Remote Sens. Environ. 64, 166–175.

Guertin, F., and Shaw, E., 1981, Definition and potential of geocoded satellite imagery products, Proceedings of the 7th Canadian Symposium on Remote Sensing, Winnipeg, Canada, September 8–11, (Winnipeg, Canada: Manitoba Remote Sensing Centre), pp. 384–394.

Gupta R.D., Mritunjay Kumar Singh, Snehmani , A. Ganju, 2014. Validation of SRTM X Band DEM over Himalayan Mountain. The International Archives of the Photogrammetry, Remote Sensing and Spatial Information Sciences, Volume XL-4, 2014 ISPRS Technical Commission IV Symposium, 14–16 May 2014, Suzhou, China

Negi, H. S., Kulkarni, A. V., and Semwal, B. S. (2009). Estimation of snow cover distribution in Beas basin, Indian Himalaya using satellite data and ground measurements. Journal of Earth System Science, 118(5), 525.

Hahn M, Baral TN and Sharma RK (2003) A study on digital orthophoto generation of Mount Everest region. Nepal. J. Geo-inf. 2, 35–43

Hilal M., Tiwari, P. S., Raghavendra, S., Pande, H., Reddy, K.S., Agarwal, S., Raju, P.L.N. (2015). Building damage assessment by integrating Terrestrial Laser scanning and Open Source Close Range Photogrammetry application National conference on open source geospatial tools in climate change research and natural resources management, 8-10th June 2015.

Jacobsen, K., 2004. DEM generation from satellite data. In:Goossens, R. (Ed.): Remote Sensing in Transition, Proc. of the 23rd EARSeL Symposium 2003, Ghent, Belgium, pp. 513–525

Jensen, J.R. (2006). Remote sensing of the environment an earth resource perspective: low price edition Pearson Education, New Delhi, India

Joshi, S. K. and Kumar, S. (2017) 'Performance of PolSAR backscatter and PolInSAR coherence for scattering characterization of forest vegetation using single pass X-band spaceborne synthetic aperture radar data', Journal of Applied Remote Sensing, 11(2), p. 26022. Joshi, S. K. and Kumar, S. (2017) 'Spaceborne PolInSAR tomography for vertical profile retrieval of forest vegetation', Journal of Applied Remote Sensing, 11(1).

Keith, R. and Wang M. (2005), "Enhancement of Image-to-Image Co-registration Accuracy Using Spectral Matching Methods", 7th International Symposium on Spatial Accuracy Assessment in Natural Resources and Environmental Sciences.

Kumar, S. (2009) Retrieval of forest parameters from Envisat ASAR data for biomass inventory in dudhwa national park, U.P.,India. ITC,International Institute for geo-information science and earth observation.

Kumar, S., Garg, R. D., Kushwaha, S. P. S., and Pandey, U. (2017b). Spaceborne SAR Tomography for Vertical Profile Retrieval of Forest Vegetation. Proceedings of the National Academy of Sciences, India Section A: Physical Sciences, 87(4), 807–816.

Kumar, S., Garg, R. D., Kushwaha, S. P. S., Jayawardhana, W. G. N. N., and Agarwal, S. (2017a). Bistatic PolInSAR Inversion Modelling for Plant Height Retrieval in a Tropical Forest. Proceedings of the National Academy of Sciences, India Section A: Physical Sciences, 87(4), 817–826.

Kumar, S., Gupta, V., Gonnuru, P., and Joshi, S. K. (2016b). PolSAR calibration and reconstruction of hybrid polarimetric RISAT-1 data for pseudo quad-pol decomposition: a comparison with quad-pol. In Earth Observing Missions and Sensors: Development, Implementation, and Characterization IV (Vol. 9881, p. 98812C). International Society for Optics and Photonics.

Kumar, S., Joshi, S. K. and Govil, H. (2017c) 'Spaceborne PolSAR Tomography for Forest Height Retrieval', IEEE Journal of Selected Topics in Applied Earth Observations and Remote Sensing, 10(12), pp. 5175–5185.

Kumar, S., Kattamuri, H. P. and Agarwal, S. (2016a) 'Dark spot detection for characterization of marine surface slicks using PolSAR remote sensing', Proc. SPIE, p. 98780K–98780K–17.

Kumar, S., Khati, U. G., Chandola, S., et al. (2017d) 'Polarimetric SAR Interferometry based modeling for tree height and aboveground biomass retrieval in a tropical deciduous forest', Advances in Space Research, 60(3), pp. 571–586.

Kumar, S., Pandey, U., Kushwaha, S. P., Chatterjee, R. S., and Bijker, W. (2012). Aboveground biomass estimation of tropical forest from Envisat advanced synthetic aperture radar data using modeling approach. Journal of applied remote sensing, 6(1), 063588.

Lefsky, M.A., Cohen, W.B., Acker, S.A., Parker, G.G., Spies, T.A. and Harding, D., 1999. LiDAR remote sensing of the canopy structure and biophysical properties of Douglas-fir western hemlock forests. Remote Sensing of Environment, 70(3): pp. 339–361

Lillesand, T. M., & Kiefer, R. W. (1994). Remote sensing and image interpretation. John Wiley and Sons.

Luckman, A. J. (1998). Correction of SAR imagery for variation in pixel scattering area caused by topography. IEEE Transactions on Geoscience and Remote Sensing.

M. N. S., R. and Kumar, S. (2016) 'Feature extraction using multi-temporal fully polarimetric SAR data', Proc. SPIE, pp. 987711–987718.

Mangla, R., & Kumar, S. (2014). DEM Construction using DInSAR. In ISPRS - International Archives of the Photogrammetry, Remote Sensing and Spatial Information Sciences (Vol. XL-8, pp. 817–820).

Mangla, R., Kumar, S. and Nandy, S. (2016) 'Random forest regression modelling for forest aboveground biomass estimation using RISAT-1 PolSAR and terrestrial LiDAR data', Proc. SPIE, p. 98790Q–98790Q–11.

Martha, T. R., Kerle, N., van Westen, C. J., Jetten, V., &Vinod Kumar, K. (2010). Effect of sun elevation angle on DSMs derived from Cartosat-1 data. Photogrammetric Engineering and Remote Sensing, 76(4), 429–438.

Mather P. M and Magaly Koch (2010), Computer Processing of Remotely-Sensed Images: an Introduction, 4th edition John Wiley & Sons.

Prajapati, R., Kumar, S. and Agrawal, S. (2016) 'Simulation of SAR backscatter for forest vegetation', Proc. SPIE, p. 98811T–98811T–12.

Qu, Z., Goetz, A.F.H, Heidebrecht, K.B., 2000. High-Accuracy Atmosphere Correction for Hyperspectral Data (HATCH). JPL AVIRIS Proceedings.

Rabus, B., Eineder, M., Roth, A., & Bamler, R. (2003). The shuttle radar topography mission—a new class of digital elevation models acquired by spaceborne radar. Journal of Photogrammetry and Remote Sensing, 57, 241–262

Raghavendra, S., Reddy, K.S., Tiwari, P. S., Pande, H, Sharma, M., Agarwal, S Raju, P.L.N., Murthy, Y.V.N.K. (2015). An open source approach for 3D street view generation. National conference on open source geospatial tools in climate change research and natural resources management, 8-10th June 2015.

Richter, R. and Schläpfer, D., 2002. Geo-atmospheric processing of airborne imaging spectrometry data. Part 2: atmospheric/topographic correction, Int. J. Remote Sensing, 23: 2631–2649.

Richter, R., 1996. A spatially adaptive fast atmospheric correction algorithm. Int. J. Remote Sensing, 17: 1201–1214.

RIEGL, 2015. RIEGL Laser Measurement Systems. , http://www.riegl.com [Accessed on 14–05-2015]

Satyabala, S. P., & Bilham, R. (2006). Surface deformation and subsurface slip of the 28 March 1999 Mw = 6.4 west Himalayan Chamoli earthquake from InSAR analysis. Geophysical Research Letters, 33(23).

Schott, J. R., Salvaggio, C. and Volchok, W. J. 1988. Radiometric scene normalization using pseudoinvariant features," Remote Sens. Environ. 26(1), 1–16.

Shimada, M. (2010). Ortho-Rectification and Slope Correction of SAR Data Using DEM and Its Accuracy Evaluation. IEEE Journal of Selected Topics in Applied Earth Observations and Remote Sensing. https://doi.org/10.1109/JSTARS.2010.2072984

Slater, J.A., Heady, B., Kroenung, G., Curtis, W., Haase, J., Hoegemann, D., Shockley, C., and Tracy, K., 2011. Global assessment of the new ASTER global digital elevation model. Photogrammetric Engineering and Remote Sensing, 77(4), pp. 335–349

Smith, J.A.; Lin, T.L.; Ranson, K.J. (1980). The Lambertian assumption and Landsat data. Photogramm. Eng. Remote Sens, 46, 1183–1189

Soenen, S. A.; Peddle D. R.; Coburn, C. A. 2005.SCS+C: a modified Sun-canopy-sensor topographic correction in forested terrain, in IEEE Transactions on Geoscience and Remote Sensing, vol. 43, no. 9, pp. 2148–2159

Sullivan, R.J. (2004). Radar foundations for imaging and advanced concepts: Scitech Publishing, INC

Tao, C.V., and Y. Hu, 2001. A comprehensive study of the rational function model for photogrammetric processing, Photogrammetric Engineering & Remote Sensing, 67(1):1347–1357.

Teillet, P.M.; Guindon B.; Goodenough, D.G. On the slope-aspect correction of multispectral scanner data. Can. J. Remote. Sens.1982, 8, 84–106.

Tomar, K. S. et al. (2016) 'Semi-empirical modelling for forest above ground biomass estimation using hybrid and fully PolSAR data', Proc. SPIE, pp. 987711–987729.

Toutin, T., 2004. Geometric processing of remote sensing images: Models algorithms and methods, International Journal of Remote Sensing, 25(10):1893–1924.

van der Sanden, J.J. (1997). Radar remote sensing to support tropical forest management: TropenbosGuyana Series 5, Guyana.

Vermote, E.F., Tanre, D., Deuzé, J.L., Herman, M., and Morcrette, J.J., 1997. Second simulation of the satellite signal in the solar spectrum, 6S: An overview. IEEE Trans. Geosc. Remote Sens. 35 (3): 675–686.

Vries, C. D., Danaher, T. and Scarth, P. 2004. Calibration of multiple Landsat sensors based on pseudo-invariant target sites in Western Queensland, Australia, in Proc. of Geoscience and Remote Sensing Symp., Alaska, pp. 3729–3732.

Xie, H., Hicks, N., Keller, G. R., Huang, H. and Kreinovich, V. (2003), "An IDL/ENVI Implementation of the FFT-based Algorithm for Automatic Image Registration, Computers & Geosciences, 29(8), 1045–1055.

Yamane N., Fujita, K., Nonaka, T., Shibayama, T., and Takagishi, S. 2008 Accuracy Evaluation of DEM derived by Terrasar-X Data in the Himalayan Region. The International Archives of the Photogrammetry, Remote Sensing and Spatial Information Sciences. Vol. XXXVII. Part B7. Beijing 2008

Yuan, D. and C.D. Elvidge, 1996. Comparison of relative radiometric normalization techniques. ISPRS J. Photogramm. Remote Sens., 51: 117–126.

Chapter 23
Geoweb Services and Open Online Data Repositories for North West Himalayas Studies Including Disaster Monitoring and Mitigation

C. M. Bhatt and Harish C. Karnatak

23.1 Introduction

With recent advancements in wireless communications and Internet technology during the last decade, it is promising to develop many citizen-centric new applications and services in various fields including spatial information technology. The users of geospatial technology have started to use online GIS (geographic information system) for a variety of applications by using web services and online data repositories. Internet today has emerged as one of the most suitable and quickest means for accessing, analyzing, displaying, and transmitting the geographical information and geographic knowledge. It is perhaps the only means which maximizes the potential of GIS data and software application for wider and easier access of geographical data to the planners and decision-makers. The World Wide Web, FTP (file transfer protocol), and HTTP programs make it convenient to access and transfer data files across the Internet. Internet technology in conjunction with GIS today in crisis situations allows geospatial information coming from multiple sources to be integrated in real time, interactively accessed and visualized to generate accurate and quick actionable information for emergency response teams coordinating the activity (Mansourian et al., 2005). Today the distributed and heterogeneous resources and data services can be accessed through a centralized and uniform interface using GIS-based web portals (Karnatak et al. 2007, 2012). Considering

C. M. Bhatt (✉)
Centre for Space Science and Technology Education in Asia and the Pacific (CSSTEAP), Dehradun, India
e-mail: cmbhatt@iirs.gov.in

H. C. Karnatak
Geoweb Services, IT & Distance Learning, Indian Institute of Remote Sensing (IIRS), Indian Space Research Organisation (ISRO), Department of Space, Government of India, Dehradun, India

R. R. Navalgund et al. (eds.), *Remote Sensing of Northwest Himalayan Ecosystems*,
https://doi.org/10.1007/978-981-13-2128-3_23

the different types of disasters varying from natural to man-made having diverse data requirements, a single GIS web service may not be sufficient enough to handle all requirements, and therefore specifically the web GIS-based portals available in the Internet are generally customized and developed for a particular theme keeping in mind a specific class of user(s). The availability of international data and information service standards published by the Open Geospatial Consortium (OGC) is playing an important role to achieve interoperability in data and information sharing. GIS service standards published by OGC are based on distributed service-oriented architectures (SOA). Such systems unify distributed services through a message-oriented architecture by using Simple Object Access Protocol (SOAP). The data and information services available through web portal applications have great scope of its massive utilization at user's end. Many GIS-based plans for a variety of applications can be developed using available GIS-based web services (also known as Geoweb services) and online data repositories. Some of the major advantages of Geoweb services and online data repositories include dynamic and updated data, real-time and multiuser access, and development of user-defined applications using mashup architecture (Karnatak et al. 2012). This approach is quite useful for the applications where real-time dynamic data is required for planning and decision-making such as disaster or emergency management.

Due to the large spatial extent of disasters affecting several people across the countries, geospatial technology today finds a wider acceptance and has become an important tool for decision-making process. The traditional desktop GIS-based disaster response to an emergency situation and relief operations using traditional maps has been transformed completely, with the evolution of Internet technology. Web-based GIS technologies today allow collaborative environment to access, process, and integrate real-time data flow from various sources and a number of users to generate actionable products and respond to an emergency situation much better than before. Today when the disasters have become more intense and severe, the online data repositories and information services can play a critical role in supporting all phases of disaster monitoring and mitigation. The increasing availability of free satellite imagery (like Sentinel data), GIS thematic maps (OpenStreetMap), GIS software (QGIS), and theme-specific hazard modeling software (HEC-RAS) to analyze the data offers considerable potential to decision-makers and planners to take appropriate informed decision in times of crisis. The present technology allows to access data and information from a variety of sources and also integrate it to one common platform for user-defined applications. This chapter presents a technological overview of Geoweb service and online data repositories and a list of popular open online resources and also demonstrates a case study for effective utilization of these services in flood disaster monitoring and mitigation.

23.2 Geoweb Services and Online Data Repositories

Advancements in information and communication technologies (ICT) have facilitated a new way for sharing and dissemination of geospatial data and information. The online data repositories and web applications are providing various means of data access by using the Internet and related technologies. Today, the users of geospatial data and information can use Internet platform for various geoscientific activities such as spatial queries, geo-visualization, and simple to complex computations for decision-making and virtual reality. Considering the importance of spatial data for humanitarian response during natural disasters, many agencies and individuals are hosting their data sets online, which has enhanced the outreach of geospatial data manyfold. These open geo-data sets can be used for various thematic applications either as a Geoweb service or as a data product(s).

The service-oriented architecture (SOA) and related technologies for distributed GIS applications are providing interoperable platform for data sharing and disemintions. The SOA-based software systems provide independent building blocks that jointly represent the software application environment (Karnatak et al. 2012). One of the unique features of SOA-based software applications is to set up complete autonomy among different service components which is important for interoperability. The most popular SOA-based web services implementations are based on XML. The Extensible Markup Language (XML) is a markup language developed by W3C to define the set of rules for encoding the documents in plain text format. The XML is used to define the data and information and is also known as data definition language. The outcomes of XML are presented as Hypertext Markup Language (HTML) document which is known as data presentation language. The SOA architecture has three basic components, i.e., a provider, a requester, and a broker (Fig. 23.1). The provider is the owner or publisher of the service, typically the client or user is a requester, and the broker is the component which maintains the registry of available services as catalogue. The interaction between the above three components is represented as three operations, i.e., publish, find, and bind.

The SOA-based web services for geospatial data, process, encoding, and other spatial operations are very important to achieve interoperability in data and information sharing. The web service standards for geospatial and data and process are designed and developed by the Open Geospatial Consortium (OGC). OGC is an international standards organization which was founded in 1994 with the objectives of development, promotion, and harmonization of open geospatial data standards.

The web services published by OGC for geospatial data and process can also be represented as Geoweb services. The Geoweb services published by OGC can be grouped into six major categories, i.e., catalogue or registry services, processing or analytical services, encoding services, data or geo-data services, portrayal or visualization services, and other services. These services have been briefly described below:

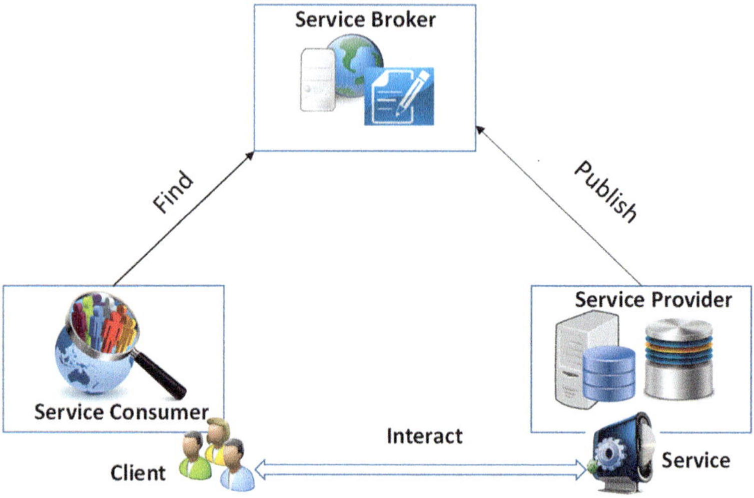

Fig. 23.1 Service-oriented architecture

23.2.1 Catalogue or Registry Services

This category of services is designed to classify, register, describe, and search the services by a user or a program. Some of the examples are CS Core, CS-WebRIM, and CS-W 19115/19119.

23.2.2 Data Processing Services

The processing services are designed to publish processes such as arithmetic operations, overlay operations, algorithms, etc. These are typically providing the capabilities of processing or transforming data-based user-defined parameters (Sahina and Gumusay 2008). The examples of WPS are Web Processing Service (WPS), Web Coverage Processing Service (WCPS), and Coordinate Transformation Service (CTS).

23.2.3 Encoding Services

Encoding is the process of putting a sequence of characters into a special format which is desired by the user for transmission or storage purposes. In GIS, encoding plays a critical role while defining the cartographic principles in spatial outputs. Typical example could be symbology to present legends in a map. Most popular encoding standards publish by OGC are Geography Markup Language (GML), Styled Layer Descriptor (SLD), KML, CityGML, etc.

23.2.4 *Geo-data Services*

The data services are designed to share the geospatial data (vector and raster) without any data loss. These services allow full or partial access of data based on user-defined area of interest (AOI). Some of the examples of OGC data services are Web Feature Service (WFS), Web Coverage Service (WCS), Sensor Observation Services (SOS), etc.

23.2.5 *Visualization Services*

The visualization services offered by OGC are typically for geo-visualization as background or overlay layer (s). These services also provide data with cartographic representations. Examples are Web Map Services (WMS) and Web Map Tile Services (WMTS).

23.2.6 *Application or Job-Oriented Services*

These services are designed for specific application or data requirements. Some of the examples include GeoXACML and GeoRSS.

The open online data repositories and Geoweb services are providing data and information by using web service standards published by OGC. The websites are available either as geo-portal or online data archive. The web portals are dynamic web applications which serve data and information to its user(s) by using database server technology. In web portal applications, the data is accessed and processed using an additional middle tier at server end using any web programming language such as PHP, C#, JAVA, Python, etc. This middle tier is also known as business logic or application server. In case of GIS data, the database servers host the raster and vector data in addition to attribute data sets in a database server. The geo-portals are developed based on GIS servers which are also known as map server(s). The GIS servers typically act as middle tier in software application architecture to make geospatial data compatible with Internet client software applications such as web browsers. Some of the popular geo-portal applications, online data repositories, early warning portals, and data analysis tools are provided in Tables 23.1, 23.2, 23.3 and 23.4. The list provided is not exhaustive, and accessing of datasets or application through these portals may require registration and proper acknowledgment. The users are therefore advised to read the data policy and data sharing policies.

Table 23.1 Popular geoportal applications

Sl. no.	Web portal	Web address	Coverage	Salient features
1.	Google maps	http://maps.google.com	Global	Satellite and aerial imageries of globe in 2D and 3D environment with many citizen applications
2.	Bing map	http://www.bing.com/maps/	Global	Global satellite imageries and infrastructure maps
3.	OpenStreetMap	http://www.openstreetmap.org	Global	Large-scale vector at city-level data created using participatory approach
4.	Wikimapia	http://wikimapia.org	Global	Identified area of interest and geolocations
5.	ISRO *Bhuvan*	http://bhuvan.nrsc.gov.in	India	Geoportal of ISRO with various thematic applications at country to village level
6.	MapMyIndia	http://www.maps.mapmyindia.com	India	Point of interest and geolocations of India
8.	Biodiversity information system	http://bis.iirs.gov.in	India	National level plant biodiversity data of India
9.	Indian bio-resource information network	http://www.ibin.gov.in	India	Distributed database bio-resource of India
10.	Visualization of earth observation data and archival system (VEDAS)	https://vedas.sac.gov.in	India	Platform for utilization of information derived over land using mainly Indian space-borne sensors
10.	India-WRIS	http://india-wris.nrsc.gov.in	India	Geoportal for water resources of India
11.	NIC GIS	http://gis.nic.in	India	Geo-enabled demography data of India

23.3 Case Study: Utilization of Online Data Repositories

This section attempts to demonstrate utilization of various online data repositories and tools which could be accessed openly and will be helpful for decision-makers and planners for taking measures to mitigate the impact of flood disaster. The main objective of the section is only to give a basic idea of how resources available from various sources in open domain can be integrated for informed decision-making. Figure 23.2 shows the various portals providing online data repositories, visualization, and analysis tools which can be accessed right from disaster watch stage to preparedness stage and finally flood disaster response phase.

For a flood disaster event, the disaster manager needs to keep a watch on the cloud persistence, heavy rainfall event, and rising river water levels which indicate the

Table 23.2 Portal providing alerts, online data repository, and natural disaster-related information

Sl. no.	Web portal	Web address	Coverage	Category	Salient features
1.	Pacific disaster center (PDC)	http://www.pdc.org/	Global	Natural disasters	Multi-hazard warning and decision support tools
2.	Global disaster alert and coordination system (GDACS)	http://www.gdacs.org/	Global	Natural disasters	Satellite and ground observation-based disaster early warning alerts in RSS and KML formats
3.	Global flood monitoring system (GFMS)	http://flood.umd.edu/	Global	Natural disasters	Experimental system using real-time TRMM multi-satellite precipitation analysis (TMPA). Streamflow, surface water storage, inundation variables at 1 km resolution
4.	Global flood detection system – version 2	http://www.gdacs.org/flooddetection/	Global	Natural disasters	Near real-time flood monitoring system using satellite data
5.	Dartmouth flood observatory	http://floodobservatory.colorado.edu/	Global	Natural disasters	Flood archive records from 1985 onward. The information contains affected area, number of people killed, the number of people displaced, the cost of damages, and a measure of the magnitude of the flood are included for each flood event. The archive is updated on an ongoing basis and new flood event are added immediately.
6.	Natural Disaster Hazards	http://www.ldeo.columbia.edu/chrr/research/hotspots/coredata.html	Global	Natural disasters	Multi-hazard information on various aspects such as frequency, loss to life & economy and other information.
7.	Fire information for resource management system (FIRMS)	https://firms.modaps.eosdis.nasa.gov/firemap/	Global	Natural disasters	Near-real time active fire data using moderate resolution imaging Spectroradiometer (MODIS) and

(continued)

Table 23.2 (continued)

Sl. no.	Web portal	Web address	Coverage	Category	Salient features
					visible infrared imaging radiometer suite (VIIRS).
8.	USGS Earthquakes Database	http://earthquake.usgs.gov/earthquakes/eqarchives/epic/kml	Global	Natural disasters	Monitoring of earthquake activities worldwide with its spatial information.
9.	Global Seismic Hazard Map	http://www.seismo.ethz.ch/static/GSHAP	Global	Natural disasters	Gridded seismic hazard risk data.
10.	NOAA/WDC Historical Tsunami Database	http://www.ngdc.noaa.gov/hazard/tsu_db.shtml	Global	Natural disasters	Spatial information on tsunami sources and run-up events.
11.	Indian National Centre for Ocean Information Services (INCOIS)	http://www.incois.gov.in/portal/index.jsp	India	Natural disasters	Near real-time tsunami information from Indian tsunami early warning Centre (ITWEC)
12.	CIMSS tropical cyclone	http://tropic.ssec.wisc.edu/#	Global	Natural disasters	Near real-time imagery, derived atmospheric analysis products.
13.	Cyclocane	https://www.cyclocane.com/	Global	Natural disasters	Data and information on cyclone active tropical storms, development potential and its tracking on map.
14.	MODIS Fire Detection Data	http://activefiremaps.fs.fed.us/gisdata.php	Global	Natural disasters	MODIS based near-real time fire location using thermal band.
15.	Lightning and Atmospheric Electricity Dataset	https://lightning.nsstc.nasa.gov/data	Global	Natural disasters	Variety of data on lightning activities such as average flashes per grid cell per year.
16.	NOAA Historical Hurricane Tracks	https://coast.noaa.gov/hurricanes/?redirect=301ocm	Global	Natural disasters	Tracking of hurricanes.
17.	Precipitation and applications viewer	https://pmm.nasa.gov/precip-apps	Global	Natural disasters	30 minutes interval updates on probable locations of landslide using landslide hazard assessment model for situational awareness (LHASA)

(continued)

Table 23.2 (continued)

Sl. no.	Web portal	Web address	Coverage	Category	Salient features
18.	Precipitation measurement missions (PMM)	https://pmm.nasa.gov/data-access	Global	Natural disasters	Precipitation data from the GPM and TRMM missions.
19.	India meteorological department (IMD)	http://www.imd.gov.in/	India	Natural disasters	Hosts information on meteorological observations, warns against severe weather phenomenon like snow, tropical cyclones, heavy rains, dust storms etc.
20.	Natural Disaster Hotspots	http://sedac.ciesin.columbia.edu/data/collection/ndh	Global	Natural disasters	Spatial data on natural hazards with frequency and economic loss.
21.	Socioeconomic and data applications center (SEDAC)	http://sedac.ciesin.columbia.edu/data/collection/ndh/sets/browse	Global	Natural disasters	Provides 209 datasets for viewing and downloading on various themes like climate, health, population, water, hazards (mortality and economic losses for flood, cyclone, drought, landslide, earthquake and volcanic hazards), etc.
22.	SEDAC's hazards mapper	http://sedac.ciesin.columbia.edu/mapping/hazards	Global	Natural disasters	Enables users to rapidly get an idea of the population, land area, dams, and nuclear power plants that could be impacted by natural events such as floods, earthquakes, fires, and volcanic eruptions
23.	International disaster database	http://www.emdat.be/	Global	Natural disasters	Core data on the occurrence and effects mass disasters in the world. The database is compiled from various sources
24.	NASA's EOSDIS	https://worldview.earthdata.nasa.gov/	Global	Natural disasters	Visualization and download of global level multi-resolution satellite

(continued)

Table 23.2 (continued)

Sl. no.	Web portal	Web address	Coverage	Category	Salient features
					imageries. The data products are updated within 3 hours of observation, essentially showing the entire earth as it looks "right now." Offer variety of applications such as forest fire management, air quality measurements, and flood monitoring
25.	Copernicus Open Access Hub	https://scihub.copernicus.eu/	Global	Natural disasters	Open access to Sentinel-1, Sentinel-2, and Sentinel-3 data products
26.	USGS earth explorer	https://earthexplorer.usgs.gov/	Global	Natural disasters	Its warehouse and source of free- and open-access satellite data from different space EO missions
27.	Humanitarian Response	https://www.humanitarianresponse.info/applications/data/country-region	Human geography	General	Information management tools and services for information exchange among operational responders during emergency. It also provides freely available datasets for many countries.
28.	Bhuvan geoportal	http://bhuvan.nrsc.gov.in/disaster/disaster/disaster.php	India	Natural disasters	Address six natural disasters such as cyclone, floods, landslides, earthquakes, and forest fire required in different phases of disaster management cycle. Most of the disaster products generated on rapid response mode based on the interpretation of Indian and international satellite

(continued)

Table 23.2 (continued)

Sl. no.	Web portal	Web address	Coverage	Category	Salient features
					datasets are made available on Bhuvan for wider public access
29.	Meteorological and Oceano-graphic Satellite Data Archival Centre (MOSDAC)	http://www.mosdac.gov.in	India	Weather and climate	Archives and disseminates data from satellites like INSAT, KALPANA-1, Oceansat, Megha-Tropiques, and SARAL. Hosts and disseminates weather-related information services and alerts over mobile devices. Provides cyclone track prediction along with intensity and cyclogenesis over Indian ocean

possibility of the event. If there is a possibility of the event to occur, disaster manager needs to have some preparedness measures in place, like which areas are likely to get inundated based on the historic inundation and identification of low-lying areas based on digital elevation model (DEM) analysis. Finally in case if the flood disaster occurs, the disaster needs to be mapped and monitored to know the spatial extent of inundation, transport network submerged, and villages marooned. In subsequent section the above-explained requirements for the three stages (watch, preparedness, and response) are explained with the help of various online data repositories and tools considering a heavy rainfall likely to occur over Srinagar, Jammu, and Kashmir which could cause floods in the Srinagar Valley.

23.3.1 Disaster Watch

Heavy continuous rainfall could trigger floods, and therefore advance information on heavy rainfall forecast for the next coming days is essential for taking appropriate mitigation measures. The Meteorological and Oceanographic Satellite Data Archival Centre (MOSDAC) at the Space Applications Centre (ISRO) provides experimental mesoscale weather forecasts in real time using WRF model (Shah et al. 2010) which can be downloaded through http://www.mosdac.gov.in/external/order-data. The

Table 23.3 Other online data repositories useful for general research and analysis including disaster mitigation

Sl. no.	Web portal	Web address	Coverage	Category	Salient features
1.	Bhuvan geoportal	http://bhuvan.nrsc.gov.in/data/download/index.php	India	General	Online geo-visualization, data download, and thematic services for a variety of EO applications. Also offers various tools and services to analyze the geo-data
2.	Natural Earth – Vector	http://www.naturalearthdata.com	Global	General	Data on natural resources such as coastline, land, oceans, islands, rivers, lakes, glaciated areas, and bathymetry is available
3.	Global map	https://www.iscgm.org/gmd	Global	General	Global geo-data at 1Km resolution on base and infrastructure layers. The data includes transportation, terrain data, drainage network, vegetation map, land cover, and land use and population centers
4.	DIVA-GIS Country Data	http://www.diva-gis.org/gdata	Global	General	Repository of data collected from various sources includes administrative areas, inland water, roads and railways, elevation, land cover, population, and climate
5.	Global self-consistent, hierarchical, high-	http://www.ngdc.noaa.gov/mgg/shorelines/gshhs.html	Global	Land and ocean boundaries	Data from two databases, i.e., World Vector Shorelines

	resolution geography database (GSHHG)				(WVS) and CIA world data Bank II (WDBII). Data can be downloaded either as shape file or in a native binary format
6.	GDEM- global digital elevation mode	http://asterweb.jpl.nasa.gov/gdem.asp	Global	Elevation	Free global DEM in 30 m posting (resolution) from ASTER satellite images
7.	CGIAR-CSI GeoPortal	http://srtm.csi.cgiar.org	Global	Elevation	Free global DEM from SRTM is available in 90 meter posting (resolution)
8.	EarthEnv-DEM90	http://www.earthenv.org/DEM	Global	Elevation	A merged global DEM in 90 meter posting (resolution) from GDEM and SRTM to fill voids and smooth data
9.	ETOPO1 global relief model	http://www.ngdc.noaa.gov/mgg/global/global.html			Global relief model of Earth's surface is available at global and regional level. Hypsographic curve of Earth's surface
10.	Global Multi-Resolution Topography (GMRT)	http://www.marine-geo.org/portals/gmrt/	Global	Elevation	Multi-beam sonar data collected by the researchers and made available as a single continuously updated global elevation data at 100 m resolution
11.	WorldClim	http://www.worldclim.org	Global	Weather and climate	Global climate gridded data at spatial resolution of about 1 km^2

(continued)

Table 23.3 (continued)

Sl. no.	Web portal	Web address	Coverage	Category	Salient features
12.	NCAR Community Climate System Model (CCSM)	http://gisclimatechange.ucar.edu	Global	Weather and climate	Climate change projected data on atmosphere and land surface in GIS format
13.	Climatic research unit (CRU)	http://www.cru.uea.ac.uk/data	Global	Weather and climate	High- and low-resolution datasets such as temperature, precipitation, pressure, drought, paleoclimate, and others, from the (infamous) climate research unit at the University of East Anglia
14.	Climate change, agriculture and food security (CCAFS)	http://www.ccafs-climate.org/data	Global	Weather and climate	Climate datasets to study biodiversity, agricultural and livestock production, and ecosystem services and hydrology
15.	Global Potential Evapotranspiration and Aridity Index	http://csi.cgiar.org/Aridity	Global	Weather and climate	Global model-driven geo-enabled raster climate data related to evapotranspiration processes and rainfall deficit for potential vegetative growth
16.	Climate Monitoring satellite application facility (CM SAF)	https://wui.cmsaf.eu/safira/action/viewProduktSearch;jsessionid=C3A056A5CE0F496B16619FAD7ACE2801.ku_2	Global	Weather and climate	Near real-time and retroactively generated datasets of cloud cover, type and temperature, surface radiation budget, temperatures, etc.
17.	Climate Analysis Indicators Tool (CAIT)	http://cait.wri.org	Global	Weather and climate	Comprehensive and comparable climate and emissions data available for free and open access

18.	International Satellite Cloud Climatology Project (ISCCP)	http://isccp.giss.nasa.gov/products/d2yeardsets.html	Global	Weather and climate	Average cloud cover and associated data like cloud top temperature, water path, optical thickness etc., on monthly basis
19.	IRI/LDEO Climate Data Library	http://iridl.ldeo.columbia.edu	Global	Weather and climate	Climate-related data download facility with various analytical tools through web browser environment
20.	World Ozone and Ultraviolet Radiation Data Center (WOUDC)	http://www.woudc.org/data_e.html	Global	Weather and climate	Data repository of ozone information and ground level ultraviolet irradiance
21.	HydroSHEDS	http://hydrosheds.cr.usgs.gov/index.php	Global	Hydrology	Hydrographic information at regional and global scale. The geo-data (raster and vector) such as river networks, watershed boundaries, drainage directions, and flow accumulations are available
22.	Water Isotopes	http://wateriso.utah.edu/waterisotopes/pages/data_access/da_main.html	Global	Hydrology	Global and regional data on grids of hydrogen and oxygen isotope composition of precipitation and environmental waters
23.	General Bathymetric Chart of the Oceans	http://www.gebco.net	Global	Hydrology	Gridded bathymetric datasets
24.	Earth Env Freshwater Ecosystems Environmental Information	http://www.earthenv.org/streams	Global	Hydrology	1 km resolution data on freshwater ecosystems which includes climate, land-cover, soil, and geology

(continued)

Table 23.3 (continued)

Sl. no.	Web portal	Web address	Coverage	Category	Salient features
25.	Randolph Glacier Inventory	http://www.glims.org/RGI	Global	Snow/ice	Multi-temporal global inventory of glacier outlines with specific focus on quality control
26.	USGS Land Cover Institute	http://landcover.usgs.gov/landcoverdata.php	Global	Land cover	Rich data and information repository of land cover data such as river observations, aquifers data, and ocean color information
27.	GLOBCOVER	http://due.esrin.esa.int/page_globcover.php	Global	Land cover	Global land cover dataset at 300 m resolution from the MERIS sensor of ENVISAT satellite
28.	MODIS Global Land Cover	http://visibleearth.nasa.gov/view.php?id=61004	Global	Land cover	Global land cover data in 1Km and 4Km from MODIS
29.	UMD GLC	http://glcf.umd.edu/data/landcover	Global	Land cover	Global land cover maps created using a classification tree approach from MODIS data
30.	Global Land Cover by National Mapping Organizations	http://www.iscgm.org/404.html	Global	Land cover	Global land cover map with classification scheme adopted from UN FAO LCCS along with other land cover products
31.	Global Lakes and Wetlands Database	http://www.worldwildlife.org/pages/conservation-science-data-and-tools	Global	Land cover	Global GIS vector layers for lakes, reservoirs, wetlands, swamps, bogs, etc.

32.	Grassland GIS	http://www.wri.org/publication/pilot-analysis-global-eco systems-grassland-ecosystems#data	Global	Land cover	Spatial data grasslands with utilizations
33.	PALSAR Forest/Non-Forest map	http://www.eorc.jaxa.jp/ALOS/en/palsar_fnf/fnf_index.htm	Global	Land cover	Global forest map in 50 meter spatial resolution generated using SAR data
34.	Global Forest Change 2000-2014	http://earthenginepartners.appspot.com/science-2013-global-forest/download_v1.2.html	Global	Land cover	Global forest tree cover change map in 30 meter spatial resolution (2000 to 2014)
35.	Atlas of the Biosphere	http://nelson.wisc.edu/sage/data-and-models/atlas/maps.php	Global	Ecology	Spatial data in raster format for environmental variables such as potential evapotranspiration, soil, pH, average snow depth, etc.
36.	Lifemapper	http://lifemapper.org/?page_id=593	Global	Ecology	Species distribution maps based on climate, terrain, and land cover
37.	Anthropogenic Biomes	http://sedac.ciesin.columbia.edu/data/collection/anthromes	Global	Ecology	Gridded anthropogenic biomes data
38.	Net Primary Productivity	http://sedac.ciesin.columbia.edu/data/collection/hanpp	Global	Ecology	Global gridded net primary productivity data (NPP)
39.	World Soil Information	https://www.soilgrids.org/#/?layer=geonode:taxnwrb_250m	Global	Ecology	Global soil data on organic carbon content, gypsum content, water capacity etc., for topsoil and subsoil. The data is available with more than 22 attributes
40.	Harmonized world soil database	http://webarchive.iiasa.ac.at/Research/LUC/External-World-soil-database/HTML/index.html?sb=1	Global	Ecology	Regional and national level soil database generated under LUC program of FAO. The spatial data is available resolution of 30 arc seconds

(continued)

Table 23.3 (continued)

Sl. no.	Web portal	Web address	Coverage	Category	Salient features
41.	ERS/MetOp soil moisture	http://www.ipf.tuwien.ac.at/404/	Global	Ecology	Soil moisture data from satellite-based scatterometer measurements
42.	Global High Resolution Soil Water Balance	http://www.cgiar-csi.org/data/global-high-resolution-soil-water-balance	Global	Ecology	Global evapotranspiration and soil water deficit data in the resolution of 30 arc seconds
43.	Carbon Dioxide Information Analysis Center	http://cdiac.ornl.gov/ftp/	Global	Ecology	Atmospheric CO2 concentrations, precipitation, and long-term modeling data
44.	UNEP WCMC	http://datadownload.unep-wcmc.org/datasets	Global	Ecology	Global data generated under United Nations Environment Programme such as wetlands, global distribution of coral reefs, mangrove distributions, etc.
45.	Terrestrial Ecoregions of the World	http://maps.tnc.org/gis_data.html	Global	Ecology	Terrestrial ecological data in vector GIS format
46.	Freshwater Ecoregions of the World	http://www.feow.org/downloads.php	Global	Ecology	Worldwide biogeographic classification of freshwater ecological regions
47.	Mineral Resources Data System	http://tin.er.usgs.gov/mrds	Global	Mineral resources/oil and gas	Global spatial data sets on mineral resources including names, locations, descriptions, geological characteristics, etc.

48.	Environmental data explorer	http://geodata.grid.unep.ch	Human geography	General	Spatial data repository at national, regional, and global levels on various themes such as freshwater, population, forests, emissions, climate, disasters, health, GDP, etc.
49.	World Bank Geo-data	https://velluminformation.com/2012/01/28/google-earth-world-bank-data-and-kml-files/	Human geography	General	World bank data available as spatial layer (KML)
50.	History Database of the Global Environment-HYDE	http://themasites.pbl.nl/tridion/en/themasites/hyde/index.html	Human geography	General	Gridded time series population and land use data of last 12,000 years. Very interesting data on GDP, value added, livestock, agricultural areas and yields, private consumption, greenhouse gas emissions, and industrial production are available for public access
51.	Natural Earth	http://www.naturalearthdata.com	Human geography	Administrative boundaries	Free- and open-access geospatial data on course resolution at global scale
52.	GADM	http://gadm.org	Human geography	Administrative boundaries	Spatial data on administrative boundaries of different countries are available in shape file, geodatabase, KMZ, and RData format
53.	World Borders	http://thematicmapping.org/downloads/world_borders.php	Human geography	Administrative boundaries	Countries boundaries with basic attributes are available in GIS format

(continued)

Table 23.3 (continued)

Sl. no.	Web portal	Web address	Coverage	Category	Salient features
54.	World Spatial Database of Protected Areas	http://free-gis-data.blogspot.in/2009/04/world-spatial-database-on-protected.html	Human geography	Environmental boundaries	The spatial data on marine and terrestrial protected areas
55.	IUCN 2013 Red List	http://www.iucnredlist.org/technical-documents/spatial-data	Human geography	Environmental boundaries	IUCN red species data with geo-locations
56.	Protected Planet	https://www.protectedplanet.net	Human geography	Environmental boundaries	Provides latest and up-to-date information on protected areas across the world
57.	Human Influence and Footprint	http://sedac.ciesin.columbia.edu/data/collection/wildareas-v2	Human geography	Land use	Provides human influence index and human footprint to assess the influence of human on terrestrial ecosystems at 30 arc second resolution
58.	Global Agricultural Lands	http://sedac.ciesin.columbia.edu/theme/sustainability	Human geography	Land use	Provides global maps depicting extent and intensity of agricultural lands in 2000 based on MODIS and SPOT images together with agricultural inventory data
59.	Global Irrigated Area and Rain fed Crops Areas	http://waterdata.iwmi.org/Applications/GIAM2000	Human geography	Land use	Provides global irrigated area map, rain-fed area map, and land use/land cover at 10 km for year 2000
60.	Crop Calendar Dataset	http://nelson.wisc.edu/sage/data-and-models/crop-calendar-dataset/index.php	Human geography	Land use	Provides gridded maps on number of parameters like planting dates, harvesting dates, etc., for about 19 crops and is available at 5 minute and 0.5 degree resolution and in netCDF and ArcINFO ASCII formats

61.	EarthStat: Agricultural Land Use and potential use	http://www.earthstat.org	Human geography	Land use	Provides geographic datasets which allow users to carry out a number of applications like understanding impact of climate change on crop yields, fertilizer, manure use, etc.
62.	Global Reservoir and Dam Database (GRanD)	http://atlas.gwsp.org/index.php	Human geography	Lakes, oceans and other water sources	Provides information on all reservoirs having storage capacity of more 0.1km^3 both in polygon and point format.
63.	Gridded Population of the World (GPW)	http://sedac.ciesin.columbia.edu/data/collection/gpw-v4	Human geography	Population	Provides population estimates for the years 2000, 2005, 2010, 2015, and 2020. The GPW data collection provides openly available, licensed under the Creative Commons Attribution 4.0
64.	WorldPop	http://www.worldpop.org.uk	Human geography	Population	Provides an open-access archive of spatial demographic datasets for Central and South America, Africa, and Asia to support development, disaster response, and health applications
65.	Large Urban Areas 1950–2050	https://nordpil.com/resources/world-database-of-large-cities	Human geography	Population	Provides database representing the historic, current, and future estimates and projections with number of inhabitants for the world's largest urban areas from 1950 to 2050

(continued)

Table 23.3 (continued)

Sl. no.	Web portal	Web address	Coverage	Category	Salient features
66.	Global Urban Extent	http://nelson.wisc.edu/sage/data-and-models/schneider.php	Human geography	Population	Consists of global maps at 500 m resolution of urban extent derived from satellite data exploiting temporal and spectral information from MODIS observations over 1 year
67.	OpenStreetMap	http://www.geofabrik.de/data/download.html	Human geography	Buildings, roads, and points of interest	OSM is a collaborative project to create a free editable map of the world. Contributors use aerial imagery, GPS devices, and low-tech field maps to verify that OSM is accurate and up to date
68.	Nuclear Power Station locations	https://fusiontables.google.com/DataSource?dsrcid=579353#rows:id=1	Human geography	Buildings, roads, and points of interest	Provides list of all nuclear power stations worldwide as per IAEA in a tabular format consisting of country, name, location, etc.
69.	Open Flights	http://openflights.org/data.html	Human geography	Transport and communications	Airport, airline, and route data across the globe. Data is provided as CSV files which can be easily processed to produce GIS outputs. Data includes all known airports and a large number of routes between airports

70.	Global Roads Open Access Data Set (gROADS)	http://sedac.ciesin.columbia.edu/data/set/groads-global-roads-open-access-v1	Human geography	Transport and communications	gROADS data set combines best available roads data by country into a global roads coverage, using the UN spatial data infrastructure transport (UNSDI-T) version 2 as a common data model
71.	NGIS Country Files	http://geonames.nga.mil/gns/html/namefiles.html	Human geography	Gazetteers (place/feature names)	Provides complete files of geographic names information covering countries or geopolitical areas in a special format amenable to input into geographic information systems, databases, and spreadsheets
72.	Global Rural-Urban Mapping Project (GRUMPv1)	http://sedac.ciesin.columbia.edu/data/set/grump-v1-settlement-points	Human geography	Gazetteers (place/feature names)	GRUMPv1 provides (a) gridded population data at 30 arc second resolution of 1990, 1995, and 2000, (b) urban extents based on night-time lights, and (c) a point's data set of all urban areas with populations of greater than 1000 persons

Table 23.4 Tools and models available in open domain for disaster analysis

Sl. no.	Model/tool	Web address	Category	Salient features
1.	ANUGA	https://anuga.anu.edu.au/	Tools/ models	ANUGA is a Free & Open Source Software (FOSS) having the capability to model the impact of hydrological disasters such as flooding, storm surge, dam breaks, and tsunamis
2.	Hydrologic Engineering Center's River Analysis System (HEC-RAS)	http://www.hec.usace.army.mil/software/hec-ras/	Tools/ models	HEC-RAS is one of the widely used free open-source software to model the hydraulics of water flow through natural rivers and other channels. It allows one-dimensional steady flow, one- and two-dimensional unsteady flow calculations, sediment transport/mobile bed computations, and water temperature/water quality modeling
3.	EPA's Stormwater Management Model (SWMM)	https://www.epa.gov/water-research/storm-water-management-model-swmm#downloads	Tools/ models	SWMM is used extensively globally in planning, analysis, and design related to storm water runoff, combined and sanitary sewers in urban areas
4.	CAPRA (Probabilistic Risk Assessment)	http://www.ecapra.org/about	Tools/ models	CAPRA Program is an initiative that aims to strengthen the institutional capacity for assessing, understanding, and communicating disaster risk, with the ultimate goal of integrating disaster risk information into development policies and programs
5.	HAZUS	https://www.fema.gov/hazus-software	Tools/ models	Hazus is a nationally applicable standardized methodology that contains models for estimating potential losses from earthquakes, floods, and hurricanes
6.	TauDEM (Terrain Analysis Using Digital Elevation Models)	http://hydrology.usu.edu/taudem/taudem5/index.html	Tools/ models	TauDEM consists of a suite of tools useful for conditioning of the DEM and the extraction and analysis of hydrological information from DEM

(continued)

Table 23.4 (continued)

Sl. no.	Model/tool	Web address	Category	Salient features
7.	Arc Hydro Tools	http://en.freedownloadmanager.org/Windows-PC/Arc-Hydro-Tools-FREE.html	Tools/models	The Arc Hydro toolset is a suite of tools which facilitate the creation, manipulation, and display of Arc Hydro features and objects within the ArcMap environment. The tools provide raster, vector, and time series functionality, and many of them populate the attributes of Arc Hydro features
8.	Grid Analysis and Display System (GrADS)	http://cola.gmu.edu/grads/	Tools/models	GrADS is an interactive desktop tool that is used for easy access, manipulation, and visualization of earth science data. GrADS has been implemented worldwide on a variety of commonly used operating systems and is freely distributed over the Internet
9.	Sentinel Application Platform (SNAP)	http://step.esa.int/main/download/	Tools/models	SNAP is an open-source common architecture for ESA toolboxes ideal for the exploitation of Earth observation data. The SNAP architecture is ideal for Earth observation processing and analysis
10.	PolSARpro	https://earth.esa.int/web/polsarpro/home	Tools/models	The polarimetric SAR data processing and educational tool aims to facilitate the accessibility and exploitation of multi-polarized SAR datasets including those from ESA (Envisat ASAR Alternating Polarization mode products and Sentinel-1) and third-party missions (ALOS-1 PALSAR, ALOS-2 PALSAR, COSMO-SkyMed, RADARSAT-2, RISAT, TerraSAR-X, and TanDEM-X)

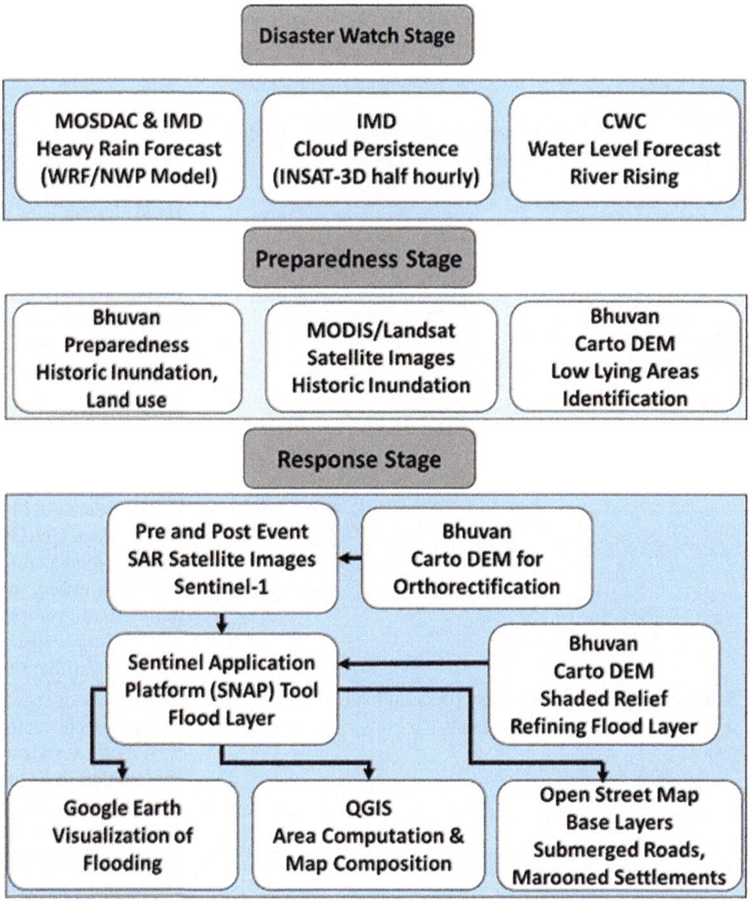

Fig. 23.2 Portals and online data repositories for flood disaster event

WRF data which is in netcdf file format can be processed, analyzed, and visualized using freely available Grid Analysis and Display System (GrADS) software accessible through http://cola.gmu.edu/grads/downloads.php. MOSDAC portal itself allows visualization of WRF-related parameters (rainfall, temperature, relative humidity, cloud fraction, etc.) in GIS environment (http://www.mosdac.gov.in/weather_forecast/index.jsp?param=hrf_24) which can be seen with base layers superimposed, the heavy rainfall and cloud fraction forecast for the next 3 days can give an understanding about areas likely to experience flood in coming days (Figs. 23.3 and 23.4). Further INSAT-3D images which are available every half hourly from MOSDAC (http://www.mosdac.gov.in/data/servlet/Image3d?imagename=3DIMG*_L1C_ASIA_MER_IR1.jpg) and also IMD (http://satellite.imd.gov.in/img/animation3d/3Dasiasec_ir1_3d.htm) as individual images and also as animation can help in monitoring the movement and persistence of cloud cover over the interested region (Fig. 23.5). Due to heavy rainfall in the upstream

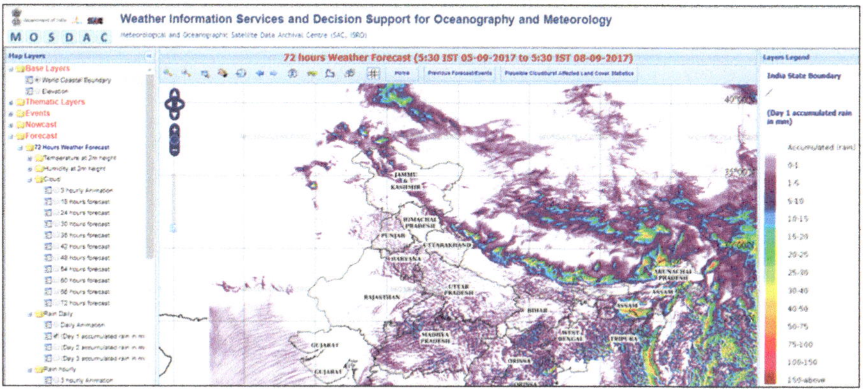

Fig. 23.3 MOSDAC portal showing daily rainfall (mm) accumulation forecast

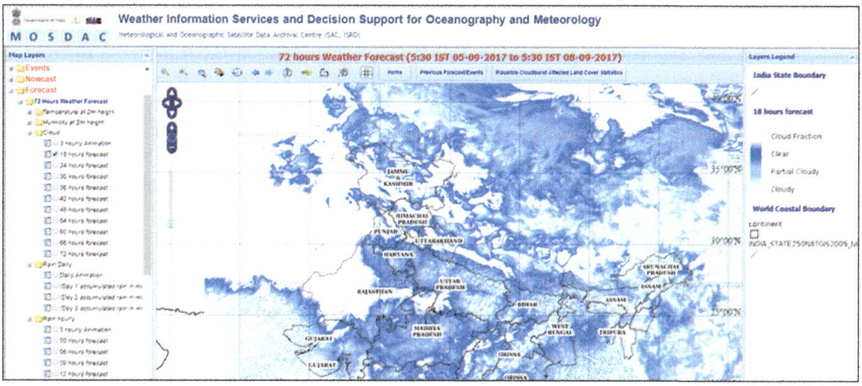

Fig. 23.4 MOSDAC portal showing daily cloud conditions forecast

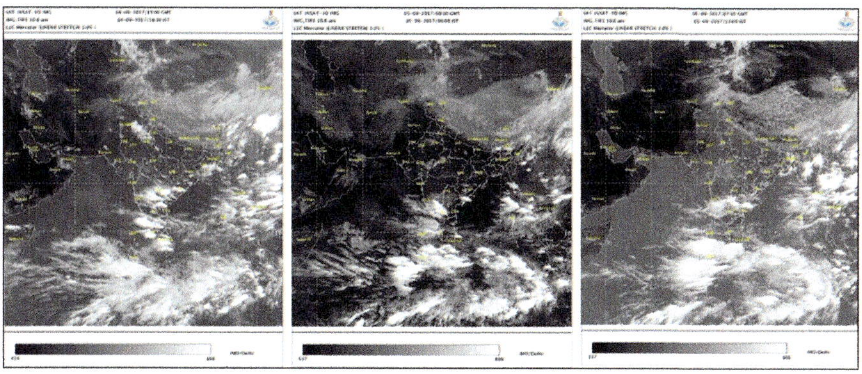

Fig. 23.5 IMD portal showing cloud persistence available from INSAT-3D images

Fig. 23.6 CWC portal showing river gauge data for Sangam, Jammu, and Kashmir station

catchment areas, the water level in the rivers also starts rising which can be monitored using the Central Water Commission (CWC) gauge station information on current level and previous water level accessed through http://www.india-water.gov.in/eSWIS-MapViewer/. The portal provides information in spatial format as well as in hydrograph format. Figure 23.6 shows that by clicking on the gauge station, information on the warning level, danger level, highest flood level (HFL), date of occurrence of HFL, present water level, and trend could be observed.

23.3.2 Disaster Preparedness

With the basic information obtained from rainfall forecast, cloud persistence, and gauge height by accessing various online portals, disaster manager gets a fair idea about the likely scenario for flood event to take place in advance can be made and can help to make plans for preparedness before the event actually happens. Knowing that there is a possibility of flood event, the decision-maker needs to have some idea about the regions that are likely to get flooded to take precautionary measures on ground for safe evacuation of the inhabitants staying in those areas. ISRO DMS services in tandem with ISRO's Geo-portal, Bhuvan, in public domain are a unique demonstration of EO data and geospatial technology utilization for visualization, understanding, effective planning, and decision-making for disaster management. Bhuvan supports management of disasters like cyclone, floods, landslides, earthquakes, forest fire, and drought, which is useful for various phases of disaster management including preparedness and response. Figure 23.7 shows historic flood inundation layers extracted from multi-temporal (08–23 Sept. 2014) satellite images during Srinagar floods in Jammu and Kashmir during September 2014 and available through Bhuvan Disaster Services (http://bhuvan-noeda.nrsc.gov.in/disaster/disaster/disaster.php#). This information can be visualized together

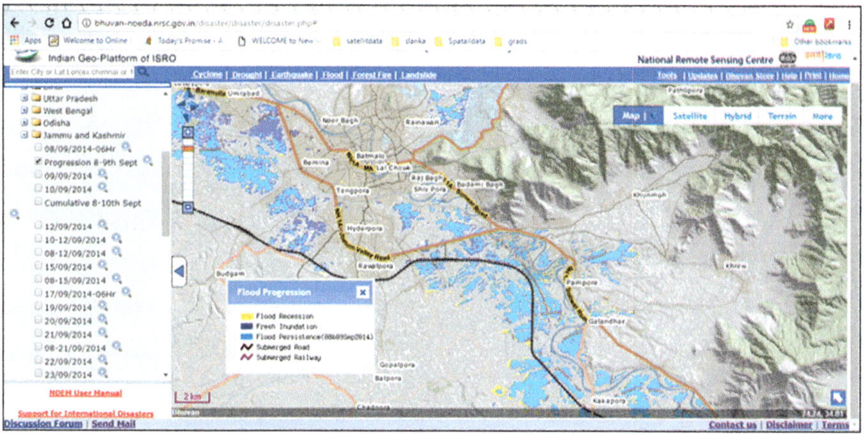

Fig. 23.7 Bhuvan portal showing historic inundation (08–23 Sept. 2014) experienced in Srinagar Valley during floods of September 2014

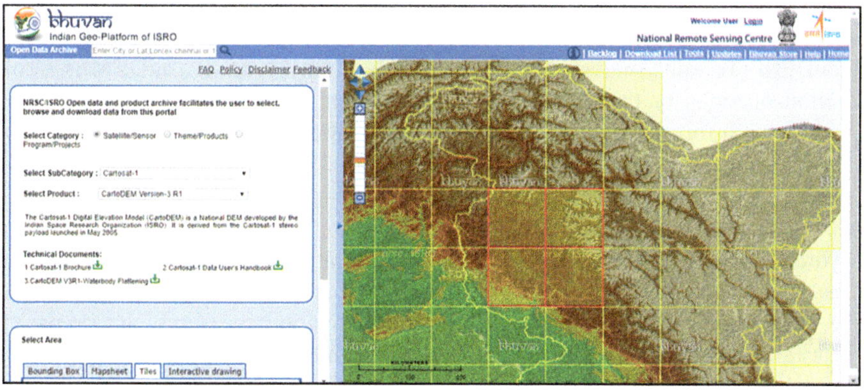

Fig. 23.8 Bhuvan portal showing CartoDEM tiles highlighted (red color) over Srinagar, Jammu, and Kashmir for downloading

with various other thematic information available with Bhuvan like land use and administrative boundaries to gain insight about areas which have experienced flood in past and also the land use that will get affected in those areas and also visualize the terrain with help of Bhuvan 3D. In addition to the available historic inundation, low-lying areas with the help of freely available digital elevation data of Cartosat DEM (~30 m) from Bhuvan (http://bhuvan.nrsc.gov.in/data/download/index.php) or SRTM elevation data (~30 m) from USGS earth explorer (https://earthexplorer.usgs.gov/) can be identified which could help in providing additional information on areas likely to get affected which may have not been affected during past floods. Figure 23.8 shows the downloading of CartoDEM through Bhuvan portal. DEM could also be utilized to derive various hydrological layers

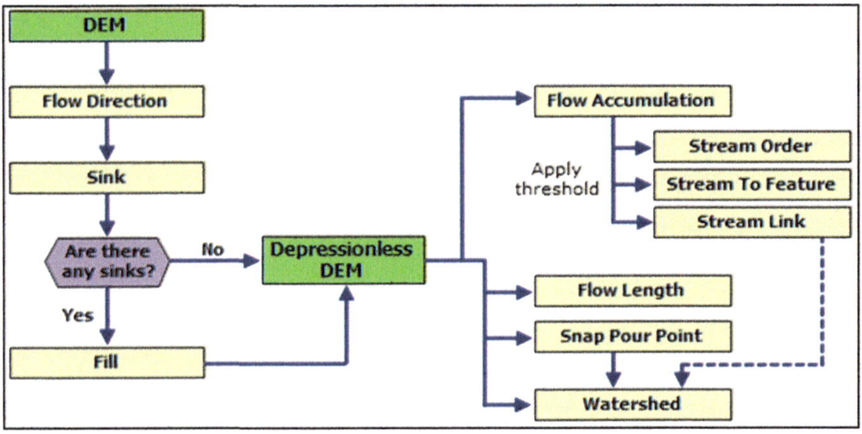

Fig. 23.9 Flowchart showing hydrological information derived from DEM

(Fig. 23.9) like flow direction, flow accumulation, slope, aspect, and basin boundaries using freely available (http://hydrology.usu.edu/taudem/taudem5/index.html) tools like TauDEM (Terrain Analysis Using Digital Elevation Models). TauDEM is a collection of tools for the extraction of hydrological information from DEM. This hydrologic information could be further utilized for advanced analysis like hydrological modeling to get extent of inundation based on varying discharge data and generate inundation scenarios using free hydrological modeling software's like HecRAS (http://www.hec.usace.army.mil/software/hec-ras/). HecRAS models the hydraulics of water flow through natural rivers and other channels. Apart from disaster preparedness during the disaster, the above information also could be utilized for planning long-term disaster mitigation measures and preparing disaster management strategies. In this effort archived satellite images freely available from MODIS (https://lance-modis.eosdis.nasa.gov/imagery/subsets/?project=fas&subset=FAS_India1), Landsat (https://earthexplorer.usgs.gov/), and Sentinel (https://vertex.daac.asf.alaska.edu/#) also could be accessed to prepare flood hazard zonation maps of the area. Figure 23.10 shows flood inundation extent captured from Modis Aqua image of 10 Sept. 2014 over Srinagar Valley. This one of the severest floods (Bhatt et al. 2016) could be very helpful in disaster management planning.

23.3.3 *Disaster Response*

Freely open data repositories and tools are very important in providing response to a disaster. For floods particularly SAR (synthetic aperture radar) data is useful because it can image the Earth through clouds and allows dynamic hydrological events like

Fig. 23.10 Flood inundation captured from Modis Aqua image of 10 Sept 2014 over Srinagar Valley (https://lance-modis.eosdis.nasa.gov/imagery/subsets/?project=fas&subset=FAS_India1.2014253.aqua.721.250m)

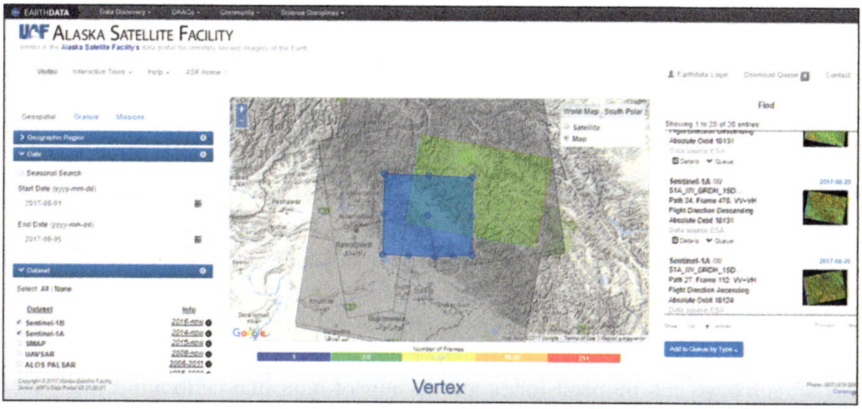

Fig. 23.11 Availability of Sentinel data over Srinagar region (area of interest) for archival and current period shown as footprint (center) and also as browse images (right side)

floods to be captured. Due to persistent cloud cover during monsoon season optical data does not provide much support in monitoring of flood events. Availability of SAR data since 2014 from Sentinel-1, a SAR mission from ESA, available free of charge, has equipped the disaster managers with a powerful dataset especially for flood disasters to respond to flood situation quickly than to wait for cloud-free optical data. Sentinel-1 data can be downloaded by registering at the Sentinels Scientific Data Hub (https://cophub.copernicus.eu/dhus/) and by specifying the area of interest, product type, sensor mode, and sensing period, among others (Fig. 23.11). The data can also be accessed by registering at Vertex (https://vertex.daac.asf.alaska.edu/#) which is the Alaska Satellite Facility's data portal for remotely sensed imagery of the Earth. Through these, portals archival (before

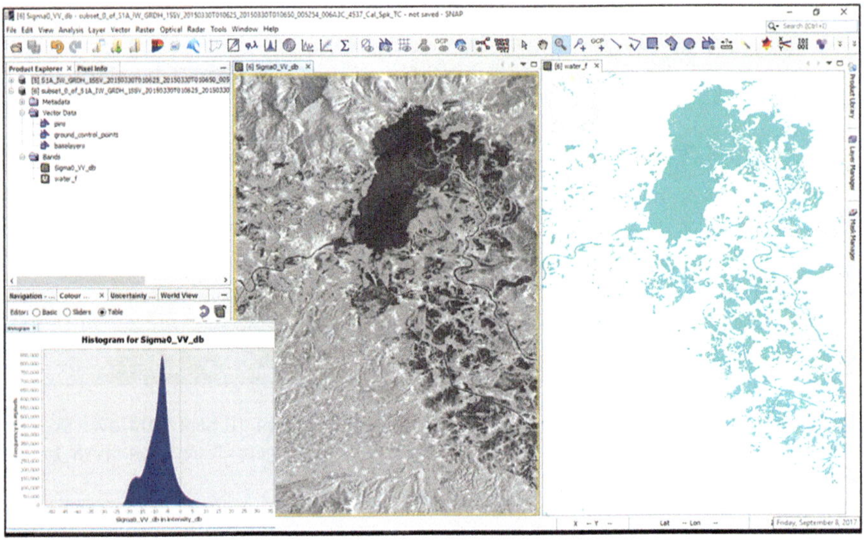

Fig. 23.12 Sentinel 1A SAR data over Srinagar region (left panel) and water layer extracted (right panel) using SNAP software

flood event) and latest (during flood event) Level-1 Ground Range Detected (GRD) Sentinel-1 data in C-band in VV and VH polarizations, which incorporates already some basic preprocessing, can be accessed for analysis. The pre- and during event Sentinel data provides (a) pre-flood river extent and waterlogged areas, and (b) present spatial extent of inundation.

For the analysis of Sentinel data, freely available tool from ESA's Sentinel Application Platform (SNAP) can be downloaded (http://step.esa.int/main/download/) and used. The user needs to have basic to intermediate knowledge of image processing and basic knowledge of SAR theory for working with the SAR images. SNAP software can be used for preprocessing of data like calibration (calibrated values of the backscatter coefficient), speckle filtering, ortho-rectification, and thresholding. To separate water from non-water histogram of the filtered backscatter coefficient image is analyzed. Low values of the backscatter in the histogram correspond to water, and high values shall correspond to the non-water class. Sentinel 1A SAR data over Srinagar region (left panel) and water layer extracted (right panel) using SNAP software (Fig. 23.12). The classified layer can be fine-tuned using shaded relief generated from CartoDEM. Pre-flood water bodies can be extracted using similar approach from the pre-event data and can be subtracted from the classified water layer to generate the flood inundation layer. This layer then can be used in a GIS software for area computation and map composition. QGIS is a cross-platform free (http://www.qgis.org/en/site/forusers/download.html) and open-source desktop geographic information system (GIS) application that supports viewing, editing, and analysis of geospatial data (Fig. 23.13). Once the flood inundation layer is available, apart from inundated area computation, a decision-

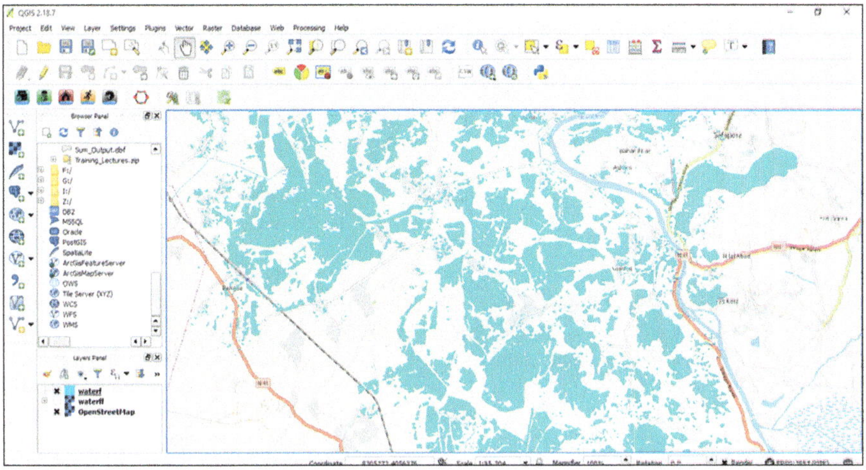

Fig. 23.13 Flood layer extracted can be visualized using OpenStreetMap as plugin using QGIS

maker is also interested in knowing the roads and rail network submerged and districts affected. OpenStreetMap (OSM) which supports and enables the development of freely reusable geospatial data can be used for downloading (https://www.openstreetmap.org/) base data like roads, rail, points of interest, and administrative boundaries and can be integrated with inundation layer for flood-related analysis and map composition.

23.4 Popular Geoweb Services and Online Data Repositories

This section provides a list of freely available resources like online data repositories, tools, and software that could help decision-makers in disaster monitoring and mitigation. The lists have been categorized into popular geoportal applications (Table 23.1), portal providing alerts, online data repository and natural disaster-related information (Table 23.2), other online data repositories useful for general research and analysis including disaster mitigation (Table 23.3), tools and models available in open domain for disaster analysis (Table 23.4), and freely available GIS software's available for geographic data analysis during disasters (Table 23.5). The information and list of websites provided for free geospatial data, software, and other datasets may not be exhaustive, and the users are requested to refer and update with more web resources and the license policy of the providers.

Table 23.5 Freely available GIS software available for geographic data analysis during disasters

Sl. no.	Software	Web address	Category	Salient features
1.	Quantum GIS (QGIS)	http://www.qgis.org/	GIS analysis	QGIS is a user-friendly and easy-to-install free- and open-source cross-platform desktop geographic information system application and is one of the most viable alternatives to proprietary desktop GIS software
2.	System for Automated Geoscientific Analyses (SAGA)	http://www.sagagis.org/en/index.html	GIS analysis	SAGA is GIS software specializing in advanced physical geography applications. There exists about over 300 modules involving various GIS operations like manipulating raster and vector data to storing, managing, and creating spatial data
3.	Geographic Resources Analysis Support System (GRASS)	http://grass.fbk.eu/	GIS analysis	GRASS is a free- and open-source geographic information system (GIS) software suite used for geospatial data management and analysis, image processing, graphics and maps production, spatial modeling, and visualization
4.	Integrated Land and Water Information System (ILWIS)	http://52north.org/communities/ilwis http://www.ilwis.org	GIS analysis	ILWIS is a window-based open-source software supporting vector and raster processing and is freely available
5.	gvSig	http://www.gvsig.org/web/	GIS analysis	gvSIG is a GIS desktop-based application which can handle a variety of vector and raster files, databases, and remote services required for planning and management
6.	MapWindow	http://mapwindow.org/	GIS analysis	MapWindow includes a free- and open-source desktop geographic information system (GIS) and includes most of GIS required operation like map viewer, identify features, processing tools, and print layout, including some higher-level tools such as TauDEM
7.	Whitebox GAT	http://www.uoguelph.ca/~hydrogeo/Whitebox/	GIS analysis	Whitebox GAT an open-source desktop GIS and remote sensing software package for general applications of geospatial analysis and data visualization

(continued)

Table 23.5 (continued)

Sl. no.	Software	Web address	Category	Salient features
8.	uDig	http://udig.refractions.net/	GIS analysis	User-friendly desktop Internet GIS (uDig) is an open-source GIS software program, built with Eclipse Rich Client (RCP) technology. It allows spatial data viewing/editing using Open GIS standards for Internet GIS, Web Map Server (WMS), and Web Feature Server (WFS) standards
9.	GeoDa	http://geodacenter.github.io/	GIS analysis	GeoDa is a free GIS software program for spatial data analysis, and its main functionality includes geostatistics, perform autocorrelation, descriptive and regression statistics
10.	LandSerf	http://www.landserf.org/	GIS analysis	LandSerf is a freely available geographic information system (GIS) for the visualization and analysis of surfaces. Applications include visualization of landscapes, geomorphological analysis, GIS file conversion, map output, archaeological mapping and analysis, surface modeling, and many more

23.5 Challenges and Gaps

Geoweb services and online open data repositories have surpassed the barrier of data availability for scientific studies in geospatial domain. Today, huge amount of geospatial data and information are available in different formats, scale, and resolution from global to local scale.

Geoweb services can bring and integrate vast amount of data from heterogeneous sources to generate effective information required to address different applications. Geospatial technology and geospatial data available in open domain today find a wider acceptance among decision-makers and planners especially for responding to disasters and other emergency events. Freely available information and tools now searchable and accessible through portals have become an important tool for decision-making process during natural disasters which have large spatial extent, affecting several people across countries and involving large number of different agencies to work in tandem. However there are still gap areas which restrict the effective use and scope of these data repositories and services. The major gaps are lack of awareness, capacity building of application user(s), and availability of network bandwidth for high-speed data transfer, online computation, algorithm development, etc. Capacity building for the development of skilled manpower who has the technical know-how about the data availability through Geoweb services, data sharing, data mining, and analysis is the major challenge that hinders the effective utilization of the technology. Internet connectivity is another major concern particularly in developing countries which prevents the full utilization of these resources. During the last decade, the data-originating organizations have come forward to share their data using geoportals and related applications, but still the data and information are available mainly for geo-visualization and querying using basic services standards such as WMS and WMTS. The data access using data service standards such WFS, WCS, WFS-T, etc. are very limited which are restricting the use of these resources by geospatial professionals. Utilization of Web Processing Services (WPS) for scientific studies is one of the exciting areas where more focus needs to be given. The WPS-based solutions provide online mode of data analysis and processing without physical download of data at user's end.

Acknowledgments This document is a compilation specifically intended for enhancing knowledge on freely available geospatial information on open-source domain which could be used by decision-makers for disaster mitigation. The data content is compiled from various web sources and all the resources considered are gratefully acknowledged.

References

Bhatt, C. M., Rao, G. S., Farooq, M., Manjusree, P., Shukla, A., Sharma, S. V. S. P., & Dadhwal, V. K. (2016). Satellite-based assessment of the catastrophic Jhelum floods of September 2014, Jammu & Kashmir, India. Geomatics, Natural Hazards and Risk, 1–19.

Harish Chandra Karnatak, Sameer Saran, Karamjit Bhatia and P.S. Roy, (2007), "Multicriteria Decision Analysis in Web GIS Environment", Geoinformatica, (2007) 11, pp: 407–429: Springer Science DOI 10.1007/s10707-006-0014-8.

K. Sahina and M. U. Gumusay, 2008, "Service Oriented Architecture (SOA) based web services for Geographic Information Systems", International Archives of the Photogrammetry, Remote Sensing and Spatial Information Sciences. Vol. XXXVII. Part B2. Beijing 2008.

Karnatak, H. C., Shukla, R., Sharma, V. K., Murthy, Y. V. S., & Bhanumurthy, V. (2012). Spatial mashup technology and real time data integration in geo-web application using open source GIS–a case study for disaster management. Geocarto International, 27(6), 499–514.

Mansourian, A.; Rajabifard, A.; Valadan Zoej, M.J. SDI Conceptual Modeling for Disaster Management. In Proceedings of the ISPRS Workshop on Service and Application of Spatial Data Infrastructure, Hangzhou, China, 14–16 October 2005.

Sahina, K. and MU Gumusay (2008). Service Oriented Architecture (SOA) based Web Services for Geographic Information Systems.

Shah, S., Rao, B. M., Kumar, P., & Pal, P. K. (2010). Verification of cloud cover forecast with INSAT observation over western India. Journal of earth system science, 119(6), 775–781.

Chapter 24
Comparison of Geostatistical and Deterministic Interpolation to Derive Climatic Surfaces for Mountain Ecosystem

Prabhakar Alok Verma, Hari Shankar, and Sameer Saran

24.1 Introduction

Spatial interpolation is the process of estimating value of continuous target variable at unknown location based on available samples. At present, there are many interpolation techniques available, and each technique has its own pros and cons. Accuracy of interpolation mainly depends on (1) sampling pattern and number of samples, (2) interpolation model adopted, and (3) presence of co-variable if a number of sample points are less. There are certain sampling techniques available, namely, regular, random, stratified, cluster, etc. Due to complex topography of mountain ecosystem like the Himalaya, stratified sampling technique is supposed to give the best prediction. But due to high variability in elevation and remote locations in mountain regions, installation of automatic weather stations (AWS) as per stratified sampling method is very difficult. So mountain regions face lack of sufficient number of samples/observations for accurate prediction (Stahl et al. 2006). There are many interpolation techniques available, but they are mainly classified into two categories: deterministic and geostatistical techniques. Deterministic techniques are based on the geometric properties of the samples, whereas geostatistical techniques are based on geometric as well as spatial autocorrelation of the target variable. Some of the deterministic techniques are inverse distance weighted (IDW), spline, Thiessen polygon, and linear regression, and geostatistical techniques are simple kriging, ordinary kriging, universal kriging, co-kriging, regression kriging, indicator kriging, etc. Stationarity, isotropy, intrinsic hypothesis, and unbiasedness are the basic assumptions of geostatistical techniques (Sluiter 2009).

P. A. Verma (✉) · H. Shankar · S. Saran
Geoinformatics Department, Indian Institute of Remote Sensing (IIRS), Indian Space Research Organisation (ISRO), Department of Space, Government of India, Dehradun, India
e-mail: prabhakar@iirs.gov.in

R. R. Navalgund et al. (eds.), *Remote Sensing of Northwest Himalayan Ecosystems*,
https://doi.org/10.1007/978-981-13-2128-3_24

Tveito et al. (2008) described stationary as “the condition that the probability distribution of the variable is constant in time and/or space, meaning that the same probability distribution function should be expected anywhere/anytime.” Sluiter (2009) described intrinsic hypothesis as “the difference between two values taken at two different locations come from a distribution which depends only on the distance (and possibly relative direction) of the two locations.” Isotropy implies that change in target variable is directionally independent. Unbiasedness implies that all sample values are having equal importance for interpolation. Since temperature decreases as per the altitude (height) increases, this variation is called the lapse rate. So in this case, elevation can be used as a co-variable to improve the prediction accuracy of primary variable (temperature).

Climatic variables have a strong relationship with the topography of the surface, and mountain ecosystem has vast variation in topography. Due to topographic variations in mountain ecosystem, climatic variables are very difficult parameters to predict. There are many deterministic and geostatistical approaches which can be used for prediction, but due to complex terrain of Himalaya, prediction is always erroneous. Guenni and Hutchinson (1998) also said that “Rainfall is most difficult to predict due to its inherent variability in space and time, prediction is even more erroneous for mountains with complex terrain.”

Air temperature is one of the important parameters related to biological processes, physical processes such as phenological development of forest (Logan and Powell 2001), and snowmelt prediction (Hock 2003). Large-scale prediction of temperature highly depends on elevation due to the lapse rate of temperature with increasing elevation. Considering the effect of lapse rate, the simplest method can be the adjustment of the nearest AWS data with the change in elevation (Stahl et al. 2006). There are other interpolation methods which give good prediction in different scenarios, e.g., inverse distance weighted (IDW) average, spline, and regression adjusted with elevation. Instead of correlation of temperature with elevation, spatial autocorrelation can also play a key role to increase prediction accuracy. There are geostatistical approaches which consider spatial variability for the prediction, e.g., simple kriging, ordinary kriging, co-kriging, universal kriging, and regression kriging.

24.2 Earlier Work

Guenni and Hutchinson (1998) developed a methodology to summarize variability of daily rainfall by a set of parameters of stochastic point rainfall model. They have used observed data from the station to calibrate the parameters and fitted a periodic function to those parameters. And the coefficients of the periodic function were interpolated using thin plate spline by considering dependence on elevation. “The reliability of the interpolated surfaces in representing the rainfall characteristics across the region was demonstrated with an independent set of locations which were not used in the fitting procedure. This shows the capability of the method in providing realistic stochastically generated rainfall at any point of the region without historical information."

Stahl et al. (2006) compared 12 variations of regression-based and weighted average-based techniques for interpolating daily maximum and minimum temperature over British Columbia, Canada, and found that techniques involving local lapse rate from the control points performed better among the rest of the 11 variations. Price et al. (2000) compared ANUSPLINE and gradient plus inverse distance squared (GIDS) for elevation-dependent spatial interpolation of climatic data from sparse weather station networks. They have used 30-year mean minimum and maximum temperature data from Eastern and Western Canada regions. It was observed that ANUSPLINE is giving lower root-mean-square error (RMSE) compared to GIDS. Di Piazza et al. (2011) compared deterministic and geostatistical techniques such as inverse distance weighting, simple linear regression, multiple regression, geographically weighted regression and artificial neural networks, and geostatistical models such as ordinary kriging and residual ordinary kriging. In some methods they have used elevation as additional information. They have used rain gauge data for Sicily (Italy). And residual ordinary kriging performed best compared to other techniques. Wagner et al. (2012) did the rainfall interpolation for Mula and Mutha river catchment of Pune city, India. He used 16 rain gauge data, Thiessen polygons, and statistical and geostatistical approach techniques with wind direction and TRMM data as co-variable. Wagner et al. (2012) found that suitable interpolation scheme should not only be based on the comparison with point measurements but should also take the representativeness of the given measurement network as well as of the interpolated spatial rainfall distribution into account. Bargaoui and Chebbi (2009) used kriging with external drift (KED) with 3D variogram where three dimensions were location, rainfall duration, and rainfall intensity. Results were compared with KED with 2D variogram. And according to Bargaoui and Chebbi (2009), "A full comparison of the accuracy of both methods (2-D, 3-D) using cross-validation scheme, shows that the 3-D kriging leads to significantly lower prediction errors than the classical 2-D kriging."

Teegavarapu et al. (2012) developed six spatial interpolation weighting methods to access their suitability for transformation of precipitation estimates in space and time. "The methods use distances and areal extents of intersection segments of the grids as weights in the interpolation schemes. Three local interpolation methods out of six methods were found to be competitive and inverse distance based on four nearest neighbors (grids) was found to be the best for the transformation of data." Goovaerts (2000) compared three multivariate geostatistical techniques for rainfall prediction named simple kriging with varying local means, kriging with external drift, and co-kriging. In these techniques, data from 36 climatic stations was used along with elevation as the co-variable. Prediction from geostatistical techniques was compared with outputs from Thiessen polygon, inverse square distance, and ordinary kriging. It was observed that multivariate geostatistical techniques performed better. Dirks et al. (1998) compared four spatial interpolation methods kriging, Thiessen polygon, areal mean, and inverse distance method for estimation of rainfall using 13 rain gauges on Norfolk Island. According to Dirks et al. (1998), "inverse distance method is recommended for interpolations using spatially dense networks." Nalder and Wein (1998) compared different forms of kriging, inverse distance squared, nearest neighbor, and gradient plus inverse distance squared (GIDS) for estimation of monthly temperature and precipitation in Western Canada. GIDS performed best for temperature as well as precipitation estimation. Hussain et al. (2010) proposed "Box–Cox transformed hierarchical Bayesian multivariate spatio-temporal interpolation method for the skewed response variable" for estimation of precipitation over a region. Results were compared with non-transformed hierarchical Bayesian interpolation and concluded that transformed hierarchical method is giving better results.

24.3 Study Area and Dataset

Uttarakhand, a state of India, is chosen as the study area in this study; the reason behind it is that this state lies in the northwest Himalayan region. Elevation of the study area varies from 400 m to 6000 m above the mean sea level, and geographical extent is 77.57°E to 81.04°E longitude and 28.71°N to 31.47°N latitude.

The India Meteorological Department (IMD) has installed few weather stations in Uttarakhand; continuous data is available for 12 stations. AWS data from ten locations is used as the primary data source for the interpolation, and AWS data from two locations is used for the accuracy assessment of the interpolated surface. Cartosat DEM and MODIS daily temperature are used as the secondary variable/co-variable for interpolation (Fig. 24.1 and Tables 24.1 and 24.2).

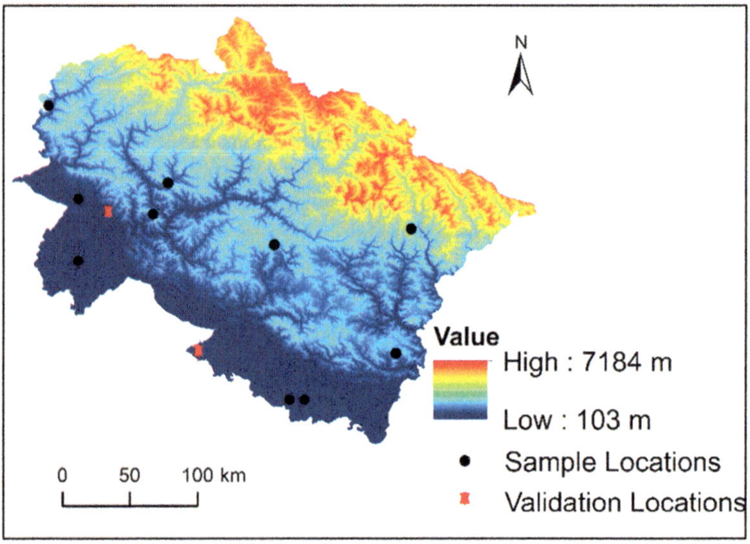

Fig. 24.1 Distribution of AWS in Uttarakhand

Table 24.1 Details of satellite data used

Satellite	Product	Date	Resolution
Cartosat	DEM version 3	–	30 m
MODIS	Daily temperature v 005	31–07–2016	5.6 km

Table 24.2 AWS data for 31–07–2016

Station	Latitude (N)	Longitude (E)	Mean temp. (K)	Elevation (m)
Ghansali	30.4	78.6	297.3375	751.89
Champawat	29.3	80.1	291.28	1659.60
Jaspur	29.3	78.8	300.51	182.3
Dhanauri	29.9	78	293.3818	211.42
Tiuni	30.9	77.8	299.65	1486.26
Munsyari	30.1	80.2	290.2211	2485.73
Dehradun	30.3	78	297.775	572.03
Jolly Grant	30.2	78.2	298.9095	548.69
Ranichauri	30.2	78.5	292.1636	2094.91
Rudrapur	29	79.4	298.37	161.04
Gairsain	30	79.3	295.2182	1503.08
Pantnagar	29	79.5	299.6667	166.58

24.4 Methodology

Spatial interpolation of daily mean temperature is carried out using deterministic as well as geostatistical techniques. And results are compared to find out the best suitable interpolation method in mountain ecosystem like Himalaya.

Deterministic and geostatistical methods are used for the downscaling of MODIS temperature product using interpolation techniques, and accuracy assessment is done to find out the best interpolation method in mountain ecosystem like Himalaya.

24.4.1 Deterministic Method

24.4.1.1 Inverse Distance Weighted (IDW)

IDW method uses the assumption that cell value for continuous variable at any location is related to its neighbors and unknown value can be determined by linear combination of neighboring cell values. Weight of each observation is determined by distance function. In IDW method, more weight is given to the nearest neighbor, and least weight is given to the farthest neighbor. If Z is the unknown value, Z_i represents observation points, and d_i represents distance observation point to the unknown value, then objective function for prediction is

$$Z = \frac{\sum_{1}^{n} \frac{Z_i}{d_i^p}}{\sum_{1}^{n} \frac{1}{d_i^p}}$$

where n represents the number of neighbors used for prediction and p denotes the power of distance.

Z_i denotes the daily mean temperature recorded from AWS.

Spline

Spline is one of the good interpolation methods where different polynomial functions are fitted between each pair of observation:

$$f(x,y) = f(x_i, y_i), \quad \text{where } x_i + 1, y_i + 1 < x, y < x_i, y_i$$

where x and y are coordinates and f is the data value.

24.4.1.2 Global Polynomial

Global polynomial interpolation method is a very coarse technique. This technique produces good result if variability of data in the study region is very less. In this technique, polynomial of degree n is fitted to get the interpolated surface:

$$z = ax + by + c$$

where z is the value of primary variable; a, b, and c are parameters; and x and y are coordinates.

24.4.2 Geostatistical Methods

Cressie (1991) and Kitanidis (1997) discussed about kriging in detail. The basis for the kriging is autocorrelation function named semivariogram. Semivariogram defines variance as a function of distance:

$$\gamma(h) = \frac{1}{2\,N(h)} \sum_{k=1}^{n} [Z(x_k + h) - Z(x_k)]^2$$

where $\gamma(h)$ is the semivariance of the variable Z with separation distance h, n is the number of samples, and $N(h)$ is the number of pair of points with h separation. Semivariance is calculated for each possible separation distance, and mean value of $\gamma(h)$ is plotted against h to produce experimental semivariogram. A semivariogram model is fitted to experimental semivariogram, and this model is used to calculate weights of kriging interpolation function by minimizing estimated variance.

$$Z = \sum_{i=1}^{n} a_i Z_i$$

24.4.2.1 Ordinary Kriging

Ordinary kriging (OK) estimates the value of climatic variable at a location as a linear combination of values from neighboring stations. And weights of linear combination are calculated using semivariogram and minimizing the estimated variance; local mean of neighbors is used for the prediction. OK falls in the category of the best linear unbiased estimator (BLUE). In the case of OK, weights are calculated in such a way that the sum of weights is 1.

24.4.2.2 Co-kriging

Co-kriging works on the principle of correlation of two or more variables. If samples of primary variable are not enough and some other variable is available which is highly correlated to primary variable, then that variable can be used as a co-variable to improve the prediction accuracy. In our case, primary variable is the daily mean temperature recorded by AWS, and co-variables are elevation and MODIS temperature product. The objective function for co-kriging is given below:

$$Z = \sum_{i=1}^{n} a_i Z_i + \sum_{j=1}^{m} b_j l_j + \ldots..$$

where Z represents the estimated value, Z_i is the primary variable, a_i denotes the weights of primary variable, l_j is representing the co-variable, and b_j denotes the weights of co-variable.

These weights have to be estimated by minimizing estimated variance with the following conditions:

In the case of ordinary co-kriging:

$$\sum_{i=1}^{n} a_i = 1$$

$$\sum_{j=1}^{m} b_j = 0$$

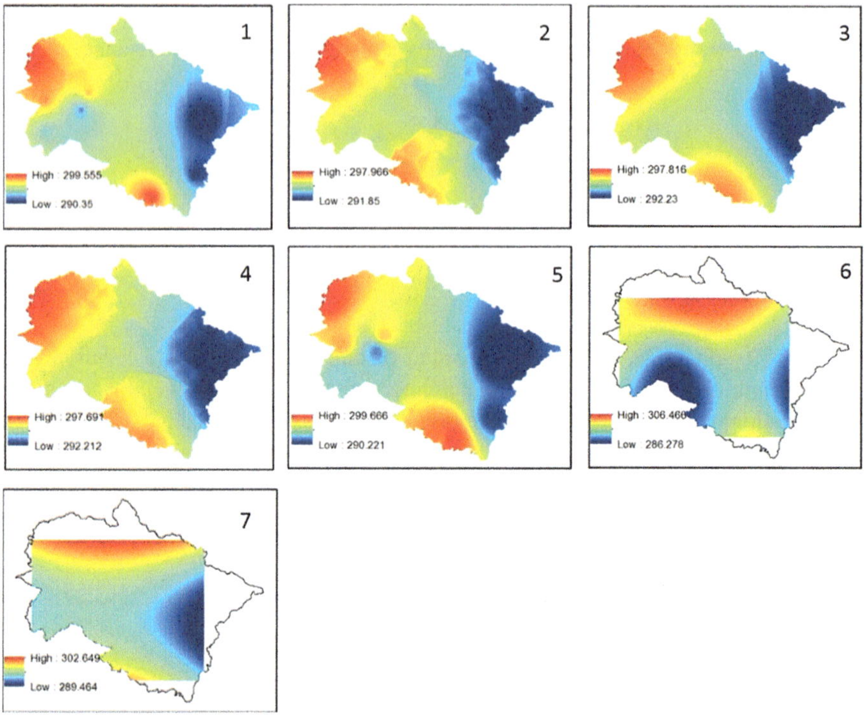

Fig. 24.2 Output of different interpolation methods. (*1*) Ordinary kriging, (*2*) co-kriging with MODIS temperature and elevation as co-variable (Co-OK-all), (*3*) co-kriging with elevation as co-variable (Co-OK elevation), (*4*) co-kriging with MODIS temperature product as co-variable (Co-OK Modis), (*5*) IDW, (*6*) spline, and (*7*) global polynomial (GP)

24.5 Results and Discussion

For deterministic methods, the only variable used is the mean temperature from AWS, whereas for geostatistical methods, the primary variable is the mean temperature from AWS, and co-variables are elevation and MODIS temperature product. For every geostatistical method, semivariogram was calculated, and spherical semivariogram model was fitted for each variable. There was a little variation in variogram models fitted for each variable, but spherical model provided best fit in the scenario of our study area.

Image 1 in Fig. 24.2 is showing the output of ordinary kriging where the mean temperature is the variable. Image 2 in Fig. 24.2 is showing the output of ordinary co-kriging where the mean temperature from AWS is the primary variable and elevation and MODIS temperature product are co-variable. Image 3 in Fig. 24.2 is

Table 24.3 Statistics of different interpolation methods

Method	Point 1		Point 2		RMSE
	Actual temp (K)	Predicted temp (K)	Actual temp (K)	Predicted temp (K)	
OK	300.51	295.58	298.91	295.44	4.27
Co-OK all	300.51	297.07	298.91	295.75	3.30
Co-OK MODIS	300.51	296.49	298.91	295.83	3.58
Co-OK elevation	300.51	296.20	298.91	295.77	3.77
IDW	300.51	296.67	298.91	295.76	3.51
Spline	300.51	288.54	298.91	294.43	9.04
GP	300.51	297.55	298.91	295.02	3.46

showing the ordinary co-kriging where the mean temperature from AWS is the primary variable and elevation is the co-variable. Image 4 in Fig. 24.2 is showing the output of ordinary co-kriging where the mean temperature from AWS is the primary variable and MODIS temperature product is the co-variable. Image 5 in Fig. 24.2 is showing the output of IDW method. Images 6 and 7 in Fig. 24.2 are the outputs of spline and global polynomial, respectively (Table 24.3).

Statistics from the table shows that interpolation of the daily mean temperature recorded by AWS is best when elevation as well as course resolution MODIS temperature product is used as co-variable in ordinary kriging method.

Due to undulating terrain of mountain ecosystem, it is very difficult to place AWS in a stratified sampling way and has a large number of observations. In such kind of scenario, it is suggestive to use one or more co-variables to improve prediction accuracy. Since temperature is highly related to elevation, elevation is used as one of the co-variables. Courser resolution MODIS temperature product is also used as co-variable to improve the prediction accuracy. And it is clear from the results that when both co-variables are used, root-mean-square error is minimum.

24.6 Conclusions

There are many interpolation methods, and they are broadly categorized into two categories, namely, deterministic and geostatistical. In this study, five methods (ordinary kriging, co-kriging, inverse distance weighted, spline, and global polynomial interpolation) were compared when interpolating climatic variable in Himalayan ecosystem. Deterministic techniques work on the geometric properties of point data but geostatistical techniques on spatial relationship between variables. In this study, the main variable which was used in interpolation is the daily mean temperature which was collected from the automatic weather stations (AWS) to produce fine-resolution temperature surface. To further improve the interpolation, elevation and course resolution MODIS temperature product were used as co-variables.

Statistical and spatial results show that co-kriging method is giving the best result when elevation and MODIS temperature product are used as co-variables.

24.7 Future Scope

There are many more interpolation techniques left to compare like regression kriging, indicator kriging, zonal kriging, etc. All the existing interpolation techniques can be explored to make a better conclusion about the best technique for mountain ecosystem.

References

Bargaoui, Z. K., Chebbi, A. (2009). Comparison of two kriging interpolation methods applied to spatiotemporal rainfall. Journal of Hydrology, 365(1–2), 56–73.

Cressie, N. (1991) Statistics for spatial data. John Wiley, London, UK, pp. 900.

Di Piazza, A., Conti, F. L., Noto, L. V., Viola, F., & La Loggia, G. (2011). Comparative analysis of different techniques for spatial interpolation of rainfall data to create a serially complete monthly time series of precipitation for Sicily, Italy. International Journal of Applied Earth Observation and Geoinformation, 13(3), 396–408.

Dirks, K. N., Hay, J. E., Stow, C. D., Harris, D. (1998). High-resolution studies of rainfall on Norfolk Island: Part II: Interpolation of rainfall data. Journal of Hydrology, 208(3–4), 187–193.

Goovaerts, P. (2000). Geostatistical approaches for incorporating elevation into the spatial interpolation of rainfall. Journal of hydrology, 228(1–2), 113–129.

Guenni, L., & Hutchinson, M. F. (1998). Spatial interpolation of the parameters of a rainfall model from ground-based data. *Journal of Hydrology*, *212*, 335–347.

Hock, R. (2003). Temperature index melt modelling in mountain areas. *Journal of Hydrology*, *282* (1), 104–115.

Hussain, I., Spöck, G., Pilz, J., Yu, H. L. (2010). Spatio-temporal interpolation of precipitation during monsoon periods in Pakistan. Advances in water resources, 33(8), 880–886.

Kitanidis, P. K. (1997). Introduction to geostatistics: applications in hydrogeology. Cambridge University Press.

Logan, J. A., & Powell, J. A. (2001). Ghost forests, global warming, and the mountain pine beetle (Coleoptera: Scolytidae). American Entomologist, 47(3), 160.

Nalder, I. A., & Wein, R. W. (1998). Spatial interpolation of climatic normals: test of a new method in the Canadian boreal forest. Agricultural and forest meteorology, 92(4), 211–225.

Price, D. T., McKenney, D. W., Nalder, I. A., Hutchinson, M. F., & Kesteven, J. L. (2000). A comparison of two statistical methods for spatial interpolation of Canadian monthly mean climate data. Agricultural and Forest meteorology, 101(2–3), 81–94.

Sluiter, R. (2009). Interpolation methods for climate data-literature review. KNMI intern rapport. De Bilt: Royal Netherlands Meteorological Institute.

Stahl, K., Moore, R. D., Floyer, J. A., Asplin, M. G., & McKendry, I. G. (2006). Comparison of approaches for spatial interpolation of daily air temperature in a large region with complex topography and highly variable station density. *Agricultural and Forest Meteorology*, *139*(3), 224–236.

Teegavarapu, R. S., Meskele, T., Pathak, C. S. (2012). Geo-spatial grid-based transformations of precipitation estimates using spatial interpolation methods. Computers & Geosciences, 40, 28–39.

Tveito, O. E., Wegehenkel, M., Wel, F. V. D. (2008). The use of geographic information systems in climatology and meteorology.

Wagner, P. D., Fiener, P., Wilken, F., Kumar, S., & Schneider, K. (2012). Comparison and evaluation of spatial interpolation schemes for daily rainfall in data scarce regions. Journal of Hydrology, 464, 388–400.

Chapter 25
Role of Citizen Science in Northwestern Himalaya: Use Case on Disaster, Bio-resource, and Governance

Kapil Oberai, Sameer Saran, Stutee Gupta, Priyanka Singh, S. K. Srivastav, and A. Senthil Kumar

25.1 Introduction

Mountain regions represent most fragile and vulnerable ecosystems (UNCED 1992) facing threats due to unpredictable socioeconomic and climate changes as well as unsustainable land use practices. The impact of unsustainable practices on the mountains is manifested in the form of depletion of bio-resources and increased occurrence of extreme events such as floods, forest fire, landslides, etc. Geospatial technology consisting of remote sensing (RS), geographic information system (GIS), and global navigation satellite system (GNSS) has a huge potential in natural resource management including governance and other environmental issues. Moreover studying the geographic phenomenon requires a large amount of data to be collected/created to model the said phenomenon to improve understanding. Citizen science these days is a valuable and important method, in which nonspecialists also called as volunteers work toward the collection of data of scientific importance in a short span of time using the power of people/volunteers. Citizen science has been

K. Oberai (✉) · S. Saran · P. Singh
Geoinformatics Department, Indian Institute of Remote Sensing (IIRS), Indian Space Research Organisation (ISRO), Department of Space, Government of India, Dehradun, India
e-mail: kapil@iirs.gov.in

S. Gupta
National Remote Sensing Centre, Indian Space Research Organisation (ISRO), Department of Space, Government of India, Hyderabad, India

S. K. Srivastav
Indian Institute of Remote Sensing (IIRS), Indian Space Research Organisation (ISRO), Department of Space, Government of India, Dehradun, India

A. Senthil Kumar
Indian Institute of Remote Sensing, Indian Space Research Organisation, Dehradun, Uttarakhand, India

R. R. Navalgund et al. (eds.), *Remote Sensing of Northwest Himalayan Ecosystems*,
https://doi.org/10.1007/978-981-13-2128-3_25

globally utilized in areas like support toward policy, electronic government, and digital democracy (Shirky 2008). Citizen science approach is very useful especially in disaster management like during recent earthquake disasters in Haiti, 2010 (Zook et al. 2010), and Nepal, 2015 (Dittus et al. 2017), assisting rescue and relief efforts. With the penetration of the Internet at grassroots levels, the advent of Web 2.0 and the availability of smartphones with good quality camera and onboard GPS receivers have supported and encouraged citizen science initiatives worldwide. These technologies enable volunteers to gather valuable dataset with geolocation (geotagged) including photographs, audio, and videos. In today's scenario, citizens have a valuable and active role as they work like "sensors" (citizens as sensors (Goodchild 2007)) assisting the state/national administrative or official initiatives through gathering and interpreting datasets. Wikipedia (www.wikipedia.org) is the classic example of citizen science approach in which a huge number of people work as volunteers by building, correcting, and revising the textual material. Likewise in the geospatial field, the popular example is the OpenStreetMap project (www.openstreetmap.org). Here volunteers work as nonspecialists toward mapping of the entire globe. Furthermore data is made available under open-access model.

25.2 Citizen Science

Citizen science basically refers to nonprofessional acting as a volunteer supporting the activities involved in a scientific project ranging from data collection to analysis including data dissemination (Arias de Reyna and Simoes 2016). Nowadays scientific activities are not limited to scientists only, and common people could contribute, for example, by collecting data about air samples (Anon. 2015). Cohn (2008) describes volunteers working in citizen science projects as "citizen scientists" working as "field assistants" in a particular scientific study. This approach enables, for example, in the case of environmental studies, to cover and collect wider geographic areas at a lower cost and in less time than may be done by professional teams (United Nations Environment Programme [UNEP] 2014). Rossiter et al. (2015) opine that "citizen-assisted science" term is a much better term to describe the field of citizen science, as citizens working as volunteers are basically "observers or experimentalist."

The concept of volunteers assisting in the scientific discovery is not a new phenomenon and can be traced in past history dating since the seventeenth century onward (Miller-Rushing et al. 2012). An early account for the same is found in China wherein tracking of locusts outbreak has been going on for over 3500 years (Tian et al. 2011). Another prominent example of citizen science used in the past is the "Christmas Bird Count" activity in the USA which dates back to 1900 and "British Trust for Ornithology Survey" in the UK wherein more than 31 million records are collected subsequently its initiation in the year 1932 (Silvertown 2009).

Alan Irwin invented the term "citizen science" during 1995 in his book *Citizen Science* wherein he highlighted "how people accumulate knowledge in order to learn about and respond to environmental threats" (Science Communication Unit, University of the West of England (UWE) 2013). Subsequently the term was modified so as to refer to the public participation in science and research activities (Bonney et al. 2009).

Some professionals working in scientific areas are susceptible toward citizen science approach due to the amateurs/nonspecialists working in such projects (Cohn 2008; Anon. 2015). But the number of research papers published in various journals suggest that such approach is highly valuable (Follett and Strezov 2015). Rossiter et al. (2015) put forward a strong case for citizen science based project wherein an amateur helps in the soil science project by mapping soil properties and types. In literature, a number of projects could be found wherein citizen science approach has been used involving the public to gather and analyze the data related to various themes like monitoring wildlife and environment, classifying images, making the written records of old records, etc. (Follett and Strezov 2015). People also question the validity of data collected by such projects; however with proper training and capacity building of the volunteers, the data collected by them can match the quality as expected from the specialists in the area (Gommerman and Monroe 2012). Kobori et al. (2016) emphasized that in order to encourage volunteer's participation in such projects, the organizers should not only keep in mind the data collection methods but also aspects such as recruiting and retaining volunteers with necessary skills and interest.

Various terms are used to describing citizen assistance/involvement working as volunteers in various endeavors related to geospatial domain. The various terms used are VGI (volunteered geographic information), crowdsourcing, and PGIS (participatory GIS). The term VGI was coined by Michael Goodchild (Goodchild 2007) to describe the involvement of public primarily nonspecialist in the GIS data creation. "Crowdsourcing" is another term used to describe the public involvement in large number in solving a complex problem through an open call. Smartphone with Internet connectivity leads to further popularity of the crowdsourcing approach (Chatzimilioudis et al. 2012). The term was first coined by Jeff Howe and Mark Robinson, the editors of the wired magazine in 2005. (Dunn 2007) describes PGIS as an approach that "seek to emphasize community involvement in the production and/or use of geographical information." Not all the citizen science projects are of geographic in nature. Hence in the citizen science, the projects which are of geographic nature wherein volunteers help in data collection/analysis, etc. come under the gamut of "geographical citizen science" (Haklay 2013).

Geographical citizen science basically represents the intersection of VGI with citizen science and therefore refers to subtype and very specific activity in citizen science domain wherein public participates in scientific research by collecting/analyzing the geospatial data only. Geographic citizen science projects could be classified in several ways. Simple classification suggested by (Haklay 2013) is on the basis of involvement of volunteer, i.e., active or passive role. In active role volunteer participates actively by contributing wholeheartedly in data collection and analysis. A volunteer acting in passive mode doesn't play active role in the project. An example of this might be in the case of the "climateprediction.net" project where

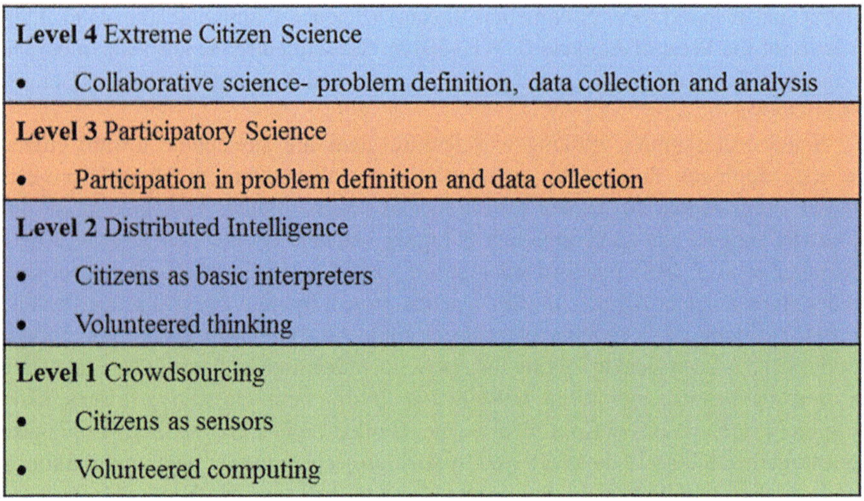

Fig. 25.1 Levels of participation in citizen science project. (Source: Haklay 2013)

volunteers contribute by sharing their computer resources to run global climate models. Haklay (2013) has also given the classification scheme and that is based on the depth of the engagement of the volunteer and their participation in the citizen science project (Fig. 25.1).

The first level is the very basic level of engagement and is called "crowdsourcing" projects, wherein the volunteer's role is to only collect the data through the sensors or to support, for example, by providing their computing resources for running a complex task having huge computing or memory requirements. In this type of project, the volunteer's cognitive ability is least used. In level 2 type of projects termed as "distributed intelligence," the cognitive ability of the volunteers is used. The examples in this category are galaxy zoo, which is a crowdsourced astronomy project wherein volunteer assists in understanding the formation of galaxy by classifying them based on shapes (https://www.galaxyzoo.org/). Here the volunteers are required to undergo elementary training before contributing toward data collection/interpretation exercise.

The level 3 type of project requires participants to also get involved in the problem definition and also discuss with the expert/scientist prior to deciding the optimal data collection methodology. The level 4 is the highest level among the level of participation wherein the volunteers are involved in all the stages in the project not only in data collection. Table 25.1 provides the list of some of the citizen science projects running worldwide in the themes related to bio-resources, disaster, and governance.

Table 25.1 Sample citizen science projects

Theme	Project	URL	Description
Bio-resources	eBird	http://ebird.org	Collecting bird observation
	SeasonWatch	http://www.seasonwatch.in	Collecting seasonal activity of plants for understanding climate affects
	eButterfly	http://www.e-butterfly.org	Collecting butterfly observations
	Golden Gate Raptor Observatory	http://www.parksconservancy.org/programs/ggro	Tracking raptors
	iNaturalist	http://www.inaturalist.org	Sharing biodiversity-related information
	EDDMapS	http://www.eddmaps.org	Documenting invasive species distribution
Disaster	GeoTag-X	http://geotagx.org	Crowdsourced disaster relief initiative
	Citizen seismology	http://www.citizenseismology.eu/index.html	Collecting seismic events
	Did You Feel It?	http://earthquake.usgs.gov/data/dyfi	Reporting earthquake occurrence data
Governance	CitiSense	http://citi-sense.nilu.no	Sensor-enabled community-based conservational governance
	Omniscientis	http://www.omniscientis.eu	Odour monitoring
	Air Quality Eggs	http://airqualityegg.com	Collecting air quality data

25.3 A Few Initiatives in Citizen Science at IIRS

25.3.1 Role of Citizen Science in Disaster: MANU Project

During June 2013, Uttarakhand state of India suffered huge devastation affecting many people including loss to property, infrastructure, etc. This was primary because of floods due to extreme rain event that happen in the higher regions of Himalaya and glacial lake outburst flood (GLOF) and events of landslides (Murthy et al. 2014). Keeping in mind the extensive destruction, harsh terrain, and urgent demand of gathering data related to damage caused by devastation in quick time, citizen science was selected. Moreover, decision was also influenced due to the limitation of RS data in identifying the damages at the local level.

Hence, an initiative involving many organizations/institutes was taken named “Map the Neighbourhood in Uttarakhand” (MANU) having the aim of primary data collection through public participation involving students of the region. A mobile app (Android) was developed for collecting data related to damage caused. Separate proformas were designed in the mobile app which enables gathering datasets related

Fig. 25.2 System architecture of MANU. (Source: Murthy et al. 2014)

to damage to either buildings and infrastructure, loss of land cover and natural resources, landslides, damage due to river bank erosion, and also points of interests. Figure 25.2 shows the system architecture of MANU App.

The data collected were visualized over Bhuvan Portal to see the extent of damage (Fig. 25.3).

A total of 18,869 points were collected by the volunteers within a short span of time. Figure 25.4 below depicts the point's distribution vis-à-vis various damage themes.

Further, geospatial analysis of damaged points was done in geospatial domain. The finding of the analysis served as inputs for restoration and developmental activities.

MANU was one of the first initiatives in India of using citizen science (level 1 – citizens as sensors) approach in disaster event for collecting valuable data for damage assessment and subsequently for future development activities.

25.3.2 *Role of Citizen Science in Governance: Citizen Science Based Swachh Bharat Turnkey Solution*

The Swachh Bharat Abhiyan (SBA) or Clean India Campaign was formerly launched by the Indian Prime Minister on the eve of October 2, 2014, the birth

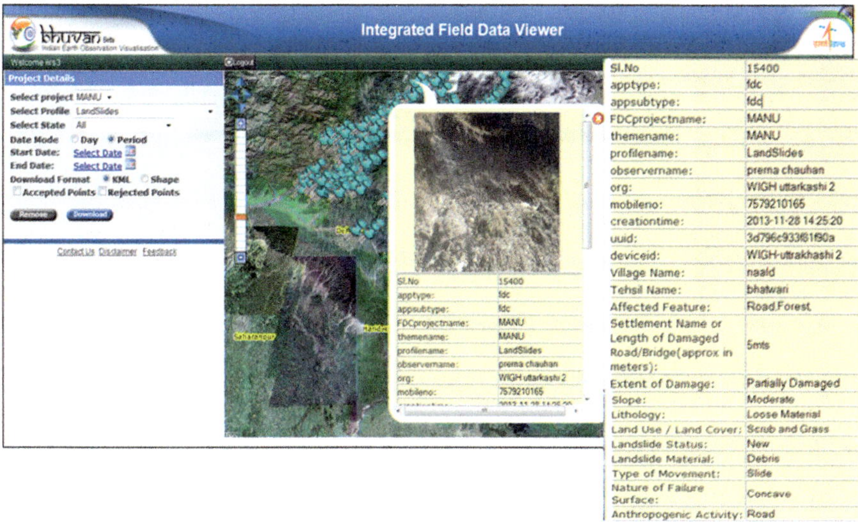

Fig. 25.3 Data visualization over Bhuvan portal. (Source: Murthy et al. 2014)

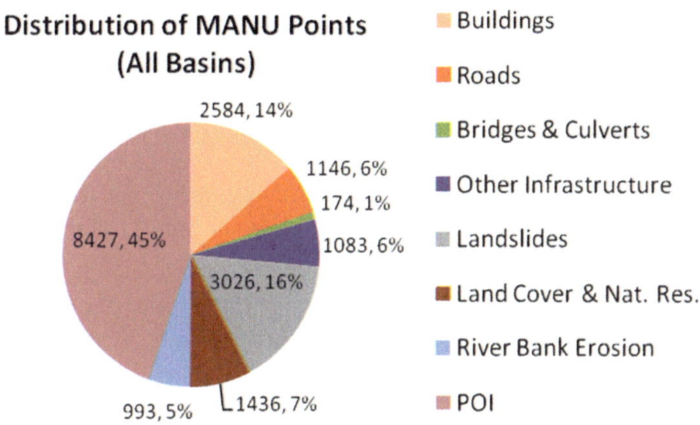

Fig. 25.4 Pie-chart depicting the theme-wise distribution of damage points

anniversary of Mahatma Gandhi. SBA reemphasized and put the focus on sanitation (the basic requirement) with the aim to achieve universal sanitation thereby achieving "Swachh Bharat" by the year 2019 which could be the best tribute on the occasion of the 150th birth anniversary of Mahatma Gandhi (Ministry of Drinking Water and Sanitation 2010). The Prime Minister urged the nation to participate with full zeal and enthusiasm to make it a grand success.

To support the above initiative, Indian Institute of Remote Sensing (ISRO) designed and developed an end-to-end geo-spatial solution utilizing citizen science

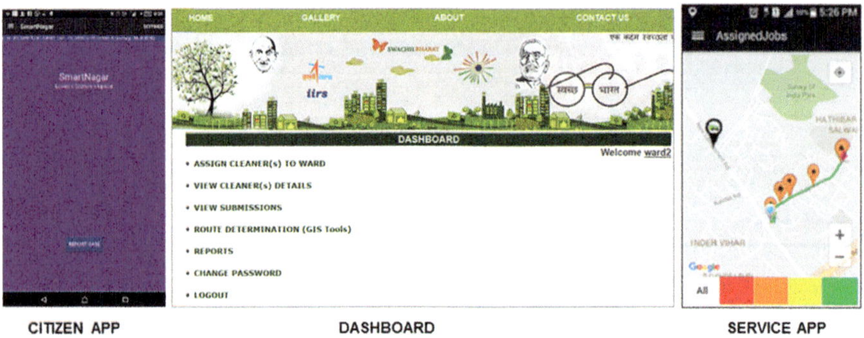

Fig. 25.5 SBA turnkey solution

method for SBA. Turnkey solution consists of three different applications, two mobile applications and one web application (Fig. 25.5).

Citizen App It is a mobile app which runs on Android OS and empowers citizens to submit complaints regarding garbage dumping in and around their locality. To report a location, the citizen needs to geotag the location using onboard GPS on the Android mobile phone, take the photograph of the garbage, and fill details like type of garbage and description and submit.

Web-Based Dashboard The data collected through the Citizen App are submitted to the centralized server and could be visualized through web-based dashboard application where data resides in the spatial database. Using the routing module in dashboard, officials can plan the optimize path that the cleaner will take to address all complaints assigned in optimal manner considering the number of complaints their locations and distance.

Service App It is an Android-based mobile application designed for the cleaners and allows them to handle the tasks in an efficient manner. Further it allows them to take pre- and post-photographs of the reported trash location.

Complete system was put to use during Oct 2, 2016 (the birth anniversary of Mahatma Gandhi), which is also celebrated as Swachh Bharat Day. Over 1000 participants from 4 universities and 9 government institutes in the Dehradun participated in this exercise. A total of 3268 garbage locations were geotagged using the citizen mobile app on August 27, 2016, within a short time of 5 h (0800–1300 h.).

RS data will have limited use in this area; hence crowdsourcing approach was a big game changer enabling collecting such a huge amount of geotagged garbage locations in a mission mode. The cleaning operation through Android-based "Service App" was carried out. For the same first, the web-based dashboard was used for job assignment and optimal path identification using the routing module. IIRS organized workshop in each university during Aug 20–26, 2016, to demonstrate the complete solution along with apprising the usage of the Citizen App which was used during Aug 27, 2016, by the participating students. During the workshop

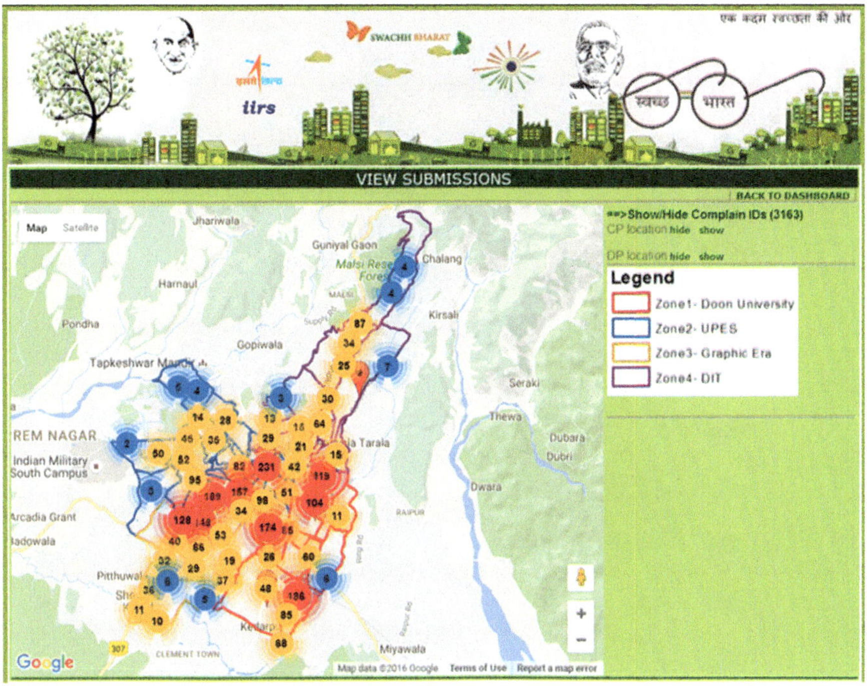

Fig. 25.6 Dashboard-visualizing complaints

participants were also briefed about the predefined routes they need to follow for geotagging the garbage locations. The screenshot of the web-based dashboard application at IIRS server showing the points reported is shown in Fig. 25.6.

25.3.2.1 Knowledge Based Classification Using Citizen Science Approach

The geotagged garbage related points contributed by the students in the above exercise were further classified using knowledge- based approach. Again using the citizen science principles (Level 2- Citizens as basic interpreters) the points collected were further classified into following categories involving the Post Graduate Students of IIRS.

- Commercial
- Natural
- Public- Open
- Public- CP/DP Spillover
- Public- Spillover
- Public- Waterbody Residential

Fig. 25.7 Web application for knowledge-based classification

A total 100 IIRS PG students participated in this activity. To help the students to classify the image, web-based application was developed (Fig. 25.7). The student by looking at the image and with the help of sample images provided, could classify each random point into one of the above mentioned categories. All images were classified into defined categories by the students within an hour.

Figure 25.8 summarizes the points classified into various categories.

The above applications clearly demonstrated the usefulness of citizen science in the national initiative like the Swachh Bharat enabling citizen's participation through garbage reporting around their locality. Further through citizen science approach, complex tasks like classifying garbage complaints based on uploaded image using knowledge based classification can be undertaken in very quick time.

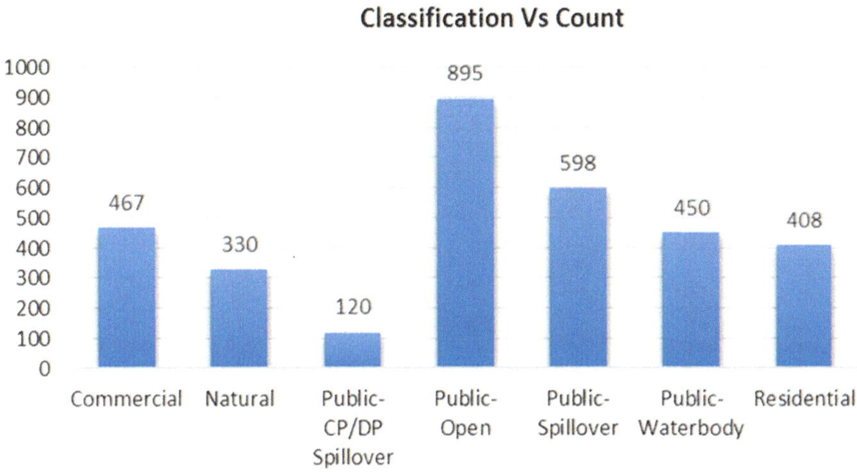

Fig. 25.8 Classification statistics of crowdsourced collected garbage points

25.4 Role of Citizen Science in Bio-resources

25.4.1 BotanyGIS: An Educative Tool

Bio-resources such as agriculture, biofuels, nutrition, water cycles and their relationships with the environmental problems depend on a comprehensive understanding of botanical principles for knowledgeable debate and decision-making (Pettit et al. 2014). However, participation in the study of botany among the students at university levels is decreasing world over for almost 100 years (Nichols 1919; Hershey 1993; Uno 1994; Drea 2011; Pettit et al. 2014). The Indian subcontinent has also witnessed a decreased interest in taxonomy for more than last three decades and regained importance only after the Convention on Biological Diversity (CBD 1992) and its effective implementation in India from 1994 (Sundaresan et al. 2001). This, indeed, can be overcome by using informal science education as provided by citizen science. BotanyGIS is a web-based GIS application (Fig. 25.9) that uses crowdsourcing capabilities for geotagging the common species in the surrounding campus environment for generating awareness among the people about the plants and hence botany education. Seventy plant species, mainly trees and shrubs belonging to 34 families, were geotagged in this study using crowdsourcing in the campus environment (Gupta et al. 2015).

The project offers capability of enhancing the tree database as well as identification of the geotagged tree based on their name and other morphological features. With this pilot study, a framework has been developed for educative citizen science project that would be integrated further with the Indian Bio-resource Information Network (IBIN) as an educative tool.

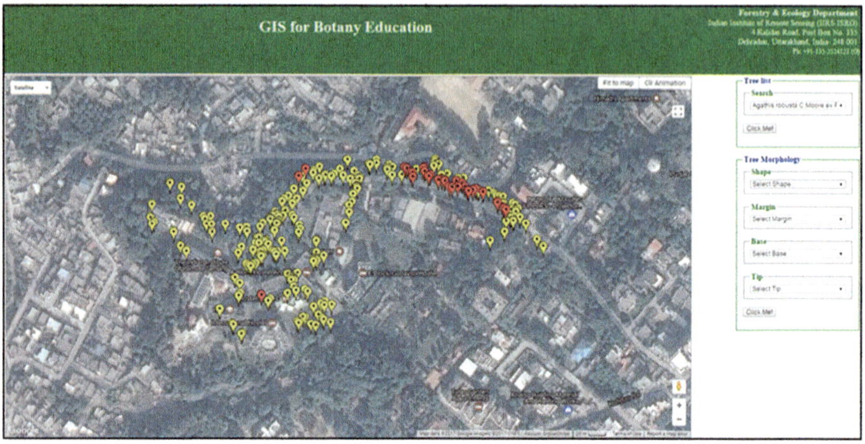

Fig. 25.9 Web application for BotanyGIS

25.4.2 IBIN Crowdsourcing Mobile App

Indian Bio-resource Information Network (IBIN) is a decentralized bio-resource database based on distributed architecture. It works on the principle of spatial data infrastructure wherein distributed databases available across the country are accessed through one single window gateway. To enable public participation in this endeavor, an Android-based mobile app is designed that enables an end user to submit datasets on species utilizing the crowdsourcing approach. Figure 25.10 depicts the Android-based app and the workflow for capturing the end-user dataset.

The application is user-friendly, encouraging nonspecialist users to contribute bio-resources datasets (species sighting). The data contributed can be visualized on the IBIN portal (www.ibin.gov.in) (Fig. 25.11). The data collected through crowdsourcing goes through the verification process; however using the Web GIS viewer, users can view both the collected data and the verified data.

Such initiatives enables a nonspecialist to contribute data and contribute toward having the species database of different categories, which otherwise may be difficult to collect and also time-consuming if the same task is left to only the identified scientist/researcher teams.

25.5 Challenges and Way Forward

Rapid technological development in the last two decades has resulted in the constant growth of processing power. This leads to generation of a prodigious flow of data which can't be handled by traditional computation systems and calls for a mechanism for handling large scale processing of datasets (Sakr et al. 2011). Citizen

Fig. 25.10 IBIN mobile app for end-user data

scientists are collectively producing big data, i.e., a high volume of high-value data. Such data collected using distributed network of citizens as data gatherers is creating new trends in data analysis and providing substantial basis of common framework to organize the findings by connecting the dots of isolated information that are otherwise fail to cumulate. Due to the availability of the data and access to all the contributors, the need for the data to be collected again and again (Parr and Cummings 2005) is also overcome, thus expanding scientific knowledge and literacy (Bonney et al. 2009). Location is a vital ingredient for the various big datasets for knowing the causal patterns and trends. GIS tools for Hadoop works with big spatial data, i.e., big data with location component, and allows spatial analysis to be performed utilizing the distributed processing capabilities.

Though crowdsourced data along with EO data is considered significant, consolidated efforts are required to gain greater acceptance and prominence of crowdsourcing (Mazumdar et al. 2017). The credibility and accuracy of VGI is

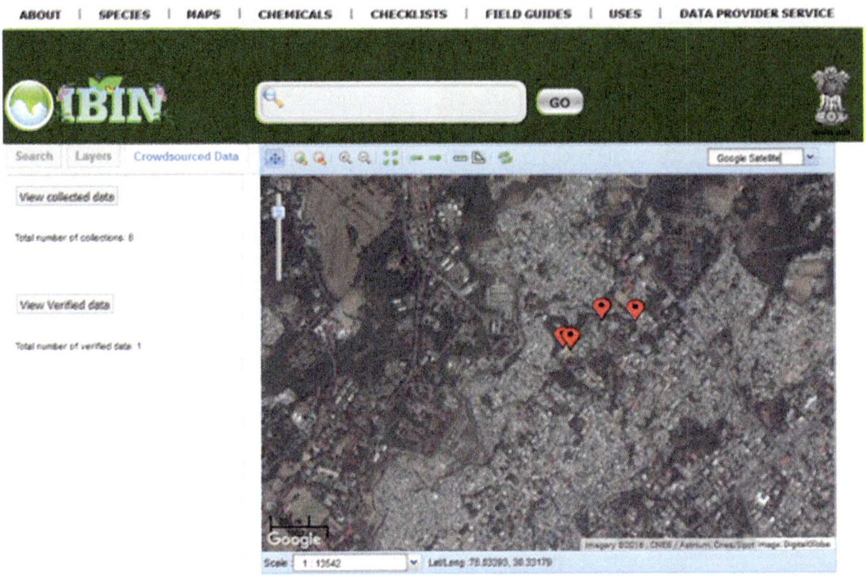

Fig. 25.11 IBIN crowdsourced data viewer

questioned by some researchers (Tüzün et al. 2009), and it is also agreed that the user-generated context approach still plays a role in helping researchers explore users' awareness and comments of certain issues (Wu 2013). With proper training and capacity building of the volunteers, the data collected by them can be comparable to that collected by the professionals. Further moderation and validation of the datasets by the experts in the domain can improve the quality of such datasets. Aspects such as recruiting and retaining volunteers with necessary skills and interest as per the project demands should also be kept in mind by the organizers prior to starting such activity.

The popularity and success of the crowdsourced projects are also guided much by Internet penetration in any region more so in the remote mountainous areas of the NW Himalayas. However the future efforts toward Digital India and other attempts in line with the Outernet (a global broadcast data company that aims in providing free access to web content using geostationary and low-Earth orbit satellites effectively to the global community) along with projects that such as Google's Project Loon and Facebook's Internet.org may contribute significantly in this direction. Funding support from the government on a routine basis is another important aspect. There is a need to make continuous efforts to reach out to the citizens as well as decision-makers and work in partnerships.

25.6 Conclusions

Mountainous region are facing serious governance issues like disasters and resource depletion. To understand the underlying patterns and processes of various phenomena affecting the mountain regions, a huge amount of primary data is prerequisite. Citizen science approach could bring revolution in these areas by assisting scientist with valuable datasets in quick possible time. This chapter discussed the role of citizen science in general and its role in the mountainous region in particular by covering some potential use cases related to disaster, governance, and bio-resources. It is believed that more and more scientific projects involving public participation through citizen science will be seen in the coming future.

References

Anon. (2015) Rise of the citizen scientist. Nature 524:265–265. doi: https://doi.org/10.1038/524265a

Arias de Reyna M, Simoes J (2016) Empowering citizen science through free and open source GIS. Open Geospatial Data, Softw Stand 1:7. doi: https://doi.org/10.1186/s40965-016-0008-x

Bonney R, Cooper CB, Dickinson J, Kelling S, Phillips T, Rosenberg K V., Shirk J (2009) Citizen Science: A Developing Tool for Expanding Science Knowledge and Scientific Literacy. Bioscience 59:977–984. doi: https://doi.org/10.1525/bio.2009.59.11.9

CBD (1992) Convention on biological diversity. Diversity 30. doi: https://doi.org/10.1146/annurev.ento.48.091801.112645

Chatzimilioudis G, Konstantinidis A, Laoudias C, Zeinalipour-yazti D (2012) Crowdsourcing with Smartphones. 1–7.

Cohn JP (2008) Citizen Science: Can Volunteers Do Real Research ? Bioscience 58:192–197. doi: https://doi.org/10.1641/B580303

Dittus M, Quattrone G, Capra L (2017) Mass participation during emergency response: event-centric crowd-sourcing in humanitarian mapping.

Drea S (2011) The End of the Botany Degree in the UK. Biosci Educ 17:1–7.

Dunn CE (2007) Participatory GIS – a people's GIS? Prog Hum Geogr 31:616–637. doi: https://doi.org/10.1177/0309132507081493

Follett R, Strezov V (2015) An analysis of citizen science based research: Usage and publication patterns. PLoS One 10:1–14. doi: https://doi.org/10.1371/journal.pone.0143687

Gommerman L, Monroe MC (2012) Lessons Learned from Evaluations of Citizen Science Are Data Collected by Citizen What Contexts Are Most. 1–5.

Goodchild MF (2007) Citizens as sensors: the world of volunteered geography. GeoJournal 69:211–221. doi: https://doi.org/10.1007/s10708-007-9111-y

Gupta S, Kandru A, Singh S (2015) Active Learning Technique for Botany Education in Campus Environment using Open Source Web GIS Application. Organ By Dep Civ Eng Indian Inst Technol (Banaras Hindu Univ Varanasi-221005 Uttar Pradesh, India 117.

Haklay M (2013) Citizen Science and Volunteered Geographic Information: Overview and Typology of Participation. In: Crowdsourcing Geographic Knowledge. Springer Netherlands, Dordrecht, pp 105–122

Hershey DR (1993) Plant neglect in biology education. Bioscience 43:418.

Kobori H, Dickinson JL, Washitani I, Sakurai R, Amano T, Komatsu N, Kitamura W, Takagawa S, Koyama K, Ogawara T, Miller-Rushing AJ (2016) Citizen science: a new approach to advance ecology, education, and conservation. Ecol Res 31:1–19. doi: https://doi.org/10.1007/s11284-015-1314-y

Mazumdar S, Wrigley S, Ciravegna F (2017) Citizen Science and Crowdsourcing for Earth Observations: An Analysis of Stakeholder Opinions on the Present and Future. Remote Sens 9:87.

Miller-Rushing A, Primack R, Bonney R (2012) The history of public participation in ecological research. Front Ecol Environ 10:285–290. doi: https://doi.org/10.1890/110278

Ministry of Drinking Water & Sanitation (2010) Swachh Bharat Mission Objectives.

Murthy YVNK, Raju PLN, Srivastav SK, Kumar P, Mitra D, Karnatak H, Saran S, Pandey K, Oberai K, Reddy KS, Gupta K, Swamy M, Deshmukh A, Dadhwal VK, Bothale V, Diwakar PG, Ravikumar M V., Leisely A, Arulraj M, Kumar S, Rao SS, Rawat RS, Pathak DN, Dutt V, Negi D, Singh J, Shukla K, Tomar A, Nadeem, Singh B, Singh AK, Kumar RS (2014) Capacity building for collecting primary data through crowdsourcing – An example of disaster affected Uttarakhand State (India). Int Arch Photogramm Remote Sens Spat Inf Sci – ISPRS Arch XL-8:1249–1252. doi: https://doi.org/10.5194/isprsarchives-XL-8-1249-2014

Nichols GE (1919) The general biology course and the teaching of elementary botany and zoology in American colleges and universities. Science (80-) 509–517.

Parr CS, Cummings MP (2005) Data sharing in ecology and evolution. Trends Ecol Evol 20:362–363.

Pettit L, Pye M, Wang X, Quinnell R (2014) Designing a bespoke app to address botanical literacy in the undergraduate science curriculum and beyond. Rhetor Real Crit Perspect Educ Technol 614–619.

Rossiter DG, Liu J, Carlisle S, Zhu A-X (2015) Can citizen science assist digital soil mapping? Geoderma 259:71–80. doi: https://doi.org/10.1016/j.geoderma.2015.05.006

Sakr S, Liu A, Batista DM, Alomari M (2011) A survey of large scale data management approaches in cloud environments. IEEE Commun Surv Tutorials 13:311–336.

Science Communication Unit, University of the West of England (UWE) B (2013) Science for Environment Policy IN-DEPTH REPORT: Environmental Citizen Science. 32.

Shirky C (2008) Here comes everybody: the power of organizing without organizations. Penguin Press

Silvertown J (2009) A new dawn for citizen science. Trends Ecol Evol 24:467–471. doi: https://doi.org/10.1016/j.tree.2009.03.017

Sundaresan V, Kathiravan K, Seshadri S, Ignacimuthu S (2001) Taxonomy in India. Curr Sci 80:912.

Tian H, Stige LC, Cazelles B, Kausrud KL, Svarverud R, Stenseth NC, Zhang Z (2011) Reconstruction of a 1,910-y-long locust series reveals consistent associations with climate fluctuations in China. Proc Natl Acad Sci 108:14521–14526. doi: https://doi.org/10.1073/pnas.1100189108

Tüzün H, Yilmaz-Soylu M, Karakus T, Inal Y, Kizilkaya G (2009) The effects of computer games on primary school students' achievement and motivation in geography learning. Comput Educ 52:68–77.

UNCED (1992) Agenda 21: The united nations programme of action from rio.

United Nations Environment Programme [UNEP] (2014) Realizing the Potential of Citizen Science. Unep Year B 2014 Emerg Issues Our Glob Environ 36–41.

Uno GE (1994) The state of precollege botanical education. Am Biol Teach 56:263–267.

Wu BS (2013) Developing an evaluation framework of spatial understanding through GIS analysis of Volunteered Geographic Information (VGI). Rev Int Geogr Educ Online 3:152.

Zook M, Graham M, Shelton T, Gorman S (2010) Volunteered Geographic Information and Crowdsourcing Disaster Relief: A Case Study of the Haitian Earthquake. World Med Heal Policy 2:6–32. doi: https://doi.org/10.2202/1948-4682.1069

Printed by Printforce, the Netherlands